CLINICAL VETERINARY MICROBIOLOGY

P. J. Quinn
MVB, PhD, MRCVS
Professor of Veterinary Microbiology & Parasitology
Faculty of Veterinary Medicine
University College Dublin

M. E. Carter
BVSc, Dip.Bact., MRCVS
Department of Veterinary Microbiology & Parasitology
Faculty of Veterinary Medicine
University College Dublin

B. K. Markey
MVB, PhD, MRCVS
Department of Veterinary Microbiology & Parasitology
Faculty of Veterinary Medicine
University College Dublin

G. R. Carter
DVM, MS, DVSc
Emeritus Professor
Division of Pathobiology
Virginia-Maryland Regional College of Veterinary Medicine
Virginia Tech, Blacksburg
Virginia
USA

Mosby
Edimburgh London New York Philadelphia St. Louis Sydney Toronto

MOSBY
An imprint of Harcourt Publishers Limited
M is a registered trademark of Harcourt Publishers Limited

Published in 1994
Reprinted 1998
Reprinted 1999
Printed in Spain by Grafos, S.A. Arte Sobre Papel
ISBN 0 7234 1711 3

A CIP catalogue record for this book is available from the British Library.

Library of Congress Cataloging-in-Publication Data: (available on request)

Contents

Section 3: Mycology

Introduction

Veterinary diagnostic microbiology is concerned with the recognition of a large number of microorganisms that either cause, or are frequently associated with infectious diseases of animals.

The contribution that a diagnostic microbiology laboratory can make to the recognition and identification of infectious disease in animals and the selection of appropriate chemotherapeutic drugs is influenced by many factors, some relating to the specimen, others relating to the laboratory. These include the type and quality of the specimen submitted, the stage of disease when it was taken, the care with which it was collected and the accuracy of the accompanying history. Other factors relate to facilities available in the laboratory for carrying out the necessary diagnostic procedures, the experience and technical proficiency of laboratory personnel and the format and content of the report furnished to the clinican. The standards prevailing in the laboratory, its organisation, work load and the training which laboratory staff have received, ultimately determine the quality of the report and its relevance to the clinical case being investigated.

This book is intended as an illustrated text to assist those with some previous microbiological experience. It is particularly intended for veterinary graduates and undergraduates, laboratory staff in veterinary diagnostic laboratories and allied areas. Throughout the book, emphasis is placed on the macroscopic and microscopic recognition of microbial pathogens, tests used to distinguish microorganisms and confirm their identification and, where appropriate, the chemotherapy, control and prevention of individual infectious diseases.

Specific sections deal with laboratory safety, collection and submission of specimens, essential laboratory equipment and standard procedures for the culture and identification of bacterial and fungal pathogens. The latter sections are organised so as to facilitate the recognition of the more important features of bacterial and fungal colonies and the application of biochemical, serological and other tests for the identification of pathogens from clinical specimens. Other sections deal with virological tests, selection of antimicrobial drugs and procedures for the control of infectious diseases. Zoonotic aspects of infectious diseases of animals are discussed under food poisoning and bacterial, viral and parasitic zoonoses. The final section (**Section 6**) summarises the infectious agents associated with the individual body systems of domestic animals. This is intended to present, in a concise manner, the more important clinical and diagnostic aspects of each disease. The information in **Section 6** is complementary to that in previous sections.

We have endeavoured to illustrate the diagnostic aspects of important pathogens with the emphasis on appropriate laboratory techniques. The magnifications given for the colour illustrations relate to the original 35 mm transparencies.

The authors would welcome notification of readers' comments and suggestions for improvements to future editions.

Dublin, 1994

Acknowledgements

Assembling material for a book usually requires forward planning, attention to detail and the means to put in place, both diagrammatically and typographically, the ideas of the authors. We were fortunate to have in our small academic department, staff who were willing to assist and advise us as we attempted to commit our ideas to print. Members of this department supported our efforts with energy and enthusiasm and we are indebted to them for their sustained efforts, especially in the final stages of preparation.

Veterinary microbiology lends itself to illustration and we have endeavoured to include illustrative material in the form of colour slides. A number of colleagues supplied slides which we would like to acknowledge:

Professor K.P. Baker, Faculty of Veterinary Medicine, University College Dublin, Ireland: **431**.

Centres for Disease Control, Atlanta, Georgia, USA: **363** and **374**.

Dr D.O. Cordes, College of Veterinary Medicine, Virginia Tech, Blacksburg, Virginia, USA: **194**, **477**, **478**, **479** and **485**.

Professor F.W.G. Hill, Faculty of Veterinary Science, University of Zimbabwe, Harare, Zimbabwe: **349**.

Dr L. Hoffmann, Iowa State University, U.S.A.: **361**.

Dr G. Joseph, Salinan dari, IPOH, Malaysia. **301**, **302** and **318**.

Professor J.F. Kazda, Institute for Experimental Biology and Medicine, Borstel, Germany: **185**, **186** and **187**.

Dr G.M. McCarthy, Faculty of Veterinary Medicine, University College Dublin, Dublin, Ireland: **76**.

Dr N. Seiranganathan, College of Veterinary Medicine, Virginia Tech, Blacksburg, Virginia, USA: **378**.

We also wish to acknowledge our predecessors in this department, particularly Mr B.T. Whitty and Mr M.A. Gallaher who left for posterity a wealth of stained smears, many of which we have used to illustrate individual chapters. We would also like to record our appreciation to colleagues who furnished material for photography:

Dr H.F. Bassett, Faculty of Veterinary Medicine, University College Dublin, Dublin, Ireland: **172**.

Mr M.J. Casey, Faculty of Veterinary Medicine, University College Dublin, Dublin, Ireland: **491**.

Mr R.P. Cooney, Faculty of Veterinary Medicine, University College Dublin, Dublin, Ireland: **192**.

Mr P. Costigan, Virus Reference Laboratory, University College Dublin, Dublin, Ireland: **98**, **99** and **100**.

Dr W.J.C. Donnelly, Faculty of Veterinary Medicine, University College Dublin, Dublin, Ireland: **200**, **201** and **489**.

Mr E. Fitzpatrick, Faculty of Veterinary Medicine, University College Dublin, Dublin, Ireland: **465**.

Dr H.A. Larkin, Faculty of Veterinary Medicine, University College Dublin, Dublin, Ireland: **373**.

Dr G.H.K. Lawson, Royal (Dick) School of Veterinary Studies, Edinburgh, Scotland, UK: **367**.
Dr F.C. Leonard, Teagasc, Moorepark, Fermoy, Co. Cork, Ireland: **356**.
Dr J.B. Power, Regional Veterinary Laboratory, Cork, Ireland: **357** and **359**.
Dr G.R. Scott, Department of Tropical Animal Health, Royal (Dick) School of Veterinary Studies, Easter Bush, Roslin, Midlothian, Scotland: **371** and **372**.
Professor B.J. Sheahan, Faculty of Veterinary Medicine, University College Dublin, Dublin, Ireland: **449**.

The manuscript was typed by our indefatigable departmental secretary, Lesley Doggett, who worked long hours converting to type our varied ideas. We made many demands on her time especially in the final weeks of preparation. Dores Maguire artistically presented our line drawings; Maurice Scanlon and Yvonne Abbott assisted with cultures and slides. Martin Loth offered expert photographic advice. Mary Gleeson provided assistance with printed material and with proof-reading. Dr W. Donnelly provided both scientific and editorial advice on a number of chapters. Professor C. Hatch, Dr G. Mulcahy and Mr M. Casey advised on matters parasitological. Mr B.A. McErlean cheerfully proof read the chapters in parallel with Joan, John and Michael Quinn.

The authors would like to record their appreciation to the publishers for their encouragement while the book was in preparation.

1 Safety in the Laboratory

Many pathogenic microorganisms encountered in veterinary medicine are potentially pathogenic for man. Staff in microbiology laboratories should be instructed on the careful handling and safe disposal of specimens submitted for laboratory investigation. Laboratory management should ensure that defined procedures are followed when handling dangerous pathogens. This book is intended as an illustrated text for the selection, application and interpretation of test procedures used in veterinary microbiology.

Laboratory Design

If the microbiology laboratory has been purpose built, it is more likely to incorporate those features necessary for good management and safe laboratory practices. Floors and walls should be impervious to liquids, work-tops should be easily cleaned and lighting should provide staff with adequate illumination at bench level. Safety cabinets appropriate for the classes of pathogens being handled by laboratory personnel should be installed. The design, layout, equipment and functioning of a diagnostic microbiology laboratory should reflect the services provided and the pathogens likely to be encountered. Standard microbiological practices may be carried out on open benches whereas dangerous pathogens may necessitate a biological safety cabinet for maximum containment, in combination with full-body protection. Sinks should be fitted with foot-operated or elbow-operated taps used in conjunction with liquid disinfectant dispensers to minimize the risk of cross-contamination via hands. Disposable paper towels or hot-air dryers are the preferred methods for hand drying.

Safety Procedures

All employees working with potentially pathogenic microorganisms should be instructed in proper safety procedures. All staff entering the laboratory should be aware of the hazards present and should meet the entry requirements. Protective clothing such as gowns, coats or uniforms must be worn when working in the laboratory and should not be taken into 'clean' areas such as libraries or dining areas. Access to laboratories handling dangerous pathogens should be restricted.

Eating, drinking, smoking or storage of food in laboratory areas should not be permitted where infectious material is used or stored. Mouth pipetting and the licking of labels should also be prohibited. Many alternatives to mouth pipetting are readily available (**1**). Because of the risk of disease transmission through infectious aerosols, all manipulations of potentially infectious clinical material should be carried out in open-fronted, negative pressure safety cabinets or other suitable equipment. Aerosols are particularly likely to be produced during homogenization or sonication procedures, when liquids are being pipetted, centrifuged or expelled from syringes, or when containers with liquids are accidentally overturned, dropped or broken.

When hazardous pathogens are likely to be present in clinical specimens, laboratory staff should take special precautions and wear disposable rubber gloves, a face mask, head cover and either work with the specimen shielded under glass or wear a visor or goggles. Laboratory personnel engaged in routine procedures with dangerous pathogens should, where possible, carry out all handling and culture procedures in a Class 1 biological safety cabinet (**Diagram 1**).

1

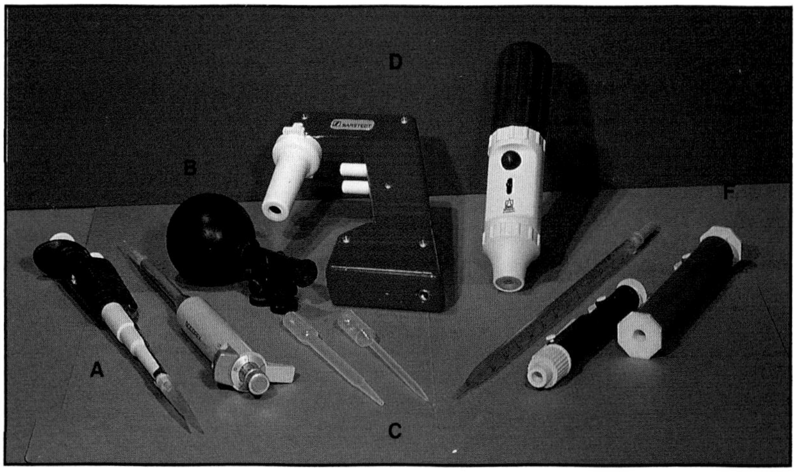

1 Micro-dispensers and macro-dispensers for liquids: from left, (**A**) two micropipettors; (**B**) filling bulb for standard pipette; (**C**) two plastic Pasteur pipettes; (**D**) rechargeable battery-operated pipettor; (**E**) manually operated pipettor; and (**F**) plastic pipette and two thumb-operated pipette fillers.

In addition to the biohazards present in a diagnostic microbiology laboratory, there are the usual safety risks arising from the use of toxic chemicals, electrical appliances, radioactive and inflammable material. Safety guidelines for the storage of inflammable material and toxic chemicals should be strictly enforced. Periodic fire drills and evacuation procedures should increase the awareness of laboratory staff to the measures appropriate for dealing with serious laboratory accidents arising from malfunctioning equipment, explosions or fires. Fire blankets and fire extinguishers should be standard equipment in each laboratory. The laboratory design should incorporate emergency exits which should not be obstructed with equipment, furniture or laboratory supplies.

Laboratory Accidents

Many of the accidents associated with diagnostic microbiology are preventable by a combination of proper training, good microbiological technique, periodic servicing of equipment, adequate supervision of laboratory staff and appropriate laboratory safety measures. Accidental penetration of the skin is a constant

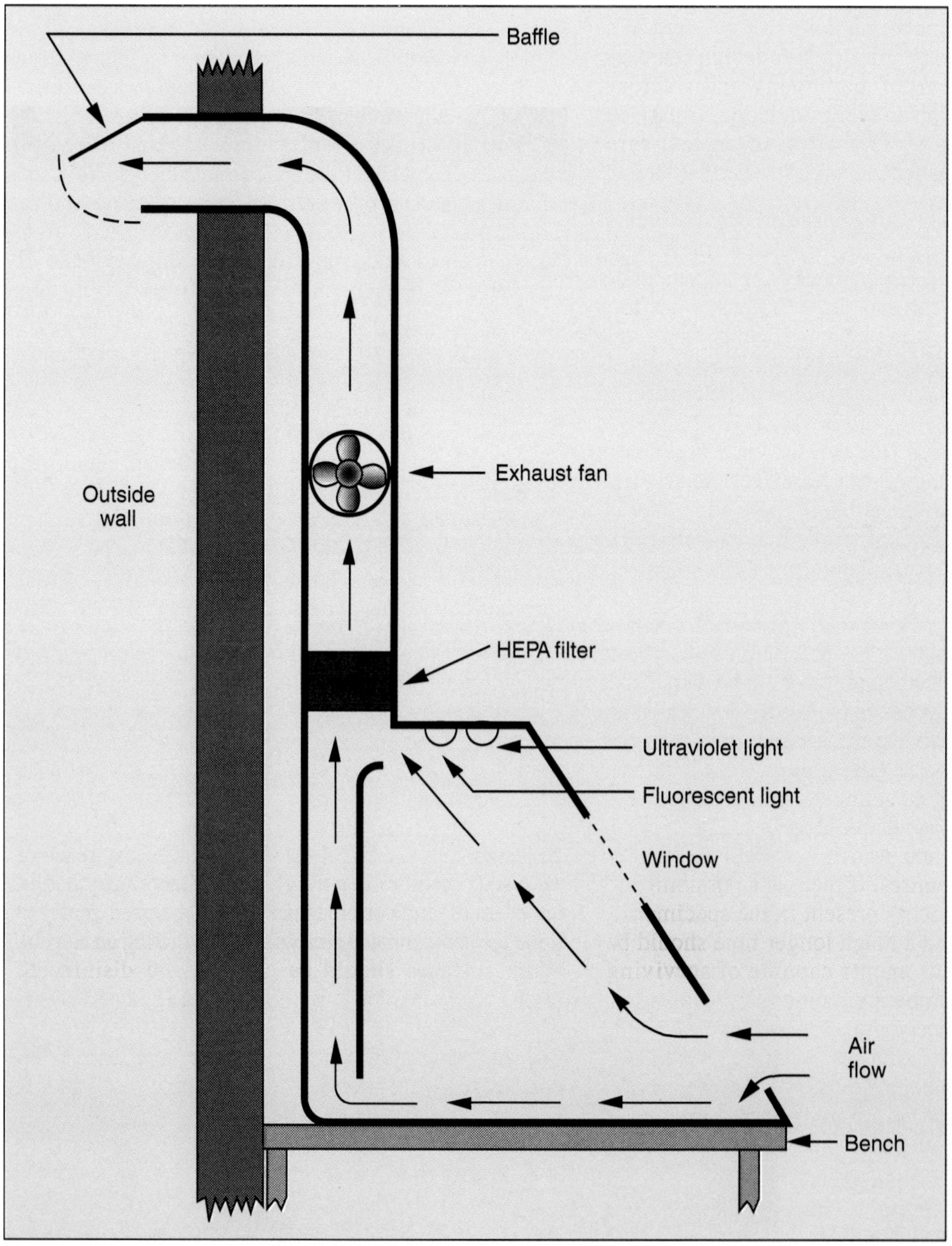

Diagram 1. Class 1 biosafety cabinet.

risk associated with laboratory breakages. Needles, blades and other sharp objects should be handled with care, disposed of in puncture-resistant containers and autoclaved promptly. Glass Pasteur pipettes should be avoided as the soft plastic varieties are much safer.

Centrifuge accidents can be avoided by proper selection of centrifuge tubes, correct balancing and adherence to appropriate acceleration and deceleration procedures. Sealed centrifuge buckets or sealed covers should be used in all procedures involving dangerous pathogens. In the absence of such safety devices, centrifugation should be carried out in a safety cabinet under negative pressure.

Facilities for laboratory animals used for diagnostic purposes should incorporate special design features ensuring containment of pathogens, satisfactory temperature, ventilation and air circulation. It may be necessary to consider vaccination of animal-care personnel against selected infectious agents transmissible to man, encountered in submitted specimens. Bite and scratch wounds should be avoided by instruction in proper animal restraint procedures and by wearing leather gloves and protective clothing when necessary.

Cleaning and support staff require special consideration in accident prevention strategies, because their awareness of the risks involved in a laboratory environment is unlikely to match that of trained laboratory personnel.

All laboratory accidents should be reported promptly to the laboratory supervisor. First Aid equipment should be at hand to deal with minor accidents. Details of the accident should be recorded and a permanent record should be retained for reference.

Decontamination

Fundamental to the safe operation of a diagnostic microbiology laboratory is the capability of effectively decontaminating potentially infectious material received by the laboratory or waste generated in the laboratory. Contaminated material including diagnostic specimens, inoculated media, viable cultures, glassware and surgical instruments must be autoclaved at 121°C for 20 minutes before leaving the premises. If there is a possibility of heat-resistant agents being present in the specimen, a higher temperature and a much longer time should be considered. Infectious agents capable of surviving standard autoclave temperatures include the agent of bovine spongiform encephalopathy (BSE).

Destruction of microorganisms by burning is practised routinely when metal loops are heated in the flame of a Bunsen burner. When loops carrying dangerous pathogens are being flamed, there may be a danger of spattering with resultant spreading of living organisms. To prevent this from occurring, a Bunsen fitted with a tube in which the flaming takes place should be used for agents such as *Mycobacterium bovis*. Alternatively, disposable plastic loops can be used which are disposed of in a disinfectant known to be active against the particular infectious agents (**2**).

Although autoclaving is the preferred method for inactivating infectious agents, ultra-violet light is commonly used in safety cabinets and chemical disinfection is extensively used for initial treatment of contaminated glassware on the work bench before autoclaving.

Disinfectant jars should be changed at frequent intervals and the date of changing the disinfectant should be marked on each jar. They should contain disinfectant at the correct concentration active against **all** the infectious agents likely to be encountered and contaminated items should be placed in the disinfectant solution without splashing. Chemical disinfectants suitable for use in disinfectant jars include sodium hypochlorite, iodophors, glutaraldehyde and some phenolic compounds. The corrosive nature of hypochlorites should be remembered if surgical instruments or metal items are being decontaminated in this manner. Washbottles filled with 70 per cent ethyl alcohol should be within easy reach of each staff member in a microbiology laboratory and in the event of spills or breakages should be used gently to flood the contaminated area without generating an aerosol. Work surfaces should be cleaned and disinfected frequently with an effective non-irritant disinfectant (*see* Chapter 50).

Heat-resistant autoclave bags are ideal for sealed containers and Petri dishes. Sharp objects such as glass Pasteur pipettes, needles and blades and broken glass should not be placed directly into these heat-resistant bags as they may puncture them with resultant release of contaminated material. Periodic checking of the efficiency of autoclaves should be carried out using commercially available non-pathogenic bacterial spores.

2

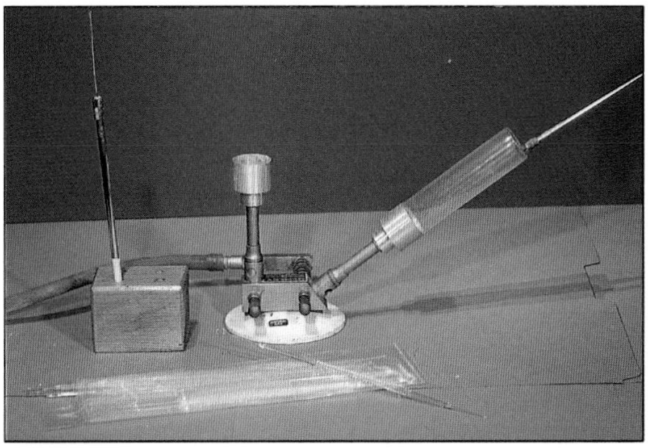

2 Safety Bunsen burner and plastic disposable inoculating loops.

Protective clothing worn by laboratory staff should be handled and laundered separately from other clothing. Laboratory coats and clothing originating in microbiology laboratories should be routinely washed, preferably on the premises, at 85°C, without a pre-wash. Clothing from laboratories dealing with dangerous pathogens should be autoclaved before routine laundering.

The authors and publishers accept no responsibility for the safety of laboratory personnel carrying out procedures described in this book. Laboratory management should ensure strict adherence to safety guidelines appropriate for the pathogens being dealt with in the laboratory. Clear labelling of containers, mandatory protective clothing, availability of safety cabinets and rigorous decontamination procedures should ensure the safety of laboratory staff and containment of infectious agents.

2 Collection and Submission of Diagnostic Specimens

The procedures used to identify infectious agents vary widely. Before a detailed laboratory investigation is undertaken, a complete case history, including a tentative diagnosis, should be submitted to the laboratory with the specimens. This will help laboratory staff to decide on the range of possible agents and so select the most appropriate tests and procedures that should identify the pathogen(s). This chapter details points of importance when collecting and submitting specimens for bacteriology, mycology or virology.

General Guidelines

- Specimens should be taken from living or **recently** dead animals.
- Samples should be taken from the affected site(s) as **early** as possible following the onset of clinical signs. This is particularly important in viral diseases as shedding of virus is usually maximal early in the infection.
- It is useful to collect samples from clinical cases and in-contact animals. In-contact animals may be at an earlier stage in the infection with a greater chance of them shedding substantial numbers of microorganisms.
- Samples should be obtained from the **edge** of lesions and include some macroscopically normal tissue. Microbial replication will be most active at the lesion's edge.
- It is important to collect specimens as aseptically as possible, otherwise the relevant pathogen may be overgrown by the numerous contaminating bacteria (**3**).
- Specimens should always be collected before the administration of any form of treatment. Samples taken from animals recently treated with antibiotics are of little value for the isolation of bacteria.
- Material sent on swabs is liable to desiccation and the relevant pathogen may not be viable on arrival at the laboratory. When possible a more generous amount of sample should be taken and submitted, such as blocks of tissue (approximately 4 cm^3), biopsy material, or several millilitres of pus, exudate or faeces.
- Specimens should be submitted that are relevant to the problem under investigation but, if appropriate, a wide range of specimens should be submitted to allow the diagnostician some flexibility in deciding the most suitable test(s) to use under the circumstances.
- Samples must be submitted individually in separate water-tight containers. Screw-capped jars that are clearly marked indicating the tissue enclosed, animal identification and the date of collection are preferable.
- If transportation to the laboratory is delayed, most samples should be refrigerated at 4°C and not frozen.
- To obtain a complete diagnosis, specimens for several disciplines should be selected, such as those for microbiology, pathology, serology, haematology and clinical biochemistry.

3

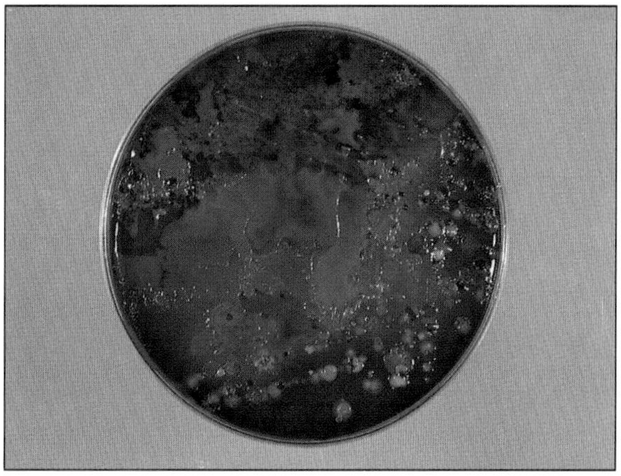

3 Grossly contaminated blood agar plate demonstrating the need to collect specimens as carefully as possible.

Specimens for Bacteriology and Mycology

This chapter deals with the selection of specimens from general pathological conditions. For specific diseases *see* 'specimens' under the relevant microorganism in Section 2 or 3 (Bacteria or Fungi) or in the Tables in Section 6.

Abortion cases

- A whole foetus should be submitted if possible. If not, foetal abomasal contents (ruminants), lung, liver and a sample of any gross lesions in or on the foetus should be sent.
- A piece of obviously affected placenta and two or more cotyledons from cattle and sheep.
- Uterine discharge (especially if no placenta is available).
- If leptospiral abortion is a possibility, 20 ml of mid-stream urine from the dam preserved with 1.5 ml of 10 per cent formalin should be submitted.
- Serum from the dam for serological tests. Acute and convalescent serum samples should be considered for endemic diseases.
- Placenta (cotyledons), foetal lesions, liver and lung in 10 per cent formalin for histopathology.

Bovine mastitic milk samples

Milk samples should be collected from cows as soon as possible after the mastitis is first noticed and not from animals treated with either intramammary or systemic antibiotics. The udder should not be rinsed with water unless very dirty. If the udder and teats are washed, they should be dried thoroughly with a paper towel. Usually it is sufficient to wipe the teats vigorously, using 70 per cent ethyl alcohol on cotton wool, paying special attention to the teat sphincters. The two teats furthest from the operator are wiped first and then the two nearest teats. The sterile narrow-necked collection bottle must be held almost **horizontally**. The first squirt of milk from each teat is discarded and then, for a composite sample, a little milk from each quarter is directed into the bottle. The milk should be collected from the two near teats first and then from the two far teats, so that an arm is less likely to brush accidentally against a cleaned teat.

Abscesses

If possible about 3 ml of pus should be collected, together with scrapings from the wall of the abscess. Pus at the centre of an abscess is often sterile. Pus from recently formed abscesses will yield the best cultural results.

Specimens for anaerobic culture

A good collection method is essential, because many anaerobes do not survive frank exposure to the oxygen in the air for more than 20 minutes. It is important not to contaminate the samples by contact with adjacent mucosal surfaces as these have a resident anaerobic flora. Specimens from animals that have been dead for more than 4 hours are usually unsuitable because of the rapid postmortem invasion of the animal body by anaerobes from the intestinal tract. Bone marrow is a good specimen to collect for the diagnosis of blackleg or malignant oedema as bone marrow appears to be one of the last tissues to be invaded by contaminating bacteria. A piece of rib stripped of the periosteum could be submitted to the laboratory for the extraction of bone marrow. Any specimens for the attempted isolation of anaerobes must arrive at the laboratory as soon as possible after collection. Collection of samples for anaerobic culture on ordinary swabs is usually of no value. Acceptable samples include blocks of tissue (4 cm^3) placed in a sterile closed container, tissues contained in commercial 'anaerobe specimen collectors', and, for liquid exudates, the sample can be collected in a disposable syringe. Air is expelled from the syringe and the needle bent back on itself or plugged.

In suspected enterotoxaemia cases, where the demonstration of a specific toxin is required, at least 20 ml of ileal contents should be submitted. A loop of ileum with contents, tied off at each end, is acceptable or the ileal contents may be drained into a screw-capped bottle.

Urine samples

Urine samples may be submitted for urinalysis, bacterial microscopy and culture or for a viable bacterial count to establish whether a clinical bacteriuria is present. For bacteriological procedures the preferred methods of collection are by cystocentesis, by catheter or mid-stream urine sample.

There are few reports of quantitative analyses for bacteriuria in dogs but the figures for humans, that could act as a guide, are:

- Clinical bacteriuria is indicated by the presence of 10^5 bacteria/ml urine
- Between 10^4 and 10^5 bacteria/ml is suggestive of infection
- Less than 10^4 bacteria/ml are not considered significant.

Samples from skin lesions

If intact pustules or vesicles are present, the surface should be disinfected with 70 per cent ethyl alcohol, allowed to dry, and material aspirated from the lesion with a sterile syringe and fine needle.

In cases where ringworm is suspected, hair should be plucked from the lesion and the **edge** scraped with a blunt scalpel blade until blood begins to ooze. Plucked hair, skin scrapings (including the scalpel blade itself) and any scab material that is present should be submitted. These specimens will also allow detection of mange or a bacterial infection, if present.

Blood cultures

These are used if a bacteraemia is suspected. Strict aseptic precautions should be taken when collecting the blood. The area over the site of venepuncture must be shaved, cleaned thoroughly with a detergent, dried and 70 per cent ethyl alcohol applied to the skin and allowed to act for at least 30 seconds. As a bacteraemia can be intermittent, it is advisable to take more than one sample within a 24-hour period. The blood should be added aseptically and without delay to one of the special commercial blood-culture bottles and then sent to the laboratory.

Samples containing bacteria that require transport media

Streptococcus spp. are particularly susceptible to desiccation especially if collected on a dry swab. Commercial swabs supplied with transport medium are satisfactory.

For the isolation of *Moraxella bovis*, ideally lacrimal secretions should be plated on blood agar immediately after collection. If this is not possible, the swab should be placed in commercial transport medium and delivered to a laboratory within 2 hours of collection.

Swabs and discharges submitted for the isolation of *Taylorella equigenitalis* should be placed in Amies transport medium containing **charcoal**. Cervical or vaginal mucus from infertile cows for the isolation of *Campylobacter fetus* subsp. *venerealis* and samples that might contain *Chlamydia psittaci* or *Mycoplasma* species should be placed in special transport media. The diagnostic laboratory should be consulted before collecting samples for the isolation of these pathogens. Formulae for transport media are given in **Appendix 2**.

Specimens for Viral Diagnosis

Samples for viral isolation such as faeces, skin scrapings, body fluids, tissue and blood with anticoagulant should be placed in bijoux bottles with viral transport medium and transported to the laboratory as quickly as possible. A formula for a suitable viral transport medium is given in **Appendix 2**.

Samples suitable for direct electron microscopy include faeces, pathological material, biopsies, skin scrapings and vesicular fluid. The latter should be smeared onto a clean glass slide, dried and stored at room temperature. Only specimens containing large concentrations of viral particles are suitable for direct electron microscopy (EM).

The fluorescent antibody technique (FAT) can be used with:

a) Cell smears fixed in acetone
b) Cryostat sections of fresh tissue.

The FAT is frequently applied for the rapid diagnosis of viral respiratory diseases. A nasopharyngeal aspirate or fluid obtained by broncho-alveolar lavage is the specimen of choice. Alternatively an extra long, absorbent, guarded swab may be used. The usual short cotton wool swabs are generally unsatisfactory for obtaining a suitable nasopharyngeal specimen of epithelial cells and mucus.

Histopathological examination can be carried out on tissues preserved in fixative. Sections may be chemically stained and examined for cellular pathology consistent with viral infection such as inclusion bodies. Alternatively, immunoperoxidase staining may be used to demonstrate the presence of specific viral antigens.

Serology may be useful to confirm a viral diagnosis. Paired serum samples are required if a disease is endemic in an area. The samples should be taken during the acute and convalescent stages of the disease, approximately 3–4 weeks apart.

Interpretation of Diagnostic Results

The following points are pertinent when interpreting reports from a diagnostic microbiology laboratory:

- A negative diagnostic report does not necessarily mean that the suspected microorganism is not the aetiological agent of the condition. There may be many reasons for the failure of the laboratory to isolate and identify the pathogen, such as a bacterium being overgrown by contaminants, a virus or other fragile microorganism having died on the way to the laboratory, or the animal may have stopped excreting the microorganisms before the sample was taken.
- Many bacteria, such as members of the *Enterobacteriaceae* are ubiquitous. Isolation of microorganisms may represent contamination of the sample by faeces or soil or their presence could be due to postmortem invasion.
- Apparently healthy animals can be subclinical shedders of microorganisms such as salmonellae, rotaviruses, enteroviruses or coronaviruses in faeces and leptospires in urine. Even if these potential pathogens are isolated and identified, the illness or death might have been due to another cause.
- Some bacteria such as salmonellae, leptospires or *Mycobacterium paratuberculosis* may be shed intermittently. A repeat sample following a negative examination report might be worthwhile.
- *Escherichia coli* in diarrhoeal faecal samples from farm animals is usually significant only if the animal is under 10 days of age and if the *E. coli* isolate possesses the K88, K99, F41 or 987P fimbrial antigens, indicating enteropathogenicity. Pigs are the exception as they are also susceptible to colibacillosis soon after weaning, and to oedema disease in the growing period.
- Diagnosis of a mycotic disease thought to be due to a ubiquitous fungus such as *Aspergillus fumigatus* should always be confirmed histopathologically. It is necessary to demonstrate the fungal hyphae actually invading the tissue.
- Two serum samples collected 2–3 weeks apart are required for serology in the case of some endemic diseases. If a four-fold increase in antibody titre in the second serum sample can be demonstrated, this indicates a recent and active infection. In the case of an exotic disease, circulating antibodies are usually an indication of exposure to that infectious agent. However, low antibody titres could be due to cross-reactivity with otherwise unrelated agents and may confuse the interpretation.

Herd or flock vaccination programmes will modify the interpretation of serological tests. For some diseases, such as brucellosis, a vaccine can be used to elicit an anamnestic response where antibody levels of an individual infected animal have dropped to a low level.

3 Essential Equipment and Reagents for a Veterinary Diagnostic Microbiology Laboratory

Suitable Laboratory Room

A separate room for microbiology is the ideal with good lighting, a washable floor and limited traffic. The fittings should be as follows:

- Benches with smooth, impervious surfaces that can be disinfected.
- A sink with foot-operated or elbow-operated taps, a liquid disinfectant dispenser and disposable paper towels.
- A gas and electricity supply.
- Easy access to an incubator and a refrigerator.

Requirements For Specimen Collection

The following items are required:

- Postmortem knife, scalpels, forceps, scissors, and a small hacksaw for bone specimens.
- Sterile, screw-capped, disposable containers: 30 ml and 100–200 ml capacity (**4**).
- Sterile plain swabs (individually wrapped), swabs with transport medium (general) and swabs in Amies transport medium with charcoal (contagious equine metritis).
- Plastic bags and waterproof marker pens.
- Wooden tongue depressors: useful for scraping mucosal surfaces and for collecting other samples such as faecal specimens. They can be sterilized by autoclaving.
- A polystyrene container, coolant pads and unbreakable containers for transport of samples (**5**).

Requirements for Staining Techniques

- A Bunsen burner and a gas supply (mains or bottled gas).
- Glass microscope slides (approx. 76 × 25 mm), preferably with frosted ends, for easy marking with a pencil, and coverslips (22 × 22mm).
- A container with 70 per cent ethyl alcohol to hold sterile forceps, scalpels and scissors. The instruments should not be stored in the alcohol for long periods or they will rust.
- A discard jar with disinfectant for used instruments and for discarding used stained smears.
- Inoculating loop holders and nichrome inoculating loops and straight wire (24 gauge).
- Sink and staining rack.
- Gram-stain kit, concentrated carbol fuchsin solution and other chemical reagents and dye powders *(see* 'preparation of staining solutions': **Appendix 1**).

4

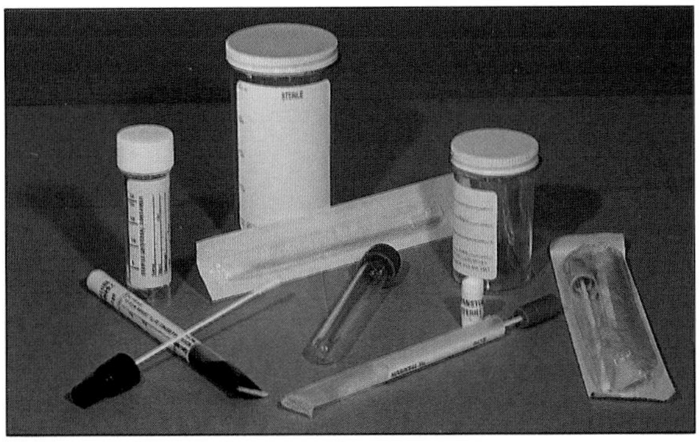

4 Sterile disposable containers and swabs for specimen collection.

5

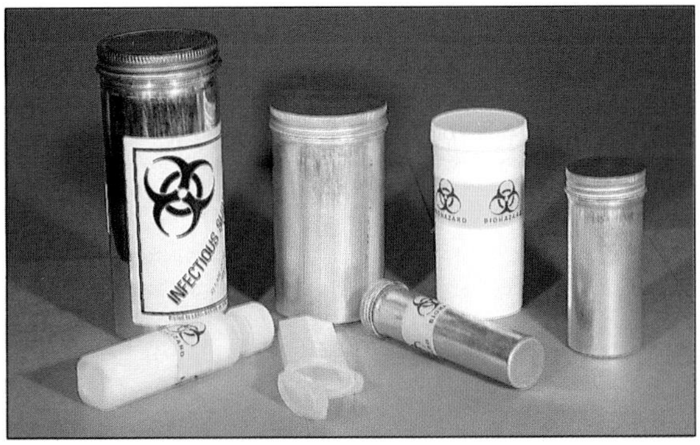

5 Unbreakable containers for transporting microbiological specimens.

- Graduated cylinders (1 litre and 100 ml), glass funnel and filter paper
- A good quality microscope with a built-in light source, low-power, high-dry and oil-immersion objectives. Microscopes for microbiology require a higher deree of resolution than those used for haematology or histopathology. **Diagram 2** indicates the size of bacteria compared to that of an erythrocyte. Viruses can be visualised only by using electron microscopy, but they are included in this diagram to demonstrate the relative sizes of bacteria and viruses. Before purchasing a microscope, a Gram-stained smear of pus from an abscess could be used to determine the optical qualities of this costly but vital piece of equipment.
- A supply of immersion oil.

Requirements for the Isolation and Identification of Bacterial Pathogens

- Refrigerator for storing media, reagents and specimens.
- Small incubator to run at 37°C ± 1°C and a maximum–minimum thermometer.
- Small autoclave, with pressure gauge and safety valve, which can be set at a temperature of 115°C or 121°C.
- Overhead pan balance to weigh in grams to at least the first decimal place.
- Water bath to run at 50–54°C and a thermometer to check the temperature.
- Hot plate and magnetic stirrer for heating media or alternatively a tripod and metal gauze for use with the Bunsen burner. A thick glass rod for stirring.
- Sterile plastic Petri dishes (90 mm) of the triple-vent type.

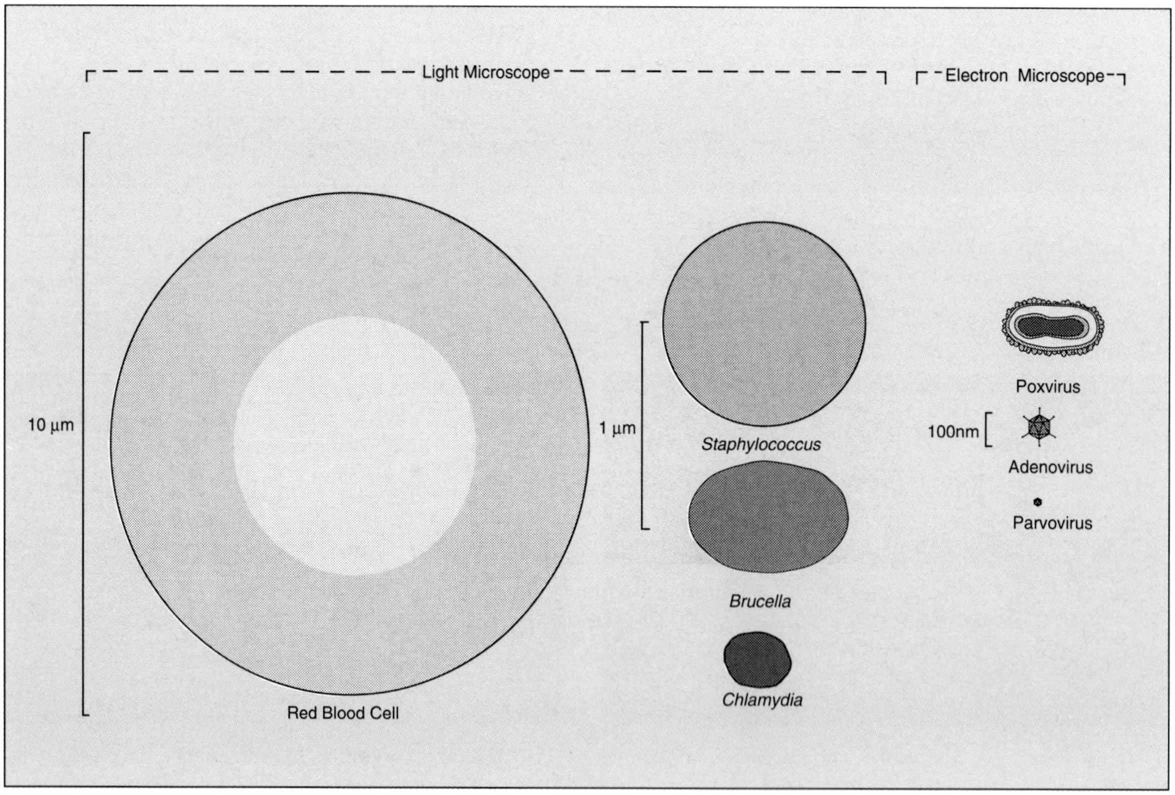

Diagram 2. Comparison of the relative sizes of a red blood cell, bacteria and viruses.

- Container with 70 per cent ethyl alcohol to store sterile scalpels, forceps (plain and rat-toothed) and scissors.
- Discard-container with disinfectant for used instruments.
- Inoculating loop holders, nichrome inoculating loops and straight wire (24 gauge).
- Sterile plain swabs (individually wrapped).
- Waterproof marker pens.
- Sterile disposable plastic Pasteur pipettes (ungraduated, multipurpose).
- Cotton-wool: absorbent for wiping down benches and non-absorbent for plugging the mouths of culture vessels.
- Screw-capped, glass, autoclavable bottles: universals (30 ml) and bijoux (5 ml).
- Pyrex test tubes (10 ml) and loose metal caps.
- Racks for bottles and test tubes.
- Flask, tubes and glass-beads (3 mm) for the collection of sterile defibrinated blood for blood agar plates.
- Dehydrated culture media (see **Table 3**).
- Pyrex flasks, with flat bottoms, for media preparation: 500 ml, 1 and 2 litre volumes. The size of flask should be selected to fit comfortably into the autoclave.
- Graduated cylinders: 100 ml and 1 litre.
- Anaerobic and CO_2 jars and envelopes that deliver the appropriate gas mixtures. A candle-jar is adequate for some capnophilic (carbon dioxide-loving) bacteria (**6**).
- Plastic squeeze bottle with 70 per cent ethyl alcohol for spillages.
- Disposal bins and plastic sacks for non-contaminated materials.
- API identification strips and other reagents for biochemical tests.

6

6 Apparatus used for culturing microorganisms under different atmospheric conditions: anaerobic and CO_2 jars with gas-generating envelopes and a candle-jar which generates approximately 2.5 per cent CO_2.

The Disposal of Culture Plates and Pathological Materials

An important consideration, when setting up a microbiology laboratory, is the safe disposal of culture plates and pathological specimens that may pose a human health hazard. Methods of destroying the pathogens present in the materials include:

- Placing the materials in an autoclavable bag, followed by autoclaving
- Enclosing the materials in a strong leak-proof plastic bag for incineration
- The least satisfactory option is to soak the culture plates and other pathological material in a suitable disinfectant for at least 48 hours before the contents can be safely discarded.

Materials containing pathogenic microorganisms should not leave the laboratory until they have been sterilised.

4 Bacterial Pathogens: Microscopy, Culture and Identification

Stained Smears from Pathological Specimens

Stained smears made from lesions can yield a considerable amount of information inexpensively and quickly. In a small laboratory, staining techniques can be used alone or in conjunction with cultural methods. However, if only stained smears are used and the pathogen is recognised as one that could be resistant to antimicrobial agents, the material should be sent to a diagnostic laboratory for culture and antibiotic susceptibility testing.

Table 1 summarises the information that can be gained from the various diagnostic staining techniques.

Preparing bacterial smears

Microscope slides are not always clean enough to use directly from the supplier. Rubbing with a clean, soft cloth and a flick through the Bunsen flame may be sufficient to remove a greasy film. If not, a mildly abrasive liquid cleaner can be used followed by rinsing the slide thoroughly and wiping it dry with a clean cloth.

A blunt scalpel and forceps should be kept in a container of 70 per cent ethyl alcohol. The instruments are flamed and cooled before use. Afterwards they should be placed into a container of disinfectant. When making a smear from tissue lesions, the specimen is held firmly with the forceps and the scalpel is used to scrape deep into the material. A small amount of the scrapings is placed on the cleaned microscope slide. Another clean slide is used with a scissor action to prepare a thin smear. With liquid or semi-liquid specimens, a little of the sample is placed on the slide with a sterile swab. The contents of the swab are smeared over the surface of the slide, with the aim of having thick and thin areas of specimen present. The smears are allowed to dry thoroughly before proceeding further.

Fixing the smears

The reasons for fixing the smears include killing the vegetative bacteria, rendering them permeable to the stain and ensuring that the material is firmly fixed to the slide. Fixed and stained smears should be handled carefully as not all bacteria, especially endospores, may have been killed. After use, the stained smears should be autoclaved or soaked in a reliable disinfectant (24–48 hours) before discarding.

For routine staining the smears are fixed by passing the slide, smear side up, quickly through the Bunsen flame two or three times, taking care not to overheat the smear. This can be tested on the back of the hand; the slide should feel warm but not hot enough to burn.

Dried smears to be stained by the Giemsa stain are first fixed in absolute methyl alcohol (Analar Grade) for 3 minutes and then dried.

Staining the smears: staining techniques

The fixed smears are placed on a staining rack over a sink. The staining solutions are flooded over the entire smear and left on the slide for the appropriate time. Between each staining reagent the smear is washed under a gently running tap, excess water tipped off and the next reagent added. Finally the stained smear is washed and air dried. The preparation method for each of the staining solutions is given in **Appendix 1**.

GRAM STAIN

Crystal violet 60 seconds
Gram's iodine (mordant) 60 seconds
Gram's decolouriser 15 seconds
Counter-stain (dilute carbol fuchsin or safronin) 60 seconds

Gram-positive bacteria retain the crystal violet-iodine complex and stain purple-blue. Gram-negative bacteria are decolourised and are stained red by the counter-stain. Gram-stained smears are illustrated (**7–11**, inclusive).

There can be slight differences in the composition of the reagents for the Gram stain. If using a commercial kit-set for Gram staining always follow the manufacturer's directions.

Table 1. Diagnostic uses of stained smears.

GRAM STAIN

Disease and species affected	*Specimen*	*Pathogens*	*Appearance in stained smears*
Abscesses or suppurative conditions Many animal species	Pus or exudate	*Staphylococcus* spp.	Gram + cocci, often in clumps
		Streptococcus spp.	Gram + cocci, usually in chains
		Actinomyces pyogenes	Gram + rods, pleomorphic
		Corynebacterium pseudotuberculosis	Gram + rods
		Pseudomonas aeruginosa	Gram - rods
		Pasteurella multocida	Gram - rods
		Fusobacterium necrophorum	Gram -, long, slender filaments, often staining irregularly
Strangles Horses	Pus	*Streptococcus equi* ssp. *equi*	Gram + cocci, often in chains
Suppurative broncho-pneumonia or superficial abscesses Foals and young horses	Pus	*Rhodococcus equi*	Gram + rods with tendency to coccal forms
Dermatophilosis (Streptothricosis) Many animals, mainly in sheep, cattle and horses	Scabs	*Dermatophilus congolensis*	Gram +, filamentous and branching with coccal zoospores arranged 2 or more across
Canine nocardiosis and **canine actinomycosis**	Pus or thoracic aspirates	*Nocardia asteroides* *Actinomyces viscosus*	Both Gram +, filamentous and branching, *Nocardia* spp. MZN +
Bovine actinobacillosis (wooden tongue)	Granules from pus	*Actinobacillus lignieresii*	Gram - rods
Bovine actinomycosis (lumpy jaw)		*Actinomyces bovis*	Gram +, filamentous and branching
Clostridial enterotoxaemia Sheep and calves mainly	Scrapings from small intestine of **recently** dead animal	*Clostridium perfringens*	Gram +, fat rods. Large numbers suggestive of enterotoxaemia
Urinary tract infections Dogs and cats	Fresh, carefully collected urine	*Escherichia coli, Proteus* spp. *Staphylococcus* spp. and *Enterococcus faecalis*	Gram - rods Gram + cocci Gram + cocci

Table 1. Diagnostic uses of stained smears (continued).

DILUTE CARBOL FUCHSIN (simple) STAIN

Disease and species affected	*Specimen*	*Pathogen*	*Appearance in stained smears*
Campylobacter infections Infertility in cattle Abortion in sheep and cattle	 Vaginal mucus Foetal stomach contents	*Campylobacter fetus*	Curved rods that can be in chains giving 'seagull' forms
Swine dysentery	Faeces or scrapings from colon	*Serpulina hyodysenteriae*	Numerous long (7.0 μm) but finely spiralled bacteria
Foot rot in sheep	Exudate from hoof	*Bacteroides nodosus* *Fusobacterium necrophorum*	Rods with a knob on one or both ends Long, slender filaments, staining irregularly

MODIFIED ZIEHL–NEELSEN (MZN) STAIN

Brucellosis in cattle, sheep, pigs and dogs	Foetal stomach contents, vaginal discharge, placenta	*Brucella* spp.	Small, red coccobacilli in clumps
Nocardiosis Canine nocardiosis Bovine mastitis	Pus and aspirates Sediment from centrifuged milk	*Nocardia asteroides*	Long-branching filaments, many staining bright red
Chlamydial infections Sheep and cattle: abortion Lambs, calves and pigs: polyarthritis Cats: feline pneumonitis	 Cotyledons Joint fluid Conjunctival scrapings	*Chlamydia psittaci*	Small, red coccobacilli in clumps. Similar to brucellae

(*continued*)

Table 1. Diagnostic uses of stained smears (continued).

ZIEHL-NEELSEN (ACID-FAST) STAIN

Disease and species affected	*Specimen*	*Pathogen*	*Appearance in stained smears*
Tuberculosis Cattle and other species	Suspect lesions	*Mycobacterium bovis*	Long, thin, bright red rods, can appear beaded. Usually not numerous in smears.
Poultry, other avian species and cervical lymph nodes of pigs	Suspect lesions	*Mycobacterium avium*	As above but larger numbers of acid-fast rods usually present
Feline leprosy	Scrapings from lesions	*Mycobacterium lepraemurium*	Large numbers of red-staining, acid-fast rods present
Paratuberculosis (Johne's disease) in cattle and sheep	Faeces, smear from ileocaecal valve area and mesenteric lymph nodes	*Mycobacterium paratuberculosis*	Fairly short, red, acid-fast rods in clumps.

GIEMSA STAIN

Disease and species affected	*Specimen*	*Pathogen*	*Appearance in stained smears*
Anthrax in cattle, sheep and pigs	Blood smear from ear vein or fluid from peritoneal cavity (pigs)	*Bacillus anthracis*	Purplish, square-ended rods in short chains surrounded by a reddish-mauve capsule
Feline infectious anaemia	Thin blood smear	*Haemobartonella felis*	Small, dark-blue coccal forms on red blood cells
Avian spirochaetosis	Blood smear taken during febrile period	*Borrelia anserina*	Helical bacteria, 8–20μm long and 0.2–0.5μm wide with 5–8 spirals
Dermatophilosis (Streptothricosis) Many animals, mainly sheep, cattle and horses	Scabs	*Dermatophilus congolensis*	Blue, filamentous and branching with coccal zoospores arranged 2 or more across

POLYCHROME METHYLENE BLUE STAIN

Disease and species affected	*Specimen*	*Pathogen*	*Appearance in stained smears*
Anthrax in cattle, sheep and pigs	Thin blood smear from ear or tail vein. Fluid from peritoneal cavity (pigs)	*Bacillus anthracis*	Blue, square-ended rods in short chains surrounded by a pinkish-red capsule

Table 1. Diagnostic uses of stained smears (continued).

WET-PREPARATIONS

Disease and species affected	*Specimen*	*Pathogen*	*Appearance in stained smears*
Leptospirosis in many animal species	Centrifuged deposit of urine or kidney tissue	*Leptospira interrogans* serovars	Helical, approx.15 µm long and 0.15µm wide. Appears beaded with hooked ends. Dark field microscopy
Ringworm in many animal species	Hair and skin scrapings in 20 per cent KOH	*Microsporum* and *Trichophyton* spp.	Chains of refractile, round arthrospores on hairs. High-dry objective
Other suspected fungal infections in many animal species	Tissue, exudates, biopsies in 20 per cent KOH	*Aspergillus fumigatus, Candida albicans* and others	Fungal mycelial elements or budding yeast cells. High-dry objective
Bovine trichomoniasis	Purulent uterine discharge in saline. Keep warm and examine within 1 hour of collection	*Trichomonas foetus*	Protozoan agent with 3 free flagella and an undulating membrane. Low/high-dry objectives

DILUTE CARBOL FUCHSIN (DCF):
A simple staining technique

Dilute carbol fuchsin 4 minutes
Wash and air dry.
The stain is used for some Gram-negative bacteria such as *Campylobacter fetus, Serpulina hyodysenteriae* and *Fusobacterium necrophorum* (**12**) where a greater depth of stain aids microscopic visualisation.

MODIFIED ZIEHL–NEELSEN (MZN)
Dilute carbol fuchsin 15 minutes
Acetic acid (0.5 per cent) 15 seconds
Methylene blue 2 minutes
Wash and air dry.

MZN-positive bacteria such as *Nocardia asteroides* (**13**), *Brucella* spp. (**14**) and *Chlamydia psittaci* stain bright red with the background and other bacteria staining blue.

ZIEHL-NEELSEN (ZN) or ACID-FAST STAIN
Strong carbol fuchsin 10 minutes with heat
Acid-alcohol decolouriser 15 minutes with several changes
Methylene blue 20 seconds
Wash and air dry.

ZN-positive or acid-fast bacteria, such as the pathogenic *Mycobacterium* sp., stain bright red with the background and other bacteria counter-stained blue (**15**). Heating the strong carbol fuchsin can be carried out in one of two ways:

- Strong carbol fuchsin is flooded onto the fixed smear with the slide on the rack over a sink. A cotton wool swab, on a metal rod, is dipped in alcohol and set alight. This is used gently to heat the smear and carbol fuchsin from below. The stain is allowed to steam for the 10 minutes but not to boil. The sink should be rinsed with water before starting the heating process in case any inflammable reagents are present.
- Heat the strong carbol fuchsin in a boiling tube to just below boiling point. Wear protective goggles when carrying out this procedure and direct the tube away from you. Add the hot stain to the smear on the staining rack over the sink. Keep topping-up the smear with hot stain for the full 10 minutes.

7

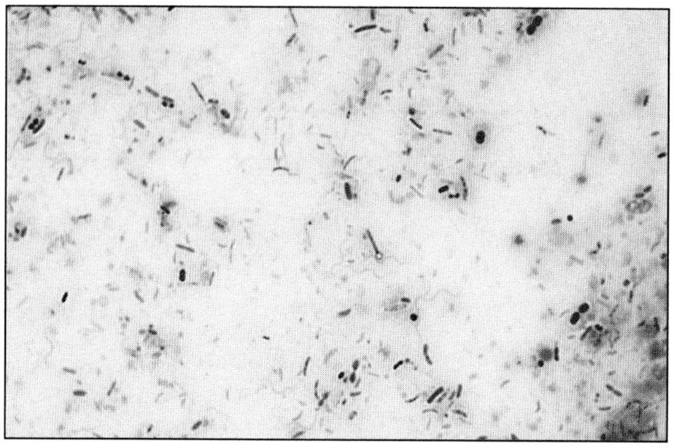

7 Gram-stained porcine faecal smear showing Gram-positive (blue) and Gram-negative (red) bacteria. Note range of morphological forms. (×1000)

8

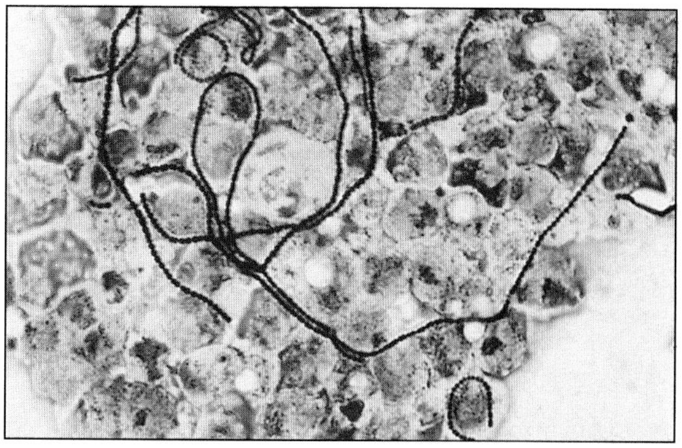

8 Gram-stained smear of mastitic milk (bovine) showing Gram-positive streptococci in chains and inflammatory cells. (×1000)

9

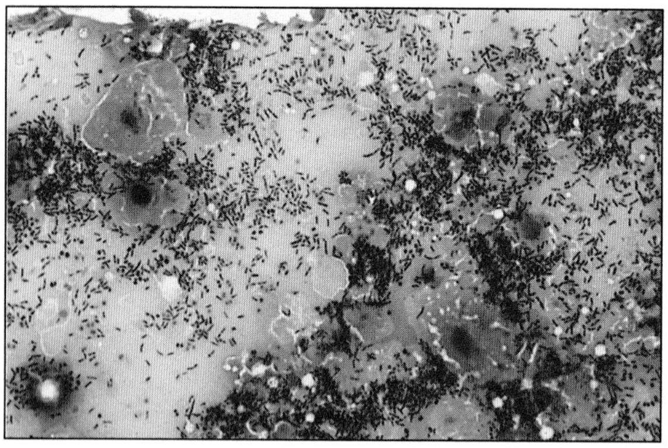

9 Gram-stained smear from an abscess with the Gram-positive, pleomorphic rods of *Actinomyces pyogenes* predominating. (×1000)

10

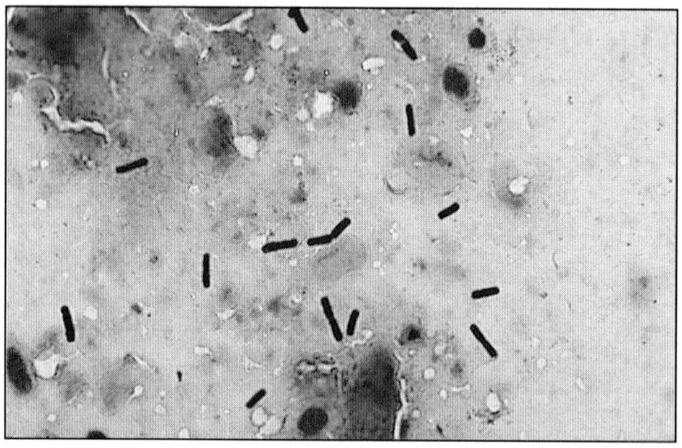

10 Gram-stained smear from the mucosa of the small intestine (lamb recently dead from enterotoxaemia): large Gram-positive rods of *Clostridium perfringens*. (×1000)

11

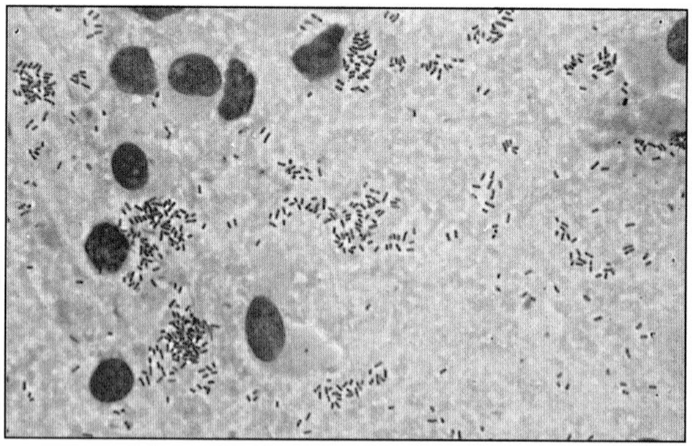

11 Gram-stained smear of urine (dog with cystitis): Gram-negative rods of *Escherichia coli*. (×1000)

12

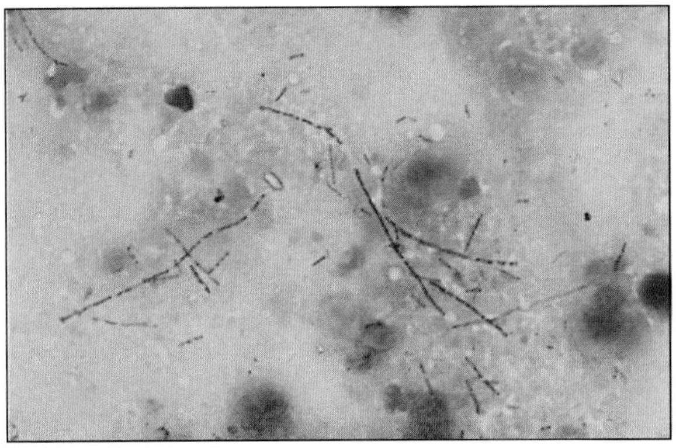

12 DCF-stained smear from a bovine abscess: red filaments of *Fusobacterium necrophorum* showing typical irregular staining. (×1000)

13

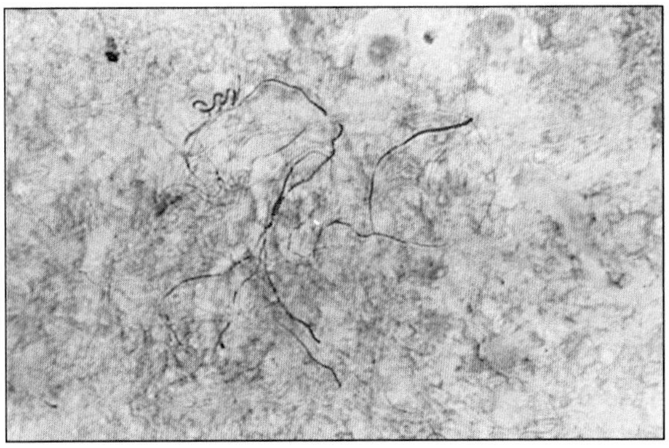

13 MZN-stained smear of a thoracic aspirate from a dog with a pleural effusion: MZN-positive branching filaments of *Nocardia asteroides*. (×1000)

14

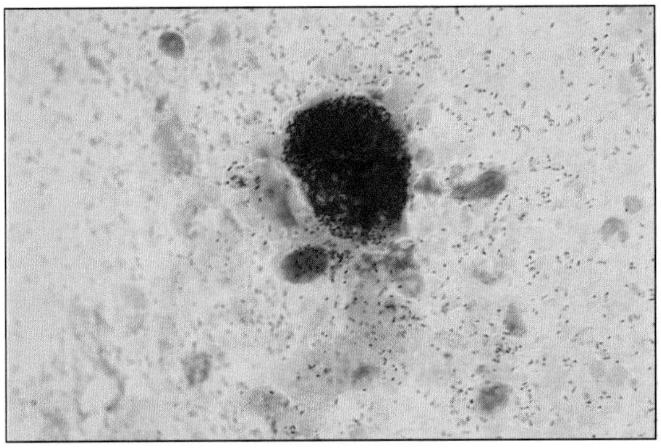

14 MZN-stained smear from a bovine placenta (an abortion case): red MZN-positive coccobacilli of *Brucella abortus* in clumps. (×1000)

15

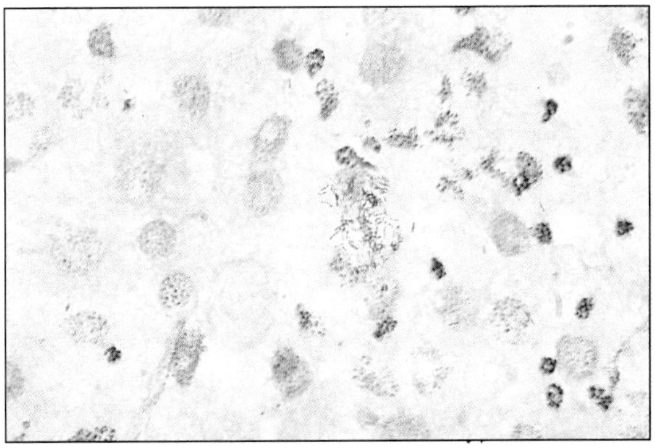

15 ZN-stained smear from a tuberculous lesion in a hen: red ZN-positive thin rods of *Mycobacterium avium*. (×1000)

Alternative Method for the Ziehl–Neelsen Stain

This method is advocated by the National Mycobacterial Laboratory, Ames, Iowa. The advantage of the method derives from using the brilliant green counter-stain with which it is almost impossible to over-do the counter-staining.

Strong carbol fuchsin 3 minutes
Wash in distilled water
Acid-alcohol decolourizer 3 minutes *exactly*
Several washes in distilled water
Alkaline brilliant green 3 minutes
Wash in distilled water and air dry.
The ZN-positive or acid-fast pathogenic *Mycobacterium* spp. are stained a bright red with the background and other microorganisms stained green.

GIEMSA STAIN

The dried smear is first fixed in absolute methyl alcohol (Analar) for 3 minutes.
1 part Giemsa stain + 9 parts buffer: 60 minutes
Wash with the buffer
Drain and air dry.
The Giemsa stain is used to stain spirochaetes such as *Borrelia anserina*, to demonstrate the capsule of *Bacillus anthracis*, to stain rickettsial organisms such as *Haemobartonella felis*, and to demonstrate the morphology of *Dermatophilus congolensis* more clearly than the Gram method (**16**).

POLYCHROME METHYLENE BLUE STAIN (M'FADYEAN'S REACTION)

Polychrome methylene blue is methylene blue solution that has been allowed to oxidise by storing it exposed to the air (loosely plugged) for several months. A thin blood or exudate smear taken from a suspect case of anthrax is air dried, flame-fixed and flooded with the stain for 2–3 minutes. The stained smear is washed and dried. The rods of *B. anthracis* stain blue and the capsular material a pale pink colour (**17**). Any suspect anthrax material should be handled with care and the stained slides autoclaved after use. Viable spores may be present on the slide after staining.

WET PREPARATIONS

Some microorganisms can be demonstrated microscopically without the use of staining techniques. Wet preparations can be examined by phase contrast or darkfield microscopy and by the use of the high-dry objective of the light microscope, with the condenser slightly lowered. Fungal structures in tissues or skin scrapings can be placed in a few drops of 10–20 per cent KOH on a microscope slide, under a coverslip, for 2–4 hours to allow clearing of the preparation to occur. The slide is then examined under the high-dry objective of the microscope. This is particularly useful to demonstrate the arthrospores of dermatophytes (**18**). Methods for microscopic examination of specimens for fungal elements are given in more detail in the mycology section.

16

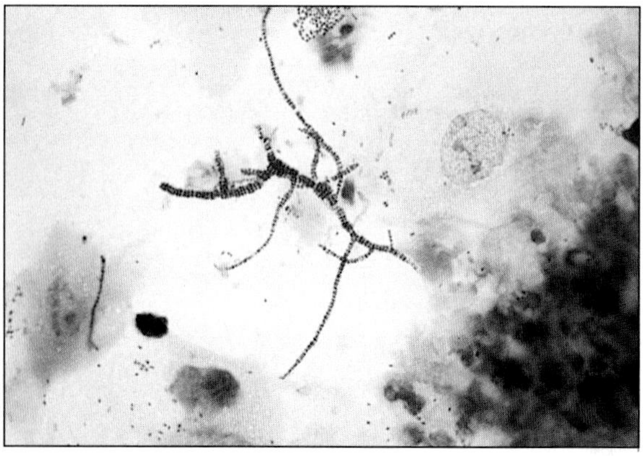

16 Giemsa-stained smear from ovine scab material showing branching filaments and zoospores of *Dermatophilus congolensis*. (×1000)

17

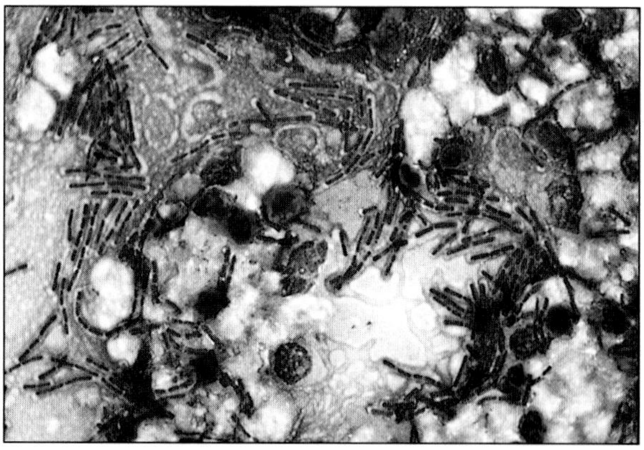

17 Polychrome methylene blue-stained bovine blood smear showing capsulated *Bacillus anthracis*. Note the square-ended rods in short chains with pink capsule surrounding the blue cytoplasm. (×1000)

18 KOH wet preparation of bovine hairs infected with *Trichophyton verrucosum* showing arthrospores. (×400)

18

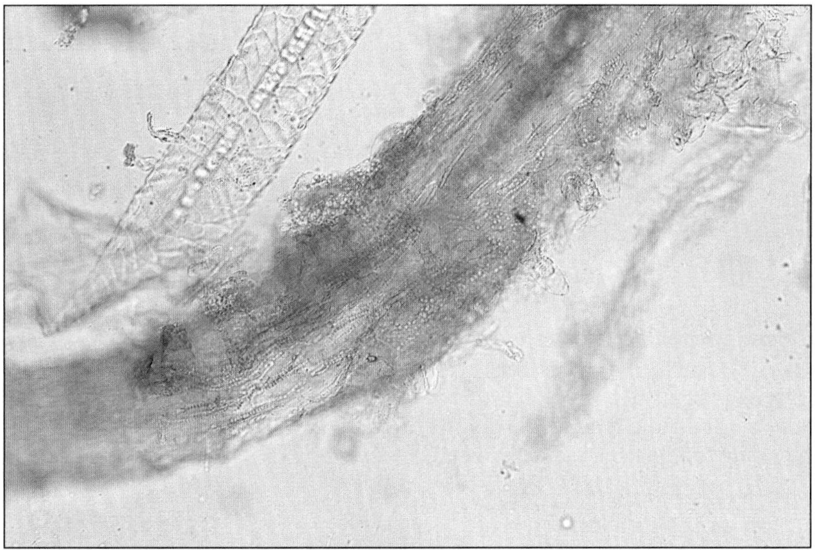

Diagrams 3 and **4** summarise the staining reactions and cellular morphology of some of the more commonly encountered Gram-positive and Gram-negative bacteria.

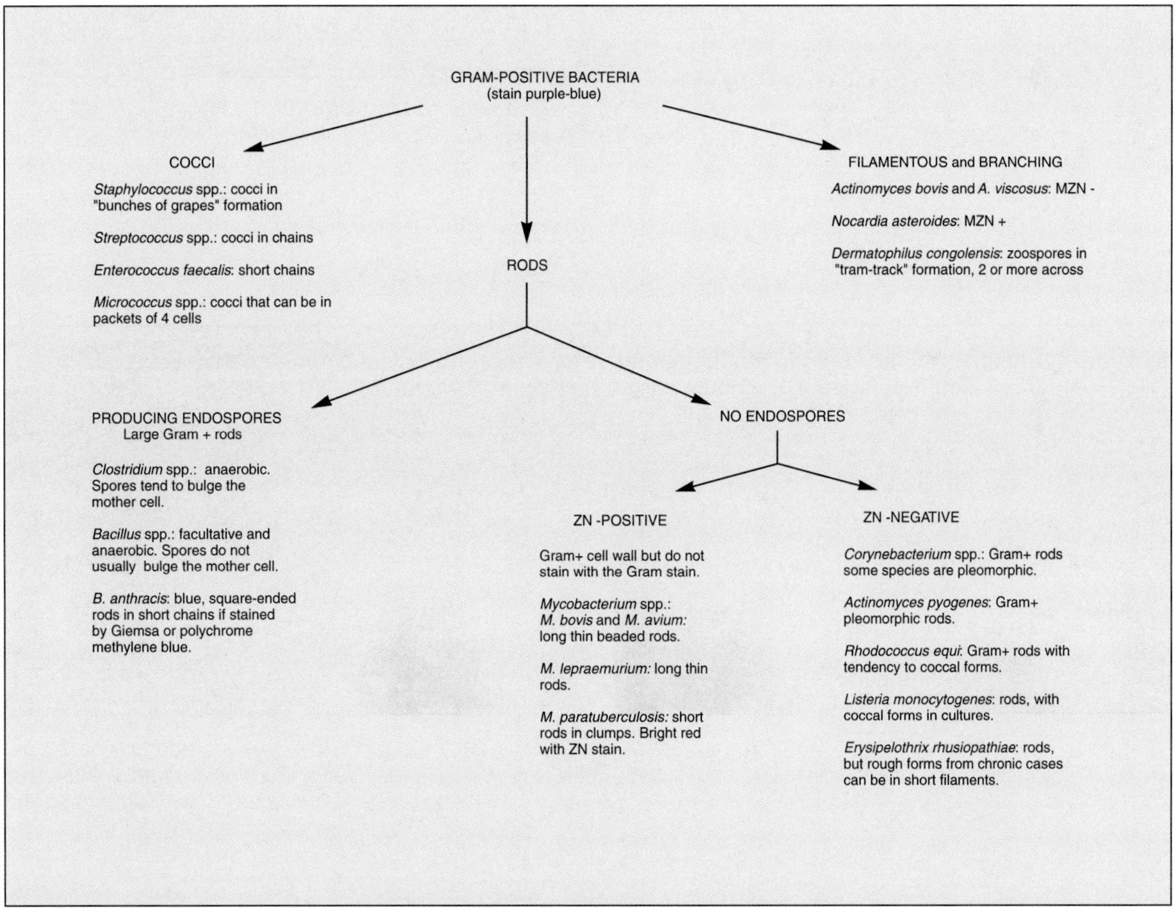

Diagram 3. Summary of the staining reactions and cellular morphology of Gram-positive bacteria. (MZN = modified Ziehl–Neelsen stain, ZN = Ziehl–Neelsen stain).

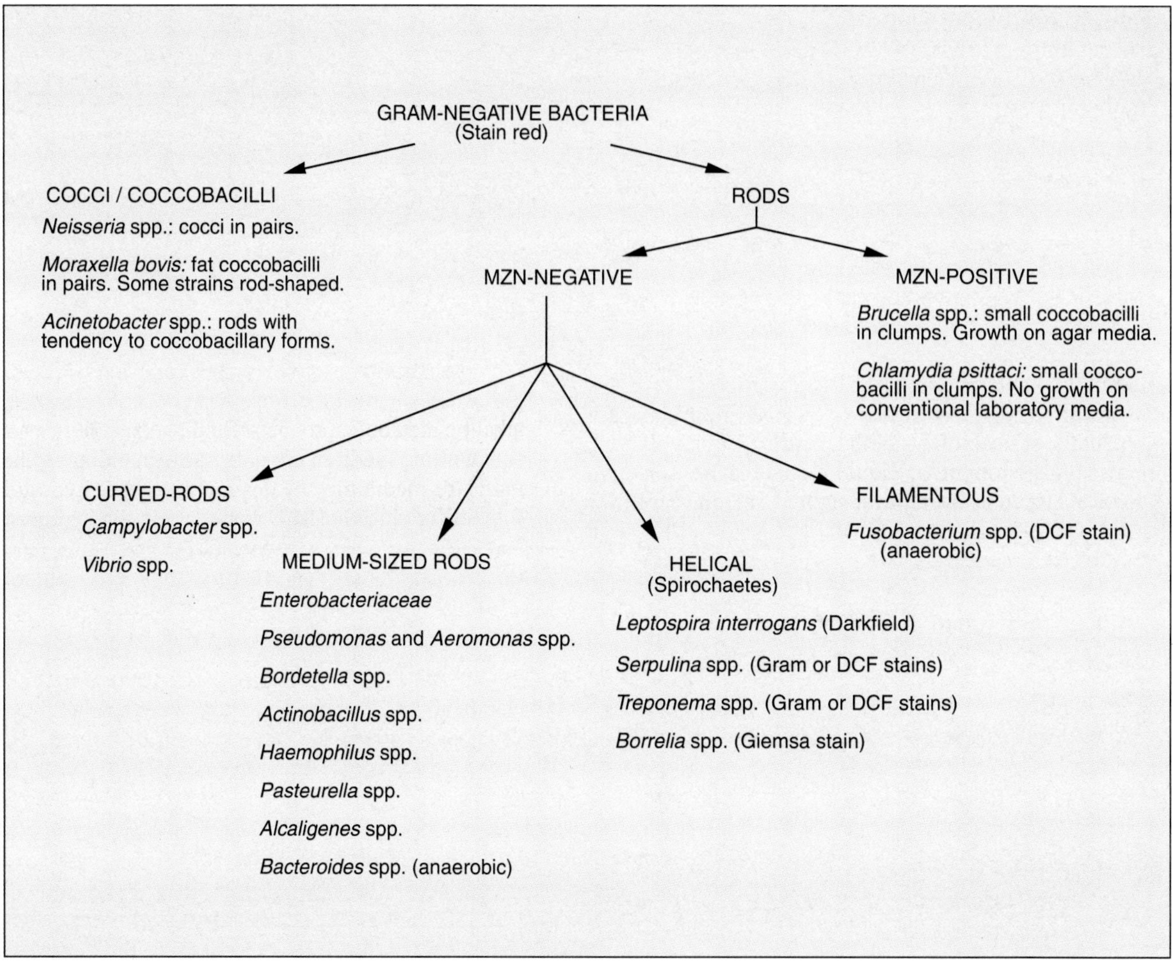

Diagram 4. Summary of the staining reactions and cellular morphology of Gram-negative bacteria. [MZN = modified Ziehl-Neelsen, DCF = dilute carbol fuchsin (simple stain)]

Bacteriological Media

Diagnostic bacteriological media can be divided into the following categories:

- **Chemically defined media:** in these, the exact amount of each ingredient is known. They are used mainly for experimental purposes but citrate broth is an example of a chemically defined medium that is used in diagnostic bacteriology.
- **Basic nutritive media:** these are capable of sustaining growth of the less fastidious bacteria. Nutrient agar is an example.
- **Enriched media:** agar media, such as blood agar, for the growth of fastidious bacteria. The media are usually enriched with blood, serum or egg yolk.
- **Enrichment broths:** liquid media that are selective for a particular bacterium, such as selenite broth for the selection of salmonellae.
- **Selective media:** these agar media have been made selective for the growth of a particular bacterium or group of bacteria and are used extensively in diagnostic bacteriology. They contain inhibitory substances that prevent the growth of unwanted bacterial species. Many selective media, such as brilliant green and MacConkey agars, can also be described as indicator media.
- **Indicator media:** these are particularly useful in diagnostic bacteriology. They are designed to give a presumptive identification of bacterial colonies due to the biochemical reactions in the media. Indicator media often contain fermentable sugars plus a pH indicator that gives a colour change in the media (**Table 2**). MacConkey agar contains the fermentable sugar lactose and neutral red as the pH indicator. Bacteria such as *Escherichia coli* that ferment lactose produce acidic metabolites that change the colonies and surrounding medium to a pink colour. Salmonellae that cannot ferment lactose will use the peptones in the medium with the production of alkaline metabolic products. Salmonella colonies and surrounding medium are pale straw in colour. Other indicator media may be designed to show hydrogen sulphide production (XLD agar) or aesculin hydrolysis (Edwards medium). Blood agar, although an enriched medium, may also be considered as an indicator medium as it shows the type of haemolysis of a particular bacterium.

Examples of media used in diagnostic bacteriology are given in **Table 3** and illustrated (**19–23**, inclusive). They can be obtained commercially as dehydrated powders or can often be purchased as pre-poured plates.

Table 2. pH Indicators used in diagnostic media and biochemical tests.

pH Indicator	*pH Range* ALKALINE ACID	*Colour change* ALKALINE ACID	*Media and biochemical tests*
Andrade's indicator	7.2–5.5	Colourless–Pink	Peptone water sugars
Bromocresol purple	6.8–5.2	Purple–Yellow	Purple agar base, lysine decarboxylase broth
Bromothymol blue	7.6–7.0–6.0	Blue–Green–Yellow	O–F medium, Simmons citrate, Smith–Baskerville medium (*Bordetella* spp.)
Litmus	8.3–4.5	Blue–Red	Litmus milk medium
Methyl red	6.4–4.4	Orange–Red	Methyl red test
Neutral red	8.0–6.8	Pale yellow–Red	MacConkey agar, Yersinia selective medium
Phenol red	8.4–6.8	Red–Yellow	XLD, brilliant green and TSI agar, peptone water sugars, Christensen urea agar

Table 3. Examples of media used in diagnostic bacteriology.

Medium	***Uses***	***Sugars and other substrates***	***pH Indicator***	***Inhibitor(s)***
Nutrient agar	Basic nutritive medium. Growth of less fastidious bacteria	–	–	–
Blood agar	Growth of most bacteria including many of the fastidious species.	Red blood cell (shows haemolysis)		
MacConkey agar	Growth of the *Enterobacteriaceae* and some other Gram-negative bacteria	Lactose	Neutral red	Bile salts
Brilliant green agar	Salmonella isolation. A few other bacteria will grow on this medium	Lactose Sucrose	Phenol red	Brilliant green dye
XLD agar*	Salmonella isolation. Some other bacteria will grow on XLD agar	Xylose, lactose sucrose, lysine, and H_2S detection	Phenol red	Bile salts
TSI agar**	Salmonella identification	Lactose, sucrose, dextrose, H_2S detection	Phenol red	–
Selenite broth	Enrichment broth for isolation of salmonellae	–	–	Selenite
EMB agar***	Presumptive identification of *Escherichia coli*	Lactose, Saccharose	Eosin and methylene blue	–
Edwards medium	Selective for the streptococci. Indicates aesculin hydrolysis and haemolysis	Aesculin Red blood cells	–	Crystal violet and thallous sulphate
Purple base agar + 1 per cent maltose	Presumptive differentiation of *Staphyloccus aureus* and *S. intermedius*	1 per cent maltose	Bromocresol purple	–
Chocolate agar	Enriched medium for *Haemophilus* spp.	Lysed red cells. X and V factors released	–	–

* = Xylose-lysine-desoxycholate agar. ** = Triple sugar iron agar. *** = Eosin-Methylene blue agar (colonies of *E. coli* have a metallic sheen).

19

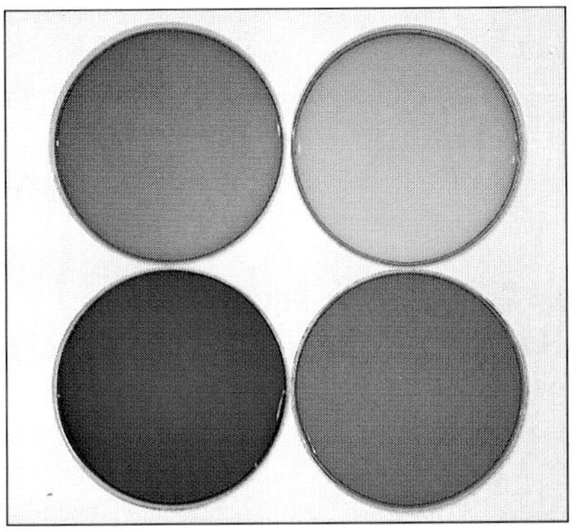

19 Indicator media: clockwise from top left, XLD, brilliant green, MacConkey and EMB agars.

20

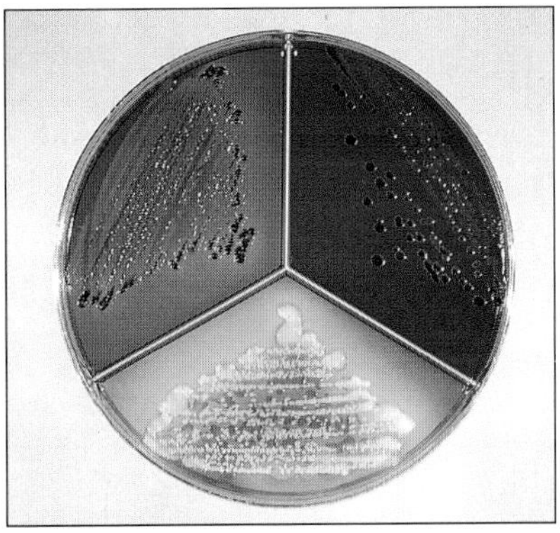

20 XLD medium with *Proteus* sp. (left), *Salmonella* sp. (right) and *Klebsiella* sp. (bottom).

21

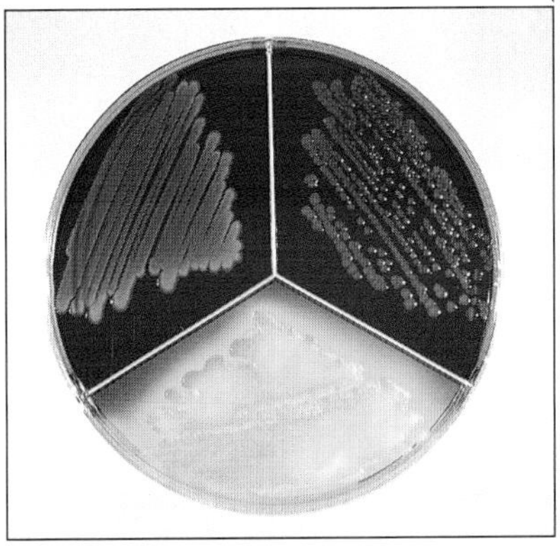

21 Brilliant green agar with *Pseudomonas aeruginosa* (left) *Salmonella* sp. (right) and *Klebsiella* sp. (bottom).

22

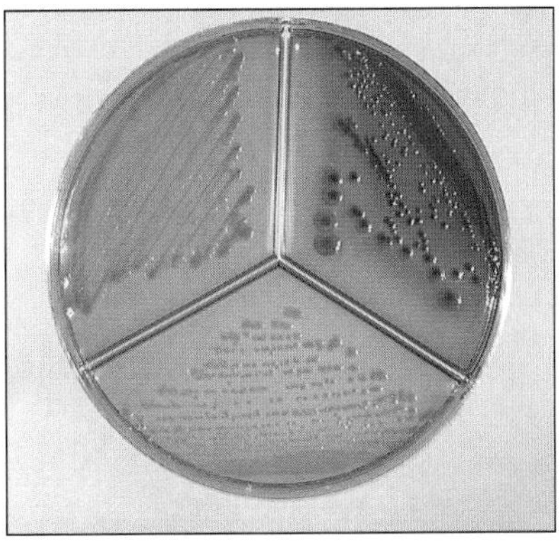

22 MacConkey agar with *Pseudomonas aeruginosa* (left), *Klebsiella* sp. (right) and *Salmonella* sp. (bottom).

23

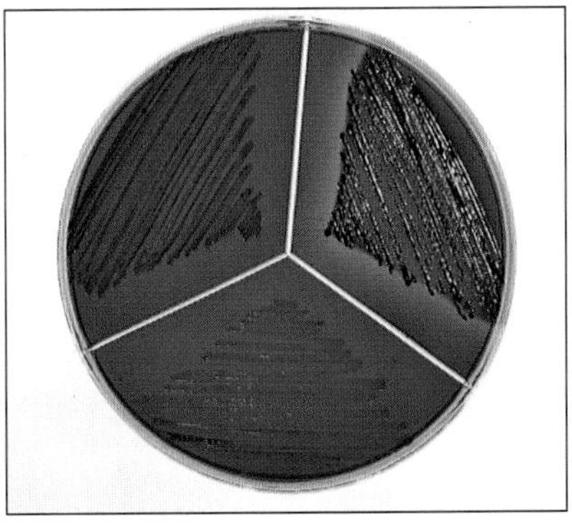

23 Eosin methylene blue agar with *Proteus* sp. (left), *Escherichia coli* (right) and *Salmonella* sp. (bottom).

Preparation of Culture Media

The manufacturer's instructions should always be followed but a few additional general points include:

- Clean glassware that has been rinsed free from detergents and other chemicals should be used.
- The glassware need not be sterile unless sterilised medium is being decanted into it.
- The appropriate amount of dehydrated medium is weighed out, placed in a flask and distilled water added to it. Glass-distilled water must be used, because this is free from chloride and heavy metal ions that can be inhibitory to bacteria.

The medium is prepared in a flask with a capacity of about twice the final volume of the medium as this will allow for adequate mixing and the frothing of the medium during heating.

Media not containing agar can usually be dissolved with gentle agitation, but dehydrated media containing agar is best dissolved by bringing to the boil with continuous stirring, using a glass rod or a hot plate that incorporates a magnetic stirrer system.

Dehydrated media once dissolved are usually sterilised in an autoclave or pressure cooker at 121°C for a holding time of 15 minutes. Some media, such as Edwards, contain ingredients that cannot tolerate this high temperature and they can be autoclaved at 115°C for a holding time of 20 minutes or in accordance with the manufacturer's instructions.

Brilliant green agar is inhibitory to many bacteria and is brought to the boil only and not autoclaved.

Media containing agar, after autoclaving, should be cooled in a water bath at 50°C before the medium is poured into Petri dishes. Agar solidifies at about 42°C. The standard (90 mm) Petri dish should contain about 15 ml of agar medium, about one-third full. Thus, one litre of medium should yield between 60–70 plates.

Some additives to the medium, such as serum or red blood cells, will not tolerate high temperatures and are added as sterile solutions or suspensions once the medium has cooled to 50°C.

After the poured plates are set, they are allowed to dry thoroughly at room temperature or for a few hours in the incubator at 37°C. For use, the surface of the agar must not contain obvious moisture. The plates are stored, agar-side upwards, in a refrigerator at 4°C.

Preparation of blood agar plates

The blood agar base is prepared from dehydrated powder, sterilized and cooled to 50°C in the usual manner.

Sterile blood at the rate of 5–10 per cent vol/vol is added to the cooled agar base and mixed well before the plates are poured. If bubbles form on the surface of the poured plates, a low Bunsen flame is quickly passed across the surface of the agar before the agar sets. If the sterile blood has been stored in the refrigerator, it should be warmed to 37°C before being added to the agar medium to avoid thermal shock to the red cells.

Collecting sterile blood

Bovine or ovine blood is most suitable for veterinary bacteriology. Sterile blood can sometimes be obtained commercially or it can be collected from a young animal that has no evidence of antibodies to the major veterinary pathogens and has not been treated with antibacterial agents. A strict aseptic technique must be used. The area over the jugular vein is clipped and shaved. The area is then saturated with 70 per cent ethyl alcohol, wiped and allowed to dry thoroughly before venepuncture. To prevent the blood from clotting one of the following methods may be used:

- Collection into a purchased human blood-donor kit.
- Collection into a pre-sterilised apparatus consisting of a tube leading into a conical flask containing glass beads (3 mm) (**24**). The flask is agitated continuously during collection and for at least 5 minutes after obtaining the blood. This defibrinated blood can be decanted into sterile bottles for storage in a refrigerator. The glass beads can be recovered from the fibrin clot and reused.

A sterile anticoagulant solution, such as 0.2 per cent sodium citrate, can be used either in the flask, in the place of the glass beads or blood can be collected in a sterile syringe and immediately added to the anticoagulant.

24

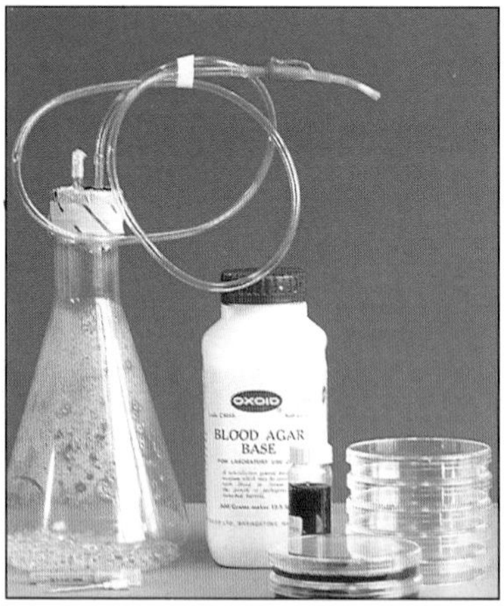

24 Sterile apparatus for the collection of blood (left) and items for blood agar preparation: blood agar base, sheep blood, Petri dishes and two poured plates.

Choice of culture media

For routine isolation of bacteria, blood agar and MacConkey agar are used. Blood agar will support the growth of most of the pathogenic bacteria and many of the Gram-negative bacteria will grow on the MacConkey agar. A comparison of the growth on the two media can indicate the types of bacteria that have been isolated. Special media used for specific pathogens are mentioned under the appropriate group of bacteria (Section 2).

Inoculation of Culture Media

If the specimen is a piece of tissue, it is easier to manipulate if it is first placed in an empty sterile Petri dish. Forceps and scalpel, previously held in 70 per cent ethyl alcohol, are flamed and allowed to cool. The tissue is scraped, paying particular attention to the edge of an observable lesion. A little of the tissue scrapings is placed at the edge or 'well' of each culture plate to be inoculated. Some of the tissue could be used to make a smear on a microscope slide for staining.

When the specimen is liquid or semi-liquid it is best applied to the well of the plate using a sterile swab.The swab is used to mix the sample uniformly. Before discarding the swab a smear could be made for microscopy.

When either inoculating or streaking culture media plates it is preferable to start with the non-inhibitory medium, such as blood agar, and then inoculate or streak any inhibitory or selective medium such as MacConkey agar.

Streaking the agar plates

The object of plate streaking is to obtain isolated bacterial colonies that are required for observing colonial morphology, antibiotic sensitivity testing and for biochemical identification. The quadrant streak method, using the whole plate, is usually employed for most diagnostic specimens (**Diagram 5**).

The plate is first labelled, on the agar side, with the type of specimen, date of inoculation and reference number. A waterproof marker pen is used and the writing should be kept as near to the edge of the plate as possible, so that after incubation the bacterial colonies are not obscured. Two inoculating loops are used so that one can be cooling while the other is in use. The loops should be flamed (**25**) before starting and then after streaks 1, 2 and 3 (**Diagram 5**). The number of times the loops are flamed will depend on the estimated number of bacteria in the original inoculum. The loop must always be flamed after streak 4 before putting it down. The inoculating loop should be kept as nearly parallel to the agar surface as possible to prevent the loop from digging into the agar (**26**).

Streaks 1, 2 and 3 should be kept as close to the edge of the plate as is practical, thus leaving plenty of room for streak 4, where it is hoped to obtain isolated colonies (**27**). This last area of the plate should be fully utilized, keeping the streak lines as close together as possible.

Half-plating or quarter-plating (**Diagram 6**) can be employed where the initial inoculum is judged to contain only a few bacterial cells, such as often occurs in bovine mastitic milk samples.

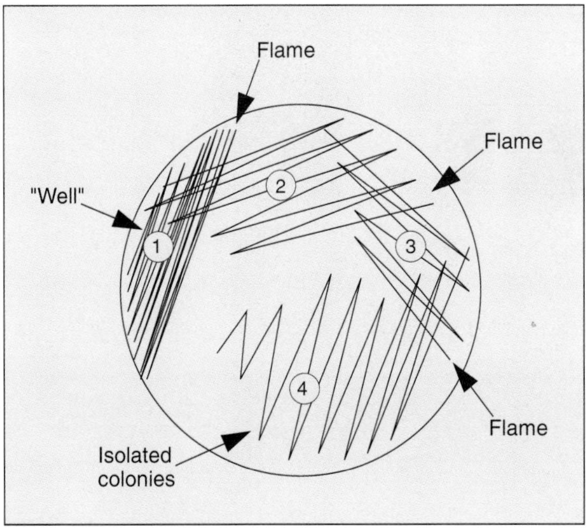

Diagram 5. Quadrant streaking method for obtaining isolated bacterial colonies on agar media. The loop should be flamed before and after streaking.

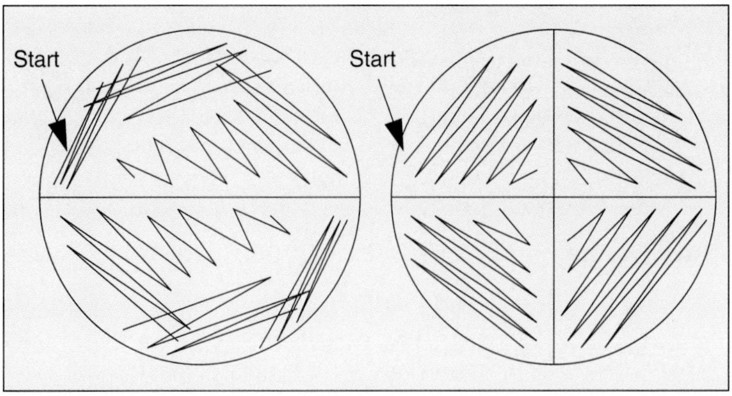

Diagram 6. Half-plating and quarter-plating on agar media. These techniques should be used only if the sample is likely to contain small numbers of bacteria or a single bacterial species, such as mastitic milk samples.

25
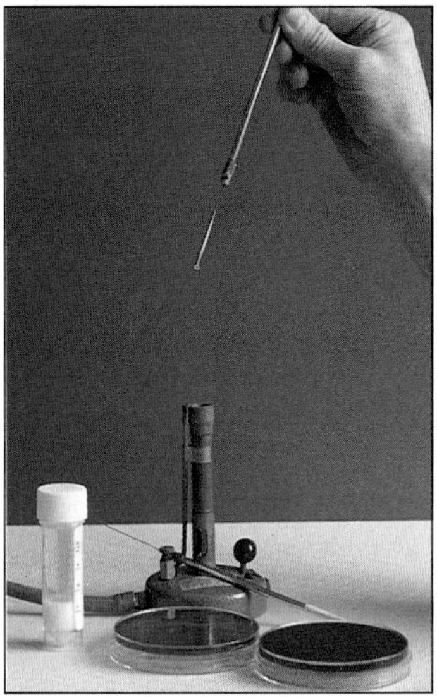

25 Correct technique for flaming an inoculating loop.

26
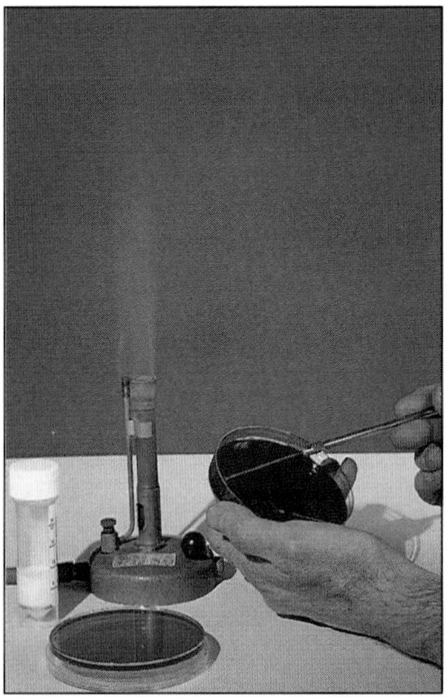

26 Plate inoculation technique: streaking culture media.

27
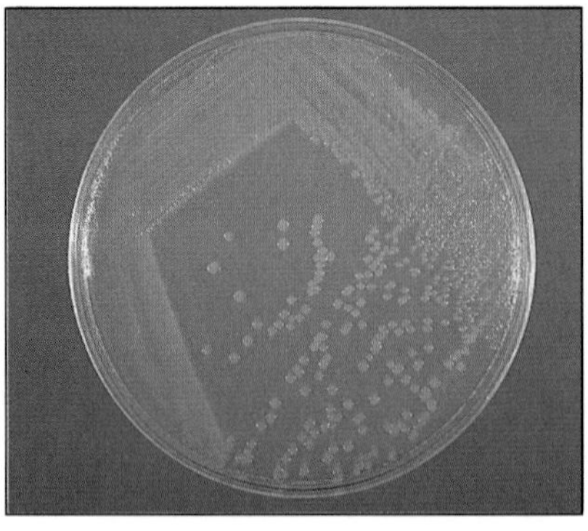

27 Nutrient agar streaked with *Enterobacter aerogenes* showing isolated colonies (*see* area 4, **Diagram 5**).

Incubation of Inoculated Culture Plates

Following selection of the media, temperature and time of incubation must be considered together with the gaseous atmosphere under which the culture plates should be incubated. These will depend upon the bacterial pathogens that are being sought. The following incubation temperatures, times and atmospheric conditions are offered as general guidelines:

Incubation atmosphere

- Normal atmosphere (aerobic) for most of the pathogenic bacteria and all the fungi.
- 5–10 per cent CO_2:
  Actinobacillus pleuropneumoniae
  Actinomyces viscosus
  Brucella species
  Campylobacter jejuni and *C. fetus* (optimally 6 per cent O_2, 10 per cent CO_2, 84 per cent N_2)
  Dermatophilus congolensis (primary isolation)
  Francisella tularensis
  Haemophilus species
  Taylorella equigenitalis
- Anaerobic:
  Actinomyces bovis
  Bacteroides species
  Campylobacter mucosalis
  Clostridium species
  Eubacterium species
  Fusobacterium species
  Peptostreptococcus species
  Serpulina hyodysenteriae

Incubation temperature

- 37°C: Most of the microorganisms pathogenic for animals including the mycoplasmas and fungi that cause systemic mycoses.
- 42°C: *Campylobacter jejuni*
  : *Serpulina hyodysenteriae*
- 28°C–30°C *Leptospira interrogans* serovars
- 25°C: The dermatophytes except *Trichophyton verrucosum* that tolerates 37°C.
- 4°C: For cold enrichment of *Listeria monocytogenes* (from brain samples) and *Yersinia enterocolitica* and *Y. pseudotuberculosis* (from faecal samples).

Incubation time

- 24–48 hours: Most of the rapidly growing bacteria.
- 48–72 hours: The rapidly growing bacteria when plated on selective media.
- 4–6 days: *Brucella* species
  Campylobacter species
  Nocardia asteroides
  Mycoplasma species
  The relatively fast-growing fungi.
- 2–3 weeks: Most of the dermatophytes (*T. verrucosum* up to 5 weeks) and *Mycobacterium avium*
- 3–8 weeks: *Mycobacterium bovis*
- 4–16 weeks: *Mycobacterium paratuberculosis*

Bacteria not yet grown on conventional agar media

These include the rickettsiae, *Chlamydia psittaci, Bacillus piliformis* and some mycobacteria. Many of these microorganism can be cultured in the yolk sac of fertile hens' eggs and/or in selected cell cultures.

Table 4 gives the features of some commonly isolated bacteria on routinely used media (blood and MacConkey agars) and their appearance in a Gram-stained smear from cultures.

Table 4. Features of some commonly isolated bacteria on routine media.

Bacterium	***Blood agar***		***MacConkey agar***		***Gram-stain***		***General comment***
	Colony*	Haemolysis	Growth	LF/NLF	Reaction	Shape	
Actinomyces pyogenes	Translucent, pin-point 0.5mm ·	+ (hazy)	–		+	R(C)	Hazy haemolysis along streak line often before small colonies can be seen. Very pleomorphic in the Gram-stained smear. Catalase –
Beta-haemolytic streptococci	Translucent, glistening and round 0.5–1.0mm •	+	–		+	C	The size of the clear zone of haemolysis varies with the Lancefield group and species. Catalase –
Listeria monocytogenes	White, smooth and round 0.5–1.0mm •	+	–		+	R(C)	Colonies very similar to the beta-haemolytic streptococci. Young colonies yield many coccal cells. Catalase +
Moraxella bovis	White, smooth and round 0.5–1.5mm ●	+ (–)	–		–	R(C)	Colonies very similar to the above two bacteria. Gram – and cells in pairs as rods or fat cocci. Some variants are non-haemolytic
Pasteurella haemolytica	White/grey, smooth and round 0.5–1.5mm ●	+	+ (Pin-point)	LF	–	R	Colonies similar to the above three bacteria but are Gram – rods and will tolerate MacConkey agar. Some strains are haemolytic only under the colonies
Enterococcus faecalis	White, smooth and round 0.5–1.0mm •	+ (alpha)	+ (Pin-point)	LF	+	C	Greenish (alpha) haemolysis. Red pin-point colonies on MacConkey agar, although Gram +
Alpha-haemolytic streptococci	White, smooth and round 0.5–1.0mm •	+ (alpha)	–		+	C	Greenish or partial haemolysis. Not usually pathogenic, can be part of the normal flora
Erysipelothrix rhusiopathiae	White, smooth and round Some strains rough 0.5–1.5mm ●	+ (alpha at 48 hours)	–		+	R	Non-haemolytic at 24 hours incubation but alpha haemolysis under the colonies occurs at 48 hours. Rough, dry colonies especially from the chronic forms of the disease. Catalase –

Table 4. Features of some commonly isolated bacteria on routine media (continued).

Bacterium	***Blood agar***		***MacConkey agar***		***Gram-stain***		***General comment***
	Colony*	Haemolysis	Growth	LF/NLF	Reaction	Shape	
Staphylococcus aureus or *S. intermedius*	White or yellow, smooth round and shiny 2.0–3.0mm ●	+ (target)	–		+	C	Human and bovine strains of *S. aureus* have a golden-yellow pigment. Hold plate to a bright light to see characteristic double-haemolysis. Catalase + (streptococci catalase –)
Clostridium perfringens	Grey, flat and often irregular edge 2.0–3.0mm ●	+ (target)	–		+	R	Anaerobic with double or target haemolysis. Colonies tend to have irregular edges
Haemolytic *Escherichia coli*	Grey, smooth, shiny and round 2.0–3.0mm ●	+	+	LF	–	R	Characteristic 'coliform' smell. Bright pink colonies on MacConkey agar about the same size as on blood agar
Aeromonas hydrophila	Grey, flat, round and shiny 2.0–3.0mm ●	+	+	V	–	R	Foul smell is characteristic but differs from that of *E. coli*. Variable lactose fermentation but good growth on MacConkey agar
Pseudomonas aeruginosa	Blue-green, flat, round. Some have metallic sheen 2.5–4.0mm ●	+	+	NLF	–	R	Amount of pyocyanin (blue-green pigment) varies between strains. Characteristic fruity-musty smell
Bacillus spp.	Grey, dry, granular with irregular edges 3.0–5.0mm ●	+ (–)	–		+	R (spores)	Many haemolytic (exception *B. anthracis).* Dry colonies as no capsular material produced on lab. media. Rhizoid *(B. mycoides)* and other unusual colony-types
Corynebacterium pseudotuberculosis	Opaque, dry, crumbling 0.5–1.0mm •	V	–		+	R	Haemolysis variable. The cells tend to be less pleomorphic than *Actinomyces pyogenes*

(*continued*)

Table 4. Features of some commonly isolated bacteria on routine media (continued).

Bacterium	***Blood agar***		***MacConkey agar***		***Gram-stain***		***General comment***
	Colony*	Haemolysis	Growth	LF/NLF	Reaction	Shape	
Coryne-bacterium renale	Grey-white, round and moist 0.5–1.0mm •	–	–		+	R	Small, moist colonies at 24–48 hours incubation, become drier later. Urease +
Brucella spp.	Translucent, convex and round 0.5mm •	–	–		–	R(C)	Some brucellae require 10 per cent CO_2 for growth. Colonies not visible until 2–3 days incubation. MZN +
Campylo-bacter fetus	Small, delicate, round and opaque 0.5mm •	–	–		–	R (curved)	Small 'dew-drop' colonies after 2–3 days incubation. Requires reduced oxygen tension for growth. Curved rods, if in pairs they have a 'seagull' appearance
Actinobacillus spp.	Grey to translucent, shiny and round 0.5–1.0mm •	–	V	V	–	R	Small, round, non-haemolytic colonies. Variable growth on MacConkey agar
Pasteurella multocida	Translucent, smooth, round and shiny 1.0–2.0mm ●	–	–		–	R	Translucent, round colonies that can appear pinkish on blood agar. Characteristic sweetish smell. Indole +
Bordetella bronchiseptica	Small, greyish-white and round 0.5–2.0mm ●	–	+	NLF	–	R	Colonies small at 24 hours, becoming considerably larger later. Grows on MacConkey. Unreactive bacterium
Rhodococcus equi	Salmon pink and mucoid. Colonies coalesce 1.0–2.0mm ●	–	–		+	R(C)	Pinkish-tan colonies, the colour becoming more definite with increased time. Mucoid colonies tend to merge
Staphylo-coccus epidermidis (coagulase –)	White, shiny, round and convex 2.0–3.0mm ●	–	–		+	C	Colonies similar to coagulase + staphylococci but are non-haemolytic and always white
Micrococcus spp.	White, yellow tan or pink. Round, convex and shiny 2.0–3.0mm ●	–	–		+	C	Shiny convex colonies, can be white but are often pigmented. Not considered pathogenic

Table 4. Features of some commonly isolated bacteria on routine media (continued).

Bacterium	***Blood agar***		***MacConkey agar***		***Gram-stain***		***General comment***
	Colony*	Haemolysis	Growth	LF/NLF	Reaction	Shape	
Salmonella spp.	Greyish, round and shiny 2.0–3.0mm	–	+	NLF	–	R	Pale colonies on MacConkey agar. No smell (unlike most of the other members of the *Enterobacteriaceae)*
Non-haemolytic *E. coli*	Greyish, round and shiny 2.0–3.0mm	–	+	LF	–	R	Non-haemolytic but otherwise similar to the haemolytic strains. Characteristic 'coliform' smell
Yersinia spp.	Greyish, round and shiny 2.0–3.0mm	–	+	NLF	–	R	Medium, round, smooth, non-haemolytic colonies. Non-lactose fermenter
Serratia marcescens and *S. rubidaea*	Red/orange, convex, round and shiny. Some strains white at 37°C 2.0–3.0 mm	–	+	NLF	–	R	Produce red pigment (prodigiosin). Some strains do not produce the pigment at 37°C
Klebsiella spp.	Grey, mucoid colonies that coalesce. 2.0–4.0mm	–	+	LF	–	R	Colonies tend to be large, mucoid and pale pink on MacConkey agar. Non-motile
Enterobacter spp.	Grey, mucoid colonies that coalesce 2.0–4.0mm	–	+	LF	–	R	Very similar to colonies of *Klebsiella* species. Motile.
Proteus spp.	Grey, swarming growth over the agar. Swarming can be in waves.	–	+	NLF	–	R	Characteristic swarming growth on non-selective medium. Turns blood agar brown. Very foul smell. Colonies pale and discreet on MacConkey but edges may be irregular.
Pseudomonas spp. other than *P. aeruginosa*	Grey or yellowish-green, flat and spreading. 2.5–4.0mm	–	+	NFL	+	R	Large , flat colonies. Some may produce the yellowish-green pigment, pyoverdin.

+ = Positive; – = Negative; V = Variable reaction; LF = Lactose-fermenter; NFL = Non-lactose fermenter; R = Rod; C = Coccus.

* Size of colonies can be variable. Size is given and described after 48 hours' incubation, as even with fast-growing bacteria the colonies are more characteristic at this stage.

Identification of Bacterial Pathogens

Identification of the bacterial pathogen(s) involved in animal disease essentially depends upon:

- Knowledge of the animal species, clinical signs of disease and/or the type of pathological lesion.
- Examination of stained smears made directly from specimens.
- The growth and colonial characteristics of the bacterial pathogen on:
  a) Blood agar: size of the colony, morphology of the colony, for example whether it is in a rough or smooth form and pigment production (**28**), whether haemolytic, and if so the type of haemolysis (**29**).
  b) Selective/indicator media: these will also demonstrate some biochemical reactions of the bacterium.
- The atmospheric conditions needed for growth can aid in the identification of the bacterium.
- Biochemical and other tests carried out on a pure culture of the suspected pathogen to confirm its identity.

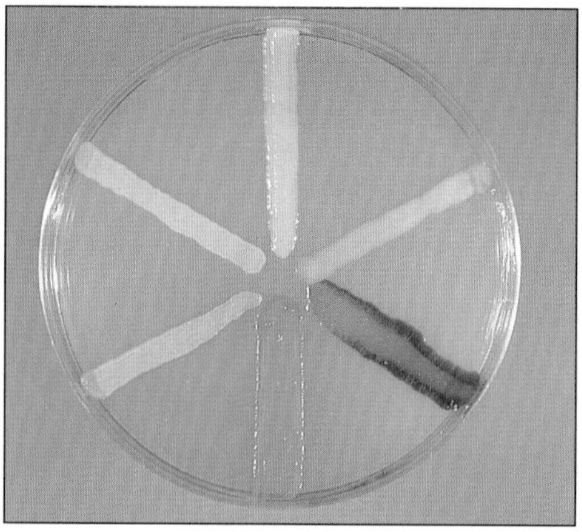

28

28 Chromogenic bacteria on nutrient agar: clockwise from top, *Micrococcus* sp.(yellow), *Rhodococcus equi, Serratia rubidaea, Pseudomonas aeruginosa* (diffusion of pigment), *Micrococcus roseus* and *Staphylococcus intermedius* (no pigment).

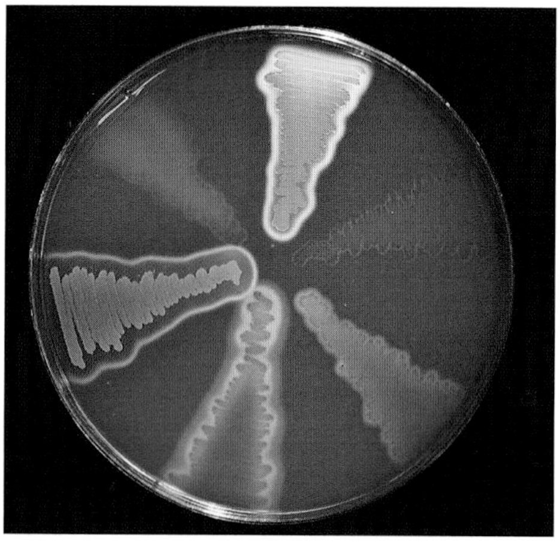

29

29 Patterns of haemolysis produced by streptococci on sheep blood agar: clockwise from top, Group A (beta-haemolytic), *S. uberis* (alpha), *S. agalactiae* (beta), *Enterococcus faecalis* (alpha), Group C (beta), *S. pneumoniae* (alpha).

Pure Culture Technique

Bacterial contaminants, representing those microorganisms from the normal flora, external environment or from postmortem invasion, may be present on the plates together with the pathogen of interest. Subculture of the bacterial colony type(s) considered most significant should be made to obtain pure cultures for use in identification tests. To obtain a pure culture, the top of two or three similar and isolated colonies from the blood agar plate should be touched with a sterile inoculating loop. Another blood agar plate and a MacConkey agar plate should be streaked out with the colonial growth. It is preferable to select colonies from a non-selective medium, such as blood agar, because suppressed microcolonies of other bacteria on selective media could inadvertently be collected together with the colony of interest.

Primary Identification of Bacteria

Once a pure culture is obtained, the results from a few comparatively simple tests can often identify the bacterium to a generic level:

- A Gram-stained smear from the culture will establish:
  a) the Gram reaction (Gram-positive or Gram-negative), and
  b) the cellular morphology (coccus or rod).
- Growth or absence of growth on MacConkey agar.
- Catalase and oxidase tests.
- Motility tests.
- An oxidation-fermentation (O-F) test in Hugh and Leifson's medium.

Gram reaction

Provided that the Gram-stained smear has been made from a pure culture, if any cells have retained the crystal violet-iodine complex then the bacterium is regarded as being Gram-positive. It is not unusual for some of the cells to decolourise, particularly older cells or those producing endospores. In some of the *Bacillus* species, such as *B. circulans*, all of the cells usually decolourise. Endospores in a Gram stain are seen as unstained areas within the mother cell. The production of endospores could indicate that the bacterium is a *Bacillus* (aerobic) or a *Clostridium* (anaerobic) species.

Other tests to distinguish Gram-positive from Gram-negative bacteria

If the results of the Gram stain are equivocal, other tests have been developed to aid in the differentiation of Gram-positive and Gram-negative bacteria.

LANA test
A swab is impregnated with L-alanine-4-nitroanilide (LANA). A colony is touched with the swab and the swab will turn yellow if the bacterium is Gram-negative (**30**).

KOH test
A loopful of the culture is taken from a non-selective medium (blood agar) and mixed with an equal amount of 3 per cent potassium hydroxide (KOH) on a clean microscope slide. After thorough mixing the loop is lifted at intervals to see whether a gel is forming. If the bacterium is Gram-negative a viscous gel forms within 60 seconds (**31**) while no gel is formed if the bacterium is Gram-positive.

Susceptibility to vancomycin
The majority of Gram-positive bacteria (except for some lactobacilli) are susceptible to vancomycin whereas Gram-negative bacteria are resistant (**32**).

30

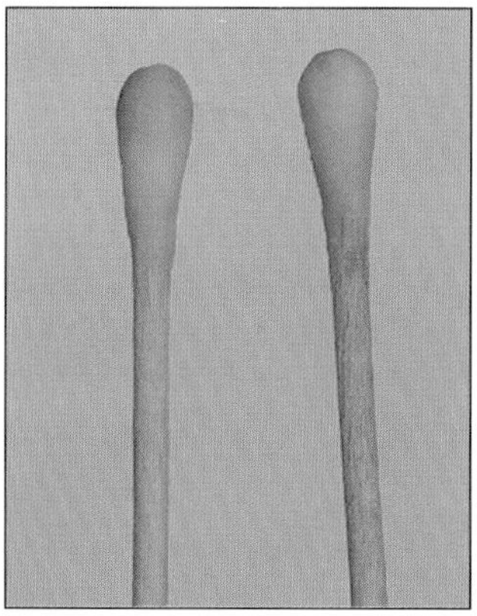

30 LANA test: Gram-positive (left) no reaction and Gram-negative (right) positive reaction.

31

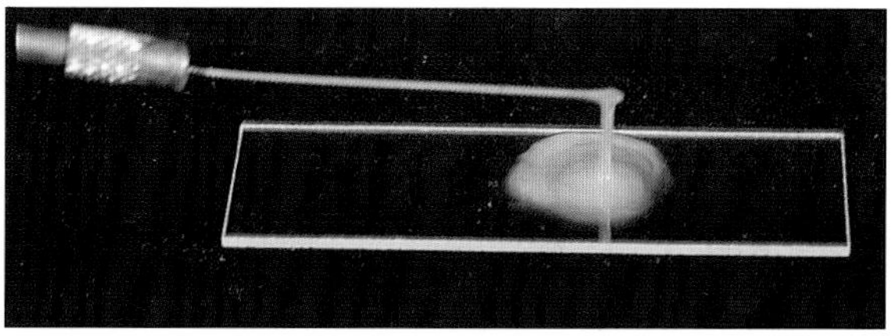

31 KOH test: showing a viscous gel formed only by Gram-negative bacteria.

32

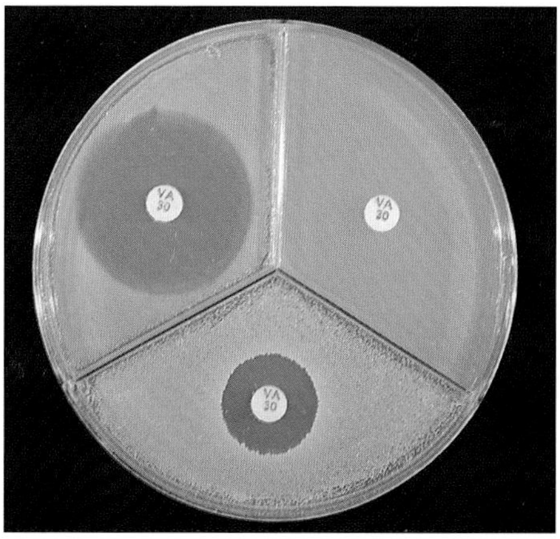

32 Differentiation of Gram-positive and Gram-negative bacteria using the susceptibility of most Gram-positive bacteria to vancomycin: *Rhodococcus equi* (left), *Escherichia coli* (right) and *Staphylococcus aureus* (bottom).

Cellular morphology (shape)

In this primary identification bacteria are regarded as being either a coccus or a rod, filamentous bacteria being viewed as elongated rods. Pleomorphic bacteria such as *Actinomyces pyogenes* and *Rhodococcus equi* can appear coccal but there are usually a few rod-shaped cells present in the Gram-stained smear, and they are classified as rods for purposes of identification. *Listeria monocytogenes* also tends to produce many coccal forms, especially from young cultures.

Growth or no-growth on MacConkey agar

Commercial firms may prepare more than one type of MacConkey agar. These differ in the composition of the bile salts and as to whether or not crystal violet (anti-Gram-positive) is present. It is advisable to use a MacConkey agar that inhibits the majority of Gram-positive bacteria, will support the growth of all members of the *Enterobacteriaceae* but is selectively inhibitory to the other Gram-negative bacteria.

33

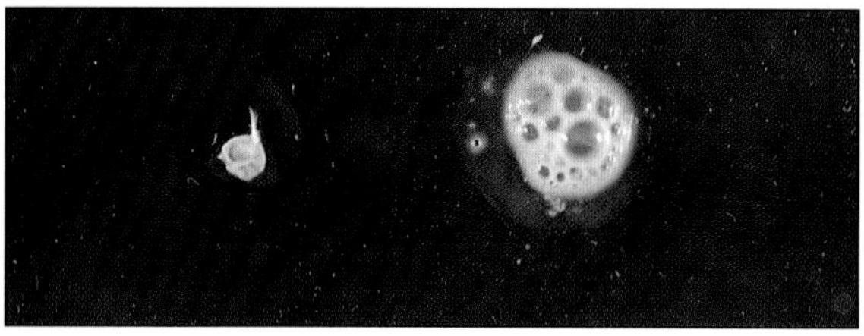

33 Catalase test on a slide: negative (left) and positive (right).

34

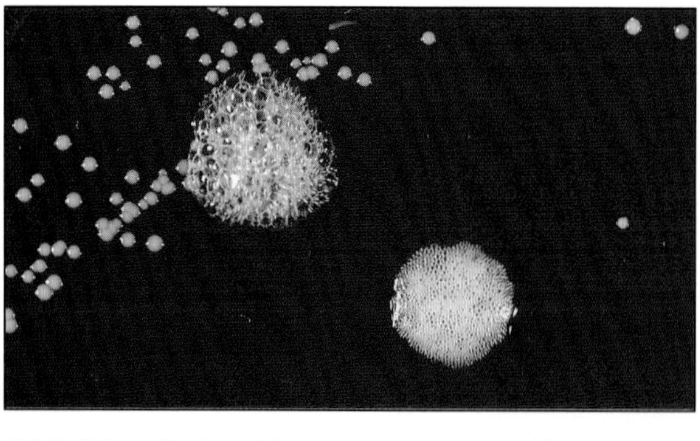

34 Catalase test on blood agar: strong positive reaction (left) from colonies and weaker reaction (right) from blood agar.

Catalase test

This test detects the enzyme catalase that converts hydrogen peroxide to water and gaseous oxygen. The reagent, 3 per cent hydrogen peroxide, should be stored at 4°C in a dark bottle. Extra care must be taken if the bacterium has been grown on blood agar, because the presence of red blood cells can lead to a false-positive reaction.

- **Method 1:** A loopful of the bacterial growth is taken from the top of the colonies avoiding the blood agar medium. The bacterial cells are placed on a clean microscope slide and a drop of 3 per cent hydrogen peroxide is added. An effervescence of oxygen gas, within a few seconds, indicates a positive reaction (**33**).
- **Method 2:** A drop of 3 per cent hydrogen peroxide is added to a colony on the plate and another drop to an area of the blood agar plate without bacterial growth. Bubbles of oxygen gas will arise from the blood agar but greater gas production from the bacterial colony indicates a positive test (**34**).

Oxidase test

This test depends on the presence of cytochrome c oxidase in a bacterial cell. Anaerobes are oxidase-negative. The reagents used in the test should be colourless and stored in a dark bottle at 4°C. The solutions must not be used if they become dark blue. Auto-oxidation of the reagents may be retarded by the addition of 1 per cent ascorbic acid. Other precautions include testing colonies on non-selective media that do not contain glucose or nitrate and the use of a glass rod or platinum loop to pick up the bacterial growth. A conventional nichrome loop contains iron and, if used, can result in a false-positive test. Several methods can be used for the oxidase test:

- **Method 1:** A piece of filter paper is moistened in a Petri dish with 1 per cent aqueous solution of tetramethyl-p-phenylenediamine dihydrochloride. The test bacterium is streaked firmly across the filter paper with a glass rod. A dark purple colour along the streak line within 10 seconds indicates a positive reaction (**35**). *Pseudomonas aeruginosa* can be used as a positive-control organism.
- **Method 2:** Equal volumes of 1 per cent alpha-naphthol in 95 per cent ethanol and 1 per cent aqueous solution of p-aminodimethylaniline oxalate are mixed. A drop of the mixture is placed on the surface of a few colonies on a blood agar plate. Colonies developing a blue colour within 10–30 seconds is interpreted as a positive reaction (**36**). Weak reactions occurring after 2 minutes are discounted. The colonies giving a positive reaction are viable only if subcultured within a few minutes of reading the test.

35

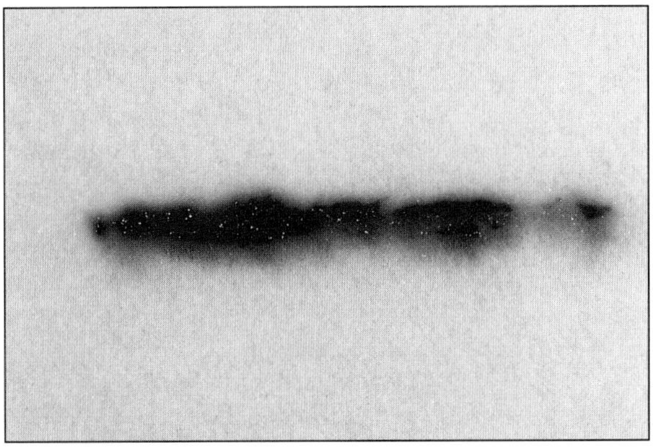

35 Oxidase test: filter paper method. A purple colour within 10 seconds indicates a positive reaction.

36

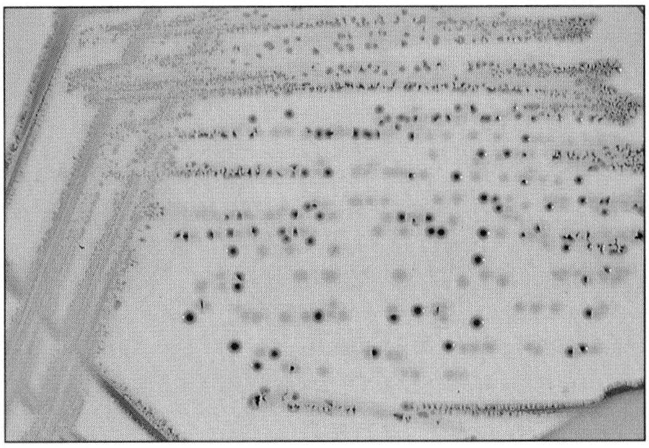

36 Oxidase test on agar medium: blue colonies are oxidase-positive.

- **Method 3:** Commercial oxidase paper strips and oxidase sticks are available. The tests should be carried out in accordance with the manufacturer's instructions.

Motility tests

The majority of bacteria are motile by means of flagella. Motility can be temperature-dependent and some bacteria tend to be motile at ambient temperatures but not at 37°C. Two main methods are used to demonstrate motility:

- **Method 1:** Direct microscopy using a young broth culture (2–4 hours' incubation) of the bacterium incubated at room temperature. A 'hanging-drop' preparation is made by placing a drop of the broth culture in the centre of a clean cover-slip and then inverting it over a transparent plastic or glass ring (about 5 mm deep) fixed to a microscope slide (**37**). The preparation is brought into focus under low-power and then examined with the high-power dry objective with reduced illumination.
  The bacterium is motile if individual cells are moving towards and away from other cells. Brownian movement is a constant and random jiggling of all bacterial cells and other small particles and must not be mistaken for true motility. If the bacterium appears to be non-motile by direct microscopy, then this negative result should be checked by the inoculation of motility media.
- **Method 2:** Semi-solid motility media are available commercially. Tetrazolium salts can be added to these media to aid in the detection of motility. Before autoclaving the motility medium, 0.05 g of 2,3,5-triphenyltetrazolium chloride (TTC) is added per litre of medium. Tetrazolium salts are colourless but as the bacterium grows the dye is incorporated into the bacterial cells where it is reduced to an insoluble red pigment, formazan. The red colour forms only in the area of medium where the bacterium is growing (**38**).

37

37 The hanging-drop method for detection of motility.

38

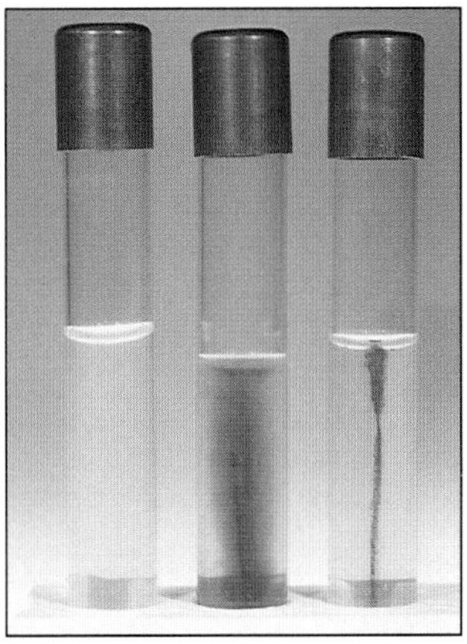

38 Motility medium (semi-solid) with TTC indicator; from left, uninoculated, motile and non-motile bacteria.

- SIM medium can be used to detect motility and will also indicate indole and hydrogen sulphide production.
- Motility media are prepared in test tubes. Two tubes of the medium are stab-inoculated using a straight wire. One tube is incubated at room temperature and the other at 37°C. The tubes are examined for motility after 24 and 48 hours. Motile bacteria migrate through the semi-solid medium which becomes turbid. If TTC has been incorporated into the medium, the motility is demonstrated by the formation of a red colour throughout the agar. The growth of a non-motile bacterium is confined to the stab line.

Oxidation-fermentation (O-F) test

This test is used to determine the oxidative or fermentative metabolism of a carbohydrate by the bacterium. The medium is semi-solid and usually contains glucose as the test sugar and bromothymol blue as the pH indicator. The uninoculated medium (pH 7.1) is green and if acid is produced by the bacterium, as a result of glucose utilization, the medium becomes yellow (pH 6.0). Bacteria that can metabolise glucose under either aerobic or anaerobic conditions are facultative anaerobes and in this test are said to be fermentative. The aerobes that require atmospheric oxygen for growth and metabolism are called oxidative. Some bacteria are unreactive in the conventional O-F medium, either because they are unable to grow in the basal medium or because they cannot attack glucose.

Two tubes of the O-F medium are heated in a beaker of boiling water immediately before use to drive off any dissolved oxygen. The tubes are then cooled rapidly under cold running water. Both tubes are stab-inoculated with the bacterium. A layer of sterile paraffin oil is layered on top of one of the tubes (sealed tube) to a depth of about 1 cm. The inoculated tubes are incubated at 37°C and examined in 24 hours and then daily for up to 14 days (**39**).

39

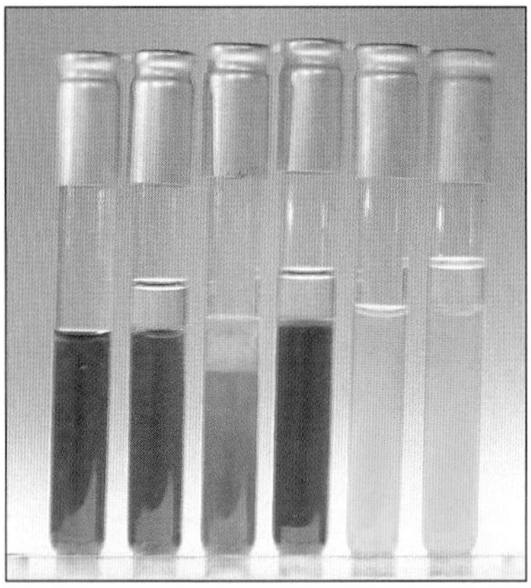

39 Oxidation-fermentation (O-F) test: results from left, unreactive (tubes 1 and 2); oxidation (tubes 3 and 4) and fermentation (tubes 5 and 6).

O-F Test reactions

Results	**Open tube**	**Sealed tube**	**Examples**
Unreactive	Green	Green	*Bordetella* spp.
Oxidation	Yellow	Green	*Pseudomonas* spp.
Fermentation	Yellow	Yellow	*Aeromonas* spp.

The conventional O-F medium is most suitable for non-fastidious Gram-negative bacteria. Modifications can be made to the medium to test for:

- Fastidious bacteria unable to grow in the medium. In this case the basal medium can be enriched with 2 per cent serum and/or 0.1 per cent yeast extract.
- Staphylococci and micrococci. The formula for the Staph. O-F test medium is:

Pancreatic digest of casein (tryptone)	10.0 g
Yeast extract	1.0 g
Agar	2.0 g
Bromocresol purple	0.001 g
Distilled water	1000 ml

The ingredients are heated gently to dissolve them and autoclaved at 121°C for 15 minutes. The medium is cooled to 50°C and sterile solutions of glucose and mannitol are added to a final concentration of 1 per cent for each sugar. The indicator is purple at pH 6.8 and yellow at pH 5.2.

Diagrams 7 and **8** give the primary identification of Gram-positive and Gram-negative bacteria, respectively, to a generic level on the basis of the previously described tests.

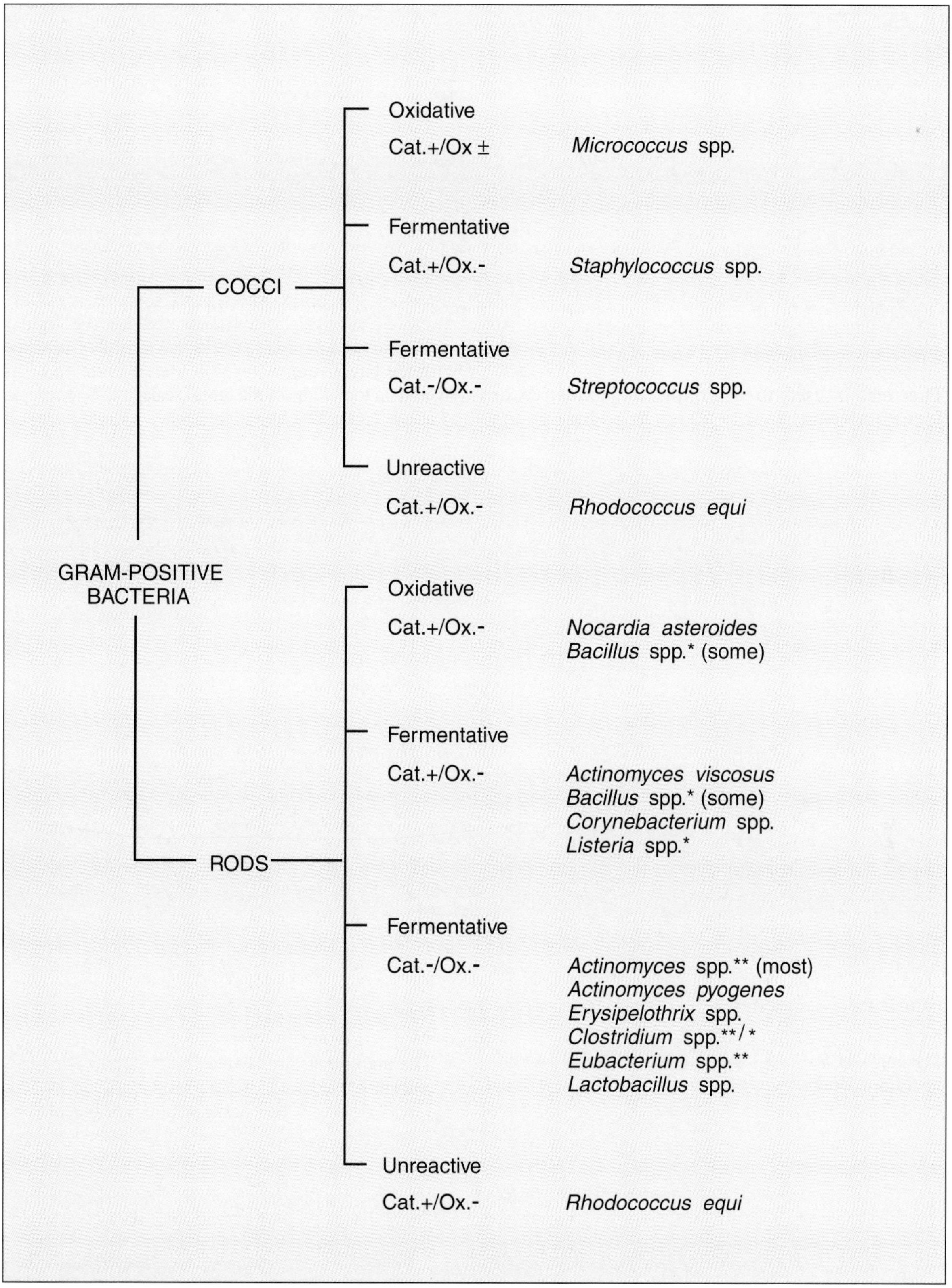

Diagram 7. Primary identification of Gram-positive bacteria. (Cat.= catalase; Ox.= oxidase; + = positive reaction; - = negative reaction; ± = variable; * = motile; ** = anaerobic)

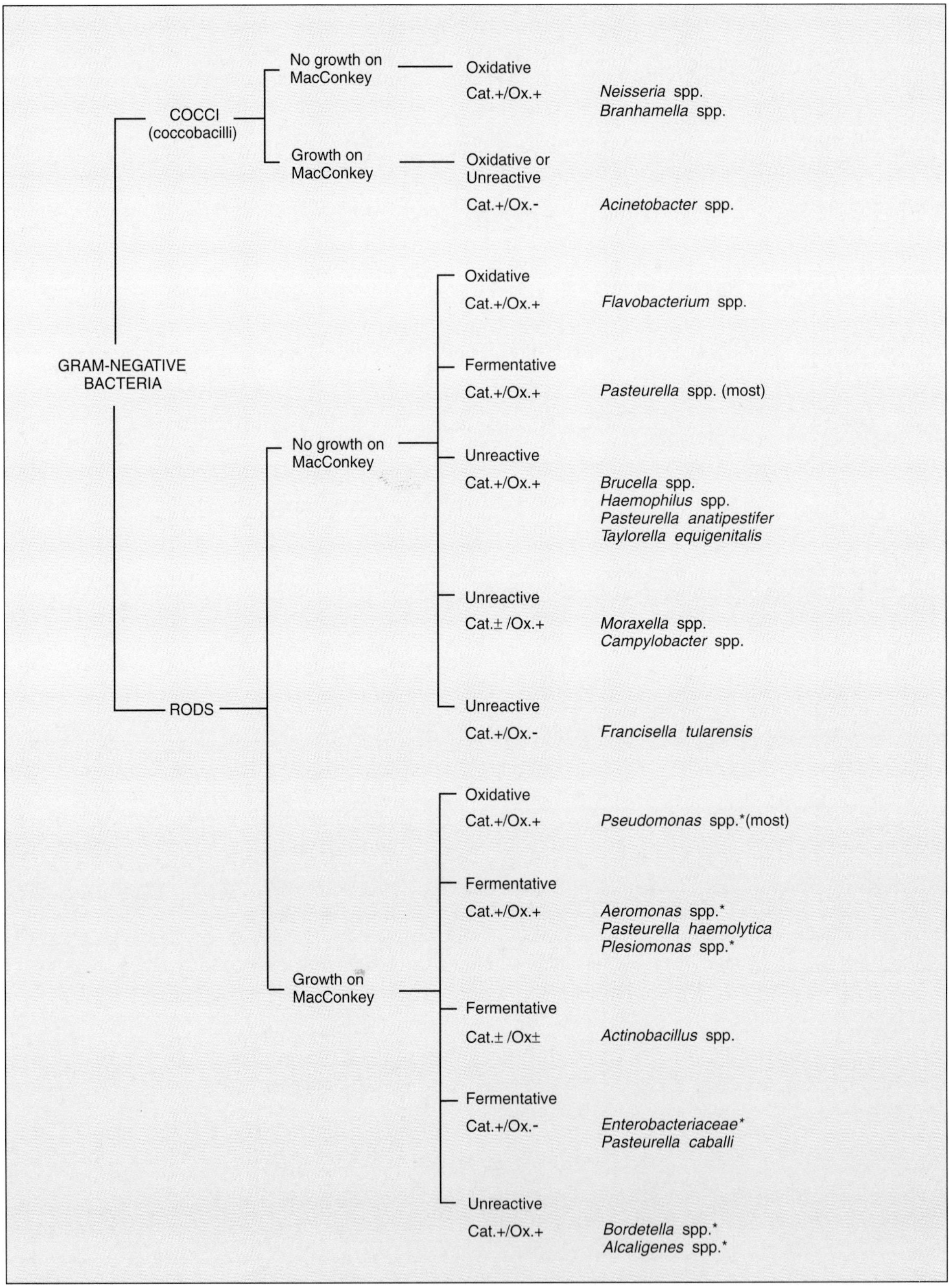

Diagram 8. Primary identification of Gram-negative bacteria. (Cat.= catalase; Ox.= oxidase; + = positive reaction; - = negative reaction; ± = variable; * = motile)

Secondary Biochemical Tests for the Identification of Bacteria

Once the bacterium has been identified to a generic level, further tests can be carried out to identify the species. Secondary biochemical tests can be grouped in the following categories:

- Commercial medium, prepared in tubes, incorporating several biochemical tests.
- Conventional biochemical tests usually carried out in small bottles or tubes. These are summarised in **Table 5** and additional methods of identification are given in Section 2 under the appropriate bacterial genus.
- Miniaturised methods that are available commercially and use small amounts of media in small chambers.
- Automated microbiology systems. These are beyond the scope of this book.

Table 5. Summary of some commonly used biochemical tests for the identification of bacteria.

Test	*Medium*	*Incubation (aerobic)*	*Product tested for*	*Test reagent*	*Result*	
					NEGATIVE (uninoculated)	*POSITIVE*
Aesculin (esculin) hydrolysis	1. Aesculin broth Aesculin 1g Peptone water 1000 ml Ferric citrate 0.5g Autoclave at 115°C for 10 minutes (**43**)**	Up to 7 days at 37°C.	Aesculin ↓ Aesculetin + iron salt ↓ Dark brown complex		Medium unaltered	Dark brown to black broth
	2. Aesculin agar e.g Edwards medium (**135**)	2–3 days at 37°C.		Wood's lamp (medium glows blue)	No browning of colonies (*Streptococcus agalactiae*)	Dark brown colonies and medium (*S.uberis*)
Carbohydrate fermentations 1. Peptone water sugars (non-fastidious bacteria)	1% sugar in peptone water (**44**)	24–48 hours at 37°C	Enzymatic attack on sugar with acid and gas (Durham tube) production	pH indicators: Phenol red → Andrade's →	Red (pH 7.4) Pale yellow (pH 7.2)	Acid produced: Yellow (pH 6.0) Pink (pH 5.5)
2. CTA medium (BBL) (fastidious bacteria)	Semi-solid medium with crystine and trypticase. Carbohydrate discs placed in medium (**45**)	18–48 hours at 37°C	Acid production and pH change	Phenol red	Red	Yellow
3. Solid agar medium such as purple agar base	1% sugar incorporated in solid agar medium (**118**)	24–48 hours at 37°C	Acid production and pH change	Bromocresol purple	Purple	Yellow colonies and medium

(*continued*)

Table 5. Summary of some commonly used biochemical tests for the identification of bacteria (continued).

Test	*Medium*	*Incubation (aerobic)*	*Product tested for*	*Test reagent*	*Result*	
					NEGATIVE (uninoculated)	***POSITIVE***
Citrate utilization	1. Koser citrate (liquid) Inoculate from broth with straight wire (**46**)	Up to 7 days at 37°C	Ability to use citrate as sole carbon source		No growth	Growth (turbidity)
	2. Simmons citrate Inoculate solid agar slant (**47**)	Up to 7 days at 37°C	As above	Bromo-thymol blue	Green (pH 6.9) (*E. coli*)	Blue (pH 7.6) (*Salmonella* spp).
Coagulated serum slant (Loeffler)	Heavy inoculum of suspect *Actinomyces pyogenes* as spot in centre of slant (**169**)	1–3 days at 37°C	Proteolytic action on coagulated serum		Slant unaltered (*Corynebacterium* spp).	Pitting of slant (*Actino-myces pyogenes*)
Decarboxy-lase (lysine and ornithine) and dehydrolase (arginine) tests	*Broth base** with: 0.5% L-arginine or 0.5% L-lysine or 0.5% L-ornithine. + 0.1% glucose (**41**)	Up to 4 days at 37°C	Arginine and ornithine to putrescine. Lysine to cadaverine. Products are alkaline.	Bromocresol purple	Yellow (acid) Glucose only attacked. (*Proteus* spp.)	Purple (alkaline) (*Salmonella* spp. (most))
	Tubed agar media: Lysine iron agar. MIO medium (ornithine).		Many bacteria first ferment the glucose with acid (yellow). If they can then attack the amino acid, the medium reverts to alkaline (purple)			
Gelatin liquefaction	1. Nutrient gelatin. Stab inoculation (**48**)	22°C for 30 days or 37°C for 14 days	Proteolytic activity (gelatinases) and gelatin liquefied		No liquefaction	Liquefaction. (Not solid at +4°C)
	2. Charcoal gelatin discs placed in a broth. (Discs do not liquefy at 37°C) (**49**)	37°C for up to 14 days	As above	Charcoal particles are released when the gelatin is liquefied	No change	Free charcoal particles
	3. X-ray film method. Small strip in heavy inoculum of bacteria in trypticase soy broth (**50**)	37°C for 48 hours	As above	Gelatin layer on X-ray film strip	No change (*Enterobacter* spp.)	Removal of gelatin layer leaving pale blue plastic film. (*Serratia marcescens*)

Table 5. Summary of some commonly used biochemical tests for the identification of bacteria (continued).

Test	*Medium*	*Incubation (aerobic)*	*Product tested for*	*Test reagent*	*Result*	
					NEGATIVE (uninoculated)	*POSITIVE*
Hippurate hydrolysis	Sodium hippurate 10 g Heart infusion broth 1 litre Sterilize at 115°C for 20 minutes (**51**)	24 hours at 37°C with an uninoculated control	Hippurate hydrolysed to benzoic acid and glycine	Centrifuge test. Add 0.2 ml ferric chloride reagent to 0.8 ml of the supernatant*	No precipitate. (*Actinobacillus equuli* and *Streptococcus pyogenes)*	Permanent preciptate. (*Actinobacillus lignieresii, Streptococcus agalactiae)*
Hydrogen sulphide	1. Iron salts in media. e.g. TSI and SIM agar (least sensitive test) (**40,42**)	16 hours at 37°C	Hydrogen sulphide gas production		No change (*E. coli)*	Blackening of the medium. (*Salmonella* spp.)
	2. Bismuth sulphite agar (plate medium) (**53**)	48 hours at 37°C	As above		No blackening	Blackening of colonies and media
	3. Lead acetate paper strip (most sensitive method).* Strip suspended over trypticase soy broth or serum glucose agar slants (*Brucella*) (**52**)	35°C for up to 7 days. Change lead acetate strip daily	Hydrogen sulphide gas production		No change on lead acetate strip	Blackening of the lead acetate strip. (*Brucella* spp.)
Indole test	1. Tryptone water (**54**)	1–2 days at 30°C	Tryptophan split to indole	Add 0.5 ml Kovac's* reagent to medium and shake. Read in 1 minute	Reagent layer yellow. (*Salmonella* spp.)	Reagent layer deep red. (*E. coli)*
	2. SIM medium in tubes (**42**)	18–24 hours at 37°C	As above	a) Kovac's reagent (0.2ml) to tube. Stand for 10 mins	No change in reagent colour	Reagent dark red
				b) Oxalic acid test paper* suspended over medium	No change in test strip	Pink colour at lower end of paper
	3. Spot test for indole	24–48 hours at 37°C on blood agar or nutrient agar	As above	Filter paper saturated with Kovac's reagent. Rub colony over filter paper with a glass rod	No reaction. *(Pasteurella haemolytica)*	Blue colour on streak within 30 seconds. (*P. multocida)*

(*continued*)

Table 5. Summary of some commonly used biochemical tests for the identification of bacteria (continued).

Test	***Medium***	***Incubation (aerobic)***	***Product tested for***	***Test reagent***	***Result***	
					NEGATIVE (uninoculated)	***POSITIVE***
Malonate utilization	Malonate broth. (0.3% sodium malonate) (**55**)	24 hours at 37°C	Utilisation of malonate as sole source of carbon	Bromothymol blue	No change. (*Salmonella* spp. (most))	Growth and a deep blue colour. (*S. arizonae)*
Methyl red (MR) test	Glucose phosphate peptone water (5 ml) (MR–VP broth) (**56**)	2 days at 37°C or 3–5 days at 30°C	Small molecular weight acids such as formic and acetic	MR reagent*: 5 drops to medium	Yellowish (*Enterobacter cloacae)*	Red (acid) pH 4.4–6.0 (*E. coli)*
Nitrate reduction	Nitrate broth (0.1% KNO_3): 5 ml (**58**)	24 hours at 37°C. (rarely up to 5 days)	Nitrate (NO_3) ↓ Nitrite (NO_2) ↓ Nitrogen gas (N_2)	Reagents for nitrite (A and B)* a. 0.5% alpha-naphthylamine in 5N acetic acid. b. 0.8% sulphanilic acid in 5N acetic acid. Add 5 drops of each reagent. Shake and wait 1–2 minutes	Colourless (no nitrite present) Add pinch of Zn dust. ↓ NO_3 converted to NO_2 ↓ Red (NO_3 not reduced)	Red (NO_3 to NO_2). Colourless (NO_3 reduced to N_2)
	2. KNO_3(40%) on filter paper*. Dry and sterilise. Place on blood agar and stab inoculate test bacterium 20mm from paper strip. Use heavy inoculum and *E. coli* as a positive control (**59**)	37°C and examine at 4 and 24 hours	Nitrate reduction		No reaction or very narrow zone of browning around inoculum	Wide zone of browning of medium between colony and strip. (*E. coli)*
ONPG test	Peptone water + 0.15% o-nitrophenyl-beta-D-galactopyranoside	24 hours at 37°C	The enzyme beta-galactosidase. Identifies potential lactose fermenters		Colourless after 24 hours. (*Salmonella* spp. (most))	Yellow colour. (*S. arizonae)*

Table 5. Summary of some commonly used biochemical tests for the identification of bacteria (continued).

Test	*Medium*	*Incubation (aerobic)*	*Product tested for*	*Test reagent*	*Result*	
					NEGATIVE (uninoculated)	***POSITIVE***
Phenylalanine deaminase test	Phenylalanine medium (BBL) (0.2% DL-phenylalanine) Tube with a slant. (**60**)	Inoculate heavily. 35°C for 4 or 18–24 hours	Phenyl-pyruvic acid formed. Acid reaction	10% aqueous ferric chloride. Add 4–5 drops direct to slant. Rotate and read in 1–5 minutes	Yellowish.	Green colour reaction in slant. (*Proteus, Morganella* and *Providencia* spp)
Phosphatase test	Nutrient agar + 0.01% phenolphthalein diphosphate (**61**)	18–24 hours at 37°C	Sufficient phosphatase to split phenol-phthalein diphosphate	Ammonia vapour. Hold colonies on the agar plate over an open bottle of ammonia	Unchanged colonies. (Coagulase -ve staphylococci)	Colonies bright pink. (Coagulase +ve staphylo-cocci)
Urease tests	1. Christensen media: a. Urea agar base + 2% urea (slant). b. Urea broth base + 2% urea. Use a heavy inoculum (**62**)	Up to 24 hours at 37°C	Urease: splits urea with formation of ammonia (alkaline)	Phenol red	Yellow (*Salmonella* spp.)	Red (alkaline) (*Proteus* spp.)
	2. Spot test: moisten filter paper with a few drops of 10% urea agar base concentrate (Difco,BBL). Rub some culture onto the paper with a glass rod test).		As above	As above	No change	Pink or red streak within 2 minutes
Voges–Proskauer (VP) test	5 ml glucose phosphate peptone water. (MR-VP broth) (**57**)	3 to 5 days at 30°C	Acetoin derived from glucose	3 ml of 5% alphanaph-thol in absolute ethyl alcohol and then 1ml of 40% KOH. Shake and leave for 5 minutes	Colourless (*E. coli*)	Red (*Entero-bacter* spp. (most))

* = further details of medium or reagent given in **Appendix 2.** ** = bold figures in brackets indicate the relevant photograph.

Commercial media incorporating several biochemical tests

- **Kligler's iron agar** contains two carbohydrates (sugars), glucose 0.1 per cent and lactose 1.0 per cent together with chemicals that indicate hydrogen sulphide production. The medium is similar to, and largely superseded by, triple sugar iron agar.
- **Triple sugar iron (TSI) agar** has three sugars, glucose 0.1 per cent, lactose 1.0 per cent and sucrose 1.0 per cent. Phenol red is the pH indicator and ferrous sulphate or ferric ammonium citrate with sodium thiosulphate detects hydrogen sulphide production. The medium is poured into test tubes and these are sloped before the agar sets to form slants. The medium is red (pH $\geq$ 7.3) if uninoculated or when an alkaline reaction occurs, and yellow (pH $\leq$ 6.8) if an acid reaction occurs (**40**). TSI agar is essentially for the presumptive identification of salmonella but is also useful for the differentiation of other members of the *Enterobacteriaceae*. This differentiation is enhanced if a tube of lysine decarboxylase broth is inoculated in parallel with the tube of TSI agar.
- **Lysine iron agar:** a solid medium, dispensed in tubes, for the detection of the decarboxylation of lysine and the production of hydrogen sulphide.
- **SIM medium:** a composite medium for the determination of hydrogen sulphide and indole production and motility (**42**). It is used mainly for the *Enterobacteriaceae*.
- **MIO medium**: this medium incorporates tests for motility, indole production and the decarboxylation of ornithine in the *Enterobacteriaceae*.

40

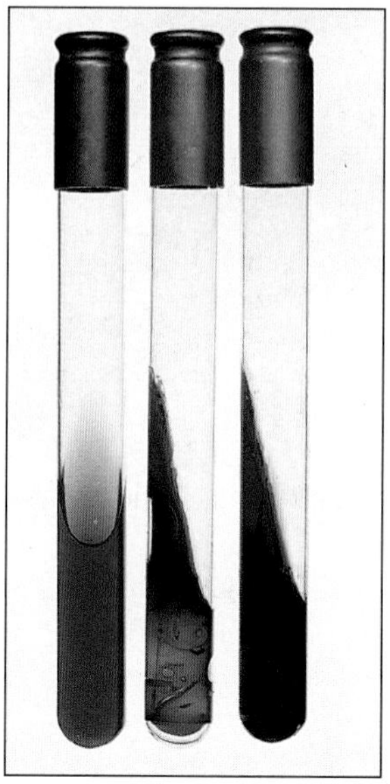

40 Triple sugar iron agar: from left, uninoculated, *Salmonella* sp. after 16 and 24 hours' incubation.

Inoculation of TSI agar and lysine decarboxylase broth

One isolated colony characteristic for *Salmonella* species from the selective/indicator media (XLD, brilliant green or MacConkey agars) is touched with a straight inoculating wire. A tube of TSI agar is stab inoculated in the middle of the agar to within 5 mm from the bottom of the tube. On the withdrawal of the straight wire, the entire slant is streaked (right to the top). The wire will still have sufficient bacterial cells to inoculate the tube of lysine broth. Both tubes are incubated at 37°C for 16 hours **with a loose cap on the TSI agar.** This is essential for the correct reaction to occur in the medium. If the incubation period is prolonged, the black colouration due to H_2S production tends to obscure the colour of the butt (**40**). The biochemical reactions in lysine decarboxylase broth (**41**) are summarised in **Table 5.** The biochemical reactions that occur in TSI agar are discussed in Chapter 18 (*Enterobacteriaceae*), the general interpretation of these reactions are as follows:

- Red (alkaline) slant and yellow (acid) butt: glucose fermentation only.
- Yellow (acid) slant and yellow (acid) butt : lactose and/or sucrose used as well as glucose.
- Blackening of the medium: hydrogen sulphide production

Most *Salmonella* species give a red slant, yellow butt and produce H_2S when inoculated into TSI agar. The notation for these reactions is R/Y/H_2S+.

41

41 Lysine decarboxylase broth: from left, uninoculated, negative and positive reactions.

42

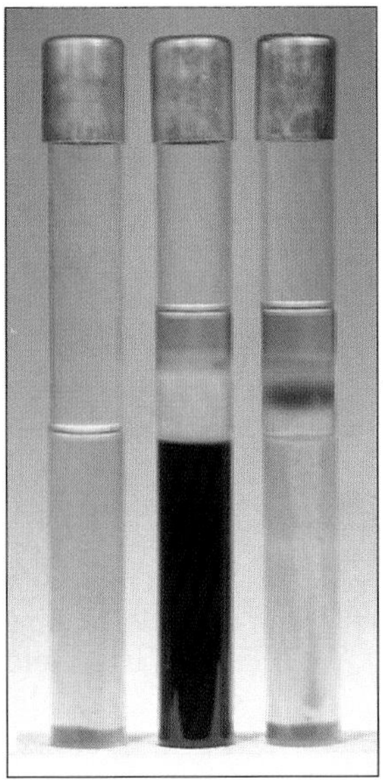

42 SIM agar: from left, uninoculated, H_2S+/Indole-/Motility+ and H_2S-/Indole+/Motility-. This medium (like TSI) uses iron salts to detect H_2S production.

Conventional biochemical tests

A summary of some of the commonly used conventional biochemical tests are given in **Table 5** and are illustrated (**43** to **62**, inclusive).

Tests to determine the range of 'sugars' that a particular bacterium can catabolise are among the more commonly used biochemical tests. Examples of these 'sugars' are:

Monosaccharides: arabinose, fructose, galactose, glucose, mannose, ribose, ribulose and xylose.
Disaccharides: sucrose (glucose and fructose molecules), maltose (glucose ×2), lactose (glucose and galactose)
Trisaccharide: raffinose (glucose, fructose and galactose)
Polysaccharide: inulin
Alcohols: adonitol, dulcitol, mannitol and sorbitol.

The fermentative or oxidative attack on the sugars by the bacterium produces acidic metabolites detected by a pH indicator. Gas may also be produced and in conventional peptone water sugars (1 per cent sugar), the gas production is demonstrated by the inclusion of a Durham tube to trap the gas (**44**). CTA medium (BBL) can be used to test the more fastidious bacteria that are unable to grow in peptone water sugars. The sugar under test is added to the inoculated medium incorporated in a paper disc (**45**). Another method to determine whether a bacterium can catabolise a certain sugar is to inoculate a solid agar medium to which the sugar has been added. An example of this type of test is purple agar with 1 per cent maltose (pH indicator bromocresol purple) used to distinguish *Staphylococcus aureus* from *S. intermedius* (**118**).

43

43 Aesculin hydrolysis: from left, uninoculated, positive and negative.

44

44 Peptone water 'sugars' (Andrade's indicator) with a Durham tube to indicate gas production. From left, uninoculated, negative, positive fermentation/no gas production and positive fermentation/gas production.

45

45 CTA medium and carbohydrate disc for fastidious bacteria: from left, uninoculated, positive and negative (phenol red indicator).

46

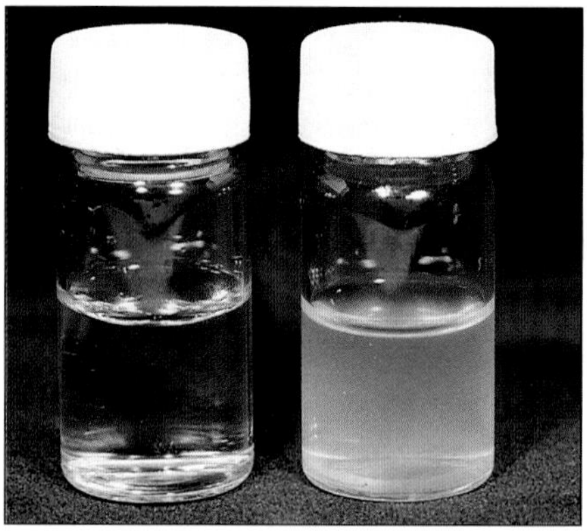

46 Koser citrate: uninoculated or negative (left), and positive (right).

47

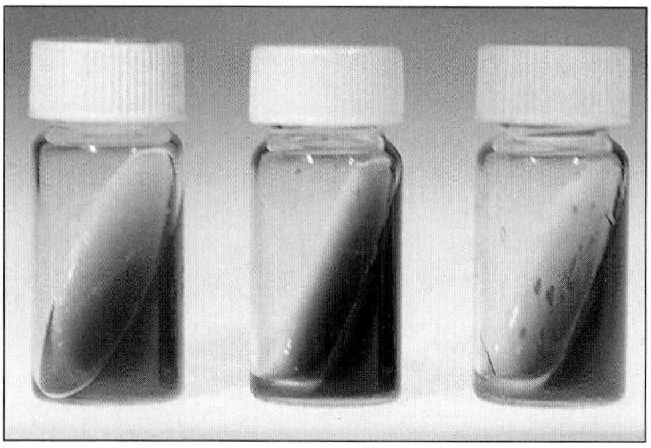

47 Simmons citrate: from left, uninoculated, positive and negative.

48

48 Gelatin liquefaction in nutrient gelatin (stab inoculation): from left, uninoculated or negative, *Serratia* sp. (positive), and *Proteus* sp. (positive).

49

49 Gelatin liquefaction using charcoal gelatin discs: uninoculated or negative (left) and positive (right).

50

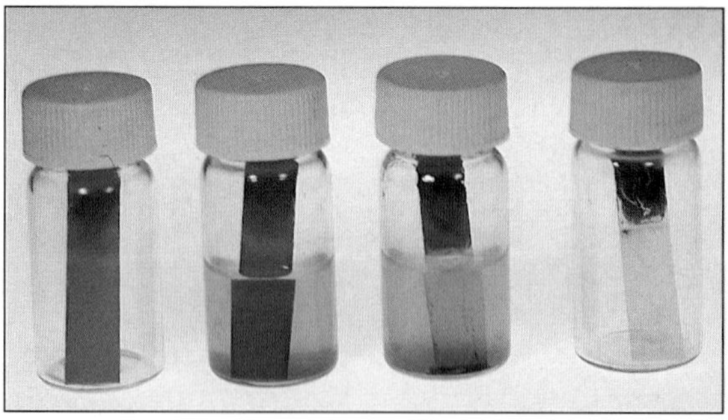

50 Gelatin liquefaction using the X-ray film method: from left, strips 1 and 2 negative, strips 3 and 4 positive.

51

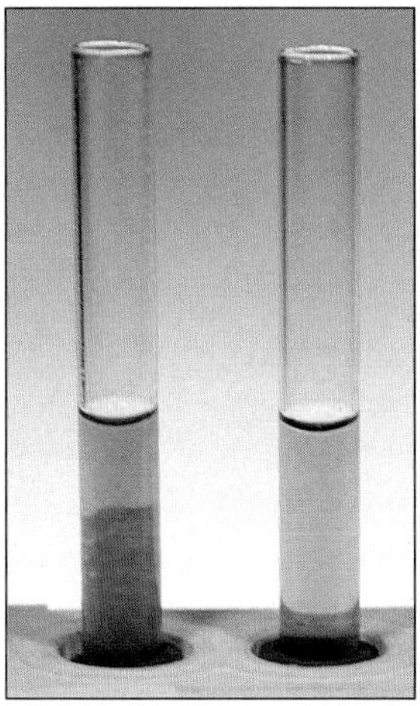

51 Hippurate hydrolysis: permanent precipitate, left (positive) and no precipitate, right (negative).

52

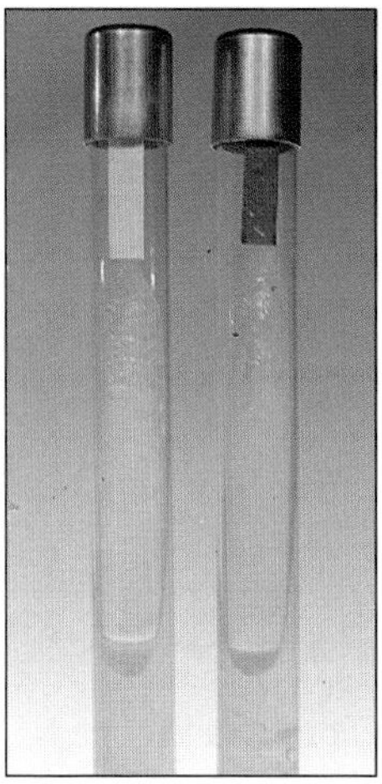

52 Lead acetate paper for detection of H_2S production: negative (left) and positive (right).

53

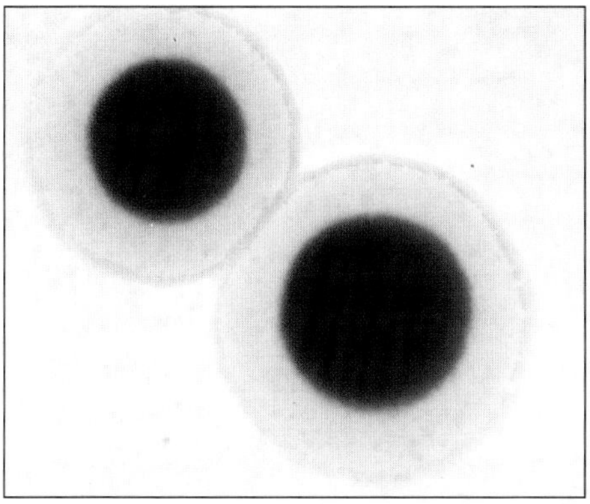

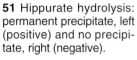

53 Close-up of *Salmonella enteriditis* colonies on bismuth sulphite medium, using bismuth salts to detect H_2S production.

54

54 Indole test: from left, uninoculated, positive and negative.

55

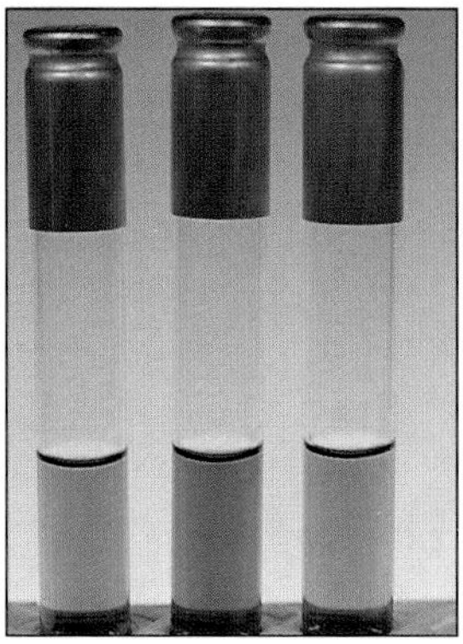

55 Malonate utilization: from left, uninoculated, positive and negative.

56

56 Methyl red (MR) test: from left, uninoculated, positive and negative.

57 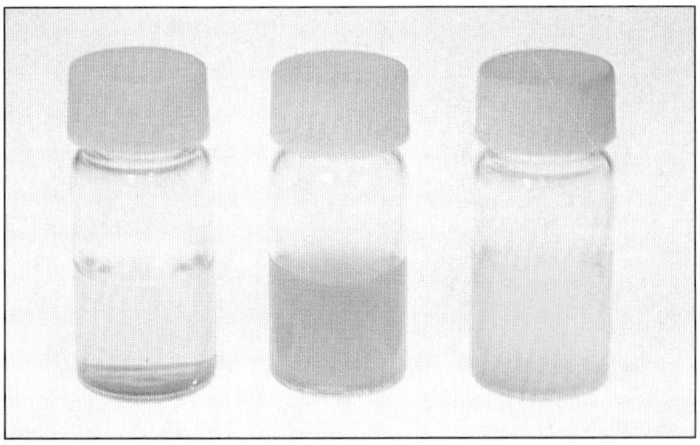

57 Voges–Proskauer (VP) test: from left, uninoculated, positive and negative.

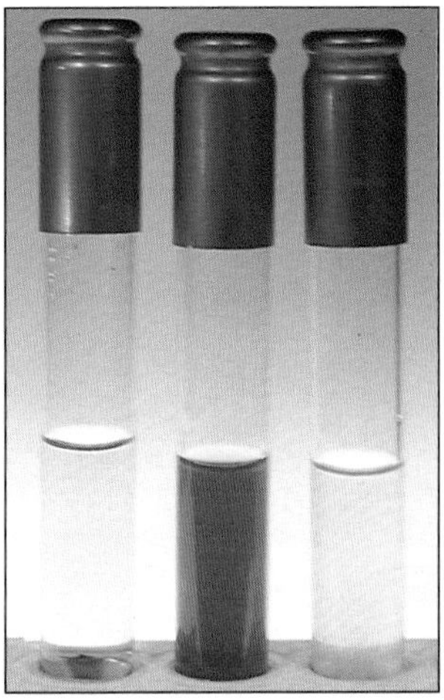 58

58 Nitrate reduction test (nitrate broth): from left, uninoculated, positive for nitrite and negative for nitrite.

59 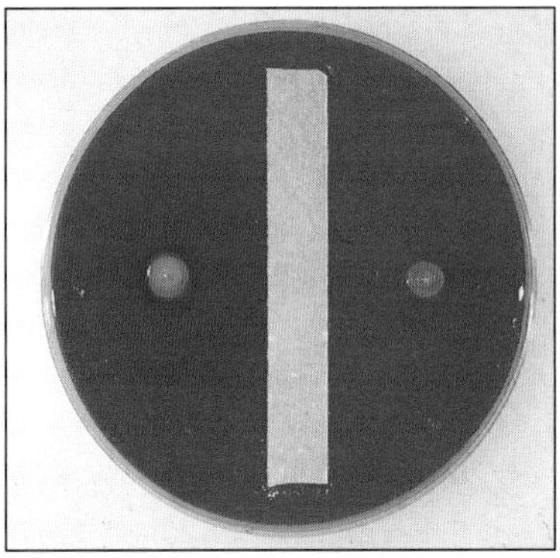

59 Nitrate reduction test (blood agar/nitrate strip): wide zone of browning around colony indicates a positive reaction (left) while absence of reaction (right) is negative.

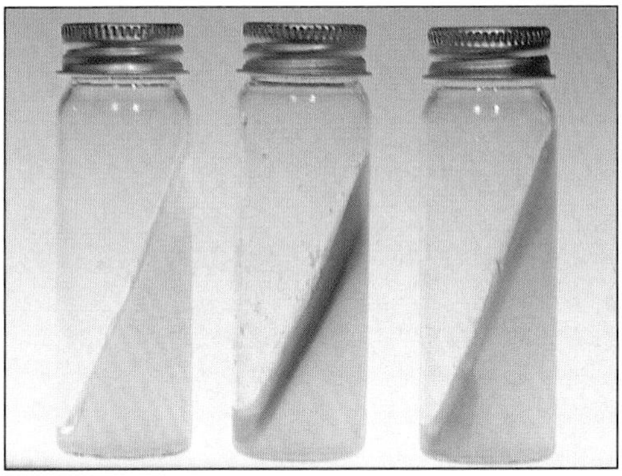 60

60 Phenylalanine deaminase test: from left, uninoculated, positive and negative.

61 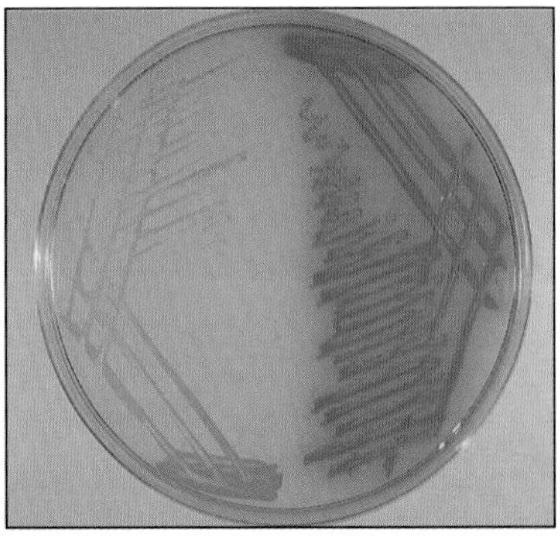

61 Phosphatase test: negative reaction left (coagulase-negative staphylococcus) and positive reaction right (coagulase-positive staphylococcus).

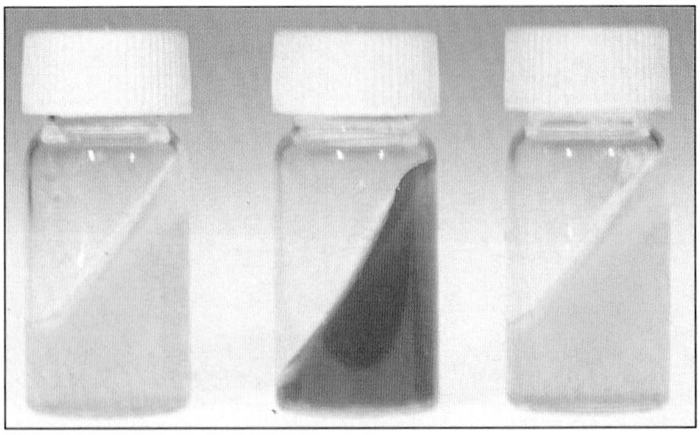 62

62 Urease test (Christensen medium): from left, uninoculated, positive and negative.

These are available from several commercial companies including Analytab Products; Roche Diagnostics; Flow Laboratories; Organon Teknika; MicroMedia Systems; BBL Microbiology Systems; Abbott Laboratories; Innovative Diagnostic Systems and Scarborough Microbiologicals. One of the most widely used systems is API (Analytab Products or bioMérieux) and individual biochemical strips are available for the identification of the *Enterobacteriaceae* and other Gram-negative bacteria. Strips are also available for anaerobes, non-fermenting bacteria, streptococci, staphylococci, and some other Gram-positive bacteria and yeasts. The results of the miniaturised biochemical tests correlate well with those obtained by the conventional methods. The advantages of the miniaturised systems include the convenience of inoculation; the media and reagents are supplied ready for use and the results are easy and quick to interpret. A diagnostic laboratory must weigh these advantages and its own particular needs against the cost of the systems.

One of the most commonly used miniaturised identification kits is the API 20E for the identification of the *Enterobacteriaceae* and some other Gram-negative bacteria. A brief description of the API 20E is given here as an example of one of the miniaturised systems. Full instructions are supplied by Analytab Products with the kit.

Inoculation and incubation

- A single, well isolated colony of the bacterium to be identified is made into a homogenous suspension in 5 ml of sterile distilled water.
- This suspension is used to inoculate the tubes of the API 20E strip using a sterile Pasteur pipette. The pipette should be held against the side of the tube when delivering the inoculum to avoid introducing bubbles into the test solution. The tube only is filled for the majority of the biochemical tests, the exceptions are:

a) CIT, VP and GEL (notated by a 3-sided box) in which both the tube and cupule are filled.
b) ADH, LDC, ODC, URE and H_2S (underlined); here the cupule is filled with sterile mineral oil.

- The API 20E strip is placed in its incubation box with a little water at the bottom to prevent dehydration and incubated between 35–37°C for 18–24 hours.
- An oxidase test is carried out separately using a conventional method. The members of the *Enterobacteriaceae* are oxidase-negative while many of the other Gram-negative bacteria are positive to the test.

Reading the biochemical reactions of the API 20E

Reagents are added to the following biochemical tubes before reading the reactions :

a) TDA (tryptophane deaminase): 1 drop of TDA reagent giving an immediate reaction.
b) IND (indole): 1 drop of indole reagent. The test is read after 2 minutes.
c) VP (acetoin production): 1 drop of reagent VP1 and 1 drop of VP2. Read after 10 minutes.

The results are read with the aid of the interpretation chart reproduced in **Table 6**.

Identification of the bacterium

The positive (+) or negative (–) reactions, read from the API 20E strip (**63**), are placed in the spaces provided on the API 20E Report Sheet (**Diagram 9**). The code numbers for the positive reactions in each triplet of biochemical reactions are added together. This gives a 7-digit 'profile number' for the bacterium. The genus and species for the bacterium can be obtained by use of the 7-digit profile number and the API 20E Analytical Profile Index or by means of the API Identification Table.

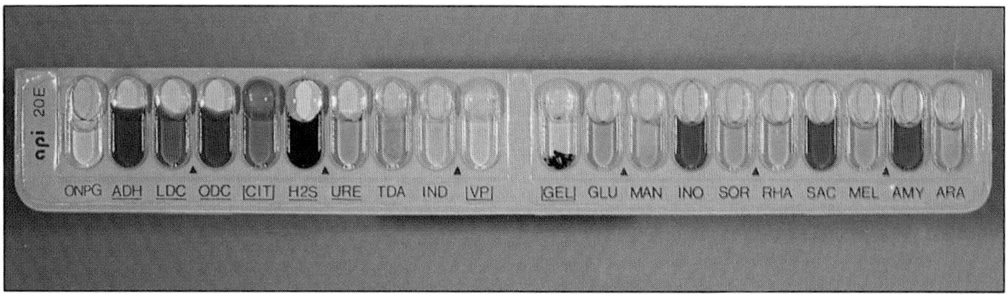

63

63 API 20E miniaturised identification system showing the reactions of a *Salmonella* sp. The results are recorded on the report sheet (**Diagram 9**).

api 20 E **Table 6.** Interpretation chart for API 20E.

TESTS	SUBSTRATES	REACTIONS/ENZYMES	RESULTS	
			NEGATIVE	POSITIVE
ONPG	ortho-nitro-phenyl-galactoside	beta-galactosidase	colourless	yellow (1)
ADH	arginine	arginine dihydrolase	yellow	red/ orange (2)
LDC	lysine	lysine decarboxylase	yellow	orange
ODC	ornithine	ornithine decarboxylase	yellow	red/ orange (2)
CIT	sodium citrate	citrate utilization	pale green/ yellow	blue-green/ green (3)
H_2S	sodium thiosulfate	H_2S production	colourless/ greyish	black deposit/ thin line
URE	urea	urease	yellow	red/ orange
TDA	tryptophane	tryptophane deaminase	**TDA** / immediate	
			yellow	dark brown
IND	tryptophane	indole production	**IND** / 2 mn maxi	
			yellow ring	red ring
VP	sodium pyruvate	acetoin production	**VP 1** + **VP 2** / 10 mn	
			colourless	pink/ red
GEL	Kohn's gelatin	gelatinase	no diffusion of black pigment	diffusion of black pigment
GLU	glucose	fermentation/ oxidation (4)	blue/ blue-green	yellow
MAN	mannitol	fermentation/ oxidation (4)	blue/ blue-green	yellow
INO	inositol	fermentation/ oxidation (4)	blue/ blue-green	yellow
SOR	sorbitol	fermentation/ oxidation (4)	blue/ blue-green	yellow
RHA	rhamnose	fermentation/ oxidation (4)	blue/ blue-green	yellow
SAC	sucrose	fermentation/ oxidation (4)	blue/ blue-green	yellow
MEL	melibiose	fermentation/ oxidation (4)	blue/ blue-green	yellow
AMY	amygdalin	fermentation/ oxidation (4)	blue/ blue-green	yellow
ARA	arabinose	fermentation/ oxidation (4)	blue/ blue-green	yellow
OX	on filter paper	cytochrome-oxidase	**OX** / 5-10 mn	
			colourless	violet ring
NO_3-NO_2	GLU tube	NO_2 production	**NIT 1** + **NIT 2** / 2-3 mn	
			yellow	red
		reduction to N_2 gas	**Zn**	
			red	yellow
MOB	**API M** or microscopic	motility	non-motile	motile
MAC	MacConkey medium	growth	absence	presence
OF	glucose (**API OF**)	fermentation : closed oxidation : open	green green	yellow yellow
CAT	in any negative sugar	catalase production	H_2O_2 / 1-2 mn	
			no effervescense	effervescense

(1) a very pale yellow should be considered positive as well
(2) an orange colour after 24 hours incubation must be considered negative
(3) reading made in the cupule (aerobic)
(4) fermentation begins in the lower portion of the tubes., oxidation begins at the cupule.

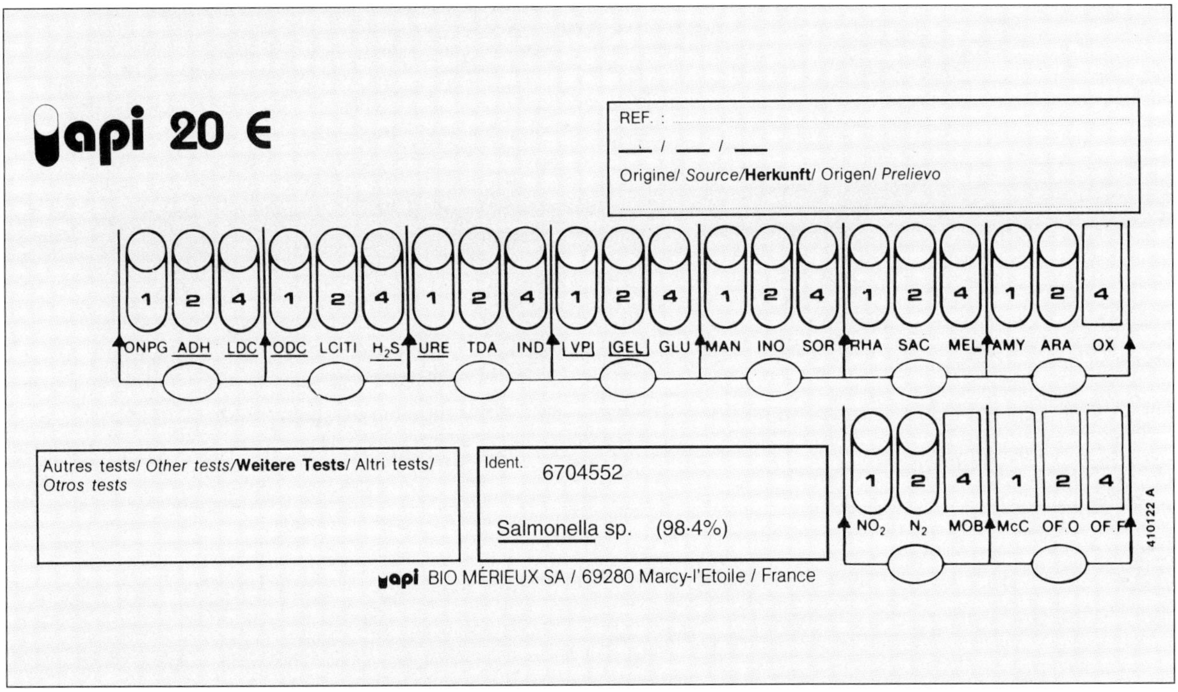

Diagram 9. An API 20E Report Sheet recording the results of the biochemical reactions and the seven digit profile number for a *Salmonella* species that is illustrated (63). (*See* Table 6 for abbreviations)

Bacterial Cell Counting Techniques

Sometimes it is necessary in diagnostic microbiology to enumerate bacterial cells in fluids such as autogenous vaccines, water, milk or urine samples. Both viable and total counts can be carried out. Viable counting techniques are more commonly used in diagnostic and food hygiene procedures. Viable bacteria are capable of multiplication with the production of visible colonies on or in agar media. In viable counting methods the assumption is made that one well-spaced, bacterial cell gives rise to one colony. Bacterial colonies, rather than the bacterial cells, are counted in most of these methods. Total counts will enumerate both viable and non-viable bacterial cells. There are inherent errors in all of these methods.

Viable Counting Methods

Serial ten-fold dilutions of the original fluid, containing bacteria, must first be made for each of the methods (**Diagram 10**). These must be carried out as accurately as possible to minimise avoidable errors and an aseptic technique should be used.

Spread plate method

A range of dilutions is used and an inoculum of 0.1 ml of each dilution is placed on the surface of an agar plate. The inoculum is spread rapidly over the entire agar surface using a thin, bent glass rod or a flame-sterilised nichrome wire, bent in a L-shape. Plate count agar, nutrient agar or even MacConkey agar can be used if a viable count of *Escherichia coli* is required. At least two, and preferably four, plates should be inoculated per dilution. The plates are incubated for 24–48 hours at 25–37°C. The incubation temperature will depend on whether environmental or pathogenic bacteria are being sought. After incubation, plates inoculated with a sample dilution yielding between 30 and 300 colonies are read, for greatest accuracy (**64**). The colony count should be an average of the two or four plates inoculated with the selected dilution. Various instruments are available to facilitate counting the colonies, including electronic counters (**65**).

64

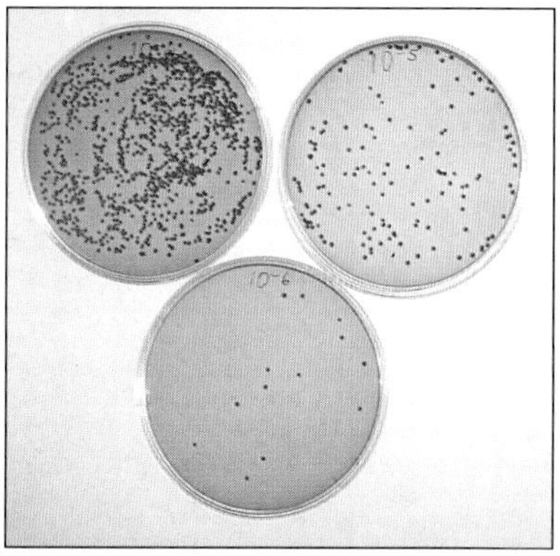

65

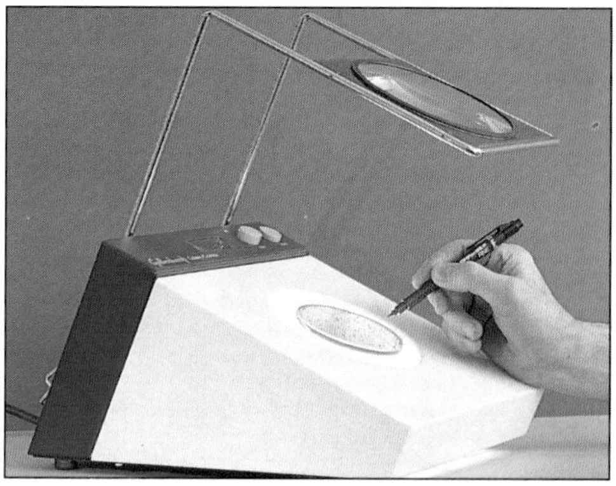

64 Colony counting, surface spread technique on MacConkey agar. The 10^{-5} dilution is suitable for counting.

65 Colony counting using an electronic counter.

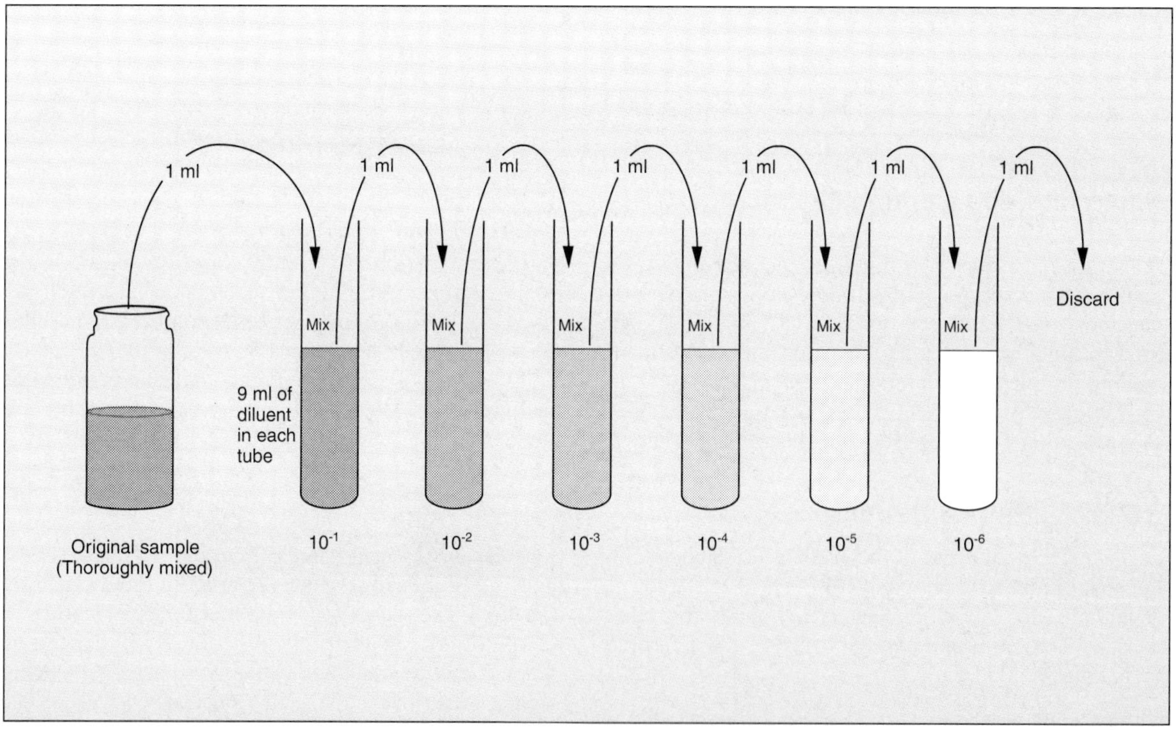

Diagram 10. Preparation of ten-fold dilutions of a bacterial suspension before conducting a viable count to find the number of bacteria/ml in the original sample. The sample should be thoroughly mixed before sampling and a separate pipette must be used for every transfer step.

Pour plate method

This method is similar to the spread-plate technique, except that the 0.1 ml inoculum is mixed thoroughly with molten agar, previously held in a water bath at 50°C. Two or four plates should be inoculated with each dilution. The agar is allowed to set and then incubated at 25–37°C for 24–48 hours. Plates inoculated with a sample dilution that yields between 30 and 300 colonies per plate should be read. The colonies will be distributed throughout the agar as well as on the surface. The subsurface colonies assume a biconvex shape.

Example of a calculation for the spread and pour plate methods:

If there is a mean count of 250 colonies per plate at the 10^{-4} dilution and as the inoculum was 0.1 ml per plate:
The number of bacteria/0.1 ml of original sample = 250×10^4
Thus, the number of bacteria/1.0 ml of original sample = $250 \times 10^4 \times 10 = 2.5 \times 10^7$

Miles–Misra technique

This method has the advantage of being economical with agar media. Lines can be drawn on the bottom of an agar plate with a waterproof marker, dividing it into 8 sectors. An inoculum of 0.02 ml, delivered as a drop, is placed on the agar in each sector. At least 4 drops per sample dilution should be used. The inocula are allowed to dry and the plates incubated at 25–37°C for 24–48 hours. A sample dilution yielding about 30 colonies per drop should be selected (**66**). An average colony count from at least 4 drops must be obtained. The calculation is similar to that for the two previous methods, but as the inoculum was 0.02 ml, the conversion factor will be 50 to obtain a figure for the bacteria/ml in the original sample.

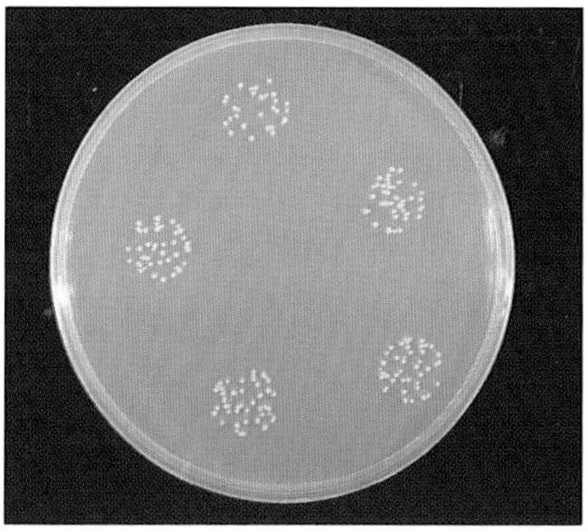

66 Miles–Misra technique: a viable bacterial counting method.

Filtration method

This is a useful method for determining the number of bacteria in a water sample or other clear fluid where the bacterial number is low. A known volume of water is passed through a membrane filter of pore size 0.22 µm. The filter will retain the bacterial cells, and is aseptically placed, bacterial-side up, on the surface of an agar plate. The medium can be selective or non-selective, depending on the bacterial species being sought. Colonies will form on the surface of the filter after incubation and can be counted (**67**). As the volume of the water or fluid is known, the bacteria/ml or per 100 ml of sample can be calculated.

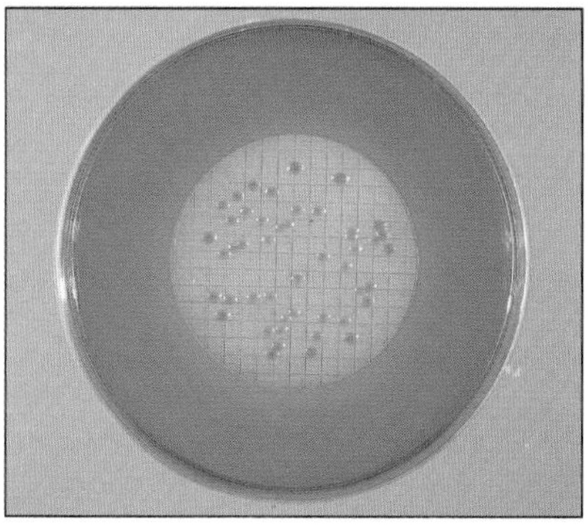

67 Filtration method for enumerating viable bacteria in water. After filtration, the membrane filter was placed on MacConkey agar with the subsequent growth of lactose-fermenting colonies.

Most probable number (MPN) techniques

These techniques are based on statistical probabilities with the assumption that there is a normal distribution of bacteria in liquid samples. If the liquid sample contains one viable bacterial cell, its growth and multiplication in a suitable broth can be detected by manifestations such as turbidity or acid and gas production. The methods can be used for most bacteria, but they are most commonly used for the detection of coliform bacteria in water supplies. MacConkey broth with bromocresol purple as the pH indicator is often used for coliform counts. Acid production is indicated by a yellow colouration of the broth and gas is trapped by a Durham tube.

One recommended method is to take one 50 ml, five 10 ml and five 1 ml quantities of the water sample. The 50 ml and 10 ml volumes are each added to their own volume of double-strength broth, while the 1 ml samples are each added to 5 ml of single-strength broth. The inoculated tubes are incubated at 35°C for 48 hours and then each is examined for acid and gas production. By referring to standard MPN probability tables (Anon, 1982) the MPN of coliforms/100 ml of water sample can be determined. For example, if one each of the tubes inoculated with 50 ml, 10 ml and 1 ml samples of water, respectively, showed acid and gas production, then from the tables the MPN of coliforms/100 ml water would be 5. For the differential coliform count specifically to detect *Escherichia coli*, tubes showing acid and gas production are subcultured into fresh MacConkey broth and incubated at 44°C. Formation of acid and gas within 48 hours at this temperature is presumptive for *E. coli* and indicative of faecal pollution of the water.

Total Counts of Bacterial Cells

These methods do not distinguish between viable and non-viable cells and thus the bacterial count will include both living and dead cells.

Breed's direct smear method

This technique is used most commonly for counting bacteria in milk. A grease-free microscope slide is placed over a template 1 cm × 1 cm (area of 100 mm^2) and a 0.01 ml of sample is carefully spread over this area. The smear is allowed to air-dry, fixed by heat and stained with methylene blue for about 1 minute. After air-drying, the stained smear is examined under the oil-immersion objective. The bacterial cells should be counted in at least 50 fields throughout the area of the smear. An average bacterial cell count per field (N) should be obtained. The radius (r) for the particular microscope's oil-immersion field can be found (in mm) using a slide and eyepiece micrometer. The area of the field will be πr^2 or approximately $3.14 \times r^2$ (mm^2).

$$\text{Bacteria/ml in sample} = \frac{N \times \text{Area of smear } (100\ mm^2)}{\text{Area of one field } (3.14 \times r^2)} \times 100 = \frac{N \times 10^4}{3.14 \times r^2}$$

(where N = average bacterial count/field and r = radius of microscope's oil immersion field in mm).

The radius of the oil immersion field is usually about 0.08 mm so that the area of the field will be 0.25 mm^2 and bacteria /ml = $N \times 4 \times 10^4$.

Counting chamber method

The Helber chamber was specially developed for counting unstained bacteria in a suspension and is supplied by Hawksley Ltd., London, but is not always readily available. An improved Neubauer haemocytometer can be used instead. The technique for counting bacterial cells and the calculation of bacteria/ml in the liquid sample is similar to that for erythrocytes.

To prevent motility of the bacteria, 2–3 drops of full-strength formalin/10 ml of bacterial suspension can be added. The chamber is filled and viewed under the low-power objective in order to orientate the marked grid. Then the high-dry objective is used to count the bacteria in the five areas marked C (**Diagram 11**) in the central region of the grid. Each of these larger squares (C), one at each corner and one central, is divided into 16 smaller squares. Thus, the bacterial cells are counted in 80 (5 ×16) of the smallest squares in the grid. The average number of bacteria (N) per small square can be calculated. The volume of fluid over each small square is:

Helber chamber: $0.0025\text{mm}^2 \times 0.02$ mm depth = 0.00005 mm^3

Neubauer haemocytometer: $0.0025\text{mm}^2 \times 0.1$ mm depth = 0.00025 mm^3

$$\text{Bacteria/ml in sample} = \frac{N \times 1 \times 1000}{0.00005} = N \times 20{,}000{,}000 \quad \text{(Helber)}$$

$$= \frac{N \times 1 \times 1000}{0.00025} = N \times 4{,}000{,}000 \quad \text{(Neubauer)}$$

Turbidity standards

Brown's or McFarland's opacity tubes are available commercially. They consist of a series of ten numbered, standard, thin glass tubes containing different dilutions of suspended barium chloride or barium sulphate, giving a range of opacities. The test bacterial suspension is placed in a 'blank' tube of similar dimensions to the standards. A visual comparison of opacity is made by rolling the test suspension across a printed page and matching it with a standard of comparable opacity. Tables are supplied with the opacity tubes that give the numerical equivalents (bacteria/ml) of the opacity standards for a certain range of bacteria. It is a convenient and simple method, but gives only an approximate total bacterial count.

Coulter counter

Coulter counters are automated, electronic counting instruments, usually used in haematology, but they can be adjusted to conduct total bacterial cell counts.

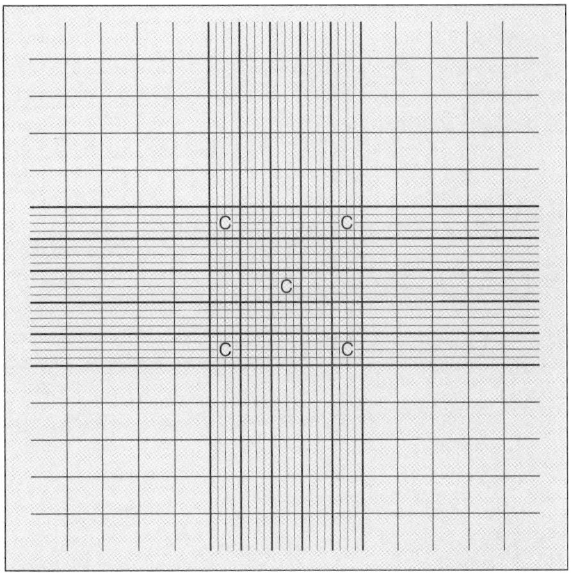

Diagram 11. Grid on a Neubauer haemocytometer. This chamber can be used to conduct a total bacterial cell count. The bacterial cells should be counted in the five areas marked 'C'. Each is divided into 16 of the smallest squares on the grid.

Surface Contact Plates

Special plastic plates are available to allow direct sampling of flat surfaces for bacteria. The technique can be used to detect a specific pathogen, such as salmonella, when a selective medium would be appropriate, or to determine the degree of contamination of a surface using a non-selective medium. An exact quantity of agar must be used to fill these plates as the agar surface should project slightly above the rim of the plate. Surfaces are sampled by placing the agar gently on the area, the plate lifted carefully and the lid replaced. The plates are incubated at 30–37°C for 24–48 hours and examined for colonial growth (**68**). If required, the number of bacteria/cm^2 of surface could be calculated, as the plastic plates incorporate a grid on the base.

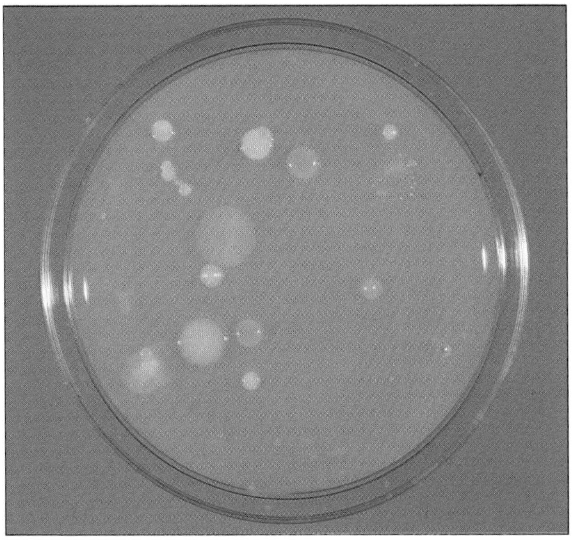

68

68 Surface contact plate for sampling surfaces: the agar projects slightly above the plastic rim to allow contact with a surface. This plate was used to sample a bench top and a mixed microbial flora was recovered.

Use of Marker Bacteria

Occasionally investigations of infectious agents may require the use of a marker organism. *Serratia rubidaea* is ideal for the purpose as it is not considered to be pathogenic, is rarely isolated from animals or the environment and its colonies have a distinctive, red pigmentation. The marker bacterium employed should have similar characteristics to the organism being investigated. Marker bacteria can be used for various purposes such as testing the efficiency of a disinfection programme, studying the dispersal of pathogenic microorganisms as an aerosol or in the determination of the efficacy of sewage treatment. A MacConkey agar plate from an investigation to determine aerosol dispersal of bacteria is illustrated (**69**).

Reference

Anon (1982). Departments of the Environment, Health, Social Security and Public Health Laboratory Service. *The Bacteriological examination of drinking water supplies*. Report No. 71, HMSO, London.

69

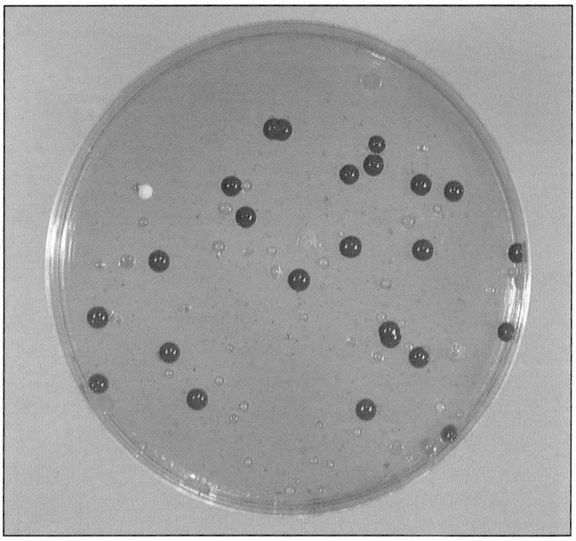

69 Demonstration of an aerosol of bacteria using *Serratia rubidaea* (red colonies) as a marker organism (MacConkey agar).

5 Diagnostic Applications of Immunological Tests

The Immune System

Most species of animals, both invertebrates and vertebrates, are capable of offering some form of resistance to infection. Among invertebrates, such resistance is based mainly on the activities of phagocytic cells and some soluble factors that have lytic or other effects on invading microorganisms. These responses are usually short-lived and relatively non-specific as they do not discriminate between the infectious agents attacking the host.

Vertebrates differ from invertebrates in that they have evolved the ability to respond in a highly specific way to foreign substances and invading microorganisms by means of a well-developed immune system. Foreign macromolecules that activate this system are called antigens and the immunoglobulins which recognise these antigens specifically are called antibodies. A parallel cellular system in which antigenic stimulation leads to the selection and activation of specific lymphocytes is called cell-mediated immunity. The immune system is a highly regulated network of mainly lymphoid cells which undergo differentiation, activation and renewal in a structured manner as the body develops full immunological competence post-natally. Non-lymphoid cells also contribute to immunological competence and likewise soluble factors, some naturally occurring, others induced by challenge, assist or amplify defence mechanisms of the host.

The body's ability to maintain itself free of infectious diseases derives not only from the highly specialised cells of the immune system capable of responding to invading pathogens and their products, but also from natural anatomical and physiological barriers to infection. The non-specific antimicrobial mechanisms that participate in innate (non-specific) immunity include the skin, mucous membranes, mechanical activities of the body, including the flushing activity of tears and urine, phagocytic cells, particularly polymorphonuclear leukocytes and mononuclear phagocytes, and secreted products such as lysozyme, complement and interferons. Commensal microorganisms in the intestine, female urogenital tract and the skin compete competitively with potential pathogens in these locations.

In contrast to non-specific immunity, which is innate in most living organisms, acquired immunity is more specialised and highly specific. Non-specific immune responses, however, play a central role in the initiation of specific immunity through the involvement of phagocytic cells and their interaction with lymphocytes which generate cell-mediated or humoral immunity. Specific immune responses also rely heavily on elements of non-specific immunity such as the complement system for amplification through the generation of chemotactic factors, promotion of phagocytosis and the development of ultrastructural lesions on target membranes. The cardinal features of specific immune responses which distinguish them from non-specific responses are specificity, immunological memory and self-discrimination.

Cells of the immune system

The cells of the immune system that participate in acquired immunity include lymphocytes, macrophages and a series of macrophage-like cells which includes dendritic cells of the spleen and lymph nodes and Langerhans' cells of the skin. The macrophage-like cells constitute a population of antigen-presenting cells (APC) which present fragments of antigenic material to lymphocytes.

Although they are indistinguishable from each other by light or electron microscopy, lymphocytes can be divided into subsets with different immunological functions. There are at least three recognisable subdivisions of lymphoid cells: T lymphocytes, B lymphocytes and a third subset, referred to as null cells, distinct from the other two lineages.

B Lymphocytes

The bone marrow is the source of stem cells from which both lymphoid cells and those cells which contribute to non-specific immunity, through phagocytic activity or other mechanisms, derive (**Diagram 12**).

B cell differentiation commences in the bone marrow and maturation in mammals probably occurs in gut-associated lymphoid tissue or in other lymphoid organs. In birds, B cells undergo maturation in the bursa of Fabricius, a lymphoid organ situated in the cloacal region. The development of B cells can be divided into four distinct phases: stem cell, pre-B cell, B cell and plasma cell. Mature B cells, each with antigen-specific receptors that have a structure and specificity identical to that of the antibody synthesised by the plasma cell, enter the circulation and migrate to special areas of residence in the spleen and other peripheral lymphoid tissue.

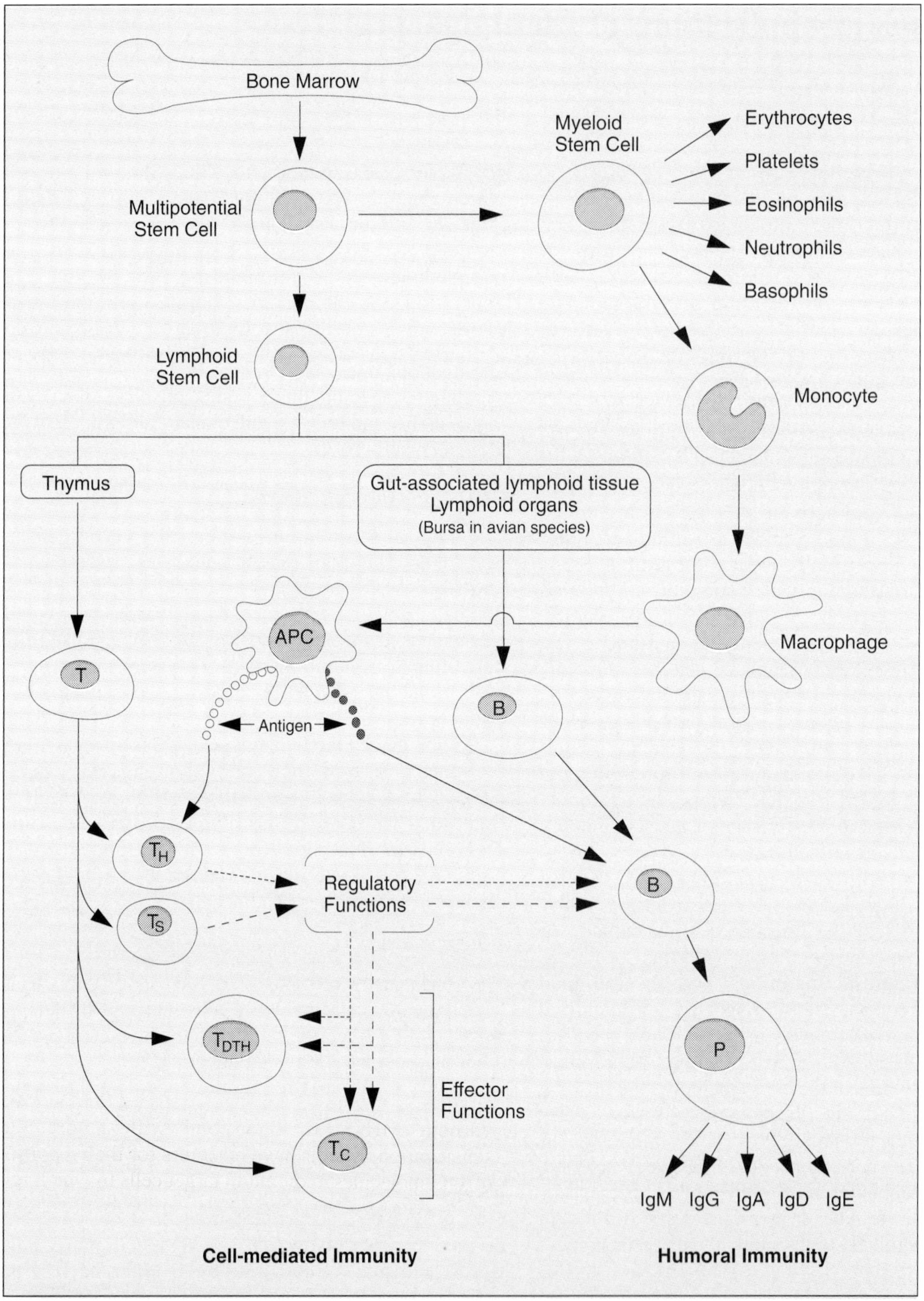

Diagram 12. Differentiation and maturation of the cells contributing to cell-mediated and humoral immunity in mammals. In birds, cells which migrate to the bursa of Fabricius differentiate into B cells.
APC: antigen-presenting cell; B: B cell; P: plasma cell; T: T cell; T_{DTH}: T cell involved in delayed-type hypersensitivity; T_H: helper T cell; T_S: suppressor T cell; T_C: cytotoxic T cell.

When B cells encounter antigens complementary to their surface immunoglobulin receptors and receive appropriate T cell co-operation, they differentiate into antibody-secreting plasma cells or into memory cells, ready to make antibodies of high affinity on later encounter with the same antigen. The antibodies secreted by plasma cells play a central role in protective immunity both directly and indirectly. Pathogenic bacteria and other infectious microorganisms may be opsonised or agglutinated by specific antibody, thereby facilitating their engulfment by phagocytes. The reaction of antibodies with the surface antigens of infectious agents may activate the complement system leading to their destruction. Antibodies can also neutralise bacterial toxins and viruses and prevent attachment of bacteria and viruses to cell receptors. The passive transfer of antibodies to newborn animals via colostrum is one of the most effective methods of conferring temporary protection on this neonatal group during a time of high susceptibility to infection.

The immunoglobulin classes IgM, IgG and IgA are associated with antibody activity against a wide range of infectious agents, and are bifunctional in that they can react with antigens on invading microorganisms and in addition initiate other biological reactions such as attachment to the membrane of phagocytic cells and activation of the complement system. IgM antibodies, the first produced in response to antigenic stimulation, are highly efficient in agglutination reactions and complement activation. Being high molecular weight molecules, they are normally confined to the blood vascular system. IgG, the predominant immunoglobulin in serum, can bind to receptors on phagocytic cells thereby facilitating destruction of antibody-coated microorganisms by intracellular killing. IgG antibodies also neutralise bacterial toxins and viruses. IgA, although present in serum, is selectively transported across mucous membranes and is the principal immunoglobulin in body secretions. Antibodies of the IgA class play an important part in protecting mucosal surfaces, particularly in the respiratory and intestinal tracts. IgE and IgD have unique roles in immune responses. IgE, associated with allergic reactions, is present in low concentrations in serum and binds with high affinity to mast cells. When cross-linked by allergens, IgE antibodies cause mast cells to degranulate and release vasoactive substances which lead to local or systemic anaphylaxis. IgD functions as an antigen receptor on B cells together with monomeric IgM. When cross-linked by antigen, IgD induces resting B cells to enlarge, preparatory to cell division.

T Lymphocytes

Following migration from the bone marrow to the thymus, lymphoid cells destined to become T lymphocytes acquire special characteristics which equip them to fulfil their role in cell-mediated reactions (**Diagram 12**). During maturation in the thymus, from which the name T cell derives, these cells acquire the ability to recognise histocompatibility antigens on host cells, gene products of the major histocompatibility complex (MHC). Unlike B cells, T cells do not recognise antigen in its native state. When antigen is presented to T cells by an APC bearing appropriate MHC molecules (class I or class II), with the antigen in physical association with the MHC gene products, recognition by the T cell receptors can occur. This 'dual' recognition for both antigen and MHC is necessary for activation of cytotoxic and immuno-regulatory T cells and is augmented by the cytokine interleukin-1 secreted by the APC. Activation leads to endogenous interleukin-2 secretion by T cells and its binding to receptors on the same cells. DNA synthesis and cell division follows with clonal proliferation of the activated cell.

T cells have both effector and regulatory roles. Functionally, they are divided into several subsets with different biological activities. Major subsets with regulatory functions include helper T cells (T_H) that assist in the development of B cells for antibody production and other T cells engaged in cytotoxic activity or delayed-type hypersensitivity reactions. Suppressor T cells (T_S) can limit or prevent immune responses by suppressing the activity of B, T_H or other subsets of T cells engaged in cell-mediated responses. T cells with effector roles include cytotoxic T cells (T_C) and delayed-type hypersensitivity cells (T_{DTH}). The former can lyse abnormal, histo-incompatible and virus-infected cells, the latter participate in delayed-type hypersensitivity reactions by secreting a variety of lymphokines which attract and activate macrophages and other T cells, thus greatly augmenting the immune response.

Other lymphocytes that are neither T nor B cells are termed natural killer cells (NK) and killer cells (K cells). Although NK and K cells are similar phenotypically, functionally they appear to be different cell populations. NK cells appear to be immunologically non-specific lymphocytes that have the ability to kill target cells such as virus infected cells and tumour cells without MHC restriction and without previous sensitisation. K cells have receptors for the Fc portion of IgG molecules and bind to target cells to which IgG antibodies are attached. This antibody-dependent reaction is referred to as antibody-dependent cell-mediated cytotoxicity. Cells other than K cells are also capable of mediating antibody-dependent killing, but to a lesser extent.

Macrophages and antigen-presenting cells

Cells of the monocyte-macrophage series originate in the bone marrow from a multipotential stem cell common to all of the haemopoietic cells and myeloid cells. Tissue macrophages arise by maturation of monocytes that have migrated from the blood. Macrophages play an essential part in non-specific immunity by ingesting and killing invading micro-organisms and by releasing many soluble factors which contribute to host defence and to inflammation. Secreted products of macrophages include a wide range of enzymes, complement components, prostaglandins and a number of cytokines with regulatory functions.

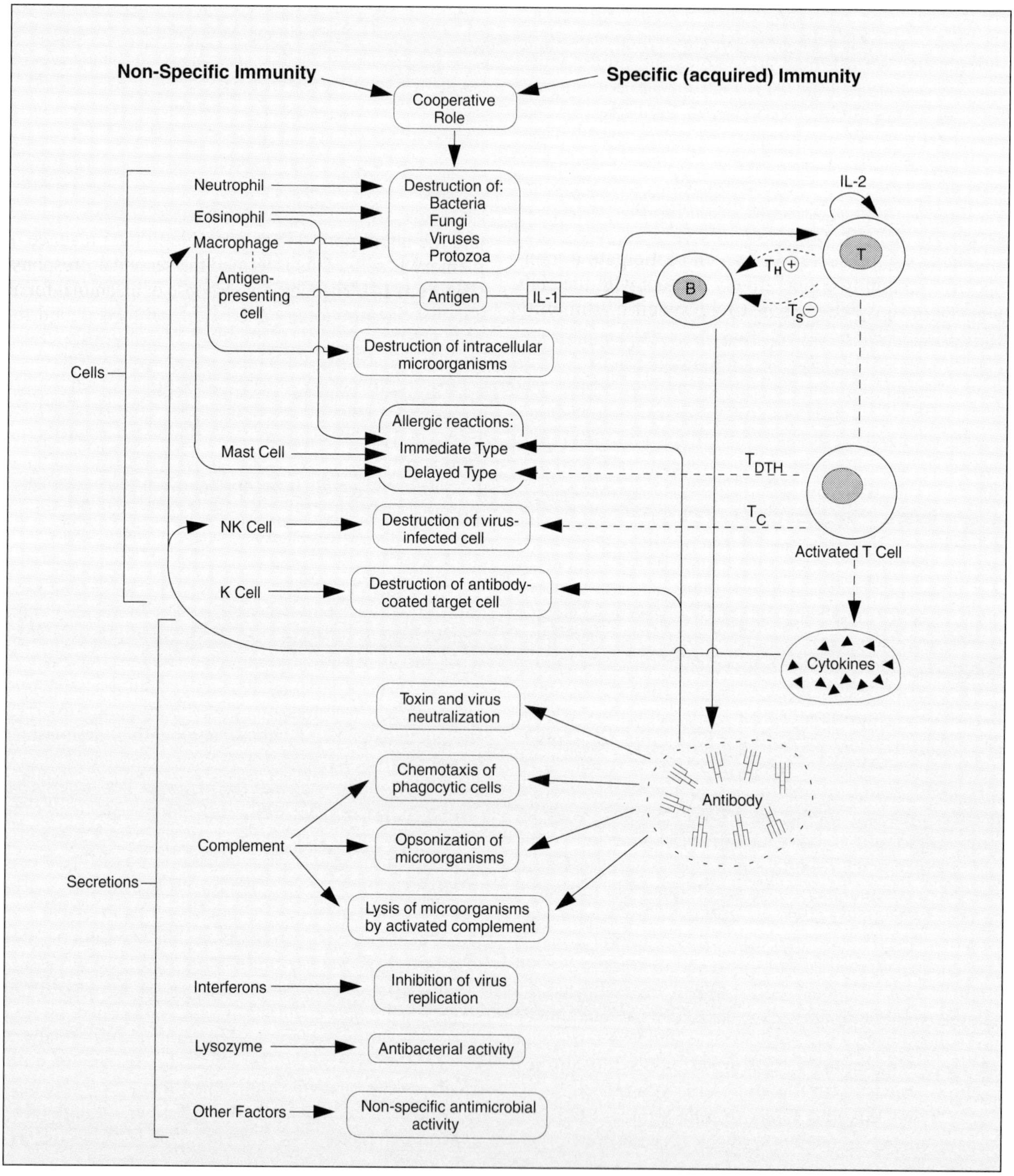

Diagram 13. The co-operative interaction of cells and secretions that contribute to non-specific immunity (left) and the cellular and humoral components of specific immunity (right).
IL-1: interleukin 1; IL-2: interleukin 2; the + and - symbols indicate the regulatory role of helper T cells (T_H) or suppressor T cells (T_S); NK: natural killer cell (lymphocyte); K: killer cell (lymphocyte); T_{DTH}: T cell involved in delayed-type hypersensitivity; T_C: cytotoxic T cell.

A feature of these cells, unusual in phagocytic cells, is their ability to become activated. Activation of macrophages results from the release of macrophage-activating lymphokines from T cells. These cells migrate more vigorously in response to chemotactic factors and deal more efficiently with intracellular parasites than normal macrophages.

Traditionally, macrophages were considered to be the major antigen-presenting cell. It is now clear that other cells, including interdigitating cells found in the paracortical region of lymph nodes and Langerhans' cells present in the epidermis, are very effective at antigen presentation. B cells, because of their ability to interact with antigen directly, are also capable of presenting antigen to T cells.

The cells and secretions which participate in non-specific immunity cooperatively interact with the cells and secretions which constitute specific immunity (**Diagram 13**). Together these two branches of the immune system protect the body against infectious agents, endeavour to prevent neoplastic changes in tissues and reject grafts from unrelated donors.

Antibody production

When an animal encounters an infectious agent or foreign substance for the first time, antibody to that agent or substance is usually detectable in the serum within 10–14 days. The time required for antibody production is influenced by the nature, amount and route of exposure to the agent, the age and immune status of the animal and the sensitivity of the assay used to demonstrate circulating antibody. The first antibody response to antigen is referred to as the primary response. The interval between exposure and production of antibody is referred to as the latent period (or lag period). If exposure to the same antigen occurs some time later, the animal responds with a stronger and more sustained response and the latent period is shorter due to the presence of memory B cells. This is termed the secondary response (**Diagram 14**). The memory response of an animal for an agent it has previously been exposed to is called the anamnestic response. When blood sampling animals for serological tests for infectious diseases, two samples may be required to demonstrate an increasing level of circulating antibody. This correlates with active infection or an anamnestic response following re-exposure to the same infectious agent.

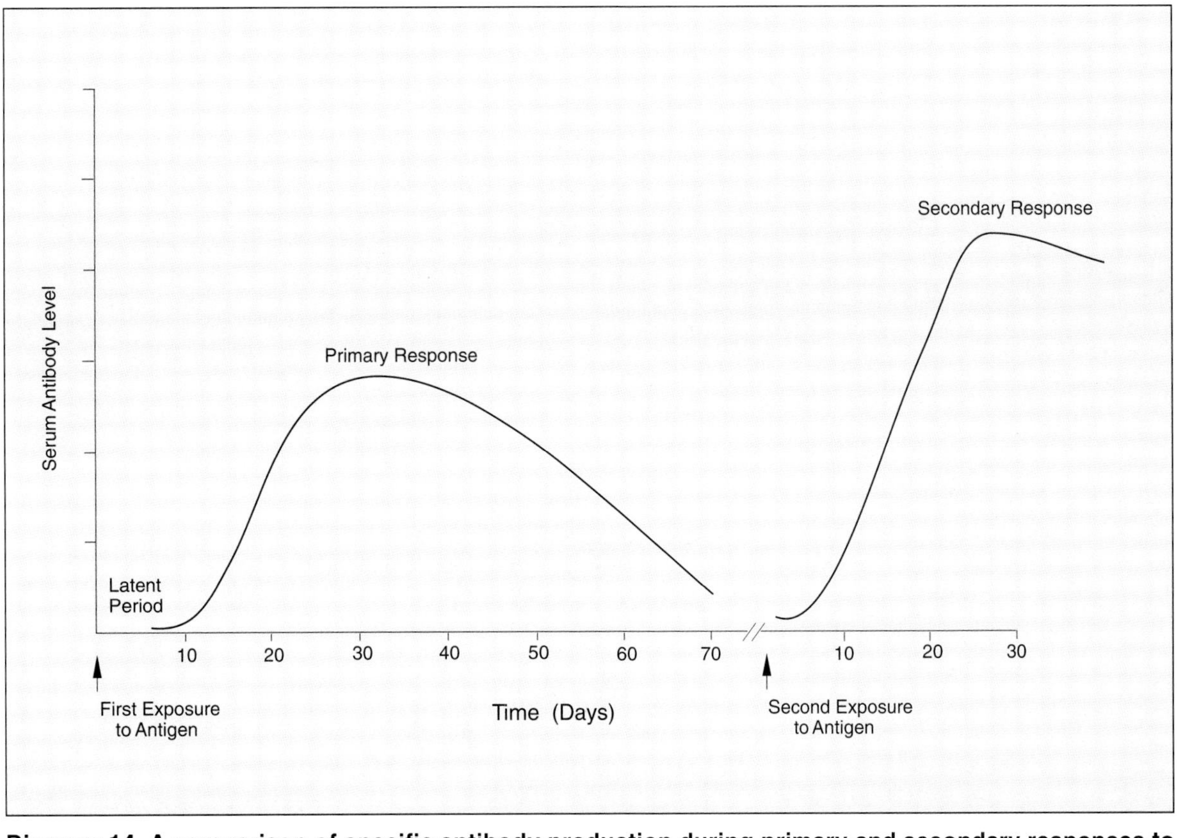

Diagram 14. A comparison of specific antibody production during primary and secondary responses to infectious agents or to vaccination. Natural infection usually results in higher levels of antibody for a longer time period.

The Complement System

The complement system consists of approximately 20 proteins, some of which are heat-labile, present in the serum of normal animals. These proteins interact with each other in sequential order once activated by antigen-antibody complexes. Complement components are numbered from Cl through to C9 and when activated they can induce membrane alteration, activation or damage. Red blood cells with fixed antibodies on their surface are lysed by complement and this lytic reaction is the basis of the haemolytic complement fixation test in diagnostic serology. Being heat-labile, complement activity is lost after heating at 56°C for 30 minutes whereas immunoglobulins are not inactivated at this temperature. Sera for titration in complement fixation tests are routinely heated at 56°C for 30 minutes to inactivate their complement, as a carefully measured amount of complement is employed in the complement fixation test procedure.

Antigen-antibody reactions

Reactions of antigens with antibodies are highly specific and because of this high specificity, these reactions can be used to identify one by means of the other. Serological tests find wide application in veterinary medicine for the diagnosis of infectious diseases. The interaction of antigen with antibody can result in a variety of immune complexes the nature of which will be determined by the state of the antigen and the type of test being carried out. In the diagnosis of infectious diseases of animals, it is usual to dilute the serum of the test animal to determine the concentration of specific antibody present and this is expressed as the **titre** of that serum. Titre may be defined as the highest dilution of a serum which gives a demonstrable reaction in a defined test procedure.

Precipitation

When soluble antigen is added to homologous antibody at the correct ratio insoluble antigen-antibody complexes are formed. If the immune complexes are of sufficient size they precipitate out of solution. This form of immunological precipitation is termed the **precipitin reaction**. When increasing amounts of antigen are added to a series of tubes containing a constant amount of antibody, variable amounts of precipitate form. If the weight of precipitate formed is plotted against the amount of antigen added, a precipitin curve similar to that shown in **Diagram 15** is formed. There are three distinct zones in this quantitative precipitin reaction: a

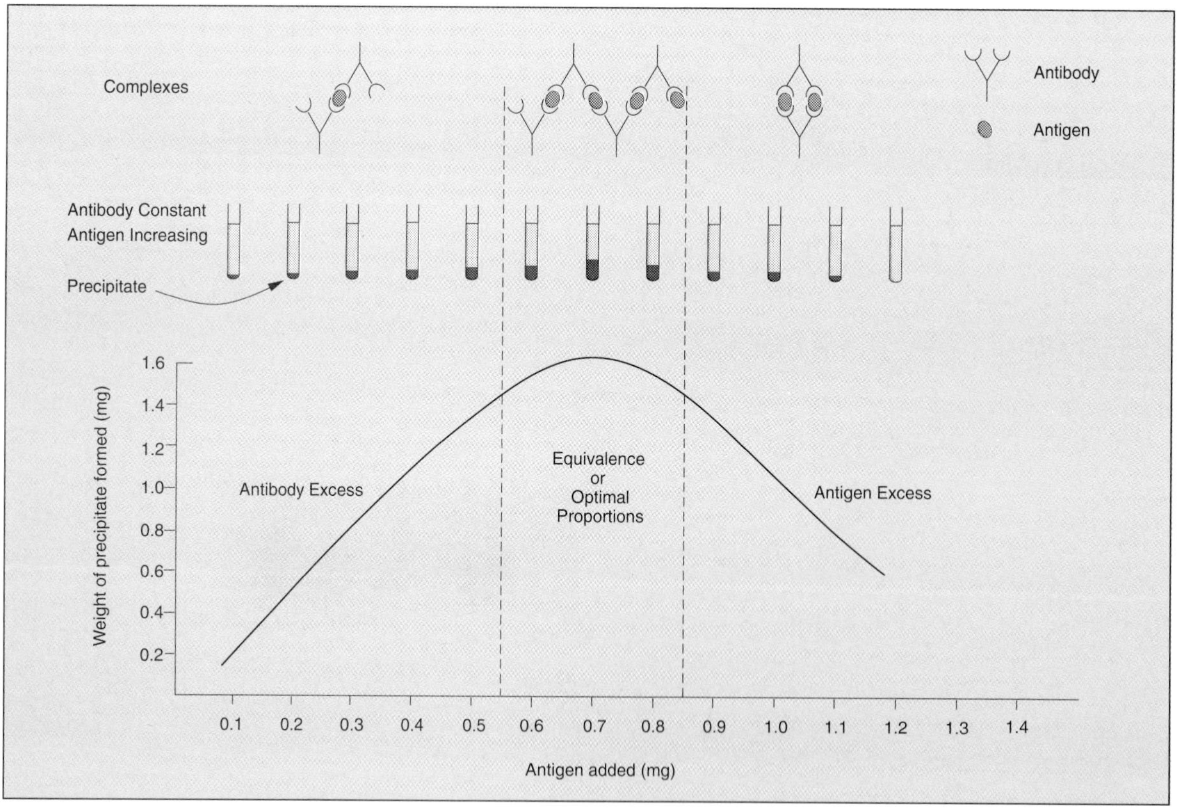

Diagram 15. The quantitative precipitin test. When an increasing amount of soluble antigen is added to a constant amount of antibody, antigen-antibody complexes of different ratios are formed. At equivalence, all of the antigen and antibody are precipitated yielding the greatest amount of insoluble precipitate.

zone of antibody excess, a zone of optimal proportions (equivalence) and a zone of antigen excess. In the equivalence zone, the proportion of antigen and antibody is optimal for maximal precipitation with neither free antigen nor antibody in the supernatant fluid. It can be seen, therefore, that the precipitin reaction is dependent on the ratio of antigen and antibody present, irrespective of the type of medium in which the reaction takes place. When diluted antigen is layered carefully over homologous antibody a band of precipitate forms where the ratio of the reagents is at optimal proportions. This method is termed the ring precipitin test (**70**) and has found application in identifying streptococci and assigning them to their appropriate groups. It is also used in forensic medicine and can be applied to many other test procedures in the laboratory. A limitation of this method is the requirement for high levels of antibody in the antiserum employed.

Precipitation reactions can take place in semi-solid media such as agar gels. When soluble antigen and antibodies are placed in wells cut in a gel, they diffuse towards each other and where they meet at or near optimal proportions, a precipitin line forms in the gel. The number of lines of precipitate formed usually corresponds to the number of different antigen-antibody systems present. This procedure is termed the Ouchterloney double diffusion method or immunodiffusion in agar. Unknown antibodies or antigens can be identified using this technique and it can also be employed to show relationship between antigens placed in adjacent wells. Patterns of reactions which can be seen by this method include identity, partial identity and non-identity. When adjacent wells contain the same antigen the precipitin lines between the antibody and the antigen wells fuse and form a continuous arc. This is referred to as a reaction of identity (**71, 72**). A reaction of partial identity occurs when adjacent wells contain molecules with some shared and some unique determinants. The presence of a spur, pointing towards the antigen which lacks some determinants, occurs on the precipitin line (**73, 74**). A pattern where the precipitin lines cross each other denotes non-identity of the two antigens (**75**). Immunodiffusion in agar is applied in diagnostic tests for infectious diseases particularly in virology and the Coggins test for equine infectious anaemia is based on this method.

Immunodiffusion usually gives a separate line for each antigen-antibody system in complex mixtures of proteins or other antigenic material, but it may be difficult to resolve all the components present in highly complex mixtures. It is possible to separate antigens present in complex mixtures by electrophoresis and then use immunodiffusion for their identification. This method is called immunoelectrophoresis. Both identification and approximate quantitation of individual proteins present in serum, urine, or other body fluids can be accomplished by this method. The individual proteins present in feline serum are shown in **76** and the absence of a particular component such as immunoglobulin can be readily detected.

Immunodiffusion can be used to identify the number of different antigen components in a biological sample, but if precise quantitation of a given antigen is required, radial immunodiffusion is used. When a soluble antigen in wells diffuses into agar containing specific antiserum, a ring of precipitate forms, the diameter of which is proportional to the concentration of antigen in each well (**77**). If the method is first standardised with known amounts of purified antigen, a standard curve can be prepared (log antigen concentration versus diameter of ring) and the amount of antigen in unknown solutions determined from this graph.

Rocket electrophoresis is a modification of radial immunodiffusion and uses direct current to migrate antigen into a gel containing antibody (**78**). The length of each rocket is proportional to the amount of antigen placed in each well. Standards containing known amounts of antigen are run simultaneously and the amount of antigen present in test samples can be extrapolated from a standard graph.

70

70 Ring precipitin test: negative (left) and positive (right).

71

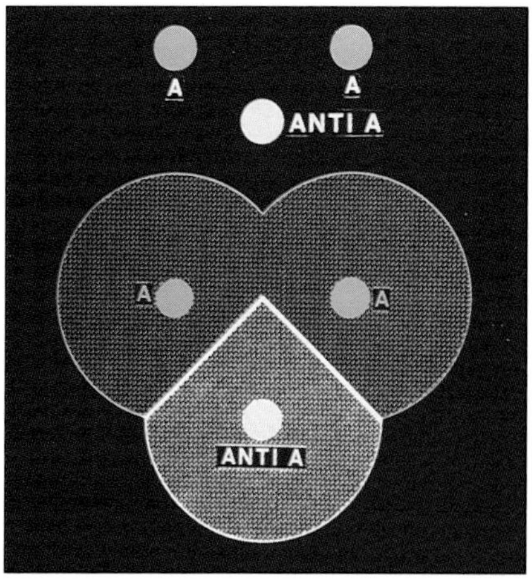

71 Agar gel immunodiffusion: diagram of a reaction of identity.

72

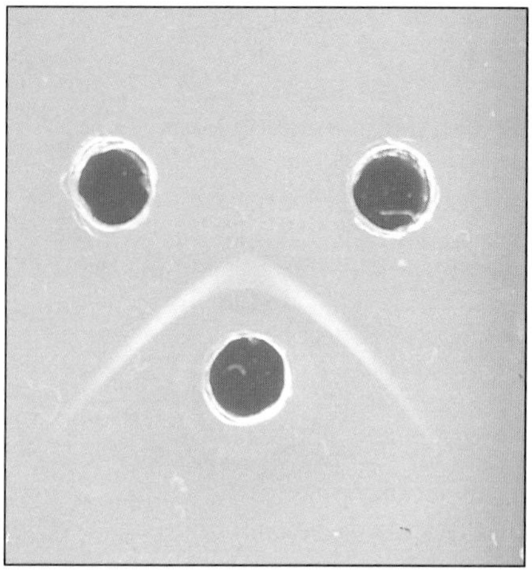

72 Agar gel immunodiffusion: reaction of identity. The well on the left contained antiserum and the wells on the right homologous antigen.

73

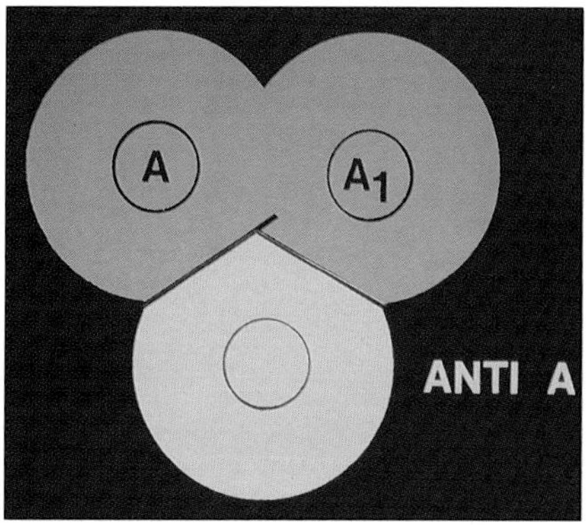

73 Agar gel immunodiffusion: diagram of a reaction of partial identity (spur formation).

74

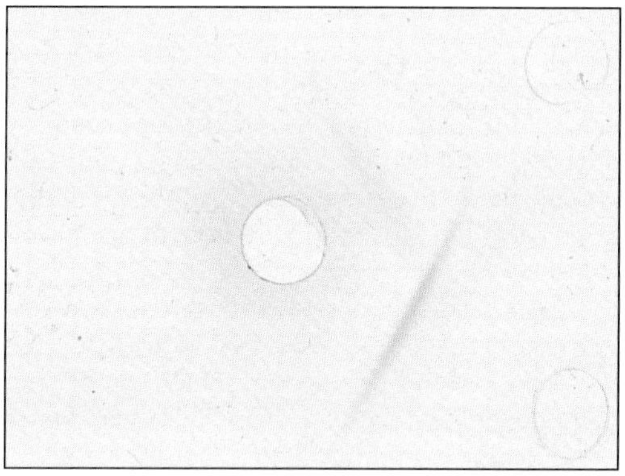

74 Agar gel immunodiffusion: reaction of partial identity showing spur formation.

75

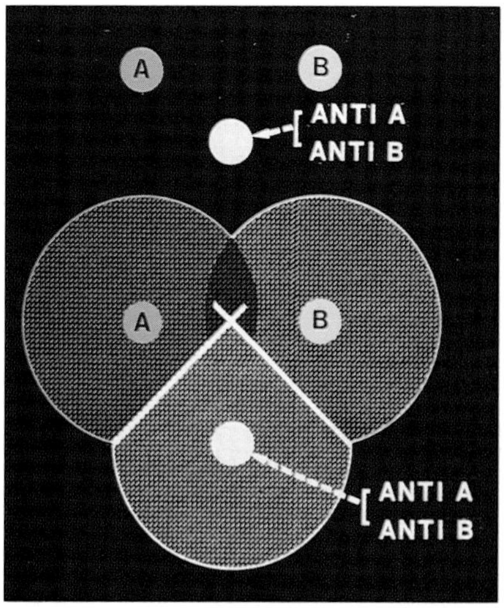

75 Agar gel immunodiffusion: diagram of a reaction of non-identity.

76

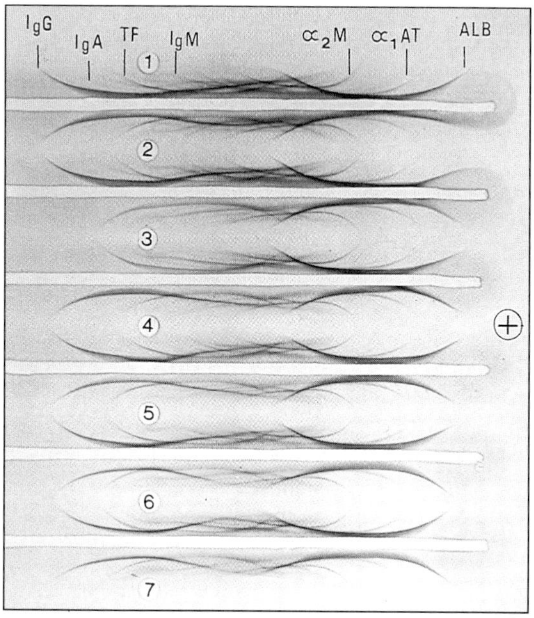

76 Immunoelectrophoresis of feline serum. Wells 2-7 contain sera of young cats of different ages, compared with adult cat serum (Well 1). In the serum of a precolostral kitten (Well 3) no immunoglobulins were detected. ALB = albumin; alpha-1 AT = alpha-1-antitrypsin; alpha-2M = alpha-2-macroglobulin; IgM = immunoglobulin M; TF = transferrin; IgA = immunoglobulin A; IgG = immunoglobulin G.

77

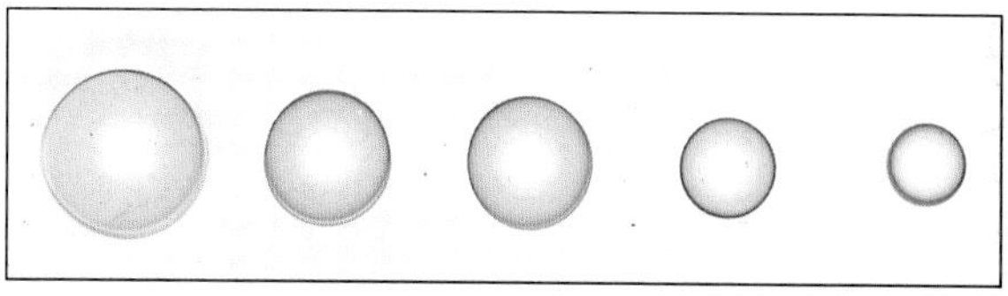

77 Radial immunodiffusion (Mancini technique) for measuring the concentration of a soluble antigen. (Amido black stain)

78

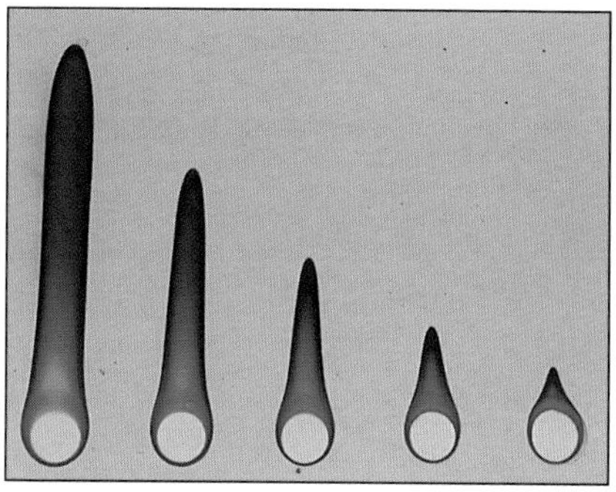

78 Rocket technique for measuring the concentration of a soluble antigen. (Amido black stain)

Agglutination

An agglutination reaction occurs when antibody reacts with particulate antigen and cross-links surface antigenic determinants. Bacteria, fungi, protozoa and red cells can be directly agglutinated by specific antibody. In diagnostic microbiology, the slide agglutination test is often used to identify unknown bacterial isolates. A drop of specific antiserum is added to a smooth suspension of bacteria in saline. A control of the bacterial suspension without antiserum is also used to eliminate autoagglutination. After the addition of antiserum the slide is rocked gently and the result is read inside 3 minutes. Clumped bacteria indicate a positive reaction. The slide agglutination test is also used to demonstrate the presence of antibody in serum. One drop of test serum is added to a standardised suspension of bacteria on a slide. After gentle rocking the result is read inside 3 minutes. In the presence of antibodies, the bacteria are agglutinated (**79**). Occasionally bacteria or other antigens may be deliberately coloured to facilitate interpretation of the result and in addition, for *Brucella abortus*, the pH of the suspending fluid is dropped to a low level to eliminate non-specific agglutination and improve the reliability of the test (**80**).

Quantitative agglutination can be performed in test tubes or microtitre plates by serially titrating the serum under test, usually in two-fold dilutions. An equal volume of standardised antigen, such as bacteria, is added to each tube and a control tube with diluent. Following careful mixing the tubes are incubated, usually at 37°C. Results are recorded after a few hours and the highest dilution of the serum giving agglutination is called the antibody titre (**Diagram 16, 81**). For diagnostic purposes, two blood samples taken several weeks apart should be used to demonstrate a rising titre in infected animals.

One practical difficulty of importance in agglutination tests is the occasional absence of agglutination in the first tubes when high titred serum is being tested. This is referred to as the prozone phenomenon. The nature of this phenomenon is not entirely clear but it is associated with high titred sera in some species, with the presence of non-agglutinating antibodies in some sera and probably with the inability of some antibodies to cross-link particulate antigen where epitopes are recessed deep in the membranes of bacterial or mammalian cells. Because of the prozone phenomenon it is imperative that test sera be checked at several dilutions to avoid errors in reporting results.

Agglutination tests can be carried out with body fluids other than serum and in the diagnosis of *Brucella abortus* infections in cattle, milk is used. In the milk ring test, stained *Brucella* organisms are added to fresh milk which is left to stand for a few hours. The stained bacteria remain dispersed throughout the milk if antibodies are absent but rise with the cream in a positive reaction (**82**).

A range of soluble antigens such as lipopolysaccharides can be passively adsorbed by red blood cells and other antigens can be chemically coupled to erythrocytes. This method is referred to as passive or indirect haemagglutination and it is a sensitive method for measuring antibodies to soluble antigens attached to carrier red blood cells (**83**).

A number of other diagnostic procedures use the principle of agglutination, and for detection of antibodies to leptospires the microscopic agglutination test is carried out using a suspension of live microorganisms and a microscope equipped with a darkfield condenser.

79

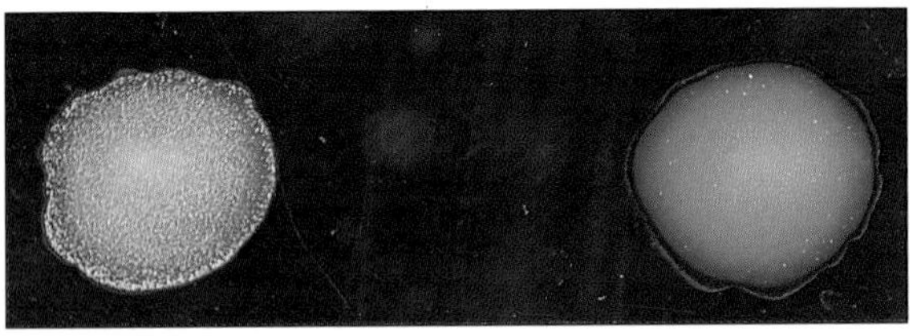

79 Slide agglutination test: positive reaction (left) and negative reaction (right).

80

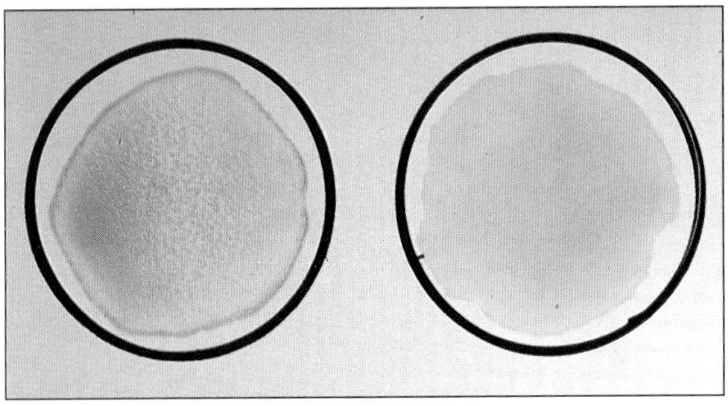

80 Rose–Bengal plate agglutination test for *Brucella abortus* (coloured antigen): positive reaction (left) and negative reaction (right).

81

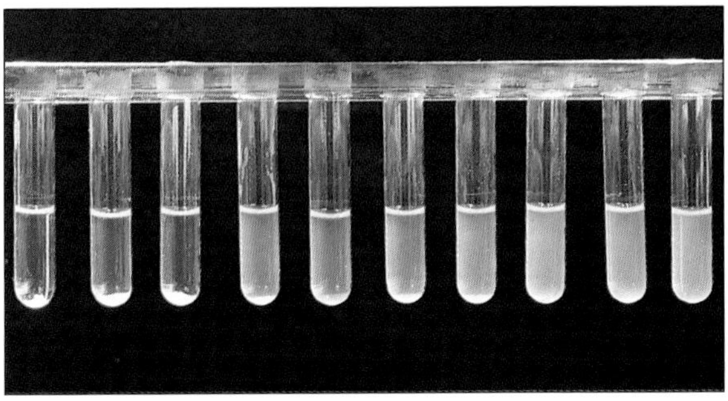

81 Tube agglutination test: doubling dilutions beginning at 1/10. The antibody titre is 1/40.

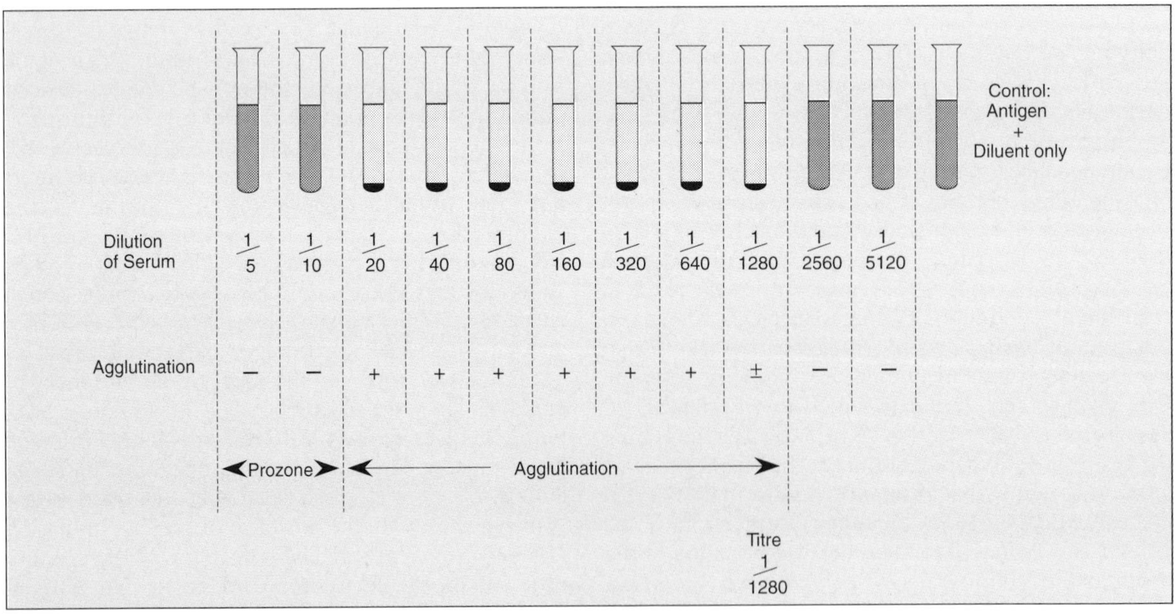

Diagram 16. Interpretation of the tube agglutination test. Tubes with high levels of antibody, where agglutination does not occur, represent a prozone effect.

82

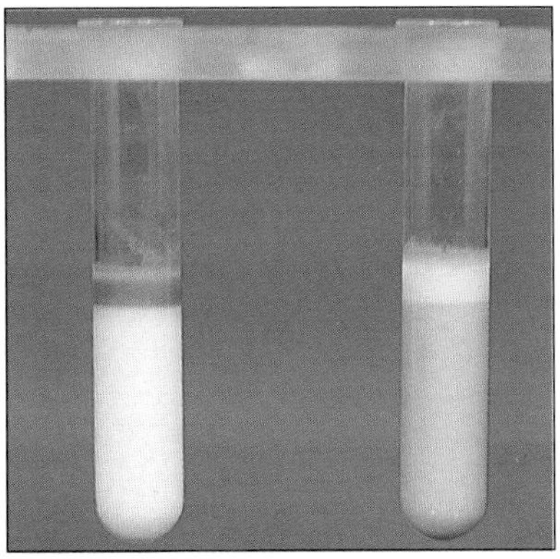

82 Brucella milk ring test (stained antigen): positive 3+ reaction (left) and negative (right).

83

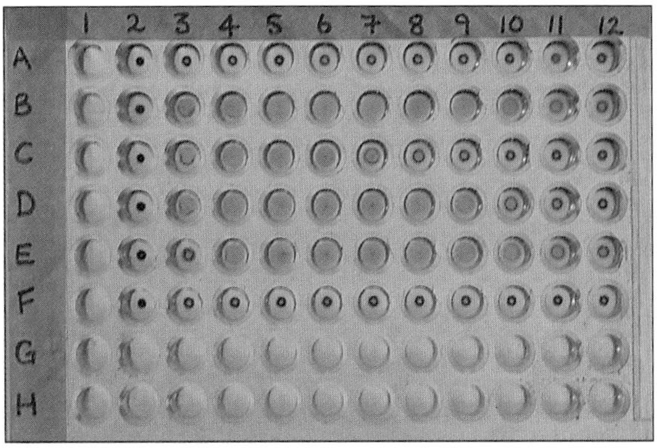

83 Indirect haemagglutination test (passive) for measuring antibody against soluble antigen (*Toxoplasma gondii* and sheep red blood cells). Column 2 downwards contains a control well for each serum and the carrier red cells. There are doubling dilutions of each serum starting at 1/8 in column 3. Interpretation: A (across), negative; B, positive 1/1024; C, positive 1/128; D, positive 1/512; E, positive 1/1024 and F, negative. U-type wells.

Complement Fixation

The fact that antibody, once it combines with antigen, activates complement, can be used for diagnostic purposes. The complement fixation test uses sheep red blood cells sensitised with rabbit antibody (haemolysin) as an indicator system. In the presence of complement, usually supplied by guinea-pig serum, the sensitised cells are lysed. The test depends on a two-stage reaction system. The test serum is heated at 56°C for 30 minutes to destroy its complement activity before titration. Antigen is then added to the titrated serum and a precise amount of guinea-pig complement is also added. After incubation at 37°C for 30 minutes, sensitised sheep red blood cells are added followed by a further incubation. Where complement is fixed by antibody in the test serum reacting with specific antigen, the sheep red cells do not lyse, but the sensitised sheep red blood cells will lyse if complement was not fixed because the test serum did not contain specific antibody (**84**).

84

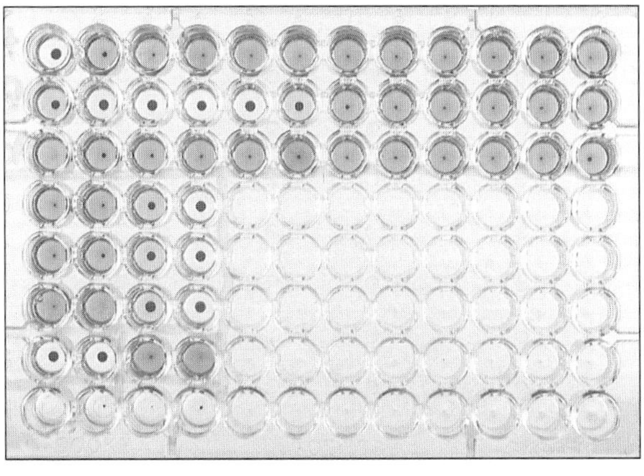

84 Complement fixation test (CFT) for enzootic abortion of ewes. Test sera in top three rows across with doubling dilutions starting at 1/8: row 1, positive 1/8; row 2, positive 1/256; row 3, negative. Control wells below. Microtitre plate with V-type wells.

Viral haemagglutination and its inhibition by antibody

Some viruses such as influenza viruses are capable of binding to red blood cells and agglutinating them. This is termed viral haemagglutination and antibodies specific for these viruses inhibit this haemagglutination by blocking their combining sites. It is usual first to screen the virus to determine its haemagglutinating activity for the red cells being used (**85**). It is common to use four haemagglutinating units (HA units) with test sera. Virus is standardised and added to dilutions of test serum. After an appropriate interval, washed red blood cells are added to each well, including control wells, and the haemagglutination inhibition titre of the serum is recorded (**86**).

85

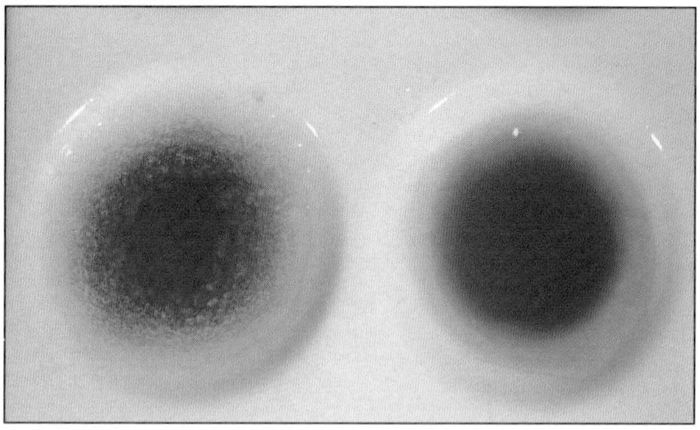

85 Screening virus haemagglutination test: positive reaction (left) and negative reaction (right) using human O red blood cells. The screening test was photographed before the red cells in the negative test had completely settled (formed a button).

86

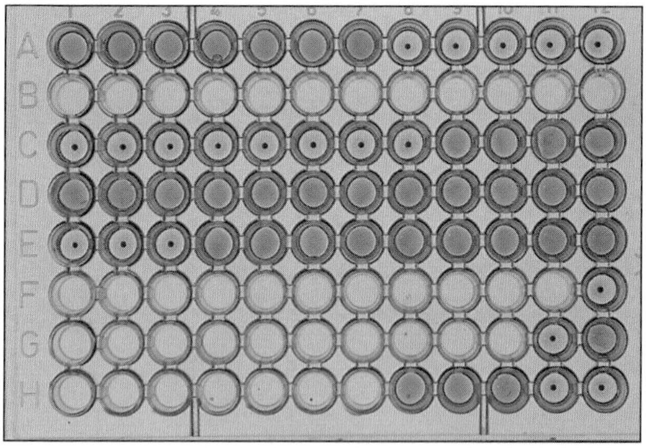

86 Haemagglutination-inhibition test (equine influenza virus and human O RBCs). Top row (A) titration of virus for haemagglutinating activity to determine 1 HA unit (well 7). Test sera are in rows C, D and E with doubling dilutions starting at 1/10. Interpretation: row C, positive 1/1280; D, negative; E, positive 1/40. Rows F, G and H contain reagent controls. V-type wells.

Enzyme-linked immunosorbent assay (ELISA)

Enzymes such as alkaline phosphatase or peroxidase can be linked to antibody without interfering with the antibody's specificity or the enzyme's activity. The enzyme is detected by assaying for enzyme activity with its substrate. Because of their simplicity and safety, ELISA methods have been widely used for the immunodiagnosis of bacterial, viral and parasitic infections. Enzyme immunoassays may be used to measure either antigen (direct, capture or sandwich ELISA) or antibody (indirect or competitive ELISA) (**Diagrams 17** and **18**).

In an indirect ELISA for antibody, known antigens are fixed to wells in polystyrene plates, or suitable membranes, to which test serum dilutions and control sera are added. Following incubation, the wells are washed with buffered diluent and enzyme-conjugated antiglobulin is added. After further incubation and washing, the specific substrate for the enzyme is added. The intensity of the colour reaction which develops may be estimated visually or spectrophotometrically, and is proportional to the amount of antibody present in the serum being tested.

Detection of antigen in serum or body fluids can be accomplished by enzyme-linked antibody methods. Polystyrene wells are coated with specific antibody, the test fluid is added and if antigen is present it binds to the antibody and remains bound after washing. An enzyme-labelled antibody specific for the antigen is added and following further washing, the addition of substrate results in colour development. The intensity of the colour reaction is directly related to the amount of bound antigen (**87**).

One commercial system combines reagents for direct and indirect ELISA into one detection system thereby detecting antigens for one infectious agent (feline leukaemia virus) and antibodies for another (feline immunodeficiency virus, **88**).

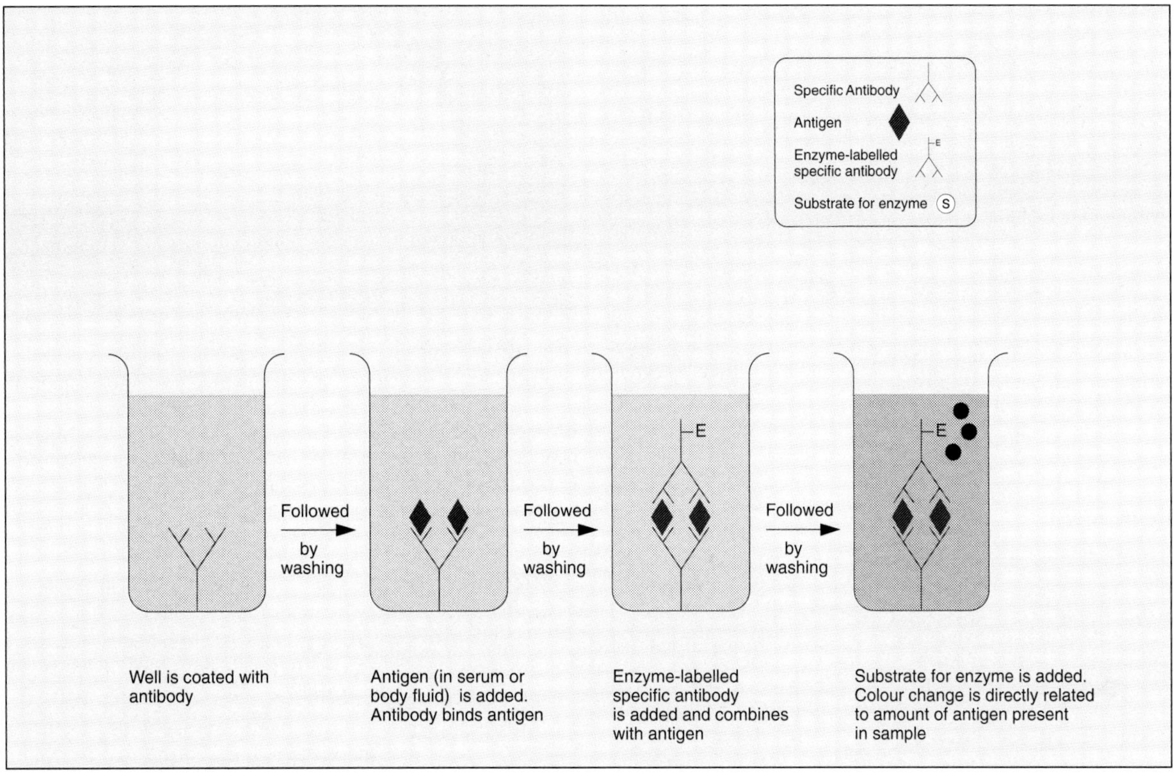

Diagram 17. ELISA technique for detection or measurement of antigen in a test sample (antigen capture, sandwich or direct method).

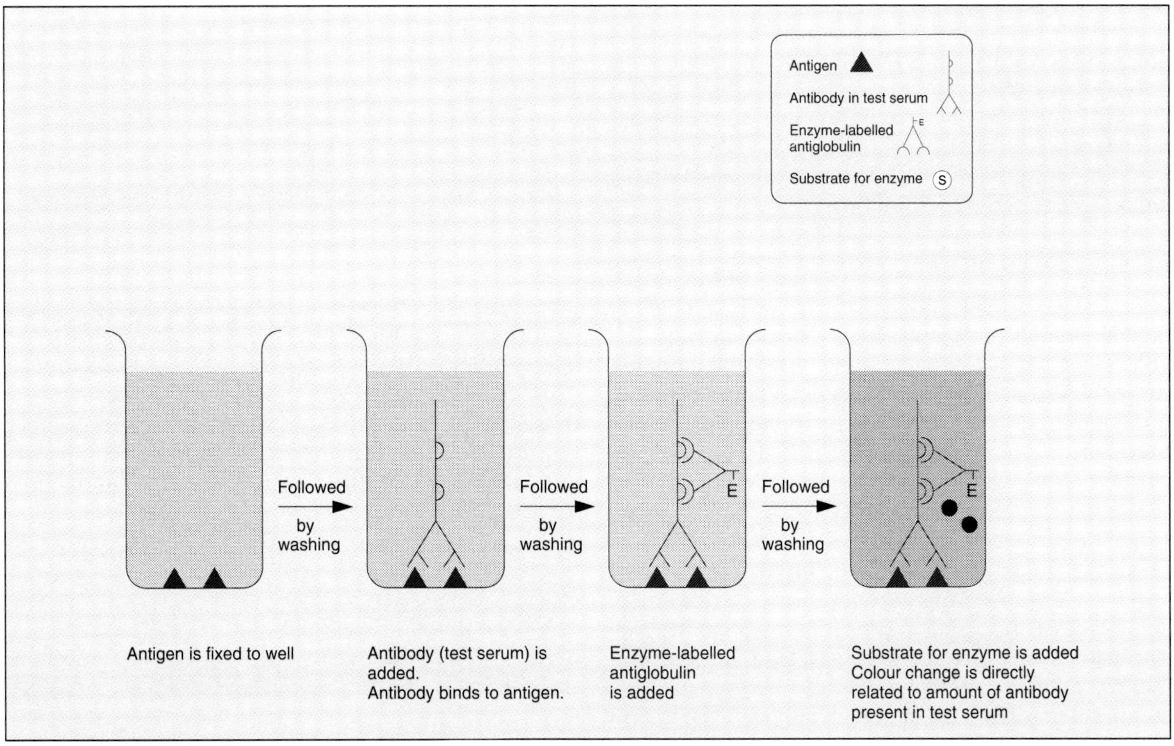

Diagram 18. ELISA technique for measuring the amount of antibody in a test serum (indirect method).

87

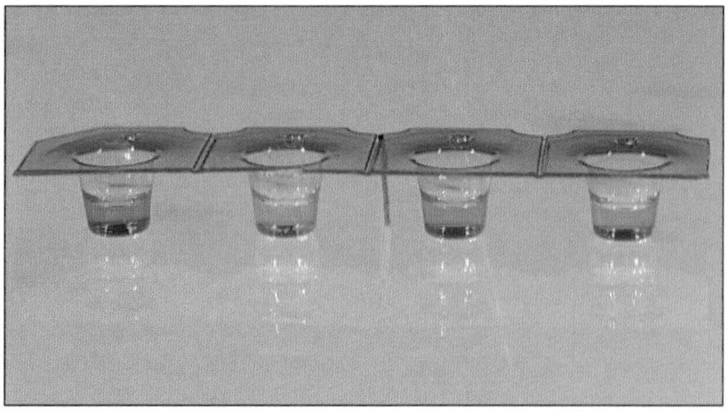

87 ELISA for the detection of feline leukaemia virus in serum: from left, positive control, negative control, positive test and negative test.

88

88 ELISA (CITE Combo, IDEXX): Feline immunodeficiency virus antibody positive (left), positive control (centre top), and feline leukaemia virus positive (right). Red dot is for orientation.

Radioimmunoassay

Radioimmunoassay (RIA) uses isotopically labelled molecules, and because of its exquisite sensitivity, extremely small amounts of antigen, antibody or immune complexes can be detected. The principle of this assay is similar in many respects to ELISA and immunofluorescence. RIA for detection of antigen is frequently performed as a competitive binding assay. A fixed concentration of radioactive antigen competes with non-radioactive antigen in the test system for attachment to solid-phase bound antibody. The radioactive label ^{125}I is often used for labelling either antigen or antibody. The short half-life of this isotope necessitates frequent standardisation of reagents. Solid-phase procedures in which antibody fixed to a polystyrene well 'captures' antigen, which is subsequently detected by a radiolabelled antibody, are often termed direct assays. Indirect assays use a radiolabelled antiglobulin to demonstrate binding of antibody to the 'captured' antigen.

Immunofluorescence

Immunofluorescence, like ELISA, uses a labelled immunoglobulin for detecting antigen or antibody. The label employed is a fluorochrome, usually fluorescein isothiocyanate (FITC) or rhodamine isothiocyanate, covalently attached to antibody molecules. When examined by ultra-violet light, fluorescein emits a characteristic green colour and rhodamine a red colour. A special fluorescence microscope with a mercury vapour light source is required for this procedure. Fluorescent antibody methods are highly sensitive but require careful interpretation to avoid errors with autofluorescence which may occur with some tissues. There are many different methods of using fluorescent-labelled antibodies in microbiology and these include the direct, indirect and sandwich methods (**Diagram 19**). Both monoclonal and polyclonal antibodies may be used in fluorescent antibody tests.

In direct immunofluorescence, conjugated antibody is added directly to a smear, tissue section, or monolayer fixed on a slide, and after incubation unbound antibody is removed by washing. The slide is examined in a fluorescence microscope and where labelled antibody binds to antigen bright fluorescence is evident (**89**).

Indirect immunofluorescence is frequently used for the detection of antibodies in serum. Test serum is added to a known antigen fixed on a slide and following incubation and washing FITC-labelled antiglobulin is added. Unbound antiglobulin is removed by washing and the slide is examined for fluorescence (**90**). The indirect test is usually more sensitive than the direct test because more labelled antibody attaches per antigenic site.

89

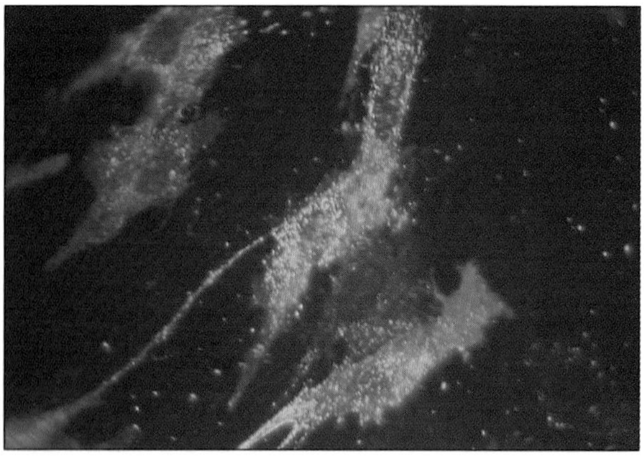

89 Direct immunofluorescence. Demonstration of parainfluenza 3 (PI3) virus in foetal calf lung cell monolayer (×400).

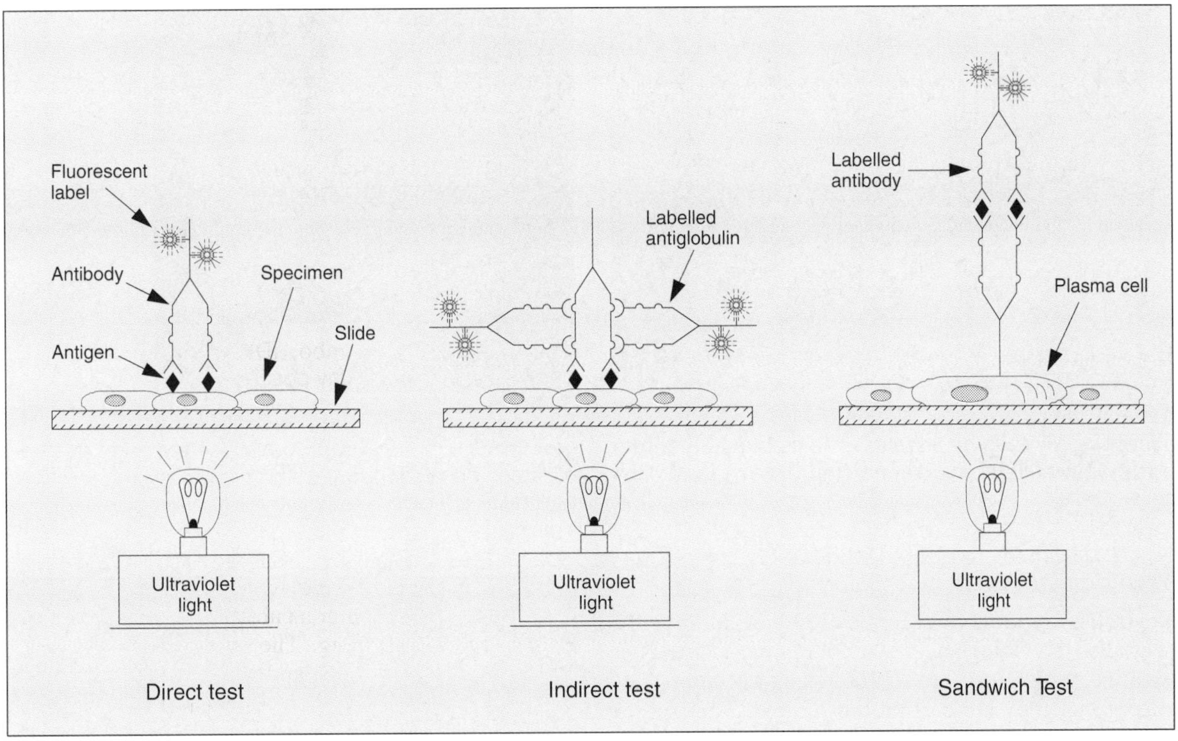

Diagram 19. Fluorescent antibody techniques (immunofluorescence). In the direct technique, specific antibody is labelled with fluorescent dye (fluorescein isothiocyanate). Labelled antiglobulin is used in the indirect test. In the 'sandwich' method, which is used to detect antibody in tissues rather than antigen, the first reagent added is specific antigen. Following washing, labelled antibody is added which reacts with the antigen bound by antibodies in the tissue section.

90

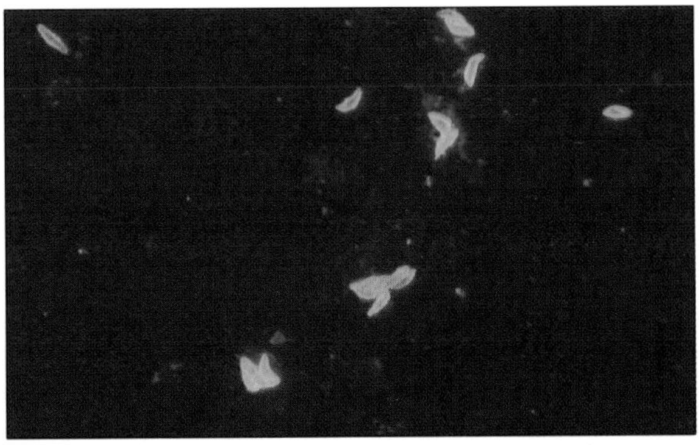

90 Tachyzoites of *Toxoplasma gondii* stained by indirect immunofluorescence. (×400)

Immunoperoxidase methods

Tissue localisation of antigens may be accomplished using enzyme-labelled reagents such as peroxidase labelled antibody. Identification of bound antibody is possible through conversion, by the enzyme, of a colourless soluble substrate to a coloured insoluble product (**91**). Horseradish peroxidase and alkaline phosphatase are the enzymes most often used for this test and the technique has significant advantages over immuno-fluorescence in that paraffin-embedded or resin-embedded sections can be used. In addition, tissues can be examined by conventional light microscopy and stained preparations can be kept for long periods without fading. In reading immunoperoxidase reactions, care must be taken to distinguish specific from non-specific staining and endogenous peroxidase staining.

91

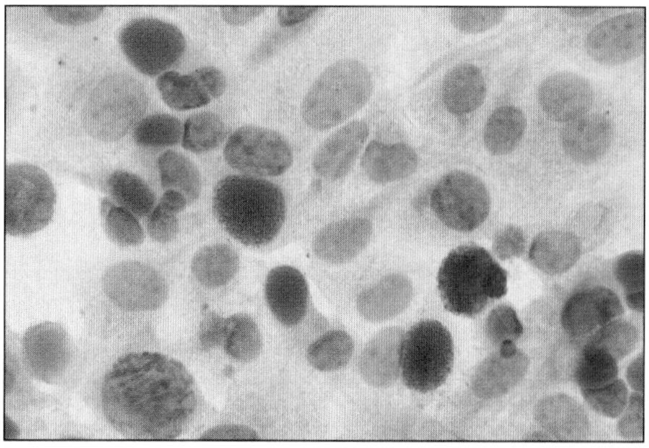

91 Immunoperoxidase staining with haematoxylin counter stain. *Chlamydia psittaci* (enzootic abortion of ewes) intracellular inclusions (red-brown) in McCoy cells. (×400)

92

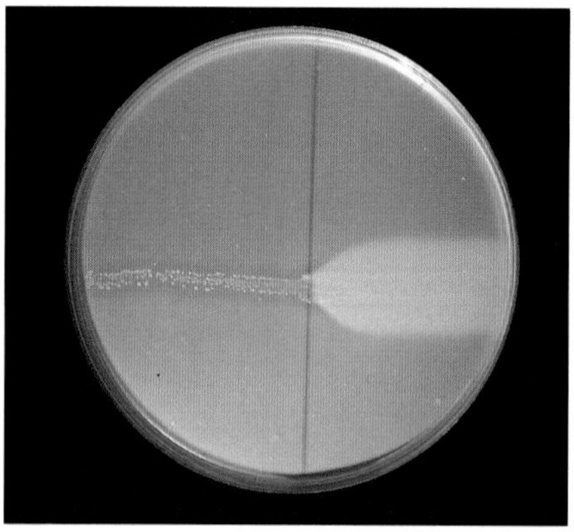

92 Nagler reaction: plate neutralisation test for *Clostridium perfringens*. Type A antitoxin on the left-half of the plate has neutralised lecithinase activity (Egg yolk agar).

Neutralisation tests

Neutralisation tests refer to the ability of antibody to neutralise the biological activity of toxins or viruses *in vitro*. Protection tests are similar to neutralisation tests except that they are carried out *in vivo*. The Nagler reaction is an example of a neutralisation test where antibody (antitoxin) to the α-toxin of *Clostridium perfringens* inhibits the lecithinase (phospholipase) activity of these bacteria growing on egg yolk agar which supplies lecithin (**92**).

Antibody, by combining with viruses, can neutralise their infectivity, thereby protecting the cells in a monolayer against virus destruction. Most changes induced by viral infections in monolayers or fertile eggs can be neutralised by specific antibody. This is the basis of virus neutralisation tests used either for identification of unknown viruses or for measurement of specific antiviral antibody. Serum-virus mixtures are inoculated into appropriate cell cultures which are incubated until the virus control, without antibody, shows evidence of cytopathic effects (**93, 94**). Virus neutralisation tests are highly specific and extremely sensitive.

93

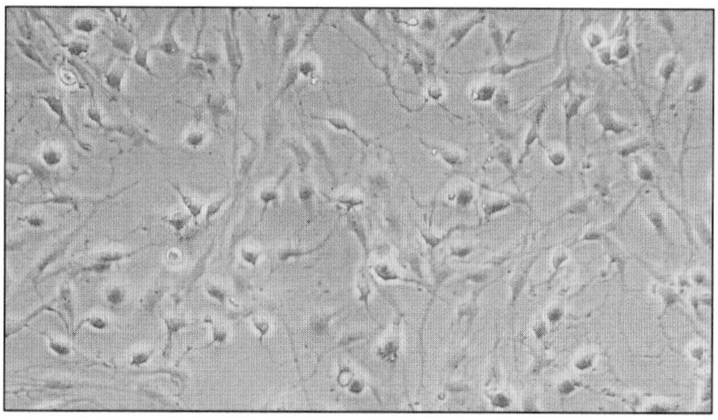

93 Virus neutralisation test in tissue culture (control): foetal calf lung (fibroblasts) cells showing evidence of cytopathic effects (CPE) due to a bovine adenovirus.

94

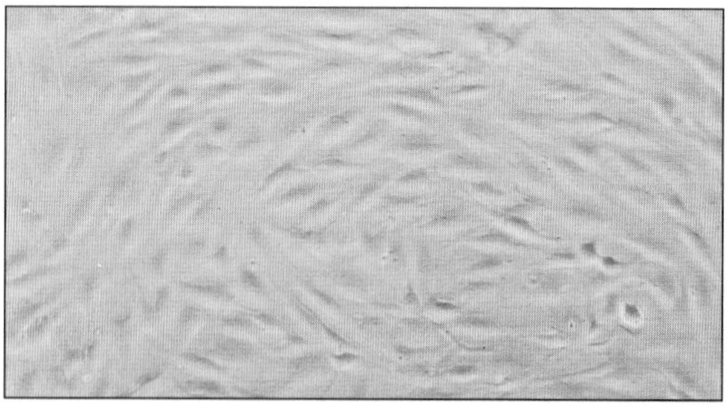

94 Virus neutralisation test in tissue culture (positive test): virus (bovine adenovirus) neutralised by specific antiserum and hence no CPE.

Bibliography

Bejamini, E. and Leskowitz, S. (1991). *Immunology, A Short Course*. Wiley-Liss, Inc. New York.

Davis, B.D., Dulbecco, R., Eisen, H.N. and Ginsberg, H.S. (1990). *Microbiology*, Fourth Edition. J.B. Lippincott Company, Philadelphia.

Halliwell, R.E.W. and Gorman, N.T. (1989). *Veterinary Clinical Immunology*. W.B. Saunders Company, Philadelphia.

Reeves, G. and Todd, I. (1991). *Lecture Notes on Immunology*. Blackwell Scientific Publications, Oxford.

Roitt, I. (1991). *Essential Immunology*. Blackwell Scientific Publications, Oxford.

Stites, D.P., and Terr, A.I. (1991). *Basic and Clinical Immunology*. Seventh edition. Appleton and Lange, Norwalk, Connecticut.

Tizard, I. (1992). *Veterinary Immunology*. Fourth edition, W.B. Saunders Company, Philadelphia.

6 The Isolation and Identification of Viral Pathogens

Viral diseases may be often diagnosed clinically, by postmortem findings or histopathologically (**95, 96**). However, in some instances laboratory confirmation of a specific viral pathogen is necessary. Tests used for laboratory diagnosis of viral infections fall into one of two categories (**Diagram 20**):

- Tests to demonstrate the presence of virus or viral antigen.
- Tests to demonstrate specific viral antibody.

95

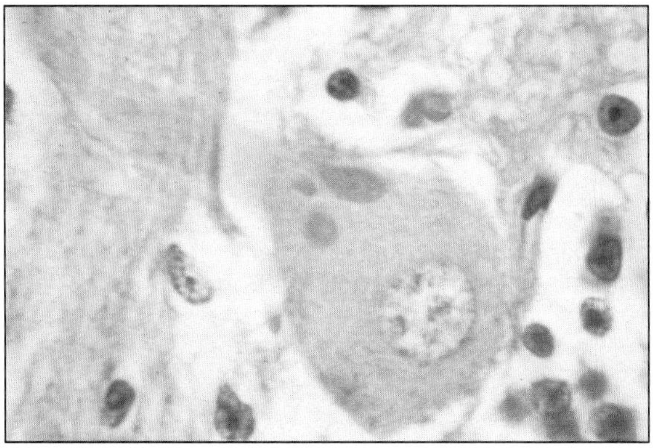

95 Three intracytoplasmic Negri bodies in a Purkinje cell from a dog with rabies showing the characteristic eosinophilic staining. (H&E stain, ×1000)

96

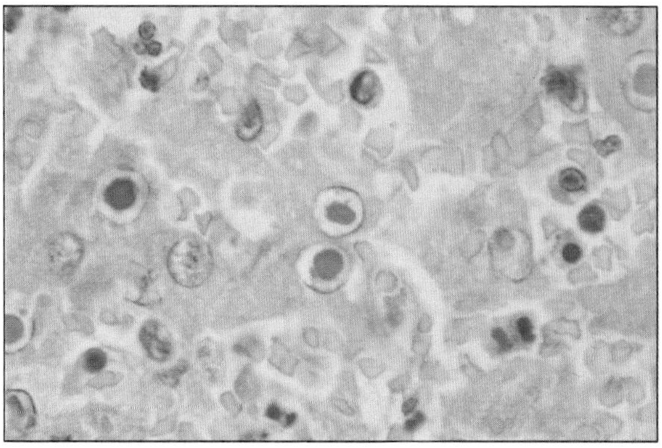

96 Numerous basophilic intranuclear inclusion bodies in hepatocytes from a dog with infectious canine hepatitis. (H&E stain, ×400)

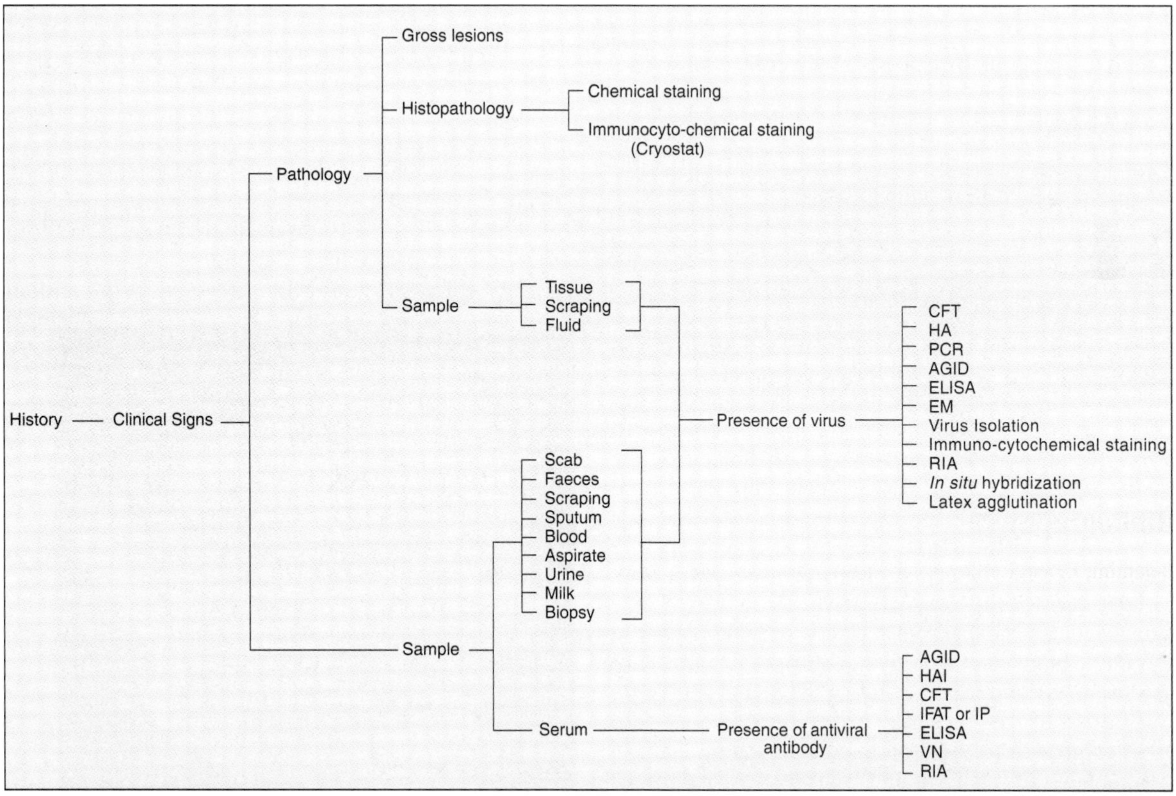

Diagram 20. Viral diagnosis. AGID = agar gel immunodiffusion, CFT = complement fixation test, ELISA = enzyme-linked immunosorbent assay, EM = electron microscopy, HA = haemagglutination, HAI = haemagglutination-inhibition, IFAT = indirect fluorescent antibody technique, IP = immunoperoxidase, PCR = polymerase chain reaction, RIA = radioimmunoassay, VN = virus neutralisation.

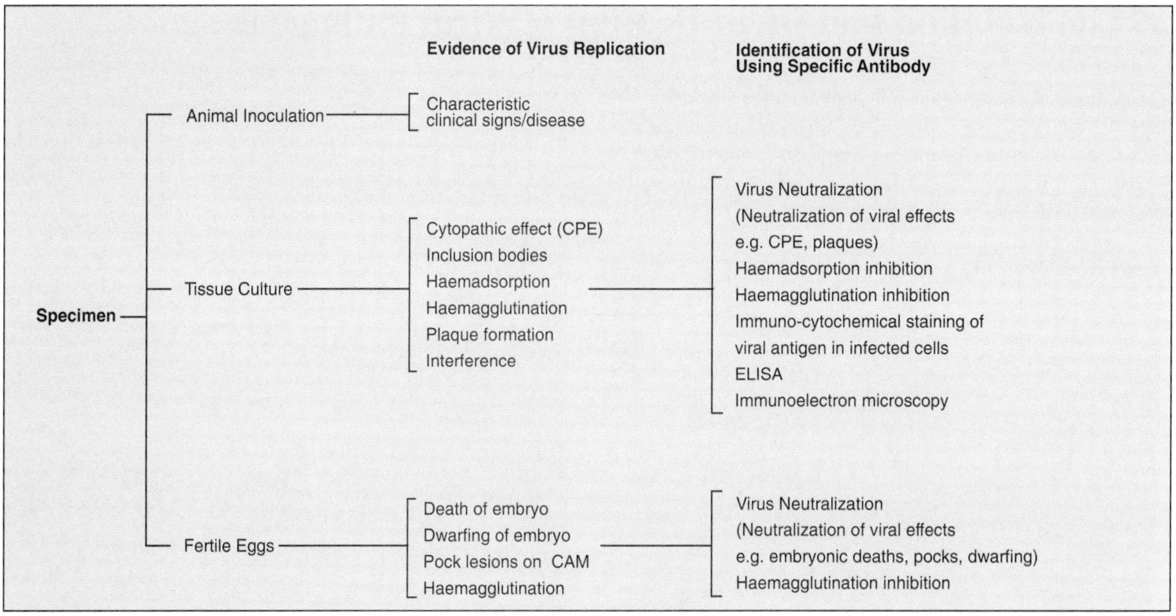

Diagram 21. Virus isolation and identification: CAM = chorioallantoic membrane, ELISA = enzyme-linked immunosorbent assay.

Tests to Demonstrate the Presence of Virus or Viral Antigen

Laboratory tests in this category include:

a) Isolation and specific identification of the virus (**Diagram 21**).
b) Direct demonstration of the virus, viral antigen or viral nucleic acid in animal tissues.

Virus isolation and identification

Virus isolation is usually the definitive criterion against which other diagnostic methods are assessed. Isolation of a virus may be carried out in cell cultures, fertile eggs (**97**) or experimental animals. The system chosen will largely depend on the suspected virus and a knowledge of the optimal method for its isolation. If optimal isolation conditions for the virus in question are provided, then isolation can be a very sensitive procedure for the detection of viruses, also permitting further study of the virus. However, virus isolation is labour intensive, expensive and slow to yield results. It should be targeted at a particular suspect virus and not used routinely as a means of ruling out a viral aetiology. A number of passages may be required to adapt the virus to growth in the laboratory and special care is required in the handling and transport of specimens (Chapter 2) to ensure the viability of viruses in the sample upon arrival at the laboratory. Cell culture involves the growth of cells *in vitro* as monolayers attached to a surface, glass or plastic, or as suspensions. The cells are no longer organized as a tissue. Three categories of cell culture are recognised. A primary cell culture is one derived from material taken directly from the animal. Foetal tissues are frequently used as they grow well and disaggregate easily. Primary cultures are the most sensitive for virus isolation but are labour intensive, requiring repeated preparation from fresh material. A secondary, diploid or semi-continuous cell culture is one that has been subcultured from a primary cell culture. The cells retain their original, diploid number of chromosomes. Secondary cultures have a finite life span related to the life span of the animal

97

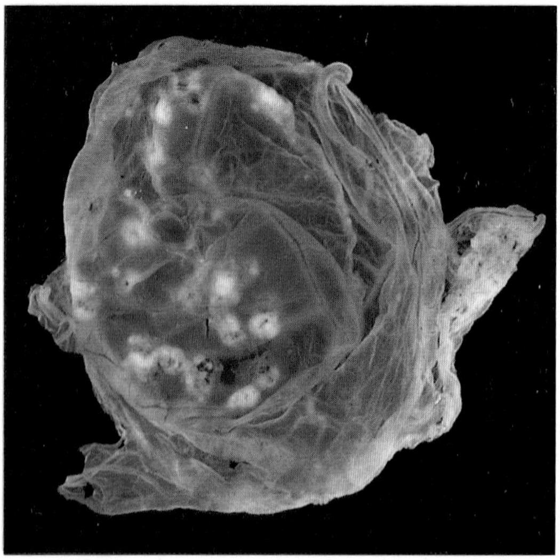

97 Chorioallantoic membrane with pock lesions caused by vaccinia virus.

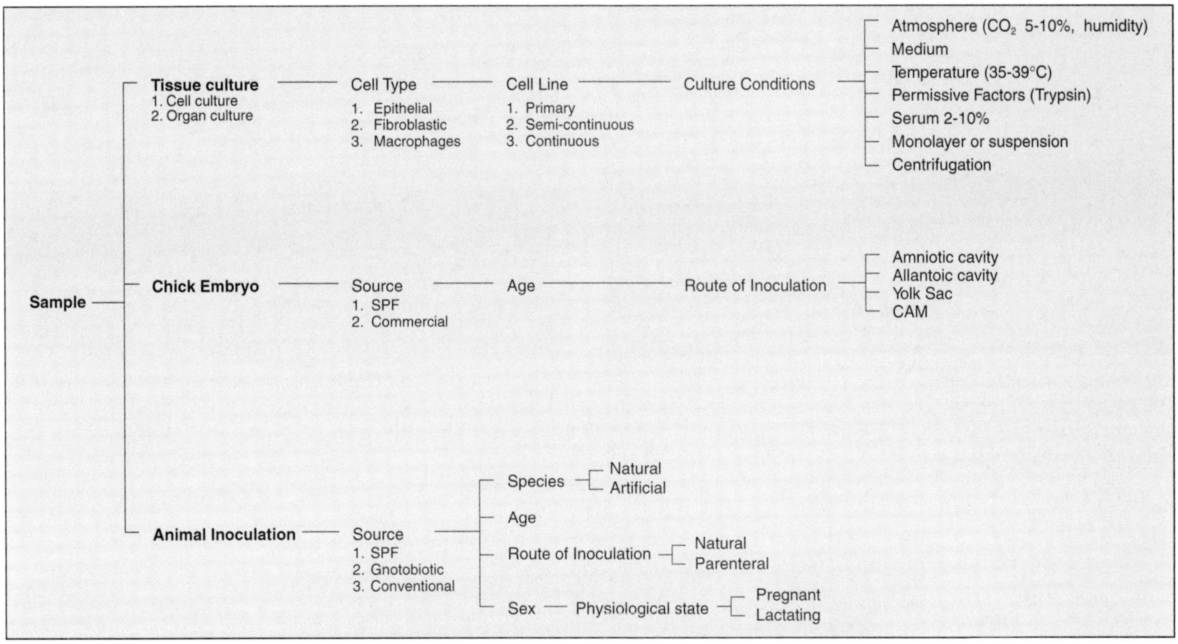

Diagram 22. Viral cultivation: SPF = specific-pathogen-free, CAM = chorioallantoic membrane.

species from which they originated. A continuous cell culture or line is capable of indefinite propagation *in vitro*. The cells are heteroploid and originate from malignant neoplasms or transformed diploid cell lines. Many continuous cell lines are well characterised and available commercially, for example Vero cells (monkey), baby hamster kidney (BHK-21) and Crandell feline kidney (CRFK). They are not as sensitive for virus isolation as primary cultures but cell culture-adapted viruses are often maintained in continuous cell lines in the laboratory.

Organ or explant cultures involve the use of a fragment of tissue or organ and its maintenance or growth *in vitro*. These cultures retain the architecture and sometimes function of the tissue. They are useful for the growth of certain fastidious viruses, such as infectious bronchitis virus *(Coronaviridae)*.

A number of cultivation factors influence the sensitivity of virus isolation (**Diagram 22**). Certain viruses, such as papillomaviruses, have not been grown in cell culture. If embryonated eggs are used they should be from a specific-pathogen-free flock to minimise problems associated with egg transmitted viruses and maternal antibody. Before inoculation, the eggs are candled to ensure the embryo is alive and to mark the positions of major blood vessels and of the air space. The eggs are inoculated between days 5–12 of incubation, depending on the route used (allantoic cavity, amniotic cavity, yolk sac, chorioallantoic membrane or intravenous). Isolation procedures avoid the use of experimental animals unless it is the only successful means of isolating or identifying a viral pathogen, for example mice inoculated intracerebrally with rabies virus. The host animal used must be free of intercurrent infection and of anti-viral antibody.

Following isolation, additional tests are usually required to identify the specific virus (**Tables 7–11**).

Direct demonstration of virus, viral antigens or viral nucleic acid in animal tissues

a) Electron microscopy

Electron microscopy is a rapid though relatively insensitive means of detecting virus. A virus concentration in excess of 10^6 per ml is required if viral particles are to be visualised (**98, 99, 100**). A negative result does not necessarily indicate the absence of virus. In certain clinical conditions such as enteric or skin infections, the numbers of viral particles frequently exceeds this figure. Alternatively, it may be necessary to isolate and successfully passage the virus in order to attain titres of this magnitude. It is possible to detect non-viable virus provided that structural morphology has been retained. Combined viral infections can be readily diagnosed. The most serious drawback is the large expense incurred in purchasing and maintaining an electron microscope. Clarification and subsequent concentration of the viral particles present in the sample, by centrifugation, is usually required for visualisation of the virus(es), particularly in the case of faecal specimens. Negative staining, using electron-dense heavy metals, such as phosphotungstic acid, is used to increase contrast. The viral particles are visualised as bright objects against a dark background. In general, members of a viral family have identical morphology and cannot be distinguished by electron

98

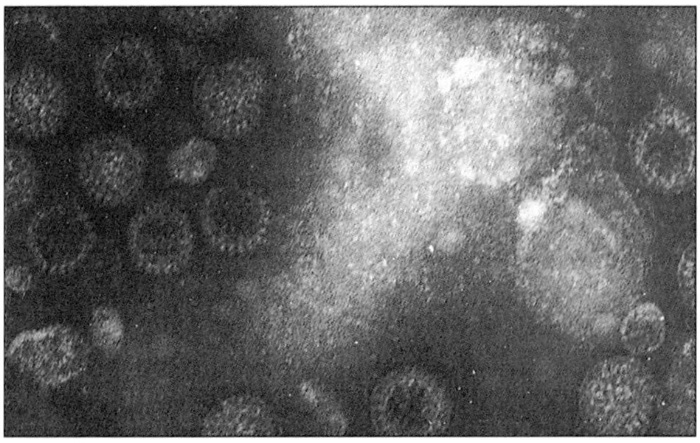

98 Electron micrograph of rotavirus in a faecal preparation showing the double capsid and typical 'wheel' effect. Negatively stained with phosphotungstic acid (PTA). (×45,000)

99

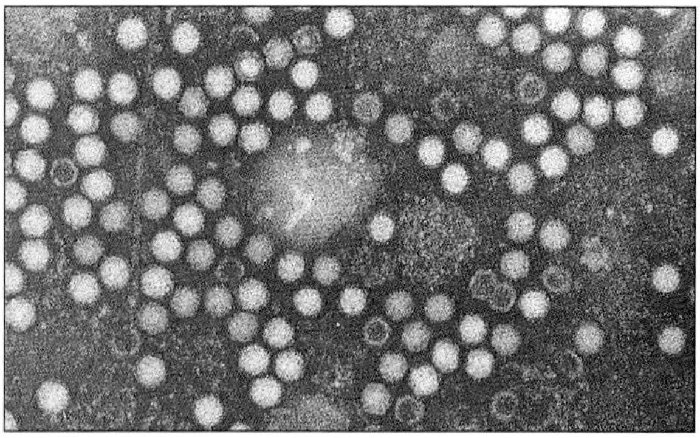

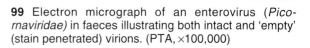

99 Electron micrograph of an enterovirus (*Picornaviridae)* in faeces illustrating both intact and 'empty' (stain penetrated) virions. (PTA, ×100,000)

100

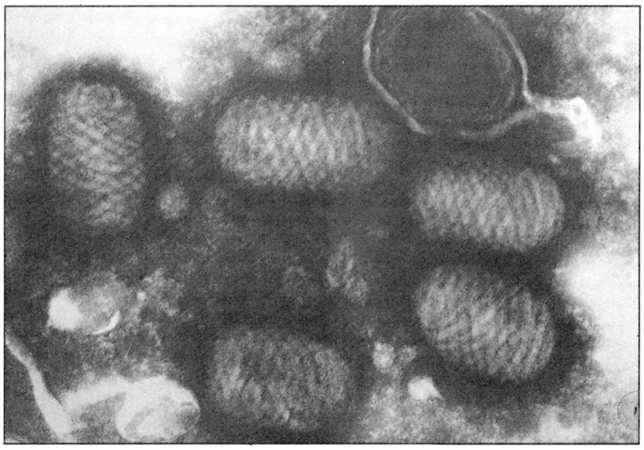

100 Electron micrograph of contagious pustular dermatitis (orf) virus from a human skin lesion acquired from sheep. Note the characteristic 'basket-weave' appearance of this parapoxvirus. (PTA, ×45,000)

101

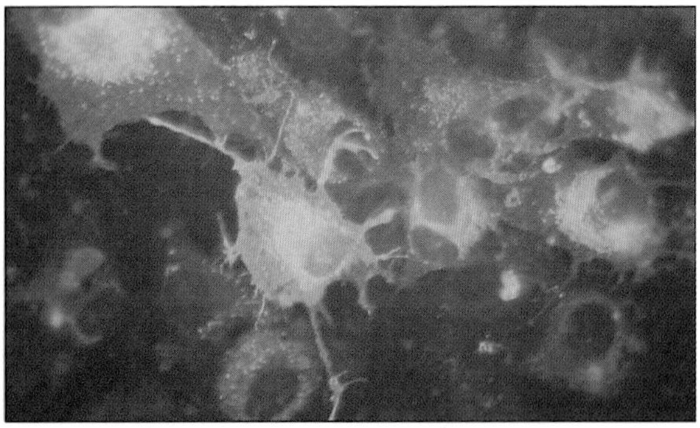

101 Bovine coronavirus in foetal calf lung cells. (Direct FA technique, ×400)

102

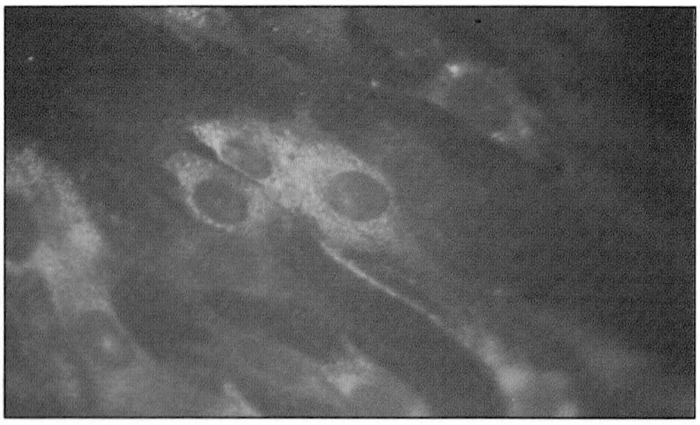

102 Bovine viral diarrhoea virus in foetal calf lung cells showing characteristic intracytoplasmic fluorescence. (Direct FA technique, ×400)

microscopy. Exceptions include the *Poxviridae*, *Papovaviridae* and *Reoviridae* where there are significant ultrastructural differences between the genera in each family. Immunoelectron microscopy is a modification of the technique capable of enhancing its sensitivity. The antiserum used may be mixed with the sample and the immune complexes pelleted by centrifugation following incubation, or the antiserum may be coated onto the copper grid and the sample applied in the usual way. The antiserum may contain polyclonal or monoclonal antibodies and be species-specific or even serotype-specific.

b) Immuno-cytochemical staining (Chapter 5)
Immunofluorescence can be used to detect viral antigen in cell monolayers, smears such as nasopharyngeal aspirates and in cryostat sections of tissues (**101, 102**). Immunoperoxidase staining may, in addition, be used to demonstrate the presence of viral antigen in histological sections. Provided the required conjugates are available these techniques can furnish a specific diagnosis within hours of the receipt of material by the laboratory. The direct labelled-antibody technique is more specific but less sensitive than the indirect method.

More recent advances in indirect staining techniques include the use of protein A and enzyme amplification techniques. These latter techniques involve the use of biotin-avidin or enzyme anti-enzyme (EAE) complexes. Avidin is a basic glycoprotein of approximately 68,000 molecular weight. Each avidin molecule is capable of binding four biotin molecules with an extraordinarily high affinity. In addition, biotin can be covalently coupled to antibody without interfering with the antigen-binding capacity of the antibody. The unlabelled antibody technique involves the use of a soluble immune complex made up of enzyme and antibodies specific for the enzyme such as peroxidase anti-peroxidase (PAP) and glucose oxidase anti-glucose oxidase (GAG). Linkage between the EAE complex and the primary anti-viral antibody is achieved by the use of an antiglobulin bridge. The primary antibody and the anti-enzyme antibody must be produced in the same or closely related species in order for the antiglobulin to form this bridge.

Immunofluorescence is extremely useful for the detection of viral antigens in samples of nasal mucus collected using a suction apparatus. For example, mixed infections of parainfluenza virus 3 (PI3) and bovine respiratory syncytial virus (RSV) are frequently involved in calf pneumonia. If virus isolation is used as the sole means of diagnosis, PI3 virus quickly outgrows RSV and the RSV may not be detected.

c) Enzyme-Linked Immunosorbent Assay (ELISA)
The ELISA is capable of great sensitivity and specificity depending on the quality of the reagents and on the controls employed. It is not technically difficult to perform, is suitable for testing large numbers of samples and can be completed in less than one working day. Increasingly, ELISAs for viral diagnosis are becoming commercially available, often in a form suitable for use in a veterinary practice (*see* Chapter 47).

d) Radioimmunoassay (RIA)
This is a rapid, specific and extremely sensitive test. However, due to the expense and potential hazards involved it is generally used as a research tool rather than as a means of routine diagnosis.

e) Immunodiffusion
Immunodiffusion in agar is easy to carry out, cheap and fairly rapid. However, it is relatively insensitive and requires high concentrations of antigen and antibody. In addition, viral antigens may not be detectable using this test after antibodies appear.

f) Complement Fixation Test (CFT)
The viral antigens detected by the CFT are often group specific. As a result the test is more broadly reactive than many other tests such as the neutralization test. It is a demanding test to perform, requiring careful standardization of several separate reagents. As a result, the ELISA has superseded the CFT for many viral diseases.

Tests based on detection of viral antigen using antisera are rapid, generally sensitive and capable of providing information on the serotype of the virus. However, only those viruses against which antibodies are present in the antiserum will be detected.

g) Haemagglutination (HA) (Chapter 5)

Many viruses have virus-coded glycoproteins on the outer surface of their envelopes which confer on the viruses the ability to attach to red blood cells and so link them together in large aggregates. This is known as haemagglutination. Poxviruses are an exception as they cause haemagglutination by lipid haemagglutinins produced during viral replication. Families of viruses containing members capable of haemagglutination include the *Orthomyxoviridae*, *Paramyxoviridae*, *Adenoviridae*, *Parvoviridae*, *Togaviridae* and *Poxviridae*. The type of erythrocyte that is agglutinated varies from virus to virus. A large number of viral particles are needed to cause visible agglutination and so the test is comparatively insensitive and unsuitable for the detection of small numbers of virions. Orthomyxoviruses (Influenzaviruses) and the paramyxoviruses carry the enzyme neuraminidase and the glycoprotein haemagglutinin as spikes which project from the lipoprotein envelope. In orthomyxoviruses the haemagglutinin and neuraminidase are separate spikes whereas in paramyxoviruses they form a joint spike. Haemagglutinin allows these viruses to attach to neuraminic acid-containing receptors on erythrocytes, while neuraminidase destroys such receptors and is responsible for elution, seen if haemagglutination is carried out at 37°C. Haemagglutinin and neuraminidase are serotype-specific in influenza viruses and form the basis of a typing system. Thirteen different haemagglutinin and nine neuraminidase influenza antigens are recognised.

h) 'In situ' hybridization

Viral DNA can be labelled with radioisotopes such as ^{32}P or ^{35}S or with biotin to produce a probe capable of detecting related DNA in infected cells. Under certain conditions double-stranded DNA will separate into single strands and subsequently reanneal to each other or to a closely related complementary strand. Hybridization between viral DNA and the DNA probe is detectable by autoradiography in the case of radioactive probes or by ELISA, immunofluorescence or immunoperoxidase staining in the case of biotinylated probes. DNA probes are extremely useful in detecting viral DNA in cells where the infection is not being expressed, for example, with endogenous retroviruses that integrate into the host cell genome as DNA provirus, and in latent herpesvirus infections. Diagnostic hybridization reactions are generally carried out by the "dot blot" technique where the clinical specimen is denatured and spotted on (fixed) to a nitrocellulose membrane. The virus specific DNA probe is added and allowed to react. Following hybridization the bound probe is detected by autoradiography or by a colorimetrical reaction. The technique can be made highly specific by using probes that are shorter than the full viral genome. It is a rapid and highly sensitive test. However, the probes are expensive to produce initially and are generally only available through research institutions.

i) Polymerase chain reaction (PCR)

This is a relatively new *in vitro* DNA amplification procedure. A typical PCR cycle consists of heat denaturation of dsDNA, the annealing of two short oligonucleotide primers, which flank the DNA segment to be amplified and the extension of the annealed primers in the presence of a thermostable DNA polymerase such as *Taq* DNA polymerase. The reaction is carried out using an excess of primers and polymerase. Typically, 30 cycles are completed giving an exponential amplification of the sequence of interest. It is possible to amplify single copy genomic sequences with great specificity by a factor of more than 10 million. The technique has been used to produce probes for virus diagnosis and also to amplify minute quantities of viral DNA present in clinical specimens, in order to facilitate detection. The PCR technique is extremely sensitive and requires only minute amounts of a clinical specimen. However, non-specific amplification of DNA may occur due to the very short sequence size, typically 20-25 base pairs, of the oligonucleotide primers. This can result in false positive results. Validation of the amplified DNA may be carried out by hybridization with a known DNA probe or by comparison of the restriction endonuclease pattern in gels with that of positive controls.

Tests to Demonstrate Antiviral Antibody

Seroconversion to a particular viral agent can be demonstrated using a variety of tests. The following serological tests are described in Chapter 5:

a) Enzyme-Linked Immunosorbent Assay

The degree of colour change is directly proportional to the amount of bound antibody. As a result, single serum dilutions are often sufficient to indicate the levels of antibody present in a sample, obviating the need for serial titrations to be carried out.

b) Complement Fixation Test

Complement fixing antibodies tend to appear before neutralizing antibody but do not persist as long. Some sera are anti-complementary and cannot be tested by the CFT.

c) Virus neutralisation (VN) Test

VN tests are highly specific and may be used to serotype viral isolates or to detect antibody. Neutralising antibodies are directed against surface antigens of the virus and are frequently type-specific or serotype-specific. They are also persistent antibodies, often lasting for the life of the animal.

d) Haemagglutination-Inhibition (HAI) Test

The HAI test can be adapted to serotype viral isolates using known antisera, or to detect antibody. The test is strain specific requiring that the antigen used be of the same virus strain as that responsible for the infection. Non-specific inhibitors of haemagglutination are frequently present in sera. Methods used to remove these inhibitors include heating at 56°C for 30 min, treatment with kaolin, trypsin, periodate or bacterial neuraminidase. The treatment used varies with the virus in question and may result in reduced antibody titres.

e) Indirect Immunofluorescence

This technique is very sensitive and rapid. However, it requires an experienced operator and may result in eye strain if a large number of tests have to be done. It is time consuming to determine the antibody titre of a sample as titres tend to be high and endpoints difficult to determine precisely.

f) Radioimmunoassay

The radioimmunoassay is similar to the ELISA, but the short half-life of the radioactive reagents requires that they be prepared and standardized at frequent intervals.

Serological testing is generally less expensive and easier to perform than the demonstration of virus or viral antigen. Serum samples are readily obtainable in most instances and some tests such as ELISAs lend themselves to automation thus allowing the screening of large numbers of samples.

Serology can also be used in retrospective studies of animal populations to establish the prevalence of virus infections. With the exception of exotic diseases, paired serum samples are required to demonstrate a rise in antibody titre for the agent suspected in a disease outbreak. Maternal antibodies in young animals, acquired passively, must be differentiated from an active immune response. Cross-reactions arising from previous infections with an antigenically related virus may occasionally render serological results difficult to interpret.

Table 7. Isolation and identification of viruses of veterinary importance: **ruminants.**

Virus	*Specimen(s)*	*Host system*	*Evidence of viral replication*	*Identification*
Bovine herpesvirus 1 (infectious bovine rhinotracheitis / infectious pustular vulvovaginitis)	Nasal and ocular swabs. Nasopharyngeal aspirate. Tracheal scraping. Foetal liver and kidney. Vaginal swab	Cell culture (bovine origin e.g. calf kidney, calf testes, foetal calf lung, etc.)	CPE I/N inclusions	VN FA
Bovine herpesvirus 2 (mammillitis)	Scabs. Swabs from teat and udder lesions	Cell culture (bovine origin)	CPE	VN
Bovine viral diarrhoea virus (bovine viral diarrhoea / mucosal disease)	Buffy coat (freshly collected heparinised blood). Spleen, abomasum, intestine. Foetal tissues	Cell culture (bovine origin)	CPE FA (non-cytopathic isolates)	VN FA
Bovine coronavirus	Faeces	Cell culture (bovine origin; trypsin enhances growth)	CPE	VN FA
Bovine parainfluenza virus 3	Nasal swabs. Nasopharyngeal aspirate. Lung	Cell culture (bovine origin)	CPE (variable) HA (guinea-pig RBCs)	VN FA HAI
Bovine respiratory syncytial virus	Nasal swabs. Nasopharyngeal aspirate. Lung	Cell culture (bovine origin; semi-continuous cell line)	CPE (slow to develop)	VN FA
Border disease virus	Foetal tissues	Cell culture (ovine or bovine origin)	CPE FA	VN FA
Louping ill virus	Brain	Cell culture (primary ovine embryo cell lines or pig kidney IB/RS-2 cell line). Suckling mice (intracerebral inoculation)	CPE (variable) Posterior paralysis Death	VN HAI (pigeon or rooster RBCs)

Table 8. Isolation and identification of viruses of veterinary importance: **pigs**.

Virus	*Specimen(s)*	*Host system*	*Evidence of viral replication*	*Identification*
Porcine herpesvirus 1 (Aujeszky's disease)	Brain, tonsil and lung	Cell culture (porcine origin e.g. PK-15 cell line)	CPE	VN FA
Porcine parvovirus	Foetal tissues	Cell culture (porcine origin; freshly seeded)	CPE	VN FA HAI (guinea-pig RBCs)
Transmissible gastroenteritis virus	Intestine	Cell culture (primary pig kidney; trypsin enhances growth)	CPE FA	VN FA
Porcine enteroviruses	Brain (Teschen disease). Foetal tissues (SMEDI)	Cell culture (porcine origin)	CPE	VN
Swine fever virus (hog cholera)	Spleen, tonsils, lymph nodes	Cell culture (PK-15 cell line)	FA	FA

Table 9. Isolation and identification of viruses of veterinary importance: **horses**.

Virus	*Specimens*	*Host system*	*Evidence of viral replication*	*Identification*
Equine herpesviruses 1 or 4 (abortion and/or rhinopneumonitis)	Nasal swabs. Foetal lung and liver	Cell culture (equine origin or PK-15 cell line)	CPE	VN
Equine influenzavirus	Nasal swabs. Lung	Chick embryo (9–11 days; allantoic cavity)	Death HA (human group O RBCs)	HAI
Equine viral arteritis virus	Nasal swabs. Blood	Cell culture (equine kidney)	CPE	VN
Equine encephalomyelitis viruses (Western, Eastern and Venezuelan)	Brain. Blood	Chick embryo (9–11 days; CAM).	Pocks, death	VN
		Cell culture (PK-15 cell line).	CPE	VN
		Weanling mice (intracerebral inoculation)	CNS signs, death	VN

Table 10. Isolation and identification of viruses of veterinary importance: **cats.**

Virus	*Specimen(s)*	*Host system*	*Evidence of viral replication*	*Identification*
Feline parvovirus (panleukopenia)	Faeces. Intestine	Cell culture (feline origin; freshly seeded)	FA Intranuclear inclusions	FA
Feline herpesvirus 1 (rhinotracheitis)	Nasal, oropharyngeal and conjunctival swabs	Cell culture (feline origin)	CPE	VN
Feline calicivirus	Nasal, oropharyngeal and conjunctival swabs. Lung	Cell culture (feline origin)	CPE	VN

Table 10a. Isolation and identification of viruses of veterinary importance: **dogs**.

Virus	*Specimen(s)*	*Host system*	*Evidence of viral replication*	*Identification*
Canine distemper virus	Lung, urinary bladder and cerebellum	Cell culture a) Canine macrophages or lymphocytes; b) Primary culture of cells from infected dog.	CPE FA	FA
		Inoculation of ferret	Clinical signs Death	FA
Canine parvovirus	Faeces	Cell culture (canine or feline origin e.g. MDCK or CRFK cell lines; freshly seeded)	FA HA (pig RBCs at 4°C)	FA VN HAI
Canine adenovirus 1 (infectious canine hepatitis)	Nasal swabs, urine, blood, lung, kidney and lymph nodes	Cell culture (canine origin)	CPE	VN
Canine adenovirus 2	Nasal or throat swabs. Lung	Cell culture (canine origin)	CPE	VN
Canine herpesvirus	Kidney, liver, lung and spleen	Cell culture (canine origin)	CPE	VN
Rabies virus	Brain	Weanling mice (intracerebral inoculation)	CNS signs, death	FA Negri bodies VN
		Cell culture (neuroblastoma cells)		FA

Table 11. Isolation and identification of viruses of veterinary importance: **poultry**.

Virus	*Specimen(s)*	*Host system*	*Evidence of viral replication*	*Identification*
Infectious bronchitis virus	Trachea, lung and faeces	Chick embryo (9–11 days; allantoic cavity)	Characteristic stunting of embryos. Curling and 'clubbing' of down.	VN
Avian herpesvirus (infectious laryngotracheitis)	Trachea and lung	Chick embryo (9–11 days; CAM). Cell culture (chick embryo kidney).	CPE I/N inclusions	VN
Avian paramyxovirus 1 (Newcastle disease)	Trachea, lung, heart, kidney, spleen and faeces	Chick embryo (9–11 days; allantoic cavity)	HA (chicken RBCs)	HAI
Avian encephalomyelitis virus (epidemic tremor)	Brain, faeces	Chick embryo (6–7 days; yolk sac). Day-old chicks (intracerebral inoculation).	Clinical signs in hatched chicks. Tremors, death.	VN VN
Avian influenzavirus (fowl plague)	Respiratory tract and faeces	Chick embryo (9–11 days; allantoic cavity)	Death HA (chicken RBCs)	VN HAI
Fowlpox virus	Scabs, scrapings from lesions	Chick embryo (9–11 days; CAM)	Pocks	EM VN
Marek's disease virus	Buffy coat (freshly collected heparinized blood).	Cell culture (chick embryo kidney)	CPE I/N inclusions	FA

CAM = Chorioallantoic membrane; CNS = Central nervous system; CPE = Cytopathic effect; CRFK = Crandell feline kidney; FA = Fluorescent antibody technique; HA = Haemagglutination; HAI = Haemagglutination-inhibition; I/N = Intranuclear; MDCK = Madin-Darby canine kidney; PK-15 = Pig kidney; RBC = Red blood cells; VN = Virus neutralization.

7 Antimicrobial Agents

Antimicrobial Susceptibility Testing

Although there are several laboratory methods for measuring the *in vitro* susceptibility of bacteria to antimicrobial drugs, the agar disc diffusion technique is used most commonly. The method described here should be used only for the rapidly growing pathogens if the results are to be clinically reliable. The technique must be standardised for consistent results. A standardised method has been proposed by the National Committee for Clinical Laboratory Standards (NCCLS, 1990, MA-A4) and is followed in this account of the Kirby-Bauer disc diffusion method (Bauer, 1966).

Factors affecting the size of the zone of inhibition

The disc diffusion method entails preparing a uniform lawn of the test bacterium on an agar plate and placing paper discs, each impregnated with an antimicrobial agent of known concentration, on the agar surface before incubation. If the bacterium is sensitive to a particular agent, then a zone of inhibition occurs around the disc after incubation. One of the reasons for the strict standardisation of the test procedures is that many factors can influence the size of the zone of inhibition and these include:

- **The size of the inoculum:** this is of great importance and the turbidity of the inoculum is adjusted to a 0.5 McFarland opacity standard. The aim is to have a dense lawn of bacterial growth with the individual colonies just touching each other (**103** and **104**).

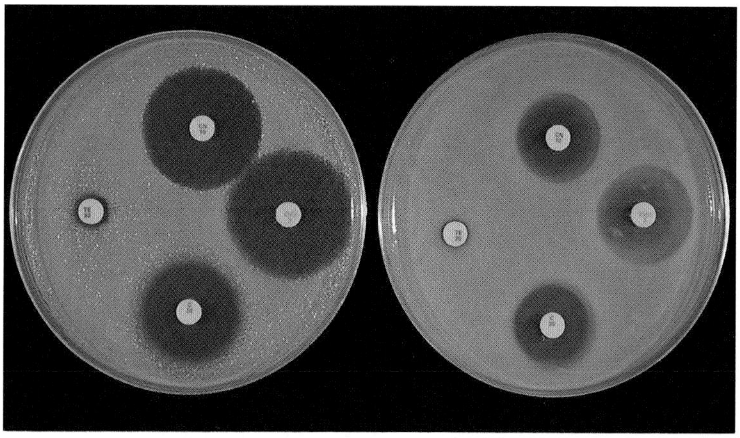

103

103 *Staphylococcus aureus* on Isosensitest agar indicating the effect of a correct inoculum (left) and a heavy inoculum (right) on the size of the zone of inhibition.

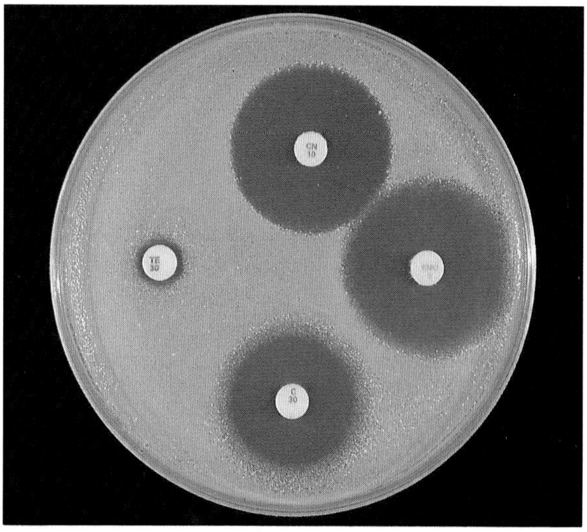

104

104 *S. aureus* on Isosensitest agar tested against gentamicin (CN), enrofloxacin (ENO), chloramphenicol (C) and tetracycline (TE) and illustrating an inoculum size that gives a lawn of the correct density.

- **The test medium:** Mueller-Hinton or a modification of this medium (Isosensitest agar, Oxoid) is usually chosen for routine susceptibility tests. It gives good batch-to-batch reproducibility, is low in sulphonamide, trimethoprim and tetracycline inhibitors, and most pathogens grow satisfactorily on the medium. However, quality control checks should be carried out on each new batch of medium. Excessive amounts of thymidine or thymine in a test medium can inhibit sulphonamides and trimethoprim, yielding smaller zones or no zones at all. Variation in divalent cations, mainly magnesium and calcium, will affect the zone size of tetracycline, polymyxin and aminoglycoside tests against *Pseudomonas aeruginosa*. For bacteria, such as streptococci, that are unable to grow on Mueller–Hinton agar, blood agar with 5–10 per cent defibrinated sheep blood can be used. But the zone size particularly for nafcillin, novobiocin and methicillin will be 2–3 mm smaller than the normal control limits (**105** and **106**). The depth of the agar and the pH of the medium must be standardised as these may also have an effect on the zone size (*see* **Quality control procedures**).
- **The antimicrobial agent and its concentration in the disc:** the ability of the agent to diffuse through the agar varies and the zone of inhibition for some drugs, such as streptomycin, is always comparatively small. The concentrations of the antimicrobial agents in the discs have been chosen to give zone sizes that correlate with achievable serum levels in the patient. The zone sizes and their interpretation are given in **Table 12**. At present there is insufficient data for the correlation of *in vitro* tests and the clinical use of topical agents for skin, eye and ear conditions. The potency of the agent in the discs must be maintained and storage and handling methods are discussed in Quality Control Methods.
- **Incubation conditions:** these have been standardised for routine susceptibility tests under aerobic incubation at 35°C for 16–18 hours and 24 hours for the staphylococci. The plates must not be incubated under an increased concentration of carbon dioxide as this can significantly alter the zones of inhibition with some antimicrobial agents. A modified method is used for *Haemophilus* species and other fastidious bacteria that require carbon dioxide for growth (NCCLS, 1990, M2-A4). Anaerobes should not be tested by the disc diffusion method and the NCCLS (1990) has published a tentative standard method for testing the susceptibility of anaerobes (M11-T2).
- **The test bacterium:** The degree of resistance or susceptibility of a bacterium to selected antimicrobial agents will vary with the species and strain of the pathogen. There are specific difficulties in detecting staphylococcal resistance to methicillin.

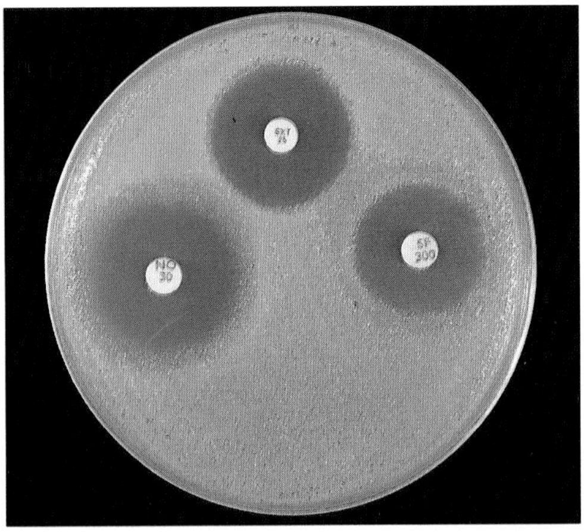

105

105 *Staphylococcus aureus* on Isosensitest agar tested against novobiocin (NO), trimethoprim-sulphamethoxazole (SXT) and sulphafurazole (SF). Compare with **106** when blood agar is used.

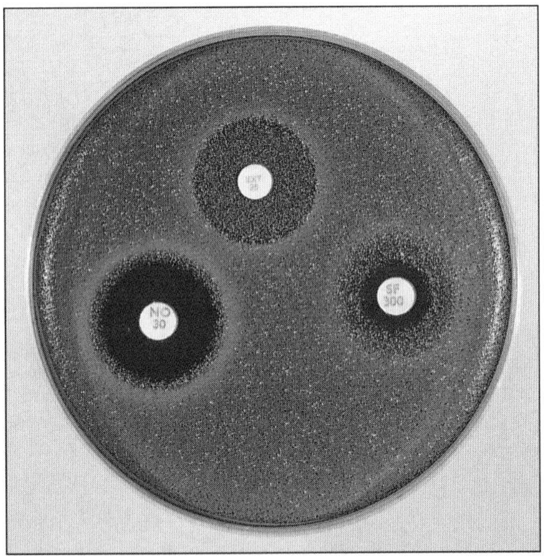

106

106 The same isolate of *S. aureus* as in **105** on blood agar tested against novobiocin (NO), trimethoprim-sulphamethoxazole (SXT) and sulphafurazole (SF). Note the smaller zone sizes compared with **105**.

Table 12. Zone size interpretation chart (modified from NCCLS (1990) M2-A4)*

Disc content in μg unless otherwise stated	Antimicrobial agents	Resistant ≤	Intermediate	Moderately Susceptible	Susceptible ≥
		Diameter of zone of inhibition to nearest mm.			
30	AMIKACIN	14	15–16	–	17
20/10	AMOXICILLIN / CLAVULANIC ACID				
	when testing staphylococci	19	–	–	20
	when testing other bacteria	13	–	14–17	18
10	AMPICILLIN				
	when testing Gram-negative enteric bacteria	13	–	14–16	17
	when testing staphylococci	28	–	–	29
	when testing enterococci	16	–	≥17	–
	when testing non-enteric streptococci	21	–	22–29	30
100	CARBENICILLIN				
	when testing *Pseudomonas* spp.	13	–	14–16	17
	when testing other Gram-negative bacteria	19	–	20–22	23
75	CEFOPERAZONE	15	–	16–20	21
30	CEFOTAXIME	14	–	15–22	23
30	CEPHALOTHIN	14	–	15–17	18
30	CHLORAMPHENICOL	12	13–17	–	18
5	CIPROFLOXACIN	15	–	16–20	21
2	CLINDAMYCIN	14	15–20	–	21
10	ENOXACIN	14	–	15–17	18
15	ERYTHROMYCIN	13	14–22	–	23
10	GENTAMICIN	12	13–14	–	15
30	KANAMYCIN	13	14–17	–	18
30	MOXALACTAM	14	–	15–22	23
1	NAFCILLIN when testing staphylococci	10	11–12	–	13
30	NALIDIXIC ACID	13	14–18	–	19
30	NETILMICIN	12	13–14	–	15
300	NITROFURANTOIN	14	15–16	–	17
10	NORFLOXACIN	12	13–16	–	17
1	OXACILLIN when testing staphylococci	10	11–12	–	13
10 units	PENICILLIN G				
	when testing staphylococci	28	–	–	29
	when testing enterococci	14	–	≥15	–
	when testing non-enteric streptococci	19	–	20–27	28
100	PIPERACILLIN				
	when testing *Pseudomonas* spp.	17	–	–	18
	when testing other Gram-negatives	17	–	18–20	21
5	RIFAMPIN	16	17–19	–	20
10	STREPTOMYCIN	11	12–14	–	15
250/300	SULPHONAMIDES	12	–	13–16	17
30	TETRACYCLINE	14	15–18	–	19
75	TICARCILLIN				
	when testing *Pseudomonas* spp.	14	–	–	15
	when testing other Gram-negatives	14	–	15–19	20
10	TOBRAMYCIN	12	13–14	–	15
1.25/23.75	TRIMETHOPRIM-SULPHAMETHOXAZOLE	10	–	11–15	16
30	VANCOMYCIN				
	when testing enterococci	14	15–16	≥17	–
	when testing other Gram-positives	9	10–11	–	12

* Updated according to the Third Informational Supplement, December 1991.

Standard Method

- At least 4–5 well-isolated colonies, of the same morphological type are selected from a non-selective agar plate. Just the top of the colonies are touched and the growth transferred to a tube containing 4–5 ml of soybean-casein digest broth or an equivalent medium such as tryptone soya broth (Oxoid).
- The inoculated broth is incubated at 35–37°C until a slight visible turbidity appears; usually within 2–8 hours.

Alternative method

- The colonies are selected as before and a suspension is made in saline or broth without the pre-incubation in broth. This is suggested as the best method for testing staphylococci, especially with suspected methicillin-resistant strains.
- The turbidity of both the pre-incubated broth and the suspension of bacteria (alternative method) is adjusted by comparison with a 0.5 McFarland turbidity standard. The standard and the test suspension are placed in similar 4–6 ml thin, glass tubes or vials. The turbidity of the test suspension is adjusted, with broth or saline and compared with the turbidity standard, against a white background with contrasting black lines, until the turbidity of the test suspension equates to that of the turbidity standard.

McFarland 0.5 Turbidity Standard

Solution A (0.048 M $BaCl_2$).
1.175 g $BaCl_2.2H_2O$
Make up to 100ml with distilled water.
Solution B (0.18 M H_2SO_4)
1.0 ml H_2SO_4 (Analar grade, sp.gr. 1.84)
Make up to 100 ml with distilled water.
For standard:
Add 0.5 ml Solution A (0.048 M $BaCl_2.2H_2O$)
to 99.5 ml Solution B (0.18 M H_2SO_4)

Shake vigorously and dispense into 4–6 ml sealed tubes or screw-capped vials. Stored in the dark at room temperature and replace 3 months after preparation. The turbidiy standard should always be agitated before use.

- A sterile, non-toxic swab on an applicator stick is dipped into the standardised suspension of bacteria and excess fluid is expressed by pressing and rotating the swab firmly against the inside of the tube above the fluid level.
- The swab is streaked in three directions over the entire surface of the agar with the objective of obtaining a uniform inoculation. A final sweep with the swab can be made against the agar around the rim of the Petri dish.
- The test agar must be Mueller-Hinton agar or a satisfactory equivalent such as Isosensitest agar (Oxoid). The exception is the use of 5–10 per cent sheep blood agar for streptococci or *Actinomyces pyogenes* that are unable to grow on Mueller-Hinton agar. The surface of the agar should be moist but no droplets of moisture should be visible on the surface of the agar.
- The inoculated plates are allowed to stand for 3–5 minutes, but no longer than 15 minutes, for any excess moisture from the inoculum to be absorbed by the agar before applying the antimicrobial discs.
- The discs are placed onto the agar surface using sterile forceps or an antibiotic disc dispenser (**107**). Each disc is gently pressed with the point of a sterile forceps to ensure complete contact with the agar surface. The discs should be placed no closer together than 24 mm (centre-to-centre). This is equivalent to 6 discs per standard 90 mm Petri dish.
- The plates are inverted and placed in a 35°C incubator within 15 minutes of applying the discs and incubated aerobically for 16–18 hours (24 hours for staphylococci).
- After incubation, the diameters of the zones of inhibition are measured to the nearest mm using a ruler or calipers. The diameters are read from the back of the plate when the test is on the comparatively clear Mueller-Hinton medium but over the surface of the agar with streptococci grown on blood agar. The diameter of the zones should be read across the centre of the discs. An interpretation of the size of the zones of inhibition is made with reference to **Table 12**. The bacterium is reported as susceptible, moderately susceptible, intermediate, or resistant to each antimicrobial agent used in the test:

- **Susceptible:** the infection may respond to the treatment at the normal dosage.
- **Moderately susceptible:** the pathogen may be inhibited by attainable concentrations of the antimicrobial agent provided a higher dosage is used or the pathogen is in a certain body site, such as urine, where the drug is physiologically concentrated.
- **Intermediate:** the result is equivocal and if the bacterium is not fully susceptible to an alternative drug, then the test should be repeated.
- **Resistant:** the bacterium is not inhibited by the usually achievable systemic concentrations of the antimicrobial agent and efficacy has not been reliable in clinical studies.

107

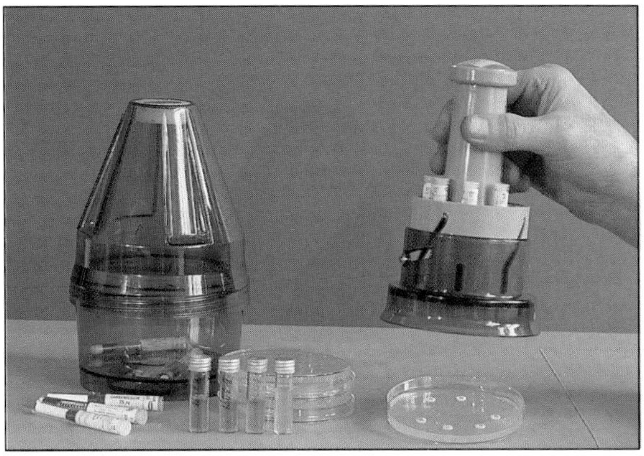

107 Susceptibility testing showing an antibiotic disc dispenser for the correct spacing of discs on the lawn of the test bacterium.

Some observations on the interpretation of zones of inhibition

The zone of inhibition is that area showing no obvious growth that can be detected by the unaided eye. The diameter of the zone is measured to each edge of this clear area. However, there are occasions when the exact extent of the zone is complicated:

a) Large colonies growing within an otherwise clear zone of inhibition should be subcultured, re-identified and retested. Faint growth of tiny colonies at the edge of the zone can be ignored.
b) *Proteus* species may swarm into areas of inhibited growth around certain antimicrobial discs. If the zones of inhibition are clearly outlined then the veil of swarming can be disregarded.
c) With trimethoprim and the sulphonamides, slight growth in the zone can be ignored and the zone measured to the margin of heavy growth.
d) For bacteria that have to be grown on blood agar plates, the zone size for nafcillin, novobiocin, oxacillin and methicillin will be 2–3 mm smaller than the normal control limits.
e) Transmitted light should be used to examine for light growth within the zone in the case of methicillin-resistant strains of staphylococci.

Methicillin-resistant staphylococci

There can be some difficulty in detecting these strains. The recommendations are as follows:

- An oxacillin disc is used as it is more stable in storage and is more likely to detect cross-resistance than a methicillin disc itself.
- The alternative method of standardising the inoculum, without preincubation in broth, is preferred for testing staphylococci.
- Resistance to methicillin or oxacillin may not be seen at 37°C. If the incubator cannot be controlled accurately at 35°C, then a separate test should be carried out with oxacillin and incubated at 30°C.
- The methicillin-resistant strains often have multiple resistance that includes the beta-lactams, aminoglycosides, macrolides, clindamycin and tetracyclines. This may act as an indication of resistance to other penicillinase-resistant penicillins.
- A film of growth within a zone of inhibition around a methicillin, oxacillin, nafcillin or cephalothin disc also indicates heteroresistance.
- Confirmation of intrinsic oxacillin-resistance in staphylococci can be made by inoculating the suspect strain onto a quadrant of Mueller–Hinton agar that has been supplemented with 4 per cent NaCl and 6 μg oxacillin per ml. The inoculation is made with a suspension adjusted to the 0.5 McFarland standard. The plate is incubated at 35°C for 24 hours. Any growth of the bacterium indicates an intrinsic oxacillin-resistance.

Selection of antimicrobial discs

The selection of the types of antimicrobial agents for use in the disc diffusion test will, to a large extent, depend on clinical considerations including the drugs that are available and in general use by the veterinarian. However, to make routine susceptibility testing relevant and practical the following guidelines may be helpful:

- A tetracycline disc will predict the result against all the other tetracyclines.
- Sulphisoxazole is a suitable representative for all the sulphonamides.
- Erythromycin will predict the result of all other macrolides.
- A clindamycin disc will predict the result for lincomycin.
- Aminoglycosides and quinolones should be tested separately. Individual members within each group are not related closely enough to assume cross-resistance.
- Chloramphenicol, vancomycin, nitrofurantoin and trimethoprim/sulphamethoxazole are tested separately as required.
- Penicillins:
  a) Staphylococci should be tested against penicillin G and oxacillin (for resistance to methicillin, cephalosporins and other beta-lactams).
  b) *Enterococcus faecalis* tested against penicillin G will predict the result against ampicillin, ampicillin-analogues, amoxicillin and the acylamino-penicillins.
  c) Streptococci should be tested against either penicillin G *or* ampicillin. Testing against both is not necessary.

- Cephalosporins:
  a) Staphylococci are usually susceptible to cephalosporins except for the methicillin-resistant strains; these should be reported as resistant even if the result of the *in vitro* test suggests otherwise.
  b) There is a lack of clinical correlation with *Enterococcus faecalis* against the cephalosporins and the results of the *in vitro* tests cannot be interpreted with accuracy.
  c) A cephalothin disc will indicate the result to cefaclor, cephapirin, cefazolin, cephradine, cephalexin and cefadroxil. Cefatozime represents ceftazidime, ceftizoxime and ceftriaxone.

Quality control procedures

Quality control procedures are needed to monitor the precision and accuracy of the test procedure including a check on the potency of the antimicrobial discs and the satisfactory performance of the test medium.

Control strains of bacteria

Stock cultures of control bacteria from the American Type Culture Collection (ATCC) should be maintained. These reference strains when tested against the antimicrobial discs should yield zones of inhibition within the control limits indicated in **Table 13**. The suggested control bacteria are:

- *Escherichia coli (E. coli)* ATCC 25922
- *Staphylococcus aureus* ATCC 25923
- *Pseudomonas aeruginosa* ATCC 27853
- *E. coli* ATCC 35218 that produces beta-lactamase.
- *Enterococcus faecalis* ATCC 29212 or 33186 to establish that the test medium is relatively free of thymidine and thymine.

The working control cultures are stored on soybean-casein digest agar or an equivalent (tryptone soya agar, Oxoid) at 4°C and subcultured weekly. These working cultures are replaced at least monthly from stock cultures maintained lyophilised, frozen below -20°C, or in liquid nitrogen. For testing, the control cultures are streaked out on an agar plate to obtain isolated colonies. The control tests are then conducted as for the routine test procedure.

Antimicrobial discs

a) Storage conditions of the discs must ensure that their potency is maintained. They should be kept under appropriate anhydrous conditions and stored at 4°C for routine use. The discs should be allowed to equilibrate with room temperature before being used in the test. Discs containing beta-lactam antibiotics should be stored for up to one week only at 4°C.Longer term storage should be at -14°C or lower. This lower temperature is also employed for long-term storage of the other antimicrobial discs.
b) Discs that have passed the manufacturer's expiry date must be discarded.
c) Quality control of discs containing combinations of beta-lactams and beta-lactamase inhibitors are monitored using both the *E. coli* control strains, one of which (ATCC 35218) is a beta-lactamase producer.
d) The potency of a new batch of antimicrobial discs should be checked against the control bacteria and the zones of inhibition compared with the control limits shown in **Table 13**.

The test medium

a) The depth of agar in a Mueller-Hinton agar plate should be approximately 4 mm. This is equivalent to:
   60–70 ml of medium in a 140 mm Petri dish,
   or 25–30 ml of medium in a 90 mm Petri dish.
b) The pH of a new batch of medium should be checked and be between 7.2 and 7.4.
c) Each batch of poured medium should be tested for sterility by incubating a few randomly selected plates at 30 to 35°C for 24 hours or longer. These plates are then discarded.
d) The medium is stored at 4°C and should be used within 7 days of preparation unless kept sealed inside a plastic sleeve.
e) To establish that a new batch of medium is free of thymidine and thymine, a control strain of *Enterococcus faecalis* (ATCC 29212 or 33186) is tested with a trimethoprim/sulphamethoxazole disc. Satisfactory medium will show a clear and distinct zone of inhibition of 20 mm or more. Unsatisfactory media will produce no distinct zone or a hazy growth within the zone.
f) The control strain *Pseudomonas aeruginosa* (ATCC 27853) is tested against tetracycline and aminoglycosides to monitor satisfactory levels of calcium and magnesium cations in the medium. The zone sizes must conform to the control limits given in **Table 13.**

Zone size limits

These zone size limits for the control bacteria against antimicrobial discs on Mueller-Hinton medium without supplements are shown in **Table 13**. Daily testing to establish the accuracy of the test should yield no more than one zone size outside the control limits in 20 consecutive tests. The zone size should be no more than four standard deviations above or below the midpoint between stated limits [midpoint + (maximum - minimum zone limits)]. Once satisfactory performance of the test has been established, testing each week and every time a new batch of medium or antimicrobial discs are introduced may be sufficient. NCCLS (1990) M2-A4 should be consulted for further details.

Table 13. Zone diameter limits for quality control of antimicrobial disc diffusion susceptibility tests on unsupplemented Mueller–Hinton medium (modified from NCCLS (1990) M2-A4)*.

Disc content	*Antimicrobial agent*	*E. coli (ATCC 25922)*	*S. aureus (ATCC 25923)*	*P. aeruginosa (ATCC 27853)*
30μg	AMIKACIN	19–26	20–26	18–26
20/10μg	AMOXICILLIN/ CLAVULANIC ACID	19–25	28–36	–
10μg	AMPICILLIN	16–22	27–35	–
100μg	CARBENICILLIN	23–29	–	18–24
75μg	CEFOPERAZONE	28–34	24–33	23–29
30μg	CEFOTAXIME	29–35	25–31	18–22
30μg	CEPHALOTHIN	15–21	29–37	–
30μg	CHLORAMPHENICOL	21–27	19–26	–
5μg	CIPROFLOXACIN	30–40	22–30	25–33
2μg	CLINDAMYCIN	–	24–30	–
10μg	ENOXACIN	28–36	22–28	22–28
15μg	ERYTHROMYCIN	–	22–30	–
10μg	GENTAMICIN	19–26	19–27	16–21
30μg	KANAMYCIN	17–25	19–26	–
30μg	MOXALACTAM	28–35	18–24	17–25
1μg	NAFCILLIN	–	16–22	–
30μg	NALIDIXIC ACID	22–28	–	–
30μg	NETILMICIN	22–30	22–31	17–23
300μg	NITROFURANTOIN	20–25	18–22	–
10μg	NORFLOXACIN	28–35	17–28	22–29
1μg	OXACILLIN	–	18–24	–
10 units	PENICILLIN G	–	26–37	–
100μg	PIPERACILLIN	24–30	–	25–33
5μg	RIFAMPIN	8–10	26–34	–
10μg	STREPTOMYCIN	12–20	14–22	–
250 or 300μg	SULPHISOXAZOLE	18–26	24–34	–
30μg	TETRACYCLINE	18–25	19–28	–
75μg	TICARCILLIN	24–30	–	22–28
10μg	TOBRAMYCIN	18–26	19–29	19–25
1.25/23.75μg	TRIMETHOPRIM-SULPHAMETHOXAZOLE	24–32	24–32	–
30μg	VANCOMYCIN	–	15–19	–

E. coli (ATCC 35218) produces a beta-lactamase and can be tested with the amoxicillin/clavulanic acid disc. The zone must be 18–22 mm.

Enterococcus faecalis (ATCC 29212 or 33186) can be tested with the trimethoprim/sulphamethoxazole disc to determine whether the Mueller–Hinton medium has sufficiently low levels of thymidine and thymine. The zone must be ≥ 20 mm.

* Updated according to the Third Informational Supplement, December 1991.

Check list of common sources of error

One or more of the following causes may be the reason for a zone size being outside the control limits given in **Table 13**.

- Clerical or reader error.
- Contamination or genetic change in the control bacterium.
- Inoculum too heavy or too light. This may result from the turbidity standard being incorrectly prepared, used later than 3 months after preparation or not agitated sufficiently before use.
- A change in the composition of the Mueller-Hinton medium.
- Loss of potency of the antimicrobial agent in the disc.

Quantitative Methods of Antibiotic Susceptibility Testing

The minimal inhibitory concentration (MIC) is the highest dilution of an antibiotic required to inhibit the growth of a bacterium. The conventional tests for determining the MIC include the broth-dilution method and agar incorporation tests.

Broth-dilution method

This test is performed by preparing two-fold dilutions of an antibiotic in a series of tubes containing a nutrient broth. Each tube is inoculated with a suspension of the test bacterium that contains between 10^4 and 10^5 bacteria/ml. The inoculated tubes of broth are incubated at 35–37°C for 24 hours. The highest dilution of the antibiotic to inhibit growth (no turbidity in the tube) is the MIC.

The minimum bactericidal concentration (MBC) can also be determined by the broth dilution method. After the MIC has been read, a standard volume of broth is taken from the tubes showing no visible growth after 24 hours' incubation and subcultured onto agar media. The MBC is arbitrarily defined as the lowest antibiotic concentration that kills 99.9 per cent of the original inoculum, or where a thousand-fold reduction in bacterial numbers has occurred.

Agar incorporation tests

These tests are similar to the broth dilution method except that the antibiotic dilutions are incorporated into an agar medium in a series of Petri dishes. These are spot-inoculated with a number of test bacteria; between 20–36 bacterial isolates can be accommodated on a 90 mm Petri dish. A series of control bacteria, of known sensitivity, should be included on each plate.

Commercial systems for determining MIC

Several commercial systems are available for determining MIC values. A novel method is illustrated (**108**). This is the E test (AB Biodisk) and consists of a thin, inert plastic test carrier with a predefined exponential gradient of antibiotic. After incubation, the MIC value is read from the scale at the point of intersection between the zone edge and the test carrier.

108

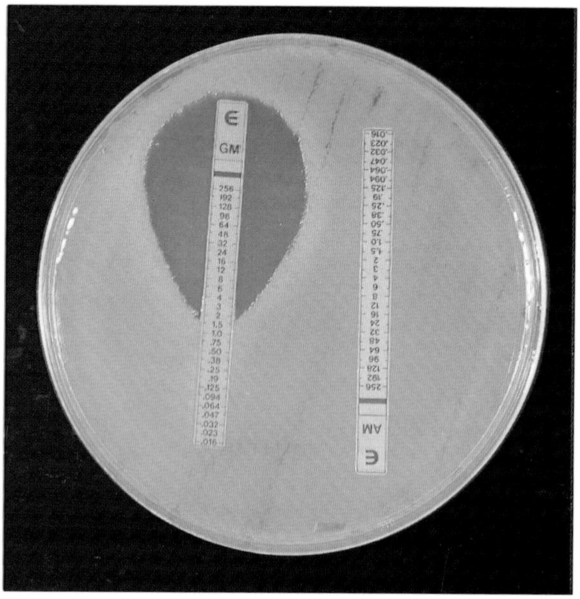

108 A commercial strip for the determination of MIC values. *Pseudomonas aeruginosa* tested against gentamicin (left) giving an MIC of 2 µg/ml and ampicillin (right) to which *P. aeruginosa* is resistant.

Antibacterial and Antifungal Chemotherapy

Successful use of antimicrobial drugs for the treatment of clinical disease depends on many factors, some relating to the sick animal and the reliability of the diagnosis, others relating to the specimen collected and its processing by the laboratory. The ultimate determining factor is the selection and administration of an appropriate drug at a frequency suitable to maintain blood levels at the optimal concentration for the duration of the treatment.

The clinical use of antimicrobial drugs begins with the clinician and the detailed examination that allows a specific clinical diagnosis to be made. With experience, it may be possible to relate clinical signs to a particular causative agent and initiate treatment with an antimicrobial drug, selected on the basis of clinical impression alone. Even with the more readily recognisable diseases, however, it may be preferable to obtain a representative specimen for diagnostic microbiology as a safeguard against error in diagnosis **before** giving antimicrobial drugs. In many instances, the relationship between causative agent and clinical picture is not constant. Chemotherapy can be initiated as soon as specimens are collected and empirical chemotherapy, can, if necessary, be modified when laboratory data are at hand.

Effective antimicrobial therapy depends on the susceptibility of the pathogen, pharmacokinetic characteristics of the drug, the amount of drug given at one time, the route, frequency of administration and the duration of treatment (**Diagram 23**). Other variables relating to chemotherapy include the drug's toxicity for the host, its half-life, concentration and persistence at the site of infection and its effect on the normal flora of the host. Because of the many species encountered in clinical practice and their individual sensitivities to drug toxicity, care should be exercised in the selection and administration of antimicrobial drugs. Apart from species variation, unexpected responses in young animals, especially in the neonatal period, may also occur. **Diagram 24** shows the factors relevant to the drug, infectious agent and the host which influence the outcome of antimicrobial chemotherapy.

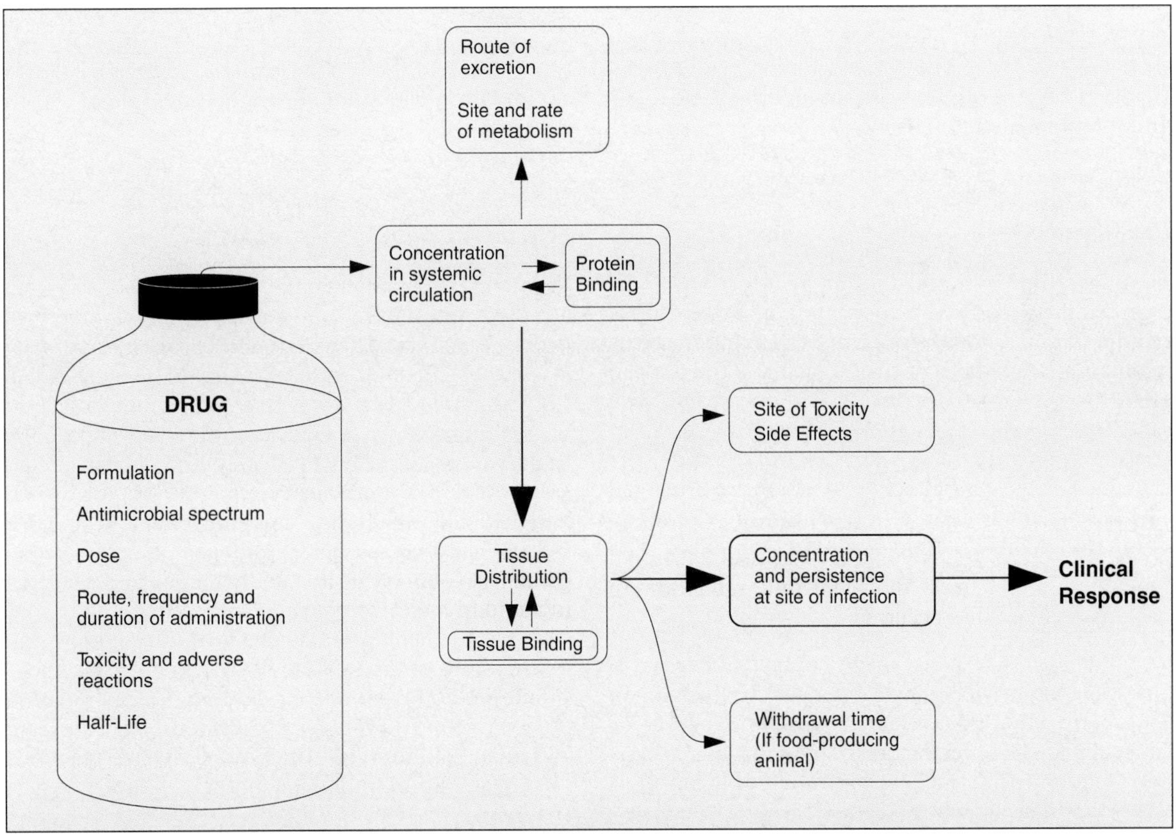

Diagram 23. Considerations involved in the selection and administration of antimicrobial drugs. Inherent toxicity or side effects arising from chemotherapy require careful consideration in some species and also in young animals. Legal regulations in some countries may prohibit the use of certain antibiotics such as chloramphenicol, in food-producing animals.

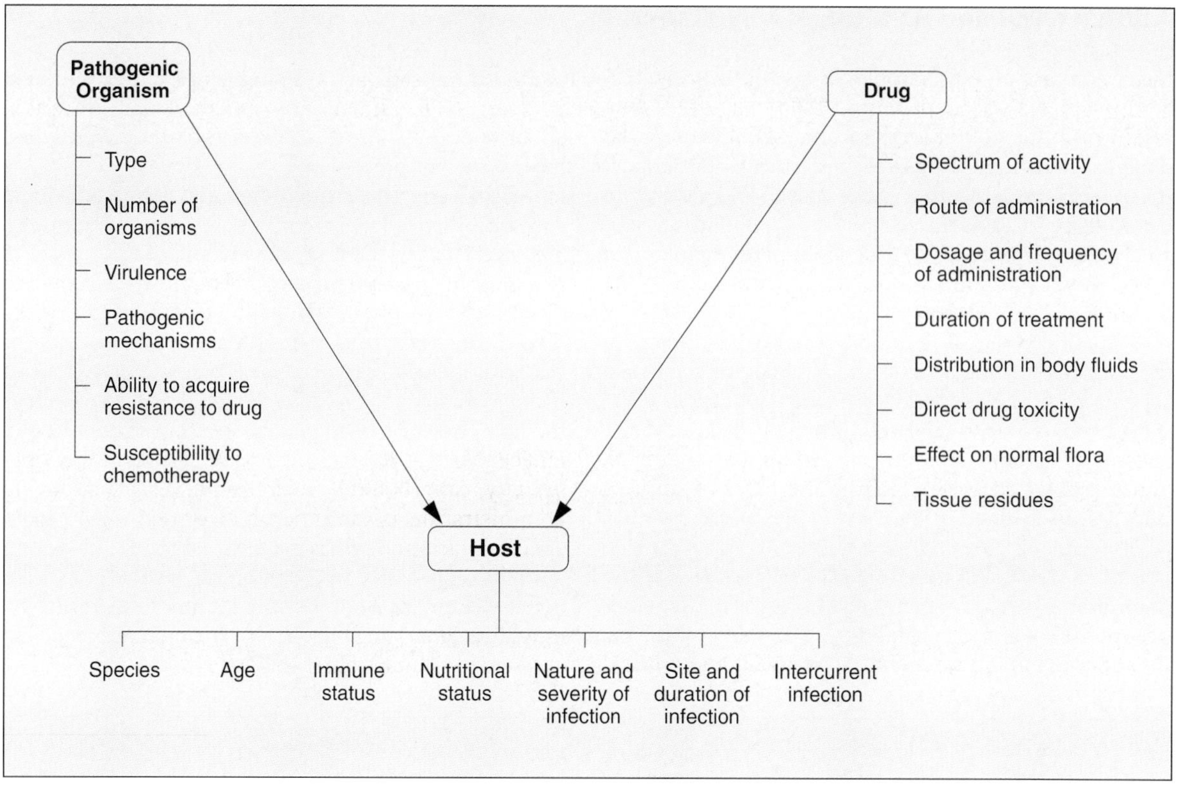

Diagram 24. Factors relevant to the infectious agent, host and drug which influence the outcome of antimicrobial chemotherapy.

Drug distribution

When treating bacterial or fungal infections, it is essential that an effective concentration of the drug is rapidly attained at the site of infection and that it is maintained for sufficient time to achieve the desired effect on the target organisms. The location of the infection can have a major influence on the drug concentration attained in a particular tissue and some sites such as the central nervous system are protected by barriers which limit or prevent the entry of many drugs except highly lipid-soluble compounds. The amount of drug, therefore, that enters each organ or tissue is determined by physiochemical properties of the drug as well as physiological factors. The effect of the inflammatory response on the structural integrity of physiological barriers may have a direct influence on drug permeation of tissues and the intracellular or extracellular site of replication of individual pathogens may determine the final concentration of the drug to which they are exposed. Necrotic tissue or accumulated pus may further hinder drug penetration into lesions caused by pyogenic bacteria, hence the need for thorough debridement and drainage if the site of infection is accessible to direct surgical intervention.

Selection of antimicrobial drugs

Before commencing chemotherapy, the body system affected, the nature and location of the infectious agent(s) and the most suitable drug likely to be effective against the pathogen should be considered. Route and frequency of administration of the drug, cost of treatment, adverse effects on the host and public health aspects of treatment (if a food-producing animal) are matters relevant to the veterinarian and the owner. Suggested antibacterial and antifungal drugs, for specific bacterial and fungal pathogens are presented in **Tables 14** and **15**. For pathogenic bacteria that commonly exhibit drug resistance such as staphylococci and pseudomonads, routine culture and susceptibility testing should be carried out. Specimens should be collected during the clinical examination procedure and before any antimicrobial drugs are administered. Sources of failure or adverse reactions following antimicrobial chemotherapy are shown in **Diagram 25**.

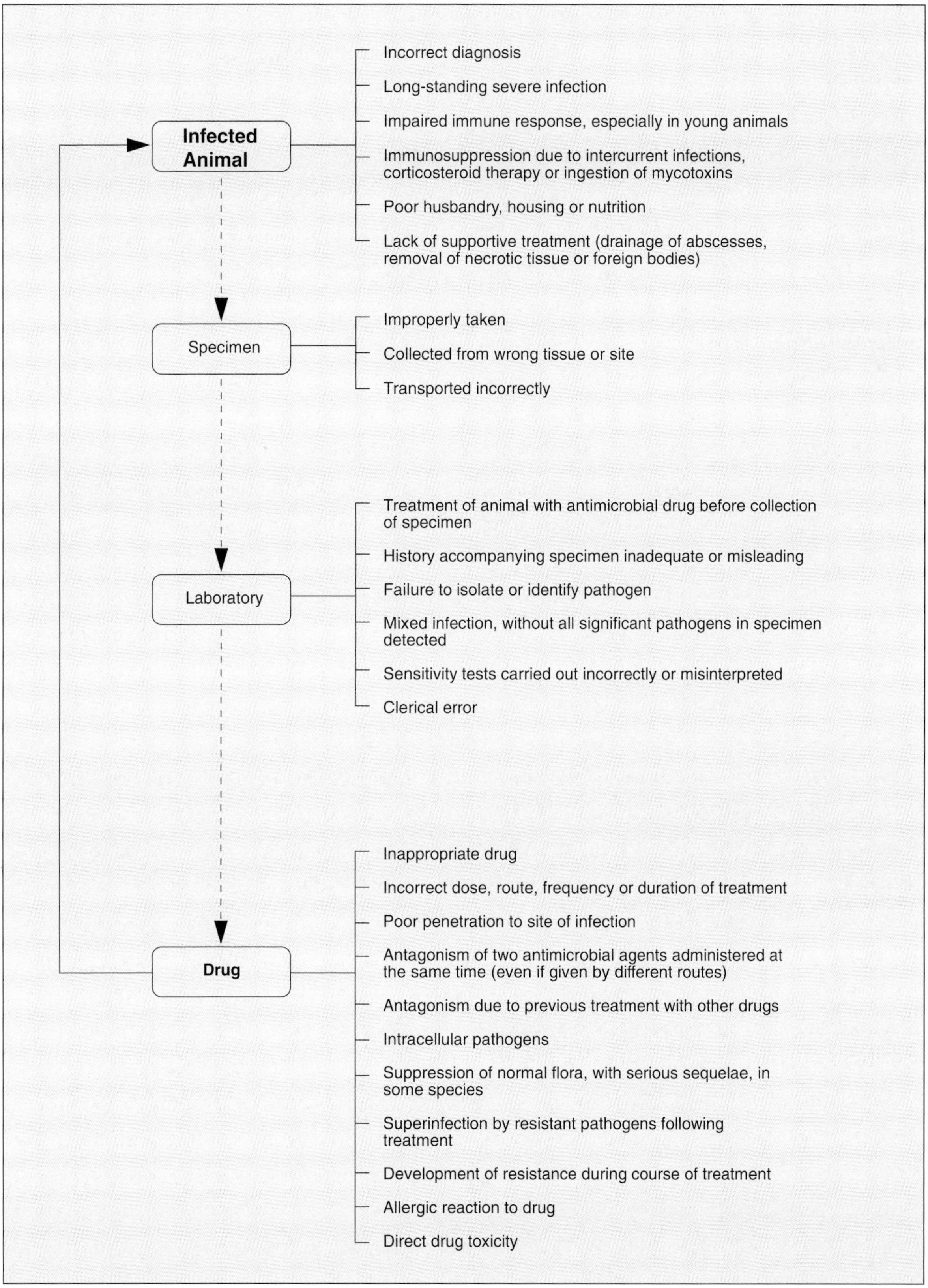

Diagram 25. Sources of failure or adverse response following prescribed antimicrobial therapy.

Table 14. Suggested antimicrobial treatment for pathogenic bacteria and related organisms*.

Pathogenic bacterium	Disease	Suggested drugs	Alternative drugs	Comments
Actinobacillus lignieresii	'Wooden or timber tongue' in cattle, superficial lesions in sheep	Potassium iodide *per os*, sodium iodide I/V, Streptomycin (+ penicillin)	Tetracycline	Drug penetration into lesion may be limited in chronic cases
A. equuli	Septicaemic and multifocal infections in foals	Streptomycin, Ampicillin	Trimethoprim-sulphameth-oxazole	Supportive treatment is necessary
A. (*Haemophilus*) ***pleuropneumoniae***	Pleuropneumonia in pigs	Trimethoprim-sulphameth-oxazole, Gentamicin	Penicillin G, Ampicillin, Tetracycline	Elimination of infected breeding animals and mass medication of in-contact pigs may eradicate infection
Actinomyces bovis	'Lumpy jaw' in cattle	Potassium iodide *per os*, sodium iodide I/V, Penicillin + streptomycin	Tetracycline	In advanced cases, treatment will not restore normal bone structure
A. pyogenes	Purulent infections of traumatic or opportunistic origin in cattle and sometimes in sheep, goats and pigs. Implicated in 'Summer mastitis' (cattle)	Penicillin G, Tetracycline	Trimethoprim-sulphameth-oxazole	Response to treatment is disappointing in 'summer mastitis' presumably due to active involvement of other bacteria
A. viscosus	Granulomatous lesions in skin or thoracic cavity in dogs	Penicillin G, Ampicillin	Erythromycin	Drainage of suppurative lesions is beneficial
Bacillus anthracis	Anthrax in cattle, sheep, pigs and other animals. Important zoonosis	Penicillin G	Erythromycin, Tetracycline	Acute disease, hence prompt treatment required; chemopro-phylaxis of in-contact animals indicated where disease is endemic
Bacteroides nodosus	Footrot in sheep and goats; implicated in infections of interdigital skin in cattle	Penicillin G + streptomycin, Clindamycin	Metronidazole	Foot trimming and topical application of disinfectants or antibiotics are essential; foot-baths beneficial in control programmes

Table 14. Suggested antimicrobial treatment for pathogenic bacteria and related organisms* (*continued*).

Pathogenic bacterium	Disease	Suggested drugs	Alternative drugs	Comments
Bordetella bronchiseptica	Kennel cough in dogs, atrophic rhinitis in pigs	Trimethoprim-sulphamethoxazole	Tetracycline	Kennel cough responds poorly to therapy and vaccination is preferable. Atrophic rhinitis is not treatable; preventative measures include vaccination, prophylactic treatment or elimination of carrier sows
Borrelia burgdorferi	Lyme disease in humans and arthritic manifestations in other animals	Tetracycline, Penicillin G		Prompt treatment is desirable
Brucella canis	Canine brucellosis	Trimethoprim-sulphamethoxazole, Tetracycline		Antimicrobial treatment is usually unsuccessful. Control is based on testing and elimination of infected dogs
***Campylobacter* species**	Diseases of the reproductive tract in cattle and sheep; enteritis in other animals	Penicillin G + streptomycin	Erythromycin	Disease is best controlled by prevention, not by treatment
Chlamydia psittaci	Wide spectrum of diseases ranging from infections in birds (psittacosis/ornithosis) to chlamydial abortion in sheep	Tetracycline, Erythromycin	Fluoroquinolones	Vaccines are beneficial for the prevention of enzootic ovine abortion
***Clostridium* species**	Gas gangrene, tetanus and enteric disease	Penicillin G	Clindamycin, Tetracycline	Clostridial diseases are best controlled by vaccination
Corynebacterium pseudotuberculosis	Caseous lymphadenitis in sheep; abscess formation in cattle and horses	Antimicrobial treatment ineffective in sheep. Prolonged penicillin therapy used in horses		Segregation and culling of affected sheep followed by thorough disinfection is the preferred control procedure
C. renale	Bovine pyelonephritis; ovine posthitis	Penicillin G	Erythromycin, Tetracycline	Antimicrobial therapy only useful in early stage of infection

(*continued*)

Table 14. Suggested antimicrobial treatment for pathogenic bacteria and related organisms* (*continued*).

Pathogenic bacterium	Disease	Suggested drugs	Alternative drugs	Comments
Dermatophilus congolensis	Streptothricosis in cattle, 'mycotic dermatitis' and 'strawberry foot rot' in sheep, 'rain scald' in horses	Penicillin G + streptomycin at high dose rates	Oxytetracycline (long-acting), Spiramycin	Treatment in tropical Africa can be disappointing. Minimising trauma to skin and control of ectoparasites is beneficial
***Enterococcus* spp.**	Opportunistic infections in many animals. Urinary tract infections in dogs and other species	Penicillin + gentamicin, Penicillin + streptomycin, Ampicillin	Vancomycin, Trimethoprim-sulphameth-oxazole	
Escherichia coli	Septicaemia or diarrhoea in calves, young pigs, lambs; oedema disease in weaned pigs; mastitis in dairy cows; urinary tract infections in many animals	Gentamicin, Trimethoprim-sulphameth-oxazole	Ampicillin, Cephalosporins	Essential to establish an antimicrobial susceptibility pattern at an early stage. Fluid-replacement therapy required in enteric infections of young animals
Eubacterium suis	Cystitis and pyelonephritis in sows	Trimethoprim-sulphameth-oxazole, Penicillin G	Tetracycline	Antimicrobial therapy is likely to be effective only in the early stages of infection
Fusobacterium necrophorum (often associated with other bacteria)	Pyonecrotic lesions in calves (calf diphtheria); bull-nose in pigs; liver abscesses and foot-rot in cattle and sheep	Penicillin G, Sulphonamides, Metronidazole	Clindamycin, Tetracycline	Surgical drainage of foot conditions may be necessary. Footbaths can be beneficial in control programmes
Haemophilus somnus	Infectious thromboembolic meningo-encephalitis	Penicillin G, Ampicillin	Tetracycline	
Klebsiella pneumoniae	Coliform mastitis in dairy cows, reproductive diseases in mares and urinary tract infections in bitches	Gentamicin, Cephalosporins	Trimethoprim-sulphameth-oxazole	Sawdust or wood-shavings used as bedding may increase the incidence of coliform mastitis in dairy cattle
Leptospira interrogans serovars	Abortion in cattle and pigs; septicaemia, hepatic and renal involvement in dogs; ocular lesions in horses	Penicillin G, Ampicillin, Streptomycin	Tetracycline	Vaccination is used in dogs, cattle and pigs. The carrier state may persist after chemotherapy

Table 14. Suggested antimicrobial treatment for pathogenic bacteria and related organisms* (*continued*).

Pathogenic bacterium	Disease	Suggested drugs	Alternative drugs	Comments
Listeria monocytogenes	Septicaemia in young animals of many species, meningo-encephalitis and abortion in cattle and sheep	Penicillin G + gentamicin, Tetracycline	Trimethoprim-sulphamethoxazole	Often associated with silage feeding in cattle and sheep; bacterium widely distributed in the farm environment
Moraxella bovis	Infectious bovine keratoconjunctivitis	Tetracycline, Penicillin G + streptomycin	Chloramphenicol	Subconjunctival administration of suitable antibiotics preferable to topical applications
***Mycoplasma* spp.**	Associated with pneumonia in many species; arthritis; occasionally mastitis and conjunctivitis	Tetracycline, Tiamulin, Fluoroquinolones	Erythromycin, Tylosin	*Mycoplasma* species vary widely in their susceptibility to antimicrobial therapy
Nocardia asteroides	Pyogranulomatous infections in dogs and other species; mastitis in dairy cattle	Trimethoprim-sulphamethoxazole, Sulphonamides	Erythromycin	Abscesses and empyema require drainage. Chemotherapy often disappointing
Pasteurella haemolytica	Bronchopneumonia in cattle; septicaemia, bronchopneumonia and mastitis in sheep	Trimethoprim-sulphamethoxazole, Fluoroquinolones	Tetracycline	Primary viral damage may facilitate proliferation of pasteurellae with subsequent pneumonia
P. multocida	Haemorrhagic septicaemia in cattle and buffaloes; pneumonic lesions in cattle; fowl cholera; atrophic rhinitis in pigs; respiratory disease in rabbits	Penicillin G, Ampicillin, Fluoroquinolones	Trimethoprim-sulphamethoxazole, Cephalosporins	Vaccination is effective in some animal species where disease seems to be independent of viral involvement
Proteus mirabilis or ***P. vulgaris***	Often associated with otitis externa and urinary tract infections in dogs and occasionally other animals	Ampicillin, Gentamicin	Cephalosporins, Trimethoprim-sulphamethoxazole	Routine susceptibility testing is required with these organisms
Pseudomonas aeruginosa	Otitis externa in dogs; mastitis in dairy cows; pneumonia in mink; opportunistic skin infections in many species	Gentamicin, Tobramycin, Fluoroquinolones	Cephalosporins	Multiple resistance commonly encountered

(*continued*)

Table 14. Suggested antimicrobial treatment for pathogenic bacteria and related organisms* (*continued*).

Pathogenic bacterium	Disease	Suggested drugs	Alternative drugs	Comments
Rhodococcus equi	Suppurative bronchopneumonia in foals. Abscesses in older horses	Erythromycin + rifampicin	Penicillin + gentamicin	Treatment should be instituted immediately the disease is recognised
***Rickettsia* species**	Febrile diseases in many species of animals. Often transmitted by ticks and occasionally by flies	Tetracycline	Chloramphenicol	Vector control should be a priority in endemic areas
***Salmonella* species**	Septicaemia and enteritis in many species, particularly ruminants where abortion may occur. Horses, pigs and poultry also affected. The infection has zoonotic implications	Trimethoprim-sulphamethoxazole, Cephalosporins	Ampicillin, Chloramphenicol	Control measures should be aimed at prevention. Vaccination is of value for some serotypes. Chemotherapy may prolong the excretion of salmonellae
***Staphylococcus* species**	Pyogenic infections in many species of animals. *S. aureus* is a major cause of mastitis in dairy cattle	**Penicillin G sensitive:** Penicillin G **Penicillin G resistant:** A penicillinase-resistant penicillin **Methicillin resistant:** Vancomycin	Cephalosporins Cephalosporins Trimethoprim-sulphamethoxazole + rifampicin, Fluoroquinolone + rifampicin	The susceptibility pattern of each isolate should be established before treatment is instituted

Table 14. Suggested antimicrobial treatment for pathogenic bacteria and related organisms* (*continued*).

Pathogenic bacterium	Disease	Suggested drugs	Alternative drugs	Comments
***Streptococcus* spp.**				
S. agalactiae* *S. dysgalactiae* *S. uberis	Mastitis in dairy cows	Penicillin G ± gentamicin Cloxacillin	Ampicillin, Cephalosporins	Intramammary preparations used. Control programmes should include teat dipping and attention to milking-machine hygiene and performance
S. equi* subsp. *equi	Strangles in horses	Penicillin G	Erythromycin	Immediate isolation of affected or suspect horses is essential
S. suis	Meningitis in pigs	Penicillin G	Trimethoprim-sulphamethoazole	
Serpulina hyodysenteriae	Swine dysentery	Tiamulin, Carbadox	Metronidazole, Tylosin, Lincomycin	Antimicrobial drugs do not eliminate the organisms
Yersinia enterocolitica	Latent infections with sporadic enteritis or generalized infection in many animals	Trimethoprim-sulphamethoxazole	Ampicillin, Tetracycline	Resistance to antimicrobial drugs is commonly encountered

* This table is intended only as a general guide for the selection of appropriate antimicrobial drugs based on the aetiological agents. Adverse reactions to antimicrobial therapy **must** be considered on a species basis. Likewise, the interaction of antimicrobial drugs with other therapeutic substances and the inherent toxicity of some antimicrobial agents in particular disease states should be considered carefully. The administration of chloramphenicol to food-producing animals is prohibited or restricted in many countries. I/V = intravenous.

Table 15. Suggested antifungal treatment for superficial and systemic mycoses.

Fungus	Diseases	Suggested drugs: Topical administration	Suggested drugs: Systemic administration	Comments
Aspergillus fumigatus	Infections of mucous membranes in dogs; lungs and air sacs in poultry; guttural pouch in horses and mastitis and abortion in cattle	Antifungal agents such as iodine or nystatin may be applied to superficial or accessible lesions	Amphotericin B, Itraconazole (orally)	Aspergillosis in poultry is preventable by efficient management. Immuno-suppression may lead to tissue invasion by aspergilli and other opportunistic fungi. Successful treatment for nasal granulomas in dogs using topical perfusion has been reported
Blastomyces dermatitidis	Disseminated disease in many species particularly dogs. Lungs, lymph nodes, skin and other tissues may be involved		Ketoconazole (orally), Itraconazole (orally), Amphotericin B	Combined ketoconazole-amphotericin B treatment may be necessary if disease is well established and treatment may be long term
Candida albicans	Mucocutaneous candidiasis occurs in many species of animals, often as a consequence of prolonged antibiotic treatment or because of immuno-suppression	Clotrimazole, Nystatin	Ketoconazole (orally), Amphotericin B ± flucytosine, Miconazole	The factors which precipitated the disease should be identified and dealt with as candidiasis is usually a secondary problem
Coccidioides immitis	Usually occurs in geographically defined areas of North and South America, affecting dogs and other species. In dogs, pulmonary involvement with dissemination to other organs occurs		Amphotericin B, Ketoconazole, Itraconazole	Long-term treatment is required. If the disease is at an advanced stage, euthanasia should be considered
Cryptococcus neoformans	Infections occur in many species, particularly cats and often involves soft tissues with a tendency to localize in the CNS. Pigeon droppings frequently contain this fungus		Amphotericin B + flucytosine (orally), Fluconazole	In advanced cases, the response to treatment is variable

Table 15. Suggested antifungal treatment for superficial and systemic mycoses *(continued).*

Fungus	Diseases	Suggested drugs: Topical administration	Suggested drugs: Systemic administration	Comments
Histoplasma capsulatum	Histoplasmosis affects many animals, particularly dogs. Primary lesions are in the lungs with dissemination to intestines, spleen, liver and lymph nodes		Ketoconazole or itraconazole (orally), Amphotericin B	Early in disease treatment may be successful but with disseminated disease the prognosis is poor
Malassezia pachydermatis	Can be present, in small numbers, in ears of normal dogs. Increased numbers in otitis externa associated with other pathogenic organisms	Nystatin, Natamycin Miconazole		*Malassezia* infections frequently recur. Repeated use of antibiotics may allow yeast proliferation. Attention to predisposing factors is helpful. Aeration and drainage hasten recovery
***Microsporum* species**	Invasion of the keratinized layers of the skin, hair, feathers, claws and nails. 'Ringworm' of many species including poultry	Clotrimazole, Tolnaftate, Natamycin, Ketoconazole	Griseofulvin (orally)	Griseofulvin is teratogenic early in gestation for dogs and cats. Thorough disinfection of premises and fittings with iodine-based or chlorine-based disinfectants essential
Sporothrix schenckii	Chronic ulcerative lesions of the skin and subcutaneous tissue, often accompanied by lymphangitis. Dissemination to viscera, bones and CNS may occur in rare cases		Potassium iodide (orally), Sodium iodide (I/V), Itraconazole, Amphotericin B	Surgery is usually contraindicated
***Trichophyton* species**	Invasion of keratinized layers of skin and hair. 'Ringworm' in many animal species	Clotrimazole, Natamycin, Tolnaftate	Griseofulvin (orally)	Thorough disinfection of premises and fittings with iodine-based or chlorine-based disinfectants is essential
Zygomycetes *(Absidia, Mortierella, Mucor, Rhizopus, Rhizomucor,* and others)	Opportunistic infections. Because of the multiple aetiology, many tissues may be invaded including respiratory, intestinal and genital tracts. Mucocutaneous granulomas occur in many animal species		Amphotericin B, Ketoconazole	Excision of lesion, where possible, followed by systemic treatment

Antimicrobial drug interactions

Antimicrobial substances, like many other classes of drugs may interact with other medication given to animals, with possible adverse results. In addition, when antimicrobial drugs are combined, their effect may be quite different from that achieved by the individual drugs used separately. When two antimicrobial drugs act simultaneously on a homogeneous microbial population the effect may be one of the following:

- Indifference, where the combined action is no greater than that of the more effective drug used alone;
- Additive, where the combined action is equivalent to the sum of the actions of each drug when used alone;
- Synergism, where the combined action is significantly greater than the sum of both effects;
- Antagonism, where the combined action is less than that of the more effective agent when used alone.

Antimicrobial synergy may take several forms.

a) Two drugs may sequentially block a microbial metabolic pathway. One of the best examples is the combination of a sulphonamide with trimethoprim. Sulphonamides compete with ρ-aminobenzoic acid, which is required by some bacteria for the synthesis of dihydrofolate. Folate antagonists such as trimethoprim inhibit the enzyme (dihydrofolic acid reductase) that reduces dihydrofolate to tetrahydrofolate. The presence of a sulphonamide and trimethoprim results in the simultaneous blocking of sequential steps leading to the synthesis of purines and nucleic acid and can achieve a greater inhibition of growth than either drug alone.

b) One drug may prevent the inactivation of a second drug by microbial enzymes. Inhibition of beta-lactamases by clavulanic acid can protect penicillin G or other susceptible antibiotics from inactivation by bacteria that produce beta-lactamase.

c) One drug may promote the uptake of a second drug thereby increasing the overall antimicrobial effect. This appears to be a widely applicable mechanism of synergism with considerable clinical importance. Penicillins enhance the uptake of aminoglycosides by enterococci and in human medicine this combination has proved highly beneficial for the treatment of enterococcal endocarditis. Cell wall inhibitors such as penicillins and cephalosporins may enhance the entry of aminoglycosides into some Gram-negative bacteria and produce synergistic effects.

d) One drug may affect the cell membrane and facilitate the entry of a second drug. Amphotericin B may be synergistic with flucytosine against *Cryptococcus neoformans*.

e) A drug combination may also prevent the emergence of resistant populations. Treatment of *Rhodococcus equi* infection in foals with erythromycin-rifampicin is used synergistically for this purpose.

Just as two antibacterial compounds can be combined in a mutually beneficial way, they can also interfere with each other's activity. A form of antagonism which has received much attention is that occurring between predominantly bacteriostatic drugs such as chloramphenicol or a tetracycline and bactericidal drugs such as penicillin or an aminoglycoside. Clearly, if bacterial growth is halted by a bacteriostatic compound, the bactericidal activity of a second drug will be abolished. Antagonism is more likely to occur if the bacteriostatic drug reaches the site of infection before the bactericidal drug. A recently recognised mechanism of drug antagonism is in the combined use of beta-lactam-stable cephalosporins and extended spectrum penicillins. The enzyme-stable cephalosporin induces the production of beta-lactamase in certain bacteria and this enzyme may then inactivate the unstable penicillin.

Administration of two antimicrobial drugs (provided that they are not antagonists) may be justified when (1) treating mixed bacterial infections, where each drug has activity against one of the pathogens; (2) treating severe infections of uncertain aetiology; (3) using synergistic combinations with documented efficacy against specific infections. There are also inherent disadvantages in combined antimicrobial treatment regimes. These include an additive toxic effect from the drugs used, an increased risk of superinfection with overgrowth of fungi or resistant bacteria and a danger of enhanced spread of R plasmids. There is the additional risk that if antimicrobial compounds are not carefully selected, antagonism may occur. Using combinations of antimicrobial drugs can lead to a more casual approach in clinical diagnosis of disease with less interest in establishing the infectious agents involved in the disease process.

Resistance to antimicrobial agents

When antibacterial or antifungal drugs are used to treat infections, the outcome is influenced by many factors. Therapeutic success depends not only on the intrinsic activity of the antimicrobial drug against the invading pathogen, but relies on the drug reaching the site of infection in sufficient concentration. The animal's response to the infectious agent often determines the course of the infection subsequently. The ability of bacteria and fungi to become resistant to antimicrobial drugs is an increasing problem in human and in veterinary medicine. Microorganisms may be resistant to certain antimicrobial drugs because the cellular mechanisms required for antimicrobial susceptibility are absent from the cell. This is sometimes referred to as constitutive or pre-existent resistance. Acquired, genetically based resistance can arise because of chromosomal mutation or through the acquisition of transferable genetic material. Mechanisms of resistance whereby bacteria either destroy antimicrobial drugs or evade their effects are presented in **Table 16**.

The development of resistance in bacteria to antibiotics usually involves a stable genetic change, heritable from generation to generation. Many mechanisms that result in the alteration of bacterial genetic composition can influence the emergence of resistance. These include mutation and transfer of genetic material from one bacterium to another by transduction, transformation or conjugation. Resistance may arise by mutation that reduces target affinity or allows the production of a drug-modifying enzyme. Sometimes the insertion of foreign DNA by recombination achieves the same result.

Frequently, resistance genes are carried on extra-chromosomal plasmids (R factors) that may be transferable from organism to organism by conjugation. Bacteriophages can carry bacterial DNA, incorporated within their protein coat. If this genetic material includes a gene for drug resistance, a newly infected bacterial cell may become resistant to an agent and be capable of passing on this resistance to its progeny. Antimicrobial resistance can be present in a bacterial population before exposure to a particular antibiotic. Treatment with a specific antimicrobial agent selects for those microorganisms that have constitutive or acquired resistance. Some genes that encode proteins that confer resistance to antimicrobial drugs have the ability to 'jump' from place to place in the bacterial genome if the gene is flanked by 'insertion' sequences. Genes with insertion sequences at each end are termed transposons, and such genes can spread widely and quickly in a bacterial population thus facilitating the spread of resistance.

Table 16. Mechanisms of resistance to antimicrobial agents.

Antimicrobial class	***Resistance mechanism***
	Alteration of target site
Aminoglycosides	Altered ribosomal protein
Beta-lactam antibiotics	Altered or new penicillin-binding proteins
Erythromycin	Ribosomal RNA methylation
Quinolones	Altered DNA gyrase
Sulphonamides	New drug-insensitive dihydropteroate synthase
Tetracyclines	Ribosomal protection
Trimethoprim	New drug-insensitive dihydrofolate reductase
Vancomycin	Altered cell-wall stem peptide
	Drug-destroying mechanisms
Aminoglycosides	Acetyltransferase, Nucleotidyltransferase, Phosphotransferase
Beta-lactamase antibiotics (penicillins, cephalosporins)	Beta-lactamase
Chloramphenicol	Acetyltransferase
	Decreased uptake (decreased permeability)
Beta-lactam antibiotics, chloramphenicol, quinolones, tetracyclines, trimethoprim	Alteration in the permeability of the bacterial cell envelope

Table 17. Adverse effects of antimicrobial drugs in different species of animals.

Drug	*Animal species*	*Toxic effect*	*Comments*
Aminoglycosides (in general)	Many species	Ototoxic, nephrotoxic and cause neuromuscular blockade	Prolonged therapy with aminoglycoside antibiotics should be avoided even in large animals
Streptomycin	Cats, dogs, mice, hamsters, guinea pigs and ferrets. Especially young animals	Vomition, salivation and ataxia in cats; vestibular damage in dogs and cats; directly toxic for mice and very young animals; toxic reactions reported in hamsters, guinea pigs and ferrets	Avoid completely in susceptible animal species
Neomycin	Cats, cattle, pigs	Nephrotoxicity, deafness and neuromuscular blockade	
Chloramphenicol	Many animal species, humans	Bone marrow depression with fatal aplastic anaemia	There is an absolute ban on the use of chloramphenicol in food-producing animals in many countries
	Many species, especially cats	Immunosuppression and bone-marrow suppression	
Fluoroquinolones	Dogs, especially greyhounds; horses	Contraindicated due to cartilage erosions in weight-bearing joints	These drugs should be administered with caution to all young animals
Griseofulvin	Cats (dogs)	Teratogenic effects in cats (dogs) early in gestation	Topical treatment with anti-fungal drugs may be preferable in pregnant small animals
Lincomycin	Horses, cattle	This drugs is **contraindicated** in the **horse** where it may cause a fatal haemorrhagic colitis and intractable diarrhoea. Oral administration in dairy cattle produces severe depression, a drop in milk production and diarrhoea, sometimes with high mortality	**Even very low doses may produce serious disease in horses and cattle**
	Rabbits, guinea-pigs, hamsters	This antibiotic is highly toxic for these animals, apparently allowing overgrowth of *Clostridium* species in the large intestine	
Penicillins	Some animals	Hypersensitivity reactions	Most likely to develop from repeated systemic administration
	Guinea-pigs, hamsters	Destroys the normal Gram-positive flora and allows overgrowth of Gram-negative bacteria and possibly other organisms, with fatal results	
Procaine penicillin	Mice	Procaine hydrochloride is toxic for mice	
Sulphonamides	Many animals	Respiratory distress and collapse if given rapidly I/V in large animals. Crystalluria, renal tubular damage and blood dyscrasias may occur in some animal species	All animals receiving sulphonamide therapy should have *ad libitum* access to water. These drugs are contraindicated in severe renal disease
Tetracyclines	Many animals	Deposited in bones and teeth of young animals with discolouration evident. Irritants if injected locally into tissues	Toxic reactions also reported in rats, hamsters and guinea pigs
	Horses	These drugs are **contraindicated** by all routes in horses as they may induce intractable diarrhoea	

Adverse reactions to antimicrobial drugs

Adverse effects of antimicrobial drugs include dose-related toxic reactions, usually involving a specific organ or system, hypersensitivity reactions, alteration in host microflora, and abnormal tissue residues. Other undesirable consequences include the risk of superinfection with overgrowth of fungi or resistant bacteria. Considerable species variation occurs and drugs which are relatively non-toxic in one species may be profoundly toxic in another species (**Table 17**). Drug administration in the presence of liver or kidney disease requires careful consideration and adjustment of dosage is required. In the presence of renal failure, aminoglycosides should be given at more widely spaced intervals and neomycin is contraindicated. Sulphonamides should be used with care and alternative drugs should be considered in severe renal failure. Severe hepatic disease usually decreases the clearance of drugs that are metabolised by the liver. Tetracyclines, chloramphenicol, lincomycin, erythromycin and oxacillin are among the antibiotics which should be avoided in the presence of severe hepatic insufficiency.

Antimicrobial drugs do not distinguish between pathogenic microorganisms and those that constitute the normal flora of the host. Some broad-spectrum antibiotics may suppress the normal flora, especially if treatment is prolonged, leading to the proliferation of drug-resistant organisms which in turn may give rise to superinfection. Even narrow-spectrum antibiotics like penicillin have a profound effect on the intestinal flora of some species such as the guinea-pig, often with fatal consequences.

References

Bauer, A.W., Kirby, W.M., Sherris, J.C., Turck, M. (1966). Antibiotic susceptibility testing by a standardised single disc method. *American Journal of Clinical Pathology*, **45**: 493–496.

National Committee for Clinical Laboratory Studies (1990). Performance standards for antimicrobial disk susceptibility tests. Fourth edition. NCCLS document M2-A4. Villanova, PA 19085, USA.

National Committee for Clinical Laboratory Studies (1990). Methods for antimicrobial susceptibility testing of anaerobic bacteria. Second edition. NCCLS document M11-T2. Villanova, PA 19085, USA.

Bibliography

Brooks, G.F., Butel, J.S., Ornston, L.N., Jawetz, E., Melnick, J.L. and Adelberg, E.A. (1991). Jawetz, Melnick and Adelberg's *Medical Microbiology*. Appleton and Lange, Norwalk, Connecticut.

Davis, B.D., Dulbecco, R., Eisen, H.N., and Ginsberg, H.S. (1990). *Microbiology*. J.B. Lippincott Company, Philadelphia.

Flecknell, P. (1983). Restraint, anaesthesia and treatment of children's pets, *In Practice*, **5**: 85-95.

Gilman, A.G., Rall, T.W., Nies, A.S. and Taylor, P. (1990). *Goodman and Gilman's The Pharmacological Basis of Therapeutics*. Eighth Edition. Pergamon Press, New York.

Greenwood, D. (1989). *Antimicrobial Chemotherapy*, Oxford University Press, Oxford.

Jacoby, G.A. and Archer, G.L. (1991). New mechanisms of bacterial resistance to antimicrobial agents, *New England Journal of Medicine*, **324**: 601-612.

Johnston, D.E. (1987). *The Bristol Veterinary Handbook of Antimicrobial Therapy*, Second Edition. Veterinary Learning Systems Co. Inc., Bristol-Myers Animal Health, Evansville, Indiana.

Katzung, B.G. (1992). *Basic and Clinical Pharmacology*, Appleton and Lange, East Norwalk, Connecticut.

Prescott, J.F. and Baggott, J.D. (1988). *Antimicrobial Therapy in Veterinary Medicine*, Blackwell Scientific Publications, Boston, Massachusetts.

Reilly, P.E.B. and Isaacs, J.P. (1983). Adverse drug reactions of importance in veterinary medicine, *Veterinary Record*, **112**: 29–33.

8 *Staphylococcus* species

The staphylococci are Gram-positive cocci that tend to be arranged in irregular clusters or 'bunches of grapes' formation (**109**). The average diameter of the cocci is 1.0 µm. They are facultative anaerobes (fermentative), catalase-positive, oxidase-negative and non-motile. Growth occurs on nutrient and blood agars but not on MacConkey agar. The pathogenic staphylococci, *Staphylococcus aureus*, *S. intermedius* and *S. hyicus* (most strains) are coagulase-positive. The coagulase test correlates well with pathogenicity. Two commonly isolated coagulase-negative staphylococci, *S. epidermidis* and *S. saprophyticus,* occur as commensals and in the environment. They cause opportunistic infections in humans and, very occasionally, in animals although they are usually regarded as non-pathogenic.

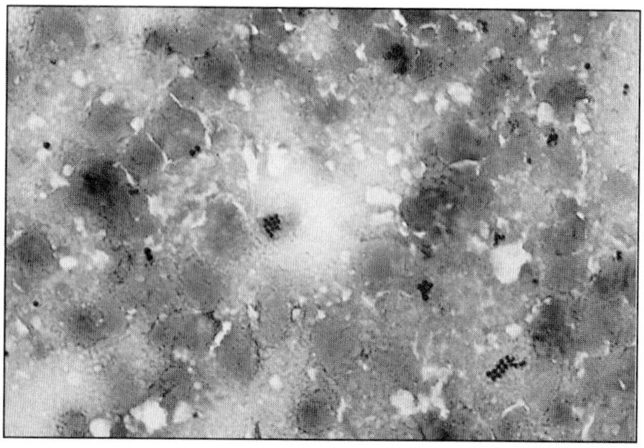
109

109 *Staphylococcus aureus* in a Gram-stained smear of a bovine mastitic milk showing the characteristic 'bunches of grapes' and other arrangements of cells. (×1000)

Changes in Nomenclature and New Species

Previous name	Present name	Significance
Staphylococcus hyicus subsp. *hyicus*	*S. hyicus*	Causes exudative epidermitis in pigs
S. hyicus subsp. *chromogenes*	*S. chromogenes*	Isolated from bovine mastitic milk but its pathogenicity is doubtful.

Relatively recently named staphylococci isolated from animals are included in the following list:

- *S. aureus* subsp. *anaerobius* is an anaerobic, catalase-negative, coagulase-positive staphylococcus that has been isolated from lesions in sheep that resemble those of caseous lymphadenitis (*Corynebacterium pseudotuberculosis*).
- *S. caprae* has been isolated from goat's milk.
- *S. gallinarum* and *S. arlettae* have been isolated from the skin of chickens.
- *S. lentus* obtained from the skin of sheep and goats.
- *S. equorum* from the skin of horses.
- *S. simulans* and *S. felis* from clinical specimens in cats.
- *S. delphini* isolated from the skin of dolphins.

Comparison of Staphylococci with other Gram-positive Cocci

Micrococci are non-pathogenic, Gram-positive cocci that could be confused with coagulase-negative staphylococci. However, micrococci are variably positive to conventional oxidase tests, oxidase-positive in a modified oxidase test (Faller and Schleifer, 1981), are oxidative in the O-F test and have a different susceptibility pattern from staphylococci to bacitracin and furazolidone. The colonies of the micrococci can be white but are often pigmented (**110**), the pigmentation ranging from a garish-yellow through cream, to buff or pink . Streptococci and enterococci are distinguished from staphylococci by the catalase test. The pertinent reactions for the commonly isolated Gram-positive cocci are summarised in **Table 18**.

110

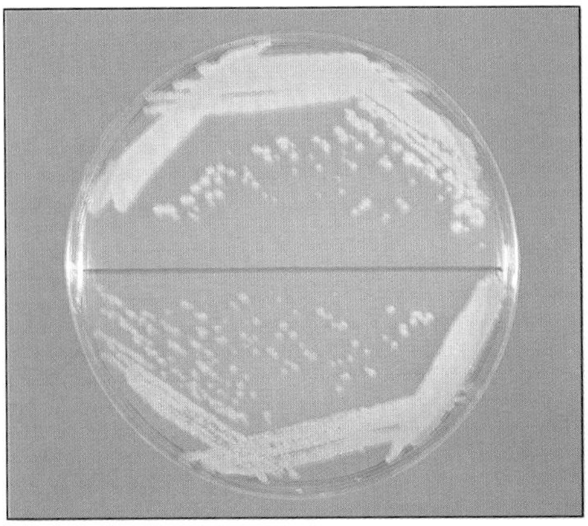

110 Two pigmented *Micrococcus* species on nutrient agar. Colonies of micrococci may be white or pigmented. They are aerobic (oxidative) and variably oxidase-positive, distinguishing them from the staphylococci.

Table 18. The main differentiating characteristics of the Gram-positive cocci.

	Coagulase	*Catalase*	*Oxidase*	*O-F glucose*	*Haemolysis*	*Bacitracin 0.04 unit disc*	*Furazolidone 100 μg disc*
Pathogenic staphylococci	+	+	–	F	+(–)	R	S
Non-pathogenic staphylococci	–	+	–	F	–(+)	R	S
Enterococci	–	–	–	F	(+)	R	S
Streptococci	–	–	–	F	(+)	R	S
Micrococci	–	+	+*	O	–(+)	S	R

+ = 90% or more strains positive; – = 90% or more strains negative; (+) = some strains positive; (–) = some strains negative; F = fermentative; O = oxidative; R = resistant; S = susceptible; * = modified oxidase test (Faller and Schleifer, 1981); susceptible to bacitracin = zone 10–25 mm; susceptible to furazolidone = zone 15–35 mm.

Natural Habitat

Staphylococci occur worldwide in mammals although the spread of staphylococcal strains between different animal species is limited. They colonise the nasal cavity (man), skin and mucous membranes and can be transient in the intestinal tract. Many infections are endogenous but prolonged survival of staphylococci in the environment permits indirect transmission.

Pathogenesis and Pathogenicity

The staphylococci are pyogenic and are associated with abscess formation and suppuration. Pus is composed of the debris of dead leukocytes and living and dead bacteria and can be surrounded by intact phagocytic cells and fibrin strands. A fibrous capsule will eventually be formed around an abscess. In chronic staphylococcal wound infections ('botryomycosis') the lesion is granulomatous with pockets of pus throughout the tissue.

The pathogenic staphylococci produce a 'battery' of toxins and enzymes, but the significance of many of them in the pathogenesis of disease is not fully understood. Enterotoxins (A-E) are involved in human food poisoning and they act by reflex stimulation of the emetic centre. Exfoliatin produces staphylococcal scalded skin syndrome (SSSS) in human infants and possibly in dogs; human toxic shock syndrome (TSS) is caused by TSS toxin-1; epidermolytic toxins are implicated in porcine exudative epidermitis and a pyrogenic exotoxin may be produced. The alpha toxin (haemolysin) is associated with gangrenous mastitis in cattle. This toxin causes lysosomal disruption in leukocytes and also affects smooth muscle, leading to constriction, paralysis and finally necrosis of the smooth muscle cells of the walls of blood vessels. A leukocidin kills neutrophils and macrophages of cattle, rabbits and humans. Protein A, a surface component of most strains of virulent *S. aureus*, binds to the Fc region of IgG and may play a part in the pathogenesis of staphylococcal diseases.

Enzymes produced by staphylococci include staphylokinase which is a plasminogen activator; coagulase which causes plasma coagulation *in vitro*; hyaluronidase ('spreading factor'); lipase; collagenase; proteases; nucleases and urease, all of which may have a role in the pathogenesis of staphylococcal infections. **Table 19** lists the main diseases caused by the pathogenic staphylococci.

Laboratory Diagnosis

Specimens

These may include exudates, pus from abscesses, mastitic milk, skin scrapings, urine and affected tissues.

Direct microscopy

Preparation and examination of Gram-stained smears of pus or exudates is often rewarding and may show the Gram-positive cocci of staphylococci in the typical 'bunches of grapes' formation.

Isolation

The routine medium for inoculation of specimens is sheep or ox blood agar. A MacConkey agar plate is inoculated in parallel to detect any Gram-negative bacteria that may also be present in the specimens.

A selective medium for Gram-positive bacteria is useful particularly if *Proteus* spp. are also present in the specimens (**111**). One such medium is sheep or ox blood agar with 15 mg nalidixic acid and 10 mg colistin sulphate (Oxoid supplement SR 70) per litre of medium. The antibiotic supplement successfully inhibits Gram-negative bacteria (**112**).

Mannitol salt agar and Baird-Parker medium are specifically selective for staphylococci but are used mainly in food microbiology.

The inoculated plates are incubated aerobically at 37°C for 24–48 hours.

Identification

Colonial characteristics

Colonies usually appear in 24 hours. After 48 hours' incubation, well-isolated colonies can reach 4 mm in diameter. They are round, smooth and glistening and on blood agar tend to appear substantial and opaque compared to the smaller, translucent colonies of beta-haemolytic streptococci.

a) **Pigmentation:** *S. aureus* strains from cattle, humans and other domestic animals usually have a golden-yellow pigment but those from dogs are almost always non-pigmented (white). The colonies of *S. intermedius* and *S. hyicus* are also non-pigmented. Some of the coagulase-negative staphylococci produce pigment, especially strains of *S. chromogenes* whose colonies are orange-yellow. Pigment enhancement in the staphylococci is said to be induced by the addition of milk, fat or glycerol monoacetate to the medium.

b) **Haemolysis:** the staphylococcal haemolysins (alpha, beta, delta and gamma) can be produced singly, in combination or not at all. The haemolysins differ antigenically, biochemically and in their effect on the red cells of various animal species. **Table 20** summarises the main properties of the haemolysins. Blood agar prepared with either ovine or bovine erythrocytes is preferable in veterinary diagnostic work as the red cells from both these animal species are susceptible to the alpha-haemolysins and beta-haemolysins that are produced commonly by staphylococcal isolates from animals. Both *S. aureus* and *S. intermedius* are usually haemolytic and often produce both the alpha-lysin and beta-lysin and so exhibit 'double-haemolysis' (**113**). The alpha-lysin is responsible for the narrow zone of clear haemolysis immediately around the colony and the beta-lysin for the broader outer zone of incomplete (partial) haemolysis. *S. hyicus* is non-haemolytic (**114**). Haemolytic activity among the coagulase-negative staphylococci is variable and often slow in appearing.

Microscopic appearance

A Gram-stained smear from colonies will reveal Gram-positive cocci randomly distributed over the field (**115**).

Tests for the pathogenicity of staphylococcal isolates

a) **Coagulase test:** this test correlates well with pathogenicity. As some pathogenic staphylococci can be negative to the slide coagulase test but positive to the tube test, some laboratories perform only the tube coagulase test (**116**). Fresh or reconstituted commercial freeze-dried rabbit plasma is the reagent used. Rabbit plasma contains fibrinogen that is converted to fibrin by the staphylococcal coagulase enzymes. 'Bound' coagulase is detected by the slide test and 'free' coagulase by the tube test.

- **Slide coagulase test:** a loopful of the staphylococcal culture is emulsified in a drop of water on a microscope slide. A loopful of rabbit plasma is added and mixed well with the bacterial suspension. The slide is gently rocked and a positive reaction is indicated by clumping within one or two minutes.
- **Tube coagulase test:** 0.5 ml of rabbit plasma is placed in a small (7 mm) test tube. Two drops of an overnight broth culture of the staphylococcus, or a heavy suspension made from the culture on an agar plate in sterile water, are added. The tube is rotated gently to mix the contents and then incubated at 37°C, preferably in a water bath. A positive test, with clotting of the plasma, can occur in 2–4 hours. However, many weak coagulase-positive strains will coagulate the plasma only after overnight incubation.

b) **DNase test:** this is carried out on commercially available DNase agar (**117**). However, this test is not entirely reliable as an indicator of pathogenicity, because it has been estimated that about 18 per cent of coagulase-negative staphylococci have DNase activity.

c) **Test for protein A:** commercial kits are available such as SeroSTAT (Scott Labs Inc., Fiskeville, RI). This test is used more commonly in medical diagnostic laboratories.

Biochemical tests

These are given in **Table 21.** The table also summarises the reactions of staphylococci to the coagulase and DNase tests and indicates which species produce haemolysins or pigment. Commercial systems are available for the identification of staphylococci such as API Staph-Ident and STaph-Trac (Analytab Products, Inc.) and Minitek Gram-Positive Set (BBL).

Purple agar base (Difco) with the addition of 1 per cent maltose (**118**) is a useful medium to differentiate the pathogenic staphylococci, particularly with coagulase-positive isolates from dogs that might be either *S. aureus* or *S. intermedius*. This presumptive identification is based on the fact that *S. aureus* strains rapidly ferment maltose and the acid metabolic products cause the pH indicator (bromocresol purple) to change the medium and colonies to yellow. *S. intermedius* gives a weak or delayed reaction and *S. hyicus* does not ferment maltose but attacks the peptone in the medium producing an alkaline reaction (a deeper purple) around the colonies. The reactions of the pathogenic staphylococci on this medium are summarised in **Table 22**.

Antibiotic susceptibility testing

This should be conducted on all coagulase-positive isolates and also when a coagulase-negative isolate appears to be significant. Resistance to the beta-lactam antibiotics is most often due to a plasmid-encoded penicillinase (beta-lactamase). Tolerance is a less common form of penicillin resistance and is thought to be due to the failure of the autolytic cell wall enzymes. Resistance to other antimicrobial agents is also common among the staphylococci.

Phage typing

Phage typing is carried out for epidemiological purposes, particularly for *S. aureus* strains from human cases of food poisoning and occasionally, in reference laboratories, for *S. aureus* isolates from bovine mastitis. Animal staphylococcal phage types are usually different from those of human origin.

111

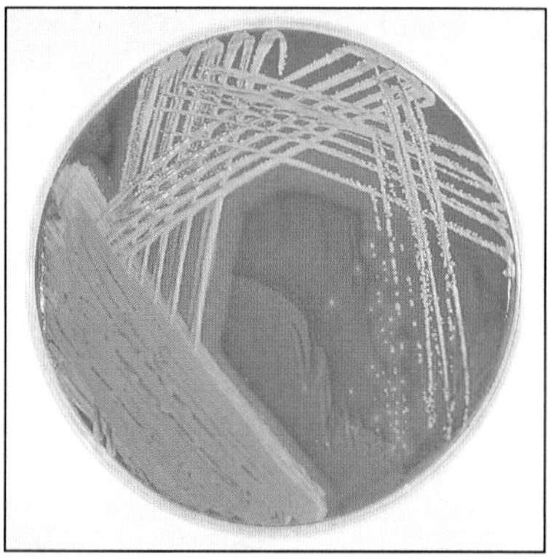

112

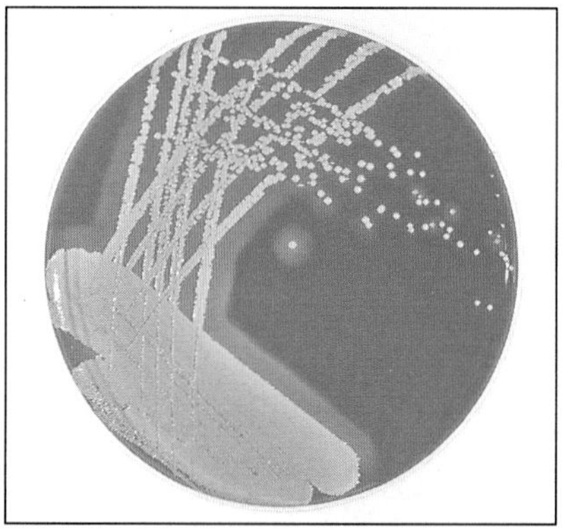

111 *Staphylococcus intermedius* on sheep blood agar overgrown by a swarming *Proteus* sp. (*see* **112** for comparison).

112 *S. intermedius* from the same sample as **111**, giving an almost pure culture, but plated on selective sheep blood agar.

113

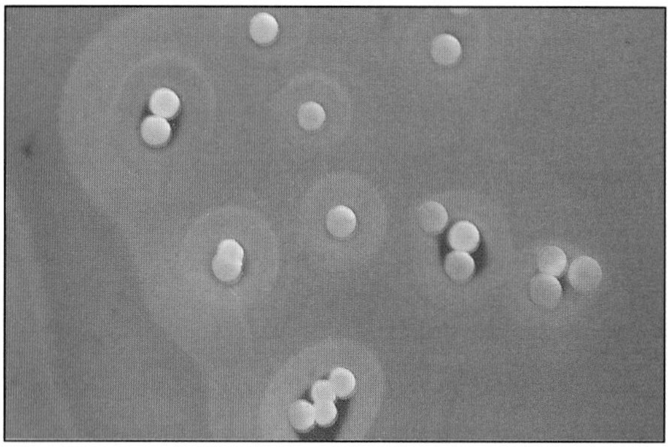

114

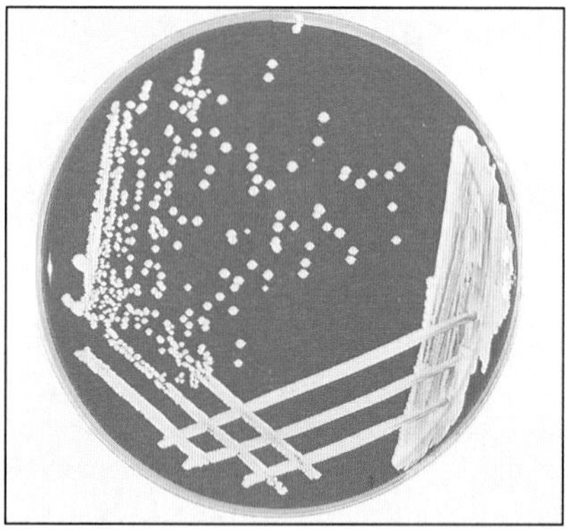

113 A close-up of the characteristic 'target-haemolysis' of a bovine isolate of *S. aureus* on sheep blood agar.

114 *S. hyicus* (non-haemolytic) on sheep blood agar.

115

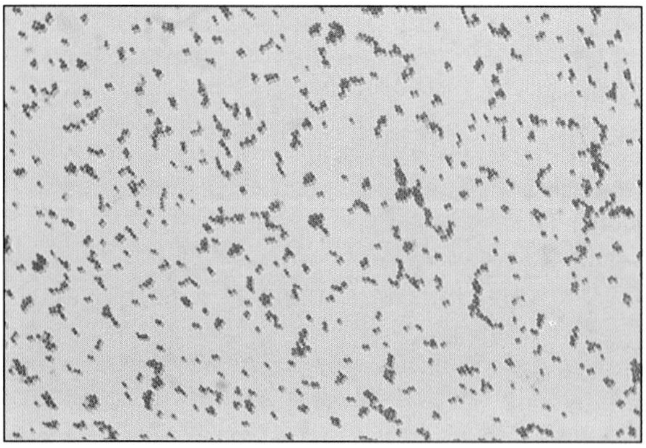

116

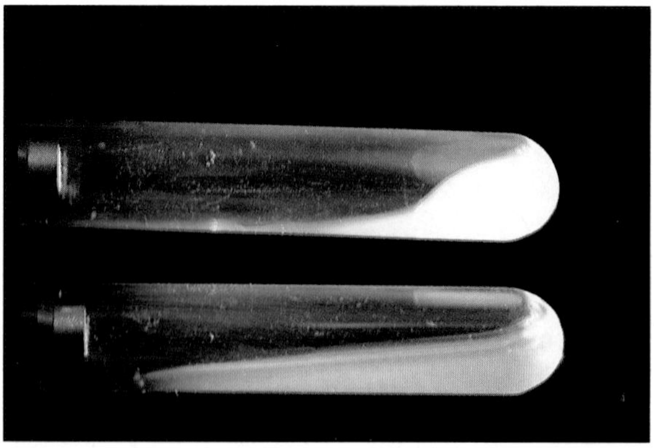

115 *S . intermedius* in a Gram-stained smear from a colony. (×1000)

116 Tube coagulase test: positive (top) and negative (bottom).

117

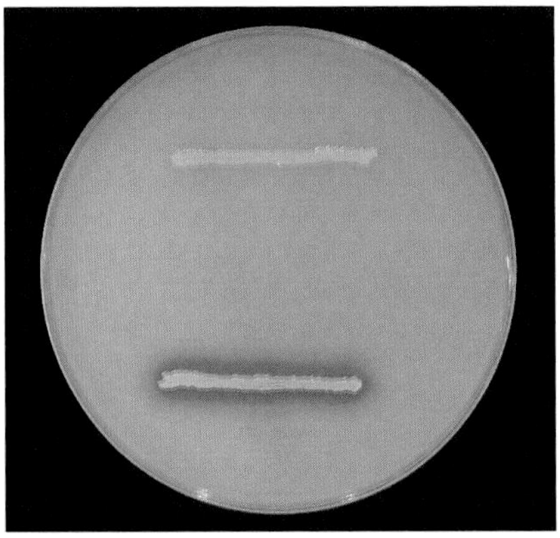

117 DNase agar: *S. epidermidis* (top) negative and *S. intermedius* (bottom) showing DNase activity.

118

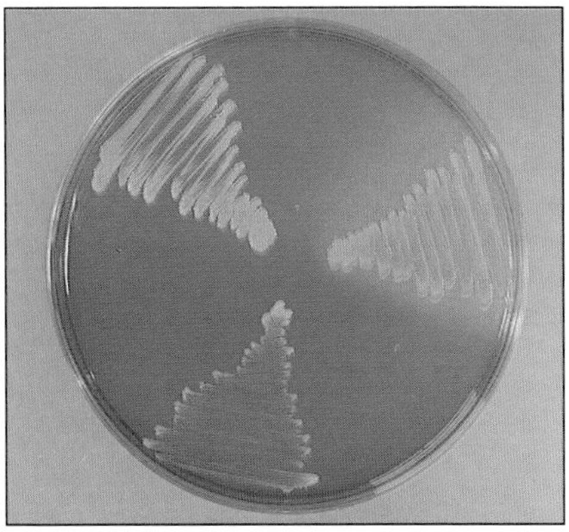

118 Purple agar base containing 1 per cent maltose: *S. intermedius* (left), *S. aureus* (right) and *S. hyicus* (bottom).

Table 19. Main diseases caused by pathogenic staphylococci.

Species	Host(s)	Diseases
Staphylococcus aureus	Many animal species	Abscesses and suppurative conditions. Infection can be systemic. Notorious for infections following surgery
	Cattle	Mastitis: subclinical, chronic, acute, peracute or gangrenous. Udder impetigo: small pustules often at base of teats
	Sheep	Mastitis: acute, peracute or gangrenous. Tick pyaemia of lambs (2–5 weeks old): associated with heavy tick infestation (*Ixodes ricinus*). Periorbital eczema (dermatitis): infections of abrasions, often due to insufficient feed-trough space. Staphylococcal dermatitis: predisposed to by scratches from vegetation such as thistles
	Goats	Mastitis: subacute or peracute. Staphylococcal dermatitis
	Pigs	Mastitis: acute, subacute and chronic (botryomycosis). Necrotising staphylococcal endometritis. Udder impetigo: after abrasions from teeth of piglets
	Horses	Mastitis: acute. Botryomycosis (spermatic cord) after castration
	Rabbits	Exudative dermatitis in neonates. Abscesses, conjunctivitis and pyaemic conditions
	Poultry	'Bumble-foot': pyogranulomatous process of subcutaneous tissue of foot that can involve the joints. Staphylococcal arthritis and septicaemia in turkeys. Omphalitis (although more commonly caused by *Escherichia coli*)
	Dogs, cats	Suppurative conditions similar to those listed for *S. intermedius*
S. aureus* subsp. *anaerobius	Sheep	Lesions similar to those of caseous lymphadenitis (*Corynebacterium pseudotuberculosis*)

(*continued*)

Table 19. Main diseases caused by pathogenic staphylococci (*continued*).

Species	Host(s)	Diseases
S. intermedius	Dogs, cats	Canine (feline) pyoderma (juvenile and adult). Chronic and recurrent pyoderma is a complex syndrome possibly involving cell mediated hypersensitivity, endocrine disorders and a genetic predisposition. Responds poorly to antibiotic therapy alone. Staphylococcal pustular dermatitis occurs in neonates or in adults under conditions of poor hygiene. Responds readily to antibiotic therapy. Pyometra. Otitis externa (together with other pathogens). Infections involving respiratory tract, bones, joints, wounds, eyelids and conjunctiva
	Horses, cattle	Rare infections in these species
S. hyicus	Pigs	Exudative epidermitis (greasy pig disease), usually in pigs under 7 weeks old. There is systemic involvement and the condition can be fatal. Septic polyarthritis
	Cattle	Rare cases of mastitis

Table 20. The staphylococcal haemolysins.

Haemolysin	*Type of haemolysis*	*Erythrocytes*		*Characteristics*
		Affected	*Not affected*	
Alpha	Complete	Rabbit Sheep Cattle	Horse Human Chicken	Cytotoxic for a large range of tissue culture cells. Lethal and necrotising in rabbits. Acts on membrane lipids and causes spasm of smooth muscle
Beta	Incomplete Becomes complete after further storage at 4–15°C ('hot-cold lysis')	Sheep and cattle at 37°C	Rabbit Chicken Horse	Unique to animal strains of staphylococci. It is a phospholipase c. May occasionally mask the effect of the alpha haemolysin Potentiation of the partial haemolysis of this haemolysin is the basis of the CAMP test
Delta	Complete	Wide spectrum including human, rabbit, sheep, horse, rat, guinea-pig and some fish erythrocytes	•	This haemolysin migrates through agar more slowly than the alpha-haemolysin so the effect takes longer to express. Lethal and necrotising effects in rabbits. It is a polypeptide and is inhibited by normal serum and phospholipids
Gamma	•	Rabbit Sheep Human Guinea-pig	•	Cytopathic for some tissue culture cells and causes oedema in rabbits. (intradermal injection). Inhibited by agar, heparin, phospholipids and cholesterol

• = data not available

Table 21. Biochemical reactions and other characteristics of staphylococci isolated from animals.

	Coagulase test	**DNase test**	**Haemolysis**	**Pigment production**	**Alkaline phosphatase**	**Urease****	**Mannitol fermented** (mannitol salt agar)	**Maltose fermented** (purple agar + 1% maltose)	**Aesculin hydrolysis**	**Novobiocin** 5µg disc	**Polymyxin B** 300 unit disc
Staphylococcus aureus	+	+	+	+*	+	d	+	+	–	S	R
S. intermedius	+	+	+	–	+	+	(d)	(±)	–	S	S
S. hyicus	d	+	–	–	+	d	–	–	–	S	R
S. epidermidis	–	d	(d)	–	d	+	–	+	–	S	R
S. saprophyticus	–	–	–	d	–	+	d	+	–	R	S
S. aureus* ssp. *anaerobius	+	+	+	–	+	•	•	+	–	S	•
S. caprae	–	–	(d)	–	(+)	+	d	(d)	–	S	S
S. gallinarum	–	–	(d)	d	(+)	+	+	+	+	R	S
S. arlettae	–	–	–	+	(+)	–	+	+	–	R	•
S. lentus	–	–	–	d	(±)	–	+	d	+	R	S
S. equorum	–	–	(d)	–	(+)	+	+	d	d	R	•
S. simulans	–	–	(d)	–	(d)	+	+	(±)	–	S	S
S. delphini	•	–	+	–	+	+	(+)	+	•	S	•
S. chromogenes	–	–	–	+	+	+	d	d	–	S	R

+ = 90% or more strains positive; ± = 90% or more weakly positive; d = 11–89% positive; – = 90% or more strains negative; () = delayed reaction; • = not known; * = strains from dogs usually white, ** = Urea R broth (Difco); R = resistant; S = susceptible; resistance to novobiocin = zone of 16 mm or less; resistance to polymyxin B = zone of less than 10 mm.

Table 22. Reactions of coagulase-positive staphylococci on purple agar with 1% maltose.

Species	Maltose fermentation	Reactions on purple agar with 1% maltose
Staphylococcus aureus	+++	Diffuse yellow colour around colonies. Rapid reaction within 24 hours' incubation.
S. intermedius	±	Little change in the medium. Slight yellowish zone under colonies and occasionally isolated colonies may have a yellowish tinge.
S. hyicus	–	Diffuse deep purple (alkaline) zone around the colonies

+++ = rapid fermentation of maltose; ± = weak and slow fermentation; – = negative.

Reference

Faller, A. and Schleifer, K.H. (1981). Modified oxidase and benzidine test for separation of staphylococci from micrococci, *Journal of Clinical Microbiology*, **13**: 1031–1035.

9 The Streptococci and Related Cocci

The streptococci and enterococci are Gram-positive cocci that occur in pairs or chains of varying lengths. Each coccus is about 1 µm in diameter. They are facultative anaerobes, catalase-negative, oxidase-negative and non-motile with the exception of some of the enterococci. The streptococci are fastidious and require the addition of blood or serum to media for growth. The enterococci tolerate the bile salts in MacConkey agar and appear as small pin-point colonies on this medium.

Natural Habitat

Streptococci are worldwide in distribution. Most of the streptococci of veterinary interest live as commensals in the mucosa of the upper respiratory and lower urinogenital tracts. They are susceptible to desiccation and do not usually survive for long away from the animal host. The enterococci are opportunists and can be found in the intestinal tract of many animals.

Recent Changes in Nomenclature

Lancefield Group	Previous Name	Present Name
C	*Streptococcus equisimilis*	*S. dysgalactiae* subsp. *equisimilis*
C	*Streptococcus equi*	*S. equi* subsp. *equi*
C	*Streptococcus zooepidemicus*	*S. equi* subsp. *zooepidemicus*
D	*Streptococcus faecalis*	*Enterococcus faecalis*
D	*Streptococcus faecium*	*Enterococcus faecium*
D	*Streptococcus durans*	*Enterococcus durans*
Q	*Streptococcus avium*	*Enterococcus avium*
N	*Streptococcus lactis*	*Lactococcus lactis*

General Differentiation of the Streptococci

- ***Haemolysis:*** the type of haemolysis produced by a streptococcal species can be variable.
  The main types of haemolysis are:
  Beta (β) -haemolysis: a clear zone of haemolysis around the colony (**119**).
  Alpha (α) -haemolysis: a zone of greening or of partial haemolysis (**120**).
  Gamma (γ) -haemolysis: no haemolysis.
  Generally the beta-haemolytic streptococci tend to be the most pathogenic for animals.
  Lancefield Group C streptococci from horses usually produce large zones of clear, beta haemolysis (**121**).
- ***Lancefield Groups***: a serological grouping determined by the C-substance that is a Group-specific cell wall polysaccharide. The methods for grouping the streptococci include:
  a) **Conventional Method**: the C-substance (antigen) is extracted either by autoclaving or by acid extraction (hydrochloric acid). A ring precipitation test is conducted by layering the extracted antigen over known antisera that can be obtained commercially for each Lancefield Group (Wellcome Diagnostics).
  (b) **Latex Agglutination Test**: kits are available commercially for identifying Lancefield Groups A, B, C, D, F and G (**122**) (Wellcome Diagnostics; Difco Laboratories; Scott Laboratories and Diagnostic Products Corporation).
  c) **Slide Coagglutination Test**: for streptococcal Groups A, B, C, F and G (Pharmacia Diagnostics).

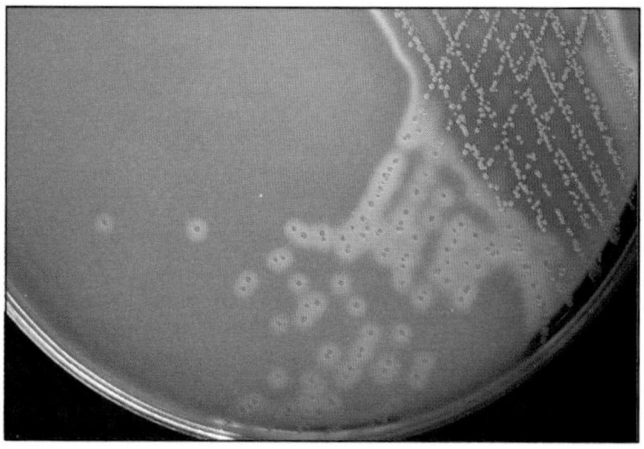
119

119 Beta-haemolytic streptococcus on sheep blood agar.

120

120 Alpha-haemolytic streptococcus on sheep blood agar.

121

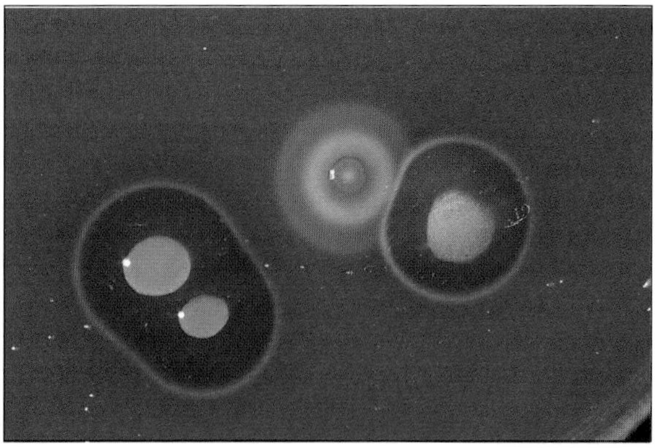

121 Close up of colonies of *Streptococcus equi* subsp. *equi* (beta-haemolytic) after 4 days incubation compared with the greening-type haemolysis of an alpha-haemolytic streptococcus (contaminant).

122

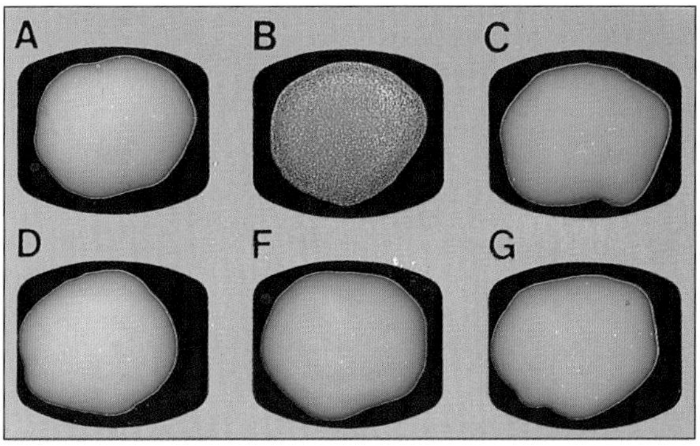

122 Latex agglutination test for Lancefield grouping of streptococci. In this instance the streptococcus under test belonged to Lancefield Group B.

Pathogenesis

Streptococci are pyogenic bacteria that are commonly associated with suppuration and abscess formation. Exotoxins include the haemolysins streptolysin O and S, a hyaluronidase, DNase, NADase, protease, streptokinase and a phage-coded pyrogenic toxin that is associated with the skin rash in scarlet fever of man. The significance of these products is poorly understood. The polysaccharide capsules of *S. pyogenes*, *S. pneumoniae* (**123**) and some strains of *S. equi* subsp. *equi* and *S. agalactiae* are antiphagocytic. The cell wall M protein of *S. pyogenes* and *S. porcinus* is also antiphagocytic and in *S. equi* subsp. *equi* it may function as an adhesin. **Table 23** briefly summarises the main streptococcal diseases of animals. The anaerobic streptococcus, *Peptostreptococcus indolicus* occurs in association with *Actinomyces pyogenes* in 'summer mastitis' of cattle in Europe.

123

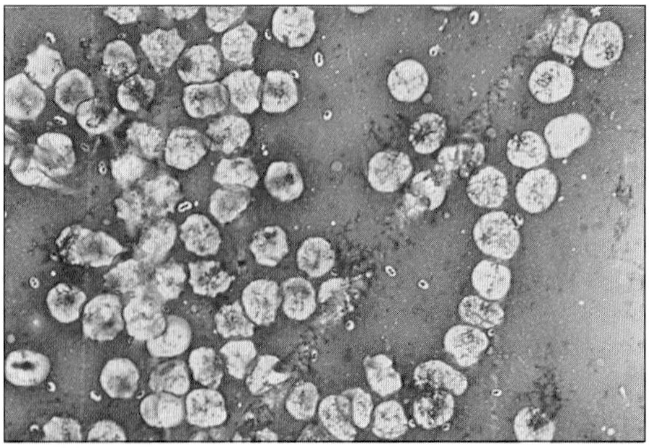

123 *S. pneumoniae* in a blood smear showing the distinctive capsule. (Nigrosin stain, ×1000)

Table 23. Streptococci of veterinary and medical significance.

Lancefield group	Species	Haemolysis	Host(s)	Disease	Natural habitat (if known)
A	*S. pyogenes*	β	Humans	Scarlet fever, septic sore throat, puerperal fever, erysipelas, abscesses and rheumatic fever	Human upper respiratory tract
			Cattle	Mastitis (rare)	
			Foals	Lymphangitis	
B	*S. agalactiae*	β (α, γ)	Cattle, sheep and goats	Chronic mastitis	Milk ducts
			Humans and dogs	Neonatal septicaemia	Maternal vagina
			Cats	Kidney and uterine infections	
C	*S. dysgalactiae*	α (β, γ)	Cattle	Acute mastitis	Buccal cavity and genitalia
			Lambs	Polyarthritis	
	S. dysgalactiae subsp. *equisimilis*	β	Horses	Abscesses, endometritis and mastitis	Skin and vagina
			Pigs, cattle, dogs and birds	Various suppurative conditions	
	S. equi subsp. *equi*	β	Horses	Strangles, genital and suppurative conditions, mastitis and purpura haemorrhagica	Equine tonsils
	S. equi subsp. *zooepidemicus*	β	Horses	Mastitis, abortion, secondary pneumonia and navel infections	Vagina and skin
			Cattle	Metritis and mastitis	
			Pigs	Septicaemia and arthritis in 1–3- week-old piglets	Skin and mucous membranes of sows
			Poultry	Septicaemia and vegetative endocarditis	
			Lambs	Pericarditis and pneumonia	

(continued)

Table 23. Streptococci of veterinary and medical significance (*continued*).

Lancefield group	Species	Haemolysis	Host(s)	Disease	Natural habitat (if known)
D	*Enterococcus faecalis* *E. faecium* *E. durans*	α (β, γ)	Many species	Opportunistic infections such as septicaemia in chickens, bovine mastitis, endocarditis in cattle and lambs, and urinary-tract infections in dogs	Intestinal tract of many animals
	S. equinus *S. bovis*	α	Many species	Opportunistic infections	Intestinal tract of many animals
E (P, U, V)	*S. porcinus*	β	Pigs	Jowl abscesses and lymphadenitis	Mucous membranes
G	*S. canis*	β	Carnivores	Neonatal septicaemia. Genital, skin and wound infections	Genital tract and anal mucosa
			Cattle	Occasional mastitis	
N	*Lactococcus lactis*	α	Cattle	Unknown pathogenicity	Milk, plants and tonsils of pigs fed on whey
Q	*Enterococcus avium*	α, γ	Many species	Unknown pathogenicity	Faeces of birds and mammals
R(D)	*S. suis* type 2	α	Pigs (weaning to 6 months)	Meningitis and arthritis	Tonsils and nasal cavity
			Humans	Meningitis and septicaemia	Pigs
S(D)	*S. suis* type 1	α (β)	Pigs (2–4 weeks old)	Meningitis, arthritis, pneumonia and septicaemia	Tonsils and nasal cavity
Ungroupable	*S. uberis*	α (γ)	Cattle	Mastitis	Skin, vagina and tonsils
	S. pneumoniae	α	Humans and primates	Pneumonia, septicaemia and meningitis	Upper respiratory tract
			Guinea-pigs and rat colonies	Pneumonia (outbreaks can occur)	

Laboratory Diagnosis

Specimens

Depending on the pathological condition, these may include exudates, pus, mastitic milk, skin scrapings, cerebrospinal fluid, urine and affected tissues. Swabs should be submitted in transport medium as streptococci are very susceptible to desiccation.

Direct microscopy

Smears from pus, exudates or centrifuged deposits of milk or urine can be fixed and stained by the Gram method (**124**). *S. pneumoniae* (the pneumococcus) occurs as pairs of cocci (**125**). Fluorescent antibody tests have been employed to identify streptococci, such as *S. suis* type 2, in tissues (Robertson and Blackmore, 1987).

Isolation

Routine media include sheep or ox blood agar, selective blood agar (as described for the staphylococci) and MacConkey agar.

Mastitic milk samples can be inoculated on blood agar, Edwards medium (Oxoid) and MacConkey agar.

Herd mastitic milk samples can be quarter-plated on blood agar containing 0.1 or 0.05 per cent aesculin to indicate aesculin hydrolysis.

Inoculated plates are incubated aerobically at 37°C for 24–48 hours.

Identification

Colonial appearance

Most streptococci produce small colonies (about 1 mm after 48 hours' incubation) and in the case of the beta-haemolytic streptococci the colonies appear translucent. Colonial variation can occur and mucoid strains of *S. equi* subsp. *equi* (**126**) and *S. pneumoniae* (**127**) are not uncommon. Although variation in the type of haemolysis within a species occurs, haemolytic activity is a useful diagnostic characteristic. The clear zones of the Group C streptococci from horses are often large. *S. pneumoniae* is alpha-haemolytic and produces either mucoid colonies or flat colonies with smooth borders and a central concavity ('draughtsman' colonies) after 48–72 hours on blood agar (**128** and **129**).

- **Gram-stained smears:** when made from colonies, Gram-positive cocci are seen. The characteristic chains occur only in broth cultures or in animal tissues.
- **Catalase test**: the streptococci are catalase-negative which helps to distinguish them from the catalase-positive staphylococci.
- **Lancefield grouping:** this can be carried out by the methods previously described.

Biochemical tests

The methods for conducting biochemical tests include:

a) Identification using the commercial kits such as API 20S, rapid STREP (Analytab Products) and the RapID STR system (Innovative Diagnostic Systems).

124

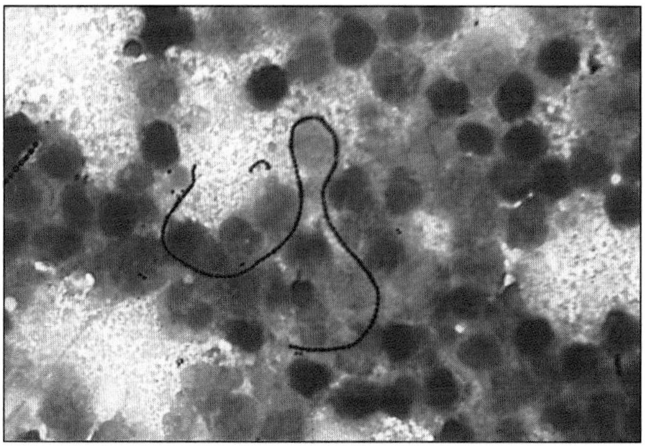

124 *S. equi* subsp.*equi* in a smear of pus from a case of strangles. The long chain of Gram-positive cocci is characteristic both of this bacterium and of the disease. (Gram stain, ×1000)

125

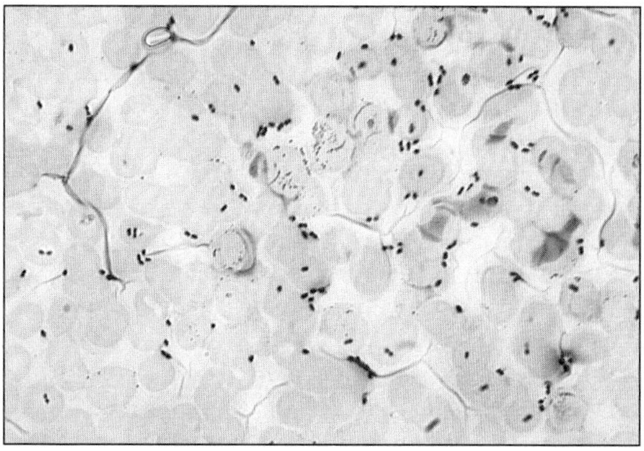

125 *S. pneumoniae* in a blood smear demonstrating pairs of cocci characteristic for this bacterium. (Leishman stain, ×1000)

126

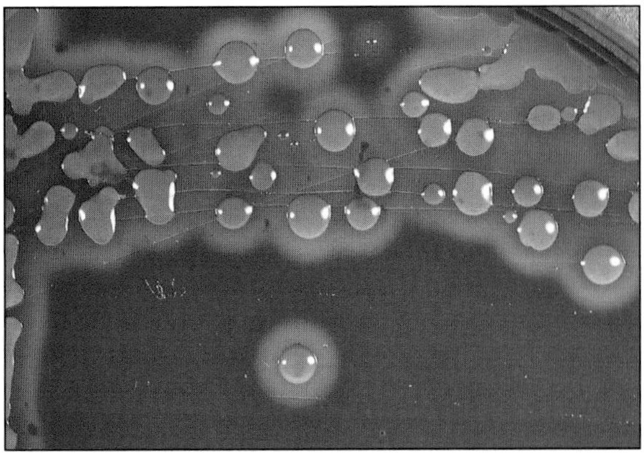

126 *S. equi* subsp. *equi* primary isolation from bastard strangles in a foal. Note the comparatively large, mucoid colonies and wide zones of haemolysis.

127

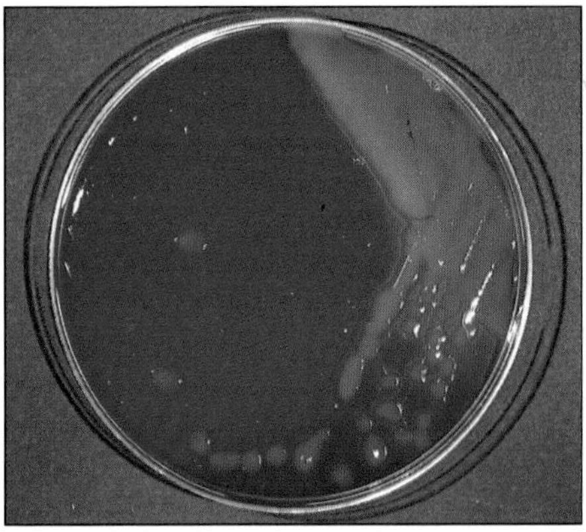

127 Mucoid-type colonies of *S. pneumoniae* on sheep blood agar showing the colonial form and alpha-haemolysis.

128

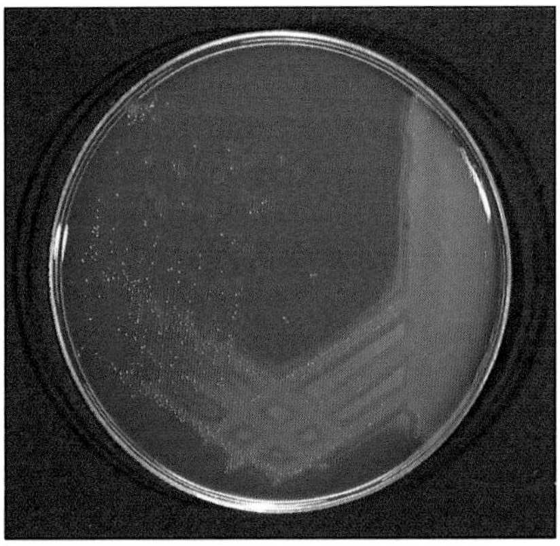

128 Non-mucoid colonies of *S. pneumoniae* on sheep blood agar. They are alpha-haemolytic.

129

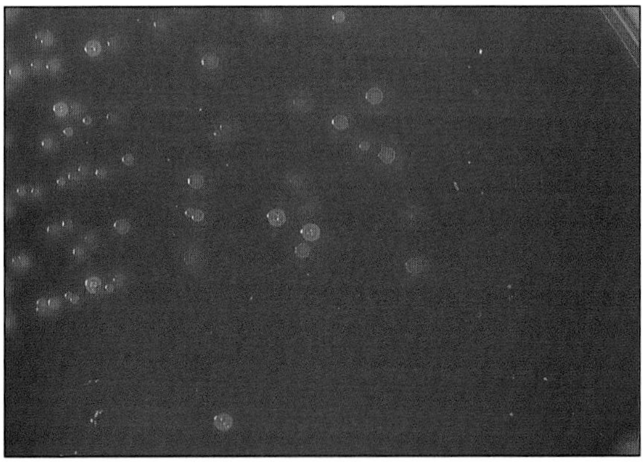

129 A close-up of the 'draughtsman' colonies characteristic of *S. pneumoniae*.

b) Conventional peptone water sugars with 4–5 drops of serum added to 2 ml of medium can be used for fastidious streptococci. An alternative is CTA medium with the appropriate carbohydrate discs. Biochemical reactions for the streptococci of veterinary significance are shown in **Table 24**. A short range of sugar fermentation tests should be carried out if a Group C streptococcus is isolated from a horse (**130**). The biochemical differentiation of Group C streptococci from horses is given in **Table 25.**

If the bacterium appears to be *S. equi* subsp. *equi*, capable of causing the highly contagious disease strangles, the result should be confirmed by a more extensive range of biochemical tests.

The enterococci and Group D streptococci are distinguished by growth at 45°C and toleration of 40 per cent bile. They all hydrolyse aesculin. A short range of tests can differentiate members of this group (**Table 26**).

130

130 *S. equi* subsp. equi in a short range of peptone water sugars (with the addition of a few drops of sterile serum in each). Andrade's pH indicator. From left, uninoculated, trehalose (-), sorbitol (-), lactose (-) and maltose (+).

Presumptive identification tests for streptococci are summarised in **Table 27**

a) *CAMP test* : a culture of a *Staphylococcus aureus*, with a wide zone of partial haemolysis (beta-haemolysin), is streaked across the centre of a sheep or ox blood agar plate. A streak of the suspect Group B streptococcus is made at right angles to, and taken to within 1 to 1.5 mm of the staphylococcal streak.

The plate is incubated at 37°C for 18–24 hours. A positive CAMP test is indicated by an arrow-head of complete haemolysis (**131**). The Group B streptococci produce a diffusible metabolite that completes the lysis of the red cells, only partially haemolysed by the beta-haemolysin of the staphylococcus.

b) *Hydrolysis of sodium hippurate*: this is another test that distinguishes the Group B streptococci from the other streptococci (**51**).

c) *Production of a carotenoid pigment by Group B streptococci under anaerobic incubation*: media such as GBS agar (Oxoid) have been designed to exploit the ability of about 97 per cent of Group B streptococci to produce an orange to red pigment when incubated anaerobically (**132**).

d) *Susceptibility to a 0.04 unit disc of bacitracin*: distinguishes the Group A streptococci from the other beta-haemolytic streptococci. The discs are available commercially and the test is carried out by the disc diffusion method (**133**).

e) *Susceptibility to optochin*: *S. pneumoniae* can be distinguished from the other alpha-haemolytic streptococci by its susceptibility to low levels of optochin (ethylhydrocuprein HCl). The test is performed as a disc diffusion test using commercially available discs. The zone of inhibition when using a 6 mm disc should be equal to or greater than 14 mm (**134**).

f) *Bile solubility*: the test is conducted by adding a suspension of the suspect *S. pneumoniae* in physiological saline or in a broth culture to an equal amount of a 10 per cent solution of sodium taurocholate. Within 10–15 minutes a pneumococcus will autolyse and the suspension will become clear.

g) *Amylase reaction and rapid VP test for S. suis*: Devriese *et al.* (1991) showed that all *S. suis* strains from pigs produced an amylase reaction on Columbia agar base with 1g/litre soluble starch and were negative to a rapid acetoin (VP) test using pelleted reagents of Rosco (Taastrup, Denmark).They proposed the use of these two tests as a rapid diagnositc method for identifying *S. suis* from pigs.

131

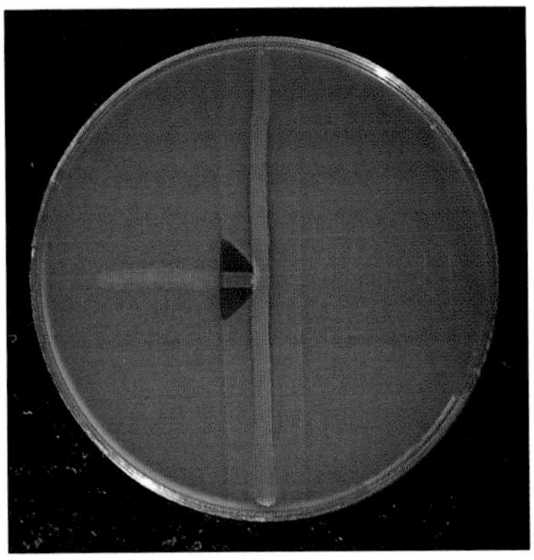

131 CAMP test for Group B streptococci. *S. agalactiae* is causing the characteristic 'arrow-head' clearing of the partial haemolysis (beta-haemolysin) of *Staphylococcus aureus* (vertical streak).

132

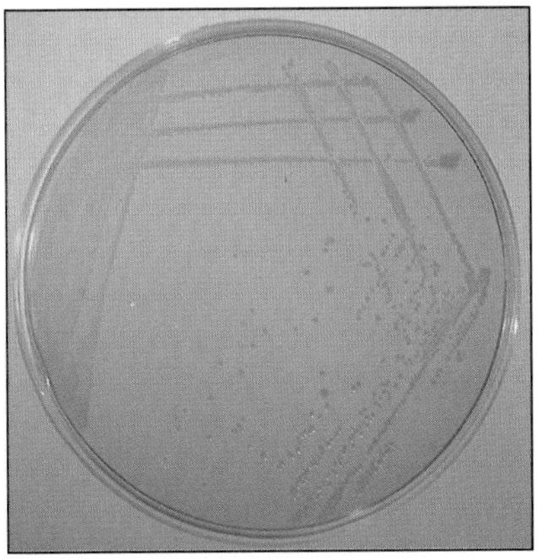

132 Group B streptococcus showing the carotenoid pigment on GBS agar under anaerobic conditions.

133

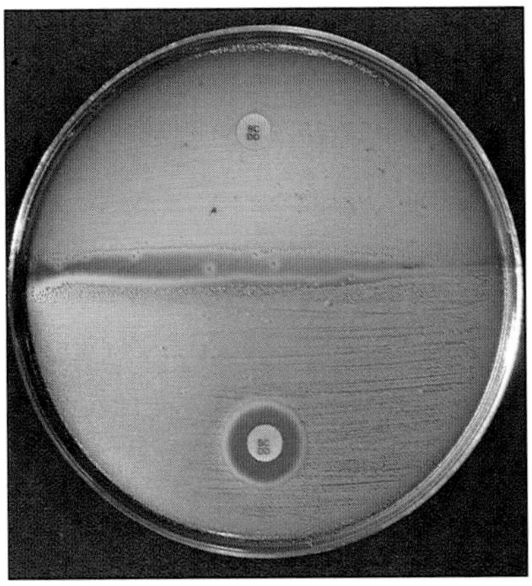

133 Bacitracin (0.04 units) susceptibility test to distinguish Group A streptococci from other beta-haemolytic streptococci. Group B (above) resistant and Group A (below) susceptible.

134 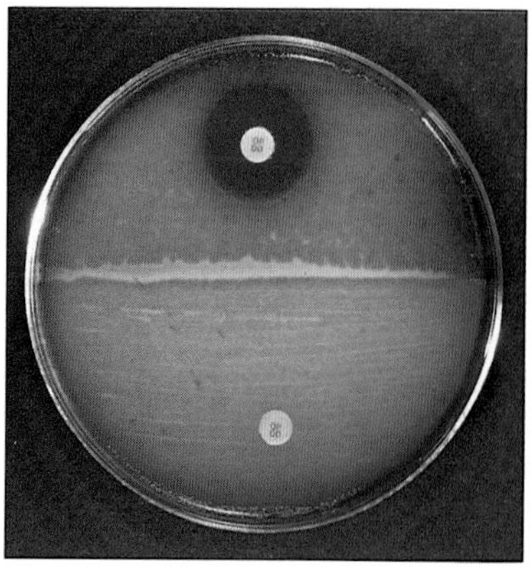

134 Optochin susceptibility test to distinguish *S. pneumoniae* from other alpha-haemolytic streptococci. *S. pneumoniae* (above) susceptible and *Enterococcus faecalis* (below) resistant.

Table 24. Biochemical reactions of important streptococci from animals.

	Lancefield group	Acid from: Inulin	Acid from: Lactose	Acid from: Mannitol	Acid from: Raffinose	Acid from: Salicin	Acid from: Sorbitol	Acid from: Trehalose	Aesculin hydrolysis	Sodium hippurate	Growth in 6.5% NaCl
Streptococcus pyogenes	A	–	+	v	–	+	–	+	–	–	–
S. agalactiae	B	–	+	–	–	(+)	–	+	–	+	–
S. dysgalactiae	C	–	+	–	–	–	–	+	–	–	–
S. dysgalactiae* subsp. *equisimilis	C	–	v	–	–	(+)	–	+	–	–	–
S. equi* subsp. *equi	C	–	–	–	–	+	–	–	–	–	–
S. equi* subsp. *zooepidemicus	C	–	+	–	–	+	+	–	–	–	–
Enterococcus faecalis	D	–	+	+	–	+	+	+	+	v	+
S. bovis	D	+	+	v	+	+	–	v	+	–	–
S. equinus	D	+	–	–	+	(+)	–	v	+	–	–
S. porcinus	E (P,U,V)	–	(+)	+	–	+	+	+	+	–	+
S. canis	G	–	(+)	–	–	•	–	(+)	v	–	–
Enterococcus avium	Q	–	+	+	–	+	+	+	+	v	(+)
***S. suis* type 2**	D(R)	(+)	+	–	(+)	+	–	+	+	–	–
***S. suis* type 1**	D(S)	+	+	–	–	•	–	+	–	–	–
S. uberis	–	+	+	+	–	+	+	+	+	+	(+)
S. pneumoniae	–	+	+	–	+	v	–	+	(+)	–	–

(+) = majority of strains positive, v = variable reactions, • = information not available.

Table 25. Differentiation of equine group C streptococci.

	Trehalose	Sorbitol	Lactose	Maltose
S. equi* subsp. *equi	–	–	–	+
S. equi* subsp. *zooepidemicus	–	+	+	+(–)
S. dysgalactiae* subsp. *equisimilis	+	–	v	+

v = variable reactions, (–) = a few strains are negative.

Table 26. Differentiation of the enterococci and group D streptococci.

	Lancefield group	Lactose	Arabinose	Sorbitol	Mannitol	Growth in 6.5% NaCl
E. faecalis	D	+	–	+	+	+
E. faecium	D	+	+	–	+	+
E. durans	D	+	–	–	–	+
E. avium	Q	+	+	+	+	+
S. equinus	D	–	–	–	–	–
S. bovis	D	+	v	–	v	–

v = variable reactions.

Table 27. Presumptive identification tests for the streptococci.

	Bacitracin (0.04 units) susceptibility	CAMP test	Sodium hippurate hydrolysis	Optochin susceptibility	Bile solubility
Group A streptococci	+	–	–	–	–
Group B streptococci	–	+	+	–	–
S. pneumoniae	–	–	–	+	+

Table 28. Differentiation of streptococci causing bovine mastitis.

	CAMP test	Aesculin hydrolysis on Edwards medium	Growth on MacConkey agar
S. agalactiae	+	–	–
S. dysgalactiae	–	–	–
S. uberis	–	+	–
E. faecalis	–	+	+

Identification of the streptococci causing bovine mastitis

These streptococci are discussed in Chapter 36 (Mastitis) and their identification is summarised in **Table 28**. Edwards medium is highly selective for streptococci. It contains both red blood cells and aesculin, thus haemolysis and aesculin hydrolysis can be observed (**135**).

135

135 Edwards medium is highly selective for the streptococci and also indicates aesculin hydrolysis and the type of haemolysis. *S. agalactiae* (left) and *S. dysgalactiae* (bottom) are non-aesculin splitters whereas *E. faecalis* (right) shows aesculin hydrolysis with darkening of the colonies and medium.

Antigen preparation for Lancefield grouping by the ring precipitation test

HOT HCl EXTRACTION

- A pure culture of the streptococcus is grown in 25 ml of Todd–Hewitt broth at 37°C for 24–48 hours.
- Centrifuge the broth to concentrate the cells. Discard the supernatant.
- Add 1 ml of the stock HCl-saline mixture (1 ml conc. HCl + 99 ml of N saline) and resuspend the cells.
- Place in a boiling water bath for 15 minutes and allow to cool.
- Add 1 drop of phenol red indicator and neutralise with N/10 NaOH until the suspension is a pale pink colour.
- Centrifuge and use the supernatant as the antigen for the test.

AUTOCLAVE EXTRACTION

- A pure culture of the streptococcus is grown in 25 ml of Todd–Hewitt broth at 37°C for 24–48 hours.
- Centrifuge the broth to concentrate the cells. Discard the supernatant.
- Add 0.5 ml of 0.85 per cent NaCl solution to the cells and shake to resuspend.
- Autoclave the suspension for 15 minutes at 121°C.
- Cool and centrifuge. Decant the supernatant into a clean tube for use as an antigen in the test.

RING PRECIPITATION TEST FOR LANCEFIELD GROUPING

To economise on commercial antisera use a capillary tube of outside diameter 1.2–1.5mm. Autoclave the capillary tubes and store in a sterile container.

The antisera chosen to test against the antigen extract will depend on the animal species and lesion from which the streptococcus was isolated. For example, the majority of the streptococci isolated from horses are Lancefield Group C.

- Dip a sterile capillary tube into the antiserum until a column about 1 cm long has been drawn into the tube. Plunge the lower end of the capillary tube into plasticine stuck on a microscope slide so that the tube is held upright.
- With a finely drawn glass Pasteur pipette carefully layer the antigen solution on top of the antiserum, taking care that there are no air bubbles and that no mixing of the antiserum and antigen occurs.
- Examine the tube in bright light against a dark background. A white ring of precipitate should appear, in a positive reaction, in 5–30 minutes. Precipitate formation after 30 minutes should be disregarded.

References

Devriese, L.A., Ceyssens, K., Hommez, J., Kilpper–Balz, R. and Schleifer, K.H. (1991). Characteristics of different *Streptococcus suis* ecovars and description of a simplified identification method. *Veterinary Microbiology*, **26**: 141–150.

Robertson, I.D. and Blackmore, D.K. (1987). The detection of pigs carrying *Streptococcus suis* type 2. *New Zealand Veterinary Journal*, **35**: 1–4.

10 *Corynebacterium* species and *Rhodococcus equi*

The corynebacteria are small pleomorphic Gram-positive rods (about 0.5 μ m in width) that occur in rod, coccoid, club and filamentous shapes. Stained smears from animal tissues often reveal groups of cells in parallel ('palisades') or cells at sharp angles to each other ('Chinese letters') (**136**). Many have metachromatic granules (high-energy phosphate stores) and these are seen best in *Corynebacterium diphtheriae* (**137**). The corynebacteria are non-spore-forming, non-acid fast, catalase-positive, oxidase-negative, usually facultatively anaerobic and the animal pathogens are non-motile. *Rhodococcus equi* can appear as a Gram-positive coccus or as a rod. It is capsulated and sometimes weakly acid-fast.

136

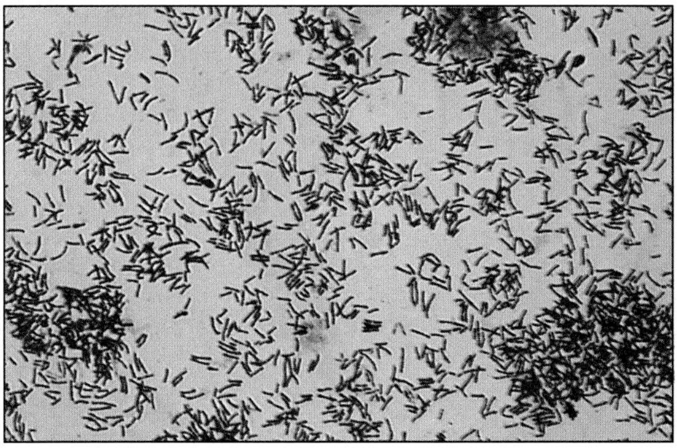

136 *Corynebacterium diphtheriae* demonstrating club-shaped cells and 'Chinese-letter' patterns characteristic of the genus. (Gram stain, ×1000)

137

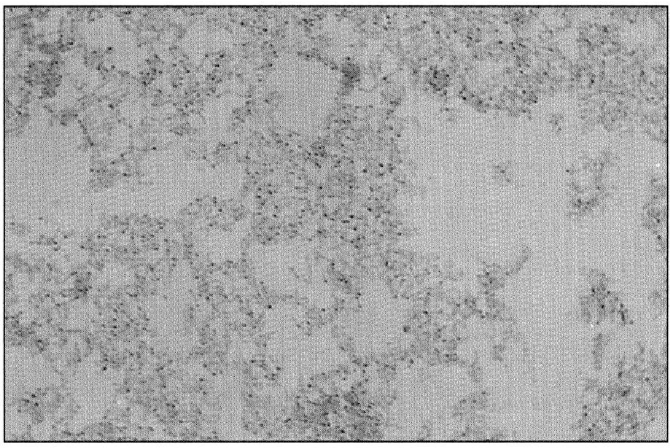

137 *C. diphtheriae* with dark-staining (reddish-purple) metachromatic granules of polyphosphate. (Methylene blue stain, ×1000)

Changes in Nomenclature

Previous Name	Present Name	
Corynebacterium equi	*Rhodococcus equi*	
Corynebacterium murium	*C. kutscheri*	
Corynebacterium ovis	*C. pseudotuberculosis*	
Corynebacterium pyogenes	*Actinomyces pyogenes* (Chapter 11)	
Corynebacterium suis	*Eubacterium suis* (Chapter 16)	
C. renale type I	*C. renale*	*Corynebacterium renale* group
C. renale type II	*C. pilosum*	
C. renale type III	*C. cystitidis*	

Pathogenesis

Corynebacteria are pyogenic bacteria causing a variety of suppurative conditions. The virulence of *C. pseudotuberculosis* is attributed to the haemolytic toxin which has phospholipase activity and to the cell wall lipids. *R. equi* can survive intracellularly through suppression of phagolysosomal fusion. It produces diffusible '*R. equi* factors' (phospholipase c and cholesterol oxidase) and these as well as the capsule and cell wall constituents probably play a part in the pathogenesis. Suppurative bronchopneumonia in foals, caused by *R. equi*, increases in prevalence where high stocking rates occur. The bacterium can multiply in soil enriched with equine faeces and may be a commensal in the intestine of horses. The disease is usually seen in 2–4 month-old foals, possibly due to the decline in maternal antibody at about 6 weeks of age. The main route of infection is by inhalation. Heavily infected sputum may be swallowed by the affected foal leading to ulcerative colitis and mesenteric lymphadenitis. **Table 29** summarises the main diseases, hosts and natural habitats of the corynebacteria and *R. equi.*

Table 29. Main hosts, diseases and natural habitats of the pathogenic corynebacteria and *Rhodococcus equi.*

Species	Main host(s)	Disease(s)	Natural habitat
Corynebacterium bovis	Cattle	Uncertain pathogenicity	Teat canal of cows
C. kutscheri	Mice, rats, (guinea-pigs)	Caseopurulent foci in liver, kidneys, lungs and lymph nodes	Mucous membranes of carrier rodents
C. pseudotuberculosis	Sheep, goats	Caseous lymphadenitis	Skin, mucous membranes and gastrointestinal tract of normal sheep and soil of sheep pens
	Horses (cattle)	Ulcerative lymphangitis. Contagious acne (Canadian horse pox)	Unknown
C. renale	Cattle	Pyelonephritis and cystitis	Prepuce and semen of asymptomatic bulls. Subclinical infection in cows
	Pigs	Kidney abscesses	Unknown
	Male sheep	Balanoposthitis ('pizzle rot')	Unknown
C. cystitidis	Cattle	Pyelonephritis	Male genital tract
C. pilosum	Cattle	Pyelonephritis	Bovine genital tract and urine
Rhodococcus equi	Foals (2–4 months old)	Suppurative bronchopneumonia	Soil and faeces of foals and other herbivores
	Older foals	Abscesses	As above
	Pigs (cattle)	Cervical lymphadenitis	Soil

Laboratory Diagnosis

Specimens

Pus or exudates are collected from suppurative conditions and mid-stream urine for attempted isolation of the members of the *C. renale* group. A tracheal wash technique, with infusion of saline, can be used for the recovery of *R. equi* from affected foals.

Direct microscopy

The corynebacteria are Gram-positive rods with varying degrees of pleomorphism (**138**). *R. equi* is usually coccal but can be rod-shaped particularly in animal tissue (**139**) and can be MZN-positive.

Isolation

For routine isolation sheep or ox blood agar is used with MacConkey agar to detect any Gram-negative contaminants that may be present. The plates are incubated at 37°C for 24–48 hours.

138 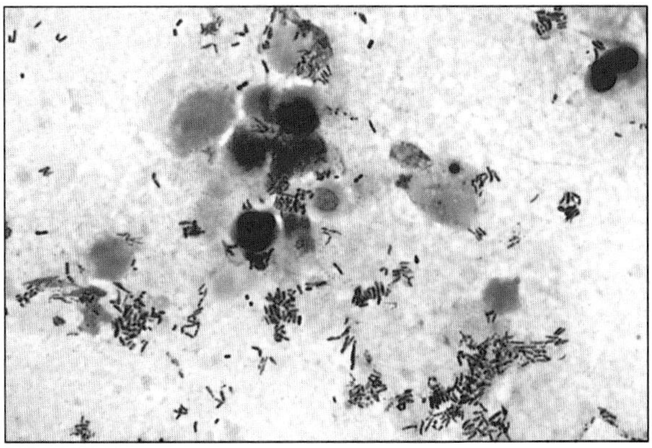

138 *C. renale* in a Gram-stained smear of bovine urine from a case of pyelonephritis. It shows extreme pleomorphism from club- shaped rods to coccal forms. (Gram stain, ×1000)

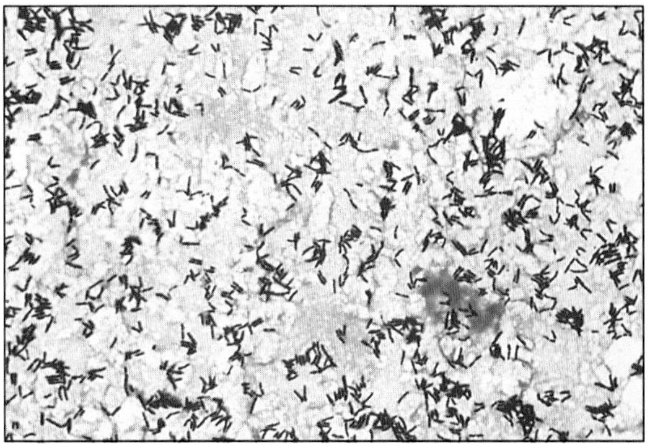 139

139 *Rhodococcus equi* in a smear of pus from a lung abscess in a case of suppurative bronchopneumonia in a foal. In this smear rod-shaped forms predominate. (Gram stain, ×1000)

Identification

Colonial morphology

- *C. bovis:* small, white, dry, non-haemolytic colonies that tend to appear in the wells of plates inoculated with a milk sample as it is a lipophilic bacterium (**140**).
- *C. kutscheri:* small, whitish colonies that bear a resemblance to those of *C. pseudotuberculosis*. Occasional strains are haemolytic (**141**).
- *C. pseudotuberculosis:* the colonies are small, white and dry (**142**). They can be surrounded by a narrow zone of haemolysis that may not appear until 48–72 hours' incubation (**143**). After several days' incubation the colonies can reach 3 mm in diameter and appear dry, crumbly and cream in colour.
- *C. renale:* colonies are non-haemolytic and very small after 24 hours' incubation. They become opaque and a dull yellow colour as they age.
- *C. pilosum:* the colonies resemble those of *C. renale* but become cream to yellow in colour with time (**144**).
- *C. cystitidis:* the colonies are similar to the above two species but are transparent to white in appearance.
- *R. equi:* the colonies are small, smooth, shiny and non-haemolytic after 24 hours' incubation but become larger, mucoid and salmon-pink in colour with age (**145** and **146**). A medium useful for enhancing the pigment of *R. equi* is formulated as follows: yeast extract 10 g; glucose 10 g; agar 15 g in 1 litre of **tap** water (**147**).

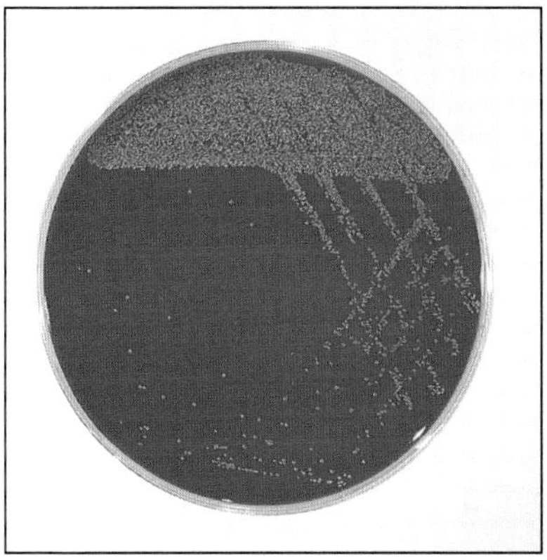 140

140 *C. bovis* on sheep blood agar: small, white, dry, non-haemolytic colonies after 48 hours' incubation.

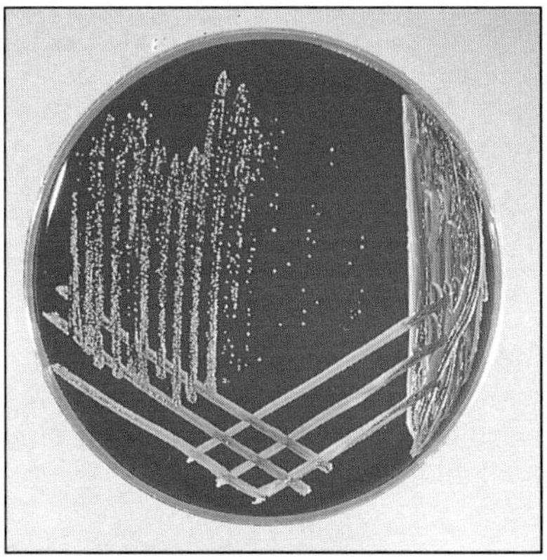 141

141 *C. kutscheri* on sheep blood agar: small, whitish colonies after 48 hours incubation. This strain, which is non-haemolytic, was isolated from a mouse colony where there were recurring deaths. Liver abscesses are a common postmortem finding in this disease.

142

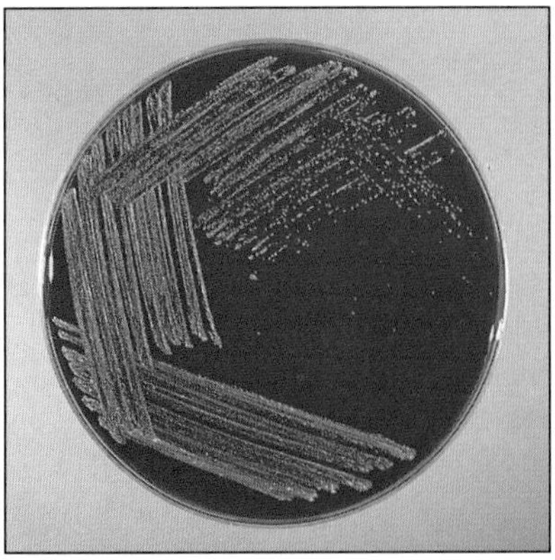

143

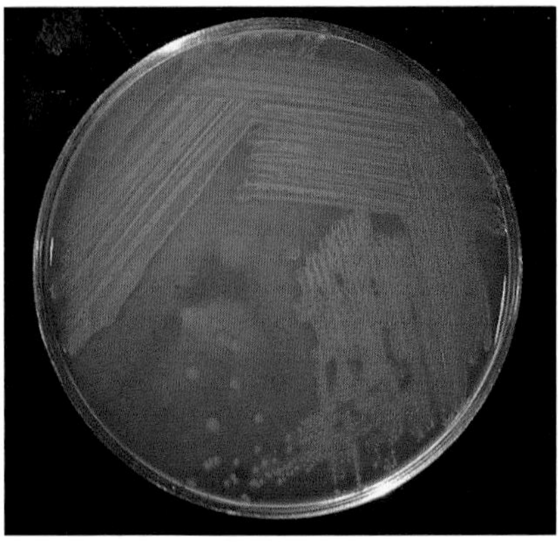

142 *C. pseudotuberculosis*: small, white and dry colonies on sheep blood agar. Non-haemolytic at 24 hours' incubation (*see* **143**).

143 *C. pseudotuberculosis* on sheep blood agar demonstrating haemolysis after 72 hours' incubation.

144

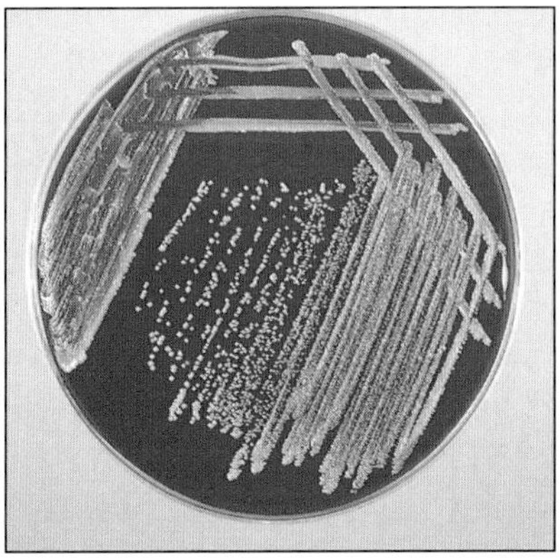

145

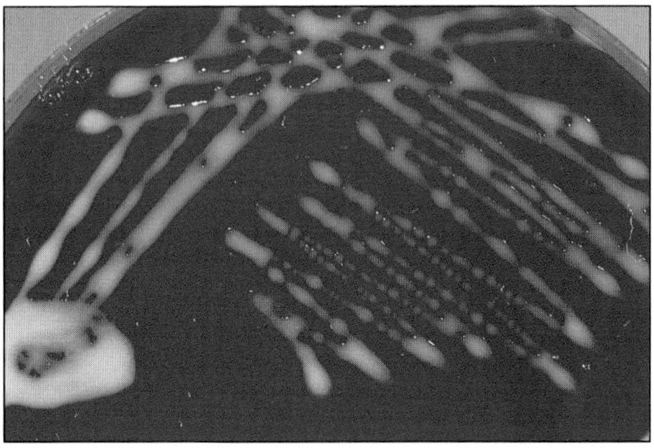

144 *C. pilosum* on sheep blood agar showing yellow pigmentation.

145 *Rhodococcus equi* on sheep blood agar showing the typical mucoid colonies (4-day culture). The salmon-pink pigmentation is not easily seen against a red background.

146

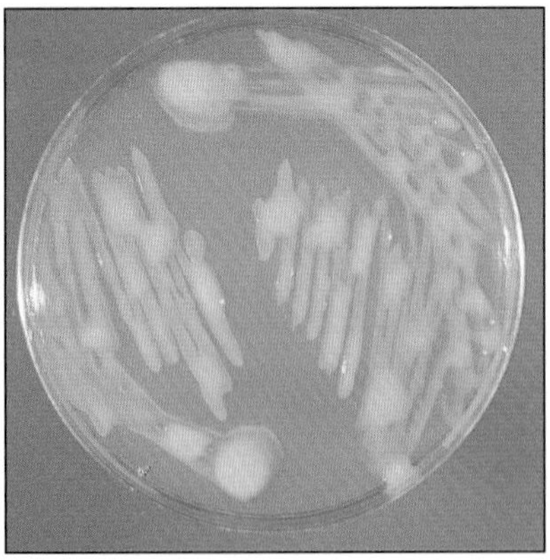

147

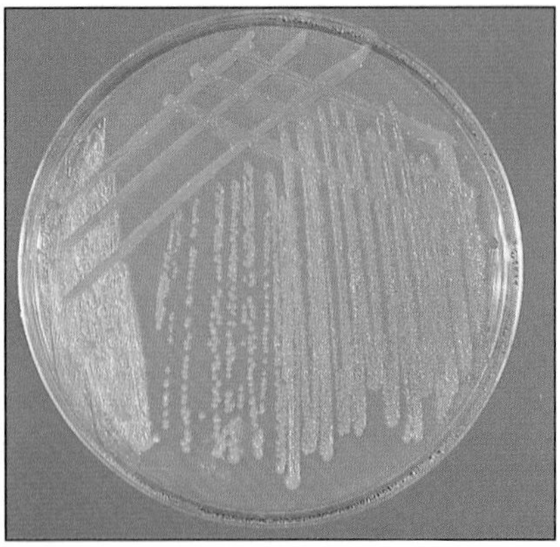

146 *Rhodococcus equi* on nutrient agar (4-day culture) demonstrating the mucoid colonies and salmon-pink pigmentation.

147 *R. equi* on pigment-enhancing medium after 48 hours' incubation, showing the typical (but enhanced) pigment and mucoid colonies.

Microscopic appearance

Gram-stained smears from colonies of *Corynebacterium* spp. will show pleomorphic Gram-positive rods. Smears of *R. equi* can have either cocci (**148**) or rods predominating.

CAMP tests

CAMP tests can be used as quick presumptive tests for *C. pseudotuberculosis*, *R. equi* and *C. renale*, interacting with the beta-haemolysin of *Staphylococcus aureus* (**149**). The results are as follows:

	Staphylococcal beta-haemolysin
C. pseudotuberculosis	Inhibition (**150**)
R. equi	Enhancement (**151**)
C. renale	Enhancement

There is a synergistic haemolytic phenomenon seen between *R. equi* ('*R. equi* factors') and *C. pseudotuberculosis* (**152**).

Biochemical tests

Definitive identification of the corynebacteria and *R. equi* is based on differential biochemical tests (**Table 30**).

148

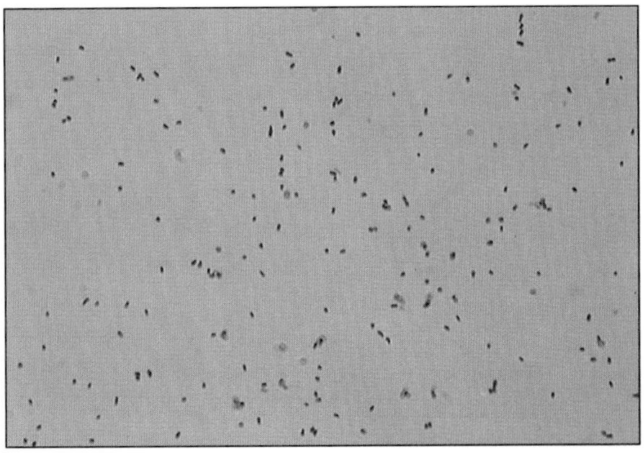

148 *R. equi* in a Gram-stained smear from a colony with coccal forms predominating. (×1000)

Table 30. Differentiation of corynebacteria and *Rhodococcus equi.*

	C. bovis	*C. kutscheri*	*C. pseudo-tuberculosis*	*C. renale* group	*R. equi*
Beta-haemolysis	–	v	+	–	–
Aesculin hydrolysis	–	+	–	–	–
Nitrate reduction	–	+	v*	v	+
Urease	–	+	+ (>18 hours)	+ (<1 hour)	+ (>18 hours)
Casein digestion	–	–	–	v	–
Glucose	+	+	+	+	–
Maltose	–	+	+	–	–
Sucrose	–	+	–	–	–

+ = positive reaction; – = negative; v = variable; v* = equine strains positive and ovine strains negative.

149

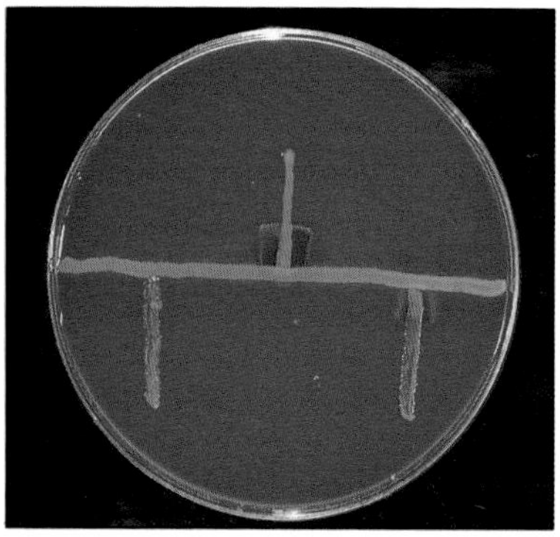

149 CAMP test with a *Staphylococcus aureus. C. pseudotuberculosis* (left), *R. equi* (centre) and *C. renale* (right). The latter two bacteria give an enhancement of the effect of the staphylococcal beta-haemolysin.

150

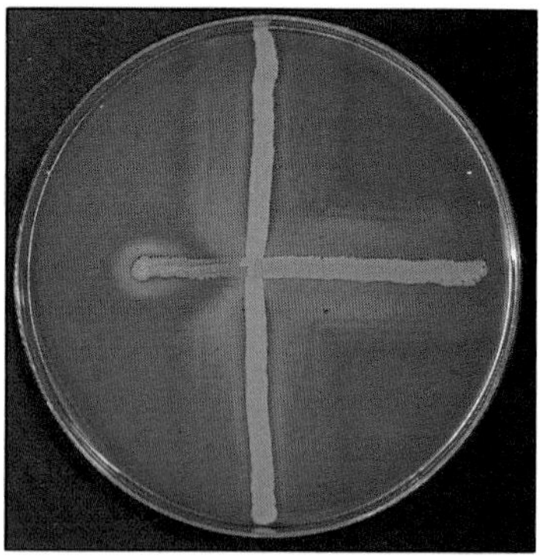

150 *C. pseudotuberculosis* drawn across (left to right) *Staphylococcus aureus,* demonstrating inhibition of the effect of the staphylococcal haemolysins.

151

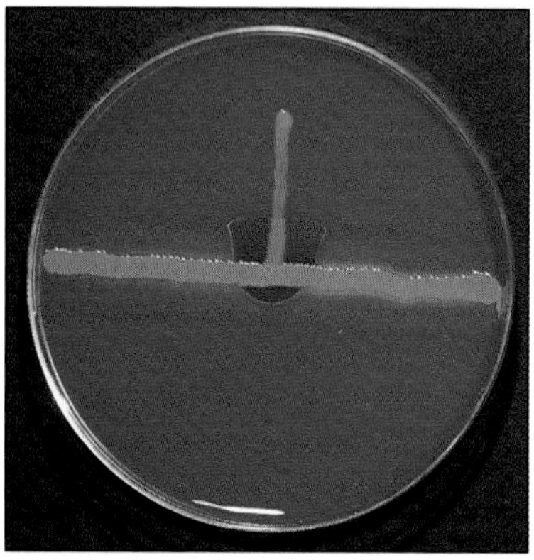

151 CAMP test with *R.equi* against *S. aureus* (horizontal) showing the typical shovel-shaped enhancement of the effect of the staphylococcal betahaemolysin that tends to extend to the opposite side of the *S. aureus* streak.

152

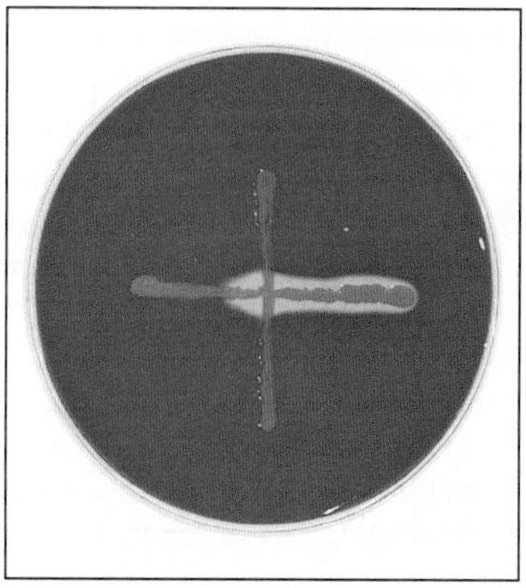

152 *C. pseudotuberculosis* (horizontal) drawn across *R. equi* (left to right) demonstrating synergistic haemolysis.

153

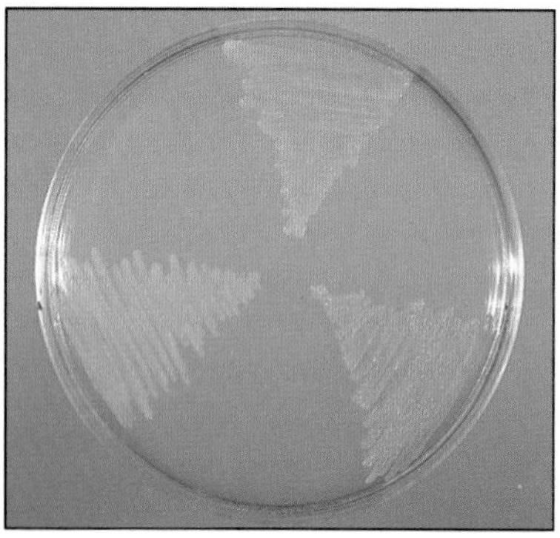

153 Pigments of the *C. renale* group: *C. pilosum* (top) distinctly yellow; *C. cystitidis* (left) white and *C. renale* (right) a dull yellow colour (48 hours' incubation).

154

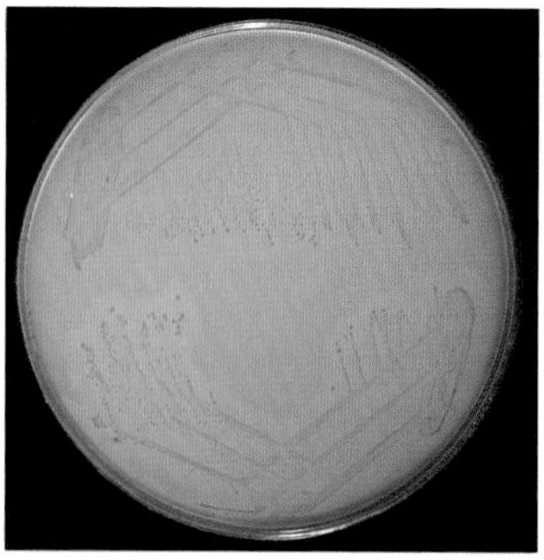

154 *C. renale* (bottom) showing casein digestion on milk agar. *C. pilosum* (top) and *C. cystitidis* do not give this reaction.

155

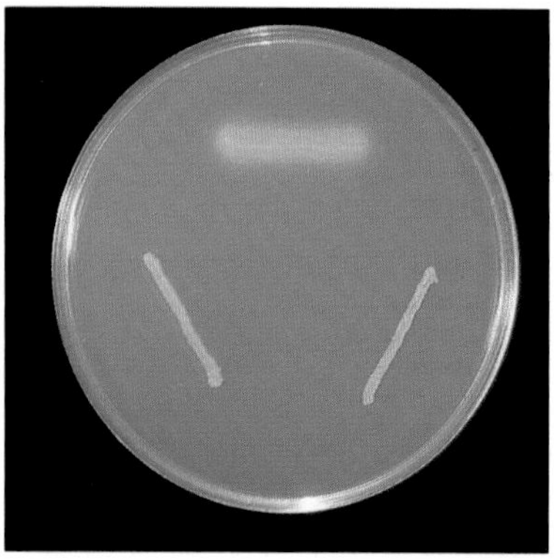

155 Hydrolysis of Tween 80 by *C. cystitidis* (top) on Tween 80 medium. Both *C. renale* (left) and *C. pilosum* (right) are negative.

Table 31. Differentiation of the *Corynebacterium renale* group.

	C. renale	*C. pilosum*	*C. cystitidis*
Colony colour (48 hours) (153*)	Yellowish	Yellow	White
Growth at pH 5.4 (broth)	+	–	–
Nitrate reduction	–	+	–
Casein digestion (154*)	+	–	–
Hydrolysis of Tween 80 (155*)	–	–	+
Acid from xylose	–	–	+
Acid from starch	–	+	+

+ = positive reaction; – = negative reaction; * denotes relevant photograph

Differentiation of members of the *C. renale* group
This is based on the colonial pigmentation (**153**), presence or absence of growth at pH 5.4 and biochemical reactions (**Table 31**). The formula of the media for demonstating casein digestion (**154**) and hydrolysis of Tween 80 (**155**) are given in **Appendix 2.**

11 The Actinomycetes

The actinomycetes comprise a heterologous group of procaryotes that have the ability to form Gram-positive, branching filaments of less than 1 µm in diameter. Fungi are eucaryotes and their filaments (hyphae) are always greater than 1 µm in width. The main animal pathogens in the actinomycetes are in the genera *Actinomyces*, *Nocardia* and *Dermatophilus*. *Nocardia* is closely related to *Corynebacterium*, *Mycobacterium* and *Rhodococcus* but *Actinomyces* differs from these in its DNA guanine/cytosine ratio and in the chemical composition of its cell wall. Other genera included in the actinomycetes are *Streptomyces* and *Actinomadura;* some species of each produce mycetomas in humans but are rarely pathogenic for animals. *Streptomyces* species are prolific producers of antimicrobial substances (**156**) and are common contaminants on laboratory agar media. Species in the genera *Micropolyspora* and *Thermoactinomyces* are not invasive but inhalation of their spores can cause allergic pulmonary disease in man and horses and possibly in other domestic animals fed or exposed to mouldy hay. The general characteristics of the genera *Actinomyces*, *Nocardia*, *Streptomyces* and *Dermatophilus* are presented in **Table 32.**

Table 32. General features of the actinomycetes.

Characteristics	*Actinomyces*	*Nocardia*	*Streptomyces (non-pathogenic)*	*Dermatophilus*
Atmospheric requirement	Anaerobic or capnophilic*	Strict aerobe	Aerobe	Aerobe/capnophilic*
Catalase	–(+)	+	+	+
Partially acid-fast (MZN-positive)	–	+	–	–
Motility	–	–	–	+ (zoospores)
Growth on Sabouraud dextrose agar	–	+	+	–
Aerial filaments	–	+	+	–
Spores	–	+ (conidia)	+ (conidia)	+ (zoospores)
Fragmentation of filaments	–	+	+	+
Odour of colonies	–	–	Pungent and earthy	–
Metabolism	Fermentative	Oxidative	Oxidative	Weakly fermentative
Reservoir	Oral mucosa and nasopharynx	Soil	Soil and common laboratory contaminant	Foci on carrier animal or within scabs in environment
Veterinary importance	Focal or systemic disease with characteristic granulomas in skin, subcutaneous tissue and internal organs		Non-pathogenic but similar to *Nocardia* in cultures. Some species produce antibiotics	Skin disease

* capnophilic (carboxyphilic) = carbon dioxide required for maximum growth, as opposed to microaerophilic, used to describe an organism that requires a reduced oxygen tension. + = positive reaction; – = negative; –(+) = most species are negative (a few positive).

156

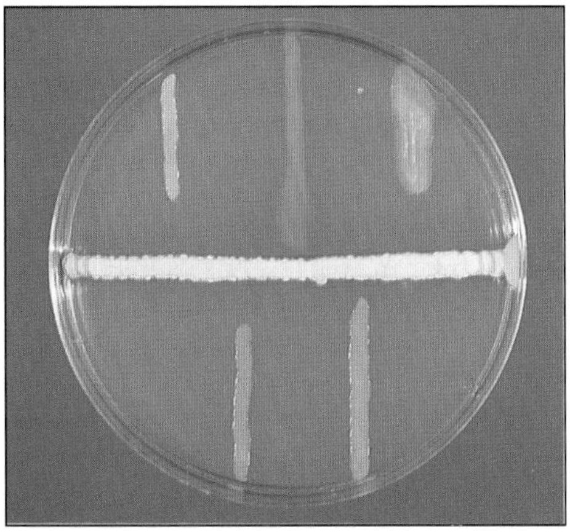

156 *Streptomyces* species (horizontal streak) on Isosensitest agar demonstrating antimicrobial activity against other bacteria that were streaked to within 2 mm of the *Streptomyces* species: *S. aureus* (top left), *Pseudomonas aeruginosa* (top centre), *Bacillus cereus* (top right), *Salmonella* species (bottom left) and *E. coli* (bottom right). *Pseudomonas aeruginosa* is the only bacterium not inhibited by the antimicrobial factor(s) from the *Streptomyces* species.

Changes in Nomenclature

Previous Name	**Present Name**
Corynebacterium pyogenes	*Actinomyces pyogenes*
Nocardia caviae	*Nocardia otitidis-caviarum*

Natural Habitat

The *Actinomyces* species are present on mucous membranes of the host animal, often in the oral cavity or nasopharynx. *Nocardia* and *Streptomyces* species are soil microorganisms. *Dermatophilus congolensis*, the only species in the genus, is thought to maintain itself in small foci of infection on a carrier animal or within scab particles in dust. It can survive in scab material for periods up to 3 years. The pathogenic actinomycetes have a worldwide distribution but severe clinical disease caused by *D. congolensis* is most common in tropical and subtropical regions.

Pathogenicity

Actinomyces species

Infections by these organisms tend to be endogenous and most of the species cause pyogranulomatous reactions in animal tissues.

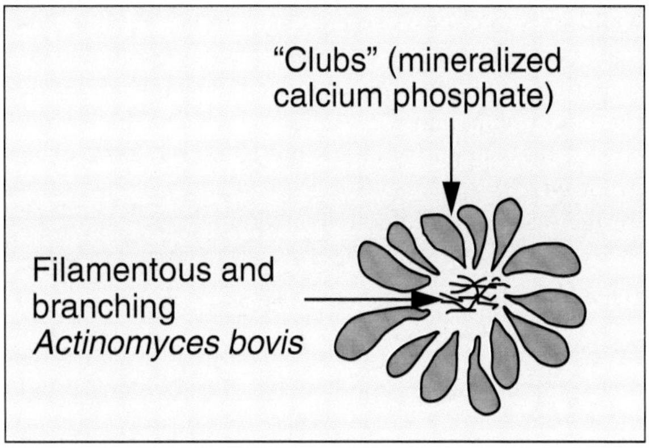

Diagram 26. A club colony.

A. bovis gains access to the alveolar region of the jaw in cattle from the oral cavity, probably through trauma to the mucosa. It initiates a rarefying osteomyelitis and soft tissue reaction, the condition being referred to as 'lumpy jaw'. Bacterial colonies form in the tissues with 'clubs' of mineralised calcium phosphate forming around them to create microscopic 'club colonies' or 'rosettes'. **Diagram 26** shows an illustration of a 'club colony'. The club formation is the result of phosphatase activity and is a host reaction to a chronic infection. Granulation, mononuclear infiltration and fibrosis occur in the lesions with sinus tracts leading to the outside. Exudate from the tracts contains pus with 'sulphur granules' that are about 1-2 mm in diameter, within which club colonies can be found if the granules are crushed and examined microscopically. The name '*Actinomyces suis*' has been proposed for the *A. bovis* strains isolated from sows with granulomatous lesions in the mammary glands. The porcine strains have minor biochemical and antigenic differences from the bovine strains.

A. viscosus causes clinical syndromes in dogs indistinguishable from that initiated by *Nocardia asteroides*. Two syndromes can occur, either separately or together. One is a localised granulomatous lesion involving skin and subcutis, the other is a pyothorax with granulomas in the thoracic cavity and often a large accumulation of sanguineopurulent pleural fluid containing soft white granules about 1 mm in diameter.

Nocardia species

These aerobic, and essentially saprophytic, bacteria cause suppurative and pyogranulomatous reactions in immunosuppressed hosts or animals that have been exposed to large doses of the bacterium. The pathogenic nocardiae survive within phagocytic vacuoles by preventing phagolysosome formation. This is probably due to the surface lipids as *Nocardia* species have a cell wall similar to the mycobacteria. Other cell wall lipids may provoke the characteristic granulomatous reaction. Exudates are sanguineopurulent and can sometimes contain soft granules consisting of bacteria, neutrophils and debris. They lack the microstructure of the sulphur granules produced by some of the *Actinomyces* species. *Nocardia asteroides* accounts for the majority of infections in animals.

Dermatophilus congolensis

D. congolensis causes skin infections that can affect many animal species and humans, but the condition is most commonly seen in cattle, sheep, goats, horses and polar bears in zoological collections. The infection is characterised by the formation of thick crusts which come away easily with a tuft of hair, leaving a moist, depressed area with bleeding points from capillaries. Infections can be localised but have a tendency to spread over large areas of the body and the morbidity and mortality can be high, especially in tropical regions. The position of the lesions varies with the predisposing conditions. In periods of high rainfall the lesions tend to occur along the backs of animals. Where there is a heavy infestation with *Amblyomma* ticks the lesions are present in the predilection sites of the ticks: dewlap, axillae, udder, scrotum and escutcheon. In the dry season, in tropical regions, when feed is scarce the lesions are on the muzzle, head and lower limbs due to the animals foraging in thorn-covered scrub.

The pathogenic actinomycetes and the diseases that they cause are summarised in **Table 33**.

Table 33. Diseases caused by the pathogenic actinomycetes.

Actinomycete	Host(s)	Disease
Actinomyces bovis	Cattle	Bovine actinomycosis ('lumpy jaw')
	Horses	Poll evil and fistulous withers (supra-atlantal and supraspinous bursitis, respectively). Can occur as a mixed infection with *Brucella* species
'A. suis'	Pigs	Pyogranulomatous mastitis
A. viscosus	Dogs	Canine actinomycosis: 1 Localised cutaneous granulomatous abscess and/or 2 Pyothorax and granulomas in the thoracic cavity
A. pyogenes	Cattle, sheep and pigs mainly	Chronic or acute suppurative mastitis, suppurative pneumonia, septic arthritis, vegetative endocarditis (cattle), endometritis, umbilical infections, wound infections and seminal vesiculitis (bulls and boars). Common in mixed infections with *Fusobacterium necrophorum.* 'Summer mastitis' (cattle), a mixed infection with an anaerobe such as *Peptostreptococcus indolicus*
A. hordeovulneris	Dogs	Localised abscesses and systemic infections such as pleuritis, peritonitis, visceral abscesses and septic arthritis. Often associated with tissue-migrating awns of the grass *Hordeum* species ('foxtails') which are common in Western USA
A. israelii	Humans and rarely pigs and cattle	Human actinomycosis. Bovine or porcine actinomycosis (rare)
Nocardia asteroides	Dogs (cats)	Canine nocardiosis: 1 Localised cutaneous granulomatous abscesses and/or 2 Pyothorax and granulomas in the thoracic cavity
	Cattle	Chronic granulomatous mastitis
	Pigs, sheep, goats and others	Less frequent infections: pneumonia, mastitis and lymphadenitis
	Whales, dolphins and birds	Uncommon infections: respiratory involvement with dissemination to other tissues

(*continued*)

Table 33. Diseases caused by the pathogenic actinomycetes (continued).

Actinomycete	Host(s)	Disease
'N. farcinica'	Cattle	Bovine farcy in tropical regions
N. brasiliensis N. otitidiscaviarum	Man and some animal species	Rare infections in animals
Dermatophilus congolensis	Cattle, horses, sheep and goats mainly, but many animal species and man can be infected	The disease has many names: 'rain-scald', streptothricosis, dermatophilosis and in the sheep: 'mycotic dermatitis' (general infection), 'lumpy wool' (wool-covered skin) and 'strawberry foot rot' (skin of lower leg and coronet). Skin infection, either focal or spreading over large areas of the body. Most common in tropical and subtropical regions
Micropolyspora faeni	Horses	The spores, in hay, of this thermophilic actinomycete have been associated with the allergic part of chronic obstructive pulmonary disease

Laboratory Diagnosis of *Actinomyces* species

Specimens

Specimens include pus, exudates, aspirates, tissue and scrapings from the walls of abscesses if they have been incised. A volume of fluid or pus should be collected and submitted, if possible, rather than just a small amount on a swab. Thin sections of granulomas in 10 per cent formalin are useful for histopathology.

Direct microscopy

Sulphur granules are the best specimens for direct examination in infections caused by *A. bovis* or *A. viscosus.* The pus or exudate is placed in a Petri dish and washed carefully with a little distilled water to expose the yellowish sulphur granules of *A. bovis* (**157**) or the softer greyish-white granules of *A. viscosus.* A granule is placed on a microscope slide in a drop of 10 per cent KOH and gently crushed by applying pressure on the coverslip. The characteristic clubs can be seen if the preparation is examined under the low power objective of a microscope. A view of a club colony, *in situ*, can be seen in stained histological sections (**158**). If smears are made from the granules and stained with the Gram stain, delicate, Gram-positive, branching filaments (**159**) can be observed. Occasionally short filaments or pleomorphic diphtheroidal forms may predominate.

Gram-stained smears of pus or mastitic milk samples in *A. pyogenes* infections usually reveal large numbers of small, highly pleomorphic, Gram-positive forms (**160**). They tend to be a mixture of cocci, rods and pear-shaped cells. Occasionally short branching forms may be seen.

157

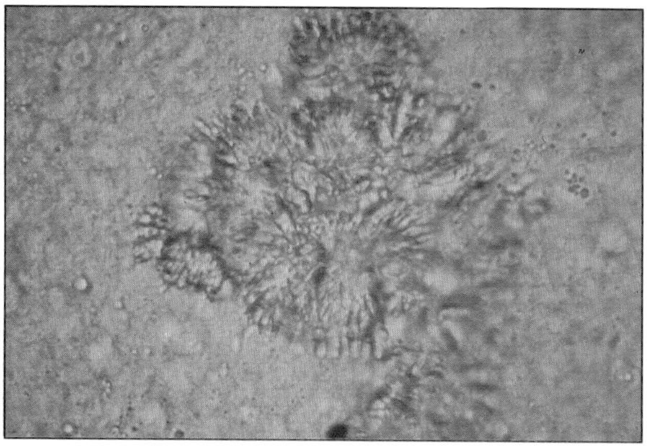

157 Sulphur granule in pus from an *Actinomyces bovis* infection. (Unstained, ×25)

158

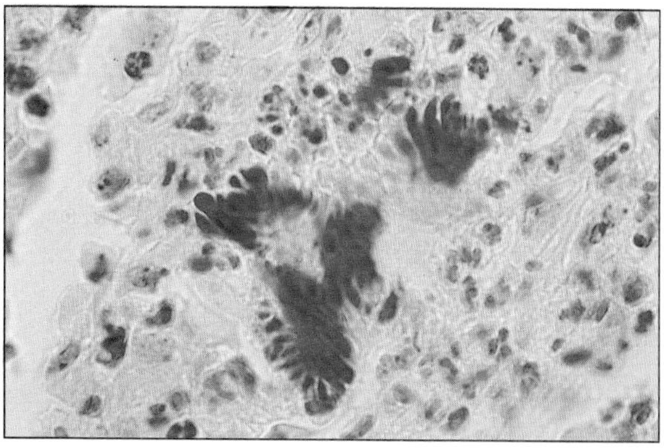

158 Club colonies in a tissue section from an *A. bovis* infection in a cow with lumpy jaw. (Plaut stain, ×400)

159 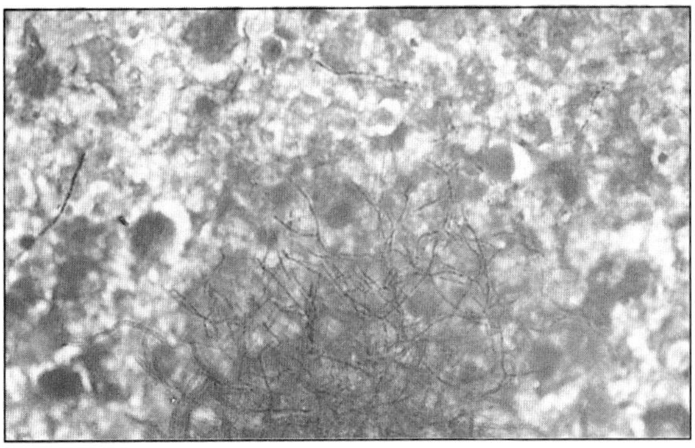

159 Gram-positive branching filaments of *A. bovis* in bovine tissue. (Gram stain, ×1000)

160 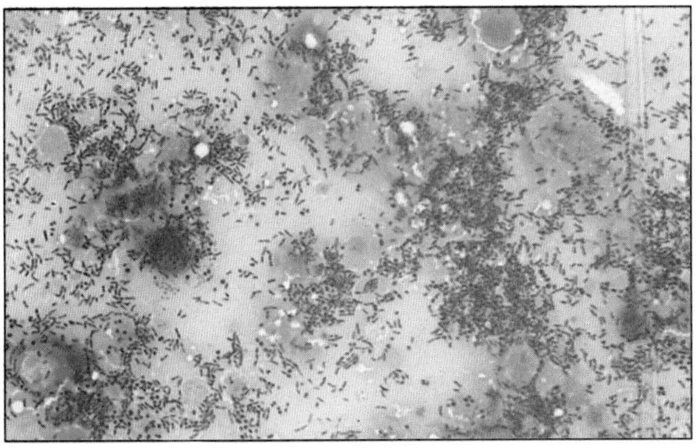

160 Gram-positive pleomorphic rods of *A. pyogenes* in a bovine mastitic milk sample. (Gram stain, ×1000)

Isolation

The *Actinomyces* species grow well on sheep or ox blood agar. *A. bovis* requires anaerobic conditions with 5–10 per cent CO_2 added ($H_2 + CO_2$ commercial envelope). *A. viscosus* and *A. pyogenes* will grow aerobically but 5–10 per cent CO_2 will enhance their growth. They are all incubated at 37°C. *A. bovis* and *A. viscosus* usually require 2–4 days but the growth of *A. pyogenes* can usually be seen in 24 hours. The isolation of *A. hordeovulneris* is greatly enhanced by 10–20 per cent foetal calf serum, although it will grow on blood agar at 37°C under 10 per cent CO_2 with either aerobic or anaerobic conditions. Colonies are visible in about 3 days.

Identification

Colonial morphology and microscopic appearance

- *A. bovis* colonies are non-haemolytic, white, rough or smooth and adhere tenaciously to solid medium (**161**). The colonies never attain a diameter of much more than 1 mm. Gram-stained smears show Gram-positive, slightly branched filaments or short forms. On subculture the bacterium may become diphtheroidal or coccobacillary (**162**). *A. bovis* grows well in thioglycollate medium giving a diffuse growth in about 7–10 days (**163**).
- *A. viscosus* commonly produces two colonial forms, one being smooth, entire, convex and glistening and the other is smaller, rough, dry and irregular (**164**). Neither are haemolytic. The larger colonial type yields Gram-positive diphtheroidal forms (**165**) and the smaller colony has short branching filaments (**166**).

161 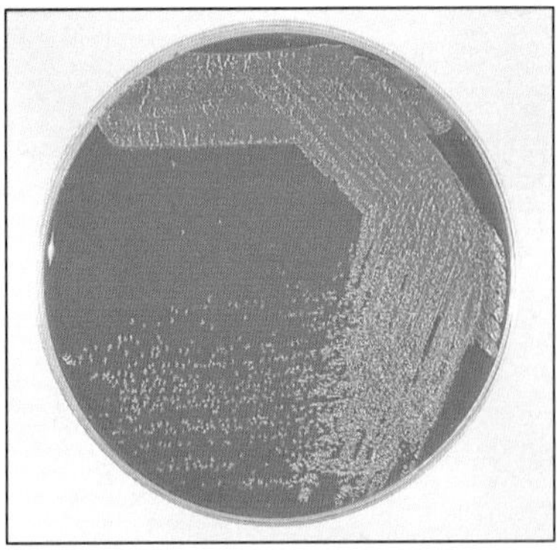

161 *A. bovis* on sheep blood agar after 4 days' incubation.

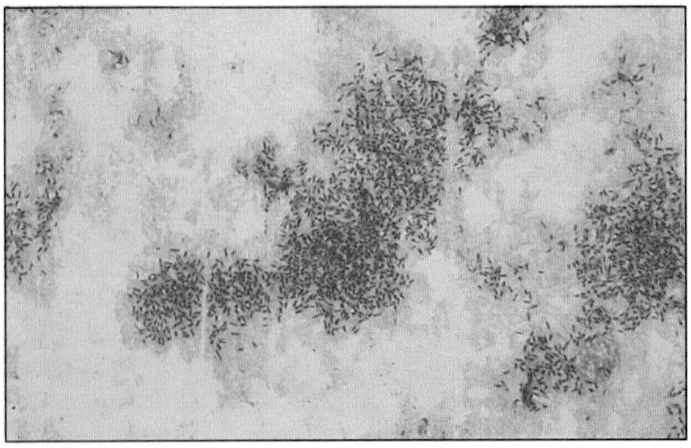 16.

162 A Gram-stained smear from an *A. bovis* culture showing diphtheroidal forms typical of repeatedly subcultured isolates. (×1000)

- *A. pyogenes:* after 24 hours' incubation a hazy haemolysis may be noticed along the streak lines before the minute colonies are visible. At 48 hours' incubation the tiny 1 mm colonies are visible surrounded by a narrow zone of complete haemolysis (**167**). A Gram-stained smear reveals the typical pleomorphic, Gram-positive rods (**168**).
- *A. hordeovulneris* colonies are about 2 mm in diameter after 72 hours' incubation. They are white, non-haemolytic, 'molar-toothed' (with ridges and valleys) but become conical and domed after further incubation. The colonies adhere firmly to the agar and a weak haemolysis may be produced after 7 days' incubation.

163

163 *A. bovis* in thioglycollate medium after 10 days' incubation showing characteristic diffuse growth.

164

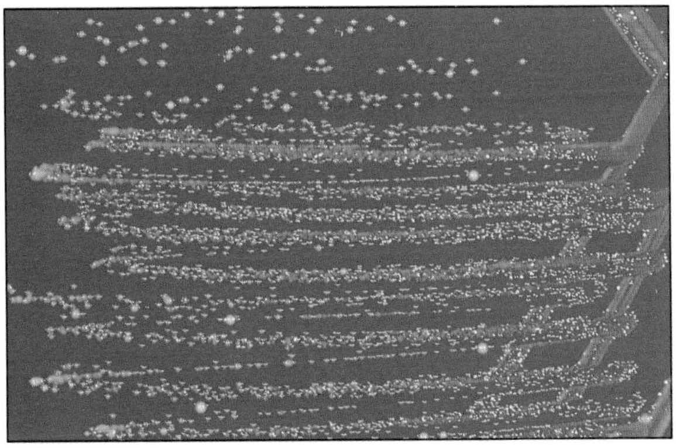

164 Close-up of *A. viscosus* on sheep blood agar showing the two colonial types. The smaller, dry, irregular type is predominating but about ten of the larger, smooth, glistening colonial variants are evident.

165

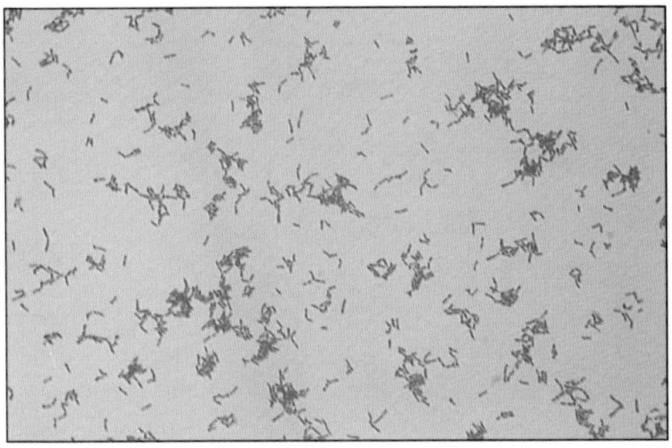

165 Gram-stained smear of the larger, smooth colonial variant of *A. viscosus* yielding Gram-positive diphtheroidal forms. (×1000)

166

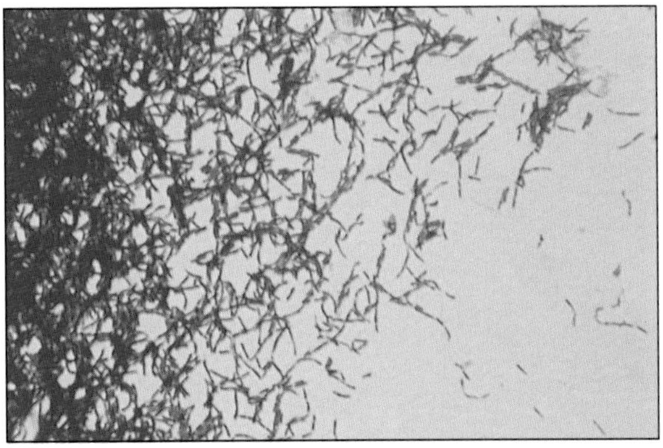

166 Gram-stained smear of the smaller, dry colonial variant of *A. viscosus* showing short branching filaments. (×1000)

167

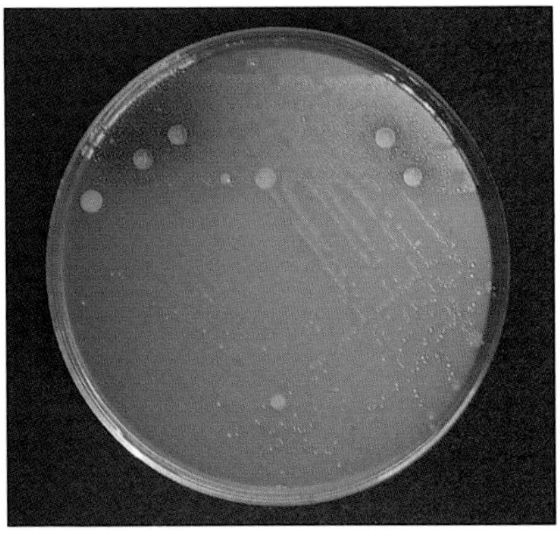

167 Small haemolytic colonies of *A. pyogenes* on a sheep blood agar diagnostic plate after 72 hours' incubation (large colonies are contaminants).

Biochemical tests

Specialised methods are required for the identification of most of the *Actinomyces* species and are usually performed only in reference laboratories. These laboratories often use a fluorescent antibody technique to differentiate the species. A rapid, presumptive test for *A. pyogenes* is to demonstrate its ability to pit a Loeffler serum slope in 24–48 hours (**169**). A loopful of a pure culture of the bacterium is taken and a heavy inoculum is made in a small area in the centre of the slope, taking care not to break the surface of the medium. The medium is incubated at 37°C for 24-48 hours. *A. pyogenes* will also give a positive CAMP test with *Staphylococcus aureus* (**170**). **Table 34** lists the main features of the *Actinomyces* species affecting animals.

168

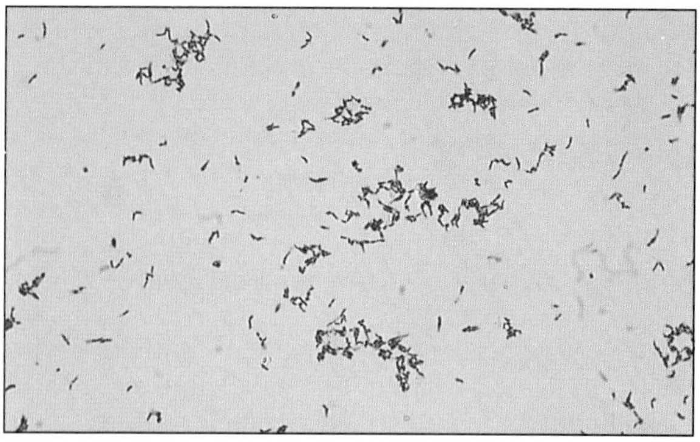

168 A Gram-stained smear from an *A. pyogenes* culture showing the typical pleomorphic appearance. (×1000)

169

169 Loeffler serum slope: 'pitting' of slope by *A. pyogenes* (left), uninoculated or negative slope (right).

170

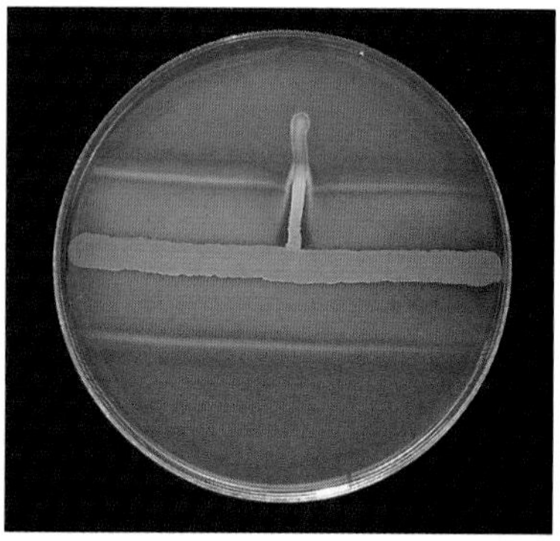

170 CAMP test with *A. pyogenes* against *Staphylococcus aureus* (horizontal) showing enhancement of the effect of the staphylococcal beta-haemolysin.

Table 34. Characteristics of *Actinomyces* species causing disease in animals.

	Actinomyces bovis	***A. viscosus***	***A. pyogenes***	***A. hordeovulneris***
Granules in pus	+ (sulphur granules)	+ (soft and whitish)	–	–(+)
Direct microscopic examination	F(D)	F/D	D	F/D
Atmospheric requirement	Anaerobic + CO_2	Aerobic (CO_2)	Aerobic (CO_2)	Aerobic + CO_2 or anaerobic + CO_2
Catalase	–	+ (weak)	–	+ (weak)
CAMP (*Staphylococcus aureus)*	–	–	+	–
Pitting of Loeffler serum slope	–	–	+	–

F = filamentous and branching form, D = shorter diphtheroidal form. CO_2 = growth enhanced by carbon dioxide.

Summary of the features allowing a presumptive identification of the Actinomyces species

- The clinical and pathological findings and the animal species affected.
- The presence of granules in pus or exudates (*A. bovis* and *A. viscosus*).
- Demonstration of fine Gram-positive, branching filaments or pleomorphic diphtheroidal forms on direct microscopic examination. To distinguish *Actinomyces* species from *Nocardia* species:
  a) The filaments of *Actinomyces* species do not fragment into bacillary forms.
  b) *Nocardia asteroides* is partially acid-fast (MZN-positive) but the *Actinomyces* species are MZN-negative.
- The presence of club-colonies in histopathological sections (*A. bovis*), although these can also occur in some other chronic infections such as bovine actinobacillosis and botryomycosis (*Staphylococcus aureus*).
- Isolation of the *Actinomyces* species on blood agar, under the appropriate atmospheric conditions, and with a consistent colonial appearance. The *Actinomyces* species cannot grow on Sabouraud dextrose agar while *Nocardia asteroides* tolerates this medium.
- The characteristic appearance in Gram-stained smears from the colonial growth that fits the *Actinomyces* species under investigation.
- Pitting of Loeffler serum slope and CAMP test for the presumptive identification of *A. pyogenes*.

Laboratory Diagnosis of *Nocardia asteroides*

Specimens

Specimens should include exudates, aspirates, mastitic milk samples, tissue from granulomas and thin sections from granulomas in 10 per cent formalin for histopathology.

Direct microscopy

Soft granules are not common in exudates from *N. asteroides* infections. Gram and MZN-stained smears are made from exudates, aspirates, granulomatous tissue and from centrifuged deposits of bovine mastitic milk. Gram-stained smears reveal Gram-positive branching filaments that often show some fragmentation into coccobacillary elements. The MZN-stained smears exhibit a similar morphology but most of the filaments retain the carbol fuchsin dye and stain red (**171**). Microcolonies of *N. asteroides* can be demonstrated in stained histopathological tissue sections (**172**).

Isolation

Blood agar plates are inoculated with the specimens and incubated aerobically at 37°C for up to 7 days, although growth should be evident in 4–5 days. Any suspect colonies could be used to heavily inoculate a Sabouraud dextrose agar plate. This is incubated at 37°C for up to 10 days.

Identification

Colonial appearance
The colonies on blood agar are often a vivid white and powdery if aerial filaments and spores are formed (**173**). Occasionally the colonies are smooth, heaped and variably pigmented. Both types of colonies are firmly adherent to the agar surface. The colonies on Sabouraud dextrose agar are dry, wrinkled and yellow, becoming deep orange colour with age (**174**). The non-pathogenic *Streptomyces* species form white, powdery colonies, embedded in the medium, on blood agar and nutrient agar (**175**). The colonies are very similar to those of *N. asteroides*. *Streptomyces* species are also able to grow on Sabouraud dextrose agar. However, *Streptomyces* species have a characteristic and powerful earthy odour.

Microscopic appearance
Gram-stained smears from colonies show Gram-positive branching filaments that characteristically break up into rods or coccobacillary elements with age. An MZN-stained smear from young cultures reveals red-staining, branching filaments (**176**). The filaments of *Streptomyces* and *Actinomyces* species are MZN-negative and stain blue with the counter-stain.

Biochemical reactions
Tests such as decomposition of casein, hypoxanthine, tyrosine, urea and xanthine are carried out in reference laboratories to differentiate between the *Nocardia* species. Seven immunotypes of *N. asteroides* have been identified.

Differentiation of *A. viscosus* and *N. asteroides*
This is important in infections in dogs as the two bacteria cause indistinguishable clinical syndromes. But although *A. viscosus* infections respond well to penicillin and other commonly used antibiotics, nocardial infections are often refractory to treatment and *N. asteroides* is susceptible only to a comparatively limited range of antimicrobial agents. Trimethoprim-sulfamethoxazole or erythromycin have been suggested as useful therapeutic agents in nocardial infections. **Table 35** summarises the differences in laboratory findings for canine nocardiosis and canine actinomycosis.

171

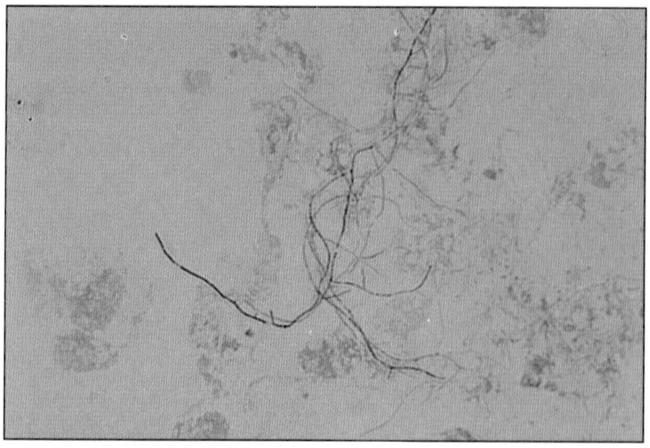

171 An MZN-stained smear of *Nocardia asteroides* in a canine thoracic aspirate: red (MZN-positive) branching filaments. (×1000)

172

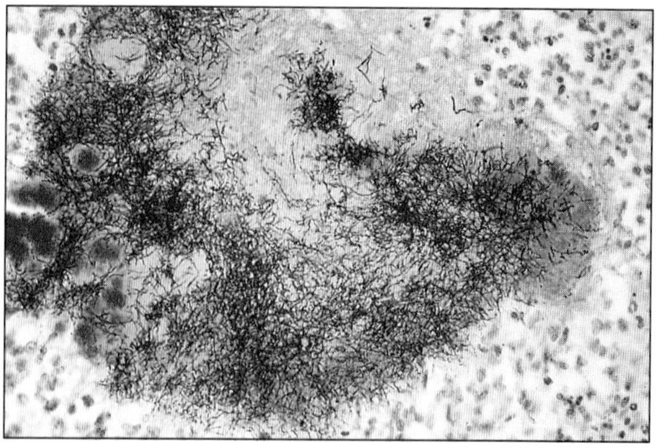

172 A Gram-stained section of a thoracic granuloma in a dog showing a microcolony of *N. asteroides*, present as branching Gram-positive filaments. (×400)

173

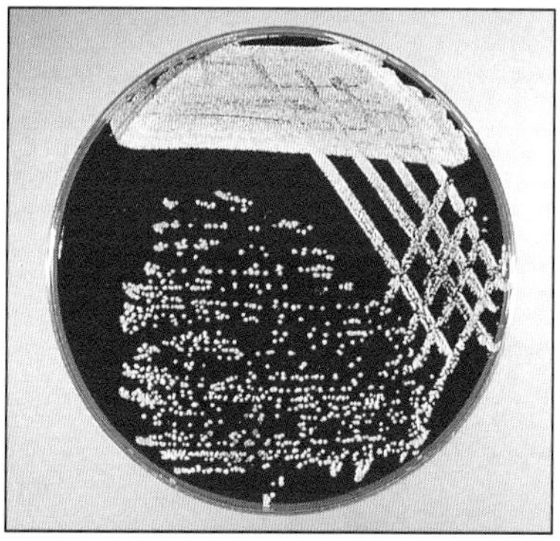

173 *N. asteroides* on sheep blood agar after 5 days' incubation. The vivid white, powdery colonies are firmly adherent to the medium.

174

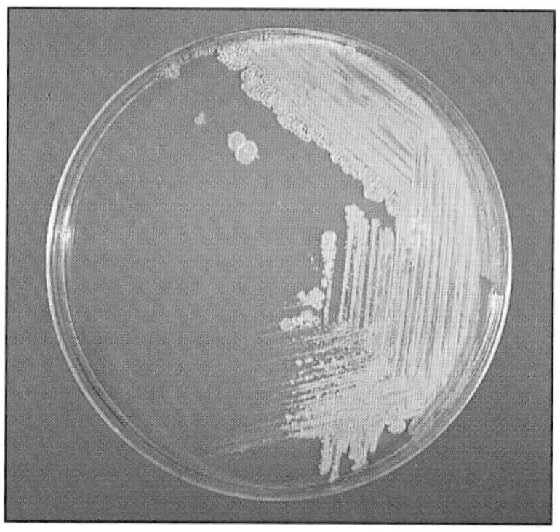

174 *N. asteroides* on Sabouraud agar after 7 days' incubation. The colonies are orange, dry and wrinkled.

175

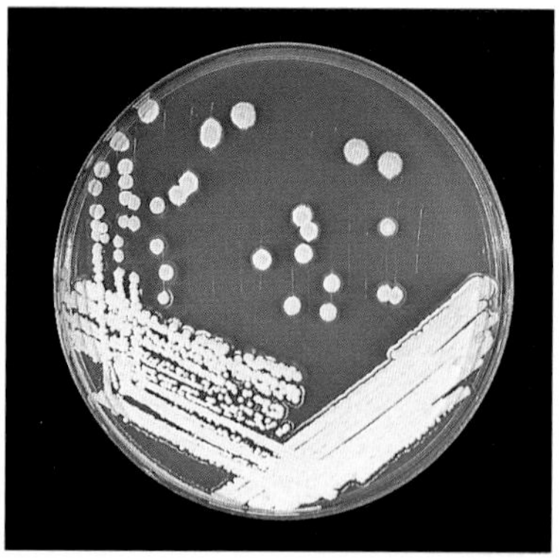

175 *Streptomyces* species on nutrient agar showing the white, powdery colonies, similar to those of *N. asteroides*. *Streptomyces* species have a characteristic earthy odour.

176

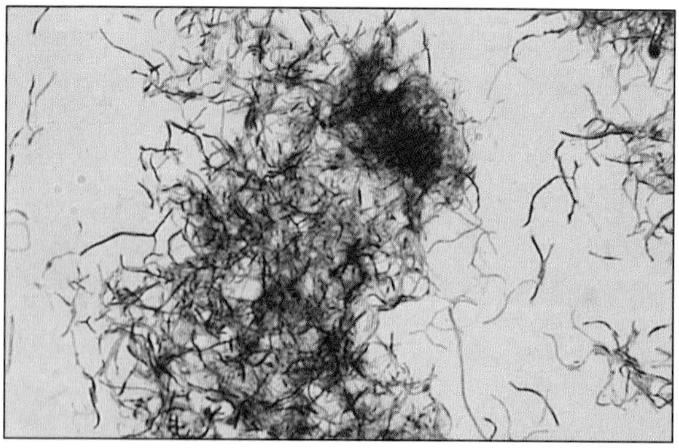

176 An MZN-stained smear from a culture of *N. asteroides* showing red (MZN-positive) filaments. (×1000)

Table 35. Differentiation of the causative agents of canine nocardiosis and canine actinomycosis.

	Canine nocardiosis	*Canine actinomycosis*
Aetiology	*Nocardia asteroides*	*Actinomyces viscosus*
Granules in exudates	Not common	Usually present
Filaments MZN-positive	+	–
Fragmentation of filaments	+	–
Growth on Sabouraud dextrose agar	+	–
Powdery, white colonies (aerial hyphae)	+	–
Susceptibility to penicillin	–*	+

* trimethoprim-sulphamethoxazole, sulphonamides or erythromycin have been suggested for treatment.

Laboratory Diagnosis of *Dermatophilus congolensis*

Specimens

A tuft of hair that is plucked from the lesion usually detaches with scab material adhering to it (**177**).

177

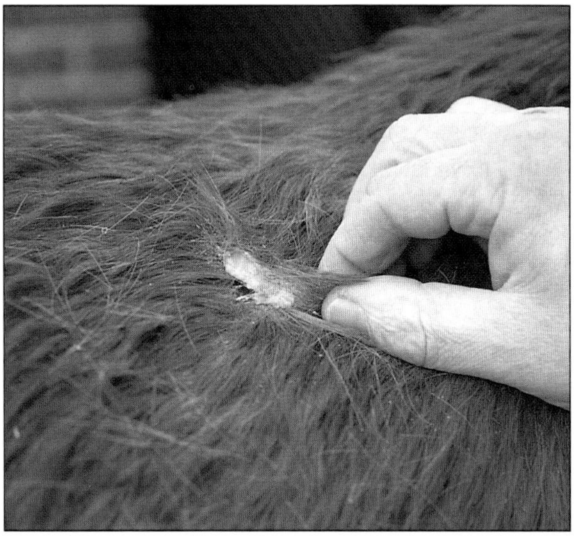

177 A tuft of hair being plucked from a horse with streptothricosis (*Dermatophilus congolensis*). The hairs came away with an adherent crust. The uneven appearance of the coat correlates with the distribution of lesions.

Direct microscopy

Small pieces of material are shaved from the scab with a scalpel and the flakes of scab are softened in a few drops of distilled water on a microscope slide. A smear is made, taking care to leave a few flakes of scab material intact. The smear can be stained by either Giemsa or Gram stains. Giemsa is the better stain to show the characteristic morphology of the bacterium (**178**). If the conventional Gram stain is used, both the cells of *D. congolensis* and debris seem to absorb the crystal violet-iodine complex avidly and stain too darkly. A modification is to leave the crystal violet on the smear for only 2–3 seconds, after which the morphology of the bacterium is easier to see (**179**). The appearance of *D. congolensis* is so unique that a strong presumptive diagnosis of streptothricosis (dermatophilosis) can be made on the direct examination of stained smears alone.

D. congolensis is filamentous and branching. Mature filaments are composed of motile, coccal zoospores, in parallel lines, at least two abreast, resulting in a 'tram-track'-like appearance. The zoospores are about 1 μm in diameter. If the flakes of scab are treated too roughly, when the smears are made, the filaments will disintegrate and only Gram-positive cocci (zoospores) will be seen. **Diagram 27** shows, diagrammatically, the developmental cycle of *D. congolensis*. Transverse, horizontal and vertical septa form in the immature filaments dividing it into coccal zoospores. When mature, these zoospores are motile by polar flagella and are infective. They can initiate an infection in macerated or traumatised skin of the host animal.

178

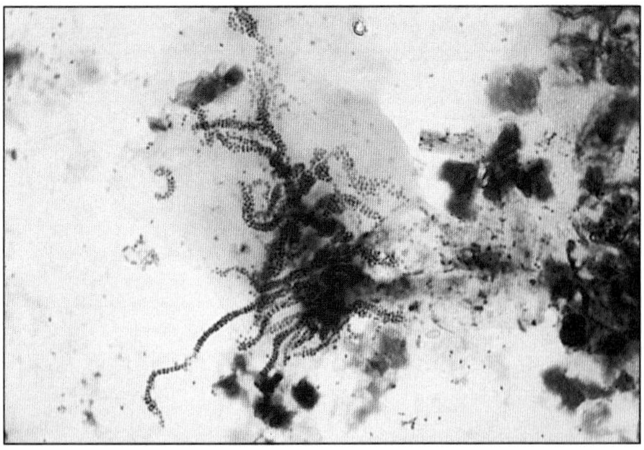

179

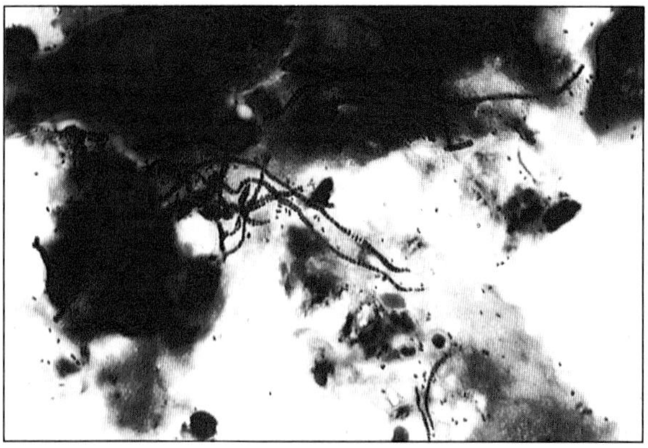

178 A Giemsa-stained smear of scab material from a case of bovine streptothricosis showing the branching filaments and zoospores of *D. congolensis*. (×1000)

179 A Gram-stained smear of *D. congolensis* in bovine scab material. The zoospores take up the crystal violet dye avidly and are sometimes less easy to visualise than in a Giemsa-stained smear. (×1000)

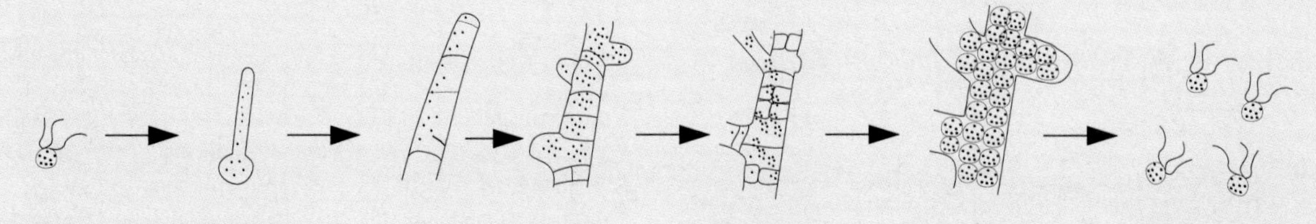

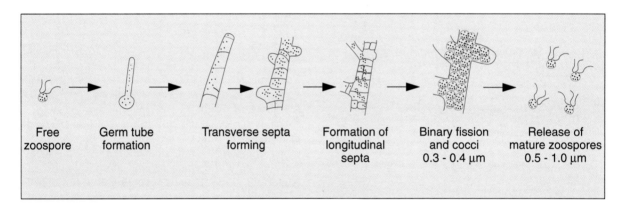

Diagram 27. Developmental cycle of *Dermatophilus congolensis.*

Isolation

Although the isolation of *D. congolensis* may not be necessary for a diagnosis of streptothricosis, the bacterium is comparatively easy to culture and grows well on sheep or ox blood agar. An atmosphere of 5–10 per cent CO_2 enhances the growth of the organism especially on primary isolation. The inoculated plates are incubated at 37°C for up to 5 days, although colonies may be seen after 24–48 hours' incubation.

Scab material contains many contaminants and Haalstra's method was developed to overcome this problem.

> Haalstra's Method for the Primary Isolation of *Dermatophilus congolensis*
>
> - Grind up a small amount of scab material and place a little in 2 ml distilled water in a bijou bottle for 3·5 hours at room temperature.
> - Place the container, with lid removed, in a candle jar at room temperature for 15 minutes.
> - The motile zoospores are chemotactically attracted to the carbon dioxide-enhanced atmosphere in the candle jar and move to the surface of the distilled water. Remove a loopful of fluid from the surface and inoculate a blood agar plate. Incubate the inoculated plate at 37°C for 72 hours under 5–10 per cent CO_2.

Identification

Colonial appearance
Small (about 1 mm) greyish-yellow, distinctly haemolytic colonies can be seen after 24–48 hours incubation. They are firmly adherent to the medium and are embedded in the agar. After 3–4 days, isolated colonies can be 3 mm in diameter and are rough, wrinkled and a golden-yellow colour (**180** and **181**). Older colonies can become mucoid. No growth occurs on Sabouraud dextrose agar.

Microscopic appearance
Gram-stained smears from colonies do not show the characteristic 'tram-track' appearance seen on direct microscopy. Usually the smears reveal uniformly staining, Gram-positive, branching filaments but sometimes coccal forms predominate.

Biochemical reactions
Biochemical tests are not usually carried out as a firm diagnosis will already have been made on the unique microscopic appearance and the characteristic colonies. *D. congolensis* is catalase-positive, urease-positive, gelatin-positive and produces acid from glucose, fructose and maltose. It is indole-negative, does not reduce nitrate and does not attack sucrose, salicin, xylose, lactose, sorbitol, mannitol or dulcitol.

180

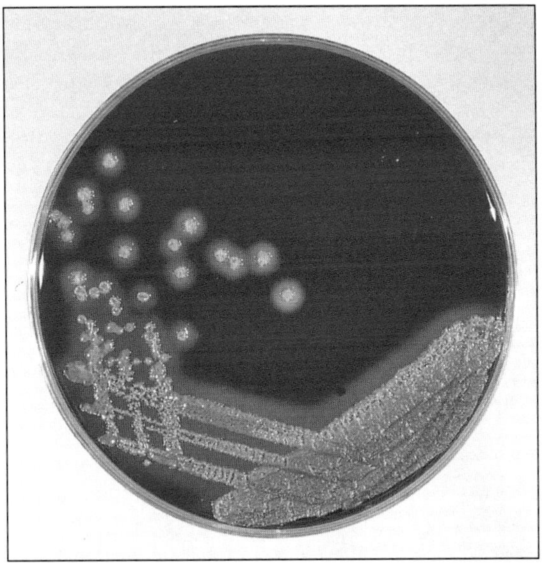

180 A culture of *D. congolensis* on sheep blood agar after 3 days' incubation in 10 per cent CO_2.

181

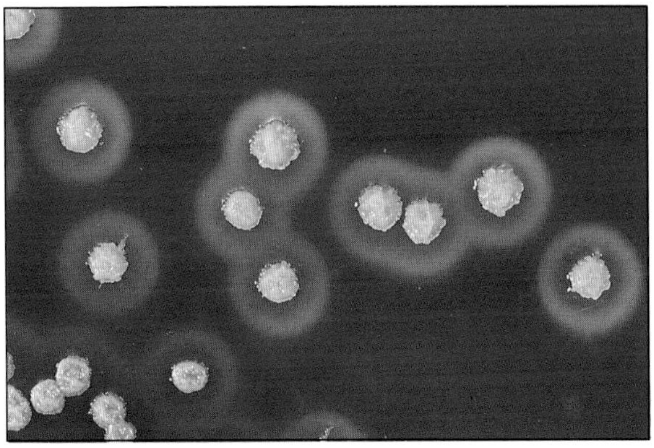

181 A close-up of the 3-day-old haemolytic colonies of *D. congolensis* on sheep blood agar showing the rough, dry, golden-yellow appearance. They are firmly embedded in the medium.

12 *Mycobacterium* species

The mycobacteria are thin rods of varying lengths (0.2–0.6 × 1.0–10.0 μm) and sometimes branching filamentous forms occur but these easily fragment into rods. They are non-motile, non-sporing, aerobic and oxidative. Although cytochemically Gram-positive, the mycobacteria do not take up the dyes of the Gram-stain because the cell walls are rich in lipids, mycolic acid forming the bulk of these. They are characteristically acid-fast, as once a dye has been taken up by the cells they are not easily decolourised, even by acid-alcohol. The rods tend to stain irregularly and often have a beaded appearance. The mycobacteria are most closely related to the genera *Nocardia* and *Rhodococcus* and all three genera have a similar cell wall type. A comparatively slow growth rate is a characteristic of the mycobacteria, with generation times ranging from 2–20 hours. Some species (chromogens) produce carotenoid pigments. The genus includes animal and human pathogens as well as saprophytic members often referred to as 'atypical', 'anonymous' or 'non-tuberculous' mycobacteria. Some of these can occasionally cause disease in animals.

Runyon's Groups

Runyon (1959) grouped the atypical mycobacteria on the basis of pigmentation, colonial morphology and growth rate. The scotochromogens are those that produce yellowish-orange pigments whether incubated in the light or in the dark. The photochromogens will produce pigment only, if exposed to light. For practical purposes, the slow-growing mycobacteria are defined as those that require over 7 days' incubation, under optimal conditons, to produce easily seen colonies and the rapid growers as those requiring less than 7 days. Although not all recently isolated species fit within the Runyon scheme, it is still a useful method of categorising the atypical mycobacteria.

Natural Habitat

The source of the pathogenic mycobacteria is usually infected animals. *Mycobacterium bovis* is excreted in respiratory discharges, faeces, milk, urine and semen. *M. avium* and *M. paratuberculosis* are shed in faeces and *M. tuberculosis* mainly in respiratory discharges. Tuberculosis is typically a disease of captivity or domestication. However, wild animal reservoirs of *M. bovis* occur such as badgers in Europe, brush-tailed opossums in New Zealand, Cape buffalo in East Africa and deer in Europe and America.

The mycobacteria are resistant to physical influences and will retain their viability in soil and particles of dried faeces for many months. The atypical mycobacteria are widespread in soil, pastures, bogs and water. A few are commensals in animals and may infect them, but such animals are not a significant source of infection for other animals.

Pathogenesis

M. bovis

Virulence appears to reside in the lipids of the cell wall. Mycosides, phospholipids and sulpholipids are thought to protect the tubercle bacilli against phagocytosis. Glycolipids cause a granulomatous response and enhance the survival of phagocytosed mycobacteria. Wax D and various tuberculoproteins induce a delayed hypersensitivity reaction detected in the tuberculin test. Infection is usually via the respiratory and intestinal tracts. In previously unexposed animals, local multiplication of the mycobacteria occurs and the resistance to phagocytic killing allows continued intracellular and extracellular replication. Infected host cells and mycobacteria can reach local lymph nodes and from there may pass to the thoracic duct with general dissemination.

After the first week, cell-mediated immune reactions begin to modify the host response and activated macrophages are able to kill some mycobacteria. The aggregation of macrophages contributes to the formation of a tubercle (**182**), and a fibrous layer may encompass the lesion. Caseous necrosis occurs at the centre of the lesion and this may proceed to calcification (cattle) or liquefaction. Once cell-mediated

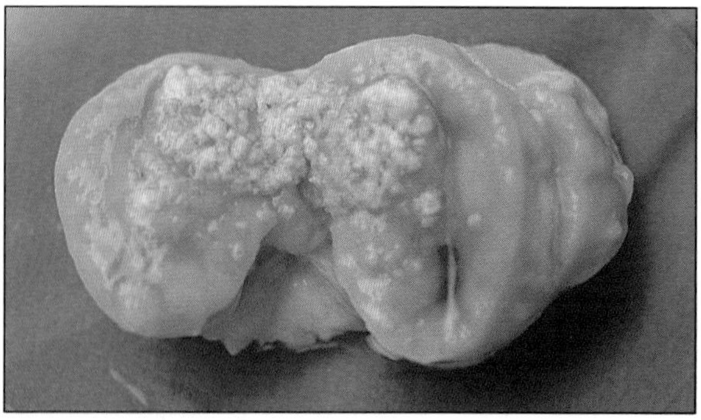

182 A tuberculous lesion in a bovine lymph node (*Mycobacterium bovis* infection).

immunity is established, lymphatic spread is retarded but occurs by contiguous extension or via the erosion of bronchi, blood vessels or viscera to new areas. T-lymphocyte-mediated reactions cause tissue damage. Haematogenous dissemination may produce miliary tuberculosis in animals such as deer. This involves multifocal tubercle formation in an organ or on the serosal surface of a cavity.

M. avium

Transmission is usually by the faecal-oral route and in birds the lesions are characteristically found in the liver. Lesions can also be present in the intestines, spleen and bone marrow. *M. avium* may sensitise cattle to the tuberculin test.

M. tuberculosis

M. tuberculosis is transmitted by aerosols or fomites and lesions are principally found in the lungs and lymph nodes. Non-human primates, dogs, canaries and psittacine birds are susceptible to human tuberculosis.

M. lepraemurium

M. lepraemurium is thought to cause both feline and murine leprosy. The disease in cats is characterised by the formation of single or multiple nodules or granulomas often with the development of non-healing ulcers. Less commonly, skin lesions in cats can also be caused by some of the atypical mycobacteria.

M. paratuberculosis

Cell-mediated immune phenomena appear to be involved in the pathogenesis of this disease. The organisms are found within macrophages, that are unable to kill them, in the submucosa of the ileocaecal area and adjacent lymph nodes. The ileum and colon are usually involved with extension to the rectum. The mucous membrane becomes thickened and permanently corrugated as a result of cellular infiltration. There is profuse diarrhoea and the disease is progressive leading to emaciation and death. Diarrhoea may not necessarily be seen in sheep and goats. Mortality is ultimately caused by malabsorption of nutrients and loss of protein into the intestine.

Large numbers of mycobacteria are present in epithelioid and giant cells in the mucosa and are shed in the faeces. Animals are thought to become infected in the neonatal period. Not all infected animals become clinical cases but they remain subclinical excretors; these animals shed small numbers of mycobacteria intermittently. The incubation period is long, about 18–24 months.

Atypical mycobacteria

The predisposing causes of comparatively rare cases of disease involving these mycobacteria may include a large infective dose and/or immunosuppression in the host. Some atypical mycobacteria can sensitise cattle to the tuberculin test.

Table 36 gives the mycobacterial diseases that occur in animals and **Table 37** indicates the susceptibility of domestic animals and birds to the mycobacteria causing tuberculosis.

Laboratory Diagnosis

Mycobacteria causing tuberculosis and atypical mycobacteria

Strict safety precautions must be enforced when working with specimens suspected of containing *M. bovis* or *M. tuberculosis.* These include preparing smears, processing specimens, inoculating media and adding reagents to biochemical tests. These procedures must be carried out within a biological safety cabinet housed in a separate room. As the mycobacteria are very resistant to disinfectants, care must be taken to use an effective one, such as a phenolic compound, for discarding contaminated slides and instruments. All contaminated materials must be autoclaved before leaving the laboratory. Any individuals who are immunocompromised in any way should be discouraged from working with these organisms. It is advisable to consult an authoritative account of safety precautions (Roberts *et al.*, 1991) before embarking on the laboratory diagnosis of the tuberculosis group of mycobacteria.

Specimens

Specimens from live animals include aspirates from cavities, lymph nodes, biopsies, tracheobronchial lavages and the centrifuged deposit from about 50 ml of milk in the case of suspected tuberculous mastitis. With dead animals, fresh and fixed (in 10 per cent formalin for histopathology) samples of lesions or a selection of lymph nodes from a tuberculin-reactor with no visible lesions are collected.

Table 36. Mycobacteria capable of causing disease in animals.

Species	Host(s)	Significance
TUBERCULOSIS-GROUP: slow-growing		
M. africanum	Humans	Human tuberculosis (Africa)
M. tuberculosis	Humans, dogs, canaries and psittacine birds	Human tuberculosis (worldwide)
M. bovis	Many animal species and humans	Bovine tuberculosis
M. microti	Voles	Vole tuberculosis. Localised lesions seen in rabbits, calves and guinea-pigs
RUNYON'S GROUPS		
I. PHOTOCHROMOGENS: slow-growing (over 7 days' incubation) saprophytes but rare disease in man and animals.		
M. kansasii	Deer, pigs and cattle	Tuberculosis-like disease. Isolated from lungs and lymph nodes
M. simiae	Humans (monkeys)	Isolated from lymph nodes of healthy monkeys. Pulmonary disease in man
M. marinum	Marine fish, aquatic mammals and amphibians	Fish tuberculosis: granulomatous and disseminated disease
M. vaccae	Saprophytic	Non-pathogenic
II. SCOTOCHROMOGENS: slow-growing, ubiquitous saprophytes found commonly in grasslands. Occasional disease in animals and humans		
M. scrofulaceum	Domestic and wild pigs, cattle and buffaloes	Tuberculous lesions in cervical and intestinal lymph nodes.
III. NON-CHROMOGENS: (slow-growing)		
M. avium	Poultry and wild birds	Avian tuberculosis. Generalised form rare in mammals
	Pigs	Lesions in cervical lymph nodes
	Horses, pigs and others	Intestinal lesions (rare)
M. intracellulare **(Battey bacillus)**	Poultry and wild birds	Avian tuberculosis. Saprophyte in soil and water
	Pigs and cattle	Can be present in intestinal lymph nodes
	Non-human primates	Granulomatous enteritis (resembles Johne's disease)
M. ulcerans	Cats	Nodulo-ulcerative skin lesions
M. xenopi	Cats	Nodulo-ulcerative skin lesions
	Pigs	Tuberculous lesions in lymph nodes of the alimentary tract
IV. RAPID-GROWING MYCOBACTERIA : need less than 7 days' incubation. Pigmentation variable. Saprophytes in soil, water and on plants. They are found regularly in intestines of pigs, ruminants and other animals. Occasionally pathogenic for animals		
M. chelonae	Fish	Disseminated granulomatous lesions
	Turtles	Tuberculosis-like lesions in lungs
	Cattle	Granulomatous lesions in lymph nodes
	Manatees, cats and pigs	Abscesses and nodulo-ulcerative lesions in various tissues
	Monkeys	Abscesses in lymph nodes or disseminated disease
M. fortuitum	Cattle	Granulomatous lesions in lymph nodes and mammary glands
	Cats	Ulcerative, pyogranulomatous lesions of skin
	Dogs	Granulomatous lesions in skin and lungs
	Pigs	Granulomas in lymph nodes, joints and lungs
M. phlei	Cats	Nodulo-ulcerative lesions of skin (rare)
M. smegmatis	Cattle	Granulomatous mastitis
	Cats	Ulcerative skin lesions

(continued)

Table 36. Mycobacteria capable of causing disease in animals (*continued*).

Species	Host(s)	Significance
OTHER MYCOBACTERIA		
M. paratuberculosis	Cattle, sheep, goats and other ruminants	Paratuberculosis (Johne's disease). Chronic, progressive, intestinal, wasting disease
M. lepraemurium	Cats and rodents	Feline and murine leprosy (respectively). Not yet isolated on conventional media
M. leprae	Humans and 9-banded armadillo	Leprosy in humans. Replication in armadillos. Not isolated *in vitro*
Unidentified acid-fast bacterium	Cattle	Skin tuberculosis (lymphangitis)

Table 37. Susceptibility of animals to the mycobacteria that cause tuberculosis.

	Mycobacterium tuberculosis	***Mycobacterium bovis***	***Mycobacterium avium***
Primates	+	+	–
Cattle	(+)	+	(+)
Sheep and goats	–	+	(+)
Pigs	(+)	+	(+)
Horses	–	+	+ (intestinal)
Dogs	+	+	–
Cats	–	+	–
Poultry	–	–	+
Canaries	+	–	+
Psittacine birds	+	–	–

+ = susceptible, (+) = slightly susceptible or may become sensitized, – = resistant to infection.

Direct microscopy

The Ziehl-Neelsen (acid-fast) stain is used to stain smears from lesions and other specimens. The mycobacteria appear as slender, often beaded, red-staining rods against a blue background (if methylene blue is the counterstain). These can be visualized only if at least 5×10^4 mycobacteria/ml of material is present. The numbers of *M. bovis* are often low in bovine specimens but avian tuberculosis in poultry (**183**) and *M. bovis* lesions in animals such as deer (**184**) and badgers usually yield large numbers of mycobacteria.

Smears stained by fluorescent dyes, such as auramine, acridine orange or fluorochrome, can be examined under a UV microscope. These stains allow the mycobacteria to be seen more easily if relatively small numbers are present.

Isolation

Several preliminary procedures are necessary in order to recover the comparatively slow-growing mycobacteria:

- Selective decontamination to reduce significantly the number of fast-growing contaminating bacteria.
- Digestion or liquefaction of mucus. Mucin-trapped mycobacteria in specimens, such as tracheobronchial exudates, may not be available for growth in cultures.
- If mycobacteria are present in small numbers they must be concentrated to allow detection by stained smears and culture. This concentration can be done by centrifugation.

Mycobacteria are comparatively resistant to acids, alkalis and quaternary ammonium compounds.

Decontaminating agents, such as 5 per cent oxalic acid, 2–4 per cent NaOH and 1 per cent quaternary ammonium compounds can be used on specimens. The mildest decontaminating procedure that gives sufficient control over contaminants is desirable. But even under optimal conditions 80–90 per cent of the mycobacteria in the specimen may be killed by the decontaminating agent. Each laboratory will need to find the best compromise between destruction of contaminants and survival of sufficient mycobacteria for isolation. Examples of the procedures for recovering mycobacteria from specimens are given in **Diagram 28**.

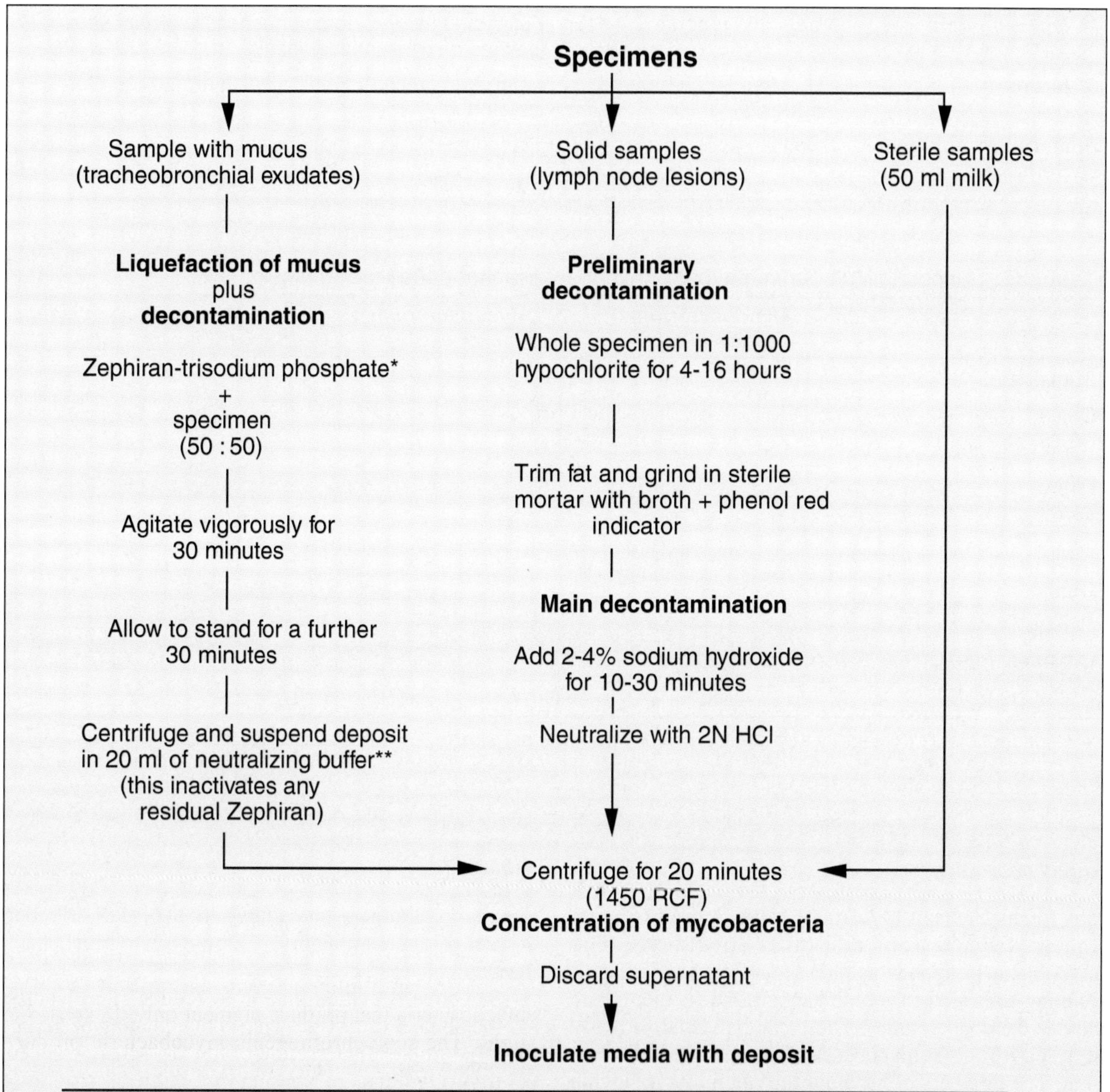

* Dissolve 1kg trisodium phosphate ($Na_3PO_4 . 12 H_2O$) in 4 litres of hot water. Add 7.5 ml of Zephiran concentrate (17% benzalkonium chloride, Winthrop Labs., N.Y.)

** Add 37.5 ml of 0.067 M disodium phosphate to 62.5ml of 0.067 M monopotassium phosphate. Final pH 6.6

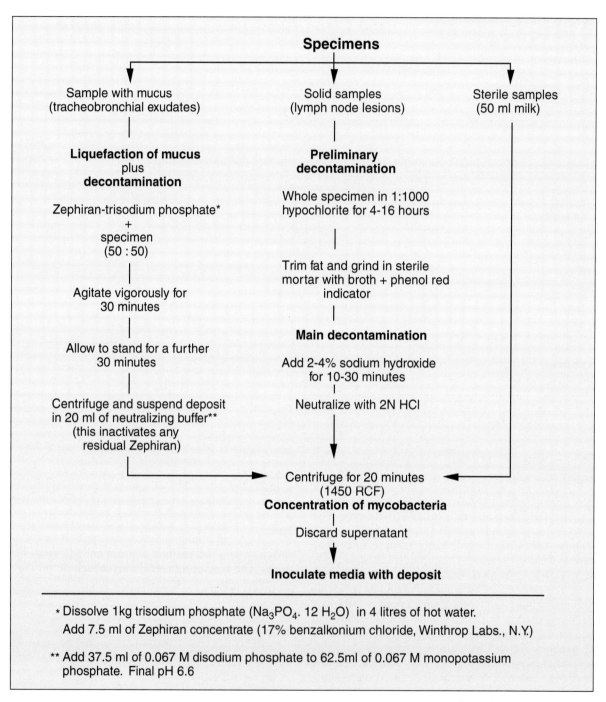

Diagram 28. Procedures for recovering mycobacteria of the tuberculosis group from clinical specimens.

183

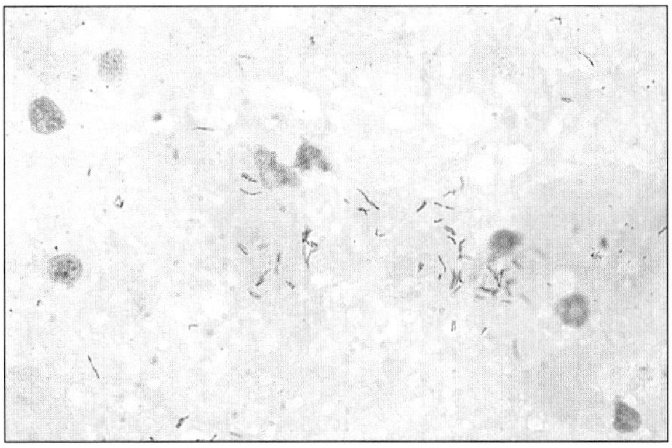

183 *M. avium* in a ZN-stained smear of material from a tubercle in a pigeon demonstrating the numerous, slender, red-staining rods characteristic of avian tuberculosis. (×1000)

184

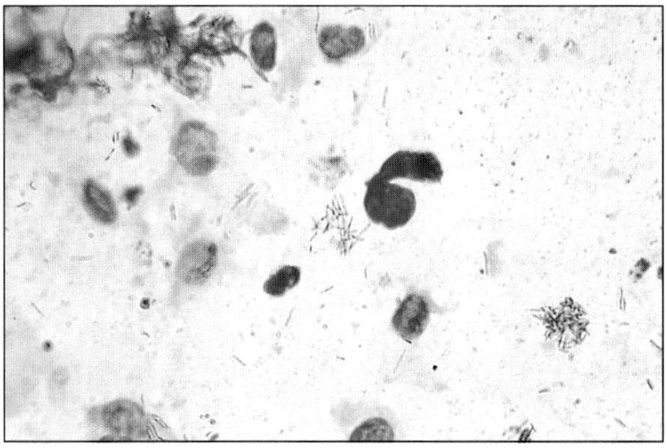

184 *M. bovis* in a ZN-stained smear of material from a tubercle in a deer. The slender, beaded, red-staining (ZN-positive) rods tend to be more numerous in lesions from deer and badgers compared to the low numbers in bovine lesions. (×1000)

Media for the mycobacteria

The egg-based Lowenstein-Jensen and Stonebrinks media are most commonly used in veterinary bacteriology. Lowenstein-Jensen medium can be obtained commercially and the formula and method of preparation of Stonebrinks medium is given in **Appendix 2**. The media are prepared as solid slants in screw-capped bottles. Malachite green dye (0.025 g/100 ml) is commonly used as the selective agent. *M. tuberculosis*, *M. avium* and many of the atypical mycobacteria require glycerol for growth. However, glycerol is inhibitory to *M. bovis* while sodium pyruvate (0.4 per cent) enhances its growth. Thus media with glycerol and without glycerol (but with sodium pyruvate) should be inoculated. The media can be made more selective by the addition of cycloheximide (400 μg/ml), lincomycin (2 μg/ml) and nalidixic acid (35 μg/ml). Each new batch of culture medium should be inoculated with stock strains of mycobacteria to ensure that the medium supports satisfactory growth. The inoculated media may have to be incubated at 37°C for up to 8 weeks for the mycobacteria in the tuberculosis group. *M. tuberculosis* and *M. avium* prefer the caps on the culture media to be loose while *M. bovis* grows best in airtight containers. Most of the atypical mycobacteria will grow on media suitable for the tuberculosis-causing mycobacteria. *M. leprae*, *M. lepraemurium* and the mycobacterium causing bovine skin tuberculosis have not yet been cultured *in vitro*.

Identification

Colonial morphology

A summary of the colonial types of the mycobacteria of the tuberculosis group is given in **Table 38**. The luxuriant growth of *M. tuberculosis* on glycerol-containing media, giving the characteristic 'rough, tough and buff' colonies (**185**), is known as eugonic. The growth of *M. avium* on media containing glycerol is also described as eugonic (**186**). *M. bovis* has sparse, thin growth on glycerol-containing media that is called dysgonic. *M. bovis*, however, grows well on pyruvate-containing media without glycerol (**187**). Cultures on Lowenstein-Jensen medium of some commonly isolated mycobacteria are illustrated (**188–191**).

Pigment production and response to light

The mycobacteria that produce yellowish-orange carotenoid pigments are called chromogenic (**192**). The term photochromogenic is applied to those mycobacteria that produce pigment only if exposed to light. The scotochromogenic mycobacteria produce pigment when incubated either in the light or in the dark. Pigment formation is tested with young, well-developed colonies on Lowenstein-Jensen medium. The cultures are exposed to a 100 watt, clear electric light bulb, at a distance of 50 cm, for at least an hour and then incubated again in darkness for a further 1–3 days. After this treatment the photochromogens will develop pigment. Older colonies of the mycobacteria in the tuberculosis group often have a yellowish hue but they are described as non-chromogenic.

Table 38. Differentiation of mycobacteria of the tuberculosis-group.

	M. tuberculosis	*M. bovis*	*M. avium*
Growth-type and colony form on media with glycerol	Eugonic 'Rough, tough and buff'. Colonies hard to break up	Dysgonic Small, moist-sheen colonies that break up easily	Eugonic Whitish, sticky colonies that break up easily
Growth time	3–8 weeks	3–8 weeks	2–6 weeks
Glycerol required for growth	+	– (inhibited by glycerol)	+
Enhanced growth with 0.4% sodium pyruvate	–	+	–
Niacin production	+	–	–
Pyrazinamidase	+	–	+
Urease	+	+	–
Nitrate reduction	+	–	–
Inhibited by TCH (10 mg/ml)	–	+	–

+ = positive reaction, – = negative reaction, TCH = thiophen-2-carbonic acid hydrazide.

185

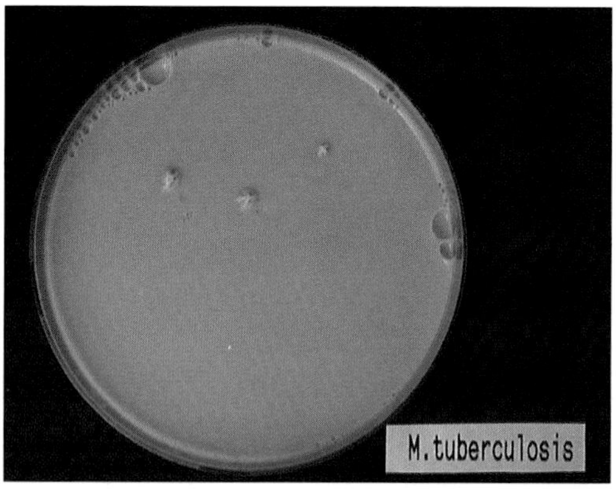

185 *M. tuberculosis* on Lowenstein-Jensen medium with glycerol showing the typical colonial morphology ('rough, tough and buff').

186

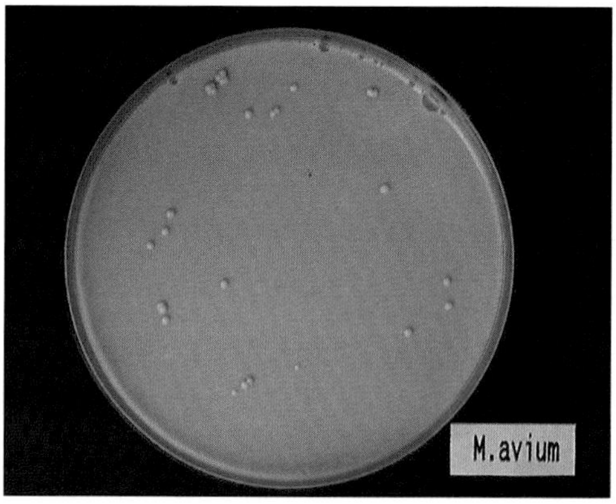

186 *M. avium* colonies on Lowenstein-Jensen medium.

187

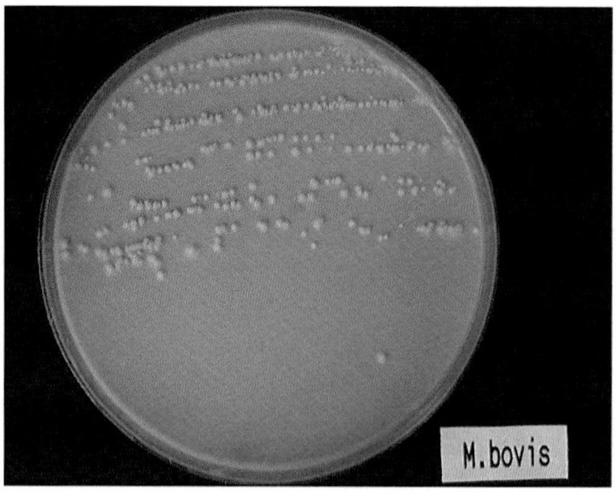

187 *M. bovis* colonies on Lowenstein-Jensen medium with pyruvate.

188 

188 Cultures of *M. tuberculosis* (left) and *M. bovis* (right) on Lowenstein-Jensen medium.

 189

189 Cultures of *M. avium* (left) and *M. fortuitum* (right) on Lowenstein-Jensen medium. *M. fortuitum* is an example of a rapidly growing atypical mycobacterium that often produces pigment.

190 

190 *M. vaccae* (left) and *M. marinum* (right) on Lowenstein-Jensen medium. They are both photochromogens.

 191

191 *M. phlei* (left) and *M. smegmatis* (right) on Lowenstein-Jensen medium. Both are rapidly growing mycobacteria that often produce pigment.

192 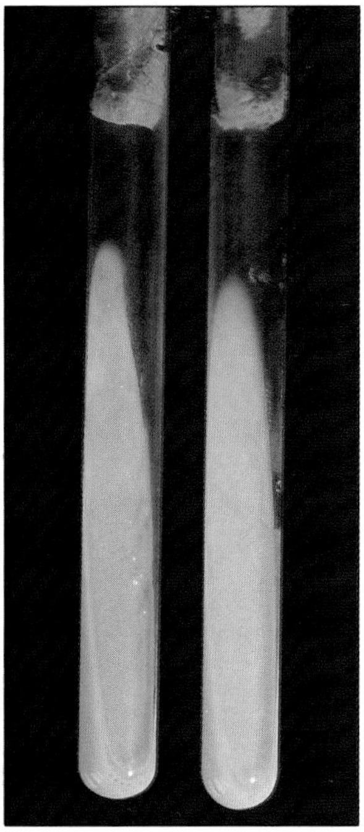

192 Two rapidly growing chromogenic mycobacteria isolated from moss in a wet pasture: *M. aurum* (left) and *M. aichiense* (right).

Microscopic appearance

The mycobacteria are acid-fast from cultures and appear as red-staining, thin, fairly long rods in ZN-stained smears.

Biochemical tests

Some of the biochemical reactions for the mycobacteria of the tuberculosis group are given in **Tables 38** and **39**, while those for the atypical mycobacteria are shown in **Table 39**. Details of some of the most commonly performed tests are given on page 165.

Table 39. *In vitro* tests for some clinically significant mycobacteria.

	Runyon group	Growth within 7 days	Inhibition by glycerol	Colonial morphology (a)	Pigmentation (b)	Niacin production	Tolerance of 5% NaCl	Deamination of pyrazinamide (4 days)	Nitrate reduction	Urease production	Growth on MacConkey agar without crystal violet
Mycobacterium tuberculosis		–	–	R	N	+	–	+	+	v	–
M. bovis		–	+	S/(R)	N	–	–	–	–	v	–
M. simiae	I	–	–	S	P	v	–	+	–	+	–
M. kansasii	I	–	–	SR/S	P	–	–	–	+	v	–
M. marinum	I	–	–	S/SR	P	–(+)	–	+	–	+	–
M. scrofulaceum	II	–	–	S	S	–	–	v	–	v	–
M. avium-intracellulare	III	–	–	S/R	N	–	–	+	–	–	–(+)
M. ulcerans	III	–	–	R	N	–	–	–	–	v	–
M. xenopi	III	–	–	S	N(S)	–	–	v	–	–	–
M. chelonae	IV	+	–	S/R	N	v	+(–)	+	–	+	+
M. fortuitum	IV	+	–	S/R	N	v	+	+	+	+	+
M. phlei	IV	+	–	R	S	•	–	•	+	•	–
M. smegmatis	IV	+	–	R/S	N	•	–	•	+	•	–

(a) R = rough, S = smooth, and SR = intermediate.
(b) P = photochromogenic (pigment produced only if culture is exposed to light)
S = scotochromogenic (pigment produced in the light and in the dark)
N = non-chromogenic (no pigment produced)
V = variable reactions, –(+) = majority negative, +(–) = most positive, • = data not available.

Biochemical Tests for Identification of Mycobacteria

- ***Niacin Production Test.:*** the commercially available niacin test strips (Difco) are easier and safer to use as this avoids employing toxic BrCN solution used in conventional tests. *M. tuberculosis* is positive and *M. avium* is negative in this test.
- ***Nitrate Reduction:*** place a few drops of sterile distilled water in a screw-capped tube (16 x 125 mm) and add a loopful of a young culture of the mycobacterium. Use an uninoculated tube as a negative control. Add 2 ml of $NaNO_3$ solution (0.01 M solution $NaNO_3$ in 0.022M phosphate buffer, pH7). Shake and incubate in a water bath at 37°C for 2 hours.
  Add: 1 drop of 1:2 dilution of conc. HCl.
  2 drops of 0.2 per cent aqueous solution of sulphanilamide.
  2 drops of 0.1 per cent aqueous N-(1-naphthyl) ethylenediamine dihydrochloride.
  Examine for the development of a pink to red colour and compare with the negative control. A strong red indicates nitrate reduced to nitrite. Add a pinch of powdered Zn to all negative tubes (converts nitrate to nitrite). The production of a red colour indicates a negative test (nitrate not reduced). The commercial paper strip method can be used but a negative result should be confirmed by the above test.
- ***Deamination of Pyrazinamide* (to pyrazinoic acid) in 4 days:** the medium is a Dubos broth base containing 0.1 g pyrazinamide, 2.0 g of pyruvic acid and 15.0 g agar per litre. Dispense in 15 ml amounts in screw-capped tubes. Autoclave at 121°C for 15 minutes and solidify in an upright position. Inoculate the agar with a heavy suspension of a young culture and incubate at 37°C for 4 days. Add 1 ml of freshly prepared 1 per cent aqueous ferrous ammonium sulphate to the tubes and place in refrigerator for 4 hours. A positive reaction is given by a pink band in the agar. Use an uninoculated tube and an *M. avium* tube as negative and positive controls, respectively.
- ***Urease Test:*** mix 1 part of urea-agar base concentrate with 9 parts of sterile water. Dispense in 4 ml amounts in screw-capped tubes (16 x 125 mm). Emulsify a loopful of young culture in the tube of substrate. Incubate at 37°C. A colour change from amber to pink or red is a positive reaction. Discard the test after 3 days.
- ***MacConkey Agar without crystal violet:*** inoculate the agar plate with a loopful of a young broth culture of the mycobacterium making a spiral streak from the centre of the agar outwards. Incubate at 37°C and examine for growth after 5 and 11 days. Only strains of *M. fortuitum* and some subspecies of *M. chelonae* will grow to the end of the spiral streak. Other mycobacteria may grow where the inoculum is heaviest.
- ***Inhibition and Tolerance Tests:*** reagents such as 5 per cent NaCl and thiophen-2-carbonic acid hydrazide (TCH) 10 μg/ml are usually incorporated into media such as Lowenstein-Jensen.

Further details of these and other biochemical tests are published (Roberts *et al.*, 1991).

Animal inoculation

This historical method of distinguishing between the tubercle bacilli was based on the variation in the pathogenicity of each for different laboratory animals (**Table 40**).

Animal inoculation is now rarely performed because of aesthetic and economic reasons as well as the risk of infection to laboratory staff.

Comparatively new methods of identifying mycobacteria

Conventional biochemical and cultural characteristics remain important in making a definite identification of mycobacterial species but these have become less important because of the newer methods now available (Roberts *et al.*, 1991).

- Rapid radiometric mycobacterial detection system (BACTEC TB system). This separates *M. tuberculosis* from other mycobacteria and can also be adapted to perform antimicrobial susceptibility testing of mycobacteria.
- Gas-liquid chromatography, a rapid and reliable method for the identification of clinically important mycobacteria, is commercially available (HP 5890A: Hewlett-Packard Co., Palo Alto, Calif., USA).
- Nucleic acid probes are available for the identification of *M. tuberculosis, M. avium, M. intracellulare* and *M. gordonae*. Other probes are being developed.
- Immunological techniques employing monoclonal antibodies.

Table 40. Inoculation of laboratory animals with mycobacteria of the tuberculosis-group.

	Mycobacterium tuberculosis	***Mycobacterium bovis***	***Mycobacterium avium***
Rabbits (intravenously)	± (pulmonary only)	++ (miliary)	++ (generalised)
Guinea-pigs (subcutaneously)	++ (generalised)	++ (generalised)	–(+) (focal)
Chicken (intravenously)	–	–	++ (generalised)

++ = systemic infection, ± = comparatively mild infection, –(+) = localised infection, – = no infection.

Field and laboratory immunological tests for tuberculosis

National eradication schemes use the delayed hypersensitivity reaction elicited in animals, infected with tubercle bacilli, after intradermal inoculation of a small amount (usually 0.1 ml) of tuberculin. Tuberculin, a purified protein derivative (PPD) prepared from *M. bovis* or *M. avium*, is a complex mixture of proteins, lipids, carbohydrates and nucleic acids.

Intradermal skin testing using PPD has proved an effective diagnostic test for identifying *M. bovis*-infected cattle. The caudal fold test uses only bovine PPD, whereas the single intradermal comparative test involves the injection of avian and bovine tuberculins simultaneously, but at different sites, into the skin of the neck. Tuberculous cattle show a delayed-type hypersensitivity reaction at the injection site, which is maximal at 72 hours' post-inoculation. The reaction is characterised by thickening of the skin due to a mononuclear cell infiltration and sometimes oedema (**193**). The tuberculin test is used mainly for cattle but it is used occasionally to test pigs, deer and poultry with appropriate modifications in technique for each species.

Cattle can be sensitised to tuberculin not only by infection with *M. bovis* but also by the acid-fast bacteria responsible for skin tuberculosis, *M. avium*, *M. tuberculosis* and saprophytic mycobacteria. This can lead to false positive reactions. False negative reactions can also occur. The term anergy has been used to describe this unresponsive state, which is not well understood. A number of laboratory-based tests have been developed in recent years for the diagnosis of tuberculosis. These include the lymphocyte transformation test, serological tests for circulating antibodies (such as the ELISA) and gamma interferon assays using whole blood. These *in vitro* tests are usually used in conjunction with the tuberculin test.

Mycobacterium lepraemurium

Feline leprosy ('acid-fast granuloma') is probably caused by *M. lepraemurium* (**194**). This mycobacterium has not yet been cultured *in vitro*. Numerous long, slender, acid-fast rods are seen in ZN-stained smears of scrapings from non-healing ulcers or from biopsies of the nodules. Histopathological sections from biopsies stained with the ZN stain will also reveal numer-

193

193 A reactor to the single intradermal comparative tuberculin test. There is no reaction to the avian PPD at the upper site but a marked reaction has occurred to the bovine tuberculin (lower site). Photographed 72 hours after injection.

194

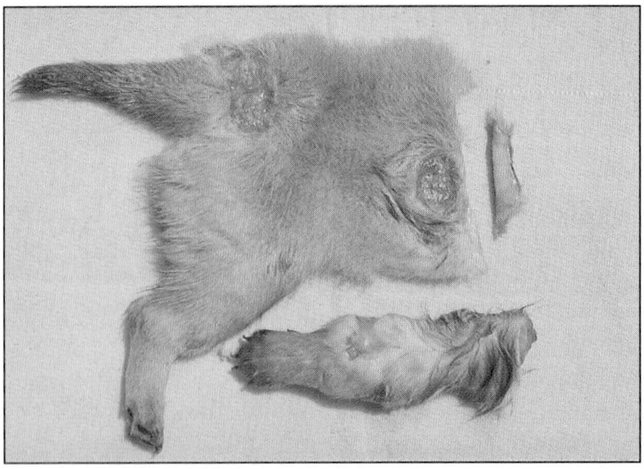

194 Lesions of feline leprosy (*M. lepraemurium)* in a young cat. They present as chronic non-healing ulcers.

ous acid-fast mycobacteria. Culture should be attempted in case the lesions have been caused by one of the atypical mycobacteria, such as *M. smegmatis*, *M. ulcerans* or *M. fortuitum*, that can cause lesions in cats.

Mycobacterium paratuberculosis

M. paratuberculosis, formerly called *M. johnei*, causes paratuberculosis (Johne's disease) in cattle (**195**), sheep, goats and other ruminants.

Specimens

In live animals with suspected Johne's disease, a small pinch biopsy from the rectum or rectal scrapings are preferred to faecal samples. A biopsy from a mesenteric lymph node, if available, would be a useful specimen. Control schemes can be based on the detection of asymptomatic shedder cattle by the isolation of *M. paratuberculosis* from faecal samples. This requires a reference laboratory willing to carry out a large number of cultures. Usually 15 g faecal samples are submitted, from every adult animal in the herd, at 6 month intervals. Specimens from dead animals include a section of ileocaecal valve, washed free of faeces, mesenteric lymph nodes (often the best specimen from sheep and goats) and sections of ileocaecal valve in 10 per cent formalin for histopathology.

Direct microscopy

Faecal or ileocaecal valve mucosal smears from advanced clinical cases, stained by the ZN-stain, usually yield large numbers of short, red-staining rods, that are characteristically in clumps indicating intracellular growth (**196**). ZN-stained histopathological sections of ileocaecal valve also reveal large numbers of acid-fast bacilli in clinical cases (**197**). Examination of stained faecal smears will detect only about 25 per cent of subclinical excretors. Field observation, however, indicates that those cattle that are excreting sufficient *M. paratuberculosis* to detect by faecal smears will eventually become clinical Johne's cases. In cattle this may be 6 months to 3 years after the animal was detected as an asymptomatic shedder.

If only a few, scattered acid-fast bacilli are seen in ZN-stained faecal smears this must be interpreted carefully, as harmless atypical mycobacteria can be present in the intestine. A repeat sample should be obtained.

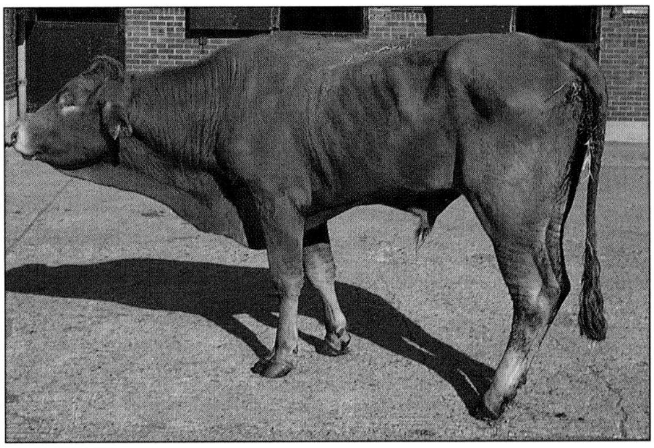

195

195 A South Devon bull with advanced Johne's disease (*M. paratuberculosis*). There is pronounced diarrhoea, emaciation and muscle atrophy.

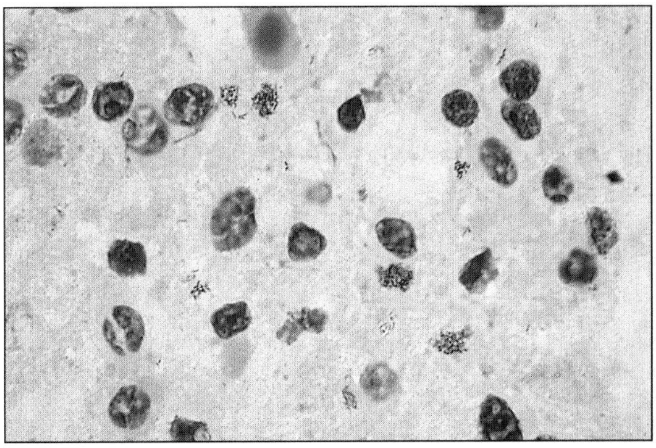

196

196 *M. paratuberculosis* in a ZN-stained smear of mucosal scrapings from a bovine ileocaecal valve. The short, acid-fast rods are in clumps indicative of intracellular growth.

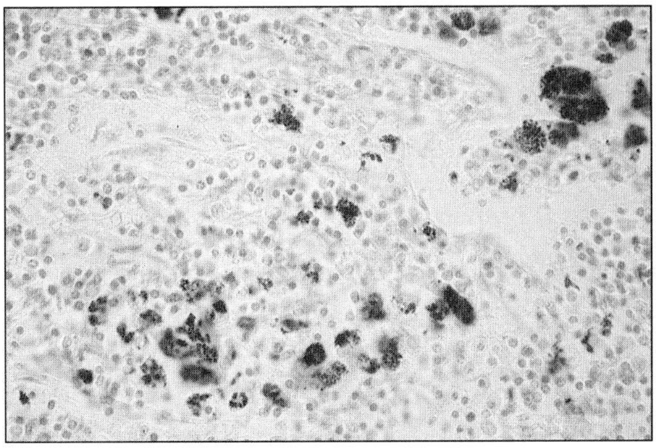

197

197 A ZN-stained tissue section from the ileocaecal valve area of a bull with Johne's disease. Large numbers of the acid-fast bacilli are present in animals with clinical disease.

Isolation

This method is far more sensitive than the examination of ZN-stained smears, but is usually undertaken only by reference laboratories. If this service is available, the culture of the faeces of all cattle, 2 years old or older, every 6 months, and the removal of all animals with positive cultures, plays an important part in a herd control or eradication programme. **Diagram 29** summarises the isolation procedures for *M. paratuberculosis.* Benzalkonium chloride (Zephiran) is used to decontaminate the specimens and Herrold's egg yolk medium with mycobactin is often used as the culture medium. If contamination by fungi is a problem, amphotericin B can be added to the medium at a final concentration of 5 µg/ml. The egg in Herrold's medium contributes sufficient phospholipids to neutralise the activity of residual Zephiran in the inoculum. The essential growth factor, mycobactin, can be extracted from *M. phlei* or preferably from a mycobactin-independent strain of *M. paratuberculosis* (mycobactin J). The method for preparation of Herrold's egg yolk medium can be obtained from manuals prepared by reference laboratories (Anon, 1974). Three slants of the Herrold's medium with mycobactin and one slant of the medium without mycobactin are inoculated per sample. The slants are incubated at 37°C and examined for growth, once a week for up to 16 weeks.

Identification

Colonial morphology

Primary colonies characteristically appear about 7 weeks after inoculation (range 5–14 weeks). The colonies are very small (1 mm diameter), colourless, translucent and hemispherical. The margins are round and even and the surfaces smooth and glistening. The colonies become more opaque and increase in size as incubation is continued, reaching a maximum diameter of about 4 mm. Roughness increases with age.

Microscopic appearance

The cells from primary cultures are highly acid-fast average 0.5 µm in width and 1.0 µm in length.

Mycobactin dependency

The primary cultures exhibit a strict dependency upon mycobactin for growth. This is somewhat modified on subculture. If growth occurs only on the slants of medium containing mycobactin and not on the slant without mycobactin, this gives evidence of mycobactin-dependency and the culture can be reported as *M. paratuberculosis* providing all other tests, such as the microscopic appearance (ZN-smear), are compatible. Occasionally if large numbers of *M. paratuberculosis* are present in the inoculum, growth may be seen on the slant without mycobactin.

Animal pathogenicity tests

M. paratuberculosis does not produce progressive disease in laboratory animals or chickens. If a susceptible sheep is inoculated with 200 mg dry weight of viable Johne's bacilli, clinical signs may appear in 8–14 months.

DNA probes

Species-specific DNA probes are being used for the rapid identification of *M. paratuberculosis* in faeces.

Field and laboratory immunological tests for paratuberculosis

- Field diagnostic tests based on delayed-hypersensitivity reactions to johnin.
  The intradermal johnin test for cattle is read 48–72 hours after inoculation. A swelling at the injection site indicates a positive reaction. The test may give up to 75 per cent false-positives.
  The intravenous johnin test is more specific. Cattle developing an elevation in temperature of 1.5°C or more are regarded as positive reactors. This test may detect about 80 per cent of clinical cases but is not useful for asymptomatic shedders. There is an antigenic relationship between *M. paratuberculosis* and *M. avium* reflected in the fact that some cattle with clinical Johne's disease react to avian tuberculin. The johnin tests are not diagnostically useful for sheep and goats.
- *In vitro* lymphocyte stimulation test.
  In this test peripheral blood lymphocytes are collected and exposed to the purified protein derivative of johnin. If the antigens of *M. paratuberculosis* have elicited a cell-mediated response, the lymphocytes will proliferate. This is measured by the addition of a radio-labelled nucleic acid precursor. The test is positive only when clinical signs are evident.
- Serological tests for antibody detection.
  Various serological tests have been developed for the diagnosis of paratuberculosis. These include an ELISA, complement fixation test (CFT), agar gel immunodiffusion (AGID), fluorescent antibody and immunoperoxidase tests. The ELISA and AGID tests show most promise. The ELISA will detect an infected animal when there are sufficient mycobacteria in the faeces for a positive culture. The AGID test becomes positive when large numbers of *M. paratuberculosis* are being excreted. The effectiveness of the ELISA and AGID tests for the diagnosis of Johne's disease in sheep and goats is not known.

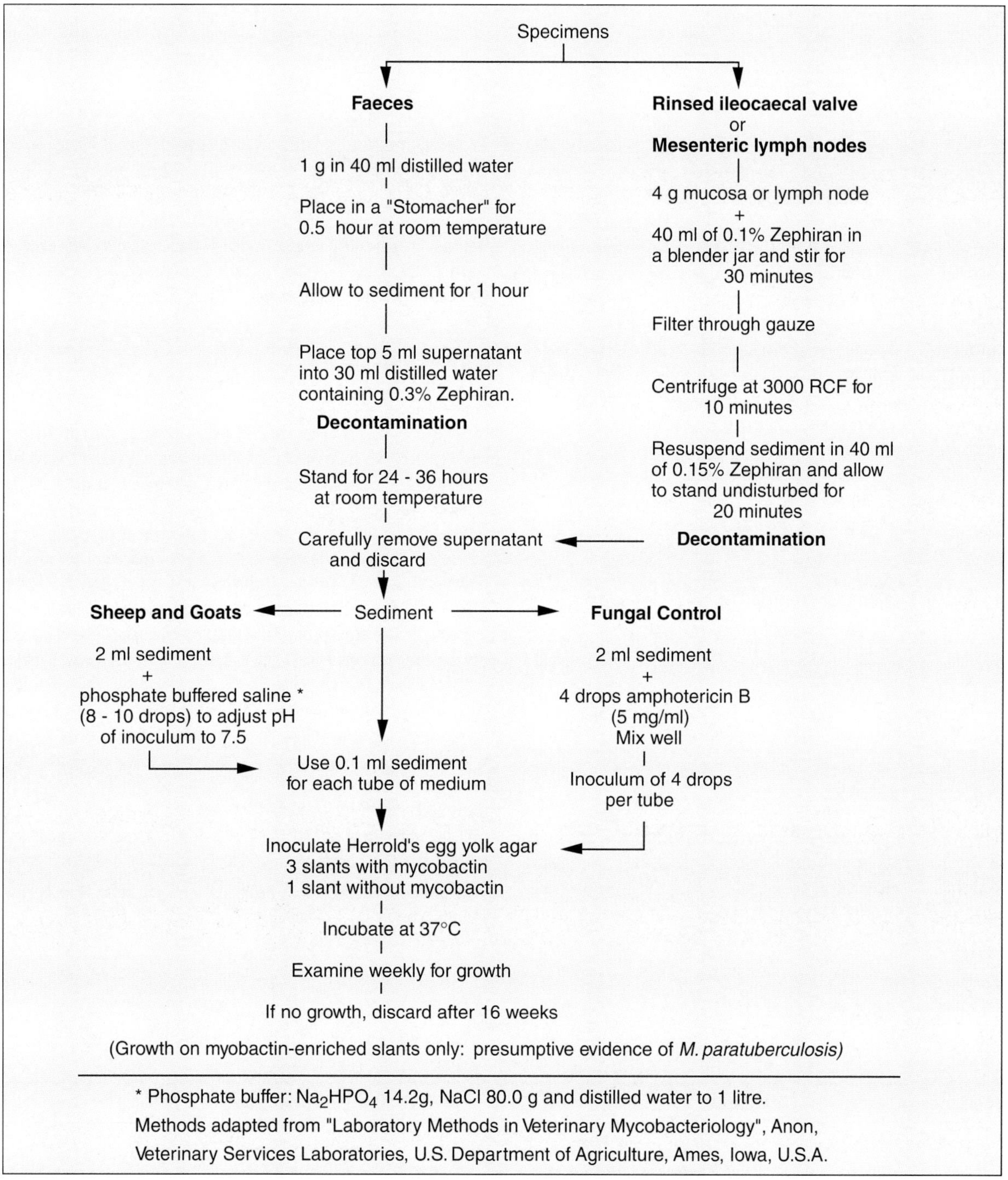

Diagram 29. Procedures for the isolation of *Mycobacterium paratuberculosis*.

References

Anon (1974). *Laboratory Methods in Veterinary Mycobacteriology*. Veterinary Service Laboratories, Animal and Plant Health Inspection Service, US Department of Agriculture, Ames, Iowa 50010, USA.

Roberts, G.D., Koneman, E.W. and Kim, Y.K., (1991). *Mycobacterium*, pp. 304–339. In *Manual of Clinical Microbiology*, 5th ed. Balows, A. (ed. in chief). American Society for Microbiology, Washington, D.C., USA.

Runyon, E.H., (1959). Anonymous mycobacteria in pulmonary disease. *Med. Clin. N. Am.*, **43**:273–290.

13 *Listeria* species

Listeria species are medium-sized Gram-positive rods, non-spore-forming and non-acid-fast, measuring about 0.4–0.5 µm in diameter by 0.5–2.0 µm in length. From rapidly growing cultures or animal tissues the cells can appear coccal. They are facultative anaerobes but growth is enhanced by 10 per cent CO_2. Listeriae are catalase-positive, oxidase-negative, hydrolyze aesculin, tolerate 10 per cent sodium chloride and are motile by a few (1–5) peritrichous flagella. They grow on nutrient agar and blood agar but not on MacConkey agar. Smooth and rough colonial variants occur with *L. monocytogenes* and the rough colonies can yield long filamentous forms.

Listeria has been divided into seven species with two distinct groups:

- *L. murrayi* and *L. grayi* are non-haemolytic, rarely isolated and considered to be non-pathogenic.
- *L. monocytogenes* and *L. ivanovii* are haemolytic and pathogenic for animals and *L. monocytogenes* also causes disease in man. *L. monocytogenes* and *L. ivanovii* are closely related to *L. seeligeri, L. innocua* and *L. welshimeri.* The latter three species are thought to be non-pathogenic. *L. seeligeri* is haemolytic.

Natural Habitat

The *Listeria* spp. are widely distributed in the environment and can be isolated from soil, plants, decaying vegetation and silage (pH over 5.5) in which the bacteria can multiply. Asymptomatic faecal carriers occur in man and many animal species. *L. monocytogenes* can be excreted in bovine milk. They can grow in a temperature range of 3–45°C and within a pH range of 5.6–9.6. Silage is commonly implicated in outbreaks of listeriosis in cattle and sheep. Human foods associated with listeriosis in man include coleslaw, soft cheeses, milk and poultry meat (*see* Chapter 37).

Pathogenesis

It is thought that the pathogenic *Listeria* spp. can penetrate the epithelial barrier in the intestine and multiply in hepatic and splenic macrophages aided by the haemolysin named listerolysin O. An alternative route may be through damaged mucosal surfaces to the central nervous system, via the neural sheath of peripheral nerve endings of the trigeminal nerve. Most pathogenic bacteria require the availability of iron in the host for metabolic activities. High iron levels in silage that lead to elevated tissue concentrations of iron may predispose cattle and sheep , fed on silage, to listeriosis (**198**). **Table 41** indicates the hosts and disease syndromes caused by the pathogenic *Listeria* spp. in animals. *L. monocytogenes* infection in humans includes an influenza-like syndrome in pregnant women that may result in infection of the foetus with abortion or premature birth. The neural form of the disease can occur in neonates and in adults with lowered cell-mediated immunity. Veterinarians and abattoir workers can acquire a primary cutaneous listeriosis and this infrequently leads to a generalised form of the disease.

Table 41. Main hosts and disease syndromes of the pathogenic *Listeria* species.

Species	Host(s)	Disease syndromes
L. monocytogenes	Young animals of many species, including lambs and calves.Birds can be affected	Visceral (septicaemic) listeriosis. Necrotic foci in liver and other abdominal organs.
	Sheep, goats and cattle. Occasionally other species	Neural listeriosis: 'circling disease' with microabscesses in the brain-stem and perivascular cuffing.
	Sheep, goats, cattle	Abortion
	Cattle	Iritis, with or without other signs. Often associated with feeding big-bale silage.
L. ivanovii	Sheep and cattle	Abortion

198 

198 *Listeria monocytogenes:* neural form of listeriosis in a silage-fed sheep showing unilateral facial paralysis.

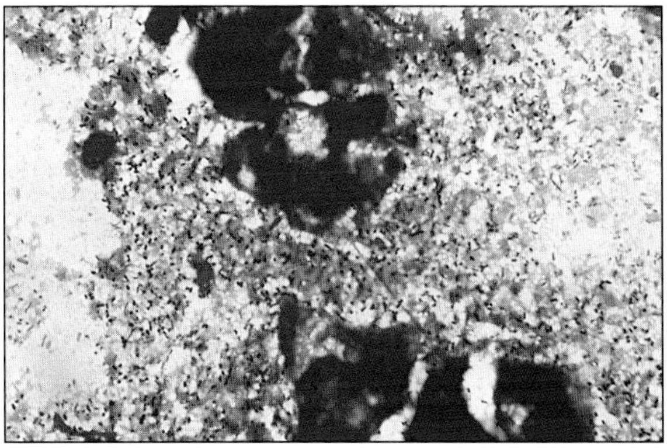 199

199 *L. monocytogenes* in a Gram-stained smear of material from a placenta (bovine abortion). (×1000)

Laboratory Diagnosis

Specimens

- Visceral form: material from lesions in liver, kidneys or spleen.
- Neural form: spinal fluid, brain stem, tissue from several sites in the medulla oblongata.
- Abortion: placenta (cotyledons), foetal abomasal contents and/or uterine discharges. A full range of specimens should be submitted so that an examination can be made for the other pathogens capable of causing abortion.

Direct microscopy

Stained smears are not as useful in listeriosis as they are in some other diseases. Smears from lesions may reveal Gram-positive rods (often coccobacillary) (**199**) but isolation should always be attempted. Histopathological examination of fixed (10 per cent formalin) brain tissue can often give a presumptive diagnosis of neural listeriosis. Microabscesses in the brain stem, usually unilateral (**200**), together with perivascular cuffing (**201**) is very characteristic of listeriosis.

Isolation

The routine medium for inoculation of specimens is ox or sheep blood agar and a MacConkey agar plate to detect any Gram-negative pathogens or contaminants. Selective media include blood agar with an antibiotic supplement (as described for *Staphylococcus* spp.) or blood agar containing 0.05 per cent potassium tellurite (inhibitory to Gram-negatives). Commercial selective and indicator media are available, such as Listeria Selective agar (Oxoid) and these are designed mainly for the isolation of listeria from human foodstuffs.

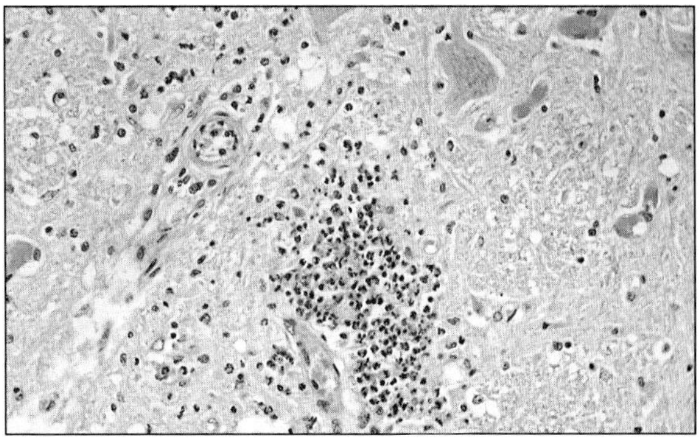 200

200 A microabscess in the medulla of a sheep with listeriosis. (H&E stain, ×400)

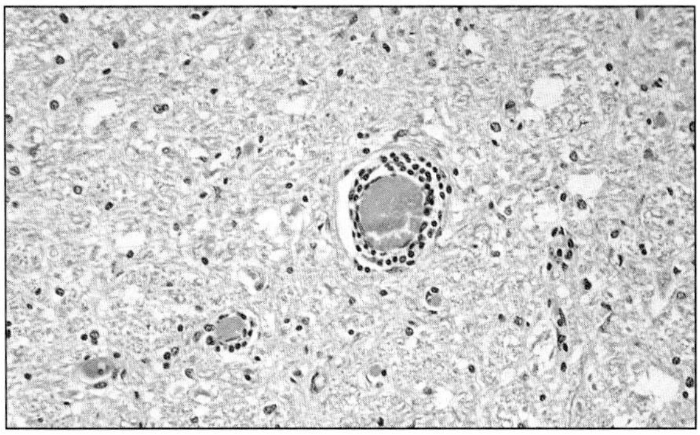 201

201 Perivascular cuffing in an ovine medulla indicative of the neural form of listeriosis. (H&E stain, ×400)

Specimens from the visceral form of the disease and from abortion cases are inoculated directly onto the laboratory media and incubated aerobically at 37°C for 24–48 hours.

A 'cold-enrichment' procedure is necessary for brain tissue from neural listeriosis. Small pieces of spinal cord and medulla are homogenized and a 10 per cent suspension is made in a nutrient broth. The broth suspension is placed in the refrigerator at 4°C and subcultured onto blood agar once weekly for up to 12 weeks. This method selects for *L. monocytogenes*, one of the few pathogens able to grow at refrigerator temperature.

Identification

Colonial appearance

Small transparent colonies with smooth borders appear on blood agar in 24 hours, becoming greyish-white and 0.5–2.0 mm in diameter in 48 hours. *L. ivanovii* produces a comparatively wide zone of haemolysis and is very similar in appearance to a beta-haemolytic streptococcus (**202**). *L. monocytogenes* and the non-pathogenic *L. seeligeri* have narrow zones of beta-haemolysis, often only under the colony itself.

Microscopic appearance

Short Gram-positive rods or coccobacilli are seen at 24 hours (**203**) with a tendency for cells from older cultures to decolourise. There are often many coccal forms in smears from young rapidly growing colonies. This, and the colonial appearance on blood agar, can lead to confusion between the pathogenic listeria and beta-haemolytic streptococci.

Biochemical and other tests

All the *Listeria* spp. hydrolyze aesculin (aesculin broth). *L. monocytogenes*, particularly, shows the characteristic 'tumbling motility' when a 2–4 hour broth culture, incubated at 25°C, is examined by the hanging-drop method. This motility is an end-over-end tumbling of individual cells with periods of quiescence. When grown in semisolid motility media the *Listeria* spp. give an unusual umbrella-shaped growth in the subsurface (**204**). Modified CAMP tests with *Staphylococcus aureus* and with *Rhodococcus equi* are useful to differentiate the two pathogenic species (**205–207**). **Table 42** gives a short list of tests for differentiating the *Listeria* spp. The API 20S (Analytab Products) also makes provision for the identification of *L. monocytogenes*.

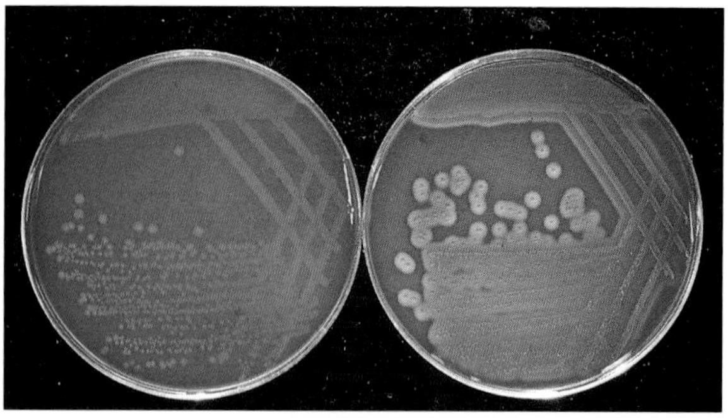

202

202 *L. monocytogenes* (left) and *L. ivanovii* (right) on sheep blood agar. The haemolysis of *L. ivanovii* tends to be more pronounced than that of *L. monocytogenes*.

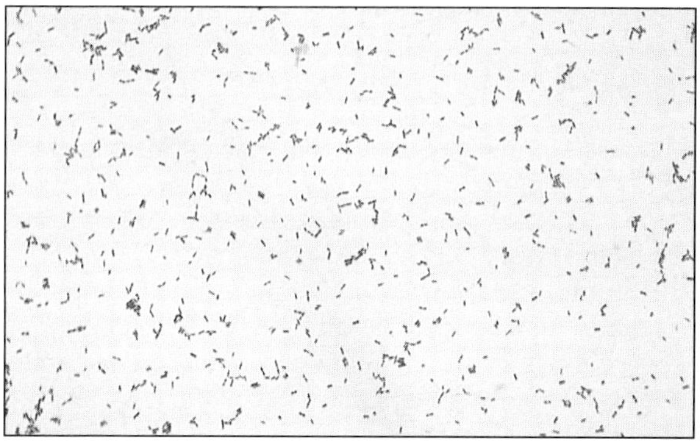

203

203 *L. monocytogenes* in a stained smear from a culture with both Gram-positive rods and cocci present. (Gram stain, ×1000)

204

204 Umbrella-shaped, sub-surface growth of *L. monocytogenes* in semisolid motility medium.

205

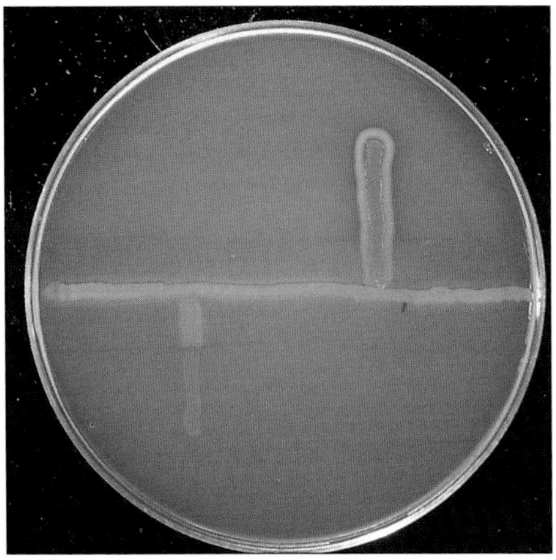

206

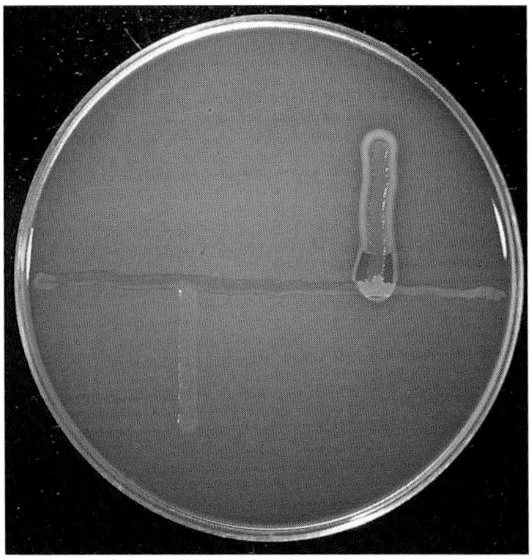

205 CAMP test with *Staphylococcus aureus* (horizontal) showing enhancement of the effect of the staphylococcal beta-haemolysin by *L. monocytogenes* (left) but not by *L. ivanovii* (right).

206 CAMP test with *Rhodococcus equi* (horizontal): no reaction by *L. monocytogenes* (left) and enhancement of haemolysis by *L. ivanovii* (right).

207 *Rhodococcus equi* streaked across (left to right) a vertical streak of *L. ivanovii* giving an enhanced haemolytic effect.

207

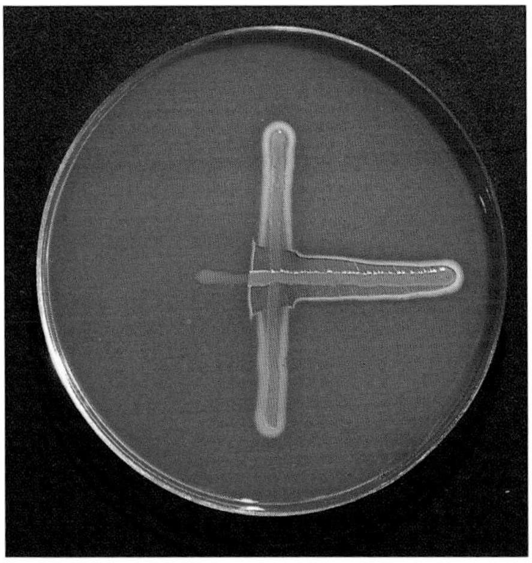

Table 42. Differentiation of the *Listeria* species.

Species	Beta-haemolysis	CAMP tests with		Fermentation of			Nitrate reduction
		*S. aureus**	*R. equi***	Mannitol	L-Rhamnose	D-Xylose	
L. monocytogenes	+	+	–	–	+	–	–
L. ivanovii	++	–	+	–	–	+	–
L. seeligeri	+	+	–	–	–	+	–
L. innocua	–	–	–	–	v	–	–
L. welshimeri	–	–	–	–	v	+	–
L. grayi	–	–	–	+	v	–	–
L. murrayi	–	–	–	+	–	–	+

v = variable reactions; * *Staphylococcus aureus*; ** *Rhodococcus equi.*

Summary of the tests for the presumptive identification of L. monocytogenes

- Small beta-haemolytic colonies at 24 hours on sheep or ox blood agar.
- Gram-positive rods tending to show many coccal forms.
- Catalase-positive (beta-haemolytic streptococci and *Erysipelothrix rhusiopathiae* are catalase-negative).
- Aesculin hydrolysis (aesculin broth).
- Tumbling motility (hanging-drop) from a 2–4 hour broth culture at 25°C.
- Subsurface umbrella-shaped growth in semisolid motility media.
- CAMP test positive with *Staphylococcus aureus*.
- Acid production from rhamnose but not from xylose.

Animal inoculation

This can be used to evaluate the virulence of *Listeria* spp., but is not employed routinely.

a) Anton test: inoculation of the conjunctiva of a rabbit or guinea-pig. Only *L. monocytogenes* causes a purulent keratoconjunctivitis within 24–36 hours of inoculation.
b) Intraperitoneal inoculation of mice with a 24-hour broth culture. Both *L. monocytogenes* and *L. ivanovii* are pathogenic for mice. They die within 5 days with necrotic lesions present in the liver.

Serotyping

This is conducted in reference laboratories. It is based on flagellar and somatic antigens and identifies 16 serovars. Except for serovar 5 (*L. ivanovii*) the serovars are not species specific. The non-pathogenic *L. innocua* and *L. seeligeri* share one or more common antigens with *L. monocytogenes*, whereas *L. grayi* and *L. murrayi* do not share antigens with the other species.

14 *Erysipelothrix rhusiopathiae*

Erysipelothrix rhusiopathiae (previously *E. insidiosa*) is at present the only species in the genus. *E. tonsillarum* has been proposed for serotype 7 of *E. rhusiopathiae* as it is avirulent and genetically distinct. *E. rhusiopathiae* from S-form (smooth) colonies and usually from acute syndromes is a Gram-positive rod 0.2–0.4 µm by 0.8–2.5 µm in size. R-form (rough) colonies and chronic forms of the disease can yield Gram-positive filaments up to 20 µm in length. The bacterium is catalase-negative and oxidase-negative, non-spore-forming, non-acid-fast, non-motile and a facultative anaerobe but growth is enhanced by 10 per cent CO_2. It is able to grow in a temperature range of 5°C–42°C, within a pH range 6.7–9.2 and at an 8.5 per cent concentration of sodium chloride. Growth occurs on nutrient agar but is improved by the addition of serum or blood. It will not grow on MacConkey agar.

Natural Habitat

E. rhusiopathiae occurs worldwide and is persistent and widespread in the environment with proliferation in organic matter. The bacterium can be isolated from many animal and bird species and from the slime layer of fish, but it is most commonly associated with pigs. It is present in the soil and slurry of piggeries and can be recovered from the faeces and tonsils of carrier pigs.

Pathogenesis

Strains of *E. rhusiopathiae* vary in virulence. The more virulent strains produce high levels of neuraminidase that can cause vascular damage and thrombus formation. Injury to articular cartilage is thought to be due to an immunological response to persistent antigen in the synovial fluid. **Table 43** indicates the main hosts and disease syndromes of the bacterium. Swine erysipelas (**208**) is a common disease.

Laboratory Diagnosis

Specimens

Liver, spleen, kidney, heart and synovial tissue can be taken from necropsy examination. Recovery of the organism from skin lesions and chronic forms of the disease may be difficult.

Direct microscopy

The morphology of *E. rhusiopathiae* may vary with the disease syndrome, but usually:

- Acute cases of the disease: Gram-positive rods (**209**).
- Chronic forms of the disease: Gram-positive filaments that tend to decolourise (**210**).

Table 43. Main hosts and diseases of *Erysipelothrix rhusiopathiae.*

Main host(s)	***Disease / disease syndrome***
Pigs	Swine erysipelas: • Acute septicaemic form (pregnant sows may abort) • Urticarial form ('diamond skin disease') • Vegetative endocarditis (chronic form) • Polyarthritis (chronic form)
Turkeys, geese and other birds	• Acute septicaemia • Vegetative endocarditis (chronic) • Arthritis (chronic)
Sheep: young	Polyarthritis (chronic) via umbilicus or wounds
: adult	Post-dipping lameness: cellulitis with extension to laminae of feet
Dolphins, cattle, dogs, horses and rabbits	Occasional infections of varying severity
Humans	Erysipeloid (localised cellulitis) and rarely endocarditis or arthritis. Occupational hazard for workers in fish, poultry and agricultural industries

Isolation

Routine isolation is made on sheep or ox blood agar with a MacConkey plate to aid in the detection of any Gram-negative pathogens or contaminants. Selective media contain sodium azide (0.1 per cent) and crystal violet (0.001 per cent). The plates are usually incubated aerobically for 24–48 hours at 37°C although the growth is enhanced by 10 per cent CO_2.

Identification

Colonial appearance
Non-haemolytic pin-point colonies (0.5 mm) appear at 24 hours' incubation (**211**). Colonial variation becomes obvious at 48 hours' incubation when a zone of greenish haemolysis often develops under and just around the colonies (**212**). The S-form colonies are 0.5–1.5 mm in diameter, convex and circular with an entire edge. The large R-form colonies are flatter, more opaque and have an irregular edge.

Microscopic appearance
The S-form colonies tend to yield medium sized Gram-positive rods while Gram-stained smears from the R-form colonies reveal Gram-positive filaments of varying lengths (**213**).

Biochemical reactions
E. rhusiopathiae is catalase-negative and oxidase-negative. It is non-motile and does not hydrolyse aesculin or produce urease. A characteristic reaction is produced when TSI agar is stab inoculated. When incubated at 37°C for 24 hours H_2S is produced as a thin, black line just along the inoculation stab (**214**). The R-forms give a 'bottle-brush' type of growth in stab cultures of nutrient gelatin incubated at room temperature. As this reaction takes about 5 days it is not very useful as a diagnostic test.

E. rhusiopathiae usually ferments lactose, glucose, levulose and dextrin, but acid production is poor or inconsistent when the bacterium is tested in 1 per cent (w/v) peptone water. Carbohydrate tests can be carried out in peptone water with added sterile horse serum (5–10 per cent) or in nutrient broth plus the test carbohydrate (0.5–1 per cent) with phenol red as the indicator. However, the fermentation pattern varies with the basal medium used. It is advisable to establish the fermentation pattern of known strains in the basal medium of choice.

Animal inoculation
This is not used routinely but both mice and pigeons will die within 4 days after intraperitoneal inoculation with 0.1–0.4 ml of a broth culture from a virulent strain. A mouse protection test, using commercial antiserum, can be used to confirm identification.

Serotyping
Serotyping is carried out only in reference laboratories. Twenty-two serovars of *E. rhusiopathiae* have been identified on the basis of heat-stable somatic antigens. Serovars 1 and 2 account for 70–80 per cent of all isolates. *E. tonsillarum* is the name proposed for the avirulent serovar 7. This serovar can be distinguished from other *E. rhusiopathiae* strains only by DNA homology comparisons.

208

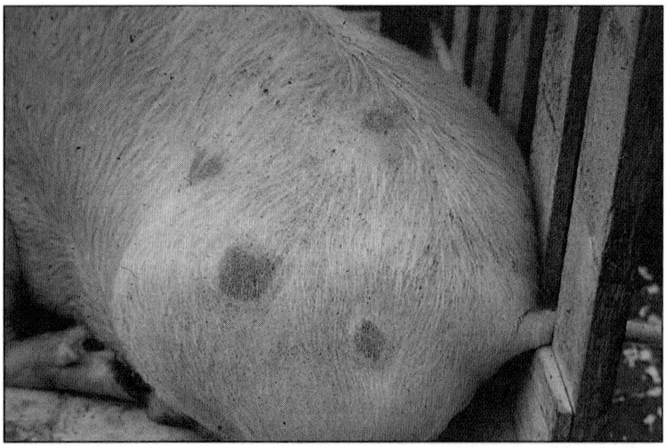

208 *Erysipelothrix rhusiopathiae:* pathognomonic diamond-shaped, red, urticarial plaques of swine erysipelas (an acute form of the disease).

209

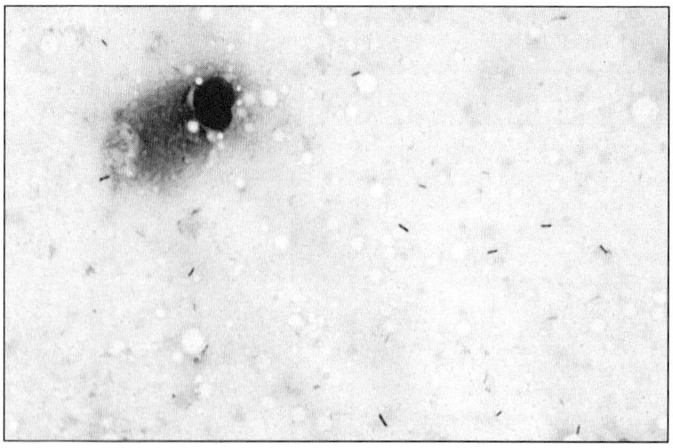

209 *E. rhusiopathiae* in an impression smear from the liver of a pig with the acute septicaemic form of swine erysipelas. The small Gram-positive rods represent the smooth form of the bacterium. (Gram stain, ×1000)

210

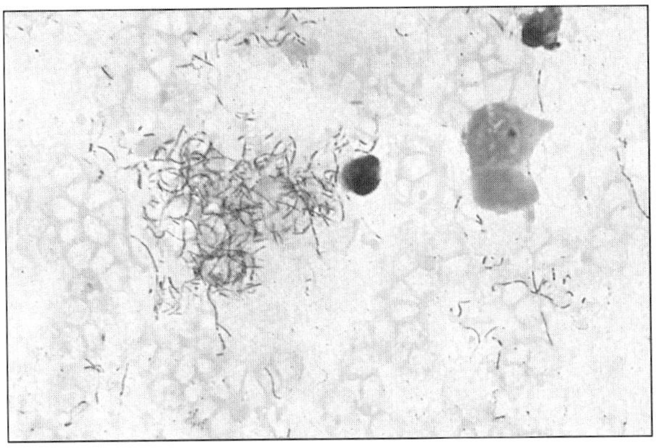

210 The rough form of *E. rhusiopathiae* showing Gram-positive filaments in vegetative lesions on the heart valves of a pig with one of the chronic forms of swine erysipelas. (Gram stain, ×1000)

211

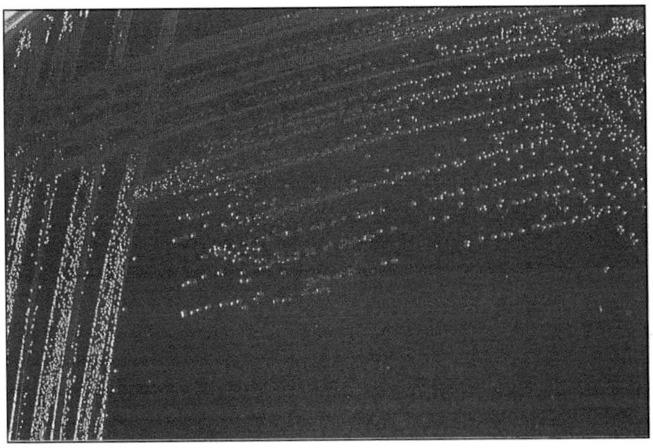

211 Close-up of *E. rhusiopathiae* showing non-haemolytic colonies after 24 hours' incubation on sheep blood agar.

212

212 *E. rhusiopathiae* colonies, showing haemolysis, after 48 hours' incubation on sheep blood agar.

213

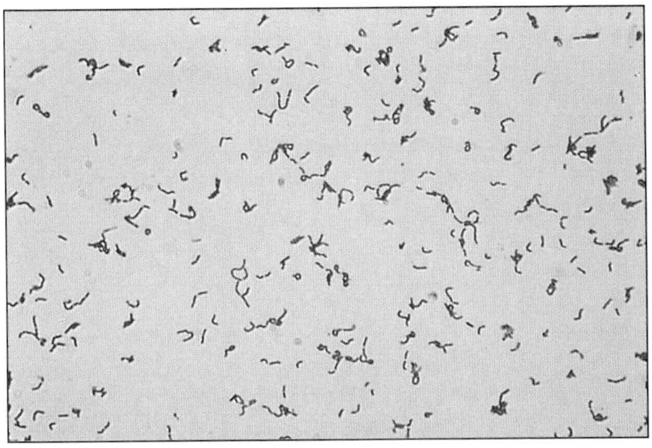

213 *E. rhusiopathiae* in a Gram-stained smear from a culture. There are small Gram-positive rods but also some short filaments indicating that the colonies are changing to the rough form. (Gram stain, × 1000)

214

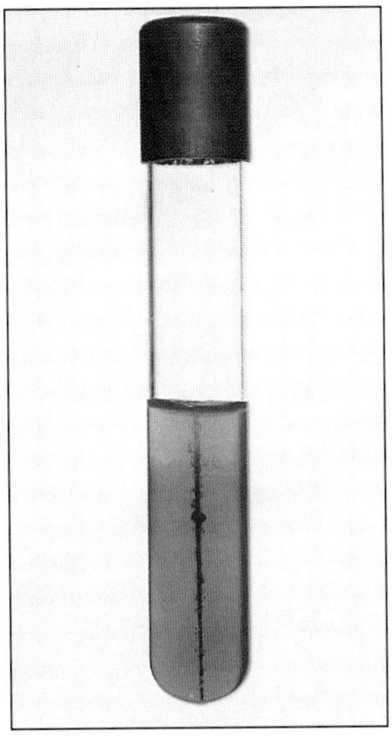

214 *E. rhusiopathiae* stab-inoculated into a tube of TSI agar. It characteristically produces a small amount of H_2S confined to the stab line.

15 *Bacillus* species

Bacillus species are large, Gram-positive or Gram-variable, endospore-forming rods. They are catalase-positive, aerobic or facultatively anaerobic and motile with the exception of *B. anthracis* and *B. mycoides*. Most will grow on nutrient agar but not on MacConkey agar. The bacilli can be divided into three main groups on the basis of endospore and mother cell (sporangium) morphology. This is illustrated in **Diagram 30**.

Natural Habitat

Most of the numerous *Bacillus* species are saprophytes widely distributed in air, soil and water. *B. anthracis* endospores can survive in soil for up to 50 years and are present in geographically limited, endemic areas. It is thought that 'incubator areas' exist where, for short periods, germination of spores and multiplication of vegetative cells can occur. These areas have a warm climate and an alkaline, calcareous soil that is subject to periodic flooding.

Pathogenicity and Pathogenesis

The majority of the *Bacillus* species have little or no pathogenic potential but can occur commonly as contaminants on laboratory media. Some species cause disease in insects, such as *B. larvae* that is associated with American foulbrood in bees. *B. anthracis* is the major animal pathogen in the genus, the cause of anthrax in both animals and man. *B. cereus* produces food poisoning in man and rare infections in animals. The taxonomic position of '*B. piliformis*' is undecided, the bacterium is responsible for Tyzzer's disease in laboratory mice and foals. *B. licheniformis* has been reported as causing abortions in cattle and sheep.

B. anthracis has a plasmid encoded 'tripartite' protein toxin with protective, lethal and oedema factors. The toxin is leukocidal, increases vascular permeability and produces capillary thrombosis causing shock. The polypeptide capsule is antiphagocytic. To be fully virulent, *B. anthracis* must produce both the tripartite toxin and the capsule.

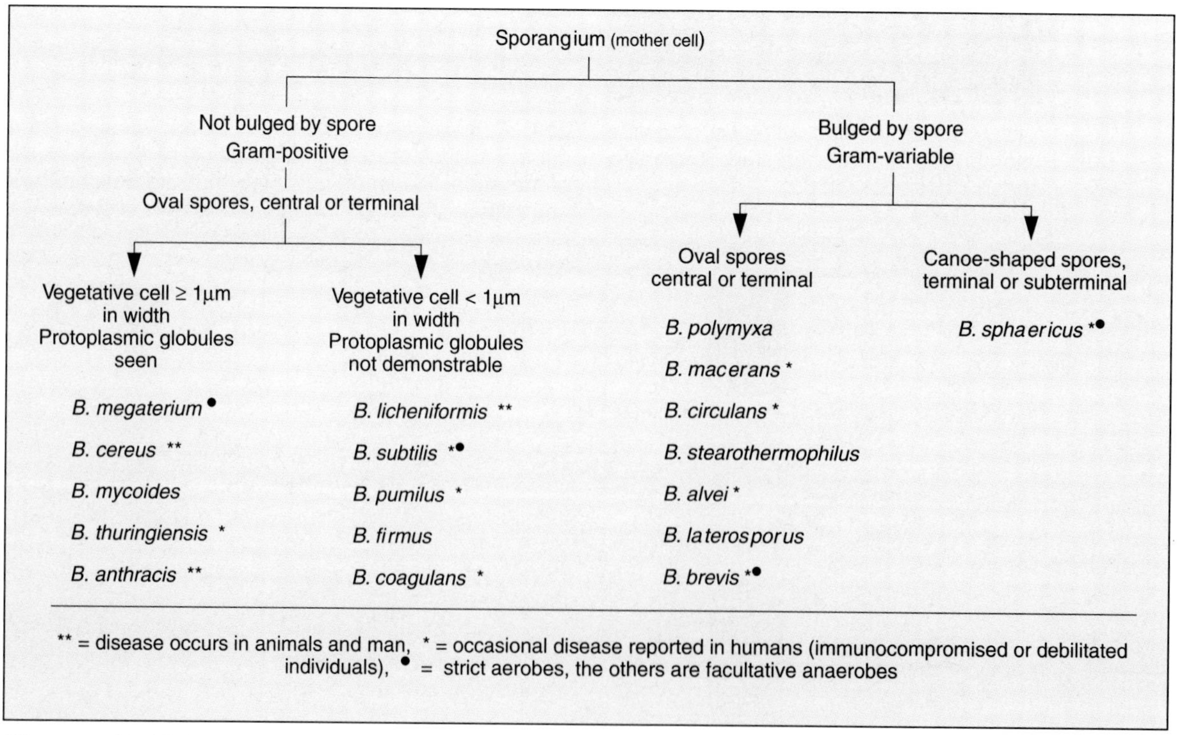

Diagram 30. Division of *Bacillus* species on the basis of endospore and mother-cell (sporangium) morphology.

Table 44. Main diseases and hosts of the *Bacillus* species.

Bacillus species	Host(s)	Disease
B. anthracis	Cattle and sheep	Septicaemic form of anthrax. Usually sudden death
	Pigs	Subacute anthrax with oedematous swelling in pharyngeal tissues and regional lymphadenitis or intestinal form with a higher mortality
	Horses	Oral route: septicaemia with colic and enteritis. Wound infections: localised oedema and lymphadenitis
	Carnivores (including mink)	Comparatively resistant. Disease pattern similar to that in pigs. A massive dose from eating anthrax-infected carcasses can lead to septicaemia
	Humans	Skin form: 'malignant pustule'. Pulmonary ('wool-sorters' disease') and intestinal forms are often fatal
B. cereus	Humans	Food poisoning
	Cattle	Rare cases of mastitis
B. licheniformis	Cattle and sheep	Reported as a cause of abortion
'B. piliformis' (taxonomy uncertain)	Laboratory mice, foals and other animals	Tyzzer's disease. An acute fatal infection causing hepatitis, enteritis and colitis

Anthrax can occur in virtually all mammalian species but birds are highly resistant. The main routes of entry of endospores are by ingestion, from soil when grazing or in contaminated food, and by infection of wounds. Inhalation of spores occurs in man but to a lesser extent in animals. Transmission by biting insects may be important especially during an outbreak of anthrax. Cattle, sheep and goats are most susceptible to infection, horses and humans occupy an intermediate position, while pigs and carnivores are comparatively resistant, but can succumb if the infective dose is high. Anthrax is a peracute disease in cattle and sheep, characterised by septicaemia and sudden death. Postmortem findings include exudation of tarry blood from body orifices, failure of the blood to clot, incomplete rigor mortis and splenomegaly in cattle. In the less susceptible species, inflammatory oedema of face, throat and neck is a common finding and colic can occur in horses and gastroenteritis in carnivores.

B. cereus produces an enterotoxin that is responsible for food poisoning in man. For further details of this bacterium *see* the 'Food poisoning' section (Chapter 37). **Table 44** summarises the main hosts and diseases of the pathogenic bacilli.

Laboratory Diagnosis

Bacillus cereus group

(B. anthracis, B. cereus, B. mycoides and *B. thuringiensis)*

Great care should be exercised when dealing with specimens from suspected cases of anthrax. All procedures should be carried out in a biohazard safety cabinet and infective and contaminated materials subsequently autoclaved. Stained smears, that have been heat-fixed, are potentially dangerous as they may contain viable spores. It should be remembered that the endospores of *B. anthracis* can remain viable for up to 50 years.

Specimens

Bacillus anthracis

A postmortem examination is usually unnecessary and should never be carried out unless the carcass can be taken to a place where, subsequently, the surrounding area can be thoroughly decontaminated. Because of the risk of human infection, personnel carrying out the postmortem should take adequate safety precautions. Endospores are not formed in the animal body but sporulation is triggered when vegetative cells are exposed to air, as happens during necropsy. If anthrax is suspected in cattle or sheep, thin blood smears should be made from blood taken from ear or tail veins. In horses and pigs, oedematous fluid can be collected from localised sites and in pigs peritoneal fluid is often more useful, diagnostically, than blood smears. Blood or homogenized spleen can be used for culture.

Direct microscopy

B. anthracis produces a capsule *in vivo* and either Giemsa or polychrome methylene blue stains are used to demonstrate the capsule which is of diagnostic importance. The capsular material is more abundant if the blood smear has been taken from a recently dead animal. Polychrome methylene blue-stained smears reveal square-ended, blue rods in short chains surrounded by pink capsular material (the M'Fadyean reaction) and is characteristic for *B. anthracis* (**215**). The capsule is reddish-mauve in a Giemsa-stained smear. Care should be taken when handling stained smears as viable spores may be present in the material. Some laboratories employ a fluorescent antibody test but cross-reactions with other *Bacillus* species may occur.

215

215 *Bacillus anthracis* in a bovine blood smear collected from a peripheral blood vessel showing square-ended, blue bacilli in short chains surrounded by a pink capsule. (Polychrome methylene blue stain, ×1000)

Isolation

Bacillus species in the *B. cereus* group grow well on sheep or ox blood agar, aerobically at 37°C in 24-48 hours. A MacConkey agar plate could also be inoculated with the specimen as a check on Gram-negative contaminants. A selective medium for *B. anthracis* has been described by Knisely (1966) and is a polymyxin-lysozyme-EDTA-thallous (PLET) acetate agar (**Appendix 2**). Contaminated specimens such as hair, bonemeal and other animal feeds should be ground finely, steeped in saline and then heated at 65°C for 10 minutes. On cooling, the suspension is strained through gauze and centrifuged. The deposit can be used for culture or animal inoculation. Commercial selective media, such as '*Bacillus cereus* selective agar base' (Oxoid) with a polymyxin supplement are available (*see* 'Food Poisoning' section Chapter 37).

Identification

Colonial morphology

B. anthracis is almost always non-haemolytic; rarely strains show weak haemolysis. After 48 hours' incubation the colonies are about 5 mm in diameter, flat, dry, greyish with a granular 'ground-glass' appearance (**216**). Under low magnification, curved and curled peripheral projections at the edge of the colonies give rise to a 'Medusa head' appearance.

B. cereus has colonies similar to those of *B. anthracis* but they tend to be slightly larger, have a slightly greenish hue and most strains are surrounded by a wide zone of complete haemolysis (**217**). The colonies of *B. thuringiensis* have a similar appearance to those of *B. cereus*.

B. mycoides has markedly rhizoid colonies that can have an almost fungal appearance. This is best seen if a nutrient agar plate is inoculated centrally and incubated at 25–30°C for a few days (**218**). Most strains are weakly haemolytic.

Microscopic appearance

All four species in the *B. cereus* group are strongly Gram-positive from young cultures and about 1 × 3–5 µm in size. *B. anthracis* often occurs in long chains (**219**). Endospores may be produced in older cultures and appear as oval, non-stained areas within the mother cell. *B. thuringiensis*, characteristically, has cuboid or diamond-shaped parasporal crystals in the cells. These glycoprotein crystals can be best seen in phase-contrast preparations from cultures over 2 days old.

Biochemical and other tests

The four species in the *B. cereus* group are closely related, however, they can be differentiated relatively easily on colonial morphology and on a few other characteristics as shown in **Table 45**. *B. anthracis* is highly susceptible to penicillin while *B. cereus* and the other two species are resistant (**220**). *B. cereus*, *B. mycoides* and *B. thuringiensis* rapidly liquefy nutrient gelatin while *B. anthracis* slowly produces an inverted fir tree type of liquefaction with side-shoots radiating from the stab line (**221**). All show lecithinase activity on egg yolk agar (**222**) but the reaction of *B. anthracis* is weak.

B. anthracis and the majority of the *Bacillus* species do not normally produce capsules in or on laboratory media and the colonies have a dry appearance. However, *B. anthracis* can be induced to produce a capsule by growing it on nutrient agar containing 0.7 per cent sodium bicarbonate under 10 per cent CO_2. The colonies are quite mucoid.

Animal inoculation

This is carried out only if any doubt remains about the identity of *B. anthracis*. Virulent *B. anthracis* strains are much more pathogenic than other *Bacillus* species and are highly invasive. A light suspension of *B. anthracis* placed on a scarified area on the base of a mouse's tail can cause death. Large doses of *B. cereus*, given to a mouse or guinea-pig subcutaneously or intraperitoneally, are needed before the bacterium proves fatal.

Ascoli test

This thermoprecipitation test is used if viable *B. anthracis* can no longer be demonstrated in tissues. About 2–3 g of homogenized material in a little saline is briefly boiled and passed through filter paper. This filtrate is used as the antigen in a ring precipitation or gel diffusion test with known *B. anthracis* precipitating antiserum.

216

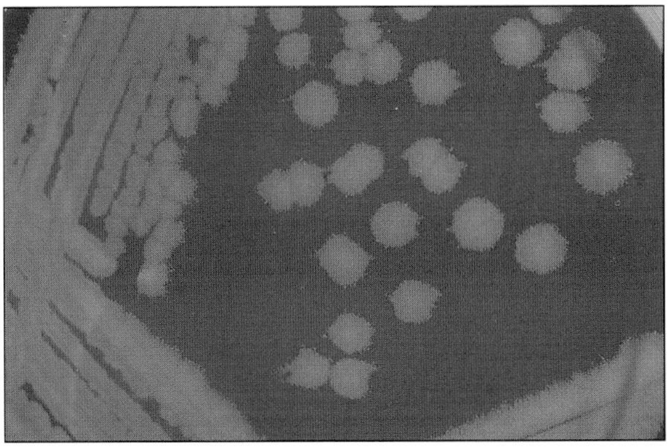

216 *B. anthracis* on sheep blood agar illustrating non-haemolytic, flat, 'ground-glass', dry colonies with irregular edges.

217

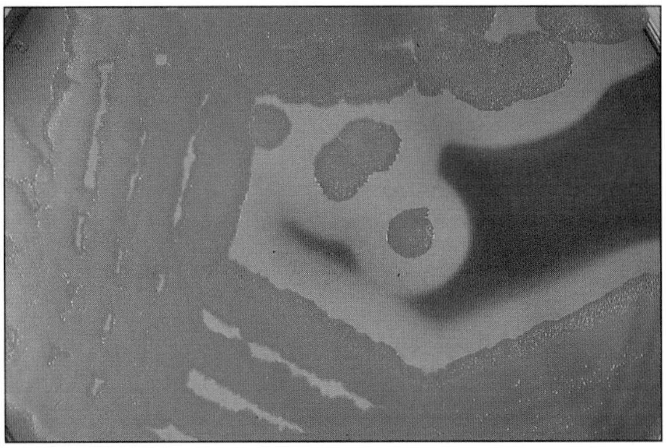

217 *B. cereus* on sheep blood agar. The morphology resembles that of *B. anthracis* but the colonies are usually strongly haemolytic.

218

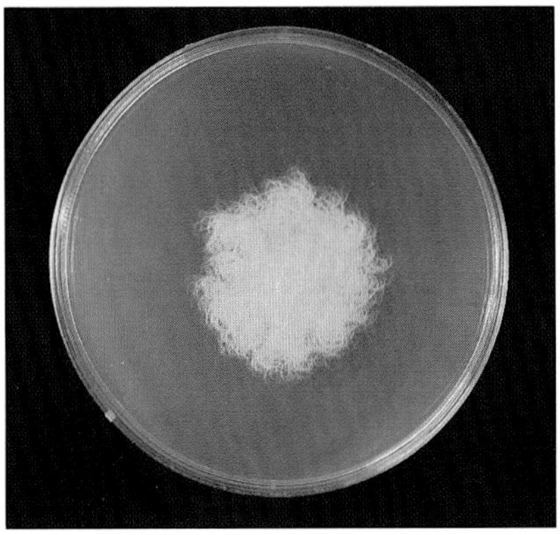

218 *B. mycoides* inoculated centrally on nutrient agar and incubated at 25°C for 3 days. It has a rhizoid-type colony.

219

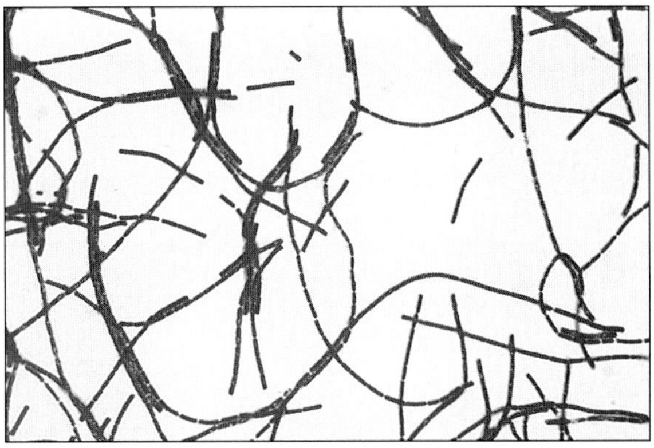

219 *B. anthracis* from a culture: Gram-positive rods in long chains with incipient sporulation. (Gram stain, ×1000)

220

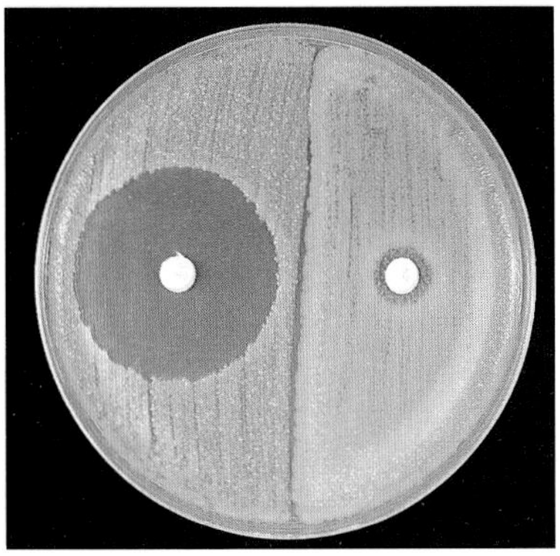

220 *B. anthracis* (left) and *B. cereus* (right) on Isosensitest agar demonstrating the susceptibility of *B. anthracis* to penicillin (10 unit disc) compared to the resistance of *B. cereus*.

221

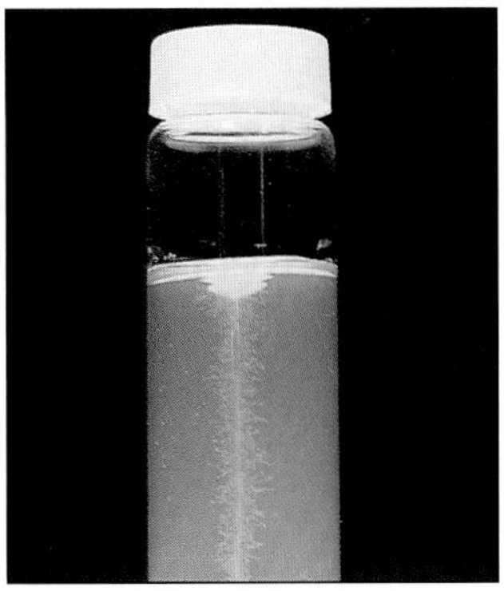

221 *B. anthracis* stab inoculated into nutrient gelatin giving the characteristic 'inverted fir-tree' type growth after 8 days at 25°C.

222

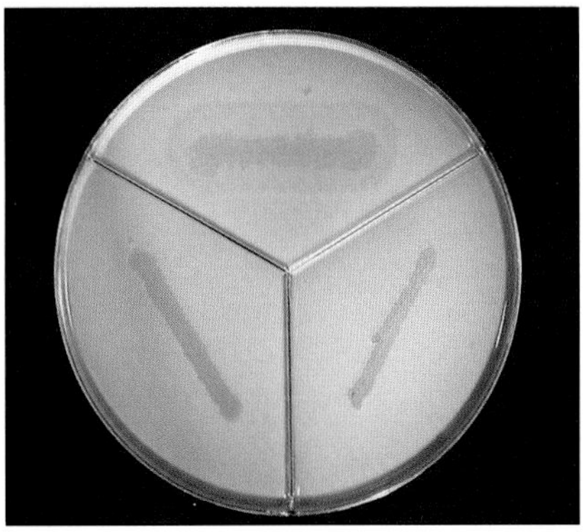

222 Strong lecithinase activity by *B. cereus* (top) on egg yolk agar after 24 hours' incubation. *B. anthracis* (left) gives a weak opaque zone after 48 hours and *B. licheniformis* (right) is unreactive on this medium.

Table 45. Summary of the differentiating characteristics of the members of the *Bacillus cereus* group.

Tests	*B. anthracis*	*B. cereus*	*B. mycoides*	*B. thuringiensis*
Motility	–	+	–	+
Crystalline, parasporal inclusions (3-day cultures)	–	–	–	+
Haemolysis	– (or weak)	+	weak	+
Penicillin susceptibility (10-unit disc)	S	R	R	R
Gelatin stab culture	'inverted fir tree' type of growth	← rapid liquefaction		→
Lecithinase activity (egg-yolk agar)	+ weak	+	+	+
Nutrient agar with 0.7% Na bicarbonate under 10% CO_2	mucoid colonies	unchanged	unchanged	unchanged
Susceptibility to cherry gamma phage*	+ (lysis)	–	(+) lysis may occur	–
Pathogenicity for mice or guinea-pigs (subcut. or i/v)	+ (death in 24-–48 hours)	+ large dose (non-invasive)	–	–

R = resistant; S = sensitive

* Available from Center for Disease Control, Atlanta, Georgia, USA.

Other *Bacillus* species

A definitive identification of the numerous *Bacillus* species, other than *B. anthracis*, requires a range of tests, as described by Turnbull and Kramer (1991), that are usually carried out in reference laboratories. The API 50CH test strip (Analytab Products) is designed to identify up to 38 *Bacillus* species and subspecies. *B. subtilis*, *B. licheniformis* and *B. circulans* are commonly seen on diagnostic culture plates as contaminants. The colonial morphology is comparatively distinctive but their identity should be confirmed by biochemical tests.

B. subtilis has round to irregular colonies with a dull, granular, cream to brownish surface (**223**). It is variably haemolytic on blood agar. Active spreading of the colonies can occur on agar with a moist surface. *B. subtilis* is Gram-positive in smears made from young colonies. The cells are 0.6–0.8 × 2–3 μm in size and the endospores, that are widespread in the environment, are ellipsoidal, central and do not bulge the sporangium.

B. licheniformis colonies are opaque, dull, rough, wrinkled, strongly adherent to the agar and hair-like outgrowths are common (**224**). It is named for the similarity of the colonies to lichen. The cells and endospores are similar to those of *B. subtilis*.

B. circulans is a species that is genetically heterogeneous. Some strains are unusual in that the colonies themselves are motile and move over the surface of an agar plate in a circular manner. As a parent colony moves, cells are left behind and these in turn form colonies that are motile (**225** and **226**). This bacillus is described as Gram-variable and the cells are often Gram-negative even in smears from young cultures. The cells are 0.5–0.7 × 2–5 μm in size and the spores are ellipsoidal, variable in position and bulge the mother cell.

223

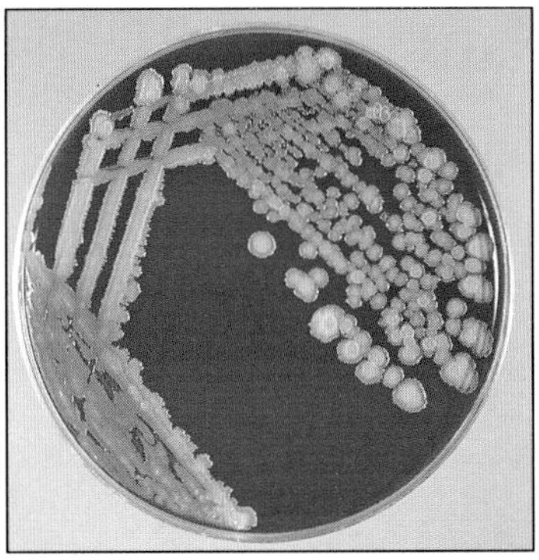

223 *B. subtilis* on sheep blood agar showing the dull, wrinkled, irregular colonies. Some strains are haemolytic.

224

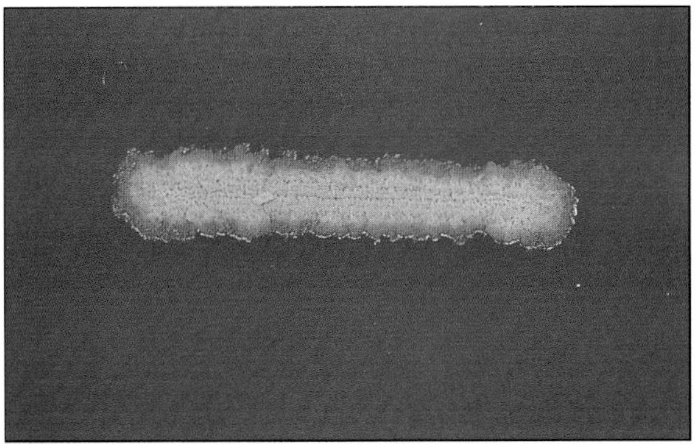

224 *B. licheniformis* on sheep blood agar inoculated as a streak to show the heaped, wrinkled, lichen-like appearance. Some strains are haemolytic.

225

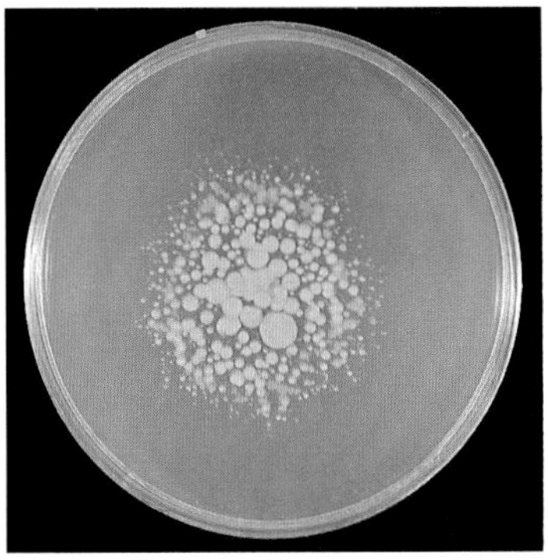

225 *B. circulans* centrally inoculated to show the motile colonies moving outwards over the agar surface. This plate was incubated at 25°C for 72 hours.

226

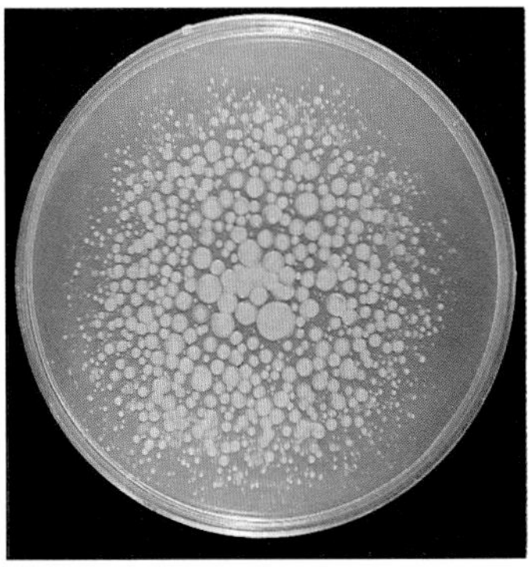

226 *B. circulans:* the same plate as **225** to show the progress of the motile colonies after a further 48 hours' incubation at 25°C.

References

Knisely, R.F. (1966). Selective medium for *Bacillus anthracis*. *Journal of Bacteriology*, **92**: 784–786.

Turnbull, P.C.B. and Kramer, J.H. (1991). *Bacillus* pp. 296–303. In *Manual of Clinical Microbiology*, 5th ed. Balows, A. (ed. in chief). American Society for Microbiology. Washington D.C., USA.

16 Non-Spore-Forming Anaerobic Bacteria

The non-sporing obligate anaerobes constitute a large group of Gram-positive and Gram-negative bacteria that exist in the environment but also as commensals on mucous membranes of animals and humans, particularly in the intestinal tract as part of the normal flora. Knowledge of these bacteria is incomplete as they can be nutritionally demanding and require strict anaerobic conditions for isolation. They are commonly implicated in necrotic and suppurative conditions, often as mixed infections with facultative anaerobic bacteria. **Diagram 31** briefly summarises the more important genera of these non-spore-forming anaerobes.

Pathogenicity

The infections are often endogenous arising from normal flora at the site or by wounds contaminated by nearby flora. For these strict anaerobes to multiply at a focus in animal tissue the redox potential of the area must be lowered. This can occur through trauma and necrosis, ischaemia, parasitic invasion or concomitant multiplication of facultative anaerobes. The conditions caused by these non-sporing anaerobes include soft- tissue abscesses and cellulitis, post-operative wound infections, periodontal abscesses, aspiration pneumonia, lung and liver abscesses, peritonitis, pleuritis, myometritis, osteomyelitis, mastitis and foot rot. Some of the more common infections are shown in **Table 46**.

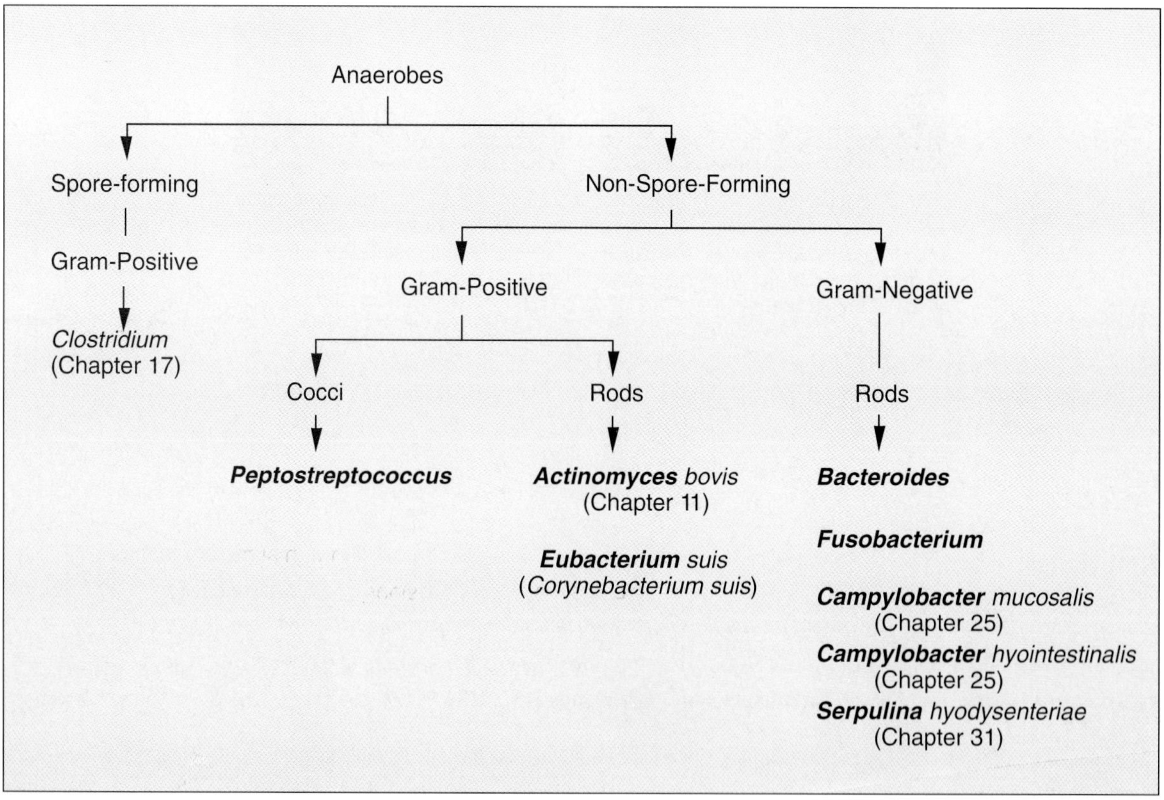

Diagram 31. Anaerobes of veterinary importance.

Table 46. Diseases caused by non-spore-forming anaerobes.

Non-spore-forming anaerobe	Associated pathogens	Host(s)	Disease
Peptostreptococcus indolicus	*Actinomyces pyogenes*	Cattle	Summer mastitis
Eubacterium suis		Sows	Pyelonephritis (normal flora in preputial diverticulum of boars)
Actinomyces bovis		Cattle	Lumpy jaw
		Horses	Fistulous withers and poll evil
		Sows	Granulomatous mastitis
Campylobacter mucosalis* and/or *C. hyointestinalis		Pigs	Proliferative intestinal adenomatosis complex
Serpulina hyodysenteriae		Pigs	Swine dysentery
Bacteroides nodosus	*Fusobacterium necrophorum, Actinomyces pyogenes, Treponema penortha*	Sheep	Contagious (virulent) foot rot
B. nodosus		Goats, cattle, pigs	Occasional infections of interdigital skin
B. melaninogenicus	*Fusobacterium necrophorum, Actinomyces pyogenes*	Cattle	Foot rot
B. melaninogenicus		Cattle, sheep, dogs and cats	Suppurative conditions
B. asaccharolyticus		Dogs, cats, horses, cattle	Osteomyelitis
B. fragilis		Calves, lambs, foals, piglets	Diarrhoeal disease (enterotoxigenic strains)
		Cattle	Mastitis
		Pigs	Abscesses
B. salivosus		Cats	Subcutaneous abscesses and emphysema
B. levii		Cattle	Associated with summer mastitis
B. heparinolyticus		Horses, cats	Lesions in the buccal cavity
Fusobacterium necrophorum	*Actinomyces pyogenes*	Cattle	Calf diphtheria (necrotic foci in larynx, trachea and buccal cavity) Liver abscesses (feed-lot cattle) Metritis, cellulitis, mastitis
		Sheep	Foot abscess Ovine interdigital dermatitis ('scald') Lip and leg ulcerations

(*continued*)

Table 46. Diseases caused by non-spore-forming anaerobes (*continued*).

Non-spore-forming anaerobe	Associated pathogens	Host(s)	Disease
Fusobacterium necrophorum		Pigs	'Bull-nose' (via injury from fitting nose rings) Necrotic enteritis Liver abscess
		Horses	'Thrush' involving the frog Necrobacillosis of lower limbs
		Chickens	Avian diphtheria (often secondary to fowl pox)
		Rabbits	Necrobacillosis of lips and mouth
F. nucleatum		Several animal species	Non-specific infections
F. russii		Cats	Soft-tissue infections

Laboratory Diagnosis: General

Choice of specimens

As the non-sporing anaerobes constitute a major portion of the normal flora, the specimens must be collected with care to avoid contamination from the normal anaerobic flora, situated mainly on mucous membranes and in the intestinal tract. Unacceptable specimens include those from the gastrointestinal tract, throat, buccal cavity, voided urine, tracheal washings, and swabs from the surface of the urogenital tract and nasopharynx. The following samples are suitable for culture of the non-spore-forming anaerobes:

- Pus from abscesses
- Discharges from wounds (surgical and traumatic)
- Direct pleural aspirates
- Peritoneal aspirates
- Joint fluids
- Urine if taken by suprapubic puncture
- Tissue specimens (biopsy, necropsy and post-operative).

Collection of specimens

Specimens for the isolation of these strict anaerobes should be placed immediately in an oxygen-free container, especially small pieces of tissue or material taken on swabs. **Appendix 2** lists the commercially available kits for anaerobic specimen collection and also the preparation of a modified Cary–Blair medium to be used with swabs sterilised and stored in an oxygen-free atmosphere.

Larger pieces of tissue (over 2 cm^3) usually maintain an anaerobic micro-environment deep in the tissue and can be placed in an air-tight jar for transportation. Fluid specimens can be collected in a sterile syringe, the air expelled and the needle bent over or plugged. However, if the specimen cannot be processed within an hour, a fluid specimen should be placed in an oxygen-free tube or vial. All specimens for anaerobic culture should be processed within a few hours of collection. It is best to keep the specimens at ambient temperature rather than in the refrigerator as oxygen absorption is greater at lower temperatures.

Direct examination

Gram-stained smears of the specimens are useful as a screening process, although many of these anaerobes are not morphologically distinctive. Dilute carbol fuchsin (4–8 minutes) stained smears are more useful for *Bacteroides* and *Fusobacterium* species as they tend to stain faintly with the Gram-stain. *Fusobacterium necrophorum* in clinical specimens is long and filamentous (about 1 µm in diameter) and characteristically stains in an irregular manner (**227**). *F. nucleatum* occurs as thin rods (3–10 µm long) with tapered ends, often in pairs. *Bacteroides nodosus* is a large rod characterised by the presence of terminal enlargements at one or both ends. Fluorescent-conjugated antiserum is available (General Diagnostics) for *Bacteroides fragilis* and is used in medical laboratories. It has been reported as being specific and sensitive. **Table 47** summarises the microscopic appearance of some of the non-spore-forming anaerobes.

Table 47. Summary of the colonial and cellular morphology of some of the non-spore-forming anaerobes.

Colonial morphology	Microscopic appearance
Eubacterium suis	
Grey, smooth, circular, 2–3 mm with a shiny centre and dull edge. Slightly raised at centre and gives poor beta-haemolysis on sheep blood agar	Pleomorphic Gram-positive rods in palisade and Chinese-letter formation. Size 0.5–1.0 × 1.0–3.0 μm
Peptostreptococcus indolicus	
Greyish to yellow, shiny, circular, entire colonies, 0.5 – 1.0 mm diameter. On freshly-prepared media the colonies are viscous; on stored media they may be friable. Some strains are surrounded by a small zone of complete haemolysis	Gram-positive cocci, 0.5–0.6 μm in diameter, occurring singly, in pairs or in short chains
Bacteroides nodosus	
Three basic colonial types are described: **B-type:** papillate or beaded (most pathogenic) from ovine foot rot. **M-type:** mucoid (less pathogenic) from non-invasive infections of sheep and cattle. **C-type:** circular (non-pathogenic) and resulting from repeated passage in media. The colonies, generally, are greyish-white and 0.5 – 3.0 mm, diameter, in 3–7 days	Gram-negative, fairly large (1.7 × 3–6 μm), slightly curved and non-motile rods. Often swollen at one or both ends. They occur singly or occasionally in pairs
Bacteroides melaninogenicus	
Circular, entire, convex and shiny colonies, 0.5 – 2.0 mm in diameter. Colonies become darker after 5 – 14 days, being black in the centre with a grey-brown periphery. Haematin pigment is seen best on media containing laked blood. A few strains are haemolytic on rabbit blood agar. The colonies fluoresce under ultra-violet light	Gram-negative rods (0.5–0.8 × 0.9–2.5 μm) with an occasional cell of 10 μm or longer
Bacteriodes asaccharolyticus	
Colonies are 0.5–1.0 mm in diameter, round, convex, opaque and light-grey after 48 hours' incubation. In 6 – 14 days the colonies may become black. Some strains are haemolytic on rabbit blood agar	Gram-negative rods (0.8–1.5 × 1.0–3.5 μm). Cells from solid media tend to be shorter than those from broth cultures
Bacteroides fragilis	
Colonies are circular, entire, low convex, translucent to semi-opaque. They tend to have concentric rings of growth. Less than 1% of strains are haemolytic. *B. fragilis* will grow on bile aesculin agar with 5% sheep blood. Aesculin is hydrolysed	Gram-negative rods (0.8–1.3 × 1.6–8.0 μm). Occur singly or in pairs and have rounded ends. Vacuoles are often present. FA antiserum is available commercially
Bacteroides levii	
Colonies are minute, circular, entire and low-convex. After 2–3 days, colonies are buff or light-brown and dark-brown after 5–7 days' incubation	Gram-negative rods (0.6–1.2 × 2.0–7.0 μm). Occur in pairs or short chains

(continued)

Table 47. Summary of the colonial and cellular morphology of some of the non-spore-forming anaerobes (*continued*).

Colonial morphology	Microscopic appearance
Fusobacterium necrophorum	
Grey to yellowish, shiny colonies, 2–3 mm in 48 hours. Haemolysis is variable. Many strains are lipase-positive on egg yolk agar. It does not produce lecithinase	Gram-negative, long and filamentous but does not branch. Filaments can be up to 100 μm in length and 0.5–0.7 μm in diameter. May have tapered or rounded ends. Irregular staining is characteristic
Fusobacterium nucleatum	
Circular to slightly irregular, convex, translucent colonies (1–2 mm) often with a 'flecked' appearance. Usually non-haemolytic except occasionally just under the colony	Long, thin, Gram-negative rods with tapered to pointed ends (0.4–0.7 × 3–10 μm). Central swellings and intracellular granules may occur
Fusobacterium russii	
Circular, smooth, shiny, entire, convex, translucent colonies, 0.5–1.0 mm in diameter. Clear haemolysis occurs on horse blood agar	Gram-negative rods (0.3–0.7 × 1.5–4.0 μm) with some thin filaments about 10–15 μm in length. Palisade arrangement of cells is often seen. Beaded forms with pointed ends are common in thioglycollate medium

227

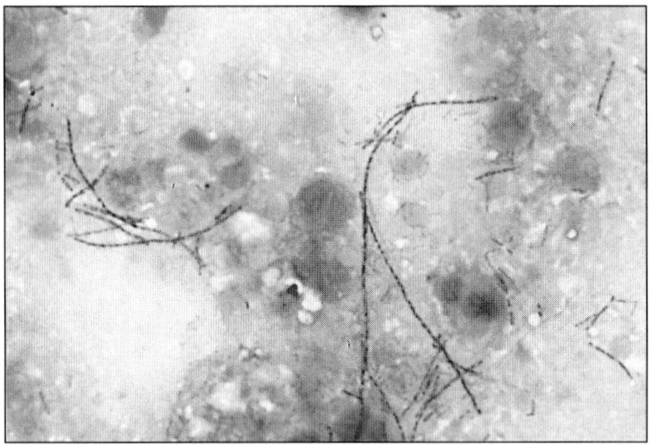

227 *Fusobacterium necrophorum* in long, non-branching filaments that characteristically stain irregularly (soft tissue abscess in cow). (DCF stain, ×1000)

Isolation

A general and brief description of the culture, media and identification of these anaerobic bacteria is presented. However, the techniques can be arduous and expensive unless a laboratory is specialising in this area. Anaerobic Laboratory Manuals by Sutter *et al.*, (1980), Dowell and Hawkins (1977) and Holeman *et al.*, (1977) are useful reference sources.

Methods for anaerobic culture

The three main methods for achieving an anaerobic atmosphere for the culture of these strict anaerobes are:

- Anaerobic jars with a catalyst, an anaerobic indicator and an atmosphere free of oxygen:
  a) Anaerobic jars with vents. These can be evacuated (to 20–24 inches of mercury), flushed twice with commercial grade nitrogen gas (N_2) and then filled with an anaerobic gas mixture [(10 per cent hydrogen (H_2), 5 per cent carbon dioxide (CO_2) and 85 per cent nitrogen (N_2)]. This mixture can be ordered in cylinders from a commercial gas supplier.
  b) Anaerobic jar without vents. These are used with commercially available envelopes that deliver an H_2–CO_2 atmosphere (Gas Pak, BBL; Gas Gendicator, Scott; Gas Generating Box 'H_2–CO_2', bioMérieux and Gas Generating Kit-anaerobic, Oxoid).
- Roll-streak tubes with pre-reduced, anaerobically sterilised medium. These large roll-tubes, with agar medium layered around the sides, are subjected to a jet of oxygen-free gas from the time they are opened for inoculation or subculture until the rubber stopper is replaced.
- Anaerobic chambers or anaerobic glove-boxes. These are usually large plastic tents kept constantly under an anaerobic atmosphere and they contain an incubator and other equipment for culturing the anaerobes. The media and specimens are introduced through a chamber lock and manipulations inside are conducted with the operator's hands and arms in gloves that are an integral part of the tent wall.

The two latter methods are usually used only in laboratories specialising in anaerobic culture work. Anaerobic jars give satisfactory results for laboratories culturing small numbers of anaerobic samples.

Media for anaerobic bacteria

a) Agar media: enriched blood agar is used for these fastidious anaerobes. To a nutritious agar base such as Eugon, Columbia, trypticase soy or Schaedler brain-heart, 0.5 per cent yeast extract, vitamin K (10 μg/ml) and haemin (5 μg/ml) is added. The preparation is given in **Appendix 2.** The media can be made selective for the Gram-negative anaerobes by the addition of an antibiotic supplement, either paromomycin (100 μg/ml) and vancomycin (7.5 μg/ml), or kanamycin (100 μg/ml) and vancomycin (7.5 μg/ml). *Bacteroides* spp. (except *B. ureolyticus*) are resistant to kanamycin but the *Fusobacterium* spp. are sensitive to this antibiotic. A 'Fastidious Anaerobe agar' (Lab M) is available commercially with various antibiotic supplements, depending on the anaerobe that is being sought.

Specific media have been recommended for *Bacteroides nodosus* such as one described by Gradin and Schmitz (1977) that consists of Eugon agar base (BBL) with 0.2 per cent (w/v) yeast extract, 10 per cent defibrinated horse blood and 1μg/ml lincomycin. Members of the *B. fragilis* group will grow on bile aesculin medium with 5 per cent sheep blood and hydrolyse the aesculin. A medium containing nalidixic acid, colistin sulphate and metronidazole for the primary isolation of *Eubacterium suis* has been described by Dagnall and Jones (1982).

Agar media should be used immediately after preparation or stored and then pre-reduced in an anaerobic jar for 6–24 hours before use. The plates are streaked with the specimens and placed as quickly as possible under an anaerobic atmosphere for incubation at 35–37°C. Plates should not be discarded until the eighth day of incubation.

b) Liquid media are useful adjuncts to agar media if the initial sample contains very small numbers of the required anaerobe and also for growing and maintaining pure cultures. Cooked meat broth with 0.4 per cent glucose or thioglycollate medium is suitable with the addition of the vitamin K-haemin supplement (**Appendix 2**).

Liquid media must be placed in a boiling water bath for 10 minutes, to expel absorbed oxygen, and rapidly cooled to 37°C immediately before inoculation. The media should be inoculated near to the bottom of the tube or bottle with as little disturbance as possible. The inoculated tubes or bottles are incubated anaerobically, with loose caps, at 35-37°C and not discarded until after 7 days of incubation. Liquid media should always be used together with agar media in plates.

Identification

Colonial morphology and microscopic appearance

The cellular morphology, and sometimes the colonial morphology, can be very variable depending on the strain, medium and cultural conditions. *F. necrophorum* produces grey to yellowish colonies on blood agar, that are about 2–3 mm in diameter after 48 hours' incubation (**228**). A Gram-stained smear from the colonies shows long Gram-negative filaments (**229**) that are less characteristic than those from direct microscopic examination of specimens. Lipase, but not lecithinase, activity is exhibited by *F. necrophorum* on egg yolk agar (**230**). *B. nodosus*, in a Gram-stained smear from enriched blood agar (**231**), appears as straight or slightly curved rods with the characteristic terminal knobs on one or both ends of the cells (**232**). **Table 47** summarises the colonial and cellular morphology of some of the pathogenic anaerobic non-sporing bacteria but, with a few exceptions, these characteristics are too variable to be relied on for identification purposes.

Commercial anaerobic identification systems

These systems have been designed for human medical microbiology and not all have been evaluated for use with veterinary isolates. However, they include many of the veterinary pathogens in the systems. Examples are API 20A and API-ZYM (Analytab Products); ATB 32A (bio Mérieux); and Minitek anaerobic II (Becton and Dickinson).

Conventional biochemical tests in tubed media

The preparation and use of conventional tubed media for biochemical tests is described by Dowell and Hawkins (1977).

Gas-liquid chromatographic analysis

This method for the detection of the end products of metabolism is most reliable and reproducible for identification of these anaerobic species. The volatile fatty acids and non-volatile fermentation products are characteristic for each species. Holeman *et al.* (1977) gives a detailed description of the gas chromatography procedures for anaerobes. The Capco Anaerobic Identification System (Capco Instruments) has proved to be a satisfactory, commercially available system.

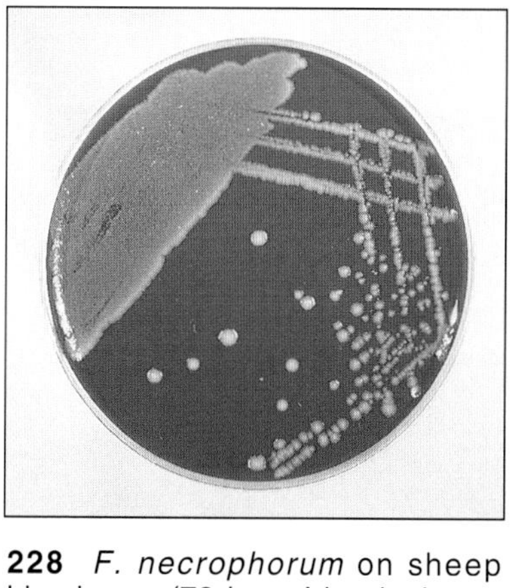

228

228 *F. necrophorum* on sheep blood agar (72 hours' incubation at 37°C).

229

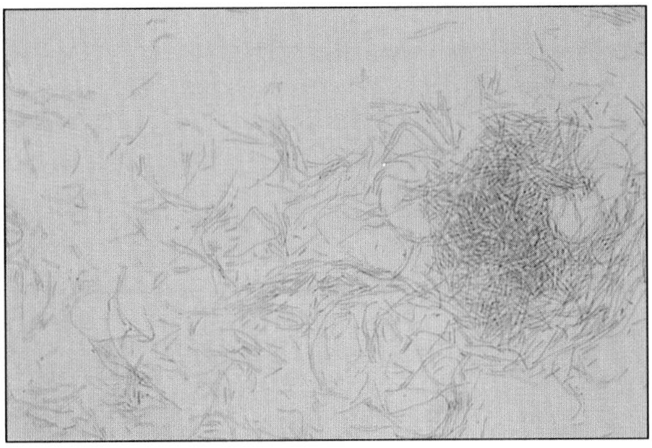

229 *F. necrophorum:* long, non-branching Gram- negative filaments from a culture. (Gram stain, ×1000)

230

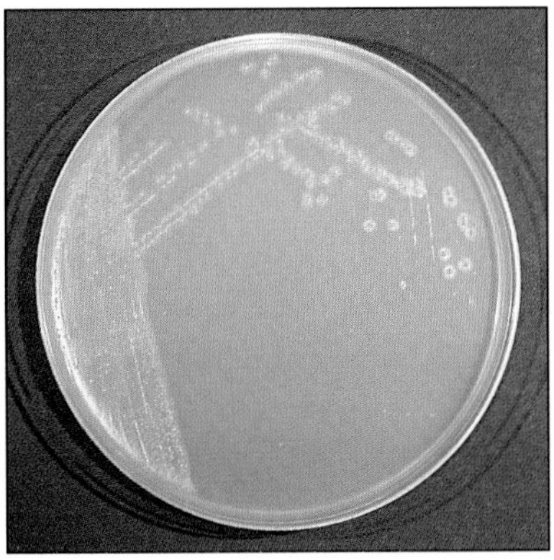

230 *F. necrophorum* on egg yolk medium with a pearly zone around the colonies due to lipase activity.

231

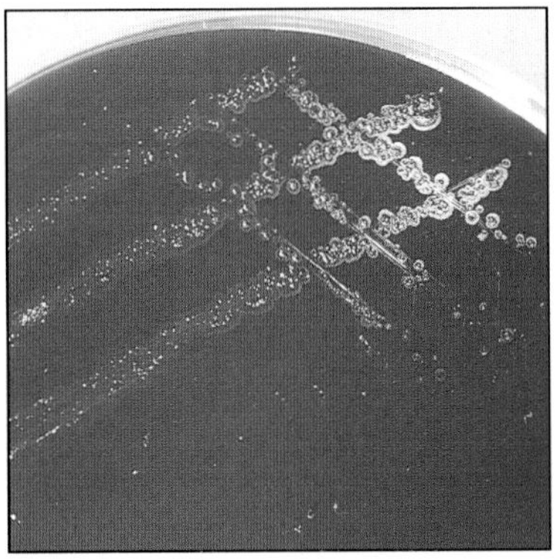

231 Close-up of *Bacteroides nodosus* on enriched sheep blood agar.

232

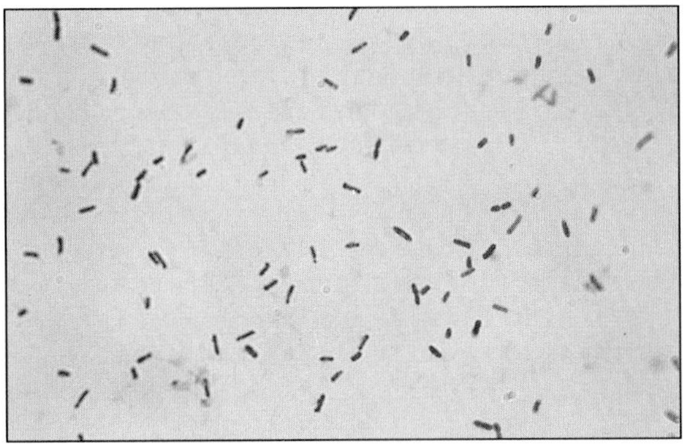

232 *B. nodosus* from a culture: straight or slightly curved large Gram-negative rods with terminal enlargements at one or both ends. (Gram stain, ×1000)

References

Dagnall, G. J. R. and Jones, J. C. T. (1982). A selective medium for the isolation of *Corynebacterium suis*, *Research in Veterinary Science*, **32**: 389–390.

Dowell, V. R. Jr. and Hawkins, T. M. (1977). *Laboratory Methods in Anaerobic Bacteriology*. CDC Laboratory Manual, DHEW Publications, No. 78-8272. Centers for Disease Control, Atlanta, Georgia, USA.

Gradin, J. L. and Schmitz, J. A. (1977). Selective medium for isolation of *Bacteroides nodosus*, *Journal of Clinical Microbiology*, **6**: 298–302.

Holeman, L. V., Cato, E. P. and Moore, W. E. C. (1977). *Anaerobe Laboratory Manual*. 4th ed. Anaerobe Laboratory, VPI, Blacksburg, Virginia 24061, USA.

Sutter, V. L., Citron, D. M. and Finegold, S. M. (1980). *Wadsworth Anaerobic Bacteriology Manual*. 3rd ed., C.V. Mosby, St. Louis, Missouri, USA.

17 *Clostridium species*

The *Clostridium* species are large (0.3–1.3 × 3–10 µm), Gram-positive, anaerobic, endospore-producing rods and the spores usually bulge the mother cell. All the pathogenic species are straight rods except *C. spiroforme* which is curved or spiral. Cells from older cultures or when producing endospores, have a tendency to decolourise. *C. perfringens* is the only species that produces a capsule in animal tissues and it is non-motile. Most of the other species are motile by peritrichate flagella. The clostridia are fermentative, oxidase-negative and catalase-negative. The strictness of anaerobic requirements varies among the species but they all prefer an atmosphere containing between 2 and 10 per cent CO_2. Most clostridia require enriched media that include amino acids, carbohydrates, vitamins and blood or serum. Optimum growth of the pathogenic clostridia occurs at 37°C. There are over 80 *Clostridium* species of which about 11 are of veterinary importance. Most of the pathogenic species produce one or more exotoxins of varying potency.

Changes in Nomenclature

Previous Name	**Present Name**
Clostridium novyi type D	*C. haemolyticum*
Clostridium botulinum type G	*C. argentinense*
Clostridium welchii	*C. perfringens*
Clostridium oedematiens	*C. novyi*

Natural Habitat

The clostridia have a wide distribution in soil, freshwater and in marine sediments throughout the world, although some species or types are present only in localised geographical areas. Many of the pathogenic clostridia are normal inhabitants of the intestinal tract of animals and man, and often cause endogenous infections. Other clostridia are more commonly present in the soil and cause exogenous infections from wound contamination or by ingestion.

Pathogenicity

Although exotoxins are important in most of the clostridial diseases, the potency of the toxin(s) produced and the invasive ability of the clostridia vary. This allows an arbitrary but convenient, division of the pathogenic *Clostridium* species into the following groups:

- Neurotropic clostridia (*C. tetani* and *C. botulinum*) that produce potent neurotoxins but are non-invasive and colonise the host to a very limited extent.
- Histotoxic clostridia (*C. chauvoei*, *C. septicum*, *C. novyi*, *C. haemolyticum*, *C. sordellii*, *C. perfringens* type A and *C. colinum*) that produce less potent toxins than the first group but are invasive. This includes the gas-gangrene-producing clostridia.
- Clostridia that produce enterotoxaemias (*C. perfringens* types A–E). Enterotoxins are formed in the intestines and absorbed into the bloodstream producing a generalised toxaemia.
- Clostridia (*C. difficile* and *C. spiroforme*) producing enteric disease that can be antibiotic-induced.

Table 48 summarises the hosts and diseases of the pathogenic clostridia.

Laboratory Diagnosis (General)

Specimens

Specimens should be taken from recently dead animals as bacteria such as *C. perfringens*, *C. septicum* and enteric facultative anaerobes are rapid postmortem invaders. For isolation, blocks of affected tissue (4 cm^3) or fluids in air-free containers should be collected when possible rather than swab-taken samples that expose the clostridia to the lethal action of atmospheric oxygen. Commercial systems are satisfactory where the swab is in an oxygen-free gas and after use the swab is placed in Cary–Blair transport medium. In some clostridial diseases, such as the enterotoxaemias, the toxin is required for diagnosis. The contents of the small intestine are collected from a recently dead animal and submitted to the laboratory as soon as possible, as the toxins are labile.

Direct microscopy

Gram-stained smears from specimens are used to observe the morphological types of organisms present.The fluorescent antibody (FA) technique is useful for specific identification.

- Gram-stained smears from affected tissues may reveal large Gram-positive rods that tend to decolourise easily when sporing. *C. spiroforme* is an exception being curved or helical. The characteristic 'drumstick' forms of *C. tetani* (**233**), due to the spherical spores being terminal and bulging the cell, may be seen in necrotic material from wounds associated with tetanus. This is suggestive, but by no means conclusive, as other clostridia, such as *C. tetanoides* and *C. tetanomorphum* have a simi-

Table 48. Summary of the hosts and diseases caused by pathogenic clostridia.

***Clostridium* species**	***Hosts***	***Diseases***
NEUROTOXIC CLOSTRIDIA		
Clostridium tetani	Horses, ruminants, humans and other animals	Tetanus
Clostridium botulinum (types A–F)	Many animal species and man	Botulism
Clostridium argentinense (*C. botulinum* type G)	Humans (Argentina)	Botulism
HISTOTOXIC CLOSTRIDIA		
Clostridium chauvoei	Cattle, sheep, (pigs)	Blackleg (Black quarter)
Clostridium septicum	Cattle, sheep and pigs	Malignant oedema
	Sheep	Braxy
	Chickens	Necrotic dermatitis
Clostridium novyi		
type A	Sheep	Big-head of rams
	Cattle and sheep	Gas gangrene
type B	Sheep, (cattle)	Black disease (necrotic hepatitis)
type C	Water buffalo	Osteomyelitis reported
Clostridium haemolyticum (*C. novyi* type D)	Cattle, (sheep)	Bacillary haemoglobinuria
Clostridium sordellii	Cattle, sheep, horses	Gas-gangrene
Clostridium colinum	Game birds, young chickens and turkey poults	Quail disease (ulcerative enteritis)
ENTEROTOXAEMIAS		
Clostridium perfringens		
type A	Humans	Food poisoning, gas gangrene
	Lambs	Enterotoxaemic jaundice
type B	Lambs (under 3 weeks old)	Lamb dysentery
	Neonatal calves and foals	Enterotoxaemia
type C	Piglets, lambs, calves, foals	Haemorrhagic enterotoxaemia
	Adult sheep	Struck
	Chickens	Necrotic enteritis
type D	Sheep (all ages except neonates) (goats, calves)	Pulpy kidney disease
type E	Calves and lambs (rare)	Enterotoxaemia
CLOSTRIDIA ASSOCIATED WITH ANTIBIOTIC-INDUCED DISEASE		
Clostridium spiroforme	Rabbits	Possible role in mucoid enteritis
	Rabbits and guinea-pigs	Spontaneous and antibiotic-induced diarrhoea
	Foals and pigs	Enterocolitis (natural)
Clostridium difficile	Humans, hamsters, rabbits, guinea-pigs	Antibiotic-induced enterocolitis
	Dogs, foals, pigs, laboratory animals	Naturally occurring diarrhoea

lar morphology. In cases of suspected enterotoxaemia, the presence of large numbers of fat Gram-positive rods in a smear of the small intestinal mucosa, from a recently dead animal (**234**), is presumptive evidence of the condition.

- FA technique is used routinely for diseases associated with *C. chauvoei* (**235**), *C. septicum*, *C. novyi* and *C. sordellii* as fluorescent labelled antisera can be obtained commercially (Wellcome Diagnostics). Affected tissue as well as a piece of rib containing bone marrow (about 14 cm long) are useful specimens. A bacteraemia usually occurs with these clostridial diseases so the bacteria would be expected to be present in bone marrow. This tissue has the added advantages of giving low background autofluorescence and being one of the last tissues to be invaded, postmortem, by bacteria such as *C. septicum*.

233

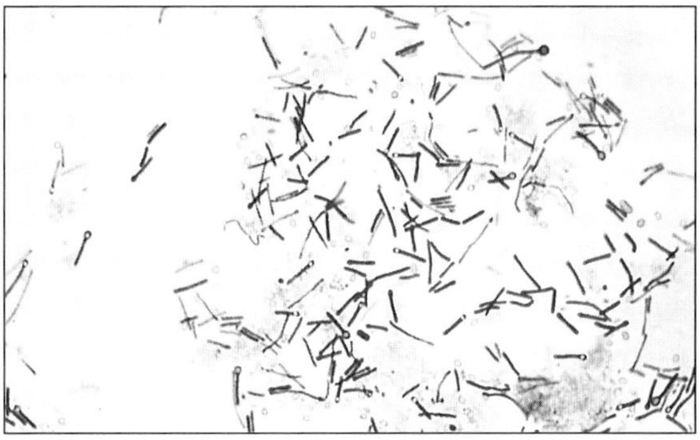

233 Sporing rods of *Clostridium tetani* in Gram-stained smear of necrotic material from a penetrating wound. The spores are spherical, terminal and bulge the mother cell giving the typical 'drum-stick' appearance. (Gram stain, × 1000)

General isolation procedures

In general, freshly prepared or pre-reduced blood agar is suitable for the isolation of clostridia. Media for the more fastidious anaerobes such as *C. chauvoei*, *C. haemolyticum* and *C. novyi* types B and C are given in **Appendix 2**. Stored agar media gradually absorb oxygen from the atmosphere, so it is important to use either freshly prepared blood agar or pre-reduced plates that have been stored under anaerobic conditions soon after preparation. A blood and a MacConkey agar plate should be inoculated and incubated aerobically. These plates will detect any aerobic pathogens that may be present and also indicate the degree of contamination of the specimen by facultative anaerobes. Liquid and semi-solid media with a low redox potential such as cooked meat broth and thioglycollate medium (**236**) can be used to grow and maintain pure cultures of the clostridia. They are of limited use for primary inoculation as any fast-growing anaerobes or facultative anaerobes will outgrow the *Clostridium* species of interest. Immediately before inoculating cooked meat broth or thioglycollate medium, boiling to expel absorbed oxygen should be undertaaken, followed by rapid cooling to 37°C.

Most of the clostridia pathogenic for animals are strict anaerobes, the exception being *C. perfringens* that is relatively aerotolerant. However, all should be grown under strict anaerobic conditions with the atmosphere containing 2–10 per cent CO_2 as this enhances their growth. An anaerobic jar with a catalyst, an anaerobic indicator and an envelope delivering $H_2 + CO_2$ is usually satisfactory.

234

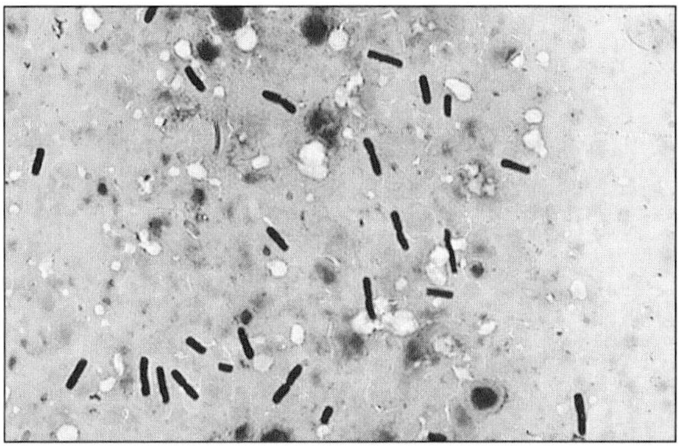

234 *C. perfringens*: large Gram-positive rods in a mucosal scraping from the small intestine of a lamb that had recently died from pulpy kidney disease. (Gram stain, ×1000)

235

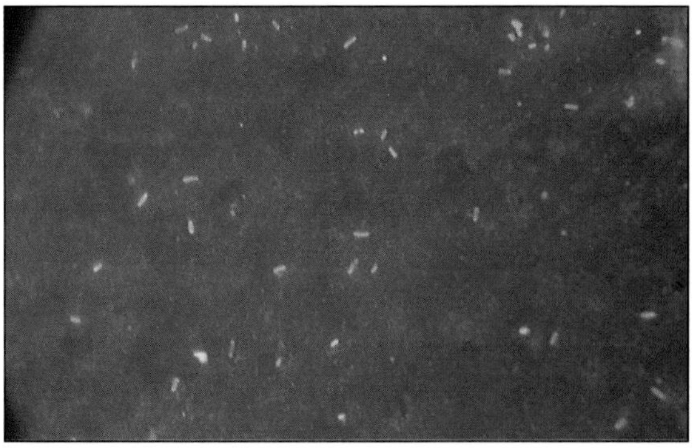

235 Direct fluorescent antibody technique showing *C. chauvoei* in muscle tissue from a case of blackleg in a heifer. (×400)

Table 49. Biochemical reactions of the clostridia pathogenic to animals.

Clostridium species		Egg Yolk agar: Lecithinase	Egg Yolk agar: Lipase	Hydrolysis of gelatin	Digestion of casein	Indole production	Acid from: Glucose	Acid from: Lactose	Acid from: Sucrose	Acid from: Maltose	Additional characteristics
C. tetani		–	–	+	–	v	–	–	–	–	Terminal, spherical endospores
C. botulinum	I	–	+	+	+	–	+	–	–	+	Toxin types A, B and F
	II	–	+	+	–	–	+	–	–	+	Toxin types B, E and F
	III	v	+	+	–	v	+	–	–	v	Toxin types C and D
	IV	–	–	+	+	–	–	•	–	•	Toxin type G
C. chauvoei		–	–	+	–	–	+	+	+	+	
C. septicum		–	–	+	+	–	+	+	–	+	
C. novyi	A	+	+	+	–	–	+	–	–	+	
	B	+	–	+	+	v	+	–	–	+	
	C	–	–	+	–	+	+	–	–	•	No toxin produced
C. haemolyticum		+	–	+	+	+	+	–	–	–	
C. sordellii		+	–	+	+	+	+	–	–	+	Urease-positive
C. colinum		–	–	–	–	–	+	–	+	+	
C. perfringens		+	–	+	+	–	+	+	+	+	Non-motile. 'Stormy-clot' in litmus milk
C. spiroforme		–	–	–	–	–	+	•	+	•	Spiral and curved
C. difficile		–	–	+	–	–	+	–	–	–	

+ = positive reaction, – = negative reaction, v = variable reaction, • = data not available.

Biochemical reactions

Some of the biochemical reactions of the pathogenic clostridia are given in **Table 49**. On egg yolk agar the clostridia with lecithinase activity produce an opalescent change around the colonies due to the enzymatic action on the lecithin in the medium. Those producing a lipase cause a pearly layer or iridescent film that can cover the colonies and in some cases extend into the surrounding agar (**237**). *C. perfringens* inoculated into litmus milk medium produces the classical 'stormy-clot' or 'stormy-fermentation' reaction (**238**). The lactose in the medium is fermented by *C. perfringens* producing acid which coagulates the casein and induces a colour change from blue to pink (litmus pH indicator). The acid clot is then broken up by gas formation. If biochemical tests are carried out by conventional methods, the tubes or bottles are placed in an anaerobic jar after inoculation. Miniaturised commercial systems such as API 20A (Analytab Products) or ATB 32A (bioMérieux) are available for the identification of many of the clostridia. The principal fermentation products can also be used to identify the *Clostridium* species by gas chromatography.

Animal inoculation

Laboratory animals, usually young guinea-pigs or mice, can be used in one of two ways:

- As 'biological filters' for contaminated specimens or for material judged to contain small numbers of the pathogenic *Clostridium* species. The pathogenicity of the clostridia can be enhanced, if equal amounts of calcium chloride solution (5 per cent w/v) and supernatant from material containing clostridia are mixed before injecting the guinea-pig or mouse intramuscularly.
- In neutralisation or protection tests to specifically identify the toxin(s) present and hence the clostridial pathogen involved in the disease. These procedures are most commonly used in tetanus, botulism and in the enterotoxaemias caused by *C. perfringens*.

236

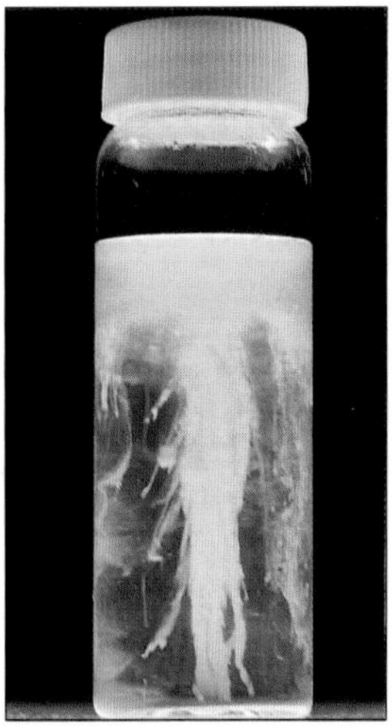

236 Growth of *C. perfringens* in thioglycollate medium.

237

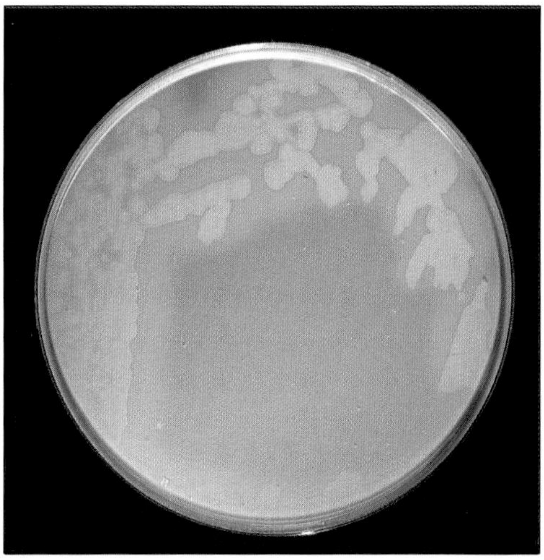

237 *C. botulinum* type C on egg yolk medium giving a pearly layer around the colonies due to lipase activity. Lecithinase is not produced by this bacterium.

238

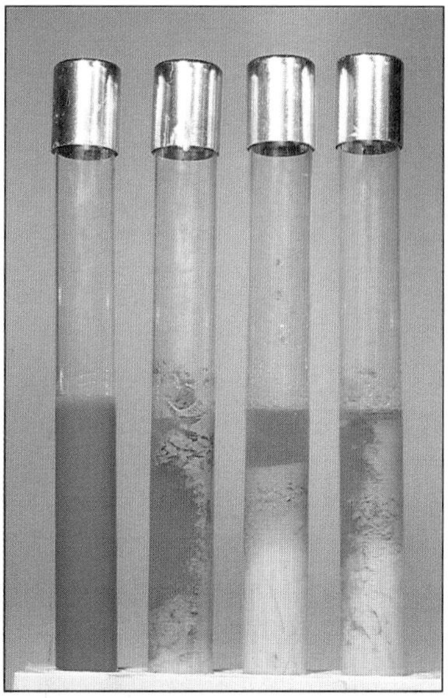

238 The 'stormy clot' reaction of three isolates of *C. perfringens* in litmus milk medium. The tube on the left is uninoculated.

Neurotoxic Clostridia

Clostridium tetani

Clostridium tetani is a straight, slender (0.4–0.6 × 2–5 μm), Gram-positive rod that characteristically produces a terminal, spherical endospore that bulges the cell giving the characteristic 'drumstick' appearance to the bacterium (**233**). The endospores are highly resistant and while boiling kills the spores of most strains in 15 minutes, autoclaving at 121°C for 15 minutes is completely sporicidal. There are 10 serological types of *C. tetani*, based on flagellar antigens, but the neurotoxin is antigenically uniform.

Natural Habitat

Soil, especially that contaminated by animal faeces, is the natural habitat as *C. tetani* is often transient in the intestines of horses and other animals.

Pathogenesis

C. tetani produces the exotoxins tetanolysin (a haemolysin) that does not appear to have pathogenic significance, and tetanospasmin (a neurotoxin) that is plasmid-coded and responsible for the clinical signs of tetanus. The endospores enter traumatized tissue or surgical wounds, especially after castration or docking, via the umbilicus (**239**) or into the uterus following dystocia in cattle and sheep. Presence of facultative anaerobes and necrotic tissue create anaerobic conditions and the *C. tetani* spores germinate. The vegetative cells multiply at the entry site and produce the potent tetanospasmin. This travels via peripheral nerves or bloodstream to ganglioside receptors of the motor nerve terminals and eventually to cells of the ventral horn of the spinal cord, thus affecting many groups of muscles at various levels. The toxin acts presynaptically on motor neurons blocking synaptic inhibition and causing a spastic paralysis and the characteristic tetanic spasms. Tetanospasmin binds specifically to gangliosides in nerve tissue and once bound cannot be neutralised by antitoxin. When toxin travels up a regional motor nerve in a limb, tetanus first develops in the muscles of that limb, then spreads to the opposite limb and moves upwards. This is known as ascending tetanus and is usually seen only in the less susceptible animals such as dogs and cats. Descending tetanus is the common form in susceptible species such as humans and horses. In this form toxin circulating in the blood stream affects the susceptible motor nerve centres that serve the head and neck first and later the limbs. Once established, signs of tetanus are similar in all animal species.

Laboratory Diagnosis

In tetanus, the diagnosis is often based on the history and on the characteristic clinical signs, without reference to laboratory tests.

Direct microscopy

Gram-stained smears of material from a wound may reveal the characteristic 'drumstick' sporing forms of *C. tetani* (**233**). This is not completely diagnostic as other clostridia such as *C. tetanomorphum* and *C. tetanoides* have a similar morphology.

Isolation

Necrotic tissue from a wound or wound exudate can be heated to 80°C for 20 minutes and used to inoculate a blood agar plate and another blood agar plate containing 3 per cent agar ('stiff agar'). A tube of thioglycollate medium or cooked meat broth could also be inoculated and subcultured onto blood agar after 2–3 days' incubation. The blood agar plates are incubated at 37°C for 3–4 days under an atmosphere of H_2 and CO_2.

Identification

Colonial morphology

C. tetani is haemolytic and on normal blood agar tends to have a spreading, swarming growth (**240** and **241**) while on 'stiff agar' (3 per cent) individual rhizoid colonies are formed (**242**).

Biochemical reactions

C. tetani liquefies gelatin but does not ferment the usual range of carbohydrates. Other reactions are given in **Table 49**. Demonstration and identification of the toxin is more important than the biochemical reactions in diagnosis.

Toxin identification

The toxin present in an animal's serum or in filtrate from cooked meat broth or thioglycollate medium can be demonstrated in laboratory animals and identified by neutralisation or protection tests using specific antitoxin. In the protection test, the animals are given antitoxin at least 2 hours before inoculation with the material containing toxin. The control mice show typical signs of tetanic spasm in the region of inoculation (**243**).

Clostridium botulinum

Clostridium botulinum is a straight rod (0.9–1.2 × 4–6 μm) and at a pH near or above neutrality produces oval, subterminal spores. The spores are very resistant but are killed at 121°C for 15 minutes while the toxins are destroyed at 100°C for 20 minutes. Eight different neurotoxins are produced by *C. botulinum* types A–G. Type G has been renamed *C. argentinense*. The toxins are identical in action but differ in potency, distribution and antigenicity. Toxin production by types C and D are known to be bacteriophage-coded. The optimum pH for *C. botulinum* is neutral to slightly alkaline (7.0–7.6) and the optimal temperature lies between 30–37°C.

239

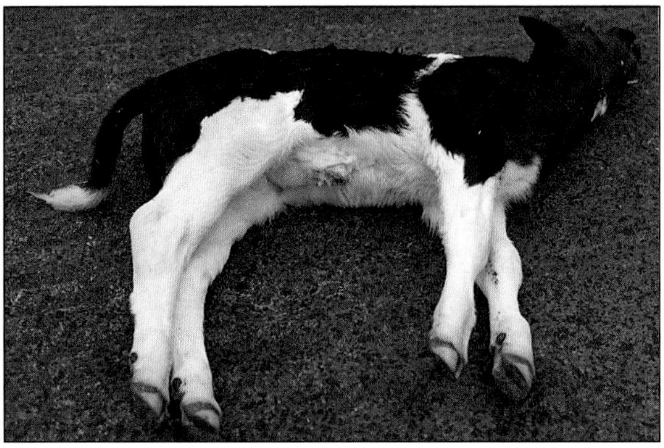

240

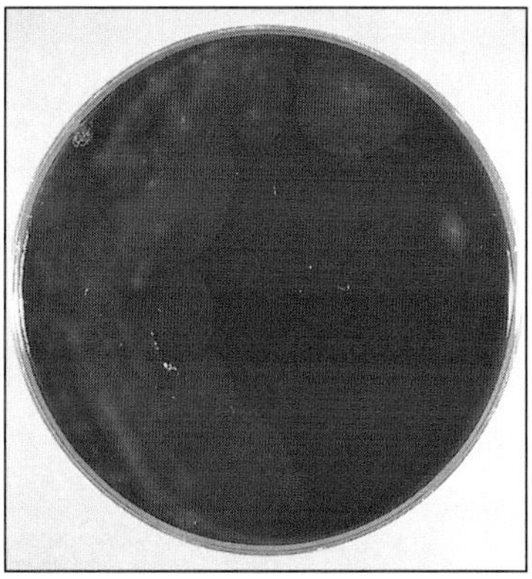

239 *C. tetani:* advanced tetanus in a young calf showing rigidity of limbs, opisthotonos and raised tail-head. Note pyogenic infection of umbilicus, the probable portal of entry of *C. tetani* in this case.

240 *C. tetani* colonies on sheep blood agar showing spreading growth and a narrow zone of beta-haemolysis.

241

241 Spot inoculation of *C. tetani* to illustrate the characteristic spreading growth on normal blood agar containing 1.5 per cent agar. Oblique illumination.

242

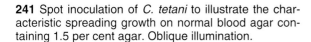

242 *C. tetani* on stiff sheep blood agar (3 per cent agar) which prevents spreading and gives individual rhizoid colonies.

243

243 Demonstration of the activity of tetanospasmin (*Clostridium tetani*) in a mouse. Injected intramuscularly into right hind leg.

Natural Habitat

The endospores are widely, but unevenly, distributed in soils and aquatic environments throughout the world. Germination of the endospores, with growth of vegetative cells and production of toxin, occurs in anaerobic situations such as contaminated cans of meat, fish or vegetables, carcases of invertebrate and vertebrate animals, rotting vegetation and baled silage. **Table 50** indicates the common sources of toxin and geographic distribution of the types of *C. botulinum*.

Pathogenesis

Botulism is an intoxication usually caused by ingestion of preformed toxin in foodstuffs. The toxin is absorbed from the intestinal tract and is transported via the bloodstream to peripheral nerve cells where it binds to susceptible cells and suppresses the release of acetylcholine at the myoneural junctions. This results in flaccid paralysis, death being caused by circulatory failure and respiratory paralysis. Less common methods of acquisition of toxin are wound botulism or toxicoinfection and infant botulism or intraintestinal toxicoinfection. In wound botulism the spores are introduced into wounds where they germinate. Toxin is formed at this localised site and spreads through the body. The 'shaker foal' syndrome is thought to be caused in this way. Infant botulism occurs when spores germinate in the intestines when the normal flora has not yet been fully established. This form is seen in human infants ('floppy baby' syndrome) and as rare epidemics of type C in broiler chickens and turkey poults. The toxin is one of the most potent known: 1 mg of the neurotoxin contains more than 120 million mouse lethal doses. A comparison of the toxins of *C. tetani* and *C. botulinum* is shown in **Table 51**. Botulism is most common in water birds (**244**), ruminants, horses, mink and poultry. Carnivores are relatively resistant to all types. Pigs are susceptible to the toxin of type A but resistant to those of B, C and D. The oral toxicity of type D toxin is high for cattle. Type C toxins are more readily absorbed through the intestinal wall of chickens and pheasants. **Table 50** indicates the toxins produced, source of toxin and animals susceptible to each of the *C. botulinum* toxins.

244

244 *C. botulinum* type C: botulism in a herring gull (*Larus argentatus*) showing flaccid paralysis of wings and legs.

Laboratory Diagnosis

The diagnosis of botulism is based on history, clinical signs and demonstration and identification of toxin in serum of moribund or recently dead animals as well as the detection of toxin and/or *C. botulinum* in the suspect foodstuff. Demonstration of toxin in animals that have been dead for some time may not be significant. *C. botulinum* spores can be transient in the intestines of normal animals and the death of the animal creates an anaerobic environment suitable for the germination of the spores and toxin production. Great care must be taken when working with materials containing *C. botulinum* toxins because of their high potency.

Toxin demonstration

Serum or centrifuged serous exudates from animals can be directly inoculated intravenously (0.3 ml) or intraperitoneally (0.5 ml) into mice. If toxin is present the characteristic 'wasp waist' appearance in the mice (**245**) will be seen in a few hours or up to 5 days. The appearance is due to abdominal breathing because of paralysis of respiratory muscles.

Extraction of toxin in foodstuffs is accomplished by macerating the product in saline overnight. The suspension is centrifuged and the supernatant filtered through a 0.45 µm bacteriological filter. As the toxin can be in a protoxin form, 9 parts of filtrate are treated with one part of 1 per cent trypsin solution and incubated at 37°C for 45 minutes. Mice or guinea-pigs are inoculated intraperitoneally.

Toxin identification

Mouse (or guinea-pig) neutralisation tests using a polyvalent antitoxin initially, followed by monovalent antitoxins, if they are available, are used to identify the toxin and the type of *C. botulinum* involved.

Isolation of C. botulinum from foodstuffs

Several samples of the foodstuffs are macerated in a small amount of physiological saline. The suspension is heated at 65–80°C for 30 minutes to kill most of the contaminanting organisms and to induce the *C. botulinum* spores to germinate. Blood agar plates are inoculated with the suspension and incubated under H_2+ CO_2 at 35°C for up to 5 days. Type E spores require treatment with lysozyme to aid germination.

Table 50. *Clostridium botulinum:* toxins, susceptible animals, sources of toxin and geographical distribution.

C. botulinum types	Toxin(s) produced	Most susceptible animals	Sources of toxin	Geographical distribution
A	A	Humans, chickens, mink	Vegetables, fruits, meat, fish	Canada, Western USA, and former USSR
B	B	Humans (cattle, horses, chicken)	Meat and meat products (often from pigs), vegetables, fish	Northern and Central Europe, Canada, Eastern USA and former USSR
Cα	$C_1(C_2)$	Waterfowl	**Limberneck** in long-necked birds. Invertebrate carcases, rotting vegetation and material on refuse dumps	Western USA, Canada, South America, Europe, Australia, New Zealand and Japan
Cβ	C_2, D (C_1)	Cattle, horses, mink, dogs, (humans)	**Forage poisoning** Carcases, baled silage, chicken manure as feed supplement, and spoiled feeds	South Africa, Australia, Europe, USA
D	C_2,D	Cattle, sheep, (horses, humans)	**Lamsiekte** Eating contaminated bones and carcases of small mammals (phosphorus-deficiency)	South Africa, former USSR, southwest USA and France
E	E	Humans, farmed fish	Humans: fish, fish-products and other foods Young fish: sludge in earth - bottomed ponds	Northern Europe, North America, Japan and former USSR
F	F	Humans	Meat (liver paste), fish	Northern Europe, USA and former USSR
C. argentinense G	G	Humans	Soil	Argentina

245

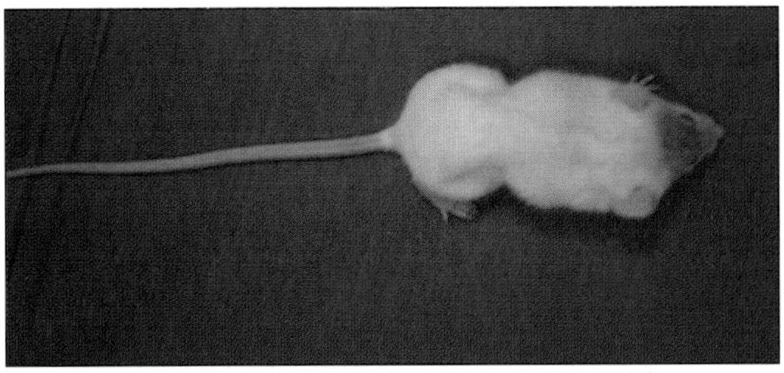

245 Mouse inoculated intravenously with serum containing the toxin of *C. botulinum.* Note the characteristic 'wasp-waist' appearance.

246

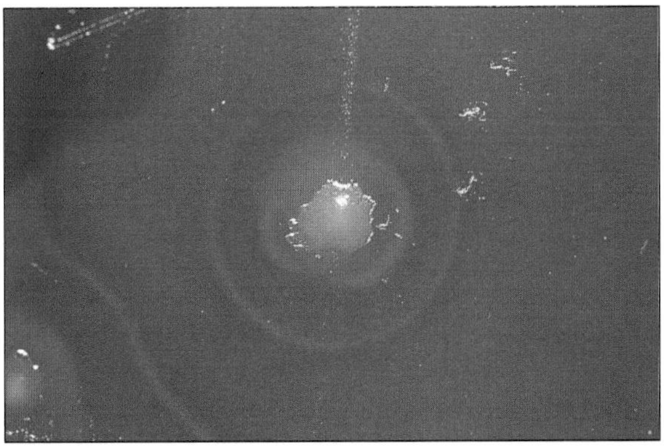

246 Close-up of *C. botulinum* type C on sheep blood agar showing beta-haemolysis and an irregular heaped colony with a granular surface.

Table 51. Comparison of the toxins of *Clostridium tetani* and *C. botulinum*.

	Clostridium tetani	***Clostridium botulinum***
Site of toxin production	Wounds	Carcases, decaying vegetation and occasionally wounds and intestine
Mode of action of toxins	Centrally by blocking synaptic inhibition	Peripherally by blocking neuromuscular transmission
Type of paralysis	Spastic paralysis	Flaccid paralysis
Antigenic types of toxin	Tetanospasmin (one antigenic type)	Eight different toxins produced by types A–G

The colonies on blood agar are usually haemolytic and vary in appearance from slightly domed with a ragged edge (**246**) to flat and rough or a film-like growth. The suspect colonies are identified by biochemical tests and, as seen in **Table 49**, there are four cultural types. To determine whether the isolate is a toxin-producing strain, a cooked meat broth is inoculated and incubated at 30°C for 5–10 days. Filtrates are prepared and laboratory animals can be used for demonstration and identification of the toxin.

Histotoxic Clostridia

Gas-Gangrene Clostridia

The clostridia commonly causing gas gangrene are summarised in **Table 52**. Occasionally other clostridia, present in soils and in the intestines of animals, are capable of causing a similar syndrome. Diseases caused by these clostridia are distributed worldwide.

Pathogenesis

The toxins produced by the gas-gangrene clostridia are not as potent as those of *C. tetani* and *C. botulinum* but the gas gangrene bacteria are invasive. The disease syndrome can vary from simple wound infections, anaerobic cellulitis to severe and fatal gas-gangrene. The infections can be either endogenous or exogenous in origin.

Endogenous infections often occur with blackleg in calves caused by *C. chauvoei*. Endospores are ingested and normally pass harmlessly through the intestinal tract but occasionally the spores pass from the intestine via the lymphatics and bloodstream to muscle masses, usually in the hindquarters but sometimes in cardiac muscle. Trauma to the area where the spores are lodged causes tissue necrosis and hence anaerobic conditions favouring germination of the spores and a supply of amino acids and other nutrients for vegetative cells. Toxin is produced followed by localised damage and finally a terminal toxaemia and bacteraemia. In exogenous infections, spores are introduced into wounds where they may germinate in the anaerobic necrotic material and toxin is produced by the vegetative cells. In braxy, the mucosa of the abomasum is damaged due to cold conditions from an adjacent rumen filled with frozen food. Any *C. septicum* spores present can germinate and replication of the bacterium leads to toxin production, toxaemia and rapid death.

Laboratory Diagnosis

• ***Fluorescent antibody technique***

Commercial fluorescein-labelled specific antisera are available for *C. chauvoei* (**235**), *C. septicum*, *C. novyi* and *C. sordellii* (Wellcome Diagnostics). The FA technique is a rapid and convenient method for identifying these clostridia. The technique is carried out on acetone-fixed smears of affected tissue or bone marrow from a rib.

• ***Gram-stained impression smears***

Gram-stained impression smears on affected tissue can yield some useful information. The morphology of the gas-gangrene organisms is given in **Table 53** and illustrated in **247–251**.

• ***Isolation and colonial appearance***

Sheep blood agar with liver extract is used to isolate the rather fastidious *C. chauvoei*. 'Stiff' blood agar (3 per cent agar) and normal blood agar are used when attempting to isolate *C. septicum* and *C. sordellii*. *C. perfringens* (**252**) and *C. novyi* type A grow well on normal blood agar. The inoculated plates are incubated under strict anaerobic conditons with 10 per cent CO_2 at 37°C for 2–4 days. The colonial appearance of each is given in **Table 53**.

• ***Biochemical reactions***

C. chauvoei ferments sucrose but rarely salicin, while *C. septicum* ferments salicin but not sucrose. Other biochemical reactions are given in **Table 49**.

247

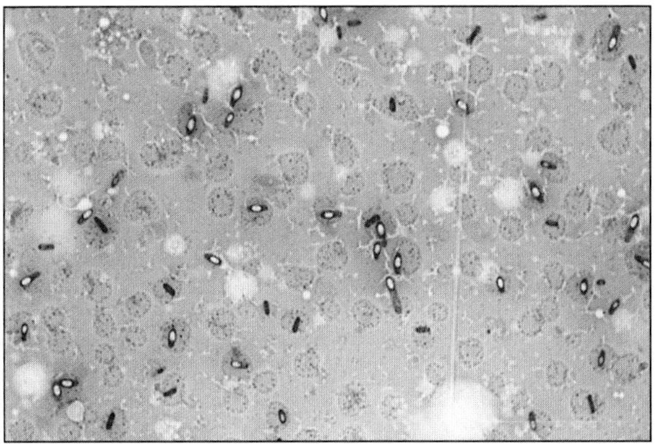

247 *C. chauvoei* spores in a tissue smear. They are oval, central to subterminal and bulge the mother cell. The citron (lemon-shaped) forms are characteristic. (Gram stain, ×1000)

248

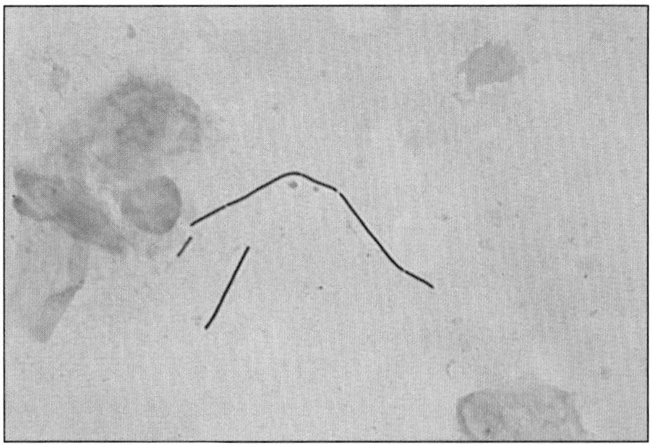

248 *C. septicum* in characteristic long forms in a Gram-stained smear of affected muscle. (×1000)

249

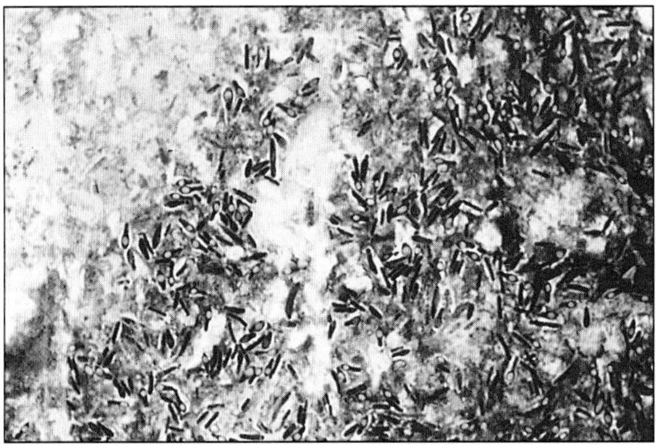

249 *C. novyi* spores in a tissue smear. They are oval, subterminal and slightly bulge the mother cell. (Gram stain, ×1000)

250

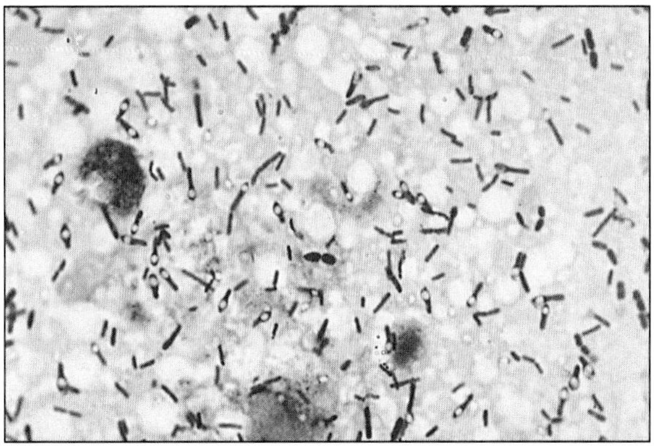

250 Spores of *C. perfringens:* not commonly seen but usually subterminal, large, oval and bulge the mother cell. (Gram stain, ×1000)

251

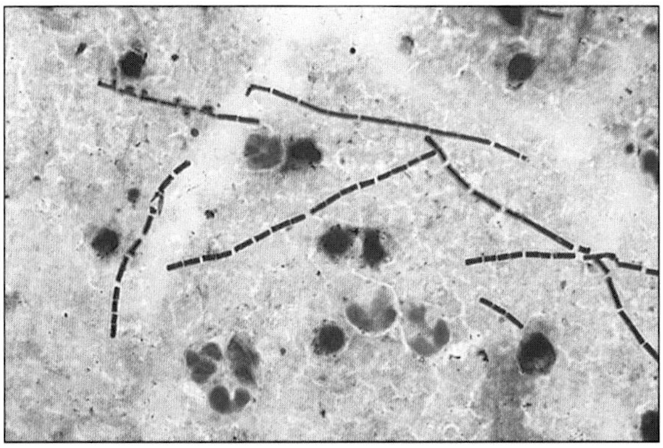

251 Chains of *C. perfringens* cells. (Methylene blue stain, ×1000)

252

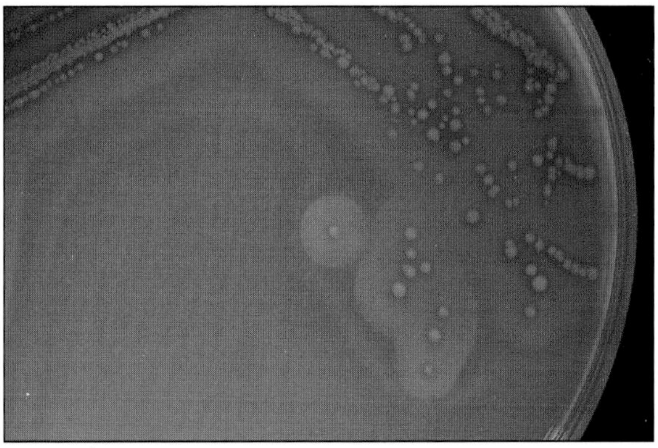

252 *C. perfringens* on sheep blood agar showing the characteristic 'target' haemolysis.

Table 52. Summary of the clostridia commonly causing gas gangrene.

Clostridium species	*Main Hosts*	*Disease*	*Route of entry*	*Clinical and postmortem signs*
C. chauvoei	Calves, 3–24 months old	Blackleg (quarter evil, or black quarter)	Endogenous, from spores in muscles	Usually sudden death, especially if heart muscle is involved. Fever, swelling of muscle masses of hind quarters. Muscles dry and spongy with small gas bubbles. Sweet rancid odour and muscles are dark red to black. Crepitation can be felt
	Sheep		Exogenous, through wounds	
C. septicum	Cattle, sheep, and pigs: all ages affected	Malignant oedema	Exogenous, through wounds	Fever, soft swelling around wound and spreading to muscles. Swelling is oedematous and wet with much exudate and gas. Muscles dark red to black colour
	Sheep	Braxy	Endogenous, from spores in abomasum	Caused by large volume of frozen food in rumen damaging localized area in abomasum. Replication of vegetative cells and toxin produced
***C. novyi* type A**	Young rams	Big-head	Wounds from fighting	Oedematous swelling over head, face and neck
	Cattle, sheep	Gas-gangrene	Wounds	Lesions similar to those of malignant oedema. Sudden death can occur
C. sordellii	Cattle, sheep	Gas-gangrene	Wounds	Similar syndrome to malignant oedema
***C. perfringens* type A**	Humans, dogs	Gas-gangrene	Wounds (road accidents)	Oedema, tissue necrosis and gangrene. Caused by the alpha toxin.

253

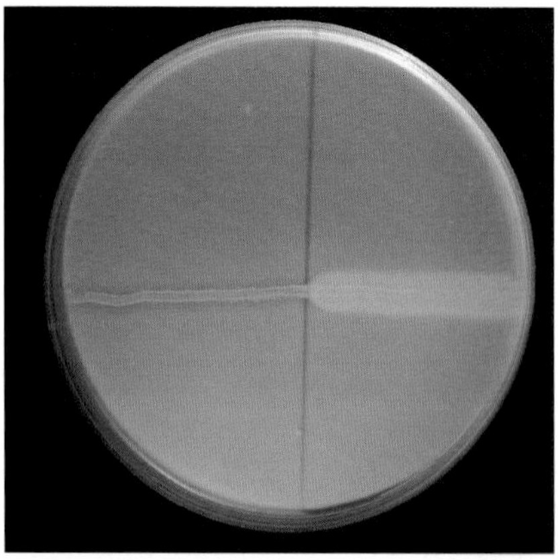

253. Nagler test for *C. perfringens* alpha toxin. The toxin is a lecithinase and attacks the lecithin in egg yolk agar (right). This reaction is neutralised on the left by specific antitoxin.

254

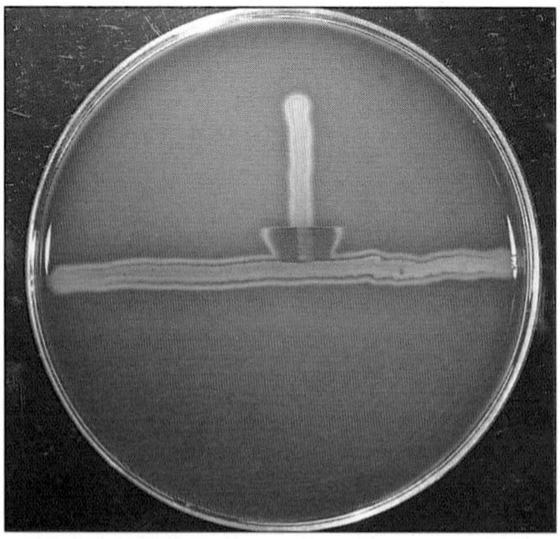

254 CAMP test with *Streptococcus agalactiae* (vertical streak) enhancing the partial haemolysis produced by the alpha toxin of *C. perfringens*.

Table 53. Microscopic and colonial appearance of the gas-gangrene clostridia.

***Clostridium* species**	**Gram-stained impression smears**	**Colonial appearance**
C. chauvoei	Oval, subterminal or central spores with typical citron (lemon-shaped) forms. Cells 0.6–0.8 x 3–8 μm	Colonies with large zone of clear haemolysis
C. septicum	Characteristic long filamentous forms. Spores oval and subterminal. Individual cells 0.6–0.8 x 3–8 μm, but filamentous forms are much longer	Swarming, spreading, haemolytic growth on normal agar. On 'stiff' agar the colonies are irregular with a rhizoid edge. Some strains produce smooth, round colonies
***C. novyi* type A**	Large Gram-positive rods with oval to cylindrical, subterminal spores. There is little or no swelling of the mother cell. Cells are 0.8–1.0 x 3–10 μm	Large, irregular colonies with a rhizoidal edge and a large zone of clear haemolysis
C. sordellii	Gram-positive rods with cylindrical spores that do not bulge the mother cell	Irregular, translucent colonies on 'stiff' agar, which become white on ageing
C. perfringens	Short, fat, Gram-positive rods that do not commonly produce spores. The spores, if present, are oval, subterminal and bulge the mother cell. Chains of cells can occur. Cells are 0.6–0.8 x 2–4 μm	Smooth, round, glistening colonies surrounded by 'target' or double-haemolysis (theta toxin giving a clear zone and partial haemolysis given by the alpha toxin)

• ***Nagler reaction of C. perfringens***

Type A antitoxin (alpha antitoxin) is spread over half of an egg yolk agar plate and allowed to dry. The suspect *C. perfringens* is streaked across both sides of the plate. All the types of *C. perfringens* produce the alpha toxin, that is a lecithinase. On the half of the plate without the antitoxin, the lecithin in the medium is attacked causing opalescence around the streak. The lecithinase reaction is neutralised on the half of the plate with the antitoxin but the growth of *C. perfringens* is unaffected (**253**).

• ***CAMP reaction of C. perfringens***

A diffusible factor produced by *Streptococcus agalactiae* enhances the partial haemolysis of the alpha toxin of *C. perfringens* (**254**). The complete zone of haemolysis seen immediately around the *C. perfringens* colonies is caused by the theta toxin.

Histotoxic Clostridia Affecting the Liver

Clostridium novyi type B is common in soil and in the normal intestinal tract of herbivores. It produces black disease (necrotic hepatitis) in sheep. *C. haemolyticum* (*C. novyi* type D) can be found in the ruminant digestive tract, liver and in the soil, and is the cause of bacillary haemoglobinuria in cattle. *C. novyi* type A is associated with gas-gangrene and type C is the reported cause of osteomyelitis of water buffalo in Southeast Asia and does not produce a toxin. *C. colinum*, the cause of quail disease, is excreted in faeces of birds with the chronic form of the condition. A toxin has not been identified for this clostridial species.

Pathogenesis

The alpha toxin is produced by *C. novyi* type A and type B. It is lethal, necrotising and phage-mediated. The beta toxin is a phospholipase and produced by *C. novyi* type B and by *C. haemolyticum* in greater amounts. This may account for the haemolytic crisis and death in bacillary haemoglobinuria.

	Toxins	
	Alpha	*Beta*
C. novyi type A	+	–
C. novyi type B	+	+
C. novyi type C	–	–
C. haemolyticum	–	+++

In black disease and bacillary haemoglobinuria, the spores, normally present in the intestine, may reach the liver and remain dormant in the Kupffer cells. Traumatic damage to the liver, especially due to migrating liver fluke, produces tissue damage and anaerobic conditions suitable for spore germination.

There is replication of the clostridia resulting in toxaemia, bacteraemia and often death. In quail disease, *C. colinum* passes from the intestine via the portal circulation and lodges in the liver where diffuse liver necrosis is produced. The intestine becomes ulcerated and in some birds extensive necrosis of the spleen occurs. Affected birds are inactive, sluggish and anorexic. They may die within 1–2 days but occasionally linger for a longer period. **Table 54** summarises the diseases caused by these clostridia.

Laboratory Diagnosis

• ***Direct Gram-stained smears***

Presence of characteristic liver lesions together with large numbers of Gram-positive rods in liver impression smears, from a recently dead animal or bird, is suggestive of the diseases. *C. novyi* type B and *C. haemolyticum* are large Gram-positive rods (0.8–1.0 × 3–10 µm) that produce oval to cylindrical, subterminal spores with little bulging of the mother cell. *C. colinum* is a Gram-positive rod, about 1 µm in diameter and 3–4 µm long. It has oval, subterminal spores but sporulation is infrequent.

• ***The fluorescent antibody technique*** is useful for the identification of *C. novyi* type B and *C. haemolyticum* in acetone-fixed liver impression smears.

• ***Isolation***

C. novyi type B and *C. haemolyticum* are very demanding in both their anaerobic and nutritional requirements. Very strict anaerobic procedures are necessary and media containing cysteine (Moore's medium), described in **Appendix 2**, should be used. These clostridia can die within 15 minutes of being exposed to atmospheric oxygen. The colonies are haemolytic, small and usually rhizoidal in nature.

C. colinum is fastidious and primary isolation is difficult. Success has been reported in tryptose-phosphate-glucose broth with 8 per cent sterile citrated horse plasma, thioglycollate broth with 3–10 per cent horse serum or in 5–8 day fertile chicken eggs. Several passages are needed after which the clostridium can be grown on blood agar. Polymyxin B (25 µg/ml) can be added to the isolation media to supress contaminants.

• ***Biochemical reactions*** are shown in **Table 49**.

• ***Animal inoculation***

Toxins in the liver can be demonstrated by intramuscular injection of homogenates into guinea-pigs. The pathogenicity is enhanced if the homogenate is added to an equal amount of 5 per cent calcium chloride solution before inoculation. The guinea-pigs die in 1–2 days. Specific antitoxin is not readily available for neutralisation tests.

The Enterotoxaemias

Clostridial enterotoxaemias are acute, highly fatal intoxications that affect sheep, lambs, calves, piglets and occasionally foals. The diseases are caused by the major exotoxins (enterotoxins) of *Clostridium perfringens* types B, C and D and occasionally types A and E.

C. perfringens is relatively aerotolerant, non-motile, has a polysaccharide capsule in tissue and is a short, fat Gram-positive rod (0.6–0.8 × 2–4 µm). The spores are oval, subterminal and bulge the mother cell. They are rarely produced although one exception is in the intestinal tract of humans in food-poisoning cases. The enterotoxin involved is identical to a component of the spore coat. Characteristic reactions are the double-zoned haemolysis on blood agar (**252**), stormy-clot (stormy-fermentation) in litmus milk medium (**238**) and the Nagler reaction (**253**). The five types (A–E) are based on the different combinations of the toxins elaborated by the organism.

Natural Habitat

Type A occurs in the intestinal tract of humans and animals and in most soils. Types B to E are more adapted to survival in the intestines but in outbreaks of disease they survive long enough in soil to infect other animals.

Pathogenesis

Minor toxins are produced such as theta (haemolysin), kappa (collagenase), mu (hyaluronidase) and nu (DNase) and these may contribute to tissue damage. However, the major toxins, alpha, beta, epsilon and iota are of greatest importance. The major toxin(s) produced by each *C. perfringens* type are shown in **Table 55**.

- ***Alpha toxin.*** This is a lecithinase (phospholipase) that attacks cell membranes causing cell death and destruction. The alpha toxin is produced by all types and gives a zone of partial haemolysis on blood agar. The Nagler reaction is based on the neutralisation of the lecithinase activity of this toxin on egg yolk medium.
- ***Beta toxin*** is lethal and necrotising. It is sensitive to trypsin and this explains the predilection of types B and C for neonates as colostrum has anti-trypsin activity. It is a labile toxin and may be destroyed if there is a delay in small intestinal contents, containing the toxin, reaching the laboratory. The beta toxin is the most important factor in the enterotoxaemias caused by type B.
- ***Epsilon toxin*** is secreted as a protoxin (prototoxin) and is activated in the intestines by proteases such as trypsin. Pulpy kidney disease is not usually seen in neonatal lambs as colostrum contains an antitrypsin factor that can prevent the epsilon toxin being activated. The toxin itself increases gut per-

Table 54. Summary of clostridia that can affect the liver.

Clostridium spp.	Main hosts	Disease	Route of entry	Clinical and postmortem signs
***C. novyi* type B**	Sheep and occasionally cattle 1–4 years old	Black disease (Infectious necrotic hepatitis)	Endogenous + liver fluke damage	Deaths are sudden. Grey-yellow foci in liver. Excess fluid in body cavities. Venous congestion occurs that darkens the skin ('black disease')
C. haemolyticum (*C. novyi* type D)	Cattle (sheep)	Bacillary haemoglobinuria (red water)	Endogenous + liver fluke damage	Sudden death with signs of abdominal pain and port-wine-coloured urine. Typical infarcts in liver: pale and raised surrounded by a blue-red zone
C. colinum	Bobwhite quail and other wild and domestic birds	Quail disease	Intestine via portal circulation to liver	Ulceration of intestinal wall and diffuse necrosis of the liver and sometimes also the spleen

Table 55. The major toxins of *Clostridium perfringens.*

Clostridium perfringens	***Major toxin***			
Type	***Alpha***	***Beta***	***Epsilon***	***Iota***
A	+	–	–	–
B	+	+	(+)	–
C	+	+	–	–
D	+	–	+	–
E	+	–	–	+

meability, assuring absorption of the toxin into the bloodstream. It damages vascular endothelium (including blood vessels in the brain) leading to fluid loss and oedema. The epsilon toxin can be regarded as an enterotoxin and a neurotoxin.

- ***Iota toxin*** is also produced as a protoxin and is not unique to *C. perfringens* type E as it is also formed by *C. spiroforme* and *C. difficile*. The action of this toxin is not fully understood.

The enterotoxaemias are often precipitated by certain husbandry and environmental factors such as abrupt changes in feeding, usually to a richer diet, and overeating and voracity on high protein and energy-rich foods. This leads to slowing of peristalsis with retention of bacteria in the intestines, absorption of toxins, inadequately digested carbohydrate and the provision of a rich medium for the proliferation of *C. perfringens*. The bacterium inhabits the large intestine in normal animals but if overgrowth occurs *C. perfringens* can spill over into the small intestine with the production of a large amount of toxin and enterotoxaemia. **Table 56** summarises the clostridial enterotoxaemias.

Laboratory Diagnosis

The definitive diagnosis of the enterotoxaemias is based on the demonstration and identification of the toxins in the small intestine using a mouse or guinea-pig neutralisation test. Other tests can be useful adjuncts particularly for pulpy kidney disease.

- *Gram-stained smears* can be made from the mucosa of the small intestine of a recently dead animal. Large numbers of fat, Gram-positive rods are highly suggestive of an enterotoxaemia as few clostridia are normally present in the small intestine.

Table 56. The clostridial enterotoxaemias.

Clostridium perfringens type	Major toxins	Hosts	Disease	Clinical and postmortem signs
A	enterotoxin	Humans	Food poisoning	Sudden onset, diarrhoea, abdominal pain and nausea, but vomiting is uncommon. Short course and rarely fatal
	alpha	Lambs	Enterotoxaemic jaundice	Occurs in California and Oregon in the spring. Depression, anaemia, icterus, haemoglobinuria and lambs die within 6–12 hours of first signs. Known as 'the yellows' or 'yellow lamb disease'
B	beta (epsilon) alpha	Lambs under 3 weeks old	Lamb dysentery	A haemorrhagic and rapidly fatal enterotoxaemia. Lambs are often found dead
		Calves and foals		Enterotoxaemia, not common
C	beta alpha	Piglets 1–3 days old Lambs, foals and calves	Haemorrhagic enterotoxaemia (clostridial enteritis)	Dysentery, collapse and death. Small intestine is dark red and has gas bubbles in mucosa. Lumen is full of bloody fluid
		Broiler chicken 2–12 weeks old	Necrotic enteritis (types A and C)	Depression, diarrhoea, death in a few hours. Mortality 2–50%. Mucosa of small intestine has a brown pseudomembrane. Most common in deep-litter units
		Adult sheep and goats	Struck	Sudden deaths due to an enterotoxaemia
D	epsilon alpha	Sheep all ages except neonates. Rare cases in calves and goats	Pulpy kidney disease	Oedema of brain, glysosuria, sudden deaths. Excess fluid in body cavities, focal symmetrical encephalomalacia in some cases. Occurs in well-grown lambs
E	iota alpha	Calves and lambs	Enterotoxaemia	Pathogenicity unclear

- *Histopathology on brain sections* to demonstrate focal symmetrical encephalomalacia in pulpy kidney disease. The lesion is not always present as its formation depends on the time from the first clinical signs to death. If present, the lesion is characteristic for the disease.
- *Glycosuria* (glucose in urine) is suggestive of pulpy kidney disease.
- *Demonstration of toxin* in the small intestine. A suitable specimen is 20–30 ml of ileal contents from a recently dead animal. The toxins are labile so the specimen should reach the laboratory as soon as possible after collection. The ileal contents are centrifuged and the clear supernatant is tested for toxin. If the ileal contents are very mucoid, it may be difficult to obtain a supernatant. Placing a small piece of cotton wool at the top of the centrifuge tube, before centrifuging, helps to take the mucus into the deposit. In ileal contents, the epsilon and iota toxins are usually in the active form. To demonstrate the toxin 0.4 ml of the clarified ileal contents can be inoculated intravenously into each of two mice. If a mouse dies within 5 minutes this is probably due to shock; deaths from toxin usually occur within 10 hours.
- *Identification of the toxin* in the clarified ileal contents is carried out by a neutralisation test. Intravenous inoculation in mice or intradermal injection into white, shaved, young guinea-pigs using with commercial antitoxins to *C. perfringens* types A–E (Wellcome Diagnostics). The mixtures and dose for inoculation are as follows:
  Test: 0.5 ml supernatant + 0.2 ml sterile saline + 0.1 ml antitoxin.
  Control: 0.5 ml supernatant + 0.3 ml sterile saline.
  Dose: 0.2 ml intradermally in guinea-pig and 0.4 ml intravenously for a mouse
  The mixtures are allowed to stand at bench temperature for 1 hour to allow neutralisation of the toxin by the antitoxin.

If pulpy kidney disease is suspected, and is the common form of enterotoxaemia in lambs in the area, it is acceptable to conduct a neutralisation test using the type D antitoxin only and omitting the initial step of demonstrating the toxin. Two mice for the test and two mice as controls should be used. The test is positive if the two control mice die in 10–12 hours but the test mice remain alive and well. If all four mice remain well, no toxin was present in the ileal contents and if all four mice die, either another toxin (other than the alpha and epsilon toxins) may have been present or the ileal contents contained an excessive amount of epsilon toxin and complete neutralisation was not attained. The test could be repeated using 0.2 ml type D antitoxin in the above mixture or the full neutralisation test carried out using the antitoxins to *C. perfringens* types A to E. Recommendations for conducting the test are given in the literature accompanying the commercial antitoxin (Wellcome Diagnostics).

Table 57 gives a simplified version of the neutralisation of toxins in ileal contents by antitoxins to *C. perfringens* types A to E. The test is often carried out in duplicate using untreated and trypsin-treated (1 per cent trypsin solution for 1 hour at 37°C) supernatant containing toxin. This is to ensure that the epsilon and iota toxins, produced in protoxin form, are converted into the active form; however, the trypsin treatment will destroy any beta toxin that may be present. This procedure is particularly necessary if a pure culture of *C. perfringens*, grown in cooked meat broth, is being typed based on the toxins that have been produced in the broth, as the epsilon and iota toxins will be in the protoxin form.

Clostridia Associated with Antibiotic-induced Disease

Clostridium spiroforme

C. spiroforme occurs as a loosely coiled, spiral Gram-positive form in smears from cultures on blood agar, but in faeces or caecal contents the spiral morphology is not so marked and it has a semicircular form. The spores are terminal. It is non-haemolytic and produces convex, circular, shiny, whitish to grey colonies on blood agar under anaerobic conditions at 37°C. *C. spiroforme* produces a cytotoxin and an exotoxin that is identical to the iota toxin of *C. perfringens* type E. It may be the cause of spontaneous diarrhoea in weanling rabbits. Diarrhoea can also be induced in adults by the administration of antibiotics, especially clindamycin. *C. spiroforme* has been frequently isolated from rabbits with 'mucoid enteritis' and may, with other microorganisms, play a part in the disease. Naturally occurring enterocolitis has been reported in foals and pigs.

The presence of semicircular Gram-positive bacteria in faeces or caecal contents is not sufficient for a diagnosis; the toxin should be demonstrated in mice or guinea-pigs and identified by a neutralisation test.

Clostridium difficile

C. difficile is a large (0.5 × 3–6 μm) Gram-positive rod that forms oval, subterminal spores. On blood agar the colonies are non-haemolytic, raised with a rhizoid edge. Special blood agar is required for isolation containing yeast extract, haemin, vitamin K, cysteine and antimicrobial agents (Borriello and Honour, 1981). It produces an enterotoxin (designated 'A') and a cytotoxin (B). *C. difficile* appears to be a cause of human, hamster, rabbit and guinea-pig enterocolitis initiated by prolonged antibiotic therapy, particularly with clindamycin. However, natural diarrhoeal diseases have been described in dogs, foals and pigs.

Both *C. difficile* and its toxins can be detected in the faeces and intestinal contents of affected animals. A definitive diagnosis is made by enzyme immunoassays for the detection of both toxins A and B in faecal specimens (Laughton *et al.*, 1984).

Table 57. Neutralization of *Clostridium perfringens* toxins by antitoxin.

	Toxins of *Clostridium perfringens* types A to E				
	A *alpha*	***B*** *alpha beta epsilon*	***C*** *alpha beta*	***D*** *alpha epsilon*	***E*** *alpha iota*
ANTITOXIN					
Type A: anti-alpha	–	x	x	x	x
Type B: anti-alpha -beta -epsilon	–	–	–	–	x
Type C: anti-alpha -beta	–	x	–	x	x
Type D: anti-alpha -epsilon	–	x	x	–	x
Type E: anti-alpha -iota	–	x	x	x	–

x = death of mouse or lesion in skin of guinea-pig.
– = antitoxin has neutralized the specific toxin and the mouse or guinea-pig is unaffected.

References

Borriello, P.S. and Honour, P. (1981). Simplified procedure for the routine isolation of *Clostridium difficile* from faeces, *Journal of Clinical Pathology*, **34**:1126–1127.

Laughton, B.E., Viscidi, R.P., Gdovin, S.L., Yolken, R.H. and Bartlett, J.G. (1984). Enzyme immunoassays for detection of *Clostridium difficile* toxins A or B in faecal specimens, *Journal of Infectious Diseases*, **149**: 781–788.

18 *Enterobacteriaceae*

Most members of the family *Enterobacteriaceae* share the following characteristics: Gram-negative, medium-sized rods (**255**) (0.4–0.6 × 2–3 μm); peritrichate arrangement of flagella, if motile; facultatively anaerobic and ferment, rather than oxidize glucose; catalase-positive and oxidase-negative; reduce nitrate to nitrite and are able to grow on non-enriched media such as nutrient agar. There are a few exceptions to these general properties, for example, *Shigella dysenteriae* is catalase-negative; *Tatumella ptyseos* is motile by polar, subpolar or lateral flagella and *Xenorhabdus* species do not regularly reduce nitrate.

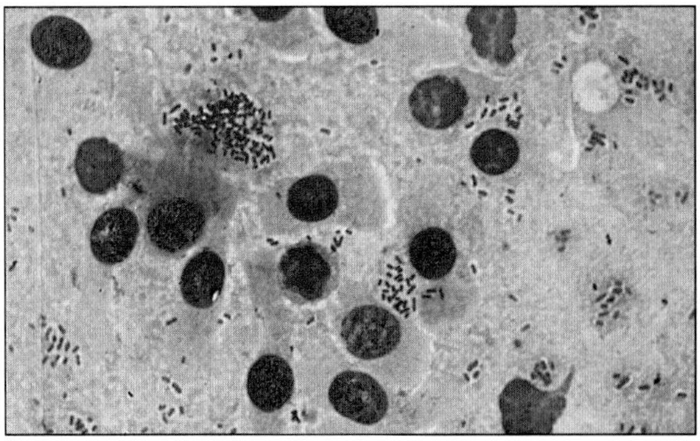
255

255 Gram-negative medium-sized rods of *Escherichia coli* in a tissue smear. The morphology is typical of most members of the *Enterobacteriaceae.* (Gram stain, ×1000)

Nomenclature

There are, at present, 28 genera and more than 80 well-defined species in the *Enterobacteriaceae,* not including the large number of *Salmonella* serotypes or 'species'. Traditionally the genera and species of the family are distinguished biochemically and this is convenient for identification of clinical isolates. However, genetic means of defining species, based on DNA-DNA homology, has led to the recognition of numerous new species, some previously regarded as aberrant biotypes, and also to regarding genetically closely related members as single genomic species. Several of the members of the *Enterobacteriaceae* have commonly used synonyms, including:

Name used in this chapter	Synonym(s)
Citrobacter diversus	*C. koseri, Levinea malonatica*
Edwardsiella tarda	*E. anguillimortifera*
Enterobacter aerogenes	*Klebsiella mobilis*
Enterobacter agglomerans	*Erwinia herbicola*
Klebsiella pneumoniae	*K. pneumoniae* subsp. *pneumoniae*
Klebsiella ozaenae	*K. pneumoniae* subsp. *ozaenae*
Klebsiella rhinoscleromatis	*K. pneumoniae* subsp. *rhinoscleromatis*
Koserella trabulsii	*Yokenella regensburgei*
Leclercia adecarboxylata	*Escherichia adecarboxylata*
Obesumbacterium proteus biogroup 1	*Hafnia alvei* biogroup 1
Serratia rubidaea	*S. marinorubra*

The terms 'coliform' or 'coliform bacteria' have no taxonomic significance but are used to refer to those members of the *Enterobacteriaceae* that usually ferment lactose, such as *Escherichia coli, Klebsiella* and *Enterobacter* species.

Habitat

The members of the *Enterobacteriaceae* are geographically widespread and many are widely distributed throughout the environment in soil, water, on plants as well as in the intestines of animals and humans. However, a few species occupy a limited ecological niche, such as *Salmonella typhi*, that causes typhoid fever and is found only in humans.

Selective and/or Indicator Media for the Enterobacteria

All enterobacteria will grow on blood and MacConkey agars and these are used routinely to isolate them in diagnostic laboratories. Although MacConkey agar is a selective medium it is relatively permissive and allows the growth of some other

Gram-negative bacteria as well as the enterobacteria. Brilliant green agar and xylose-lysine-deoxycholate (XLD) medium are more selective and used for the isolation of salmonellae, although the media will support the growth of some other enterobacteria. **Table 58** gives the reactions of some members of the *Enterobacteriaceae* and *Pseudomonas aeruginosa* (for comparison) on MacConkey agar, brilliant green agar and XLD medium. The uninoculated media (**256**) are illustrated together with the appearance of 11 enterobacteria and *P. aeruginosa* on these selective/indicator media (**257–268** inclusive).

Table 58. Reactions of some members of the *Enterobacteriaceae* and *Pseudomonas aeruginosa* on selective/indicator media.

Photograph number	*Bacterium*	Lactose	Sucrose	Xylose	H_2S	Lysine	*MacConkey agar (lactose, neutral red)*	*Brilliant green agar (lactose, sucrose, phenol red)*	*XLD agar (lactose, sucrose, xylose, lysine, H_2S, phenol red)*
257	*Salmonella enteritidis*	–	–	+	+	+	Pale colonies (alk)	Red (alk)	Red/black centre (alk)
258	*Proteus mirabilis*	–	(–)	+	+	–	Pale (alk)	Red (alk)	Yellowish/black centre (alk)
259	*Edwardsiella tarda*	–	–	–	+	+	Pale (alk)	No growth	Reddish/black centre (alk)
260	*Escherichia coli*	+	d	+	–	(+)	Bright pink (acid)	[Yellow-green] (acid)	Yellow (acid)
261	*Klebsiella pneumoniae*	+	+	+	–	+	Pink (acid)	Yellow-green (acid)	Yellow (acid)
262	*Enterobacter aerogenes*	+	+	+	–	+	Pink (acid)	Yellow-green (acid)	Yellow (acid)
263	*Providencia stuartii*	–	d	–	–	–	Pale (alk)	[Yellow-green] (acid)	Yellow (acid)
264	*Citrobacter diversus*	d	(–)	+	–	–	Pink (acid)	[Green] (acid)	Yellow (acid)
265	*Serratia marcescens*	–	+	–	–	+	Pale + pigment	Red-yellow (acid)	Red (alk)
266	*Yersinia enterocolitica*	–	+	d	–	–	Pale pink	[Green] (acid)	Yellow (acid)
267	*Yersinia pseudo-tuberculosis*	–	–	+	–	–	Pale (alk)	[Red] (alk)	Red (alk)
268	*Pseudomonas aeruginosa**	–	–	–	–	+	Pale + pigment	Red + pigment	Red (alk)

[ ] = poor growth, alk = alkaline reaction, H_2S = hydrogen sulphide, + = 90–100% strains positive, (+) = 76–89% positive, d = 26–75% positive, (–) = 11–25% positive, – = 0–10% positive. * = not a member of the *Enterobacteriaceae*, included for comparison.

256

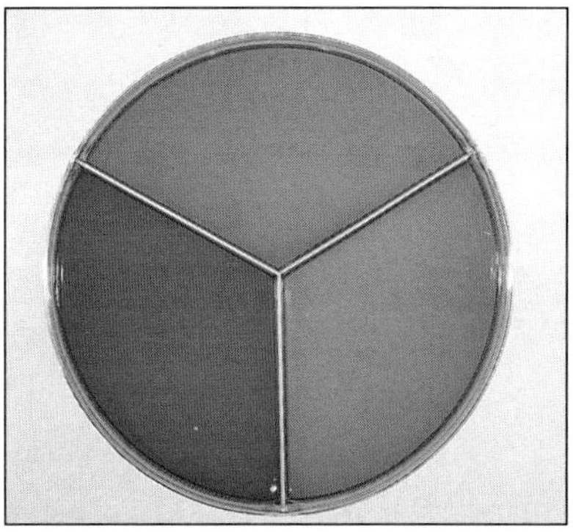

257–267 Reactions of some members of the *Enterobacteriaceae* on XLD medium (left), brilliant green agar (top) and MacConkey agar (right).

256 Uninoculated media.

257 *Salmonella enteritidis.*

257

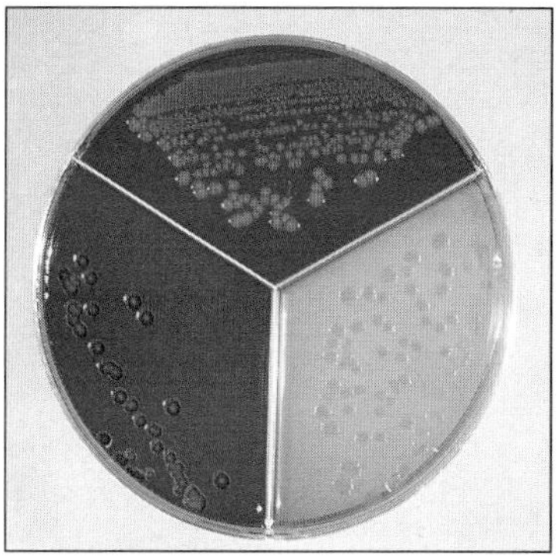

258

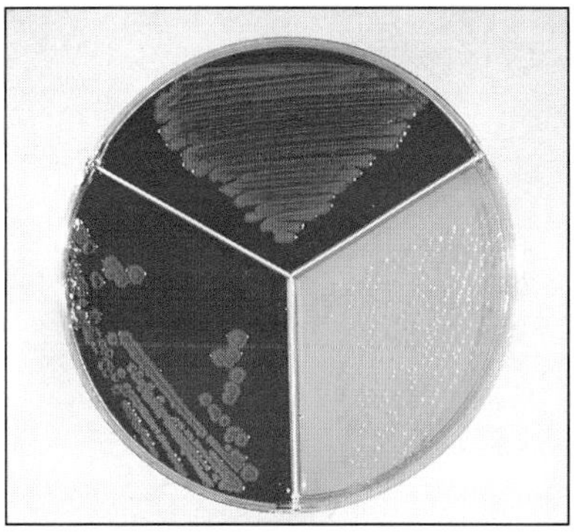

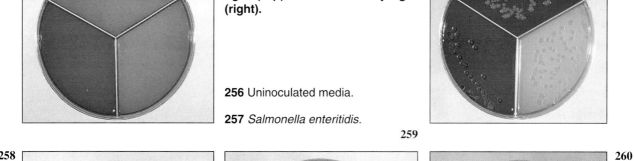

258 *Proteus mirabilis.*

259

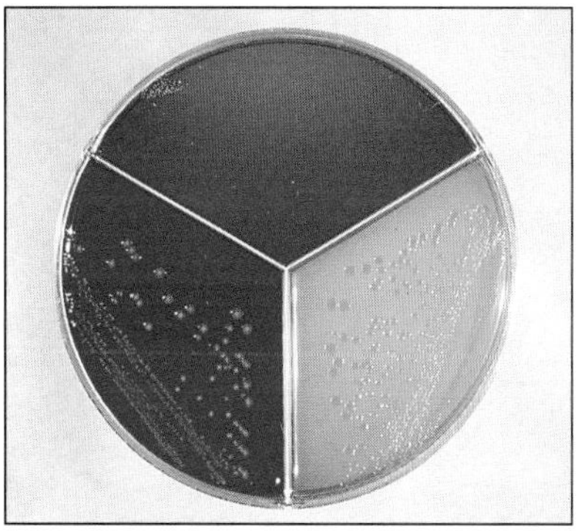

259 *Edwardsiella tarda.*

260

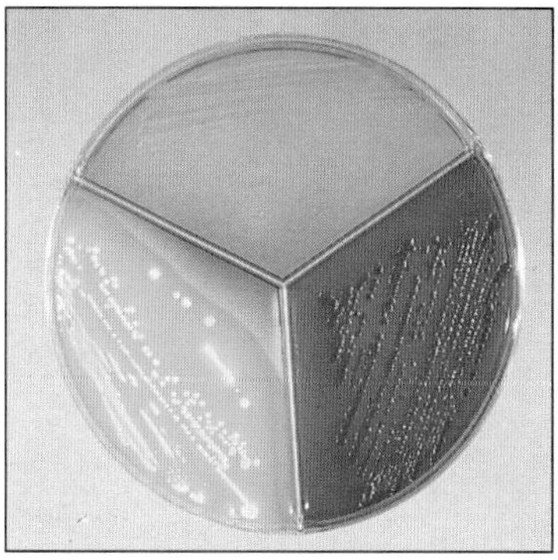

260 *Escherichia coli.*

261

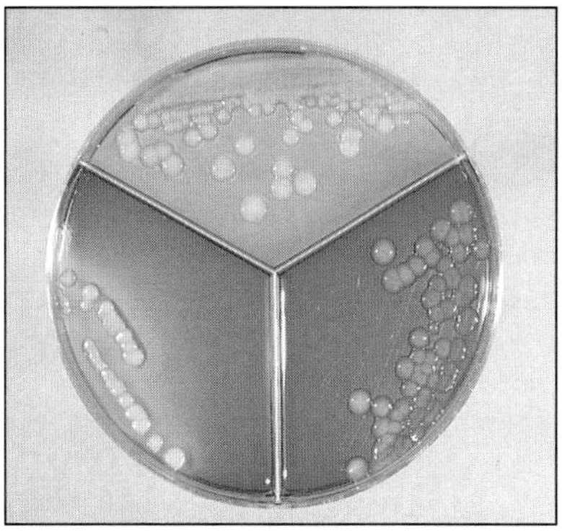

261 *Klebsiella pneumoniae.*

262

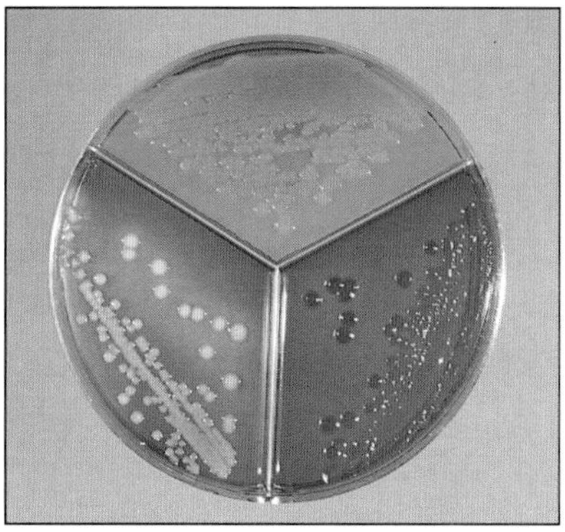

262 *Enterobacter aerogenes.*

263

263 *Providencia stuartii.*

264

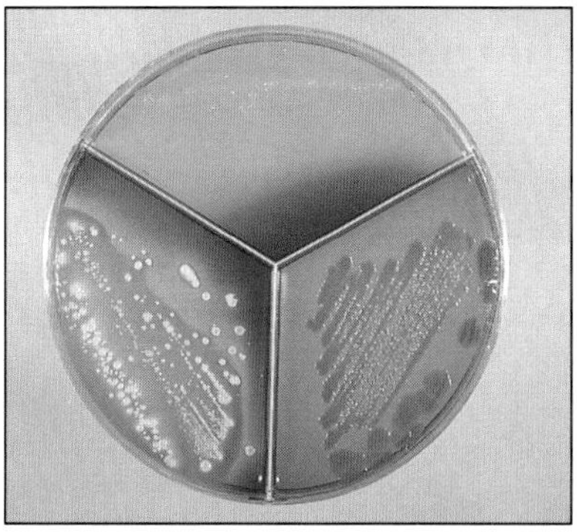

264 *Citrobacter diversus.*

265

265 *Serratia marcescens.*

266

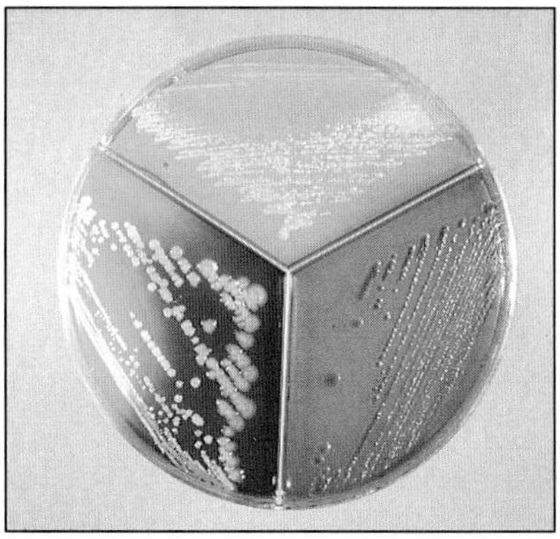

266 *Yersinia enterocolitica.*

267

267 *Yersinia pseudotuberculosis.*

268

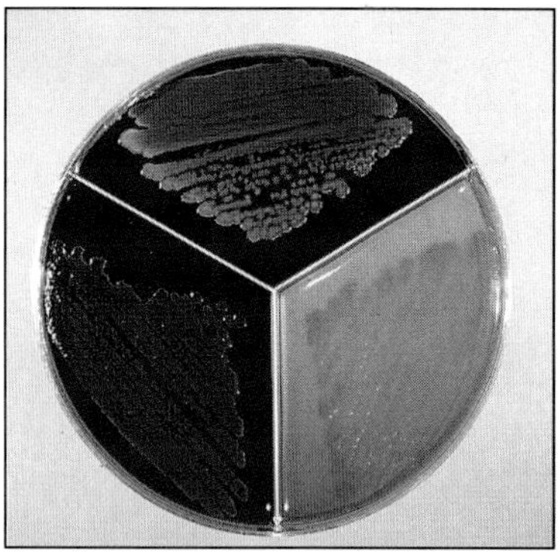

268 *Pseudomonas aeruginosa* for comparison with the reactions of the *Enterobacteriaceae.*

MacConkey agar

Fermentable sugar: lactose.
pH indicator: neutral red (pale straw at pH 8 and pink at pH 6.8).
Inhibitors: bile salts and crystal violet (anti-Gram-positive bacteria).
Reactions: if the bacterium can ferment the lactose, acidic metabolic products are produced and the medium and colonies are pink (lactose-fermenter). If the organism is unable to utilize the lactose, then it attacks the peptone (nitrogen source) in the medium with resulting alkaline metabolic products and the medium and colonies are pale straw-coloured (non-lactose-fermenter)

Brilliant green agar

Fermentable sugars: lactose and sucrose.
pH indicator: phenol red (red at pH 8.2 and yellow at pH 6.4).
Inhibitor: brilliant green dye that to some extent inhibits the growth of most enterobacteria, except *Salmonella* species.
Reactions: similar to those occurring on MacConkey agar except that the bacteria may ferment one or both of the sugars with an acid reaction (yellowish-green) or be unable to ferment either sugar and attack the peptone instead, with an alkaline reaction (red colonies and medium).

XLD medium

Fermentable sugars: lactose, sucrose and xylose.
pH indicator: phenol red (red at pH 8.2 and yellow at pH 6.4).
Other substrates: lysine and chemicals for detecting hydrogen sulphide (H_2S) production.
Inhibitor: bile salts (sodium deoxycholate).
Reactions: salmonellae will first ferment the xylose creating a temporary acid reaction but this is reversed by the subsequent decarboxylation of lysine with alkaline metabolic products. Superimposed on the red (alkaline) colonies is the production of hydrogen sulphide, so most salmonellae have red colonies with a black centre (**257**). *Edwardsiella tarda* also gives this reaction although the H_2S production is less marked and the periphery of the colonies tends to be a yellowish-red colour. The large amount of acid produced by enterobacteria that can ferment either lactose or sucrose, or both, prevents the reversion to alkaline conditions even if the bacterium is able to decarboxylate the lysine.

Triple sugar iron (TSI) agar

This is an indicator medium only and does not contain an inhibitor. A brief description of the medium and technique for inoculation is given in Chapter 4. It is prepared as slants in tubes.
Fermentable sugars: glucose 0.1 per cent, lactose 1.0 per cent and sucrose (or saccharose) 1.0 per cent.
Other substrates: chemicals to indicate hydrogen sulphide (H_2S) production.
pH indicator: phenol red (red at pH 8.2 and yellow at pH 6.4).
Reactions: all members of the *Enterobacteriaceae* are capable of fermenting glucose and the small amount (0.1 per cent) will be attacked preferentially and rapidly. At this early stage both the butt and slant will be yellow due to acid production from the glucose fermentation. Some enterobacteria attack the lactose and/or sucrose (each at a 1.0 per cent concentration) in the medium and in this case sufficient acid is produced to maintain both the butt and the slant in an acid (yellow) condition.
Bacteria that are unable to ferment either lactose or sucrose, after the depletion of the limited amount of glucose, will utilize the peptones in the medium. This is a less efficient method of producing energy and occurs mainly at the surface of the slant in the presence of atmospheric oxygen. The metabolites of peptones are alkaline and this causes the slant to revert back to the original red colour. Some members of the

Enterobacteriaceae, including most *Salmonella* spp., are able to produce hydrogen sulphide. This reaction is superimposed over the sugar fermentations and is seen as a blackening of the medium.

The general interpretation of the reactions is as follows:

- Alkaline (red) slant and acid (yellow) butt: glucose fermentation only.
- Acid (yellow) slant and acid (yellow) butt: lactose and/or sucrose attacked as well as the glucose.
- Blackening of the medium: hydrogen sulphide production.

The reactions are illustrated (**269**) and **Diagram 32** gives the differentiation of some of the enterobacteria by their reactions in TSI agar and lysine decarboxylase broth.

269

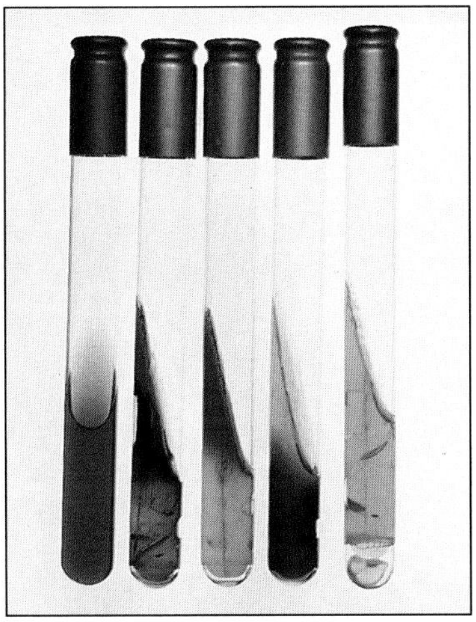

269 TSI agar slopes showing the range of reactions from the left, uninoculated, R/Y/H_2S+, R/Y/H_2S-, Y/Y/H_2S+, Y/Y/H_2S-. *See* **Diagram 32** for notation.

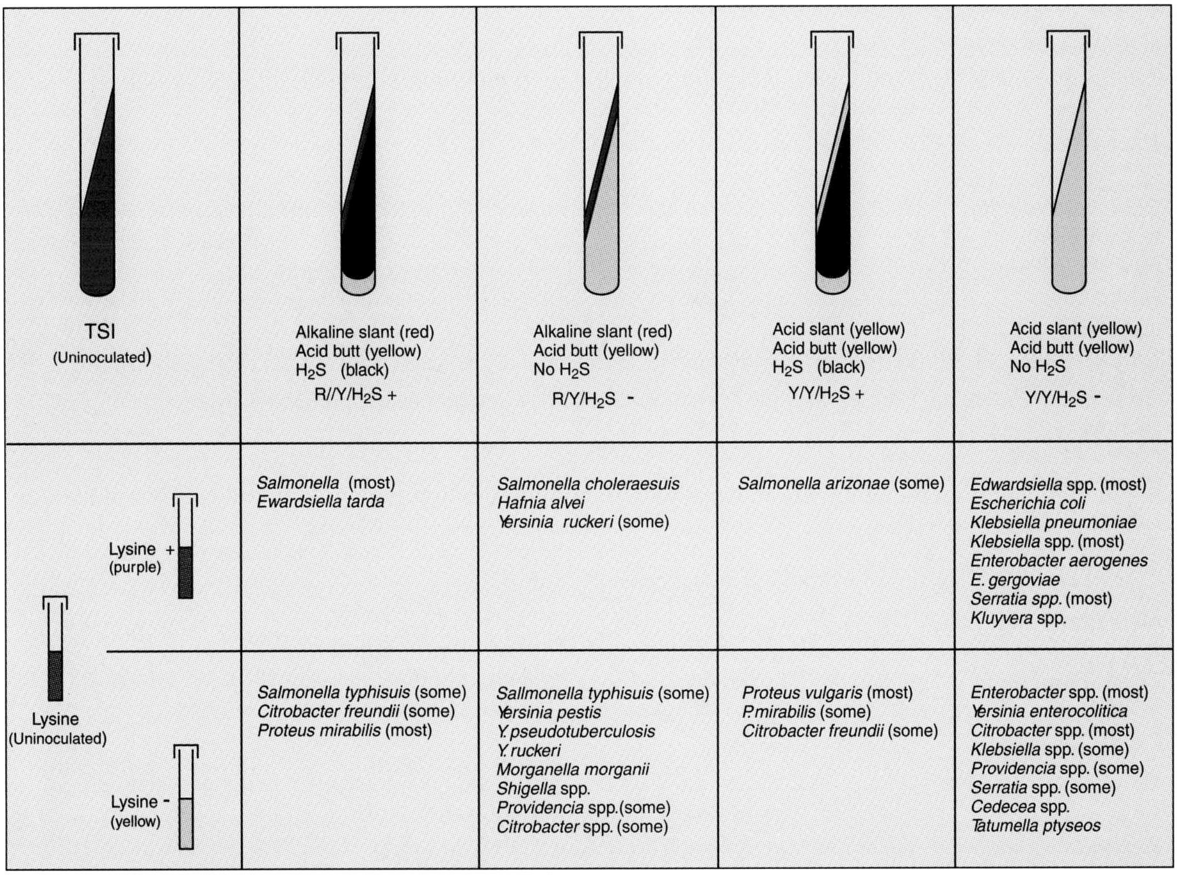

Diagram 32. Reactions of the *Enterobacteriaceae* in triple sugar iron agar and lysine decarboxylase broth. R= red (alkaline), Y= yellow (acid), H_2S+ = hydrogen sulphide produced, H_2S- = hydrogen sulphide not produced.

Pathogenicity

The *Enterobacteriaceae* can be divided into three groups based on their pathogenicity for animals:

- Uncertain significance for animals. These include species from 17 genera of the *Enterobacteriaceae* and they are summarised in **Table 59**. As some of them may be isolated from clinical specimens a range of their biochemical reactions is given in **Table 60**.
- Opportunistic pathogens that are known occasionally to cause infections in animals. These include species within the genera *Klebsiella, Enterobacter, Proteus, Serratia, Edwardsiella, Citrobacter, Morganella,* and *Shigella* which is a human pathogen that also causes disease in other primates.
- Major pathogens of animals such as *Salmonella* species, *Escherichia coli* and three of the *Yersinia* species.

All Gram-negative bacteria, including the members of the *Enterobacteriaceae*, have lipopolysaccharides in the outer membrane of the cell wall that are potent endotoxins, the main endotoxic principle being lipid A. Endotoxin is released when bacteria die and lyse. The effects of endotoxin in the animal body include fever, leukopaenia followed by leukocytosis and hyperglycaemia with a subsequent fall in blood sugar and lethal shock after a latent period. The more pathogenic members of the *Enterobacteriaceae* have other virulence factors such as adhesins for attachment to host cells, capsules that are antiphagocytic, siderophores that aid the bacterium in its competition with the host for iron and exotoxins that include enterotoxins and cytotoxins.

Table 59. Genera of the *Enterobacteriaceae* whose species are of uncertain significance for animals.

Genus (species)	Site of isolation and possible pathogenicity
Budvicia aquatica	Water and human faeces
Buttiauxella agrestis	Water
***Cedecea* species**	Rare isolates from human clinical specimens (most commonly from the respiratory tract). Bacteraemia in humans has been reported
Ewingella americana	Rare isolate from human clinical specimens
***Erwinia herbicola* (= *Enterobacter agglomerans*)**	This and other *Erwinia* species are associated with plants as pathogens or saprophytes
Hafnia alvei	Faeces of humans and animals, sewage, soil, water and dairy products. Rarely, if ever, pathogenic
***Kluyvera* species**	Water, sewage, soil and milk. Occasionally isolated from human clinical specimens
***Koserella trabulsii* (= *Yokenella regensburgei*)**	Human respiratory tract, wounds, urine and faeces but pathogenicity uncertain
Leclercia* (*Escherichia*) *adecarboxylata	Environment, food and water. It has been isolated from human clinical specimens
***Leminorella* species**	Isolated from human faeces and urine
Moellerella wisconsensis	Human faeces
Obesumbacterium proteus	Found only in contaminated beer. Biogroup 1 is thought to be a brewery-adapted biochemical variant of *Hafnia alvei*. Unlikely to be pathogenic for animals
Pragia fontium	Isolated from water
***Providencia* species**	Urinary tract of compromised or catheterised human patients, burn infections and diarrhoea. Rarely isolated from faeces of healthy humans
Rahnella aquatilis	Water and from a human burn wound
Tatumella ptyseos	Occasionally isolated from human clinical specimens, mainly from the respiratory tract
***Xenorhabdus* species**	Pathogenic for nematodes

Table 60. Biochemical reactions: members of the *Enterobacteriaceae* of uncertain significance.

	Indole production	Methyl red	Voges–Proskauer	Citrate	Urease	Phenylalanine deaminase	Hydrogen sulphide	Lysine decarboxylase	Ornithine decarboxylase	Motiliy (36°C)	Gelatin liquefaction	Growth in KCN broth	ONPG (beta-galactosidase)	Acid from: Dulcitol	Inositol	Lactose	Maltose	Mannitol	Mannose	Rhamnose	Sorbitol	Sucrose	Xylose	Yellow pigment
Budvicia aquatica	–	+	–	–	d	–	(+)	–	–	d	–	–	+	–	–	(+)	–	d	–	+	–	–	+	–
Buttiauxella agrestis	–	+	–	+	–	–	–	–	+	+	–	(+)	+	–	–	+	+	+	+	+	–	–	+	–
Cedecea species	–	+	d	+	–	–	–	–	d	+	–	(+)	+	–	–	v	+	+	+	–	v	v	(+)	–
Ewingella americana	–	(+)	+	+	–	–	–	–	–	d	–	–	(+)	–	–	d	(–)	+	+	(–)	–	–	(–)	–
Erwinia herbicola (*Enterobacter agglomerans*)	(–)	d	d	d	(–)	(–)	–	–	–	(+)	–	d	+	(–)	(–)	d	(+)	+	+	(+)	d	(+)	+	(+)
Hafnia alvei	–	d	(+)	–	–	–	–	+	+	(+)	–	+	+	–	–	–	+	+	+	+	–	–	+	–
Kluyvera species	+	+	–	+	–	–	–	v	+	+	–	+	+	(–)	–	+	+	+	+	+	d	+	+	–
Koserella trabulsii	–	+	–	+	–	–	–	+	+	+	–	+	+	–	–	–	+	+	+	+	–	–	+	–
Leclercia adecarboxylata	+	+	–	–	d	–	–	–	–	(+)	–	+	+	(+)	–	+	+	+	+	+	–	d	+	d
Leminorella species	–	v	–	v	–	–	+	–	–	–	–	–	–	v	–	–	–	–	–	–	–	–	+	–
Moellerella wisconsensis	–	+	–	(+)	–	–	–	–	–	–	–	d	+	–	–	+	d	d	+	–	–	+	–	–
Obesumbacterium proteus																								
biogroup 1	–	(+)	d	–	–	–	–	+	d	–	–	–	d	–	–	–	–	d	+	–	–	–	–	–
biogroup 2	–	(–)	–	–	–	–	–	+	+	–	–	–	–	–	–	–	d	–	(+)	(–)	–	–	(–)	–
Pragia fontium	–	+	–	(+)	–	(–)	(+)	–	–	+	–	–	–	–	–	–	–	–	–	–	–	–	–	–
Providencia species	+	+	–	+	v	+	–	–	–	+	–	+	–	–	v	(–)	–	v	+	v	–	v	–	–
Rahnella aquatilis	–	(+)	+	+	–	+	–	–	–	–	–	–	+	(+)	–	+	+	+	+	+	+	+	+	–
Tatumella ptyseos	–	–	–	–	–	+	–	–	–	–	–	–	–	–	–	–	–	–	+	–	–	+	–	–
Xenorhabdus species	v	–	–	v	v	–	–	–	–	+	d	(–)	–	–	–	–	v	–	+	–	–	–	–	d

+ = 90–100% strains positive, (+) = 76–89% positive, d = 26–75% positive, (–) = 11–25% positive, – = 0–10% positive, v = reaction variable among species.

Enterobacteria that are Opportunistic Pathogens: Natural Habitat and Pathogenicity

The diseases and natural habitat of these members of the *Enterobacteriaceae* are summarised in **Table 61**.

Table 61. Diseases caused by the opportunistic pathogens of the *Enterobacteriaceae*.

Pathogen	Habitat	Disease(s)
Citrobacter diversus	Faeces of man, animals and in soil and sewage	Meningitis in human neonates and mastitis in cattle have been reported. It is probably capable of opportunistic infections in other mammals
Edwardsiella tarda	Water, mud and reptilian intestines	Fish (eels and catfish) and marine mammals: abscesses in muscle, liver and kidneys Mild diarrhoea reported in pigs, calves and dogs Latent intestinal infections in tortoises
Enterobacter aerogenes	Water, soil, sewage and faeces	Coliform mastitis in cattle Uterine infections in mares Occasionally part of the mastitis-metritis-agalactia (MMA) syndrome in sows
Klebsiella pneumoniae	Intestinal tract of animals and man, soil and sawdust	Coliform mastitis in cattle Cervicitis and metritis in mares Urinary tract infections in dogs Pneumonia and suppurative conditions in foals
Morganella morganii	Faeces of animals	Ear and urinary tract infections in dogs and cats
Proteus mirabilis* and *P. vulgaris	Faeces of mammals and environment	Urinary tract infections of dogs and horses Associated with otitis externa in dogs and cats Diarrhoea in young mink, lambs, calves, goats and pups
Serratia marcescens	Environmental organism	Bovine mastitis Septicaemia in chickens and immunosuppressed mammals Infections in geckos and tortoises
***Shigella* species**	Intestinal tract of man and other primates	Dysentery/diarrhoea in man and other primates. Very few reports of infections in other animals. Dogs can be infected from human owners and may excrete the bacteria for short periods, without clinical signs

Laboratory Diagnosis

Direct microscopy

As the enterobacteria share the property of being Gram-negative, medium-sized rods with many other bacterial genera, direct microscopy is not usually helpful. However, in urinary tract infections a bacterial count (Chapter 4) could be carried out on freshly taken, mid-stream urine. In dogs, 10^5 bacteria/ml urine is taken to indicate a clinical bacteriuria.

Isolation

All the members of the *Enterobacteriaceae* will grow on the routine diagnostic media, blood and MacConkey agars. **Diagram 33** gives the routine culture methods as well as tests for the presumptive identification of some of the enterobacteria.

Identification

Colonial morphology and reactions on selective/indicator media

Reactions on MacConkey agar indicate whether or not the bacterium ferments the lactose in the medium. In several genera, lactose fermentation is late or irregular and in these cases the ONPG test for β-galactosidase (Chapter 4) reveals a potential ability to attack lactose.

Generally, on blood agar, the colonies of most of the members of the *Enterobacteriaceae* are similar. They are usually relatively large, 2–3 mm after 24 hours' incubation, non-haemolytic, shiny, round and greyish in colour. However, a few enterobacteria have distinctive colonial characteristics. Most *Proteus mirabilis* and *P. vulgaris* strains will swarm on blood agar. *Proteus* species grow in a colonial fashion for a period causing a build-up of toxic metabolic products. Long, flagellated swarm cells are formed (**270**) that move quickly across the contaminated agar and colonial growth occurs on the fresh agar. Metabolic products again accumulate to a critical level and the swarming cycle is repeated. If a spot inoculation of a *Proteus* species is made on a blood agar or nutrient agar plate, rings of heavy growth interspersed with thin areas of swarming growth will be seen (**271**). Normally the bile salts in MacConkey agar prevent the swarming of *Proteus* species, but if the surface of the agar is moist, swarming may be seen even on this medium (**272**). On blood agar, particularly, the powerful and foul odour of *Proteus* species will be noticed and the bacteria tend to turn the blood agar a chocolate-brown colour. Most *P. vulgaris* and *P. mirabilis* strains produce hydrogen sulphide (H_2S) in TSI and XLD media. As they are also lactose-negative, *P. mirabilis* can give a reaction similar to most of the salmonellae (R/Y/H_2S+) in TSI. However, *Proteus* species are almost always lysine-decarboxylase-negative. Similarly on XLD medium (**273**) some strains of *Proteus* can mimic salmonella colonies by having a black centre (H_2S production) but the periphery of the colony tends to have a yellowish tinge (**258**).

Klebsiella pneumoniae (**274**) and *Enterobacter aerogenes* have very mucoid colonies on primary isolation indicative of the presence of a large capsule around individual cells. Both are lactose-fermenters but the colonies are pale pink on MacConkey agar. The rare strains of *Escherichia coli* that are mucoid are usually a more vivid pink (**275**).

A few members of the *Enterobacteriaceae* produce pigments. Both *Serratia marcescens* and *S. rubidaea* form red pigment called prodigiosin (**276**). This pigment is produced best at 25°C but some strains can form the pigment at 37°C. The red pigmentation is superimposed on any colour reaction that occurs in MacConkey agar due to the pH indicator, neutral red (**277**). A number of the enterobacteria produce a yellow pigment as demonstrated by *Enterobacter agglomerans* (**278**).

Microscopy and tests for primary identification

All the enterobacteria are medium-sized, Gram-negative rods (**279**) although occasionally some members show coccobacillary or long forms, such as the swarm cells of *Proteus* species. An O-F test demonstrates that they are facultative anaerobes. The fact that all the enterobacteria are oxidase-negative is the most useful characteristic as most of the other Gram-negative bacteria are oxidase-positive.

Biochemical tests

Diagram 33 indicates the short range of biochemical tests that, together with the reaction on MacConkey agar and the colonial morphology, can give a presumptive identification of the opportunistic enterobacteria. The combination of a TSI agar slope and lysine decarboxylase broth, while designed for the presumptive identification of salmonellae, can also be useful for other enterobacteria (**Diagram 32**). If there is doubt about the identification of an isolate, a full range of biochemical tests should be carried out (**Table 62**). The methods and interpretation of the tests are given in Chapter 4. A more expensive, but less time consuming, method of identification is to use a commercial strip such as the API 20E (Analytab Products) for the enterobacteria and other Gram-negative organisms. The method of inoculation and interpretation of the API 20E is also given in Chapter 4.

Serotyping for antigen detection

Klebsiella species are sometimes serogrouped by their capsular (K) antigens in reference laboratories. At least 77 capsular types of *Klebsiella* have been described. K1, K5 and K7 are the predominant types in the isolates from metritis in mares.

Antibiotic sensitivity tests

Included in the R-factor plasmid, which is common in the enterobacteria, is the resistance transfer factor and genes for resistance to several antibiotics. Antibiotic sensitivity testing should be carried out with any isolate that is considered to be significant, before the treatment of an animal with antimicrobial drugs.

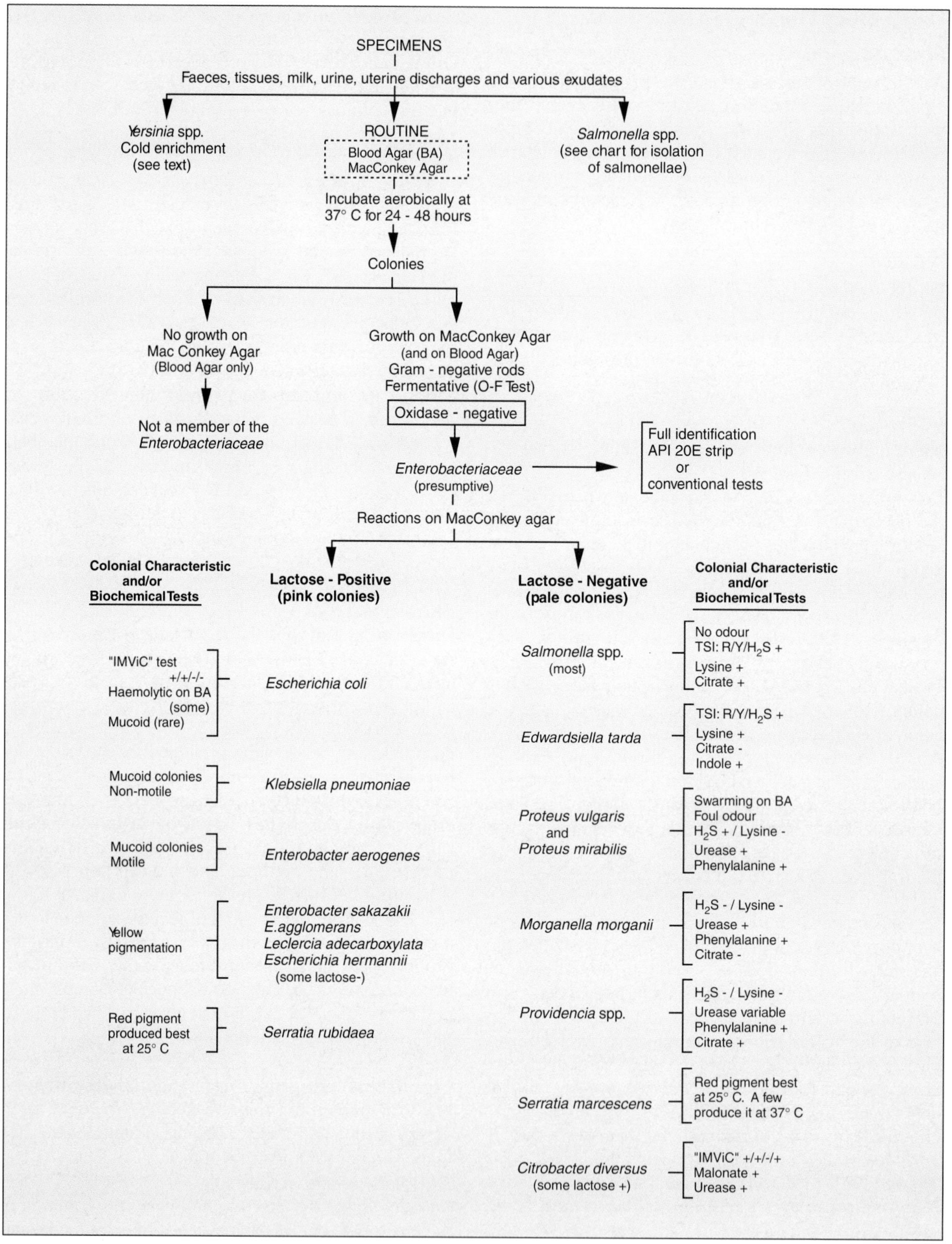

Diagram 33. Routine isolation of important members of the *Enterobacteriaceae* and their presumptive identification on colonial morphology and/or biochemical tests.
+ = positive reaction, - = negative, 'IMViC' = indole, methyl red, Voges–Proskauer and citrate tests, TSI = triple sugar iron agar, lysine = lysine decarboxylase test.

270

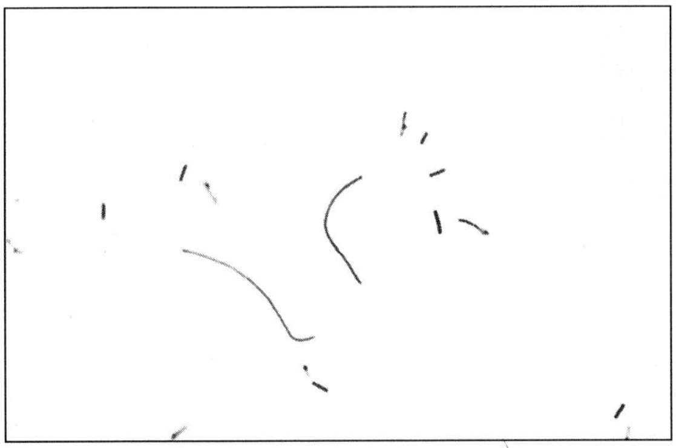

270 Gram-stained smear from the swarming growth of *Proteus mirabilis* showing the long 'swarm cells'. (Gram stain, ×1000)

271

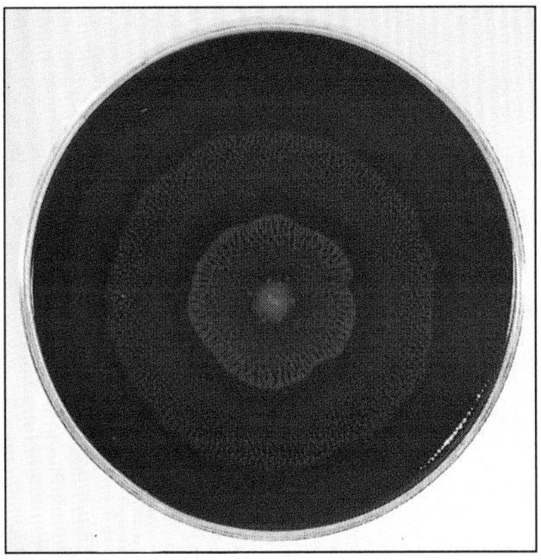

271 Characteristic swarming of a *Proteus* species following spot inoculation in the centre of a blood agar plate.

272

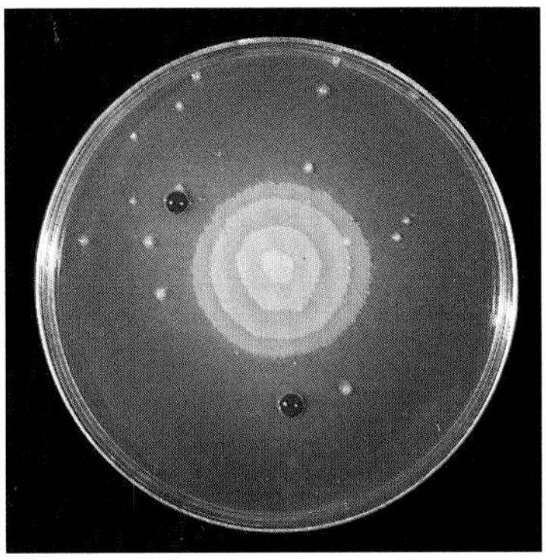

272 A MacConkey agar plate exposed to an aerosol during the spreading of slurry. It shows a *Proteus* species swarming, red-pigmented colonies of *Serratia rubidaea* and mucoid, pale pink colonies of a *Klebsiella* species.

273

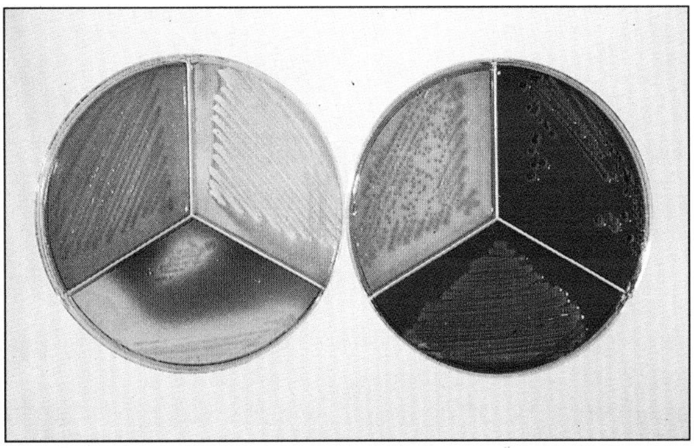

273 Strains of *P. vulgaris* (left) and *P. mirabilis* (right) on two separate plates: MacConkey agar (left), XLD medium (right) and brilliant green agar (bottom). Note that the *P. mirabilis* strain on the right is giving similar reactions to that of most salmonellae except that with this strain, the periphery of the colonies on the XLD medium tends to be yellowish.

274

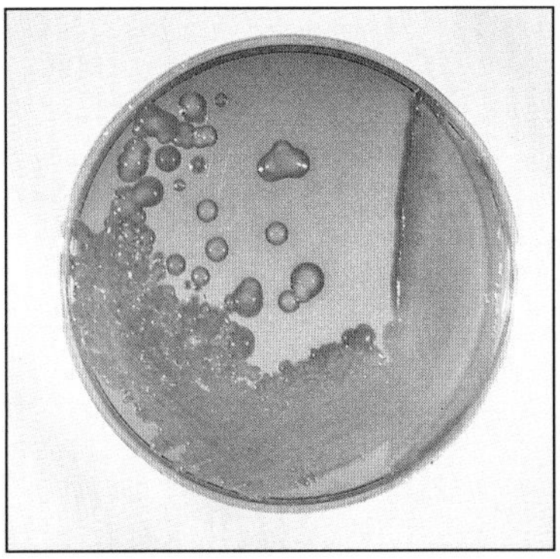

274 *Klebsiella pneumoniae* on MacConkey agar. The bacterium is a lactose-fermenter but the characteristic large, mucoid colonies always tend to be pale pink.

275 A mucoid strain of *E. coli* (left) compared to a non-mucoid *E. coli* strain (right) on MacConkey agar. *K. pneumoniae* (bottom) is normally mucoid and tends to be pale pink.

275

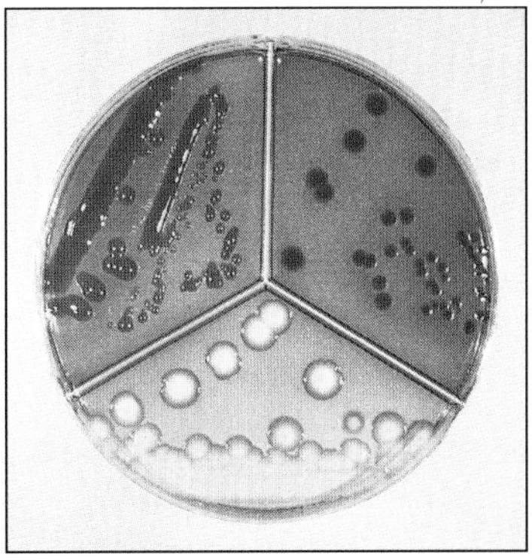

276

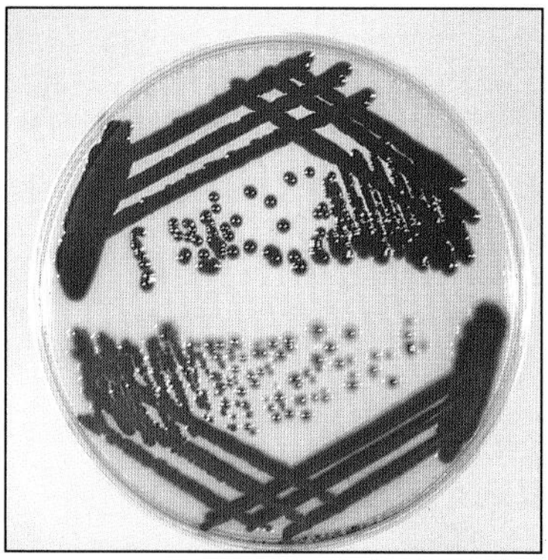

276 *Serratia rubidaea* (top) and *Serratia marcescens* (bottom) on nutrient agar showing the production of red pigment (prodigiosin).

277

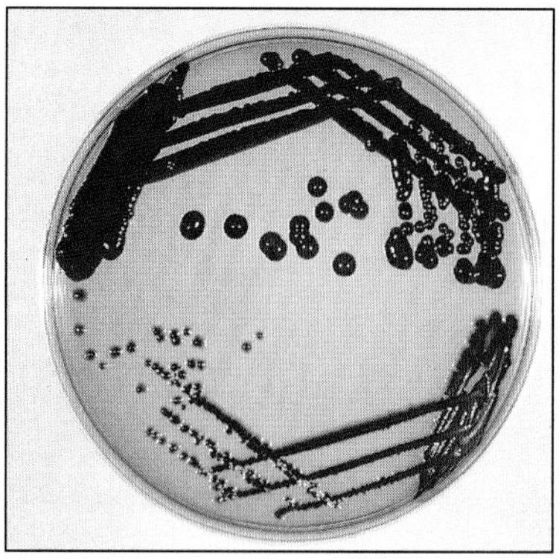

277 *Serratia rubidaea* (top) and *S. marcescens* (bottom) on MacConkey agar. Colony colour is due to the production of red pigment, not a pH change.

278

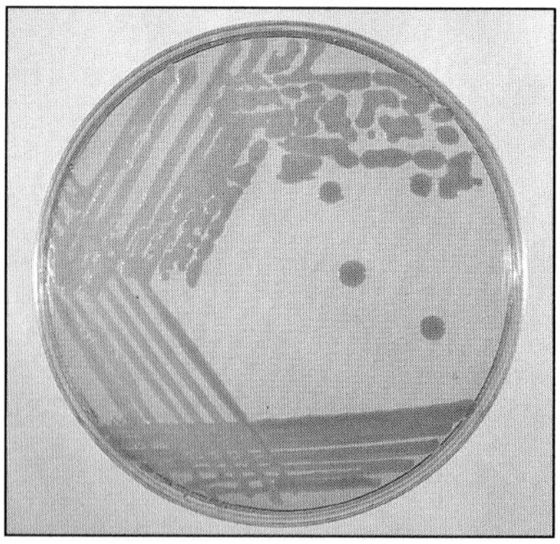

278 *Enterobacter agglomerans* on nutrient agar showing mucoid colonies and yellow pigmentation.

Escherichia coli

Natural Habitat

Escherichia coli is a natural inhabitant of the large intestine and lower small intestine of all mammals. It is usually present in larger numbers in carnivores and omnivores than in herbivores. *E. coli* is excreted in faeces and can survive in faecal particles, dust and water for weeks or months. The presence of *E. coli* in water samples, being tested for potability, is taken as evidence of faecal pollution.

Surface Antigens of *E. coli*

The capsular (K) antigens are polysaccharides and the cell wall or somatic (O) antigens are determined by the sugar side-chains on the lipopolysaccharide molecules of the outer membrane. The flagellar (H) and fimbrial (F) antigens are proteins. Some of the well-known fimbrial antigens, K88(F4) and K99(F5) are adhesins that allow pathogenic *E. coli* strains to adhere to intestinal cells and colonise the small intestine. The O, H and K antigens can be used to serotype strains of *E. coli*; each serotype is designated by the numbers of the antigens that it bears, for example O157:K85:H19.

Pathogenesis and Pathogenicity

Predisposing causes

The predisposing causes are of paramount importance and to a large extent determine whether or not clinical signs of illness will occur:

- Neonates obtaining insufficient passive immunity (antibodies) from colostrum. This might be due to either a quantitative or qualitative deficiency.
- Intensive husbandry practices lend themselves to a rapid transmission of the pathogenic *E. coli* strains.

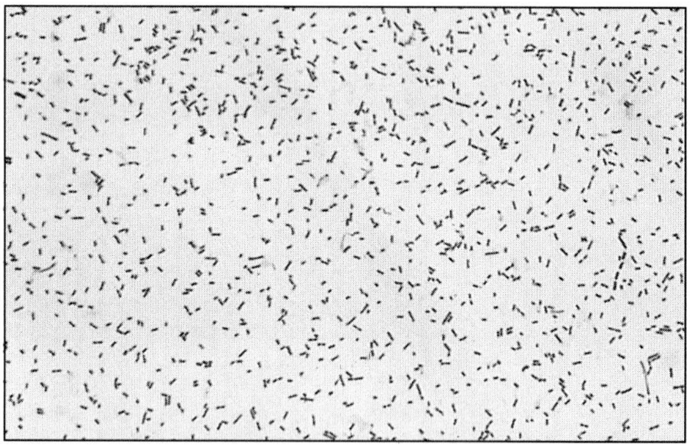

279

279 Medium-sized Gram-negative rods of *E. coli* from a culture. The morphology is representative of most members of the *Enterobacteriaceae*. (Gram stain, ×1000)

- Poor hygiene often allows a build-up of pathogenic strains in the young animal's environment. A large dose of pathogenic *E. coli* may overcome colostral immunity.
- Young neonates, under one week of age, are particularly susceptible because:
  - a) The normal flora of the intestines is not fully established.
  - b) They have a naive immune system.
  - c) Receptors for the adhesins of *E. coli* are present for the first week of life only in calves and for the first 6 weeks of life in piglets.
- In recently weaned pigs there are stress factors such as different surroundings, companions and diet. Heavy grain diets particularly, can lead to a massive colonisation of the anterior small intestine by enterotoxigenic, K88 and K99 strains of *E. coli*.

Table 62. Biochemical reactions for some clinically significant members of the *Enterobacteriaceae*.

	Indole production	Methyl red	Voges–Proskauer	Citrate	Urease	Phenylalanine deaminase	Hydrogen sulphide	Lysine decarboxylase	Ornithine decarboxylase	Motility (36°C)	Gelatin liquefaction	Growth in KCN broth	ONPG (beta-galactosidase)	ACID FROM: Dulcitol	Inositol	Lactose	Maltose	Mannitol	Mannose	Rhamnose	Sorbitol	Sucrose	Xylose	Red pigment	Swarming (blood agar)	Mucoid colonies
Citrobacter diversus	+	+	–	+	(+)	–	–	–	+	+	–	–	+	d	–	d	+	+	+	+	+	(–)	+	–	–	–
Edwardsiella tarda	+	+	–	–	–	–	+	+	+	+	–	–	–	–	–	–	+	–	+	–	–	–	–	–	–	–
Enterobacter aerogenes	–	–	+	+	–	–	–	+	+	+	–	+	+	–	+	+	+	+	+	+	+	+	+	–	–	+
Enterobacter cloacae	–	–	+	+	d	–	–	–	+	+	–	+	+	(–)	(–)	+	+	+	+	+	+	+	+	–	–	–
Escherichia coli	+	+	–	–	–	–	–	(+)	d	(+)	–	–	+	d	–	+	+	+	+	(+)	+	d	+	–	–	(–)
Klebsiella pneumoniae	–	(–)	+	+	+	–	–	+	–	–	–	+	+	d	+	+	+	+	+	+	+	+	+	–	–	+
Morganella morganii	+	+	–	–	+	+	–	–	+	+	–	+	–	–	–	–	–	–	+	–	–	–	–	–	–	–
Proteus mirabilis	–	+	(–)	d	+	+	+	–	+	+	+	+	–	–	–	–	–	–	–	–	–	(–)	+	–	+	–
Proteus vulgaris	+	+	–	(–)	+	+	+	–	–	+	+	+	–	–	–	–	+	–	–	–	–	+	+	–	+	–
Salmonella subgenus I	–	+	–	+	–	–	+	+	+	+	–	–	–	+	d	–	+	+	+	+	+	–	+	–	–	–
(‘Arizona’) subgenus III	–	+	–	+	–	–	+	+	+	+	–	–	+	–	+	d	+	+	+	+	+	–	+	–	–	–
Serratia marcescens	–	(–)	+	+	(–)	–	–	+	+	+	+	+	+	–	(+)	–	+	+	+	–	+	+	–	+	–	–
Serratia rubidaea	–	(–)	+	+	–	–	–	d	–	(+)	+	(–)	+	–	(–)	+	+	+	+	–	–	+	+	+	–	–
Shigella species	v	+	–	–	–	–	–	–	v	–	–	–	v	–	–	–	v	v	+	v	v	–	–	–	–	–
Yersinia enterocolitica	d	+	–	–	(+)	–	–	–	+	–	–	–	+	–	d	–	d	+	+	–	+	+	d	–	–	–
Y. pestis	–	(+)	–	–	–	–	–	–	–	–	–	–	(+)	–	–	–	(+)	+	+	–	–	–	+	–	–	–
Y. pseudo - tuberculosis	–	+	–	–	+	–	–	–	–	–	–	–	d	–	–	–	+	+	+	+	–	–	+	–	–	–

+ = 90–100% strains positive, (+) = 76–89% positive, d = 26–75% positive, (–) = 11–25% positive, – = 0–10% positive, v = reaction variable between species. Tests read after 48 hours at 37°C.

- Oedema disease occurs most commonly in young, weanling pigs but the disease can occur in older pigs.

The following factors are often present prior to the occurrence of oedema disease in pigs:

a) Recent change in feed.
b) The pigs are thriving and growing rapidly.
c) Mild diarrhoea noticed a few days before the signs of oedema disease appear.

Virulence factors of pathogenic E. coli strains

- The capsular polysaccharide (K antigen) is antiphagocytic and also protects the cell wall from the damaging effects of complement.
- The endotoxin (lipid A) is the toxin associated with *E. coli* septicaemias (colisepticaemias) and the toxaemia in coliform mastitis. Lipid A also interferes with the complement components that are responsible for the attack on the outer membrane of *E. coli.*
- Certain fimbriae are protein adhesins. They allow the attachment of *E. coli* to glycoproteins on the surface of epithelial cells of the jejunum and ileum with consequent colonisation. Otherwise peristalsis would move *E. coli* into the large intestine where the cells are not susceptible to enterotoxin, while the cells of the jejunum and ileum are highly sensitive to the enterotoxin. The production of K88 (F4), K99 (F5), 987P (F6), F41 and F165 is closely linked with the ability of the *E. coli* strain to form enterotoxin. K99 and K88 are plasmid-encoded and receptors for K88 are inherited by pigs as a dominant Mendelian trait. *E. coli* with K88 adhesins is the most common type in pigs. All the major fimbrial types have been isolated from pigs, whereas K99 and F41 occur in calves and K99 in lambs.
- *E. coli* strains produce both alpha and beta haemolysins. The alpha haemolysin is a protein that damages host cell membranes. Loss of the ability to produce this alpha haemolysin causes a decrease in virulence.
- Siderophores are particularly important for the invasive strains. Siderophores are concerned with removing iron from iron-binding proteins of the host. Iron is an absolute requirement for the growth of almost all bacteria, thus pathogenic bacteria have to compete with the host for iron.
- Both heat-labile (LT) and heat-stable (ST) enterotoxins are produced by the enterotoxigenic *E. coli* strains that also have the K88, K99 or other colonising antigens. The LT enterotoxin is a protein (91,000 M.W.) and is antigenically related to the cholera toxin (*Vibrio cholerae*). The toxin affects the adenylate cyclase system leading to diarrhoea, hypovolaemia, metabolic acidosis and hyperkalaemia if the acidosis is severe. The ST enterotoxin is also a protein (1,500–2,000 M.W.) and there are two types. ST_a is plasmid-encoded and affects the guanylate cyclase system, but the method by which the ST_b enterotoxin causes diarrhoea is not known. ST_a causes fluid accumulation in the intestines of suckling mice and piglets and ST_b in the intestines of weaned pigs only.
- Verotoxin or Shiga-like toxins are similar in activity to the Shiga toxin (cytotoxin) of *Shigella* species. Both the Shiga and Shiga-like toxins inhibit protein synthesis in host cells following interaction with the 60S ribosomal subunit. There are two types of Shiga-like toxins, SLT-1 that is neutralised by antibody specific for Shiga toxin and SLT-2,which causes haemorrhagic colitis in man, and is not neutralised by Shiga antitoxin. SLT-2 causes death in mice and induces changes in Vero cells in tissue culture. The oedema disease toxin is thought to be a variant of SLT-2. Shiga-like toxins can cause destruction of intestinal epithelial cells where there is a dense adherence of the pathogenic *E. coli.* The O groups 26 and 111, in neonatal calves and piglets, produce Shiga-like toxins.

Types of pathogenic E. coli

E. coli strains, normally regarded as non-pathogenic, can cause opportunistic infections in sites of the body such as mammary glands (mastitis) and uterus (metritis). *E. coli* strains that cause enteritis have been classified as:

- Enterotoxigenic *E. coli* (ETEC) which has the fimbrial adhesins K88, K99 or others. The production of these colonisation factors correlates with enterotoxin production. These strains cause the majority of cases of neonatal colibacillosis.
- Enteropathogenic (EPEC) strains do not appear to produce enterotoxins or Shiga-like toxins but they can cause enteritis and diarrhoea by other mechanisms. These strains have been recovered from lambs with diarrhoea.
- Enteroinvasive (EIEC) strains adhere to cells of the distal small intestine, invade the enterocytes and deeper layers of the intestinal mucosa. They reach the lymphatic system where there is multiplication. The death of some *E. coli* cells occurs and endotoxin is released. The virulence factors such as capsules, adhesins, siderophores and alpha-haemolysin are important as survival factors for these invasive strains which are responsible for colisepticaemia.
- Attaching and effacing *E. coli* (AEEC) strains colonise the small intestine, attach to target cells and kill them. The Shiga-like toxins (verotoxins) destroy the microvilli by unknown means. These strains have been isolated from calves and rabbits with enteric disease.

Oedema disease in pigs is often associated with *E. coli* O139 and O141 and these strains are usually haemolytic. A verotoxin, which is probably a variant of the SLT-2 toxin, is produced by these *E. coli*. The toxin is produced in the intestine but is absorbed and carried via the bloodstream to the target cells, usually endothelial cells of the small arteries. These oedema disease strains of *E. coli* are normally present in the large intestine of pigs, but they appear to multiply rapidly under conditions of stress, particularly a change of diet. The diseases caused by *E. coli* in domestic animals are summarised in **Table 63**.

Laboratory Diagnosis

- ***Diagnosis of the opportunistic infections caused by E. coli***

  In these infections it is sufficient to isolate *E. coli* in an almost pure growth from carefully taken samples such as cervical swabs, mastitic milk samples and mid-stream urine. The culture and presumptive identification methods are shown in **Diagram 33**. Pathogenic strains of *E. coli* are quite often haemolytic (**280**) and as they are a strong lactose-fermenter the colonies on MacConkey agar are bright pink (**281**). Eosin methylene blue (EMB) agar is occasionally used in diagnostic laboratories and on this medium *E. coli* colonies have a unique and characteristic metallic sheen (**282**). The 'IMViC' test (indole+/ MR+/ VP–/ citrate–) is a quick presumptive method of identifying *E. coli* (**283**) as almost no other lactose-positive member of the *Enterobacteriaceae* gives this combination of results for the tests.
- ***Diagnosis of the enteroinvasive strains***

  The diagnosis, in this case, must be based on the isolation and identification of *E. coli* from normally sterile sites in the body such as bone marrow, joints, spleen or blood. Liver specimens should be avoided as there can be movement of enteric bacteria to this organ in the agonal stage of a disease.
- ***Demonstration of the enterotoxigenic strains***

  The enterotoxigenic strains of *E. coli* are present in large numbers in the small intestine, but in this case it is insufficient merely to isolate and identify the *E. coli*. Demonstration of the significant fimbrial antigens (K88, K99, F41, 987P or F165) or the enterotoxin itself is necessary.

  a) Fimbrial antigens: Fimbriae are expressed poorly on selective and some types of non-selective laboratory medium. E medium (Francis *et al.*, 1982) is advised for K88, K99 and F41. Minca medium (BBL) has been found satisfactory for 987P, K99 and F41. Commercial test kits are available for the detection of fimbrial antigens such as a latex agglutination test (Fimbrex K88, K99, F41 and 987P, CVL, New Haw, UK) that is illustrated (**284**). Specific antiserum (Rijks Institut) can be obtained for use in a slide agglutination test. An enzyme-linked immunosorbent assay (ELISA) is available for directly measuring the presence of K99 fimbriae-expressing *E. coli* in faecal extracts (Beldico, Belgium). The fluorescent antibody technique using conjungates prepared against each of the common colonising antigens can be used on smears made from scrapings from the ileum of a fresh carcase.

  b) Demonstration of enterotoxins: The most sensitive of the methods being developed for the ST and LT toxins is an ELISA which employs monoclonal antibodies. Carroll *et al.* (1990) found that two Oxoid test kits, for ST and for LT in culture filtrates or supernatants, were useful for veterinary diagnosis. A competitive enzyme immunoassay (EIA) is employed for the ST (ST EIA, code TD700) and a reverse passive latex agglutination

280

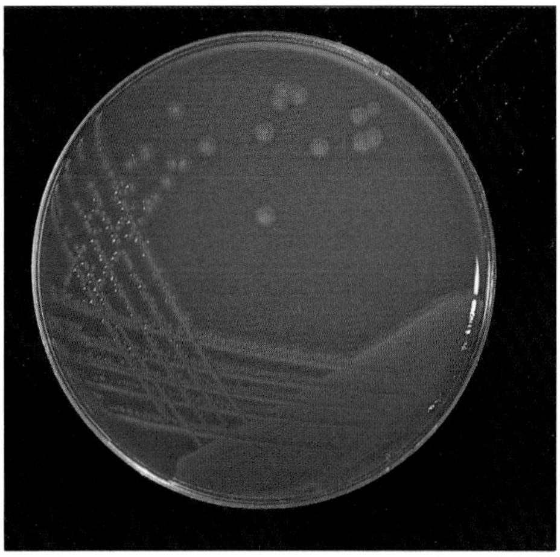

280 Haemolytic *Escherichia coli* on sheep blood agar.

281

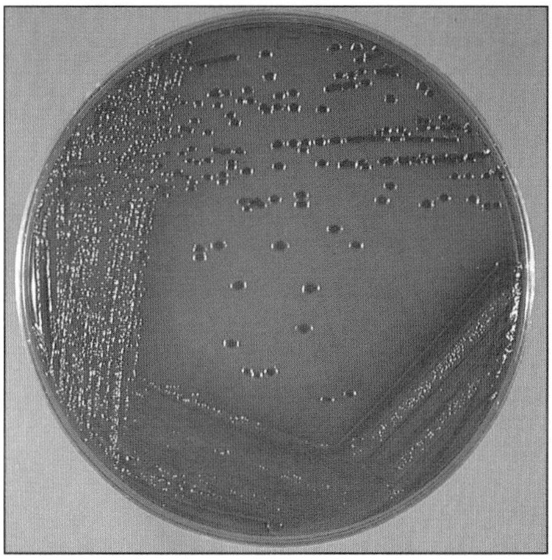

281 *E. coli* on MacConkey agar. Bright pink colonies indicating acid production as a result of the fermentation of lactose. (Neutral red indicator)

282

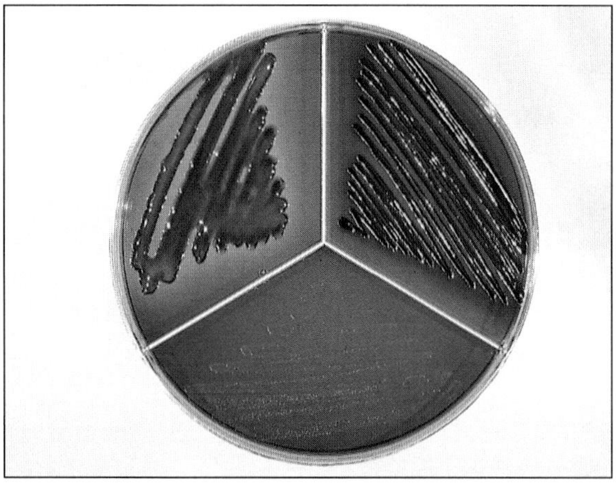

282 *E. coli* (right) giving a distinctive metallic sheen on EMB agar distinguishing it from other members of the *Enterobacteriaceae* such as *Salmonella* sp. (bottom) and *Klebsiella pneumoniae* (left).

283

283 The 'IMViC' test for *E. coli:* from left: Indole + / MR + / VP - / Citrate -.

test for the LT (LT VET-RPLA, code TD920). DNA probes specific for the base sequences of *E. coli* genes encoding enterotoxin (LT and ST) are available (Molecular Biosystems, San Diego, California). Such probes have been used to detect enterotoxigenic strains in cultures and in faecal extracts. The LT has been assayed in ligated loops and its effect on adrenal and Vero cell monolayers observed. ST_a enterotoxin can be assayed in suckling mice and the ST_b toxin in weaned pig and rabbit intestinal loops.

- ***Diagnosis of the attaching and effacing strains of E. coli***

  Most of these *E. coli* isolates have been shown to produce urease. This is an unusual characteristic as generally less than 1 per cent of *E. coli* strains produce this enzyme. The Shiga-like toxin in culture supernatants can be tested for cytotoxicity for Vero cells. DNA probes for genes encoding the Shiga-like toxins 1 and 2 are used in medical laboratories. Histopathological examination of sections of the ileum should demonstrate the characteristic distortion of the microvilli and the effacement of the mucosal surface.

- ***Diagnosis of oedema disease***

  The disease is usually diagnosed on clinical and postmortem findings. These *E. coli* isolates are usually haemolytic and the strains commonly have O antigens 139 and 141. The Shiga-like toxin, which is a variant of SLT-2, can be detected in a Vero cell culture and in a mouse assay.

Table 63. Diseases caused by *Escherichia coli.*

Animals involved	Disease	Clinical signs and pathogenesis
PIGS		
Piglets less than 1 week old	Neonatal diarrhoea (colibacillosis)	Profuse watery diarrhoea and severe dehydration, mortality 90–100%
	Colisepticaemia	Occasionally septicaemia with invasive strains and death of a piglet within 48 hours of birth
	Piglet meningitis	Acute meningitis and fibrinous polyserositis in piglets has been reported
Pigs about 2 weeks after weaning	Weanling enteritis (colibacillosis)	Diarrhoea, anorexia and fever. Mortality lower than in neonatal pigs Enterotoxins involved (ST_b and LT)
Weaned pigs	Oedema disease	Often sudden death. Oedema of forehead, eyelids, stomach wall and larynx (hoarse squeal). Nervous signs such as ataxia, convulsions and paralysis may be seen Verotoxin (variant of SLT-2 toxin) involved often associated with *E. coli* 0139 and 0141
Sows after farrowing	Coliform mastitis	One or more mammary glands affected
Gilts after farrowing	Mastitis-metritis-agalactia (MMA) syndrome	Complex syndrome involving hysteria, hormonal imbalance and coliform infection (often *E. coli*)

(continued)

Table 63. Diseases caused by *Escherichia coli.* (*continued*)

Animals involved	Disease	Clinical signs and pathogenesis
CATTLE		
Calves less than 1-week old	'White scours' (colibacillosis)	The appetite is normal at first but decreases as the faeces become more fluid. White pasty faeces around rectum. Dehydration and emaciation occur. Death usually within 4–5 days if untreated. Enterotoxins involved (St_a) and possibly a Shiga-like toxin in some cases
Calves less than 1-week old	Colisepticaemia	Sudden death due to endotoxic shock or diarrhoea, depression, respiratory distress and death. Endotoxin mainly involved from invasive *E. coli* strains
Calves surviving a colisepticaemia	Joint ill	*E. coli* localised in joints and/or kidneys ('white spot'). Entry can be via the umbilicus
Dairy cows soon after parturition	Coliform mastitis	Peracute disease: fever, anorexia, depression and sunken eyes. Death due to endotoxic shock. Most common in housed cows
SHEEP		
Neonatal lambs	Colibacillosis and colisepticaemia	Syndromes similar to those that occur in calves but less common. Enterotoxigenic and enteropathogenic strains involved in colibacillosis and *E. coli* 078 is common in colisepticaemia
Neonatal lambs	'Watery mouth'	The lamb is dull and anorectic, saliva drools over the muzzle and there is abdominal tympany. There are splashing sounds within the abomasum and death within 6–24 hours. Associated with *E. coli* endotoxaemia
Ewes	Coliform mastitis	Peracute and similar to the condition in cows. Most commonly seen in ewes housed during lambing
DOGS (CATS)		
Neonatal pups	Colisepticaemia	Septicaemia, often fatal
Bitches	Pyometra	Associated with the progesterone-stimulated endometrium. A vaginal discharge may occur 4–8 weeks after oestrus. In a closed-cervix pyometra the bitch is toxaemic and ill. Endotoxin involved
Adult dogs	Urinary tract infection	Cystitis (often in bitches) is most common but *E. coli* can ascend higher in the urinary tract. A bacteriuria occurs with greater than 10^5 *E. coli* /ml urine
POULTRY		
Young chicks	Omphalitis	Infection of the vestigal yolk sac. The contents are dark, fluid and evil-smelling. Common name is 'mushy-yolk disease'
All ages	Colisepticaemia	Primary or secondary infection via the intestines or respiratory tract. Many body organs are affected with air-sacculitis, peritonitis and ovarian infection
All ages	Coligranuloma	Chronic condition, possibly following a colisepticaemia. There are nodular lesions in the liver and intestines
OTHER ANIMALS		
Neonatal animals such as foals and rabbits	Colibacillosis and colisepticaemia	Diarrhoea and septicaemia. Less common than in calves and piglets

284

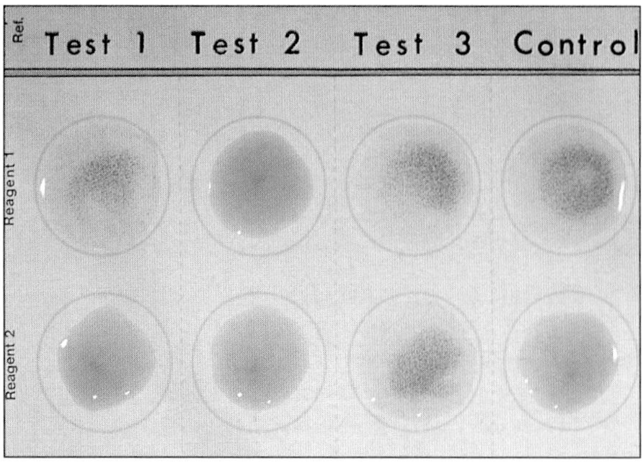

284 A latex plate agglutination test for detecting the K99 pili antigens of enteropathogenic *E. coli*. Suspensions of three *E. coli* isolates (tests 1, 2 and 3) and a control antigen have each been placed in a top and a bottom well. Reagent 1 is latex particles coated with monospecific antibody to the K99 antigen and has been placed in all the wells in the top row. Reagent 2 is a suspension of latex particles only and has been placed in all the wells in the bottom row. *E. coli* (test 1) is positive for the K99 antigen; *E. coli* (test 2) is negative and *E. coli* (test 3) has autoagglutinated in the bottom well indicating that the test is invalid for this isolate.

Antibiotic sensitivity test

This should be carried out on all *E. coli* isolates considered to be significant because of the R-factor plasmids, common in the enteric bacteria, whose resistance transfer factor has genes that code for multiple drug resistance.

Salmonella

The genus *Salmonella* comprises a single species that has been divided into over 2,000 serotypes in the Kauffmann–White Schema, based on the O (somatic), H (flagellar) and occasionally capsular (Vi) antigens. More recently the genus has been divided into 7 subgroups. Subgroup I contains most of the salmonellae that are significant animal pathogens and most have been given names, for example *dublin* or *typhimurium.* Subgroups IIIa and IIIb contain the bacteria once known as 'Arizona' and now called 'arizonae' if monophasic (IIIa) or 'diarizonae' if diphasic (IIIb). There are nearly 400 serotypes in subgroups IIIa and IIIb. **Table 64** indicates the subspecies name of each group and the properties of each. A recent proposal (Le Minor and Popnoff, 1987) is that the salmonellae in subgroup I be named *Salmonella enterica* subsp. *enterica* and the serotypes be presented as, for example, Dublin or Typhimurium. A simplified nomenclature is often preferred in diagnostic laboratories with the named serotypes of *Salmonella* being regarded as 'species', such as *S. dublin* or *S. typhimurium.* Salmonellae in subgroups IIIa and IIIb occasionally cause disease in animals and birds; these will be referred to as '*S. arizonae*' in this chapter.

Natural Habitat

The reservoir for salmonellae is the intestinal tract of warm-blooded and cold-blooded animals. The majority of infected animals become subclinical excretors. However, salmonellae can survive for 9 months or more in the environment in sites such as moist soil, water, faecal particles and animal feeds, especially in blood-and-bone and fish meals.

Pathogenesis and Pathogenicity

Transmission of salmonellae is usually by the faecal-oral route but infection via mucous membranes of the conjunctivae or upper respiratory tract is suspected.

Colonisation of the intestinal tract and enteric disease

Salmonellae need to colonise the distal small intestine or colon to initiate enteric disease. Volatile organic acids produced by the indigenous normal anaerobic flora inhibit the growth of salmonellae and the normal flora usually block access to attachment sites required by the *Salmonella* species. Disruption of the normal intestinal flora by factors such as antibiotic therapy, diet and water deprivation increases the host's susceptibility to infection. Reduced peristalsis, stress due to transportation and overcrowding also predispose to colonisation of the intestine by the salmonellae. The attachment of salmonellae is usually by fimbriae. Some strains producing enteritis and diarrhoea appear capable of forming LT-like and ST-like enterotoxins and a cytotoxin.

Invasion and septicaemia

The invasive strains that produce septicaemia are able to escape destruction by the host and to multiply within the macrophages of the liver and spleen as well as intravascularly. The invasive abilities of some strains of *S. typhimurium* are increased by the presence of genes carried on a plasmid. Destruction within the bloodstream is prevented by O-repeat units of the lipopolysaccharide. It is thought that they may mask determinants on the bacterial cell surface that would normally bind complement and activate it by means of the alternate pathway. This would reduce the chances of chemotaxis, opsonisation and phagocytosis. Any salmonellae, in non-immune animals, that are phagocytosed tend to survive within the phagocyte. Siderophores, which remove iron from the iron-binding proteins of the host, are secreted by these invasive salmonellae. Multiplication of the organisms in the body leads to a severe endotoxaemia. Some serotypes

Table 64. Characteristics of the seven *Salmonella* subgroups.

Salmonella subgroup	*I*	*II*	*IIIa*	*IIIb*	*IV*	*V*	*VI*
Subspecies name	*enterica**	*salamae*	*arizonae*	*diarizonae*	*houtenae*	*bongori*	*indica*
Flagella usually monophasic or diphasic (Di)	Di	Di	Mono	Di	Mono	Mono	Di
Habitat of majority of strains:							
Warm-blooded animals	+	–	–	–	–	–	–
Cold-blooded animals and environment	–	+	+	+	+	+	+
Differential tests:							
Beta-galactosidase (ONPG)	–	–	+	+	–	+	d
Lactose	–	–	(–)	(+)	–	–	(–)
Dulcitol	+	+	–	–	–	+	d
Malonate utilization	–	+	+	+	–	–	–
Gelatin hydrolysis**	–	+	+	+	+	–	+
Growth in KCN medium	–	–	–	–	+	+	–

* Le Minor and Popoff (1987), ** = rapid X-ray film method.

seem to be more commonly invasive than others. Invasive strains occur frequently in *S. typhi*, *S. dublin* and *S. typhimurium*.

The diseases and conditions caused by some of the salmonella serotypes are summarised in **Table 65**. *S. dublin* is host-adapted to cattle, although infections can occur in other animal species. This serotype occurs in Europe, western USA and South Africa. Both subclinical excretors and latent adult carriers occur. The latent carriers may excrete *S. dublin* in faeces if stressed by factors such as parturition or a concurrent infection. *S. dublin* can cause an unusually wide range of clinical syndromes including ischaemic necrosis of the tips of the ears, tail and limbs (**285**). This terminal dry gangrene usually follows a few weeks after the recovery of calves from acute disease. There is evidence that endotoxin damages the endothelium of blood vessels and also activates the alternate pathway of complement as well as the blood- clotting mechanism. This probably leads to a localised form of disseminated intravascular coagulation (DIC) that causes the terminal ischaemia.

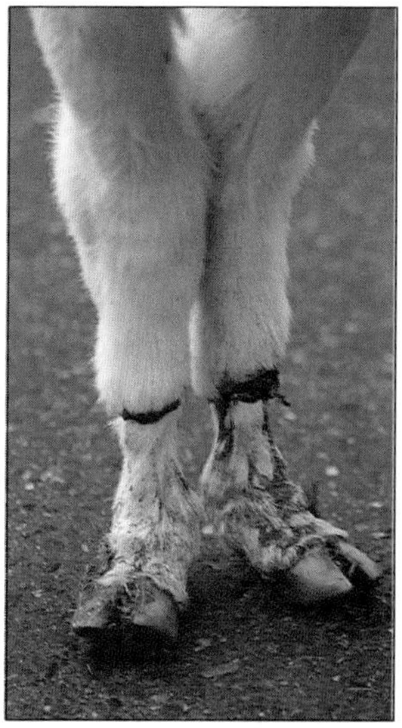

285

285 Terminal dry gangrene in a 3-week-old calf following a *Salmonella dublin* infection. Typically, one or both hind limbs, tip of tail and ears are affected (compare with **484**, ergotism in a cow).

Laboratory Diagnosis

Isolation

Diagram 34 indicates the steps for the isolation of salmonellae from clinical specimens. The host-adapted serotypes from pigs and poultry are more fastidious than most of the other commonly isolated serotypes. They do not tolerate selenite broth, tetrathionate broth or brilliant green agar, although most strains of *S. choleraesuis* will grow on modified brilliant green agar (Oxoid). If *Proteus* species are a problem, the enrichment broths can be incubated at 43°C, or sodium sulphathiazole added to the broths at 0.125 mg/100 ml. Some laboratories add sodium sulphadiazine to brilliant green agar (80 mg/l) to make it more selective.

Table 65. Diseases caused by selected *Salmonella* serotypes.

Host	*Salmonella* serotypes	Disease
Humans	*S. typhi**	Typhoid fever
	S. paratyphi A*	Paratyphoid fever
	*S. schottmuelleri**	Paratyphoid fever
	S. enteritidis, S. typhimurium and others	Food poisoning
Cattle	*S. dublin**	Subclinical excreters, latent carriers, enteritis, septicaemia, meningitis in calves, abortion (with or without other apparent clinical signs), osteomyelitis, joint ill, terminal dry gangrene in calves
	S. typhimurium, S. bovismorbificans and others	Enteritis or septicaemia
Pigs	*S. choleraesuis** and *S. choleraesuis* biotype Kunzendorf*	Severe outbreaks clinically similar to swine fever (hog cholera), and swine fever can be followed by a secondary infection with *S. choleraesuis*, earning it the name of 'hog cholera bacillus'. Rectal stricture is sometimes a sequel of the disease
	*S. typhisuis**	Chronic enteritis in young pigs. Far less virulent than *S. choleraesuis*
	S. typhimurium and others	Enteritis or septicaemia
Sheep	*S. abortusovis* *S. montevideo* *S. dublin*	Abortion in ewes. *S. abortusovis* is present in Britain, Europe and the Middle East
	S. typhimurium, S. anatum and others	Enteritis or septicaemia
Horses	*S. abortusequi**	Abortion in mares. The serovar is present in Europe, South Africa and South America but is now rare in the USA
	S. typhimurium and others	Enteritis or septicaemia, especially in foals and stressed adults
Poultry and other birds	*S. pullorum**	Pullorum disease (bacillary white diarrhoea) in chicks. Transovarian transmission
	*S. gallinarum**	Fowl typhoid in all ages, mainly adults. Egg transmitted
	S. arizonae	Severe infections (enteritis and septicaemia) in chicks and turkey poults. Egg transmitted. Serovar associated with reptiles. Occasional infections in other animals
	S. enteritidis, S. typhimurium and many other serotypes	Collectively known as 'fowl paratyphoid'. Inapparent infections, enteritis and septicaemia. *S. enteritidis* may be egg transmitted *S. typhimurium* can cause sudden deaths (septicaemia) in pigeon squabs, or if they survive, swollen wing joints

* = host-adapted *Salmonella* serotypes.

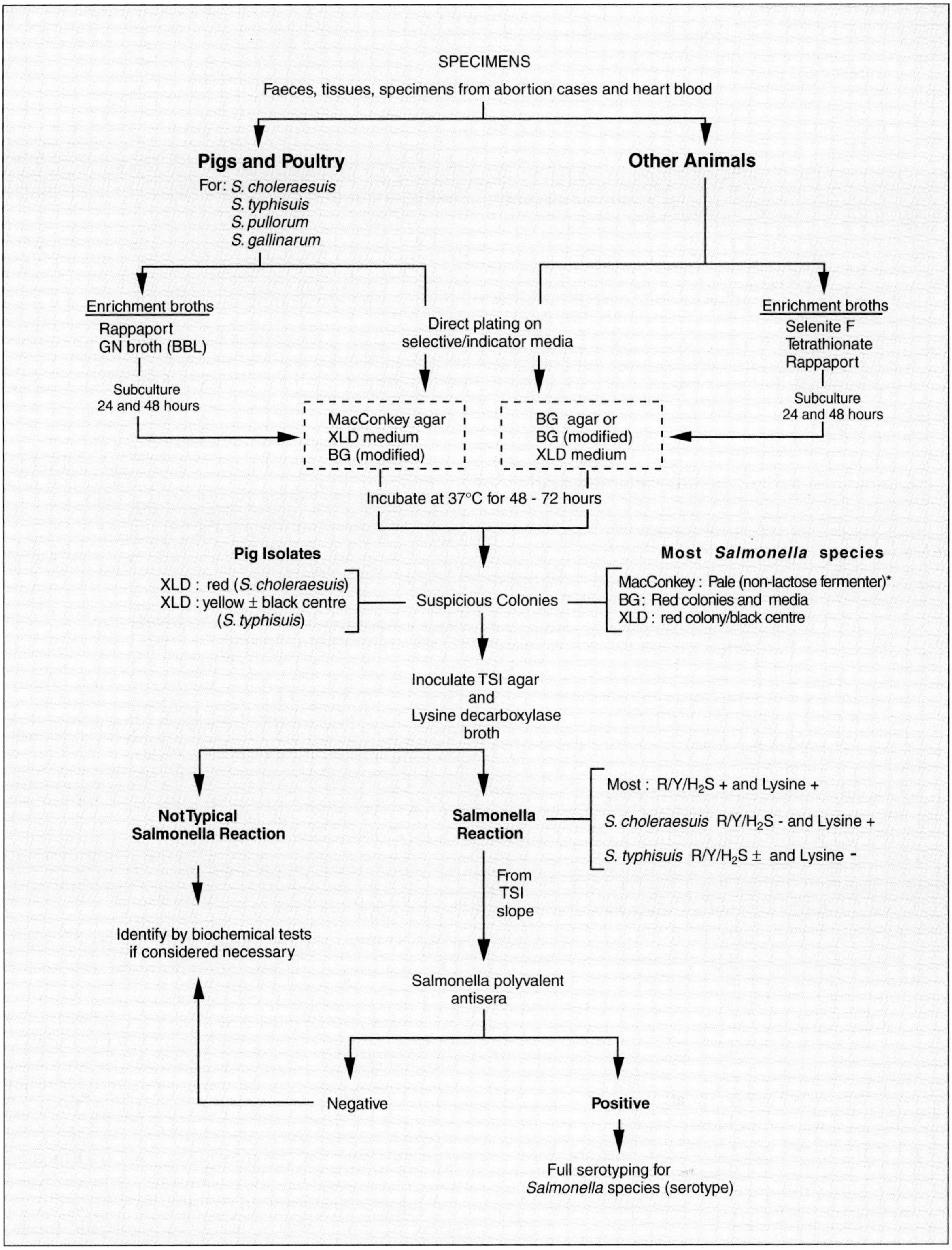

Diagram 34. Isolation of salmonellae from clinical specimens.
BG = brilliant green agar, BG (modified) = brilliant green agar, modified (Oxoid), XLD = xylose-lysine-deoxycholate medium, *= very occasionally salmonella strains are lactose-positive, TSI = triple sugar iron agar.

Water and feed samples

Occasionally veterinary diagnostic laboratories are asked to examine water and animal feedstuffs for the presence of salmonellae.

a) Water samples

The bacterium can be isolated from water in one of three ways:

- About 100 ml of a water sample is added to an equal amount of double-strength enrichment broth. The broth is incubated for 48 hours and subcultured onto selective/indicator media.
- If the water sample does not contain particulate matter about 100 ml can be passed through a sterile 0.45 µm membrane filter. Any salmonellae present will be concentrated on the surface of the filter. The filter can be placed on a selective medium (**286**) or if few salmonellae are present, the membrane filter can be placed in 10 ml of enrichment broth to increase the number of salmonellae during incubation at 37°C for 24–48 hours. Subsequently a subculture can be made on selective/indicator media.
- A very effective method for isolating salmonellae from running water, such as a stream, involves the immersion of a sterile pad in the water for 48 hours. The pad consists of cotton wool wrapped in a square of surgical gauze and securely tied with one end of a long piece of strong string (**287**). The device can be placed in a paper envelope and sterilised in an autoclave. The string is held at the free end and the pad thrown out into the water to be sampled. The end of the string is secured to an object on the bank. After 48 hours the pad is retrieved and carefully placed in a jar containing 50 ml of enrichment broth and the string cut below the point where it was handled. After 24–48 hours' incubation at 37°C, subcultures are made on selective/indicator media.

b) Feed samples

The salmonellae in feed samples may have been subjected to heat-treatment and will be in a desiccated condition. A representative 30 g sample is broken up and placed in 100–200 ml of a non-selective broth such as nutrient broth or lactose broth. Six ml of 10 per cent Tergitol No. 7 is added to samples with a high fat content. The sample is shaken vigorously and incubated at 37°C for 48 hours. The broth can be streaked directly onto brilliant green agar or XLD medium but a 1 ml aliquot should be placed in 10 ml of a selective enrichment broth, such as tetrathionate. This is incubated at 37°C for 24 hours and then subcultures are made onto selective/indicator media.

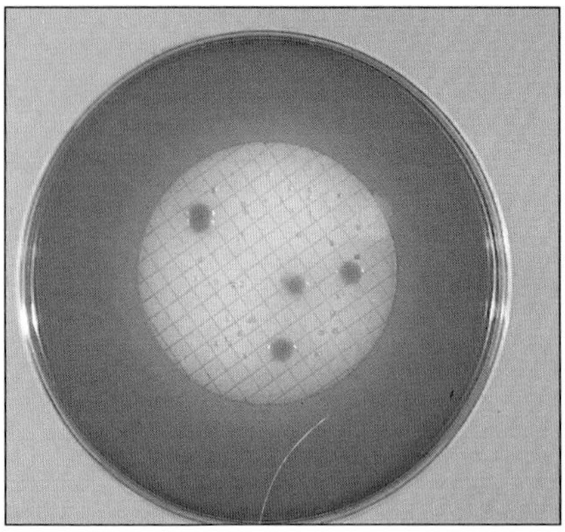

286

286 Filtration method for isolating salmonellae from water. The membrane filter on MacConkey agar shows lactose-fermenting and non-lactose-fermenting colonies.

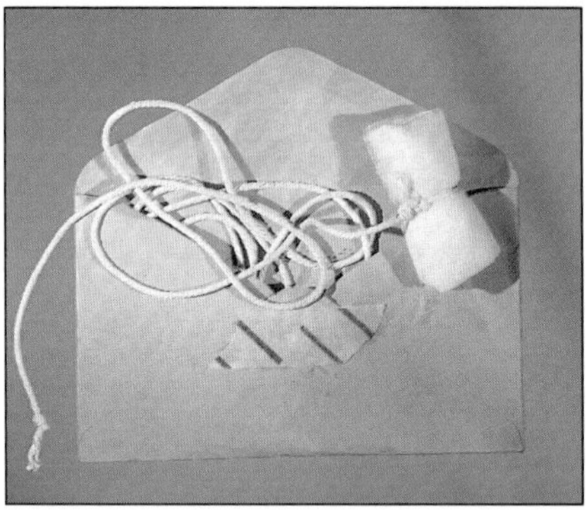

287

287 An improvised device for recovering salmonellae from ponds or streams.

Identification

Colonial morphology on selective/indicator media

The majority of salmonellae are non-lactose-fermenters and produce pale colonies on MacConkey agar and an alkaline reaction in the medium (**288**). However, it must be remembered that some strains of *S. arizonae* are lactose-positive and strains of *S. typhimurium* have been encountered carrying plasmids with genes coding for lactose fermentation. Most salmonellae give an alkaline reaction in brilliant green agar and have red colonies (**289**). On XLD medium the majority of *Salmonella* serotypes produce hydrogen sulphide and have red colonies with a black (H_2S) centre (**290** and **291**). Colonies characteristic for salmonella on the selective/indicator media are inoculated, singly, into a TSI agar slope and lysine decarboxylase broth. The typical reaction for salmonella in TSI agar is a red (alkaline) slant, yellow (acid) butt and superimposed (black) H_2S production (R/Y/H_2S+). The test for lysine decarboxylation is positive. However, *S. choleraesuis* does not produce H_2S (**292**) although *S. choleraesuis* biotype Kunzendorf is H_2S-positive. The fastidious *S. typhisuis* is variable in H_2S production and is lysine-negative (Blessman *et al.*, 1981). If the reaction in TSI agar and lysine decarboxylase broth is equivocal, further biochemical tests should be carried out (**Table 62**) or an identification system used such as API 20E (Analytab Products).

288

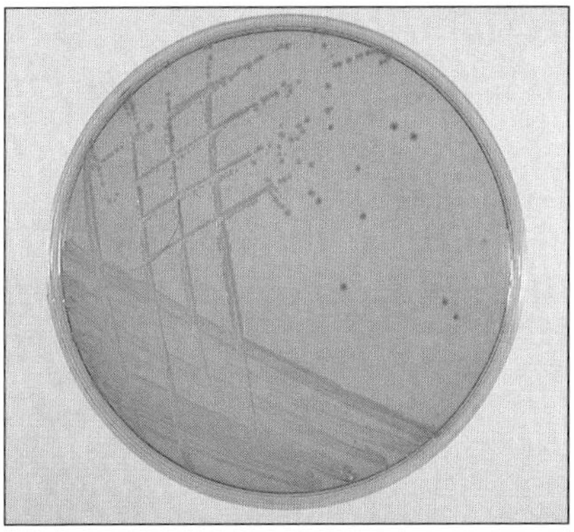

288 Pale, non-lactose-fermenting colonies of *Salmonella* sp. on MacConkey agar.

289

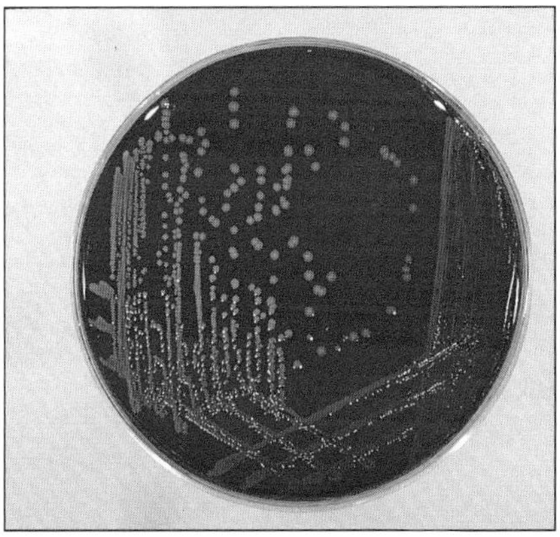

289 *Salmonella typhimurium* on brilliant green agar. Most salmonellae are unable to ferment either the lactose or sucrose in the medium but instead use the peptone with a consequent alkaline reaction. (Phenol red indicator).

290

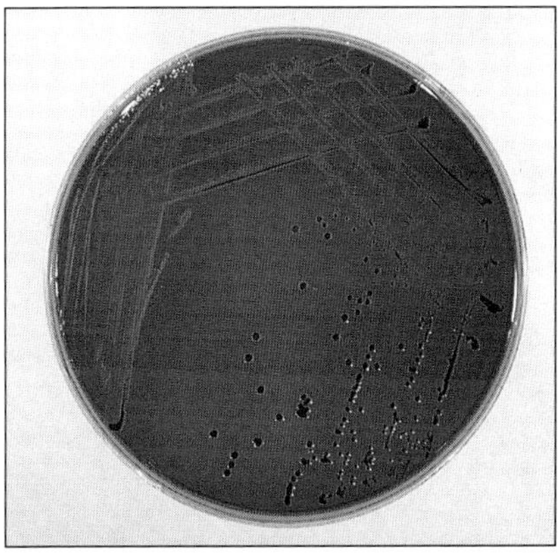

290 *Salmonella dublin* on XLD medium showing the H_2S production (black centre) and alkaline (red) reaction in the medium and periphery of colony. (Phenol red indicator).

291

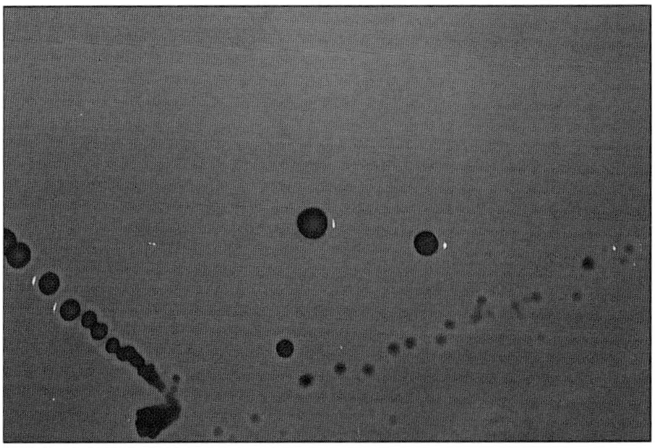

291 Close-up of *S. dublin* colonies on XLD medium. The colonial appearance is characteristic of most salmonellae on this medium (black centre (H_2S) and red 'skirt').

292

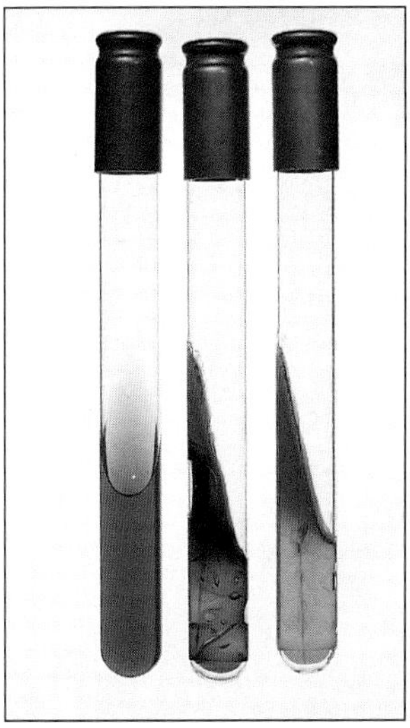

292 Triple sugar iron (TSI) agar: from left, uninoculated, *S. typhimurium* with H_2S production (representative of most salmonellae) and *S. choleraesuis* with no H_2S production.

Salmonella serotyping

Bacterial growth for serotyping should be taken from a TSI agar slant or from nutrient agar as culture from selective media is often unsuitable for typing. Serotyping is based on the O (somatic) and H (flagellar) antigens (**Table 66**) and a slide agglutination test is used (**293**). Rare strains of *S. dublin* have a Vi (virulence) capsular antigen that can mask the cell wall (O) antigens. Boiling a suspension of *S. dublin* for 10–20 minutes will destroy the Vi antigen. A loopful of culture of the salmonella to be serotyped should be suspended in a drop of saline on a microscope slide and examined for autoagglutination. This can occur with rough strains and will invalidate the serotyping. Smooth-rough (S→R) dissociation occurs after subculture and most frequently from media containing carbohydrates.

A smooth salmonella to be serotyped is emulsified in a drop of 0.85 per cent saline on a clean microscope slide. A drop of antiserum (Wellcome Diagnostics or Behring) is added to, and mixed well with, the salmonella suspension. The slide is rocked gently for about 30 seconds and the antigen-antibody mixture examined for agglutination (**293**). The salmonella is first tested against antisera to the O (somatic) antigens and then the H (flagella) antigens.

Cultures of a salmonella that are motile and diphasic will contain cells that have either phase 1 (specific) or phase 2 (non-specific) flagellar antigens. Usually the majority of cells have flagella in one phase but there will be a very few cells with the alternative flagellar antigen. The salmonella will agglutinate only with antisera to the flagellar antigen that predominates. To obtain a complete antigenic formula for the salmonella, the phase must be 'changed'. In reality this involves the selection of the few cells that have the alternative flagellar antigens. This 'phase-changing' can be carried out by the Craigie tube method (**294**) or by the ditch-plate method (**295**). Some *Salmonella* serotypes are monophasic with the flagellar antigens in one phase. *S. pullorum* and *S. gallinarum* are unusual in being non-motile and lacking flagellar antigens.

A serogroup comprises serotypes with similar O antigens. Lysogenisation by certain converting bacteriophages may produce changes in the O antigens. In serogroup E, a phage can alter the O-antigen 3,10 to 3,15 thus changing *S. anatum* to *S. newington*. Certain serotypes share an almost identical antigenic formula. These serotypes must be distinguished by biochemical tests (**Table 67**) and are known as biotypes or biovars.

A rapid latex agglutination test is available as a commercial test kit for the specific identification of *S. enteritidis* from cultures ('Sefex', CVL, New Haw, UK). The latex particles are coated with a monoclonal antibody to a novel fimbrial antigen of *S. enteritidis*.

293

293 Salmonella slide agglutination test for serotyping salmonellae showing agglutination with homologous antiserum (left) and a negative reaction (right).

294

294 The Craigie tube method for 'phase-changing' salmonellae. The semi-solid agar contains antiserum specific to the predominating H phase of the salmonella under test. The salmonella culture is inoculated on the agar surface within the central tube. Salmonella cells in the alternative H phase are able to move through the agar uninhibited and can be collected, after 24 hours incubation, from the surface outside the central tube.

295

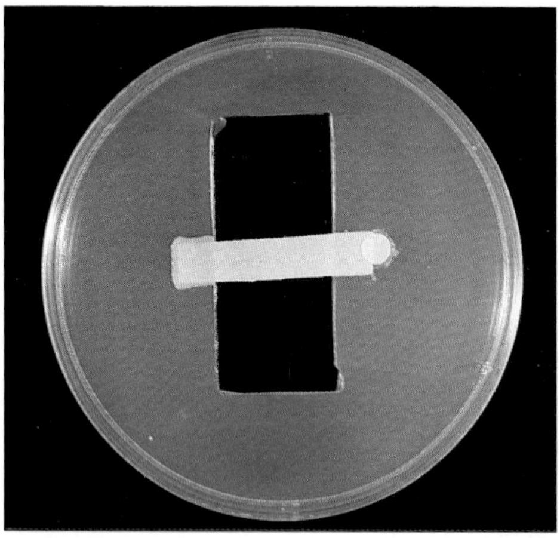

295 The ditch-plate method for 'phase-changing' salmonella. The filter paper strip contains antiserum specific for the predominating H phase of the isolate. The salmonella under test has been inoculated on the left side of the filter paper bridge. Salmonella cells in the alternative H phase move across the bridge and can be collected on the filter paper disc on the right.

Table 66. Antigens of some *Salmonella* serotypes.

Serotype	Serogroup	Somatic (O) antigens	Flagella (H) antigens	
			Phase 1	Phase 2
S. paratyphi A	A	$\underline{1}$, 2, 12	a	[1, 5]
S. typhimurium	B	$\underline{1}$, 4, [5], 12	i	1, 2
S. derby	B	$\underline{1}$, 4, [5], 12	f, g	[1, 2]
S. agona	B	4, 12	f, g, s	–
S. saint paul	B	$\underline{1}$, 4, [5], 12	e, h	1, 2
S. heidelberg	B	$\underline{1}$, 4, [5], 12	r	1, 2
S. abortusovis	B	4, 12	c	1, 6
S. abortusequi	B	4, 12	–	e, n, x
S. typhisuis	C_1	6, 7	c	1, 5
S. choleraesuis	C_1	6, 7	c	1, 5
S. choleraesuis biotype Kunzendorf	C_1	6, 7	[c]	1, 5
S. montevideo	C_1	6, 7, $\underline{14}$	g, m, [p], s	–
S. oranienburg	C_1	6, 7	m, t	–
S. newport	C_2	6, 8	e, h	1, 2
S. bovismorbificans	C_2	6, 8	r	1, 5
S. kentucky	C_3	8, $\underline{20}$	i	z_6
S. typhi	D_1	9, 12, [Vi]	d	–
S. enteritidis	D_1	$\underline{1}$, 9, 12	g, m	[1, 7]
S. dublin	D_1	$\underline{1}$, 9, 12, [Vi]	g, p	–
S. gallinarum	D_1	$\underline{1}$, 9, 12	–	–
S. pullorum	D_1	9, 12	–	–
S. anatum	E_1	3, 10	e, h	1, 6
S. newington	E_2	3, $\underline{15}$	e, h	1, 6
S. senftenberg	E_4	1, 3, 19	g, [s], t	–
S. worthington	G_2	$\underline{1}$, 13, 23	z	l, w

[ ] = antigen may be present or absent, $\underline{1}$ = O factor whose presence is due to phage conversion.

Table 67. Biochemical differentiation of *Salmonella* biotypes.

	S. typhisuis	*S. choleraesuis*	*S. choleraesuis* biotype Kunzendorf	*Salmonella* (most serotypes)
Hydrogen sulphide (TSI)	v (58% +)	–	+	+
Lysine decarboxylase	–	+	+	+
Citrate (Simmons)	–	+	+	+
Mannitol	–	+	+	+
Inositol	+	–	–	v
Sorbitol	–	(+)	(+)	+
Trehalose	–	–	–	+
Maltose	–	+	+	+

	S. pullorum	*S. gallinarum*	*Salmonella* (most serotypes)
Glucose (gas)	(+)	–	+
Dulcitol	–	+	+
Maltose	–	+	+
Ornithine decarboxylase	+	–	+
Rhamnose	+	–	+
Motility	–	–	+

v = variable reactions, (+) = most strains positive.

Phage typing salmonella isolates

Phage typing is based on the sensitivity of a particular isolate to a series of bacteriophages at appropriate dilutions. Phage typing of *S. typhi*, *S. typhimurium* and *S. enteritidis* is carried out at reference laboratories. *S. enteritidis* phage type (PT) 4 has recently been the cause of a large proportion of the human food-poisoning cases in Britain. Since 1987 this phage type, has become common in broiler and laying flocks.

Serology for the detection of salmonella antibodies

Agglutination tests, ELISA, antiglobulin and complement fixation tests have been used to detect antibody responses to salmonella infections. For cattle, serological methods of diagnosis are more useful on a herd basis, rather than for individual animals. Paired acute and convalescent sera should be tested in order to detect a rising antibody titre. Plate agglutination tests, using serum or whole blood, with a stained antigen, have been used in national eradication schemes for *S. pullorum* in chickens. An ELISA has been developed for the detection of *S. enteritidis* antibodies in chickens and found to be sensitive and specific (Kim *et al.*, 1991).

Yersinia species

Yersinia pestis and *Y. pseudotuberculosis* were once classified as *Pasteurella* species. There are now 11 species in the genus *Yersinia. Y. pestis*, *Y. pseudotuberculosis* and *Y. enterocolitica* are responsible for zoonotic infections and *Y. ruckeri* is a pathogen of fish. *Y. enterocolitica* is a significant enteric pathogen for humans.The 'biochemically atypical' strains of *Y. enterocolitica* have now been classified into seven additional species. They are rarely pathogenic for animals or man.

Natural Habitat

Y. pestis, the cause of bubonic plague in man and a sylvatic cycle in animals, is transmitted mainly by fleas from tolerant rodents. Human infections through cuts, bites, scratches and aerosols can also occur. Cats are susceptible to *Y. pestis* and naturally infected cats can pose a health hazard for humans in endemic areas. *Y. pseudotuberculosis* persists in wild rodents and birds as well as in the environment. The intestinal tract of wild and domestic animals appears to be the reservoir for *Y. enterocolitica*. Pigs, particularly, are carriers of *Y. enterocolitica* strains pathogenic for humans.

Pathogenesis

The virulence factors of *Yersinia* species include anti-phagocytic outer membrane proteins, a plasmid-encoded exotoxin, a bacteriocin, a coagulase and a fibrinolytic factor, all of which correlate with virulence. The virulent strains of *Y. pestis* resist phagocytosis and are able to grow within macrophages. Endotoxin is also thought to contribute to tissue damage and clinical signs in plague. *Y. enterocolitica* produces a heat-stable enterotoxin that is associated with food-poisoning strains in man. The diseases caused by the three *Yersinia* species pathogenic for animals and man are summarised in **Table 68**.

Antigens of *Yersinia* species

Y. pseudotuberculosis can be divided into 6 serogroups based on the thermostable O antigens (I-VI) and five H (flagella) antigens (a-e). There are shared antigens between the closely related *Y. pestis* and *Y. pseudotuberculosis*. An antigenic relationship exists between *Y. pseudotuberculosis* and salmonella O antigens in serogroups B, D and E and also with some *E. coli* O antigens. *Y. enterocolitica* shares an antigen with *Brucella* species that can give rise to false brucella agglutination test results. *Y. enterocolitica* is divided into more than 37 serotypes, many of which do not appear to be pathogenic.

Laboratory Diagnosis

Great care must be exercised if a live or dead animal is presented that might be infected with *Y. pestis*. The public health authorities should be notified immediately. The animal, whether alive or dead, should be treated promptly to kill any fleas. It is advisable to wear a gown, mask and gloves when handling the animal. All bacteriological culture work should be conducted in a biohazard cabinet.

Specimens

For *Y. pestis*, samples could include oedematous tissue, lymph nodes, nasopharyngeal swabs, transtracheal aspirates, cerebrospinal fluid and blood for culture and serology. Specimens for *Y. pseudotuberculosis* comprise any necrotic internal lesions, lymph nodes and faeces. Faecal samples are usually taken for *Y. enterocolitica*.

Isolation

Yersinia species grow on nutrient, blood and MacConkey agars but the colonies, after 24 hours' incubation, tend to be smaller than those of the other members of the *Enterobacteriaceae*. *Y. pestis* grows poorly on agars containing desoxycholate whereas *Y. enterocolitica* and *Y. pseudotuberculosis* grow well on these media. Yersinia selective medium (CIN agar) containing the antibiotic supplement cefsulodin (15 mg/litre), irgasin (4 mg/litre) and novobiocin (2.5 mg/litre) is designed for the isolation of *Y. enterocolitica* from faeces. A cold-enrichment procedure may be necessary for the isolation of both *Y. enterocolitica* and *Y. pseudotuberculosis* from faecal specimens. A faecal specimen (approximately 5 per cent by volume) is placed in 1/15 M phosphate buffered saline (Oxoid) and held in the refrigerator (4°C) for 3 weeks. Subcultures, at weekly intervals, can be made on MacConkey and Yersinia selective medium. *Yersinia* species usually grow faster at 37°C but prefer lower

Table 68. Diseases caused by *Yersinia* species.

Yersinia species	Host	Disease(s)	Transmission	Reservoir
Y. pestis	Humans	Bubonic plague ('black death')	Flea and rat bites	Many rodent species
	Rodents	Sylvatic plague. Infection in rodents usually latent with occasional outbreaks of disease	Occasionally cat bites and scratches	
	Cats	Cats in endemic areas may show mandibular lymphadenitis, fever, depression, anorexia, sneezing and occasionally nervous disturbances. Most infections are fatal		
Y. pseudo-tuberculosis	Guinea-pigs, other rodents, rabbits, wild and captive birds	Pseudotuberculosis: 1 Septicaemic syndrome. 2 Classical syndrome with nodules in internal organs Seen most commonly in guinea-pigs and canaries	Ingestion	Faeces of carrier rodents and wild birds
	Farm animals	Latent infections, Occasional disease such as in captive deer		
	Sheep	Orchitis and epididymitis reported		
	Humans	Mesenteric lymphadenitis, acute terminal ileitis and rare cases of septicaemia. Mainly children and young adults		
Y. enterocolitica	Farm animals	Latent infections with sporadic cases of enteritis or generalised infections. Captive deer particularly susceptible	Ingestion	Carrier state in many animal species, especially pigs
	Humans	Food poisoning (enteritis), mesenteric lymphadenitis (pseudo-appendicitis). Most common in children		Pigs are carriers of strains pathogenic for humans

incubation temperatures, particularly on primary isolation. Sometimes additional culture plates, incubated at 22–25°C, can be useful for initial isolation.

Identification

Colonial morphology

The *Yersinia* species are lactose-negative (**296**) although lactose-positive strains of *Y. enterocolitica* occur and this property is thought to be plasmid-mediated. *Y. enterocolitica* will grow well on media, such as brilliant green and XLD agars, intended for salmonella isolation but *Y. pseudotuberculosis* is less tolerant. *Y. enterocolitica* and *Y. pseudotuberculosis* (**297**) are non-haemolytic on blood agar. The colonies of *Y. enterocolitica* on *Yersinia* selective medium (Oxoid) have dark red centres with a transparent periphery.

Biochemical tests

The yersiniae are identified by conventional biochemical tests (**Table 62**) or by means of a commercial test strip such as API 20E (Analytab Products). *Y. pestis*, *Y. pseudotuberculosis* and *Y. enterocolitica* are non-motile at 37°C but *Y. pseudotuberculosis* and *Y. enterocolitica* strains can be motile at 28°C. *Y. pseudotuberculosis* is almost always urease-positive as are most strains of *Y. enterocolitica*, but *Y. pestis* does not produce this enzyme. Only certain strains of *Y. enterocolitica* cause human infections and these pathogenic serotypes can be detected biochemically (Farmer and Kelly, 1991). They are negative for the pyrazinamidase test, salicin fermentation and aesculin hydrolysis, whereas the non-pathogenic strains are positive to these tests. The pathogenic strains grow as small red colonies on congo red-magnesium oxalate (CR-MOX) agar but non-pathogenic strains are unable to grow on this medium. *Y. pestis* can be confirmed in reference laboratories by the fluorescent antibody technique, bacteriophage-susceptibility and by mouse or guinea-pig inoculation. The infected animals usually die in 3–8 days.

Serological tests to demonstrate antibodies

Microagglutination and complement fixation tests or enzyme immunoassays are used in some medical reference laboratories for the detection of antibodies to *Y. enterocolitica* and *Y. pseudotuberculosis.*

296

297

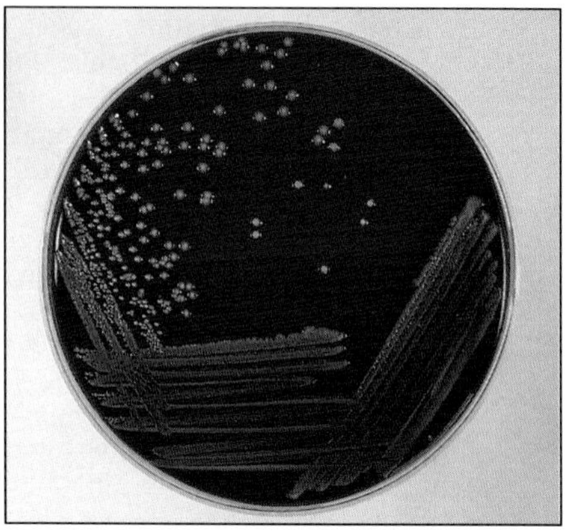

296 Comparison of the reactions of *Y. enterocolitica* (left plate) and *Y. pseudotuberculosis* (right plate) on MacConkey agar (top), brilliant green agar (left) and XLD medium (right).

297 *Yersinia pseudotuberculosis* on sheep blood agar. The non-haemolytic, greyish, shiny, discrete colonies are similar to those of many other members of the *Enterobacteriaceae.*

References

Blessman, B.H., Morse, E.V., Midla, D.A. and Swaminathan, B. (1981). Culture and identification of *Salmonella typhisuis* and related serotypes. American Association of Veterinary Laboratory Diagnosticians, 24th Ann. Proc. pp 1–10.

Carroll, P.J., Woodward, M.J. and Wray, C (1990). Detection of LT and $ST1_a$ toxins by latex and EIA tests, *Veterinary Record,* **127**: 335–336.

Farmer, J.J. and Kelly, M.T. (1991). *Enterobacteriaceae* pp. 360–383 In Manual of Clinical Microbiology, 5th ed., Balows, A., Hausler, W.J., Herrmann, K.L., Isenberg, H.D. and Shadomy, H.J. (eds), American Society for Microbiology, Washington, DC.

Francis, D.H., Remmers, G.A. and De Zeeuw, P.S. (1982). Production of K88, K99, and 987P antigens by *Escherichia coli* cultured on synthetic and complex media, *Journal of Clinical Microbiology,* **15**: 181–183.

Kim, C.J., Nagaraja K.V., Pomeroy B.S. (1991). Enzyme-linked immunosorbent assay for the detection of *S. enteritidis* infection in chicken, *American Journal of Veterinary Research,* **52**: 1069.

Le Minor, L and Popoff, M.Y. (1987). *International Journal of Systematic Bacteriology,* **37**: 465.

19 *Pseudomonas* species

Pseudomonas species are medium-sized (0.5–1.0 × 1.5–5.0 μm) Gram-negative rods. They are strict aerobes, oxidative, catalase-positive, oxidase-positive and most are motile by one or several polar flagella, an exception being *P. mallei* which is non-motile. Some species produce soluble pigments and most will grow on MacConkey agar.

Natural Habitat

The numerous species in the genus are almost exclusively saprophytes including two of the species of importance in animals, *P. aeruginosa* and *P. pseudomallei*. Infected equidae are the reservoir for *P. mallei*. *P. aeruginosa* and *P. pseudomallei* are present in soil and water, *P. aeruginosa* is found worldwide and *P. pseudomallei* mainly in tropical regions. *P. aeruginosa* can be found on skin, mucous membranes and in the faeces of animals. *P. fluorescens*, present in soil and water, is associated with food spoilage and can cause lesions in reptiles and fish.

Pathogenesis

P. aeruginosa produces a number of protein exotoxins, an enterotoxin that is responsible for diarrhoea during initial infections, an endotoxin and numerous extracellular products, such as proteases and haemolysins, that may play a role in the pathogenesis. *P. aeruginosa* possesses pili which facilitate adherence to epithelial cells and some strains have a capsule that is antiphagocytic. The bacteriocins (pyocins) and pigments exhibit antimicrobial activities. The blue-green pigment (pyocyanin) can colour pus and stain wool a greenish hue. However, *P. aeruginosa* is opportunistic and is involved in primary disease only rarely. Predisposing causes include trauma to tissue (burns and wounds), debilitation due to malignancy or immunodeficiency, and reduced numbers of normal flora, often caused by antibiotic therapy. *P. aeruginosa* is resistant to many commonly used antibiotics.

P. pseudomallei has a wide host range, including humans, and causes melioidosis or pseudoglanders. Infections are usually systemic and the manifestations depend on the extent and distribution of the lesions. The lesions are nodules that may suppurate and can form in any tissue, including the brain. Most infections are chronic but acute disease with terminal septicaemia may occur. The toxins include a lethal factor with anticoagulant activity and a skin-necrotising proteolytic agent. Melioidosis usually occurs in tropical regions between 20° northern and southern latitudes but has been reported in localised areas of France, Iran, China and the USA.

P. mallei causes glanders or farcy (the skin form) in the equidae. Humans and members of the cat family are susceptible with occasional infections in dogs, goats, sheep and camels. Cattle, pigs, rats and birds are resistant to infection. Transmission occurs from infected animals via contaminated food and water and less commonly from aerosols and infection of wounds. Toxins are suspected in the pathogenesis but the mode of action is uncertain. Primary lesions occur at the point of entry with dissemination via the lymphatic system and the bloodstream. The disease can be acute or chronic and many infections are fatal if not treated at an early stage. Infection is characterised by the formation of tubercle-like nodules that frequently ulcerate. Strain variations may determine whether suppurative or granulomatous lesions predominate. Glanders once had a wide geographical distribution but now is seen only in China and Mongolia with pockets of infection in India, Iraq, the Philippines and Eastern Europe.

Other saprophytic *Pseudomonas* spp. have been incriminated in infrequent infections of humans. Common contaminants in clinical specimens include *P. maltophilia*, *P. putida*, *P. fluorescens*, *P. stutzeri* and *P. cepacia*. Little is known of their involvement in animal disease.

Table 69 summarises the diseases and conditions caused by the main pathogens of the genus *Pseudomonas*.

Laboratory Diagnosis

P. mallei and *P. pseudomallei* are among the most dangerous bacteria to work with in a laboratory. A biohazard cabinet must be used and all necessary safety procedures taken.

Specimens

These will be varied and depend on the clinical signs and site of lesions.

Direct microscopy

Direct microscopy from specimens is of little diagnostic use as the pseudomonads are medium-sized, Gram-negative rods with no other distinctive characteristics. A fluorescent antibody technique can be useful for *P. mallei* and *P. pseudomallei*.

Isolation

The *Pseudomonas* species are non-fastidious and will grow on trypticase soy agar, blood agar and on less complex media. The growth of *P. mallei* is enhanced by 1 per cent glycerol. A selective medium for *P. mallei* can be made by adding 1000 units polymyxin

Table 69. Diseases and main hosts of the pathogenic *Pseudomonas* species.

Species	Host(s)	Disease
P. aeruginosa	Cattle	Mastitis, uterine infections, skin infections, abscesses, enteritis and arthritis
	Sheep and goats	Mastitis, pneumonia, lung abscesses and 'green wool' (a skin infection in sheep)
	Pigs	Enteritis, respiratory infections and otitis
	Horses	Metritis, lung abscesses and eye infections
	Dogs and cats	Otitis externa, cystitis, endocarditis, dermatitis, wound infections and conjunctivitis
	Mink	Septicaemia and pneumonia
	Chinchilla	Generalised infection with conjunctivitis, otitis, pneumonia, enteritis and infection of the genital organs
	Reptiles	Necrotic stomatitis and other necrotic lesions, especially in captive snakes
	Many animal species	Infection of burns and other wounds, diarrhoea, genital and nosocomial infections
P. pseudomallei	Many animal species	Melioidosis (pseudoglanders):
	Horses	The disease can mimic glanders
	Cattle	Acute and chronic forms with localisation of lesions in lungs, joints and uterus
	Sheep	Arthritis and lymphangitis predominate
	Goats	Loss of condition, respiratory and central nervous disturbances, arthritis and mastitis
	Pigs	As for goats but in addition diarrhoea and abortion
	Dogs	Febrile disease with localising suppurative foci
P. mallei	Horses and other equids	Glanders: acute form with high fever, mucopurulent nasal discharge, respiratory signs, septicaemia and death within 2 weeks Chronic forms of glanders • Pulmonary: small nodules in lungs that break down and discharge *P. mallei* into the bronchioles • Cutaneous form: Farcy, which is a lymphangitis with ulcers along lymphatic vessels of the limbs and chest. The ulcers eventually heal leaving 'star-shaped' scars
	Humans, cats and other animals	Acute, septicaemic disease

E, 1250 units bacitracin and 0.25 mg actidione to 100 ml of trypticase soy agar. Commercial selective media are available for *P. aeruginosa* and usually contain 0.03 per cent cetrimide (cetyl trimethyl ammonium bromide). *P. aeruginosa* will also grow on many of the selective media intended for the *Enterobacteriaceae* such as MacConkey, brilliant green and XLD agars (**268**).

The cultures for *P. aeruginosa*, *P. pseudomallei* and *P. mallei* are incubated aerobically at 37°C for 24–48 hours. Some of the saprophytic pseudomonads, such as *P. fluorescens,* grow extremely poorly, or not at all, at 37°C as 30°C is often the upper temperature limit of their growth range.

Identification

Colonial morphology

• ***P. aeruginosa*:** the colonies are large (3–4 mm), flat, greyish-blue with a characteristic fruity, grape-like odour of aminoacetophenone. Most strains give a clear zone of haemolysis on blood agar (**298**). Pyocyanin, a bluish pigment unique to *P. aeruginosa*, gives the blue colour associated with many cultures. Colonial variation includes S-forms (soft and shiny), R-forms (dry and granular) that are not unlike the colonies of some *Bacillus* species, and mucoid M-forms that are frequently biochemically atypical. Some strains have colonies with a distinctive metallic sheen (**299**). *P. aeruginosa* produces large, pale colonies on MacConkey agar (unable to utilise lactose) with greenish-blue pigment superimposed (**300**). Red colonies and medium, indicative of an alkaline reaction, are seen on brilliant green (**299**) and XLD agars. No H_2S is produced on XLD medium.

• ***P. pseudomallei***: colonial growth varies from smooth and mucoid to rough with a dull, wrinkled, corrugated surface. In the smooth form the colonies are round, low-convex, entire, shiny and greyish-yellow. After several days the colonies become opaque, yellowish-brown and umbonate (**301**). The growth has a characteristic earthy or musty odour. Partial and, later, complete haemolysis occurs on sheep blood agar. *P. pseudomallei* grows on MacConkey agar, utilising lactose (**302**), but there is no growth on deoxycholate or Salmonella-Shigella (SS) agars.

• ***P. mallei***: growth is slower than that of *P. aeruginosa* and *P. pseudomallei* but in 24–48 hours the colonies are 1–2 mm in diameter, smooth and white to cream. As they age, they become granular and yellowish or brown in colour. *P. mallei* is unable to grow on MacConkey agar.

298

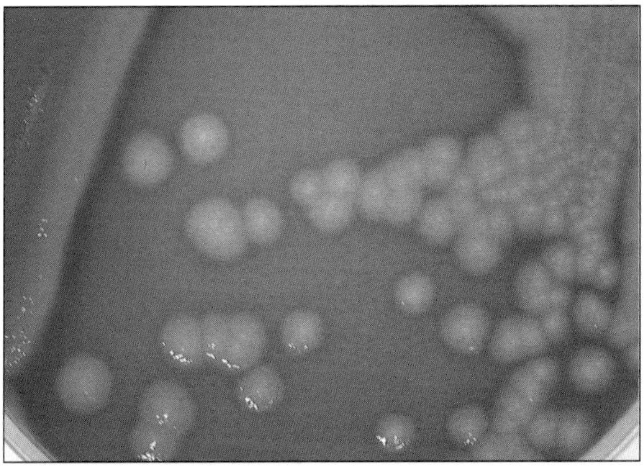

298 *Pseudomonas aeruginosa* on sheep blood agar showing large, flat, irregular-edged colonies resembling those of some *Bacillus* species. The green-blue pyocyanin pigment is most obvious in areas of heaviest growth.

299

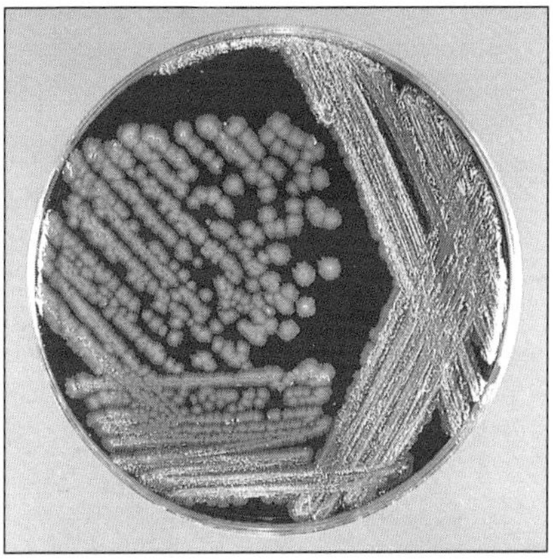

299 *P. aeruginosa* on brilliant green agar where the alkaline reaction is similar to that given by *Salmonella* species. The metallic sheen displayed by this strain is a feature of some isolates.

300

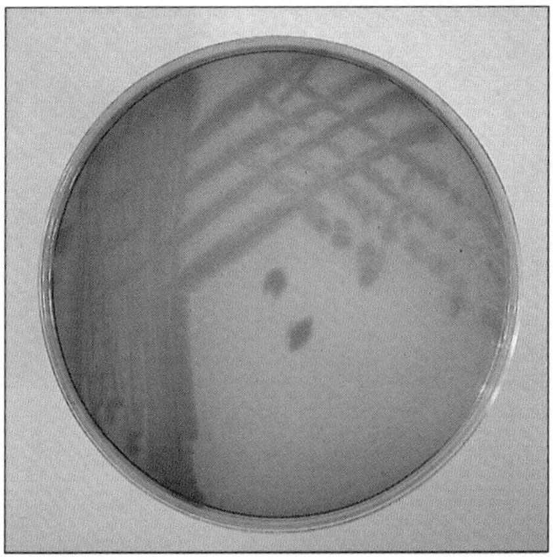

300 *P. aeruginosa* on MacConkey agar. It has the pale colonies of a non-lactose-fermenter with green-blue, pyocyanin pigment superimposed.

301

301 *Pseudomonas pseudomallei:* smooth colonial form on sheep blood agar after several days' incubation. The colonies are smooth, glistening, opaque, yellowish-brown and umbonate with a zone of clear haemolysis.

302

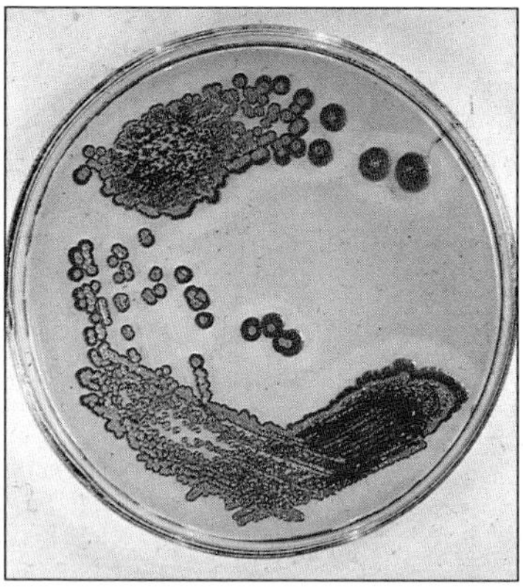

302 *P. pseudomallei* on MacConkey agar (*P. mallei* does not grow on this medium).

The pigments of *P. aeruginosa*

Strains of *P. aeruginosa* produce the diffusible pigments pyocyanin (blue), pyoverdin (yellow), pyorubin (red) and pyomelanin (dark brown) (**303**) in varying combinations and amounts. Some strains produce all four pigments. Pyorubin and pyomelanin are less commonly produced, develop slowly and are seen best by growing the strains on nutrient agar slants at room temperature for up to 2 weeks. As pyocyanin is unique to *P. aeruginosa* this is an important diagnostic characteristic although strains vary in the amount of the pigment they produce. Media such as Pseudomonas agar P (Difco) (**304**) will enhance pyocyanin production and Pseudomonas agar F (Difco) enhances pyoverdin production (**305**). Pyoverdin, once called 'fluorescein', will fluoresce under ultra-violet light. Some strains of *P. aeruginosa* do not produce pyocyanin and these may also be atypical in certain biochemical reactions, making them difficult to identify.

303

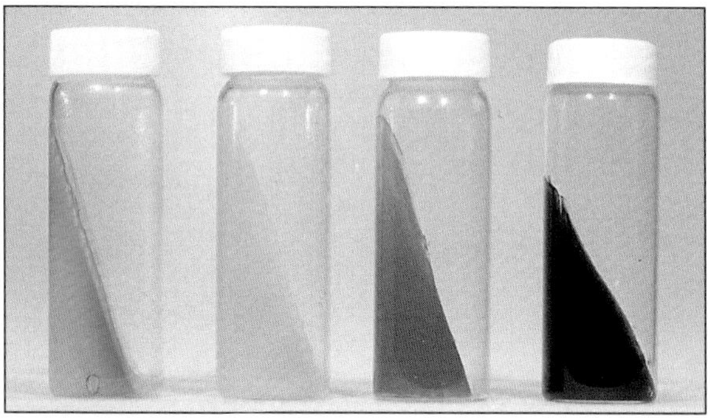

303 Pigments produced by *P. aeruginosa*. The nutrient agar slopes show from left: pyocyanin (blue-green), pyoverdin (greenish-yellow), pyorubin (red) and pyomelanin (dark brown).

304

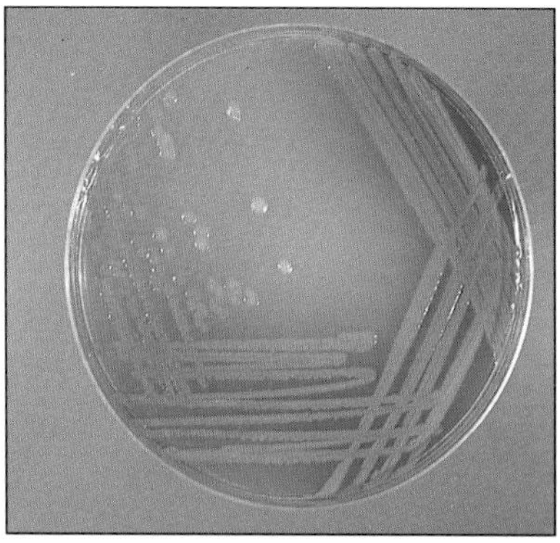

304 *P. aeruginosa* on 'Pseudomonas agar P'. This medium enhances the production of pyocyanin.

305

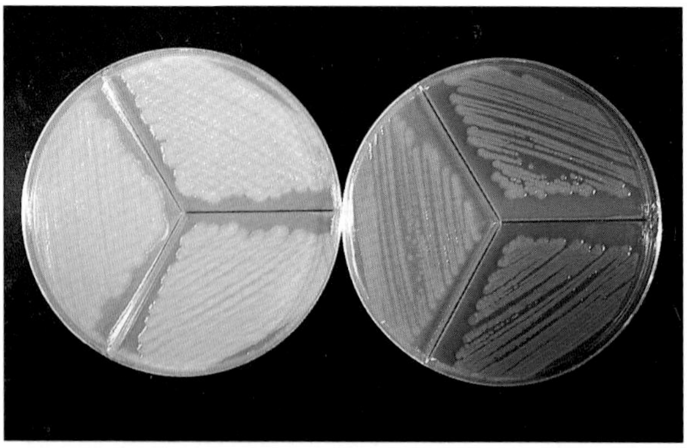

305 Three strains of *P. aeruginosa* on 'Pseudomonas agar F' (left) and on 'Pseudomonas agar P' (right). These media are used to enhance pigment production.

Microscopic appearance

All the pseudomonads are medium-sized Gram-negative rods.

Biochemical characteristics

The characteristic colonial appearance, including pyocyanin production and odour, and a strong oxidase reaction (the members of the *Enterobacteriaceae* are oxidase-negative), will give a presumptive identification of *P. aeruginosa*. Other characteristics of *P. aeruginosa* and those of *P. pseudomallei* and *P. mallei* are given in **Table 70**. Commercial identification systems such as API 20E (Analytab Products) can be used to identify some of the *Pseudomonas* species. **Table 71** contrasts the reactions of *P. aeruginosa* with other saprophytic pseudomonads that can be isolated from clinical specimens, although their significance is not known.

Immunological tests

- Melioidosis: complement-fixing and indirect haemagglutinating antibodies are produced after infection with *P. pseudomallei*. However, the diagnosis of melioidosis depends more on the isolation and identification of the bacterium than on clinical findings and serological tests.
- Glanders: both cell-mediated and antibody-mediated responses are elicited by infection with *P. mallei*. Complement fixation, agglutination, indirect haemagglutination and counter-immunoelectrophoresis tests are used in the diagnosis of glanders. False-positive reactions may occur in areas where melioidoisis is endemic as the serological tests may detect antibodies that cross-react with those of *P. pseudomallei*.

Mallein tests are used to demonstrate the hypersensitivity developed after infection with *P. mallei*.

Table 70. Main characteristics of the pathogenic *Pseudomonas* species.

Characteristic	*P. aeruginosa*	*P. pseudomallei*	*P. mallei*
Pigment produced	++	– but colonies become orange to cream	– but colonies are yellow to brown
Odour	'fruity', grape-like	putrid becoming earthy	–
Growth on MacConkey agar	+	+	–
Growth at 5°C	–	–	–
Growth at 42°C	+	+	–
Oxidation of: glucose	+	+	+
lactose	–	+	–
Arginine dihydrolase	+	+	(+)
Reduction of nitrate to nitrite	+	+	+
Reduction of nitrate to N_2 gas	v	+	–
Motility	+	+	–

+ = positive reaction, (+) = most strains positive, v = variable reaction, – = negative.

Table 71. Differentiation of *Pseudomonas aeruginosa* and other saprophytic pseudomonads.

	Growth on MacConkey agar	Pyoverdin produced	Oxidase	Growth at 5°C	Growth at 42°C	Gelatin	Urease	Oxidation of Glu.	Oxidation of Lact.	Oxidation of Malt.	Arginine dihydrolase
P. aeruginosa	+	+	+	–	+	+	+	+	–	–	+
P. fluorescens	+	+	+	+	–	+	(+)	+	–	–	+
P. putida	+	+	+	(+)	–	–	(+)	+	–	(–)	+
P. cepacia	+	–*	+	–	(+)	+	+	+	+	+	–
P. maltophilia	+	–	(–)	–	–	+	–	(–)	–	+	–
P. stutzeri	+	–	+	(+)	(+)	–	(+)	+	–	(+)	(+)

+ = positive reaction, (+) = most strains positive, (–) = most strains negative, – = negative reaction, * = some strains produce a yellowish non-fluorescent pigment, Glu. = glucose, Lact. = lactose, Malt. = maltose.

Mallein is a glycoprotein extracted from the bacterium. In infected animals, subcutaneous inoculation of mallein (subcutaneous test) results in swelling at the injection site and fever. Instillation of mallein into the conjunctival sac (ophthalmic test) is followed in 6–12 hours by an inflammatory and purulent reaction in the eye, whereas inoculation of a small amount of mallein into the skin of the lower eyelid (intrapalpebral test) gives a localised, oedematous swelling and purulent conjunctivitis. Healed lesions of the nasal mucosa are activated in glanderous animals after mallein tests and this can be a useful diagnostic feature.

Animal inoculation

The Straus reaction is seen in male guinea-pigs inoculated intraperitoneally with infective material containing either *P. pseudomallei* or *P. mallei*. A localised peritonitis and a purulent inflammation of the testicular tunica vaginalis develops in 2–3 days.

Antibiotic susceptibility tests

These need to be carried out with *P. aeruginosa* isolates as multiple drug resistance (**306**), associated with R factors, is frequently encountered with this bacterium.

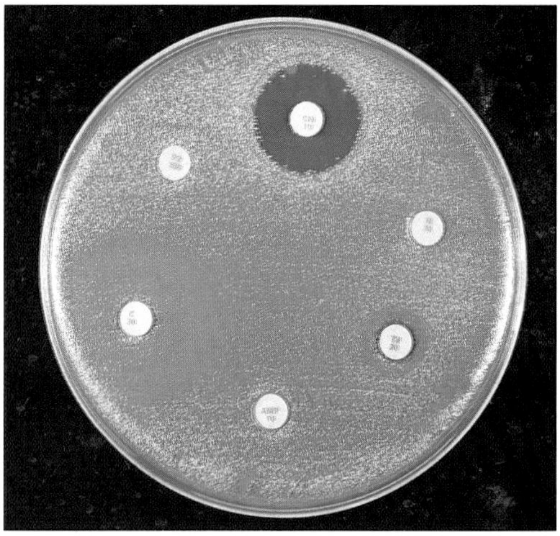

306

306 *P. aeruginosa* on Isosensitest agar demonstrating characteristic multiple resistance to antibiotics.

20 *Aeromonas, Plesiomonas* and *Vibrio* species

Members of the *Aeromonas*, *Plesiomonas* and *Vibrio* genera are Gram-negative rods (0.5–0.8 × 3.0–4.0 μm) that are either straight or curved. They are facultative anaerobes, catalase-positive and motile by polar flagella (except *A. salmonicida* which is non-motile). All ferment glucose with acid production and a few *Aeromonas* spp. also produce gas. Many of the *Vibrio* spp. require sodium chloride for growth. *Plesiomonas* lacks exoenzymes whereas *Vibrio* and *Aeromonas* spp. produce diastase, lipase, DNase and various proteinases such as gelatinase. Most of the species in the three genera grow on common laboratory media at 35–37°C. However, many of the saprophytic *Vibrio* and *Aeromonas* species have an optimum temperature for growth lower than 35°C.

Changes In Nomenclature

Previous Name	**Present Name**
Aeromonas shigelloides	*Plesiomonas shigelloides*
The microaerophilic *Vibrio* spp	*Campylobacter* spp.

Natural Habitat

Many of the species in the three genera are free-living saprophytes although some are associated with reptiles, fish and animals.

Aeromonas spp. are widespread in freshwater, sewage and soil. The numbers increase with the amount of organic matter present. *A. hydrophila* is part of the normal flora of freshwater fish and is commonly present in fish ponds and tanks. Animals can be faecal carriers of *Aeromonas* spp.

Plesiomonas shigelloides (only species in genus) is present in freshwater but distribution is limited by its 8°C minimum temperature for growth and lack of halophilism. It has been isolated from a wide host range that includes freshwater fish, shellfish, oysters, toads, snakes, monkeys, dogs, cats, goats, pigs, cattle and poultry.

Vibrio spp. can be present in both freshwater and seawater as well as in the alimentary tracts of animals and man.

Pathogenicity

A. hydrophila is an opportunistic pathogen causing disease in fish and reptiles with rare reports of infections in mammals. *A. hydrophila* can occasionally cause infections in humans that range from wound infections, septicaemia to self-limiting diarrhoea in children and food poisoning. *A. salmonicida* is an obligate parasite of salmonid fish.

P. shigelloides has been reported as a cause of gastroenteritis in man, cases occurring mainly in tropical and subtropical regions. Virulent strains have been found to produce heat-stable and heat-labile enterotoxins. It can be isolated from diagnostic specimens but its role in animal disease is uncertain .

At least five *Vibrio* spp. are human pathogens including *V. cholerae*, the cholera bacillus, and *V. parahaemolyticus* which causes food poisoning. Only *V. metschnikovii* is associated with disease in domestic animals. It causes a cholera-like disease in chickens and other birds but its geographical distribution is very limited.

V. anguillarum causes infections in many species of fish especially in salt or brackish water. It causes high mortality in salt water eels.

A summary of the diseases of veterinary importance caused by *Aeromonas*, *Plesiomonas* and *Vibrio* species is given in **Table 72**.

Laboratory Diagnosis

Specimens

Specimens include swabs and scrapings, affected tissue, faeces and mastitic milk. As *A. hydrophila* is associated with fish, great care must be taken with specimens from freshwater fish to ensure the validity of the cultural findings.

Direct microscopy

Many of the species are Gram-negative, straight rods without characteristic morphology, although some of the *Vibrio* spp. are distinctively curved and the findings from Gram-stained or DCF-smears from specimens may suggest the genus. A fluorescent antibody test has been developed for *A. salmonicida*.

Isolation (all cultures are incubated aerobically):

- *A. hydrophila:* blood agar and MacConkey agar, at 37°C for 24 hours.
- *A. salmonicida:* blood agar, Furunculosis agar (Difco) at 25°C for 48 hours.
- *V. anguillarum:* blood agar with nutrient agar base and 2.0 per cent NaCl at 20°C for 48 hours.
- *P. shigelloides*, *V. metschnikovii* and *V. parahaemolyticus*: nutrient agar or blood agar at 37°C for 24–48 hours.

Table 72. Diseases and hosts of *Aeromonas, Plesiomonas* and *Vibrio* species.

Genus and species	Host(s)	Disease
Aeromonas hydrophila	Frogs	'Red-leg disease'
	Reptiles	Necrotic stomatitis in snakes, septicaemias
	Eels, cyprinids, pike	Skin lesions
	Pike, grass carp	Swim bladder inflammation
	Cyprinids	Infectious ascites (following a viral infection) and haemorrhagic septicaemia
	Eels	'Fresh-water eel disease'
	Mammals (rare infections):	
	Dogs	Neonatal septicaemia
	Cattle	Mastitis
	Turkeys	Septicaemia
	Pigs	Diarrhoea
	Humans	Food poisoning
A. salmonicida* subsp. *salmonicida	Salmonids	Furunculosis
	Goldfish (carp)	'Ulcer disease'
Plesiomonas shigelloides	Humans	Gastroenteritis
	Animals and birds	Pathogenicity uncertain. It is isolated from clinical specimens
Vibrio parahaemolyticus	Humans	Food poisoning, associated with seafoods
V. metschnikovii	Chickens and other birds	Cholera-like enteric disease
V. anguillarum	Salt-water eels and other fish	Skin necrosis, red areas on skin and generalisation with high mortality

Selective media

- *A. hydrophila:* blood agar with 10 mg/litre ampicillin.
- *V. parahaemolyticus* will grow on TCBS agar (BBL) formulated for *V. cholerae* and other enteric vibrios. Being halophilic, it also grows on mannitol salt agar (BBL) containing 7.5 per cent NaCl intended for the pathogenic staphylococci.

Identification

Colonial morphology

- *A. hydrophila:* colonies are large (2–3 mm), flat, greyish and surrounded by a large zone of beta-haemolysis (**307**). Newly isolated strains have a pungent, foul odour. It grows well on MacConkey agar (**308**), often with pale colonies (non-lactose fermenting), but a minority of strains yield lactose-fermenting colonies.

307

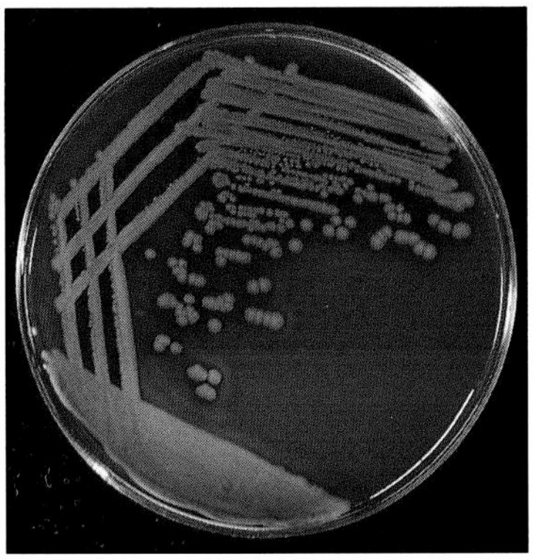

308

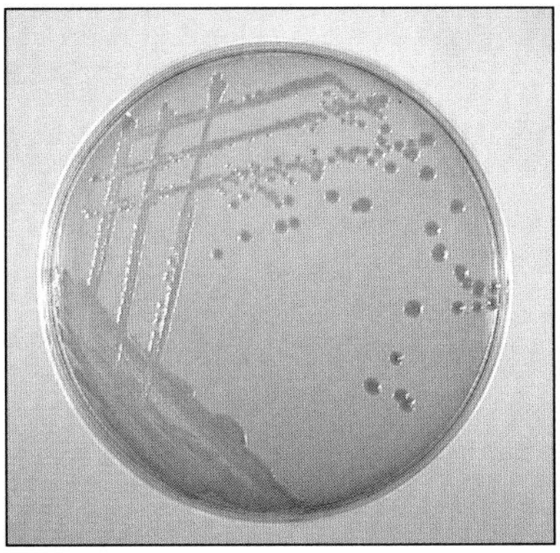

309

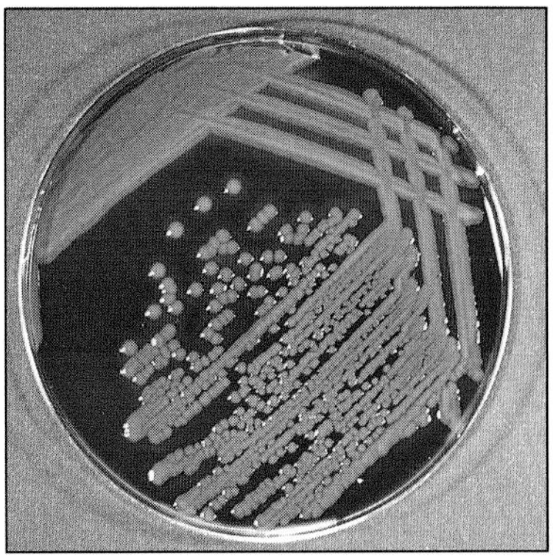

307 *Aeromonas hydrophila* on sheep blood agar, usually markedly haemolytic with large colonies after 48 hours. A putrid odour is characteristic of recent isolates.

308 *A. hydrophila* on MacConkey agar. The ability to ferment lactose is variable.

309 *Vibrio metschnikovii* on sheep blood agar showing large, glistening, haemolytic colonies.

- *A. salmonicida*: forms small colonies on blood agar that produce haemolysis after 48 hours. Brown pigment develops on Furunculosis agar (Difco) and often also on nutrient agar.
- *P. shigelloides* is non-haemolytic on blood agar and the colonies resemble those of the *Enterobacteriaceae.* It is a non-lactose fermenter on enteric media.
- *V. metschnikovii*: smooth, transparent colonies that are 2–4 mm in diameter at 48 hours. It may be haemolytic on blood agar (**309**) and grows poorly on MacConkey agar.
- *V. anguillarum*: small, smooth colonies within 48 hours. It is haemolytic on blood agar.
- *V. parahaemolyticus*: moderate-sized (about 2 mm diameter) colonies in 24 hours, non-haemolytic on sheep blood agar. It forms greenish colonies on TCBS agar (BBL) and ferments mannitol causing an acid reaction (yellow) on mannitol salt agar.

Microscopic appearance

The *Aeromonas* and *Plesiomonas* species are medium-sized, straight, Gram-negative rods although *A. salmonicida* tends to be coccoid, in pairs, chains or clusters. The *Vibrio* spp. are Gram-negative rods, curved (**310**) to a greater or lesser extent.

Biochemical and other characteristics

A presumptive diagnosis of *Aeromonas* species is based initially on a positive oxidase reaction, growth on MacConkey and fermentation of carbohydrates. *A. hydrophila* produces acid and gas from glucose, but not all *Aeromonas* spp. are able to produce gas.

P. shigelloides is DNase-negative and fails to attack gelatin. It ferments glucose, inositol, maltose and trehalose but is otherwise not very reactive in 'sugars'. *A. hydrophila* is resistant to the vibriostat O/129 (Oxoid)

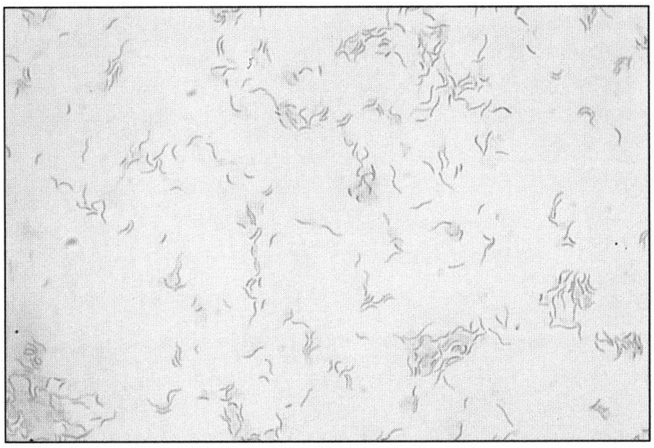
310

310 *V. metschnikovii* in a Gram-stained smear from culture showing strongly curved Gram-negative rods. (×1000)

whereas *P. shigelloides* and most *Vibrio* species are sensitive, one of the exceptions being *V. parahaemolyticus.*

V. cholerae and *V. mimicus* have only a slight requirement for Na^+(NaCl), but most of the halophilic *Vibrio* spp. require the supplementation of biochemical tests with 1 per cent NaCl. *V. parahaemolyticus* belongs to the lysine-decarboxylase-positive, arginine-dihydrolase-negative group of *Vibrio* spp. It is distinguished from other members of the group by negative reactions for sucrose, salicin and cellobiose fermentation but a positive reaction for arabinose. *V. metschnikovii* is the only clinically significant *Vibrio* sp. that is oxidase-negative and nitrate-reductase-negative.

Biochemical and other characteristics for identification of *A. hydrophila, A. salmonicida* subsp. *salmonicida, P. shigelloides, V. parahaemolyticus* and *V. metschnikovii* are given in **Table 73**.

Table 73. Characteristics of *Aeromonas, Plesiomonas* and *Vibrio* species.

Characteristics	*A. hydrophila*	*A. salmonicida* ss. *salmonicida*	*P. shigelloides*	*V. parahaemo-lyticus*	*V. metschni-kovii*
Beta-haemolysis (BA)	+	+	–	–	(+)
Motility	+	–	+	+	+
Growth with 6.5% NaCl	–	–	–	+	(+)
Exoenzymes produced	+	+	–	+	+
DNase	+	+	–	+	(+)
Gelatin	+	+	–	+	(+)
Oxidase	+	+	+	+	–
Catalase	+	+	+	+	+
Sensitive to 0/129 (150 µg)*	–	–	+	(–)	+
Indole production	+	–	+	+	(–)
Nitrate reduction	+	–	+	+	–
Urease	–	–	–	(–)	–
Aesculin hydrolysis	+	(–)	–	–	(+)
Lysine decarboxylase	+	•	+	+	(–)
Ornithine decarboxylase	–	–	+	+	–
Arginine dihydrolase	+	+	+	–	(+)
Glucose (gas)	+	+	–	–	–
Inositol	–	•	+	–	(–)
Arabinose	+	+	–	+	–
Mannitol	+	+	–	+	+
Sucrose	+	–	–	–	+
Lactose	v	–	v	–	v
Growth on MacConkey agar	+	•	+	–	–

* = 2,4-diamino-6,7 diisopropylpteridine phosphate (0/129, Oxoid), a vibriostat.
+ = positive reaction, (+) = most strains positive, v = variable reactions, (–) = most strains negative,
– = negative reaction, BA = blood agar, • = data not available.

V. anguillarum is very sensitive to the vibriostat O/129 (Oxoid) and requires a high salt concentration for growth, the optimum being between 1.5 and 3.5 per cent NaCl. Several biotypes and serotypes can be distinguished within this species. The biochemical reactions of the biotypes are shown in **Table 74**.

A. salmonicida is non-motile and four subspecies have been described. These subspecies vary in biochemical and other characteristics, some of these are given in **Table 75**. *A. salmonicida* subsp. *salmonicida* can be identified, presumptively, by the brown pigment formed on Furunculosis agar (Difco), by lack of motility, absence of gas production from glucose, a negative result in the indole test (subsp. *masoucida* is indole-positive) and positive reactions in the oxidase and catalase tests.

Table 74. Biotypes of *Vibrio anguillarum*.

	V. anguillarum biotypes		
	A	**B**	**C**
Indole	+	–	–
Glucose	+	+	+
Trehalose	+	+	+
Glycerol	+	+	–
Arabinose	+	+	–
Mannitol	+	–	+
Saccharose	+	–	+

\+ = positive reaction, – = negative reaction.

Table 75. Subspecies of *Aeromonas salmonicida*.

***A. salmonicida* subspecies**	**Brown pigment**	**Gas from glucose**	**Catalase**	**Remarks**
subsp. *salmonicida*	+	+	+	Furunculosis: typical strains
subsp. *nova*	–	–	–	Erythrodermatitis of carp
subsp. *achromogenes*	–	–	+	Isolated from salmonids and they represent atypical strains
subsp. *masoucida*	–	+	+	

\+ = positive reaction, – = negative reactions

21 *Actinobacillus* species

The *Actinobacillus* species are Gram-negative, medium-sized rods (0.3–0.5 × 0.6–1.4 μm) that can produce coccal forms. They are non-motile, non-spore-forming and non-acid-fast. A surface slime is present in the three major species (*A. lignieresii*, *A. equuli* and *A. suis*) and may be related to the stickiness of their colonies on agar. Surface cultures have low viability and die in 5–7 days. The actinobacilli ferment carbohydrates, without the production of gas, within 24 hours and most species produce urease and grow on MacConkey agar. The reactions in the oxidase and catalase tests are variable. The genus is still in a state of flux, mainly because of the close similarities between actinobacilli and species in the genera *Pasteurella* and *Haemophilus*.

Changes in Nomenclature

Previous Name	Present Name
Haemophilus pleuropneumoniae	*Actinobacillus pleuropneumoniae*
Pasteurella ureae	*Actinobacillus ureae*

Species of Uncertain Classification

'Actinobacillus seminis', the cause of ovine epididymitis in Australia, New Zealand, South Africa, the USA and more recently in the UK (Heath *et al.*, 1991), is unreactive in conventional biochemical tests and is not closely related to other actinobacilli.

Actinobacillus salpingitidis, isolated from the reproductive tract of hens, has been shown by DNA hybridisation studies not to be closely related to the other *Actinobacillus* species.

Actinobacillus actinoides, which produces pneumonia in calves and seminal vesiculitis in bulls, has similarities to *Haemophilus somnus* and has now been excluded from the actinobacilli.

Natural Habitat

The actinobacilli are commensals on the mucous membranes of their hosts. With the exception of *A. seminis* the geographical distribution of the *Actinobacillus* species is worldwide.

Pathogenesis

The actual pathogenic mechanisms of actinobacilli are unknown. Pyogranulomatous lesions are formed by *A. lignieresii* in cattle, and less commonly in sheep, that resemble those of *Actinomyces bovis*. But unlike actinomycosis, bovine actinobacillosis is spread by the lymphatics. The specific disease is wooden (timber) tongue but granulomatous lesions can also involve the skin and underlying tissues, usually of the head, neck and limbs. Less commonly, the lungs and other internal organs are affected. Small greyish-white granules (about 1 mm in diameter) are present in exudates from lesions. If these granules are crushed on a slide and stained, club colonies are seen consisting of club-like processes of calcium phosphate with the Gram-negative rods of *A. lignieresii* in the centre. If *A. lignieresii* produces a granulomatous lesion in the soft tissues of the jaw area this, clinically, can be difficult to distinguish from lumpy jaw (*A. bovis*). **Table 76** summarises the differential features of the two conditions.

A. capsulatus causes arthritis in rabbits. *A. equuli* and *A. suis* are responsible for enteritis, suppurative, multifocal nephritis and arthritis in both horses and pigs. *A. pleuropneumoniae* is the cause of acute, subacute and chronic respiratory infections in pigs. It is thought that immune complexes may damage the endothelium of blood vessels and this results in vasculitis and thrombosis with consequent oedema, necrosis, infarction and haemorrhage. A thermolabile cytotoxin is produced that damages porcine macrophages and blood monocytes.

Table 77 summarises the hosts and diseases of the actinobacilli affecting animals.

Laboratory Diagnosis

Specimens

Specimens should include pus, exudates from lesions, tissue biopsies, pneumonic lesions and biopsies of granulomatous material fresh and in 10 per cent formalin for histopathology.

Direct microscopy

This is worth while only in *A. lignieresii* infections where exudates are available. Pus or exudates are washed with distilled water in a Petri dish to reveal the small greyish-white granules. A few of these granules are placed in a drop of 10 per cent KOH on a microscope slide and crushed gently with a coverslip. The structures of the clubs that surround the club colonies can be seen under the low-power objective. Stained histopathological sections (**311**) also demonstrate the characteristic clubs and club colonies. If a Gram-stained smear is made from the crushed granules, the presence of medium-sized Gram-negative rods would suggest actinobacillosis rather than actinomycosis.

Table 76. Differentiation of *Actinobacillus lignieresii* and *Actinomyces bovis* infections in cattle.

Characteristics	***Actinobacillus lignieresii***	***Actinomyces bovis***
Specific disease	Bovine actinobacillosis Wooden (timber) tongue	Bovine actinomycosis Lumpy jaw
Granulomatous abscesses	Jaw, head, neck and limbs	Jaw region
Granules in exudates	Greyish-white, about 1 mm	Yellow 'sulphur granules' about 1–3 mm
Club colonies	+	+
Spread via lymphatics	+	–
Bone affected (osteomyelitis)	Uncommon	Common
Gram-stain reaction	Gram-negative rods	Gram-positive branching filaments or diphtheroidal forms
Atmospheric requirements	Growth in air (facultative anaerobe)	Anaerobic (H_2+ CO_2)

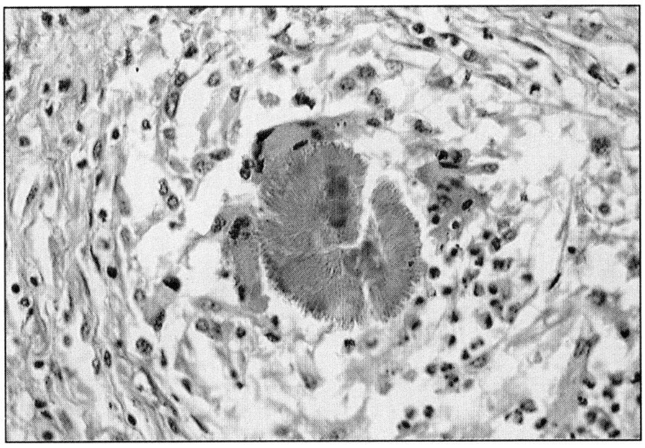

311

311 Club colonies (yellow) in a histopathological section from a case of bovine actinobacillosis (*Actinobacillus lignieresii*). (H&E stain, ×400)

Isolation

Most of the actinobacilli can be cultured on sheep or ox blood agar and some will grow on MacConkey agar. However, *A. pleuropneumoniae* benefits from factor V that can be provided by chocolate agar or a staphylococcal streak on blood agar. The growth of all the actinobacilli, and particularly *A. pleuropneumoniae*, is improved by 5–10 per cent CO_2 (a candle-jar is satisfactory). The inoculated plates are incubated at 37°C for 24–72 hours.

Table 77. Diseases caused by the pathogenic actinobacilli.

Species	***Host(s)***	***Disease***
Actinobacillus lignieresii	Cattle	Bovine actinobacillosis: 'wooden (timber) tongue' and pyogranulomatous lesions around the head, neck and limbs and occasionally in the internal organs
	Sheep	Pyogranulomatous lesions usually in head and neck region
	Pigs	Rare granulomatous abscesses in mammary glands
A. equuli	Neonatal foals	Sleepy foal disease (shigellosis or viscosum infection): septicaemia and those foals that survive for a few days develop purulent nephritis, pneumonia and arthritis
	Postnatal foals	Purulent arthritis, enteritis and infection of aneurysms
	Mares	Occasional abortion and/or septicaemia
	Pigs	Arthritis in piglets and endocarditis in older pigs
	Calves, dogs, rabbits and rats	Occasional infections
A. suis	Pigs under 3 months-old	Septicaemia
	Older pigs	Arthritis, pneumonia and pericarditis
	Neonatal and postnatal foals	Some strains cause syndromes similar to those in foals associated with *A. equuli*
A. pleuropneumoniae	Pigs	Acute: severe fibrinous pleuropneumonia. Chronic: pleuritis and pulmonary sequestration and abscessation. Morbidity and mortality high in non-immune pigs
A. capsulatus	Rabbits	Arthritis
A. actinomycetem-comitans	Humans	Role in periodontal disease and endocarditis
	Rams	Associated with epididymitis
'A. seminis'	Rams	Epididymitis: relatively common in New Zealand, Australia and South Africa; cases reported in Britain and USA
A. ureae	Humans	Upper respiratory tract infections
	Sows	Reported as a cause of abortion

Identification

Colonial appearance

- *A. lignieresii*: small, glistening colonies develop in 24 hours (**312**). They are usually slightly sticky (viscid) on primary isolation but lose this characteristic on subculture. The colonies are non-haemolytic and develop to about 2 mm in diameter in 48 hours. The organism grows well on MacConkey agar, the colonies are at first pale but become pinkish as *A. lignieresii* is a late lactose-fermenter (**313**).
- *A. equuli*: some strains are haemolytic and the colonies are sticky with this feature remaining on subculture (**314**). It is a lactose-fermenter on MacConkey agar (**313**).
- *A. suis*: all strains are haemolytic with colonies similar to those of *A. lignieresii* but more sticky. It grows well on MacConkey agar.
- *A. pleuropneumoniae*: small (1 mm) colonies surrounded by a zone of beta-haemolysis, which somewhat resemble those of a beta-haemolytic streptococcus (**315**). Two distinct colony forms are possible, one waxy and the other a soft glistening type. No growth occurs on MacConkey agar.
- *A. capsulatus*: the cells are capsulated and very sticky colonies are produced on blood agar. It grows well on MacConkey agar.
- *A. actinomycetemcomitans*: the colonies are small (1 mm after 2–3 days), adherent to the medium and difficult to break-up. Colonies are described as 'star-like'. No growth occurs on MacConkey agar.
- *'A. seminis'*: small, round, pinpoint colonies that are greyish-white, non-haemolytic and appear after 24–48 hours on blood agar. No growth occurs on MacConkey agar.
- *A. ureae*: colonies on blood agar are mucoid and are usually accompanied by some greening of the medium.

312

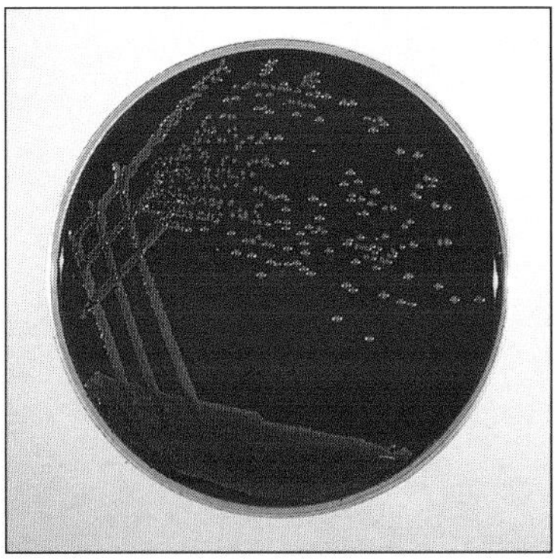

312 *A. lignieresii* on sheep blood agar. On primary isolation the colonies are non-haemolytic, shiny and slightly sticky but this property is lost on subculture.

313

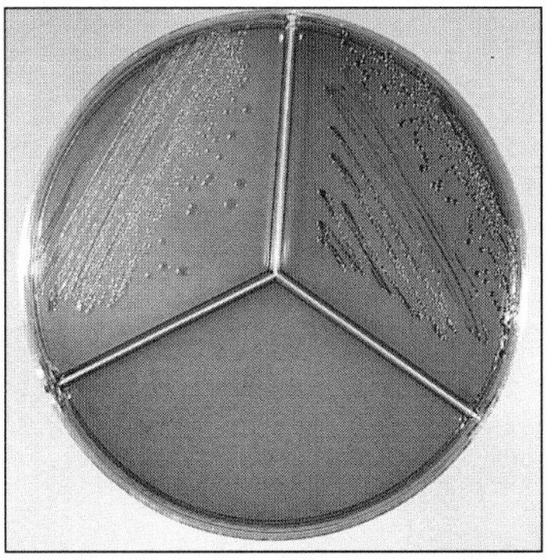

313 *A. lignieresii* (left) and *A. equuli* (right) on MacConkey agar. *A. pleuropneumoniae* (inoculated, bottom) is unable to grow on this medium. *A. equuli* ferments lactose but *A. lignieresii* gives a late reaction.

314

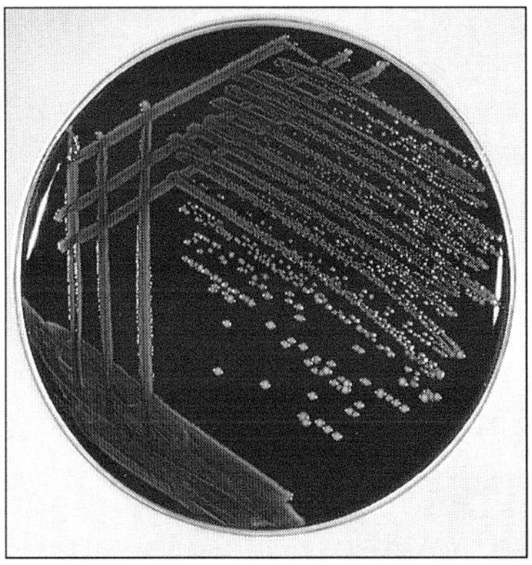

314 *A. equuli* on sheep blood agar. Colonies are viscid and remain so on subculture.

315

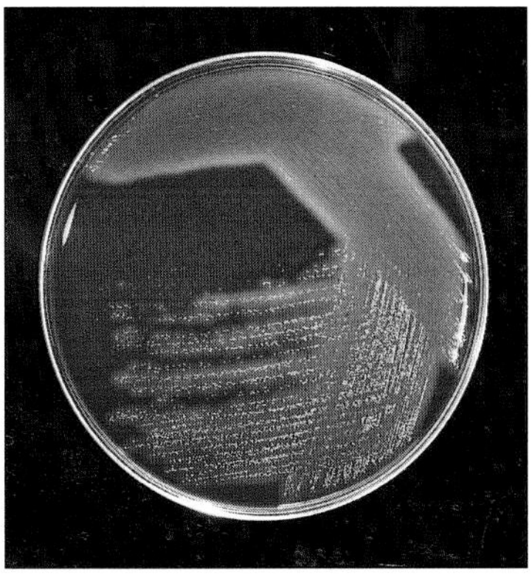

315 *A. pleuropneumoniae* on sheep blood agar. The small colonies are surrounded by a narrow zone of beta-haemolysis.

Microscopic appearance

All actinobacilli are Gram-negative rods or coccobacilli (**316**).

Biochemical reactions

A. pleuropneumoniae enhances the haemolysis caused by the beta-haemolysin of *Staphylococcus aureus* in a CAMP test (**317**). The differential characteristics of the actinobacilli are shown in **Table 78**. '*A. seminis*' is usually catalase-positive but oxidase-negative, and almost completely unreactive in conventionally performed biochemical tests (Phillips, 1984). A few strains show slight acid production, after incubation periods of up to 28 days, in arabinose, fructose, mannose, trehalose, glucose, maltose, mannitol and xylose. Most strains are reported as being nitrate, H_2S, MR, VP, indole, urease, phosphatase, gelatinase and citrate negative. Heath *et al.* (1991) used a combination of API 20NE, API 20E and API 50CHS (Analytab Products) identification systems to test four field isolates of *A. seminis* from rams and *A. seminis* NCTC 10851. Normal inocula were used for the API 20NE strip but heavy inocula, with or without supplementation with serum, were used for the other two systems. In this case the field isolates and type strain of *A. seminis* were apparently positive to a considerable number of biochemical tests.

316

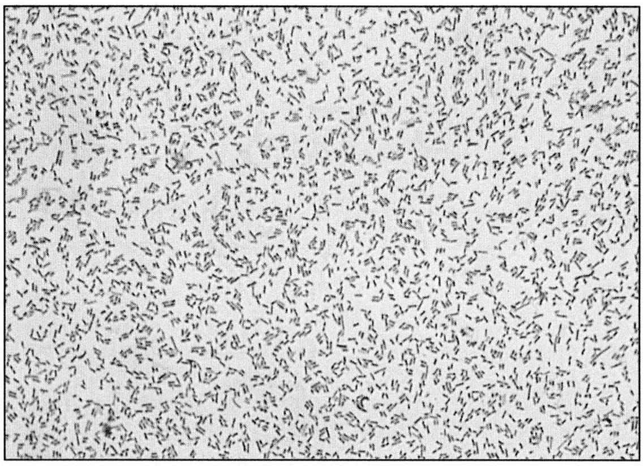

316 A Gram-stained smear showing the medium-sized Gram-negative rods of *A. lignieresii*, representative of the genus. (×1000)

317

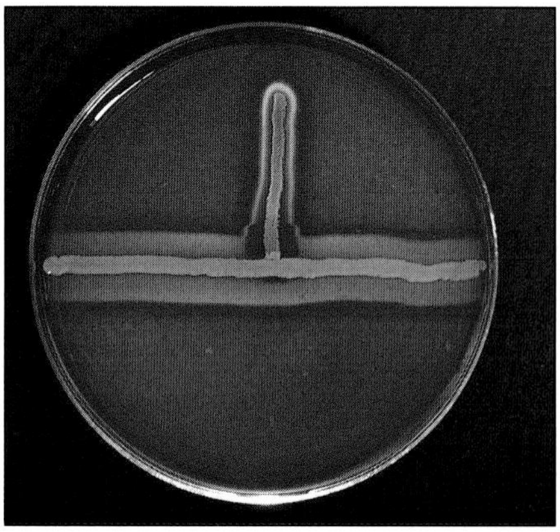

317 CAMP test with *A. pleuropneumoniae* against *Staphylococcus aureus* (horizontal streak) showing enhancement of the haemolytic effect of the staphylococcal beta-haemolysin.

Table 78. Differential characteristics of the actinobacilli.

		A. lignieresii	*A. equuli*	*A. suis*	*A. pleuropneumoniae*	*A. capsulatus*	*A. actinomycetemcomitans*	*A. ureae*	'*A. seminis*'
Haemolysis (sheep blood agar)		–	v	+	+	–	–	+(α)	–
CAMP (*Staphylococcus aureus*)		–	–	–	+	–	–	–	–
Growth on MacConkey agar		+	+	+	–	+	–	–	–
Catalase		v+	v–	+	v+	+	+	+	–
Oxidase		+	v	v	v	+	+	+	–
Hydrolysis of aesculin		–	–	+	–	(+)	–	–	–
Urease		+	+	+	+	+	+	+	–
Acid from:	Arabinose	(+)	–	+	–	+	–	–	(+)
	Lactose	(+)	+	+	–	+	–	–	–
	Maltose	+	+	+	+	+	+	+	–
	Mannitol	+	+	–*	–	+	v	+	–
	Melibiose	–	+	+	–	+	–	–	–
	Sucrose	+	+	+	+	+	–	+	–
	Trehalose	–	+	+	–	+	–	–	–

v = variable reactions, v+ = variable, most positive, v– = variable, most negative, (+) = late reaction, * = *A. suis* strains from horses can be mannitol positive.

References

Heath, P.J., Davies, I.H., Morgan, J.H. and Aitken, I.A. (1991). Isolation of *Actinobacillus seminis* from rams in the United Kingdom, *Veterinary Record*, **129**: 304–307.

Phillips, J.E. (1984). *Actinobacillus*, pp. 574–575, In Krieg, N.R. and Holt, J.G. (ed.), *Bergey's Manual of Systematic Bacteriology*, Vol. 1. Williams and Wilkins Co., Baltimore, USA.

22 *Pasteurella* species

The pasteurellae are small (0.2 μm by up to 2.0 μm), Gram-negative rods or coccobacilli. They are non-motile, non-sporing, facultatively anaerobic, fermentative (except for *P. anatipestifer*), oxidase-positive and catalase-positive (except for *P. caballi*). Although unenriched media support their growth, they grow best on media supplemented with serum or blood. The principal hosts and diseases associated with *Pasteurella* spp. are listed in **Table 79**.

Recent Changes in Nomenclature

Previous name	**Present name**
Pasteurella ureae	*Actinobacillus ureae*
Haemophilus avium	*Pasteurella avium*
Pasteurella pneumotropica (Henriksen biotype)	*Pasteurella dagmatis*

Natural Habitat

Pasteurella species are worldwide in distribution with a wide spectrum of hosts. Most are commensals on the mucous membranes of the upper respiratory and intestinal tracts of animals. The carrier rate for different species varies greatly.

Pathogenesis

The mechanisms of disease production by the pasteurellae are not fully understood. Endotoxins are particularly important in the septicaemic diseases such as fowl cholera and bovine haemorrhagic septicaemia. Infections may be endogenous or exogenous. The portal of entry is usually via the respiratory tract and virulence is enhanced by animal-to-animal transmission as in pneumonic pasteurellosis. All *Pasteurella* spp. are probably extracellular parasites with various stresses, including concurrent viral infections, predisposing to infection as in 'shipping fever'. The thermolabile dermonecrotoxin produced by some of the type D strains of *P. multocida* (AR^+ strains) have an important role in atrophic rhinitis of pigs. *Bordetella bronchiseptica* infection of the turbinates appears to facilitate colonisation by the type D strains. *P. haemolytica* produces a soluble cytotoxin (leukotoxin) that has a role in breaching the lung's primary defence mechanism by its action on the alveolar macrophages and other leukocytes of ruminants. In some chronic *Pasteurella* infections it is thought that immune complexes may contribute to the lesions.

Serogroups or Types of *Pasteurella* spp.

As indicated in **Table 79**, types or serogroups of *P. multocida* have been identified based on differences in capsular substances (polysaccharides). They have been designated A, B, D, E and F. Somatic types (lipopolysaccharides) have also been determined and given numbers. A serotype is identified by its serogroup followed by its somatic type, for example the cause of bovine haemorrhagic septicaemia is B:6 in Southeast Asia and E:6 in Africa. *P. haemolytica* has analogous capsular types that are identified by numbers. The two biotypes of *P. haemolytica* are based upon differences in a number of characteristics including pathogenicity, antigenic nature and biochemical activity (for example, biotype A ferments arabinose and biotype T ferments trehalose).

Laboratory Diagnosis

Specimens

The specimens required depend on the animal species and on the disease syndrome. Portions of pneumonic lung should be taken from the edge of the lesion. In septicaemias, pieces of liver, spleen, kidney and lymph nodes could be submitted. From live animals the specimens might include pus, exudates, nasal swabs, bronchial lavages and mastitic milk.

Direct microscopy

The small, Gram-negative rods or coccobacilli are not always readily discernible in Gram-stained smears from affected tissues. In septicaemias, such as fowl cholera, distinctive bipolar-staining pasteurellae can be seen in Giemsa-stained or Leishman-stained smears (**318**).

Isolation

The routine medium for the isolation of *Pasteurella* spp. is ox or sheep blood agar. Clinical materials should be inoculated on both blood and MacConkey agars. Selective medium containing clindamycin (2 μg/ml) should be used for the isolation of *P. multocida* from porcine nasal swabs. The plates are incubated aerobically at 37°C for 24–48 hours. *P. anatipestifer* grows best on blood or serum agar under 5–10 per cent CO_2 (a candle-jar is satisfactory).

Animal inoculation

Intraperitoneal inoculation of mice is sometimes necessary to recover *P. multocida* from clinical specimens that contain large numbers of other bacteria.

Table 79. Principal hosts and diseases of the *Pasturella* species.

Species	***Principal host(s)***	***Disease and commensal status***
P. multocida		
type A	Cattle	Part of 'shipping fever' complex Part of the 'enzootic pneumonia' complex in calves Occasional but severe mastitis
	Sheep	Pleuropneumonia Mastitis
	Pigs	Pneumonia (often secondary)
	Rabbits	One cause of 'snuffles', pleuropneumonia, abscesses, otitis media, conjunctivitis and genital infections
	Poultry	Fowl cholera (primary infection)
	Many domestic and wild animals	Pneumonia and other infections in stressed animals Commensals in respiratory and digestive tract
type B	Cattle, water buffalo, bison, yak and other ruminants	Epizootic haemorrhagic septicaemia (primary infection) Nasopharynx of carrier animals South-East Asia and other countries
type D	Pigs	Atrophic rhinitis (with or without *Bordetella bronchiseptica*)
	Pigs, less commonly other domestic animals and poultry	Pneumonia (usually secondary)
type E	Cattle and water buffalo	Epizootic haemorrhagic septicaemia (primary infection) Africa only
type F	Turkeys mainly	Role in disease not clear
P. haemolytica		
biotype A	Cattle	Part of 'shipping fever' complex Pneumonia (primary or secondary)
	Sheep	Enzootic pneumonia (primary or secondary). Septicaemia in lambs under 3-months-old. Gangrenous mastitis ('blue bag')
biotype T	Sheep	Septicaemia in lambs 5–12 months-old
P. pneumotropica	Rodents	Pneumonia (secondary) and abscesses (possibly from bites) Commensal in nasopharynx
	Dogs and cats	Present in the nasopharynx of some animals. Not a significant pathogen
P. canis	Dogs (man)	Recovered from the oral cavity of dogs and from dog bites in humans
P. dagmatis	Dogs, cats (man)	Present in the oral cavity and intestinal tract of dogs and cats and recovered from dog and cat bites in humans
P. stomatis	Dogs, cats	Isolated from the respiratory tract but not usually pathogenic
P. caballi	Horses	Respiratory infections including pneumonia
P. aerogenes	Pigs (man)	Commensal in intestine of pigs. Rarely pathogenic but one isolate associated with abortion. Abscesses from pig bites in humans
P. anatipestifer	Ducklings mainly but also chickens, turkeys, pheasants, water fowl	'New duck disease': severe fibrinous polyserositis in ducklings 1–8 weeks old
P. anatis	Ducks	Recovered from intestines. Pathogenicity not demonstrated
P. gallinarum	Poultry	Commensal in respiratory mucosa. Occasional low grade infections
P. avium, *P. langaa* *P. volantium*	Chickens	Isolated from the respiratory tract of healthy birds Pathogenicity not demonstrated for any of the three species
P. testudinis	Turtles, tortoises	Associated with abscesses. Possibly opportunistic
P. granulomatis	Cattle (Brazil)	Fibrogranulomatous disease

318

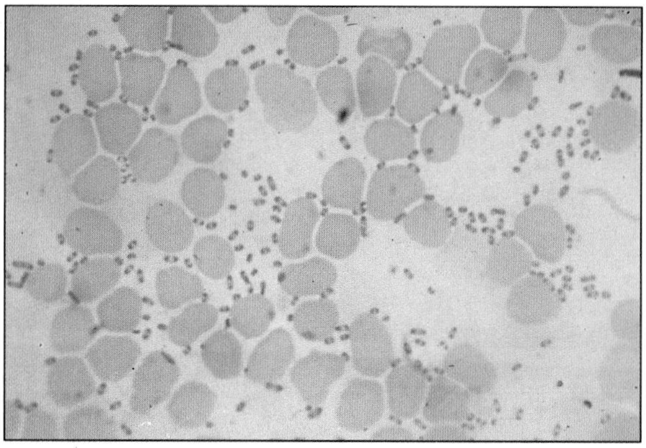

318 *Pasteurella multocida* in a bovine blood smear from a case of haemorrhagic septicaemia showing the characteristic bipolar staining. (Leishman stain, ×1000)

319

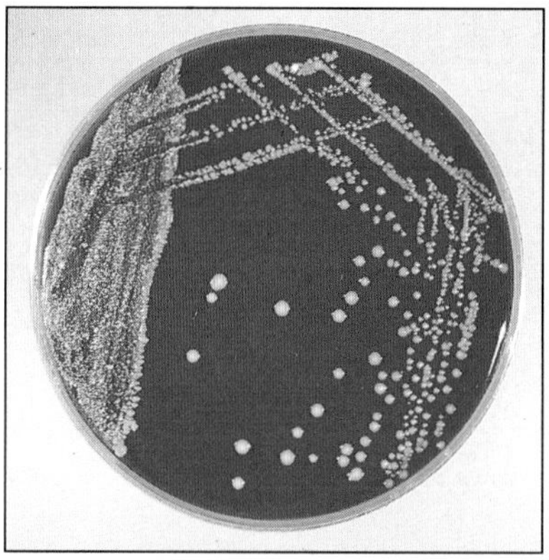

319 *P. pneumotropica* on sheep blood agar.

Identification

Colonial morphology

The colonies of all species are usually evident in 24 hours. They are of moderate size, round and greyish. *P. anatipestifer* produces small 'dewdrop' colonies within 48 hours. The colonies of *P. pneumotropica* are non-haemolytic (**319**) and somewhat similar to those of *P. multocida* (**320** and **321**). Type A strains of *P. multocida* often produce relatively large, mucoid colonies due to their large capsules of hyaluronic acid (**322**). *P. multocida* has a characteristic 'sweetish' odour, is non-haemolytic, does not grow on MacConkey agar and is a good indole producer. *P. haemolytica* is beta-haemolytic (**323**) and usually tolerates the bile salts in MacConkey agar to grow as pinpoint red colonies (**324**). It has no odour and does not produce indole. *P. testudinis* and *P. granulomatis* are also haemolytic on blood agar.

320

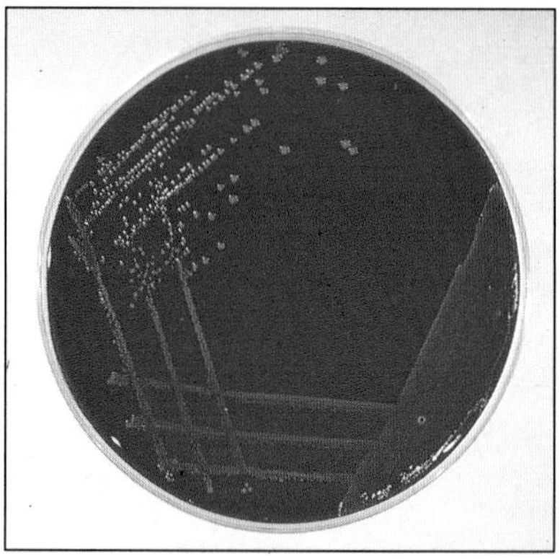

320 *P. multocida* on sheep blood agar. The colonies are non-haemolytic and have a characteristic sweetish odour.

321

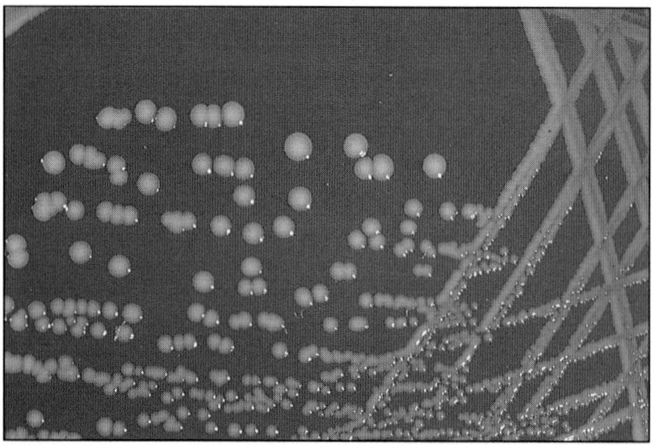

321 A close-up of the colonies of *P. multocida* shown in **320**.

322

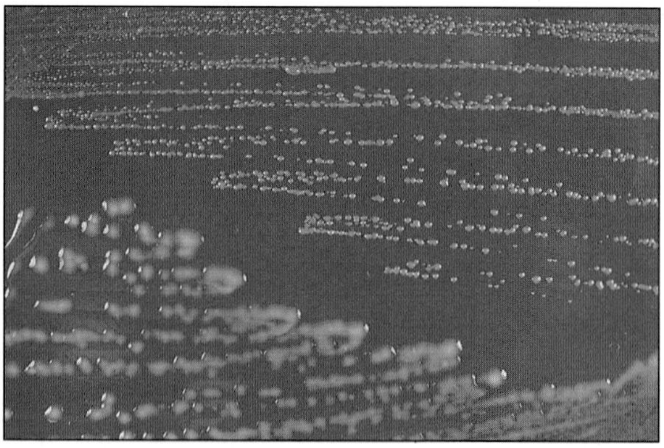

322 Comparison of the colonial types of *P. multocida*. The non-mucoid strain (top) of low virulence was isolated from a dog, while the mucoid colonies (bottom), are those of a virulent type A strain from a pig.

323

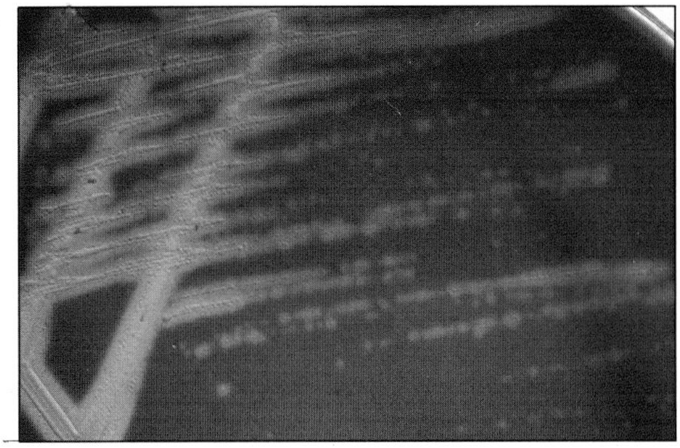

323 *P. haemolytica* on sheep blood agar isolated from the pneumonic lung of a lamb showing small colonies surrounded by a narrow zone of beta-haemolysis.

324

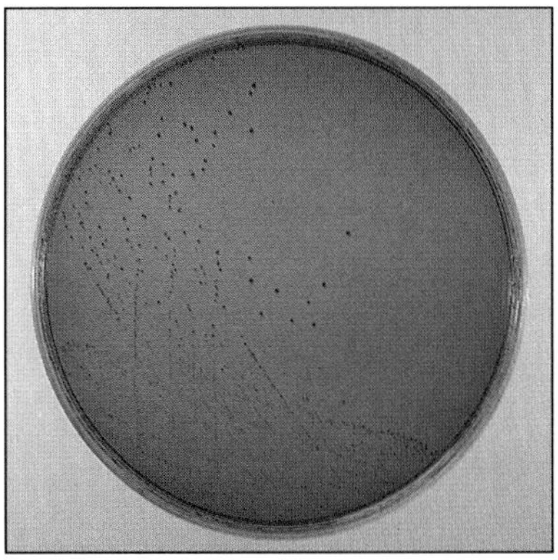

324 *P. haemolytica* on MacConkey agar: the small, red, pin-point colonies indicate a tolerance of the bile salts in the medium.

Microscopic appearance

Smears from colonies reveal small, Gram-negative rods or coccobacilli (**325**). *Pasteurella* spp. are non-motile.

Biochemical reactions

Characteristic colonies yielding small, Gram-negative rods or coccobacilli that are oxidase-positive (the enterobacteria are oxidase-negative) and catalase-positive (except for *P. caballi*) are inoculated into TSI slopes. The usual reaction is a yellow slant and butt with no gas or H_2S production. The suspected *Pasteurella* sp. can then be inoculated into differential media to determine the identifying characteristics listed in **Table 80**. This table gives the reactions of the most important species. *P. anatipestifer* is a non-fermenter and may eventually be reclassified in another genus. It is non-haemolytic, does not grow on MacConkey agar and is indole and urease-negative.

Identification of the proposed subtypes of *P. multocida*

- *P. multocida* subsp. *multocida:* strains that cause significant disease in domestic animals.
- *P. multocida* subsp. *septica:* strains recovered from various sources including dogs, cats, birds and man.
- *P. multocida* subsp. *gallicida:* strains recovered from birds that may occasionally cause fowl cholera.

The three subspecies are differentiated by minor differences in their fermentation of carbohydrates (**Table 81**). The identification of the subspecies may be of use in epidemiological studies rather than for routine diagnostic purposes.

325

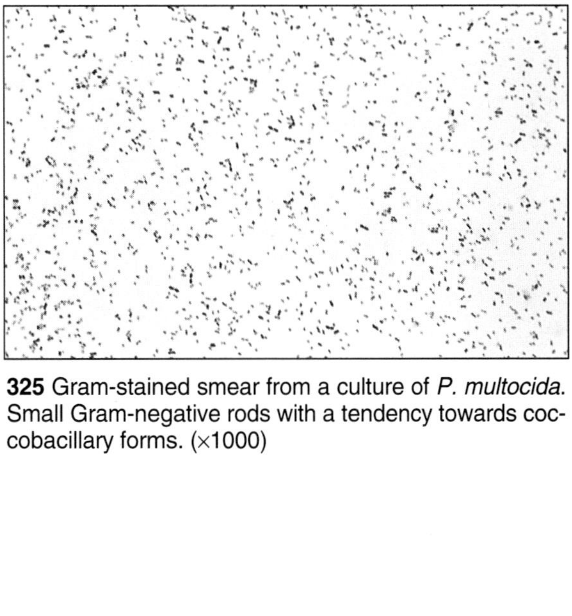

325 Gram-stained smear from a culture of *P. multocida.* Small Gram-negative rods with a tendency towards coccobacillary forms. (×1000)

Serotyping of *P. multocida* and *P. haemolytica*

Serotypes of these two pasteurellae can be identified by some reference laboratories.

Table 80. Differentiation of important *Pasteurella* species.

Species	Beta-haemolysis	Growth on MacConkey agar	Indole	Urease	Ornithine decarboxylase	Acid (24–48 hrs) from: Glucose	Lactose	Sucrose	Maltose	Mannitol
P. multocida	–	–	+	–	(+)	+	(–)	+	(–)	(+)
P. haemolytica	+	(+)	–	–	(–)	+	(+)	+	+	+
P. pneumo - tropica	–	(–)	(+)	+	+	+	(+)	+	+	–
P. canis	–	–	+/–	–	+	+	NA	NA	–	–
P. dagmatis	–	–	+	+	–	+*	–	–	+	–
P. caballi	–	–	–	–	+/–	+*	+	NA	(+)	+
P. aerogenes	–	+	–	+	(+)	+*	(–)	+	+	–
P. gallinarum	–	(–)	–	–	+/–	+	–	+	+	–

(+) = most positive; (–) = most negative; +/– = positive and negative strains; NA = data not available; +* = gas and acid from glucose.

Table 81. Differentiation of the *Pasteurella multocida* subspecies.

Subspecies	Fermentation of: Trehalose	D-Xylose	L-Arabinose	Sorbitol	Dulcitol
P. multocida subsp. *multocida*	v	v	–	+	–
P. multocida subsp. *septica*	+	+	–	–	–
P. multocida subsp. *gallicida*	–	+	v	+	+

v = variable reactions.

Antibiotic susceptibility tests

Most strains of *P. multocida* are susceptible to penicillin (**326**). However, antibiotic susceptibility tests should be carried out on strains of *P. multocida* and *P. haemolytica* as plasmid-based resistance to sulphonamides and some commonly used antibiotics has been widely encountered.

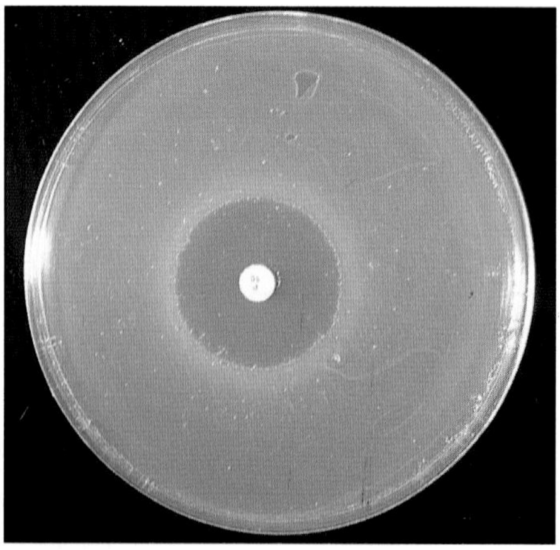

326

326 *P. multocida* on Isosensitest agar showing susceptibility to penicillin (uncommon for Gram-negative bacteria).

23 *Francisella tularensis*

Francisella tularensis consists of two biotypes, the more virulent type A, referred to as *tularensis* and the milder type B known as *palaeartica*. Both are small, pleomorphic, non-motile, Gram-negative rods or coccobacilli. Individual cells are less than 1 µm in any dimension. *F. tularensis* is a strict aerobe that grows best on blood agar supplemented with cystine. It is oxidase-negative, catalase-positive and non-fermentative. The biotypes differ in host specificity, biochemical activity and geographical distribution.

Natural Habitat

F. tularensis occurs mainly in the northern hemisphere with biotype A, *tularensis*, predominating in North America and biotype B, *palaeartica*, in Eurasia. Wild animals are the reservoir of infection, especially rabbits and hares but also beavers, muskrats, squirrels, woodchucks, opossums, skunks, deer and foxes. *F. tularensis* is most frequently transmitted by any one of a large range of biting arthropods including flies, mosquitoes, lice and ticks. The methods of transmission to domestic animals and to man are summarised in **Table 82**.

Pathogenesis

An appreciable number of infections in farm animals can occur in some regions, especially in sheep. Tularaemia is rarely seen in dogs but cats are more susceptible and there are reports of transmission from cats to humans. Domestic fowl can act as a reservoir of infection. *F. tularensis* is highly invasive and after infection a bacteraemia develops with localisation and granuloma formation in parenchymatous organs and lymph nodes. The characteristic gross lesions seen in rabbits and other wild animals are small necrotic granulomatous foci in the spleen, liver and lymph nodes.

Laboratory Diagnosis

F. tularensis is highly infectious for humans. All work with material from suspected tularaemia cases must be carried out in a biological safety cabinet or alternatively sent to a laboratory equipped to handle this dangerous pathogen.

Specimens

Portions of liver, spleen and lymph nodes are selected. Representative portions should be frozen at –30°C or –70°C for possible future reference. Blood samples for serology are taken from living animals.

Direct microscopy

The bacterium is very small and does not show up well in Gram-stained smears. Direct or indirect fluorescent antibody (FA) staining of smears from lesions can be carried out in reference laboratories.

Isolation

Material from specimens is plated on blood agar with and without cystine. MacConkey agar plates are inoculated in parallel in order to detect any Gram-negative pathogens or contaminants. The inoculated plates are incubated at 37°C, aerobically for at least 7 days. *F. tularensis* is indifferent to the presence or absence of carbon dioxide.

Identification

Colonial morphology

F. tularensis grows slowly requiring 2–4 days for maximum colony size. After 48 hours, the colonies are minute and dew-drop in appearance (**327**). On further incubation the colonies tend to coalesce and a greenish discoloration is evident around them.

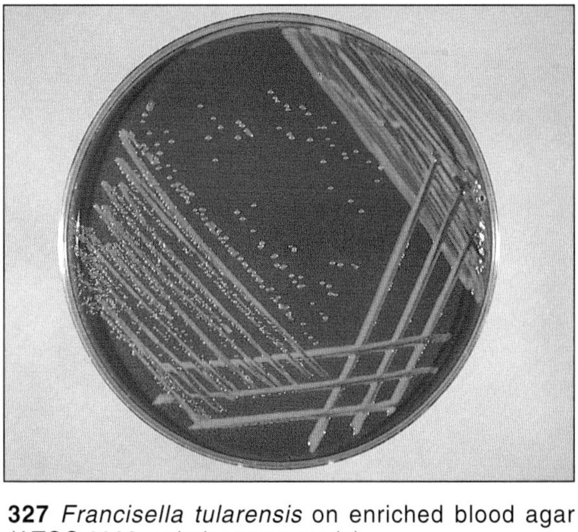

327

327 *Francisella tularensis* on enriched blood agar (ATCC 6223, avirulent type strain).

Table 82. Transmission of *Francisella tularensis.*

Host(s)	Transmission
Sheep and other farm animals	Biting arthropods including *Dermacentor andersoni* (wood tick)
Cats	• Infected rabbits and rodents • Bites from other cats
Man	• Skin inoculation from handling infected wild animals, especially rabbits. This is the most common method • Ingestion of contaminated meat or water • Bites from arthropods • Bites from infected dogs or cats

Microscopic appearance
Gram-stained smears from colonies reveal small Gram-negative rods and coccobacilli.

Biochemical characteristics
F. tularensis is oxidase-negative and weakly catalase-positive. Biochemical characterisation is difficult and not necessary for identification. Differentiation of the two biotypes is carried out in reference laboratories.

Immunological techniques for antigen
Immunological techniques include:
a) FA technique on smears from the cultures;
b) A slide agglutination test on cultures using known specific antiserum.

Serology
Agglutination antibody titres of 1:40–1:60 indicate infection in dogs and cats. These are attained about 2 weeks after the initial infection.

Animal inoculation
Guinea-pig inoculation with tissue homogenates is recommended only if specimens are so heavily contaminated that isolation cannot be made on culture media. Serum can be obtained from the guinea-pig about 2 weeks after inoculation. The agglutination titres and interpretation are similar to those for dogs and cats.

24 *Brucella* species

Brucella species are small Gram-negative rods (0.5–0.7 × 0.6–1.5 μm) that often appear coccobacillary. They are non-motile, non-spore-forming and partially acid-fast in that they are not decolourised by 0.5 per cent acetic acid in the modified Ziehl–Neelsen (MZN) stain. The carbol fuchsin is retained and the brucellae appear as red-staining coccobacilli. The brucellae are aerobic and capnophilic (carboxyphilic), catalase-positive, oxidase-positive (except *B. ovis* and *B. neotomae*), urease-positive (except *B. ovis*) and will not grow on MacConkey agar. *B. ovis* and *B. abortus* biotype 2 require media enriched with blood or serum. The growth of the other brucellae is enhanced on enriched media but they are able to grow on nutrient agar.

Natural Habitat

Brucella species are obligate parasites and each species has a preferred natural host that serves as a reservoir of infection. Brucellae have a predilection for ungulate placentas, foetal fluids and testes of bulls, rams, boars and dogs. *B. abortus* is excreted in bovine milk and can remain viable in milk, water and damp soil for up to 4 months.

Pathogenesis

Transmission is by direct or indirect contact with infective excretors. The route of infection is often by ingestion but venereal transmission may occur and is the main route for *B. ovis.* Less commonly infection may occur *in utero,* via conjunctiva or by inhalation. The brucellae possess an endotoxin that contributes to the pathogenesis, as does a surface cell wall carbohydrate that is responsible for binding to B-lymphocytes. Soon after entry into the host the brucellae are engulfed by phagocytic cells in which they survive, multiply and are transported to the regional lymph nodes. The organisms pass to the thoracic duct and then via the bloodstream to parenchymatous organs and other tissues such as joints. Granulomatous foci can develop in tissue with occasional suppuration and caseation.

Brucellosis is essentially a disease of the sexually mature animal, the predilection sites being the reproductive tracts of males and females, especially the pregnant uterus. Allantoic factors stimulate the growth of most brucellae. These factors include erythritol, possibly steroid hormones and other substances. Erythritol is present in the placenta and male genital tract of cattle, sheep, goats and pigs but not humans. Erythritol does not stimulate the growth of *B. ovis* and inhibits *B. abortus* strain 19, the attenuated vaccinal strain. A pyogranulomatous reaction occurs in affected placentae and abortion occurs from mid-gestation onwards. Apparently normal, but infected neonates can be born but the infection is of limited duration in these animals. Females usually abort only once, after which a degree of immunity develops, and the animals remain infected and large numbers of brucellae can be excreted in foetal fluids at subsequent parturitions. Permanent infertility may occur in male dogs infected with *B. canis.*

Humans can be infected by all the *Brucella* species, except *B. ovis* and the non-pathogenic *B. neotomae,* and develop undulant fever. The manifestations are an undulating pyrexia, malaise, fatigue, night sweats, muscle and joint pains, but not abortion. Osteomyelitis is the most common complication. Humans can also develop a hypersensitivity to the antigens of both virulent *B. abortus* and the vaccinal strain 19.

Table 83 lists the *Brucella* species and indicates the main hosts, diseases and geographical distribution.

Laboratory Diagnosis

Extreme care must be exercised when working with brucellae as humans are highly susceptible to brucellosis and laboratory infections are not uncommon.

Specimens

In abortion cases a full range of specimens should be collected and submitted for a differential diagnosis. A whole foetus should be sent, if feasible. Alternatively, foetal stomach contents, any foetal lesions, cotyledons, uterine discharges, urine (for leptospirosis), colostrum, paired serum samples, and sections of cotyledon and foetal lesions in 10 per cent formalin for histopathology. Semen and tissue from epididymides or testes from males could be examined.

Table 83. Diseases and principal hosts of the *Brucella* species.

Species	Host(s)	Diseases	Geographical distribution
B. abortus	CATTLE* Sheep, goats and pigs Horses Humans	Abortion and orchitis Sporadic abortion Associated with bursitis (poll evil and fistulous withers) Undulant fever	Biotypes: 1: Worldwide (common) 2: Worldwide (not common) 3: India, Egypt, East Africa 5: Britain and Germany Other biotypes are infrequently isolated
B. melitensis	GOATS, sheep Cattle Humans	Abortion Occasional abortion and excretion in milk Malta fever	Many sheep- and goat-raising regions except New Zealand, Australia and North America
B. suis	PIGS Humans	Abortion, orchitis, arthritis, spondylitis and herd infertility Undulant fever	Biotypes: 1: Worldwide 2: Western and Central Europe 3: USA, Argentina and Singapore 4: Arctic Circle (Canada, Alaska and Siberia) in reindeer and caribou
B. ovis	SHEEP	Epididymitis in rams and sporadic abortion in ewes	New Zealand, Australia and some other sheep-raising countries: USA, Romania, Czechoslovakia, South Africa, South America
B.canis	DOGS Humans	Abortion, epididymitis, disco-spondylitis and permanent infertility in males Undulant fever	North America and parts of Europe Becoming worldwide but not common
B. neotomae	Desert wood rat (*Neotoma lepida*)	Non-pathogenic for the wood rat and has not been recovered from any other animal species	USA (Utah)

* Natural host given in capital letters.

Direct examination

Smears are made from specimens and stained by the modified Ziehl–Neelsen (MZN) stain (**Appendix 1**). Brucellae appear as small, red-staining coccobacilli in clumps (**328**) because of their intracellular growth.

Isolation

The brucellae grow well on 5–10 per cent blood agar. However, other than foetal abomasal contents and colostrum, the specimens are likely to contain many contaminating bacteria and fungi, so selective media are usually required. The selective media contain a nutritive blood agar base with 5 per cent sterile sero-negative equine or bovine serum and an antibiotic supplement. The antibiotic supplement used in selective media for *B. ovis* usually differs from that for *B. abortus*. Terzolo *et al.* (1991) have suggested that Skirrow agar is a satisfactory medium for both the *Campylobacter fetus* subspecies and for brucellae, including the most fastidious species such as *B. abortus* biotype 2, *B. canis* and *B. ovis*. Formulae for the media that have been mentioned are given in **Appendix 2.**

328

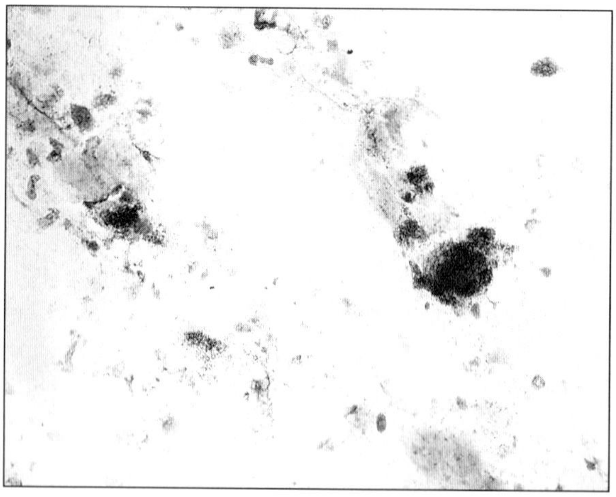

328 *Brucella abortus* in an MZN-stained smear of a cotyledon from a case of bovine abortion. The small, red (MZN-positive) coccobacilli characteristically occur in clumps reflecting their intracellular growth. (×1000)

Liquid specimens can be inoculated straight onto the plates. Scrapings from cotyledons are used, tissue samples are homogenized and aliquots used for culture. Milk or colostrum can be centrifuged at 2000 *g* for 20 minutes and loopfuls of both cream and sediment used to inoculate the plates. A loopful of cream from a positive brucella milk ring test often yields *B. abortus.*

The inoculated plates are incubated at 37°C under 5–10 per cent CO_2 for up to 15 days.

Identification

Colonial morphology

After 3–5 days' incubation on selective serum agar, pinpoint, smooth, glistening, bluish, translucent colonies appear. As they age the colonies become opaque and about 2–3 mm in diameter (**329**). Strains of *B. abortus, B. suis, B. melitensis* and *B. neotomae* are usually in the smooth form when first isolated. Colonies of rough morphology occur in each of these species on subculture. *B. ovis* and *B. canis* are always in the rough form. The rough colonies are dull, yellowish, opaque and when touched with an inoculation loop are found to be friable. Brucellae are non-haemolytic on blood agar (**330**).

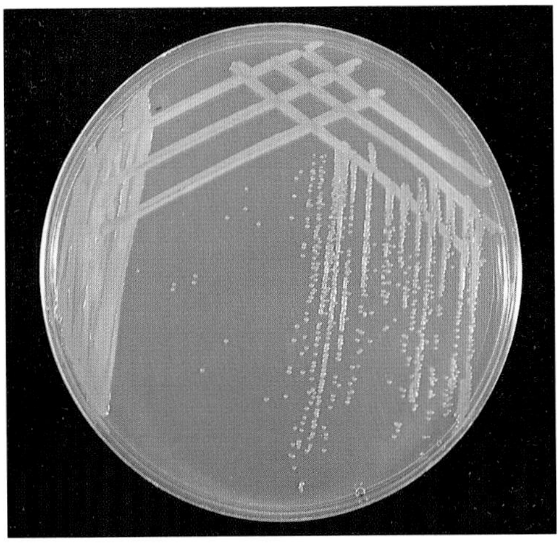

329 *B. abortus* on selective serum agar after 4 days at 37°C.

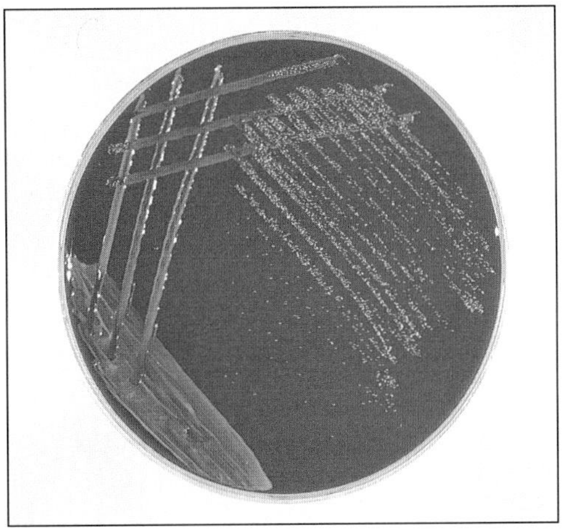

330 *B. abortus* on sheep blood agar: small, glistening, non-haemolytic colonies becoming opaque with age.

Microscopic appearance

MZN-stained smears from suspect colonies show small red-staining (MZN-positive) coccobacilli. Some laboratories use a fluorescent antibody (FA) test that identifies the organism to the generic level.

Biochemical tests

For routine identification, a few biochemical tests, together with colonial morphology and staining properties, will presumptively identify the isolate as a *Brucella* sp. In summary, the brucellae are non-motile, catalase-positive, oxidase-positive (except *B. ovis* and *B. neotomae*), give rapid urease activity (except *B. ovis* and some strains of *B. melitensis*), reduce nitrate and are indole-negative. Using known positive antisera a rapid slide agglutination test can be used for *B. abortus.*

Brucella biotyping

This is usually carried out in reference laboratories and the tests involved definitively identify the species and biotype (**Table 84**).

- *CO_2 requirement for primary isolation*: this test must be carried out immediately after isolation as the brucellae can quickly adapt to growth without CO_2. Duplicate, non-selective medium plates are streaked and incubated aerobically and under CO_2.
- *H_2S production:* the brucellae are not strong producers of H_2S so the sensitive lead-acetate-strip method is used. The brucella isolate is grown on a trypticase soy (tryptose) agar slant with the test strip suspended over the slant. The strip is examined daily for 4 days and replaced each day. Blackening of the strip indicates a positive test.
- *Urease activity*: a Christensen urea slope is inoculated with a loopful of a culture. The test is incubated at room temperature and examined at half-hour intervals. All species are positive except *B. ovis* and some strains of *B. melitensis*. Rapid urease activity is usually shown by *B. suis, B. canis* and *B. neotomae* with a positve reaction (slope turning pink) within half an hour, and *B. abortus* biotypes within 2 hours. The test is regarded as negative if there is no reaction after 24 hours.

Table 84. Differential characteristics of the species and biotypes in the genus *Brucella*.

	Biotype	CO_2 required	H_2S production	Urease activity (hours)	Growth in presence of dyes: Thionin (20 µg/ml)	Growth in presence of dyes: Basic fuchsin (20 µg/ml)	Agglutination in monospecific sera: A	Agglutination in monospecific sera: M	Agglutination in monospecific sera: R	Lysis by phage Tbilisi: RTD	Lysis by phage Tbilisi: 10^4 x RTD	Main host(s)
B. abortus	1	(+)	+	1–2 +	–	+	+	–	–	+	+	Cattle
	2	(+)	+	1–2 +	–	–	+	–	–	+	+	Cattle
	3	(+)	+	1–2 +	+	+	+	–	–	+	+	Cattle
	4	(+)	+	1–2 +	–	(+)	–	+	–	+	+	Cattle
	5	–	–	1–2 +	+*	+	–	+	–	+	+	Cattle
	6	–	(+)	1–2 +	+*	+	+	–	–	+	+	Cattle
	7[a]	(–)	+	1–2 +	+*	+	–	+	–	+	+	Cattle
Strain	19	–	+	1–2 +	–	+	+	–	–	+	+	Vaccine
B. melitensis	1	–	–	v	+*	+	–	+	–	–	–	Goats, sheep
	2	–	–	v	+*	+	+	–	–	–	–	Goats, sheep
	3	–	–	v	+*	+	+	+	–	–	–	Goats, sheep
B. suis	1	–	+	0-0.5 +	+	(–)	+	–	–	–	+	Pigs
	2	–	–	0-0.5 +	+*	–	+	–	–	–	+	Pigs, hares
	3	–	–	0-0.5 +	+	+	+	–	–	–	+	Pigs
	4	–	–	0-0.5 +	+	(–)	+	+	–	–	+	Rein-deer, caribou
	5	–	–	0-0.5 +	+	–	–	+	–	–	+	Rodents
B. ovis		+	–	–	+*	(–)	–	–	+	–	–	Sheep
B. canis		–	–	0-0.5 +	+	–	–	–	+	–	–	Dogs
B. neotomae		–	+	0-0.5 +	–	–	+	–	–	–	+	Desert wood rat

+ = positive, (+) = most strains positive, (–) = most strains negative, – = negative, v = variable reactions, * = inhibited by 40 µg/ml thionin, a = formerly *B. abortus* biotype 9, biotypes 7 and 8 were deleted by International Committee on Bacterial Taxonomy, Subcommittee on Taxonomy of *Brucella* (1988), *Int. J. Syst. Bacteriol.,* **38**: 450–452.
A = *B. abortus* antigen, M = *B. melitensis* antigen, R = rough, RTD = routine test dilution.

- *Growth in the presence of dyes*: the conventional test is carried out by incorporating the dyes thionin (blue) or basic fuchsin (red), separately in trypticase soy agar at the concentration of 20 µg/ml (1:50,000) or 40 µg/ml (1:25,000). The medium is prepared by heating a 0.1 per cent solution of either dye in a boiling water bath for 20 minutes and then adding it to the required amount of autoclaved agar. The dye is mixed with the agar and poured into Petri dishes. A sterile swab is used to inoculate the dye media with a suspension of the test strain and a reference strain as a control. Six cultures, including the control, may be tested per agar plate. The inoculated plates are incubated at 37°C under 5–10 per cent CO_2 for 3–4 days and then examined for growth. An example of the growth patterns on dye plates is illustrated (**331**).
- *Agglutination with monospecific sera*: *B. abortus, B. melitensis* and *B. suis* possess two important surface antigens named A and M, which are present on the lipopolysaccharide-protein complex. The basis of the test is that the biotypes of the three species have the two antigens in different proportions. Only the permanently rough species, *B. ovis* and *B. canis,* will agglutinate with the R, anti-rough, monospecific serum.

  A dense suspension of the test brucella is prepared in 0.5 per cent phenol-saline and heated at 60°C for 1 hour. A drop of the suspension is added to a drop of each monospecific antiserum and mixed. Agglutination should occur within 1 minute. Control cultures of *B. abortus* biotype 1, *B. melitensis* biotype 1, *B. ovis* and *B. canis* are recommended to monitor the test.

331

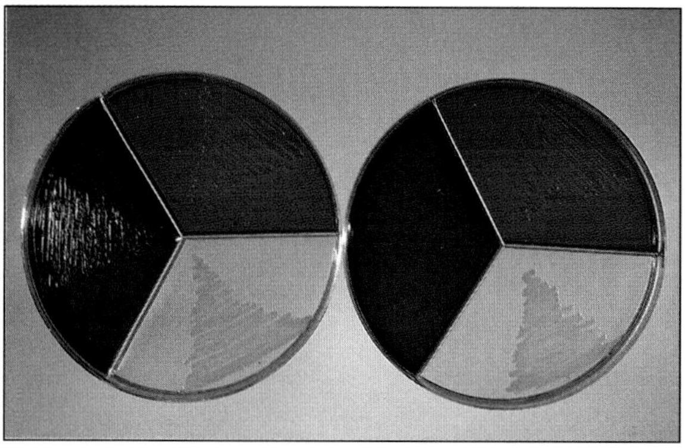

331 Growth of *B. abortus* biotype 3 (left plate) and biotype 1 (right plate) on serum agar containing the dyes thionin (left) and basic fuchsin (right). A control segment, without dye, is included for each biotype. Biotype 3 is able to grow in the presence of both dyes at a 20µg/ml concentration but biotype 1 will tolerate only basic fuchsin (20µg/ml).

- *Phage typing* : a number of bacteriophages are available to aid in the identification of the brucellae. The reactions of the Tbilisi (Tb) phage are given in **Table 84.** All biotypes of *B. abortus* will be lysed at the routine test dilution (RTD) and both *B. abortus* and *B. suis* biotypes will lyse at a phage concentration of $10^4 \times$ RTD.
- *Oxidative metabolic rates* on selected substrates are conducted in reference laboratories. These rather involved tests need be carried out only if the results of the previously mentioned tests are equivocal.

Full descriptions of the tests for biotyping brucellae are given by Alton *et al.* (1975) and Corbel and Morgan (1975).

Differentiation of *B. abortus* biotype 1 and strain 19 (S19)

The vaccinal strain 19, derived from *B. abortus* biotype 1, gives similar reactions in the biotyping tests to CO_2-independent biotype 1. Strain 19 is inhibited by erythritol. Recovery of strain 19 from diagnostic samples is uncommon but has been recorded. **Table 85** gives the tests to distinguish strain 19 from *B. abortus* biotype 1 isolates.

Animal inoculation

Guinea-pig inoculation is not commonly used but it is the most sensitive test for isolating pathogenic brucellae. Guinea-pigs are inoculated intramuscularly with 0.5–1.0 ml of suspect tissue homogenate. Two animals per sample are usually used and euthanised at 3 and 6 weeks after inoculation. Serum is taken for serology and the spleen, together with any abnormal tissue, is collected for bacteriological examination.

Immunological tests for detecting antibodies to *Brucella abortus*

A wide range of immunological tests has been developed, in conjunction with state controlled National Eradication Schemes, for bovine brucellosis. Some of the most commonly used serological tests are summarised in **Table 86**. The techniques can be obtained by reference to the texts listed in the bibliography.

In cattle, naturally infected with *B. abortus,* there is a subsequent rise in both IgM and IgG antibodies. However, the IgM titre declines and later the predominant immunoglobulin present is IgG. In some cases of chronic brucellosis the animal can have high levels of IgG1 that agglutinate poorly themselves and can also mask the normally efficient agglutinating properties of any IgM antibodies that may be present.

In adults vaccinated with strain 19, IgM antibody reaches peak values at about 13 days post-inoculation and the highest IgG values are reached in 28–42 days. Unlike natural infections, the IgM antibody titres remain high after vaccination.

Shared antigens occur between the brucellae and

other Gram-negative bacteria such as *Francisella, Campylobacter, Salmonella, Pasteurella* and *Yersinia enterocolitica* (especially serotype O9) involving the somatic antigens. In Europe, the pig is most frequently infected with *Y. enterocolitica* and this can cause cross-reactions, particularly in the agglutination test. Cross-reactions have also been recorded in cattle, dogs and other species.

Immunological tests for detecting antibodies to *B. melitensis, B. ovis, B. suis* and *B. canis*

- *B. melitensis*: serological tests such as the complement fixation test, ELISA, agar-gel immunodiffusion and the Rose Bengal plate test are available and give comparable results to the serum agglutination test. Milk ring tests are unreliable with sheep milk.

Table 85. Differentiation of *Brucella abortus* biotype 1 and strain 19.

	Brucella abortus	***Strain 19***
Growth on media containing:		
Thionin blue 2 μg/ml	+	–
Erythritol 1 mg/ml	+	–
Penicillin (5 unit disc)	R	S

+ = growth, – = no growth, R = resistant, S = susceptible.

Table 86. Summary of the serological tests for bovine brucellosis.

Serological test	***Principal immunoglobulin class identified***	***Remarks***
HERD TEST		
Brucella milk ring test	IgM, IgGl, IgA	Conducted on bulk milk from a herd. If positive, sera are taken from individual cows in the herd and subjected to one or more of the other tests
INDIVIDUAL TESTS		
Plate agglutination tests: Rose-Bengal plate test and the Card test	IgGl, IgM	Useful screening tests. Antigens buffered to pH 3.65–4.0 and this allows IgGl to cause agglutination
Brucella serum agglutination test (SAT) (tube agglutination)	IgM, IgG2, (IgGl)	Widely used test but often IgGl antibodies fail to agglutinate, so false negatives may occur
Brucella complement fixation test (CFT)	IgGl, IgM	One of the most specific of the serological tests
Enzyme-linked immuno-sorbent assay (ELISA)	Has capacity to detect all immunoglobulins	Comparatively new test for brucella antibodies but is proving reliable and is easily automated
SUPPLEMENTARY TESTS		
Coombs antiglobulin test	IgM, IgGl, IgG2	Very sensitive test, will detect 'incomplete antibodies' that do not react in the brucella SAT
Heat inactivation Rivanol precipitation Mercaptoethanol treatment	IgG IgG IgG	These are designed to differentiate the 'non-specific' reactions by destroying IgM, the antibody most commonly produced by adult vaccination

Ig = immunoglobulin

- *B. ovis*: the complement fixation test and ELISA have been found to be sensitive and specific. An indirect haemagglutination test is sometimes used but the tube agglutination test is unsatisfactory because autoagglutination may occur with this antigen.
- *B. suis*: the tube agglutination test has been used on a herd basis but is not reliable for individual animals because of low titres to *B. suis* and nonspecific reactions. The Rose Bengal test and complement fixation tests give satisfactory results and an ELISA has been developed that will probably become the main test.
- *B. canis* : the serological tests include a commercially available rapid slide agglutination test (Pitman–Moore, Inc.), a mercaptoethanol tube agglutination test (TAT), complement fixation test and an agar-gel immunodiffusion (AGID) test. The tests are subject to the occasional false-positive reaction. The slide agglutination test should be used only as a screening test and a definite diagnosis must be based on the results of a TAT or AGID test together with the culture of uterine discharges.

References

Alton, G.G., Jones, L.M. and Pietz, D.E. (1975). *Laboratory techniques in brucellosis*, 2nd ed., Monogr. Ser. No. 55. Geneva, World Health Organization.

Corbel, M.J. and Morgan, W.J.B. (1975). Proposal for minimal standards for descriptions of new species and biotypes of the genus *Brucella*, *International Journal of Systematic Bacteriology*, **25**: 83–89.

Terzolo, H.R., Paolicchi, F.A., Moreira, A.R. and Homse, A. (1991). Skirrow agar for simultaneous isolation of *Brucella* and *Campylobacter* species, *Veterinary Record*, **129**: 531–532.

Bibliography

Nicoletti, P. (1990). Serological diagnosis of canine brucellosis, pp. 102–104, In *Diagnostic Procedures in Veterinary Bacteriology and Mycology,* 5th ed., Carter, G.R. and Cole, J.R. Jn. (ed.). Academic Press, Inc., New York.

United States Department of Agriculture (1965). Manual Nos. 64A, B, C and D, Ames, Iowa, Agricultural Research Service, National Animal Disease Center, Diagnostic Reagents Division.

25 *Campylobacter* species

The *Campylobacter* species are thin, curved, Gram-negative, motile rods. The cells are 0.2–0.5 µm in width and when daughter cells remain joined, S-shaped, seagull-shaped and sometimes long spiral forms may be seen (**332**). They are motile by a single flagellum, non-fermentative, oxidase-positive, catalase-variable and most are microaerophilic. They grow best on nutritious basal media supplemented with 5–10 per cent blood under reduced oxygen tension. The most important animal pathogens are *C. fetus* subsp. *fetus, C. fetus* subsp. *venerealis, C. jejuni* and possibly *C. mucosalis* and *C. hyointestinalis.* The *Campylobacter* species were once placed in the genus *Vibrio* and some of the diseases are still occasionally referred to as 'vibriosis'.

Natural Habitat

The *Campylobacter* species are worldwide in distribution. Many of the animal species are commensals on the mucosa of the oral cavity and intestinal tract. *C. fetus* subsp. *venerealis* occurs in the prepuce of bulls and in the genital tract of cows in herds where bovine genital campylobacteriosis is, or has been, present. There are also some nonpathogenic species that are saprophytes in the environment.

Pathogenicity

Little is known of the pathogenic mechanisms of most *Campylobacter* species. *C. jejuni* produces an adhesin, a cytotoxin and a heat-labile toxin similar to that of *Escherichia coli.* Transmission of many of the *Campylobacter* species, including *C. fetus* subsp. *fetus,* is by the faecal-oral route. *C. fetus* subsp. *venerealis* is transmitted by coitus and infection of the female genital tract may lead to metritis with resulting death and resorption of the embryo (infertility), or occasionally to abortion. The diseases associated with the *Campylobacter* species and/or their commensal status are given in **Table 87**.

Laboratory Diagnosis

Specimens

Table 88 summarises the specimens required from the various clinical conditions for the diagnosis of *Campylobacter* species. Transport medium should be used for the collection of specimens for the isolation of *C. fetus* (**Appendix 2**).

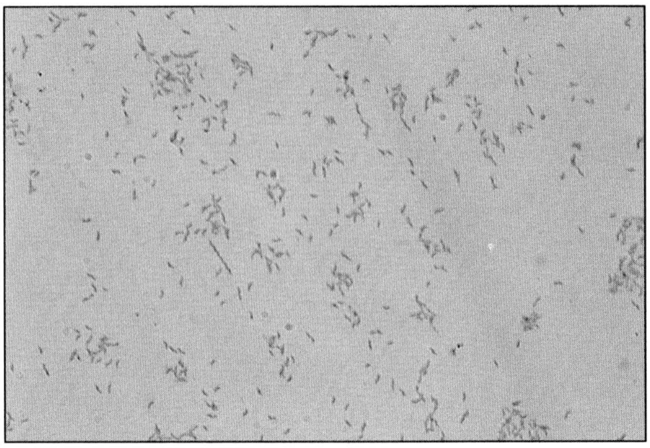

332

332 *Campylobacter fetus* subsp. *fetus* in a DCF-stained smear from culture showing the curved rods and 'seagull' forms characteristic of the genus. (×1000)

Direct microscopy

Both subspecies of *Campylobacter fetus* can be demonstrated in foetal abomasal contents using the DCF stain (dilute carbol fuchsin for 4 minutes) but they tend to stain poorly in Gram-stained smears. Fluorescent antibody staining of smears from foetal abomasal contents, cervical mucus and preputial washings is most reliable, especially when small numbers of the bacterium are present.

C. jejuni can be seen in wet mounts of faeces by phase contrast or darkfield microscopy. The typical darting motility of corkscrew-like organisms is suggestive of *Campylobacter* species. Characteristic slender, curved rods can be demonstrated in DCF-stained smears, or by phase contrast of ovine foetal abomasal contents and in bile from chickens with hepatitis.

C. mucosalis can be visualised using the modified Ziehl–Neelsen (MZN) stain on heat-fixed smears from mucosal scrapings. Large numbers of pink-staining, slender, curved rods, located intracellularly, are seen.

Other *Campylobacter* species can be demonstrated by methods similar to those described above.

Isolation procedures

- *Campylobacter fetus* (both subspecies): cervical mucus and preputial washings can be passed through a 0.65 µm membrane filter to reduce

Table 87. Pathogenic and nonpathogenic *Campylobacter* species.

Species	Principal host(s)	Disease and/or commensal status
C. fetus subsp. *venerealis*	Cattle	Bovine genital campylobacteriosis (epizootic bovine infertility): infertility, early embryonic death and occasional abortion. Prepuce of asymptomatic bulls
C. fetus subsp. *fetus*	Sheep	Ovine genital campylobacteriosis: outbreaks of abortion
	Cattle	Occasional abortions
	Man	Occasional infections
	Cattle, sheep	Commensal in the intestinal tract
C. jejuni	Sheep	Outbreaks of abortion
	Dogs, cats, other animals and man	Enteritis with diarrhoea
	Poultry	Avian vibrionic hepatitis
	Many domestic and wild animals and birds	Commensal in the intestinal tract
C. mucosalis	Pigs	Associated with the porcine intestinal adenomatosis complex. Present in intestinal tract of normal pigs
C. hyointestinalis	Pigs	Associated with the porcine intestinal adenomatosis complex. Commensal in the intestine of normal pigs
C. coli	Pigs, man	May cause mild diarrhoea in pigs and enteritis in man. Commensal in the intestine of pigs. Tends to increase in numbers in pigs with swine dysentery caused by *Serpulina hyodysenteriae*
C. cryaerophila	Cattle, pigs, sheep, horses	Isolated from faeces of normal animals and infrequently from aborted foetuses. Significance unknown
C. laridis	Dogs, horses, birds	Isolated from faeces. Disease status uncertain
C. sputorum biovar *sputorum*	Man	Commensal in oral cavity. Considered nonpathogenic
C. sputorum biovar *bubulus*	Cattle	Commensal in preputial cavity of bulls and genital tract of cows. Considered to be nonpathogenic
C. sputorum biovar *fecalis*	Sheep, cattle	Present in the intestinal tract and has been isolated from semen and vagina of cattle. Considered to be nonpathogenic
C. upsaliensis	Dogs, man	Isolated from diarrhoeic and normal individuals. Disease status uncertain

Table 88. Specimens for *Campylobacter* spp.

Clinical entity	Specimens
Bovine infertility (*C. fetus* subsp. *venerealis*)	Anoestrous mucus from females collected on a herd basis (10 to 20 animals). Culture within 4 hours of collection Preputial washings from bulls. Transport medium is recommended (*see* Appendix 2)
Bovine and ovine abortion (*C. fetus* subsp. *venerealis, C. fetus* subsp. *fetus* and *C. jejuni*)	Foetal abomasal contents preferred. Further samples from dam and foetus to exclude other infectious agents (*see* Chapter 2)
Avian vibrionic hepatitis (*C. jejuni*)	Bile
Diarrhoea (*C. jejuni* and others)	Rectal scrapings or faeces. If culture of the specimens will be delayed, store in Cary Blair medium (available commercially, also *see* Appendix 2) at 4°C
Porcine intestinal adenomatosis complex (*C. mucosalis* and/or *C. hyointestinalis*)	Smears of mucosal scrapings of lesions (heat fixed) for staining by the modified Ziehl–Neelsen stain. Sections of affected ileum and caecum in 10% formalin for histopathology (Warthin-Starry silver stain). Sections of affected intestine (fresh) for attempted isolation. These *Campylobacter* species are cell-associated and faecal samples are of no diagnostic value

contamination. Foetal abomasal contents and filtrates are inoculated onto a nutritious base (Brucella, Columbia or brain heart infusion agars) supplemented with 5–10 per cent blood. The medium can be made selective by the addition of polymyxin B sulphate (2 units/ml), novobiocin (2 µg/ml) and cycloheximide (20 µg/ml). Non-selective blood agar should be inoculated simultaneously for the possible isolation of other pathogens. The plates are incubated at 37°C for 4–6 days in a microaerophilic atmosphere containing 6 per cent O_2, 10 per cent CO_2 and 84 per cent N_2. These atmospheric conditions must be adhered to strictly and can be attained by one of the following methods:

a) Filling an anaerobic/CO_2 jar with the specified ready-mixed gas in a cylinder obtained from a commercial gas supplier.

b) Using special gas-generating envelopes such as CampyPak II (BBL) or BR56/BR60 (Oxoid) with a palladium catalyst in an anaerobic/CO_2 jar.

c) An anaerobic gas-generating envelope in the jar without a catalyst.

d) Incubating a plate inoculated with a facultative anaerobe together with the plate for isolation of *Campylobacter* spp. Both plates are placed in an air-tight plastic bag.

One of the first two methods is preferable.

- *C. jejuni*: rectal swabs or faeces are inoculated onto one of several selective media, which can be obtained commercially, such as charcoal-cefoperazole-deoxycholate agar (Oxoid) (**333**) or Blaser's Campy-BAP medium. This latter medium is Brucella agar with 5 per cent sheep blood and vancomycin (10 µg/ml), polymyxin B sulphate (2–5 units/ml), trimethoprim lactate (5 µg/ml), cephalothin (15 µg/ml), and amphotericin B (2 µg/ml). The plates should be incubated under the atmosphere described above for 2–3 days at 42°C as 37°C is suboptimal for this bacterium.
- *C. mucosalis*: homogenates of mucosal scrapings from the affected intestine are plated on Columbia blood agar with and without 1:60,000 brilliant green. The inoculated plates are incubated anaerobically at 37°C for 2–5 days. After primary isolation the bacterium will grow well microaerophilically.
- Other *Campylobacter* species: those of faecal origin can be isolated similarly to *C. jejuni* and those from the reproductive tract in the same manner as *C. fetus*. *C. cryaerophila* is isolated in a two-stage procedure by reference laboratories.

Identification

Colonial morphology

Both subspecies of *C. fetus* have small (1 mm), round, slightly raised, smooth, translucent colonies said to have a 'dewdrop' appearance (**334**). The colonies of *C. jejuni* are usually flat, grey, larger than those of *C. fetus* and can be spreading and watery on moist plates (**335**). Other *Campylobacter* species vary somewhat in colonial appearance. *C. coli* produces a pink-tan pigment and the growth of *C. mucosalis* may appear as a dirty-yellow colour on the inoculating loop.

Microscopic appearance

a) DCF-stained smears from colonies show small, curved or seagull-shaped rods.
b) Wet mounts under phase contrast or darkfield microscopy reveal the characteristic curved forms with darting motility.
c) Fluorescent antibody-stained smears can be used to identify *C. fetus,* but the technique does not distinguish between the subtypes, so this must be done by biochemical tests (**Table 89**).

Biochemical and other tests

Susceptibility or resistance to nalidixic acid or cephalothin (**336**), hydrogen sulphide production, nitrate reduction, growth at 25°C or 45°C and the catalase reaction are some of the criteria on which a definitive identification of the *Campylobacter* species is based (**Table 89**).

C. jejuni is the only species that hydrolyses sodium hippurate. For this test a large loopful of a 24–48 hour culture of *C. jejuni* is emulsified in 0.4 ml of a 1 per cent aqueous solution of sodium hippurate and incubated at 37°C for 2 hours. Then 0.2 ml of a ninhydrin solution is added to the tubes at 37°C. A positive reaction is given by a deep purple colour developing after 10 minutes. The ninhydrin solution is prepared by adding 3.5 g of ninhydrin to 100 ml of a 1:1 mixture of acetone and butanol.

Serology

The cervical mucus agglutination test for *C. fetus* subsp. *venerealis* is accurate if carried out 2–7 months post-infection. A vaginal mucus agglutination test has been found useful but the serum agglutination test is unreliable.

As *C. jejuni* is present in the intestines of many normal animals, its isolation from faeces may not necessarily be significant. A four-fold increase in an agglutinating antibody titre to the bacterium would suggest involvement of the organism in the diarrhoea.

Antibiotic susceptibility tests

These tests are not usually performed with *Campylobacter* species. Many *Campylobacter* species produce beta-lactamase, thus accounting for their resistance to penicillin and ampicillin.

333

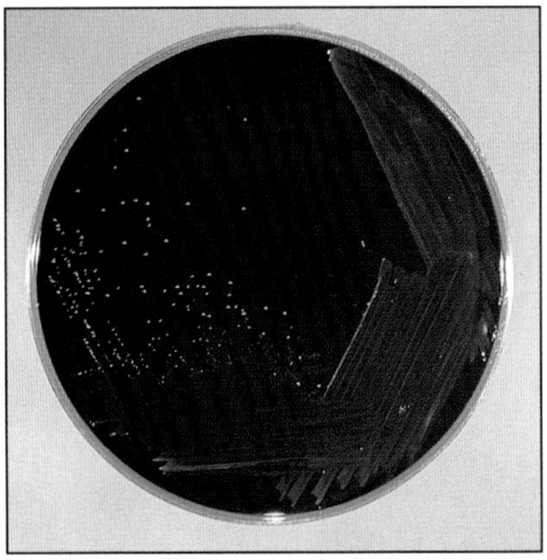

333 *C. jejuni* on charcoal-cefoperazone-deoxycholate agar.

334

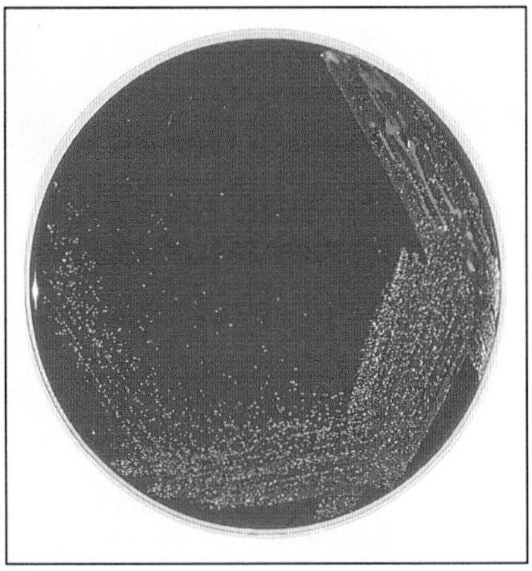

334 *C. fetus* subsp. *fetus* on sheep blood agar demonstrating the small colonies characteristic of both subspecies *fetus* and *venerealis*.

335

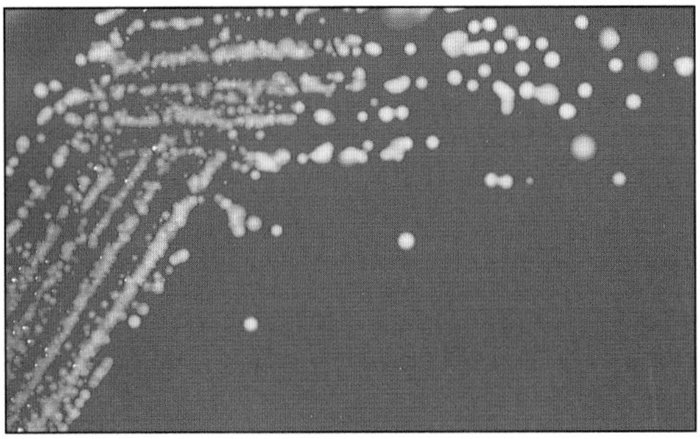

335 A close-up of *C. jejuni* on sheep blood agar. The colonies have a spreading, watery appearance on slightly moist plates.

336 Susceptibility or resistance to 30μg discs of nalidixic acid (NA) or cephalothin (KF) as an aid to the identification of *Campylobacter* species. *C. jejuni*, susceptible to nalidixic acid (right), is shown above and *C. fetus* subsp. *venerealis*, susceptible to cephalothin (left), is shown below.

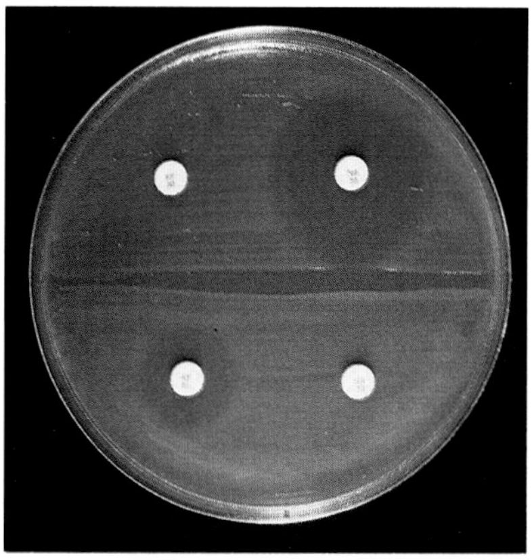

336

Table 89. Differentiation of the principal *Campylobacter* species.

Species	Growth at*		Catalase	Nitrate reduction	H_2S production		Susceptibility to (30 μg disc)	
	25°C	42°C			Lead acetate**	TSI	Nalidixic acid	Cephalothin
C. fetus subsp. *venerealis*	+	–	+	–	–	–	R	S
C. fetus subsp. *fetus*	+	–	+	–	+	–	R	S
C. jejuni	–	+	+	–	+	–	S	R
C. mucosalis	–	•	–	+	+	+	v	S
C. hyointestinalis	+	poor	+	+	+	+	R	S
C. coli	–	+	+	–	+	–	S	R
C. cryaerophila	+	–	+	+	–	–	S	R
C. laridis	–	+	+	+	+	–	R	R
C. sputorum biovar *sputorum*	–	+	–	+	+	+	v	S
C. sputorum biovar *bubulus*	v	–	–	+	+	+	v	S
C. sputorum biovar *fecalis*	–	+	+	+	+	+	R	S
C. upsaliensis	–	+	v	+	•	–	S	S

* = Tests carried out in thioglycollate medium.
** = Lead acetate strips with inoculation of semi-solid Brucella or brain-heart infusion broth with 0.02% cysteine for 4–6 days.
v = variable reaction; • = data not available ; R = resistant; S = susceptible; TSI = triple sugar iron agar.

26 *Haemophilus* species

Haemophilus species are small Gram-negative rods, less than 1 µm wide by 1–3 µm, but can be coccobacillary or produce short filaments. Capsules can be produced by *H. influenzae* and *H. paragallinarum*. Traditionally *Haemophilus* species had to have, by definition, an absolute requirement for one or both of the growth factors, X (haemin) which is heat-stable and V [nicotinamide adenine dinucleotide (NAD)] which is heat-labile. However, nucleic acid hybridisation studies have shown that this classification embraced genetically heterogeneous bacteria. Some have been reclassified but the genus still contains species not closely related to the type-species, *H. influenzae*. The haemophili are motile, facultative anaerobes, produce acid from glucose, reduce nitrates, and are variable in the oxidase and catalase tests. They are nutritionally fastidious, will not grow on MacConkey agar and grow best on chocolate agar (supplying the X and V factors) under 5–10 per cent CO_2 at 37°C.

Changes in Nomenclature

Previous Name	Present Name
Haemophilus equigenitalis	*Taylorella equigenitalis*
Haemophilus pleuropneumoniae	*Actinobacillus pleuropneumoniae*
Haemophilus avium	*Pasteurella avium*

From DNA studies, *Histophilus ovis, Haemophilus agni* and *Haemophilus somnus* are indistinguishable.

Natural Habitat

Haemophilus species are commensals or parasites of the mucous membranes of humans and animals, most commonly of the upper respiratory and lower genital tracts. *H. parasuis* inhabits the nasopharynx of normal pigs and *H. somnus* is present in the respiratory tract of healthy cattle, but *H. paragallinarum* is more closely associated with the respiratory tract of sick or recovered birds. *H. agni* is a commensal in the genital tract of sheep and *H. haemoglobinophilus* in that of dogs.

Pathogenesis

The capsule and cytotoxic factor of *H. paragallinarum* are thought to be virulence factors and endotoxin may play a role in the disease process. *H. somnus* is resistant to the lethal effects of phagocytes and serum, can adhere to epithelium and is toxic to endothelial cells to which it also adheres. *H. somnus* can bind to immunoglobulins like the staphylococcal protein A. Lesions of lungs, body cavities and joints are serofibrinous and/or suppurative. Thrombotic vasculitis leading to encephalitis and meningitis as well as haemorrhagic necrotising processes are caused by *H. somnus*. Young or previously unexposed animals are most susceptible to *Haemophilus* infections with stress factors contributing to the development of signs of disease. **Table 90** summarises the diseases or significance of *Haemophilus* species associated with animals.

Laboratory Diagnosis

Specimens

Haemophilus species are fragile and the specimens should be protected from drying and cultured as soon as possible (within 24 hours) after collection. Refrigeration and transport media do not appear to be beneficial and deep freezing, below –60°C, is the only definite method for the preservation of these bacteria. The type of specimen required will depend on the disease or lesions present.

Direct microscopy

Demonstration of these small Gram-negative rods in tissues is often difficult and specific fluorescent antibody staining is a sensitive and specific method.

Isolation

X and V factors must be supplied for all the *Haemophilus* species except *H. somnus*. The X factor (haemin) is heat-stable and present in adequate amounts in 5 per cent blood agar. V factor is present mainly intracellularly in red cells and is susceptible to NADases present in most bloods. In chocolate agar the V factor is released from the red cells, the NADases are destroyed, and the heat-stable X factor is still present. *Staphylococcus aureus* grown as a streak across a blood agar plate will provide the V factor. V-factor-requiring haemophili will grow as satellite colonies near the streak. Commercially available media, with supplements, are available for *Haemophilus* species but chocolate agar is the most satisfactory medium for the haemophili isolated from animals. Selective media have been designed for *H. somnus* but their performance has not been consistently successful. The

Table 90. Diseases of the *Haemophilus* species.

Species	Host(s)	Disease or significance
H. influenzae	Humans	Variety of diseases ranging from respiratory infections to meningitis
H. parainfluenzae	Humans	Normal flora of upper respiratory tract and implicated in urethritis
H. somnus	Cattle	• Infectious thromboembolic meningoencephalitis (TEME): septicaemia with infarcts in cerebellum • Respiratory disease: pneumonia and pleurisy, often in mixed infections with other agents • Genital infections such as endometritis and abortion. It can occur in semen and genital tract of bulls
	Sheep	Commensal of genital tract of sheep and reported as a cause of epididymitis and orchitis in rams. May also cause pneumonia, mastitis, polyarthritis, meningitis and septicaemia
H. parasuis	Pigs	• Primary agent of Glasser's disease: polyserositis and meningitis in young pigs • Arthritis and pneumonia in older pigs • Secondary invader in swine influenza or in enzootic pneumonia (*Mycoplasma hyopneumoniae*)
H. paragallinarum	Poultry	Infectious coryza of chicken: nasal discharge, sneezing and oedema of the face, reduction in egg production
	Japanese quail	Highly susceptible to the infection
H. haemoglobinophilus	Dogs	Commensal of lower genital tract and sometimes causes cystitis and neonatal infections. Isolated from balanoposthitis and vaginitis but the role of the bacterium in these conditions is uncertain
H. aphrophilus	Humans	Part of the oral flora, found in dental plaque, and can be an occasional cause of endocarditis
	Dogs	Recovered from the pharynx of dogs
H. ovis	Sheep	Rare reports of bronchopneumonia
H. paracuniculus	Rabbits	Significance not known. Isolated from the intestine of rabbits with mucoid enteritis
H. influenzaemurium	Mice	Respiratory infections and conjunctivitis
H. piscium	Trout	Ulcer disease: ulcerating inflammation of gills and mouth

chances of isolating *H. somnus* from contaminated specimens is increased if the specimens are first incubated in an infusion broth.

The growth of many of the *Haemophilus* species is enhanced by 10 per cent CO_2. As this is not inhibitory for any of them, CO_2 should be used for routine isolation. The inoculated chocolate agar plates are incubated under 10 per cent CO_2 at 35–37°C for 3–4 days, although some growth may be seen after 24 hours.

Identification

Colonial morphology

Small dewdrop-like colonies may appear after 24–48 hours' incubation (**337**) and none are consistently haemolytic. A few strains of *H. somnus* may show a frank clearing around the colonies especially on Columbia-base sheep blood agar. *H. somnus* colonies may appear yellowish (**338**) especially in a loopful of growth or on a confluent lawn.

337 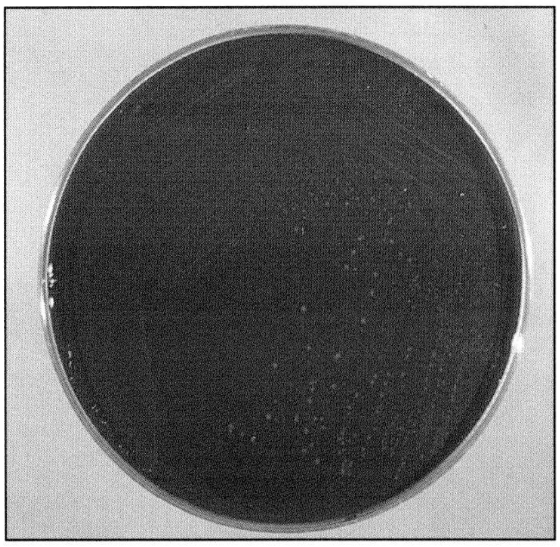

337 *Haemophilus paragallinarum* on chocolate agar demonstrating the typical dew-drop type colonies of many *Haemophilus* species.

338 *H. somnus* on chocolate agar showing the characteristic yellowish tinge of the colonies.

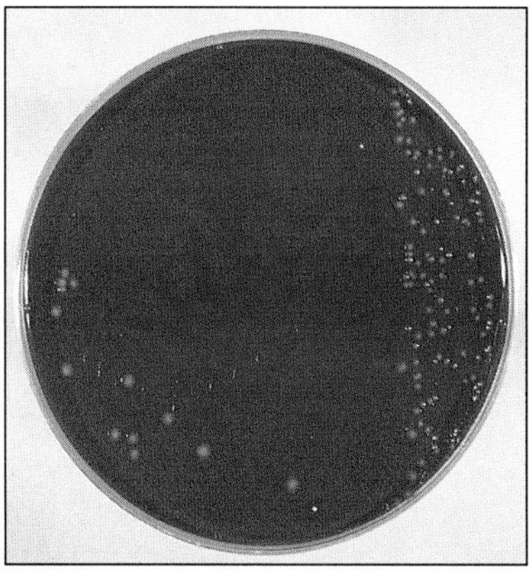 338

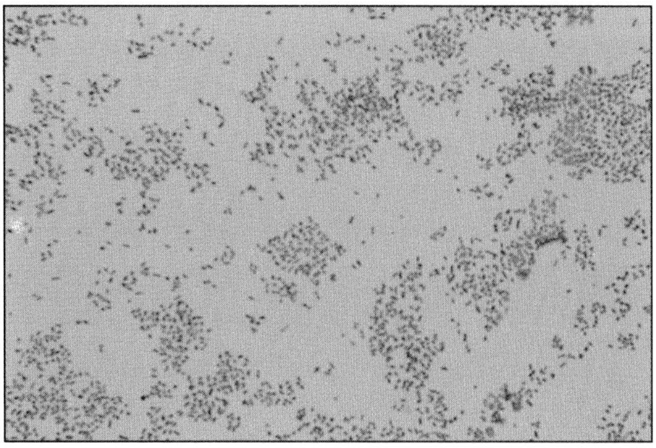 339

339 Microscopic appearance of *Haemophilus* species: small Gram-negative rods, often coccobacillary. (×1000)

Microscopic appearance

Haemophili are small Gram-negative rods that can be coccobacillary in form (**339**). More rarely short filaments occur.

Biochemical reactions

Some differential biochemical reactions for the *Haemophilus* species isolated from animals are given in **Table 91**. In non-specialist laboratories a presumptive identification of the fastidious *Haemophilus* species is based on the host species, clinical signs and lesions, colonial and microscopic characteristics, X and V factor requirements, oxidase and catalase reactions and whether or not CO_2 enhances growth. *H. somnus* is rather variable in biochemical activities and the most reliable reactions are oxidase-positive, catalase-negative with CO_2 giving a considerable enhancement of growth. If the indole test is positive, this is useful diagnostically.

- *Indole test*: the substrate is 0.1 per cent L-tryptophan in 0.05 M phosphate buffer at pH 6.8. This is inoculated with a loopful of the suspect *Haemophilus* species and incubated for 4 hours. The test is shaken with 0.5 ml of Kovac's reagent. A red colour in the upper alcohol phase indicates the presence of indole.
- *Urease test*: the test medium contains:
  0.1 g KH_2PO_4
  0.1 g K_2HPO_4
  0.5 g NaCl
  0.5 ml of 0.2 per cent phenol red
  100 ml distilled water

  The 0.2 per cent phenol red is prepared by dissolving 0.2 g phenol red crystals in 8.0 ml 1N NaOH and 92.0 ml distilled water.

  The test medium pH is adjusted to 7.0 with 5N NaOH. This is autoclaved and then 10.4 ml of a 20 per cent filter-sterilised urea solution is added.

  The medium is inoculated with the test *Haemophilus* species and incubated for 4 hours at 37°C. A red colour indicates urease activity.
- *Ornithine and Arginine tests*: commercial Moller's medium can be used but should be inoculated with a heavy loopful of the suspect *Haemophilus* culture. A purple colour that develops after 4–24 hours' incubation indicates a positive reaction.
- *Carbohydrate fermentation*: a phenol red broth (Difco) containing 1 per cent of the sugar and filter-sterilised X and V factors (10 mg/litre of each) is used. Other additions should be made for some of the species:

  1 per cent serum in the medium for *H. parasuis* and *H. paragallinarum*. A drop of defibrinated blood/ml for *H. somnus*.

  H. paracuniculus is tested in bromocresol purple broth as it may not grow in the phenol red broth.

Tests for X and V factor requirements

- *V factor* : the need for the V factor can be demonstrated by satellitism around a V factor-producing bacterium such as *Staphylococcus aureus* (**340**). The test is carried out on tryptose agar (Difco) which does not contain either the X or V factor.
- *Disc Method for X and V factors* : three commercial discs impregnated with V factor, X factor and

XV factors, respectively, are placed on a lawn of the test bacterium on a tryptose agar plate. Colonies will cluster around the disc(s) supplying the required growth factor(s) (**341**). However, the results of this test are invalidated:

a) If there is a carry-over of, particularly, the X factor from a previous richer medium.
b) If a contaminating colony is present on the plate, this may act as a feeder-organism.
c) If the test medium contains traces of X or V factors.

- *Porphyrin test*: this is the most satisfactory method for testing the requirement for the X factor. A loopful of growth from a young culture is suspended in 0.5 ml of a 2 mM solution of delta-aminolevulinic acid (ALA) hydrochloride (Sigma) and 0.8 mM $MgSO_4$ in 0.1 M phosphate buffer at pH 6.9. It is incubated for at least 4 hours at 37°C and exposed to a Wood's UV lamp in a dark room. A red fluorescence indicates that porphyrin is present and the X factor is not required. The test is based on X factor-independent strains being able to convert ALA, a porphyrin precursor, to porphyrin (an intermediate in the the haemin biosynthetic pathway). Haemin-dependent strains do not have the appropriate enzymes. Filter paper discs impregnated with ALA are available commercially.

For further information on biochemical or X and V factor requirement tests Killian and Biberstein (1984) should be consulted.

Table 91. Differential characteristics for *Haemophilus* species isolated from animals (the reactions for *Taylorella equigenitalis* are given for purposes of comparison).

	Requirement for factors		Catalase	Oxidase	CO_2 enhances growth	Haemolysis	Indole	Urease	Glucose (acid)	Nitrate reduction	Sucrose	Lactose	D-Xylose	Mannitol	Ornithine decarboxylase	Arginine dihydrolase
	X	***V***														
Haemophilus somnus (*H. agni*)	–	–	–	+	+	v	v	–	+	+	–	–	+	+	+	–
H. parasuis	–	+	+	–	v	–	–	–	+	+	+	v	–	–	–	–
H. paragallinarum	–	+	–	–	+	–	–	–	+	+	+	v	+	+	–	–
H. haemoglobinophilus	+	–	+	+	–	–	+	–	+	+	+	–	+	+	–	–
H. aphrophilus	+	–	–	–	+	–	–	–	+	+	+	+	–	–	–	–
H. paracuniculus	–	+	+	+	+	–	+	+	+	+	+	–	–	–	+	+
H. influenzaemurium	+	–	+	–	–	–	–	–	+	+	+	–	–	–	•	•
H. ovis	+	–	v	+	–	–	–	–	+	+	–	+	+	v	•	•
Taylorella equigenitalis	–	–	+	+	+	–	–	–	–	–	–	–	–	–	–	–

+ = over 90% strains positive, – = less than 90% strains positive, v = variable, • = data not available.

340

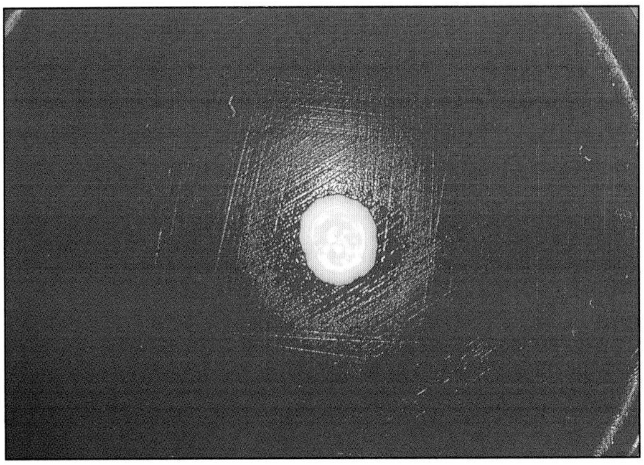

340 Satellitism of *H. paragallinarum* around a V factor-producing bacterium such as *Staphylococcus aureus*.

341

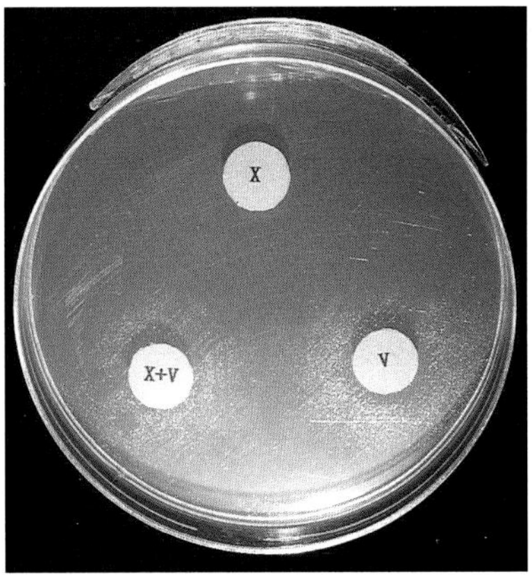

341 Conventional test for the X and V factor requirements on a nutrient agar plate. *H. paragallinarum* shows satellitism around the V and XV discs indicating an absolute requirement for the V factor.

Serology

Surveys have shown that antibody to *H. somnus* is widespread in cattle populations. So far there is no test that is used for the diagnosis of clinical cases.

In poultry, antibody to *H. paragallinarum* is demonstrable after 1–2 weeks of infection and can be detected for over a year. Serological procedures are used to identify potential carrier birds; these include slide and tube agglutination tests, agar gel precipitation, latex agglutination and haemagglutination and haemagglutination-inhibition tests.

Haemophilus piscium

H. piscium, responsible for 'ulcer disease' in trout, is probably not a true *Haemophilus* species as it does not require either the X or V factors for growth. It is a Gram-negative rod (0.5–0.7 × 2–3 μm), non-motile and will not grow at 37°C, the optimum temperature for growth being 20–25°C. Media for isolation should be supplemented with a peptic digest of fish tissue. Colonies develop to 1–3 mm in diameter and are circular, convex and cream-coloured. On blood agar complete haemolysis is produced.

Reference

Killian, M. and Biberstein, E.L. (1984). *Haemophilus*, pp. 558–596, In Krieg, N.R. and Holt, J.G. (ed.), *Bergey's Manual of Systematic Bacteriology*, vol. 1, Williams and Wilkins Co., Baltimore, USA.

27 *Taylorella equigenitalis*

Taylorella equigenitalis (formerly *Haemophilus equigenitalis*) is a Gram-negative rod about 0.8 × 5.0 µm in size. It is a facultative anaerobe, non-motile, oxidase-positive, catalase-positive, phosphatase-positive and produces no acid from carbohydrates. *T. equigenitalis* is a fastidious and slow-growing bacterium and optimal growth is obtained on chocolate agar with a rich base (Eugon or Columbia agar) at 37°C under 5–10 per cent CO_2. It does not grow on MacConkey agar.

Natural Habitat

T. equigenitalis is the causal agent of contagious equine metritis (CEM). It resides exclusively in the equine genital tract. Stallions develop no signs of disease but both sexes can remain carriers indefinitely. The disease is highly contagious. The geographical distribution is limited, but from Europe the disease spread to Japan, Australia and the USA.

Pathogenicity

Transmission is essentially venereal, but the mares can also be infected by attendants and via veterinary instruments. The organisms can be isolated from neonatal and virgin animals. A purulent metritis develops within a few days of infection and the mare often has a copious mucopurulent uterine discharge. The infectious process is limited to the mucous membranes of the uterus, cervix and vagina. There is erosion and degenerative change in the endometrium. After endometrial repair is complete, within a few weeks, the organism may still be present in the clitoral sinuses and fossa and can remain there for long periods. No clinical signs occur in the stallion but *T. equigenitalis* can be found on the surface of the penis, in preputial smegma and in the urethral fossa. The infection in mares causes a temporary infertility and occasionally abortion within the first 60 days of pregnancy.

Laboratory diagnosis

Specimens

In many countries contagious equine metritis is controlled by the Department of Agriculture or by the Thoroughbred Breeders' Association. These bodies lay down the method of sample taking, the type of samples to be taken and the culture media to be used, when examining asymptomatic stallions and mares. The requirements may be amended periodically. Often only approved laboratories are licensed to process and culture the specimens. In general, acceptable samples are swabs or biopsies from:

- Mares: cervix, uterus, clitoral fossa and clitoral sinuses.
- Stallions: urethra, urethral fossa and diverticulum, prepuce and pre-ejaculatory fluid.

In some cases it is specified that a stallion serves two maiden mares and these are sampled instead of the stallion. The specimens are collected on sterile swabs and these are placed into Amies transport medium with charcoal. They must reach the laboratory, under refrigeration, within 48 hours of collection.

Direct microscopy

Gram-stained smears are of use only on uterine exudates from a mare with clinical disease. *T. equigenitalis* can appear as Gram-negative rods, coccobacilli or short filaments.

Isolation

Routine medium is chocolate agar with a highly nutritive base such as Eugon or Columbia agar and preferably equine blood. The inoculated plates are incubated at 37°C under 10 per cent CO_2. Growth may be seen at 48 hours but negative plates should be examined daily for up to 7 days before discarding them. However, selective media are required to suppress contaminating bacteria. If streptomycin is used as one of the selective agents, two plates should be inoculated in parallel, with and without streptomycin, as some strains of *T. equigenitalis* are susceptible to this antibiotic.

Examples of this type of medium are:

- Plate 1: Eugon chocolate agar with 10 per cent horse blood and 200 µg/ml streptomycin.
- Plate 2: Eugon chocolate agar with 10 per cent horse blood and
  5 µg/ml amphotericin B (Fungizone, Squibb)
  1 µg/ml crystal violet.

Timoney *et al.* (1982) suggested a selective medium that proved very effective in controlling the bacterial and fungal flora in material on swabs taken from the external genitalia of mares and stallions. As the medium does not contain streptomycin, it is suitable for the isolation of both streptomycin-sensitive and streptomycin-resistant strains of *T. equigenitalis*. The formula for the CEM Selective Medium (Timoney) is given in **Appendix 2**.

Identification

Colonial morphology
After 48 hours' incubation the colonies are under 1 mm in diameter, shiny, smooth and greyish-white. They may attain a size of 1.5 mm on further incubation (**342**).

Microscopic appearance
Gram-negative pleomorphic coccobacilli are seen in smears from the colonies.

342

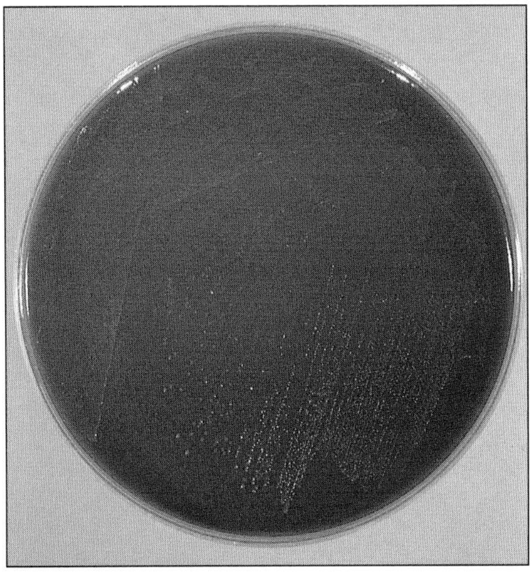

342 *Taylorella equigenitalis* on chocolate agar after 3 days at 37°C.

Biochemical reactions
Colonies with the correct macroscopic and microscopic appearance that are catalase-positive and oxidase-positive are subcultured onto Eugon chocolate agar without antibiotics and subjected to further tests:

- Inability to grow in air.
- Agglutination with *T. equigenitalis* specific antiserum in a slide test. Weak spontaneous agglutination may sometimes occur in the saline control.
- Phosphatase activity: 0.5 ml of *p*-nitrophenyl phosphate solution (1 mg/ml) is added to a suspension of the suspect colonies in 0.5 ml of Tris buffer (pH 8.0). The mixture is incubated at 37°C for up to 2 hours. A yellow colour indicates a positive result.
- Other biochemical test results, mainly negative, are given in **Table 91** (Chapter 26) in the section on the *Haemophilus* species.

Serology
Complement-fixing antibodies are consistently detectable from the third to seventh week post-infection in mares. However, this is rather late for the test to be useful diagnostically and the CFT titres do not correlate sufficiently well with the carrier state. Demonstration of CFT antibodies may be useful, in retrospect, to confirm a past infection.

Antibiotic susceptibility
Conventional antibiotic susceptibility tests are difficult with this slow-growing, fastidious bacterium. Topical treatment of the clitoral fossa in mares and external genitalia of stallions, on 5 consecutive days, with 2 per cent chlorhexidine followed by 0.2 per cent nitrofurazone ointment is often used.

Reference

Timoney, P.J., Shin, S.J. and Jacobson, R.H. (1982). Improved selective medium for isolation of the contagious metritis organism, *Veterinary Record*, **111**: 107–108.

28 *Bordetella* species

The *bordetellae* are small (0.2–0.5 × 0.5–1.0 µm). Gram-negative rods that tend to be coccobacillary. They are strict aerobes and do not attack carbohydrates but derive energy by the oxidation of amino acids. *B. avium* and *B. bronchiseptica* are motile by peritrichous flagella but *B. pertussis* and *B. parapertussis* are non-motile. All are catalase-positive and oxidase-positive. *B. bronchiseptica* and *B. avium* will grow on MacConkey agar.

Change in Nomenclature

Previous Name	**Present Name**
Alcaligenes faecalis (strains causing turkey coryza)	*Bordetella aium*

Natural Habitat

The bordetellae are inhabitants primarily of the upper respiratory tract of healthy and diseased humans, animals and birds. *B. pertussis* and *B. parapertussis* are human pathogens causing whooping cough and a mild form of whooping cough, respectively. *B. bronchiseptica* can be present in the upper respiratory tract of pigs, dogs, cats, rabbits, guinea-pigs, rats, horses and possibly other animals. *B. avium* inhabits the respiratory tract of infected poultry, principally turkeys. Mammalian infections are mainly transmitted by aerosols but in turkeys indirect spread can occur via water and litter.

Pathogenesis

B. bronchiseptica attaches firmly to the ciliated respiratory epithelium and this is followed by rapid proliferation, ciliary paralysis and an inflammatory response. Virulent strains produce pili and an extracellular enzyme, adenylate cyclase. The pili aid adherence and the adenylate cyclase has antiphagocytic activity, protects the bacterium from intracellular destruction and causes immobility of respiratory cilia. A dermonecrotising toxin is also formed, which is primarily responsible for nasal turbinate atrophy and may play a role in pneumonia and other respiratory infections. The effect on the turbinate bones is most serious in young pigs under 3 weeks of age when osteogenesis is most active.

Bordetella infections depress the respiratory clearance mechanisms, facilitating invasion by other organisms. Atrophic rhinitis in pigs is transient and self-limiting when caused by *B. bronchiseptica* alone but the bacterium aids the establishment of the piliated and toxigenic (AR+) strains of *Pasteurella multocida* and the combined infection causes more serious and permanent lessons (**343**). *B. bronchiseptica* also produces proteases, a haemolysin and haemagglutinins that may play a part in the pathogenesis.

The infectivity of *B. avium* is associated with a plasmid. The bacterium can depress some cell-mediated immune reactions and produces a histamine-sensitizing factor resembling the toxin of *B. pertussis*. A cytotoxin, a haemagglutinin and a dermonecrotising toxin, all of which are probably involved in the disease process in birds, are also produced. **Table 92** summarises the main diseases and hosts of the bordetellae.

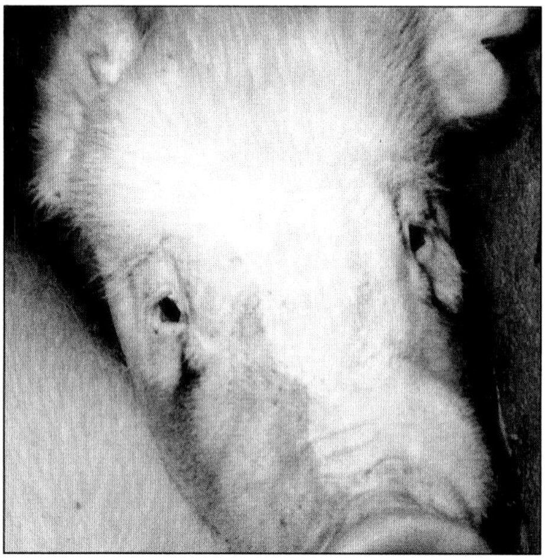

343

343 Pig with early signs of atrophic rhinitis. Note the lacrimation and the shortened, deeply wrinkled snout.

Laboratory Diagnosis

Specimens

Specimens may include nasal swabs, tracheal washings and pneumonic lungs. If nasal swabs are to be taken from animals where the nasal orifice is small, such as in young pigs, dogs and laboratory animals, the narrow gauge, flexible swabs designed for human infants (such as Mini-Tip Culturette swabs, Marion Scientific, USA) should be used.

Table 92. Diseases and hosts of the bordetellae.

Species	Host(s)	Disease
Bordetella pertussis	Humans	Classical form of whooping cough
	Chimpanzees	Rare case of whooping cough-like disease in captive animals
B. parapertussis	Humans	Mild form of whooping cough
	Lambs	Isolated from healthy and pneumonic lambs. Significance not known
B. bronchiseptica	Pigs	Atrophic rhinitis (with or without AR^+ strains of *Pasteurella multocida*) Bronchopneumonia seen in young pigs
	Dogs	Canine infectious tracheobronchitis ('kennel cough') with or without concurrent respiratory viruses. Secondary invader in canine distemper
	Rabbits	'Snuffles'-like syndrome with upper respiratory tract infection, bronchopneumonia or septicaemia
	Guinea pigs and rats	Similar to the disease in rabbits
	Horses, cats	Respiratory infections (not common)
	Humans	Occasionally isolated from wounds and body fluids (presumed to be zoonotic infections)
B. avium	Turkeys and less commonly other birds	Turkey coryza: rhinotracheitis and sinusitis in young poults. Morbidity high but mortality low

Direct microscopy

As the bordetellae are small Gram-negative coccobacilli, smears directly from specimens are not very useful. A fluorescent antibody technique would be useful.

Culture

The routine media used are sheep blood and MacConkey agars. *B. avium* and *B. bronchiseptica* grow well on both media. The plates are incubated aerobically at 37°C for 24–48 hours.

If isolations are to be attempted from specimens containing a large number of bacterial contaminants, such as nasal swabs, a selective medium is required. The reason is two-fold – to prevent overgrowth and to maintain the alkaline to neutral conditions for the bordetellae. Even a few fermentative bacteria on a medium containing carbohydrates can produce sufficient acid to inhibit the *Bordetella* species.

Several selective media have been described, such as MacConkey agar with 1 per cent glucose and 20 µg/ml furaltadone or blood agar with 2 µg/ml clindamycin and 4 µg/ml neomycin but the Smith–Baskerville (SB) medium (Smith and Baskerville, 1979) gives a high isolation rate and is also an indicator medium (**Appendix 2**). The SB medium was designed for the isolation of *B. bronchiseptica* from pigs. If it is used for the isolation of strains from dogs or rabbits, the gentamicin should be omitted as some isolates have been reported as being gentamicin-susceptible. *B. avium* will grow well on SB medium, with or without the antibiotic supplement. The inoculated SB medium is incubated aerobically at 37°C for 48 hours.

Identification

Colonial morphology

On sheep or horse blood agar *B. bronchiseptica* forms very small, convex, smooth colonies with an entire edge after 24 hours (**344**). Some strains may be haemolytic. The colonies of *B. avium* are similar (**345**) but are non-haemolytic. Phase modulation occurs in both species and this is thought to be due to loss of a capsule-like structure on subculture. The virulent, encapsulated phase I colonies are convex and shiny, those of phase II are larger, circular and convex with a smooth surface and the avirulent phase III colonies are large, flat and granular with an irregular edge.

The colonies on MacConkey agar are small, pale with a pinkish hue and amber discolouration of the underlying medium. *B. avium* (**346**) and *B. bronchiseptica* (**347**) have colonies of similar appearance on MacConkey agar.

344

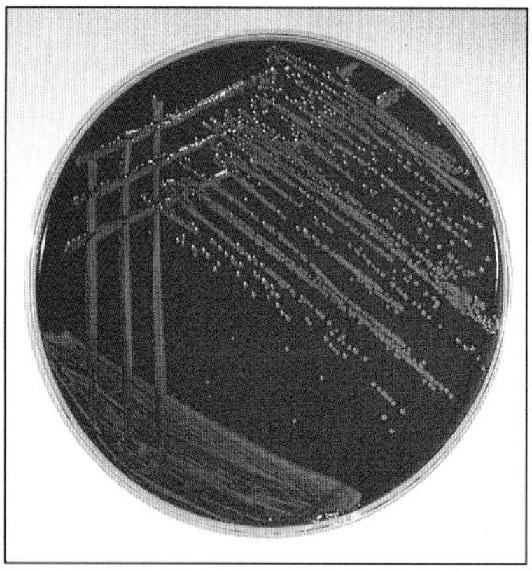

344 *Bordetella bronchiseptica* on sheep blood agar.

3

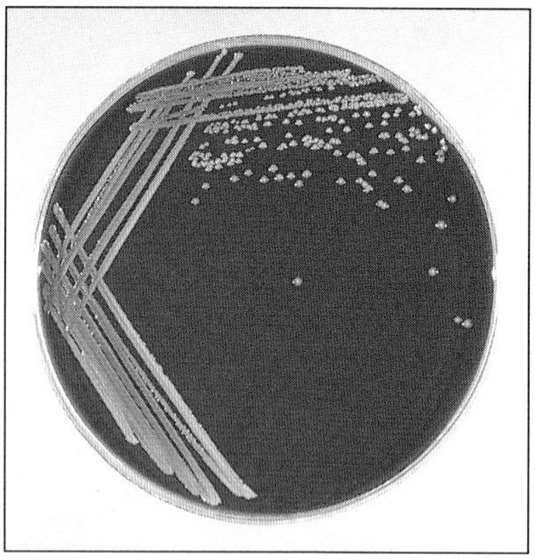

345 *B. avium* on sheep blood agar.

346

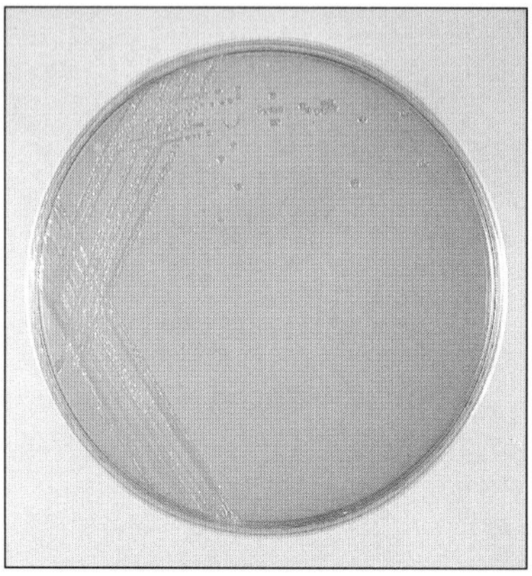

346 *B. avium* on MacConkey agar.

3

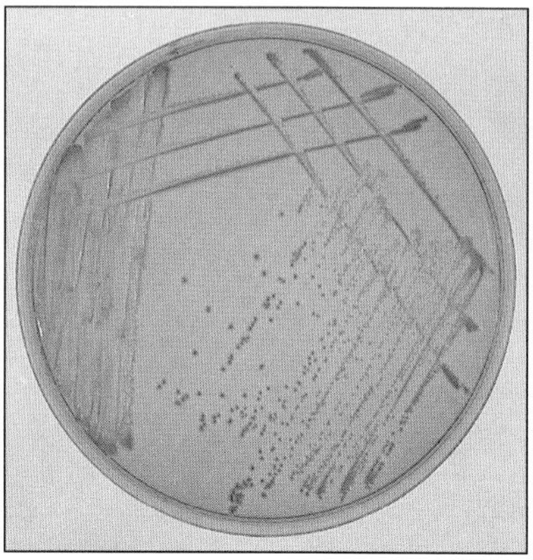

347 *B. bronchiseptica* on MacConkey agar showing the pale, slightly tan colonies.

Smith–Baskerville (SB) medium contains the pH indicator bromothymol blue and the agar is green at pH 6.8. The colonies of *B. avium* and *B. bronchiseptica*, after 24 hours incubation, are small (0.5 mm diameter or less), blue colonies with a lighter blue (alkaline) reaction in the medium around them. After 48 hours' incubation, the colonies are 1.0–2.0 mm diameter, blue or blue with a green centre and the surrounding medium is blue. The non-fermentative contaminants such as *Alcaligenes, Pseudomonas* and *Flavobacterium* species, at 24 and 48 hours' incubation, tend to be larger and more greenish than those of the bordetellae. Any fermentative contaminants give an acid reaction and the colonies and surrounding medium become yellow (**348**).

Microscopic appearance

The bordetellae are small Gram-negative coccobacilli.

Biochemical reactions

B. bronchiseptica is positive to the oxidase, catalase, citrate, urease and nitrate tests. It is motile and carbohydrates are not utilized. *B. avium* and *Alcaligenes faecalis* have similar reactions to those of *B. bronchiseptica* but are urease-negative and nitrate-negative. *A. faecalis* is present in soil, water and faeces and because of its ubiquity it can occasionally be present as a contaminant in clinical specimens. It has many

3

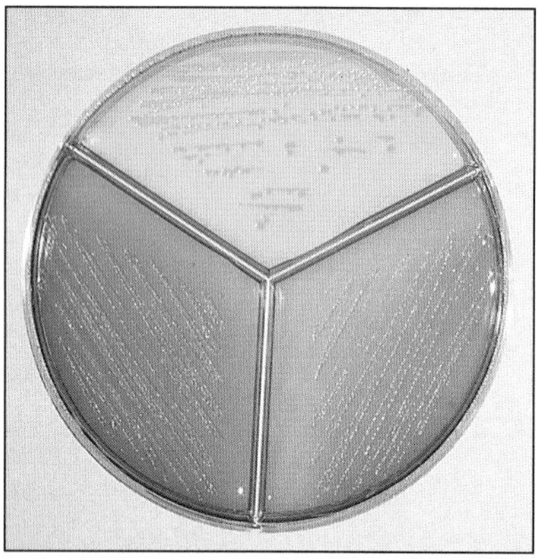

348 Smith–Baskerville medium with *B. bronchiseptica* (left), a lactose and/or glucose-fermenting bacterium (top) and *B. avium* (right). The uninoculated medium, with bromothymol blue as the pH indicator, is green.

Table 93. Differentiation of *Bordetella* species and *Alcaligenes faecalis*.

Characteristic	*Bordetella bronchiseptica*	*B. avium*	*Alcaligenes faecalis*
Beta-haemolysis (blood agar)	(+)	–	–
Growth on MacConkey agar	+	+	+
Growth on SS agar	–	+	(+)
Colonial morphology on SB medium	Small, blue	Small, blue	Larger, greenish
Nitrate reduction	+	–	–
Urease	+	–	–
Sole source of carbon for growth:			
Citrate	+	+	+
Malonate	–	–	+
Glycolate	–	–	+
Assimilation of adipate*	•	+	–
Assimilation of caprate*	•	–	+

+ = positive reaction, (+) = most strains positive, – = negative, • = data not available
SB medium = Smith-Baskerville medium, * = using API Rapid NFT System (Analytab Products).

properties in common with *B. avium* from which it must be distinguished (**Table 93**).

Both automated and miniaturised commercial identification systems are available for glucose-non-fermenting bacteria that include *B. bronchiseptica* and *A. faecalis*. These are listed in Chapter 30 (glucose-non-fermenting, Gram-negative bacteria).

Haemagglutination test

B. bronchiseptica possesses a haemagglutinin and will haemagglutinate washed sheep erythrocytes. A young 24-hour culture should be used as older colonies tend to lose their haemagglutinating ability. Two colonies of a suspected *B. bronchiseptica* culture are suspended in a drop of physiological saline on a slide. An equal volume of a 3 per cent suspension of washed sheep red cells is added and mixed. To check for autoagglutination, controls should include a suspension of colonies without erythrocytes and a suspension of erythrocytes alone. *B. bronchiseptica* will haemagglutinate the red cells within 1–2 minutes.

Serology

Tube agglutination, microagglutination and ELISA procedures have been developed for *B. avium* and *B. bronchiseptica.*

Animal inoculation

The dermonecrotising toxins of *B. avium* and *B. bronchiseptica* are thought to be important virulence factors. There appears to be no cross-reactivity between the two toxins. They are intracellular, heat-labile toxins. The dermonecrotising toxin of *B. bronchiseptica* is lethal if inoculated intraperitoneally into mice and produces skin necrosis when injected intradermally into guinea-pigs. Fatal infections can also be produced in guinea-pigs by injection of young, intact cells given intraperitoneally.

Reference

Smith, I.M. and Baskerville, A.J. (1979). A selective medium facilitating the isolation and recognition of *Bordetella bronchiseptica* in pigs. *Research in Veterinary Science*, **27**:187–192.

29 *Moraxella* species

The moraxellae are short, plump, Gram-negative rods (1.0–1.5 × 1.5–2.5 μm), characteristically in pairs. Some strains approach a completely coccal shape. They are strict aerobes, oxidative, oxidase-positive, usually catalase-positive, non-motile and do not attack carbohydrates. Although they will grow on non-enriched media, their growth is enhanced by the addition of blood or serum. The optimal temperature for growth is 33–35°C. Most *M. phenylpyruvica* strains will grow on MacConkey agar, but *M. bovis* and *M. lacunata* are unable to do so.

Natural Habitat

The moraxellae are commensals on the mucous membranes of man and other mammals. The reservoir of *M. bovis* is thought to be the conjunctiva or naso-pharynx of asymptomatic cattle over 2 years of age. *Moraxella* species are susceptible to desiccation and do not survive well away from the animal host. Transmission is by direct contact or via flying insects.

349

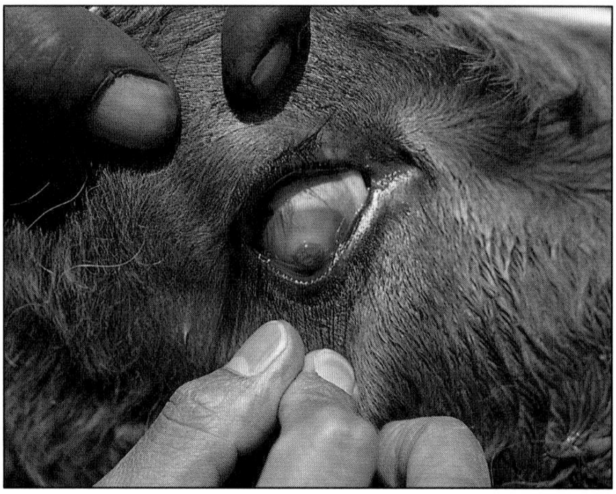

349 Young steer with infectious keratoconjunctivitis showing the healing stage with the characteristic red cone of granulation tissue projecting from the cornea.

Pathogenesis and Pathogenicity

M. bovis, the cause of infectious bovine keratoconjunctivitis, is the major animal pathogen in the genus. *M. lacunata* and *M. phenylpyruvica* are only occasionally associated with disease. *Moraxella* species, isolated from the eyes of horses with conjunctivitis, were referred to as '*M. equi*' but they are now considered to be strains of *M. bovis*.

Virulent *M. bovis* strains produce a haemolysin, pili and a cytotoxin that damages bovine neutrophils. Lipopolysaccharides, a collagenase and a hyaluronidase, produced by *M. bovis*, may also contribute to virulence. Predisposing environmental factors are implicated in infectious bovine keratocojunctivitis. These include irritation to the eyes by ultraviolet light (in sunlight), dust, long vegetation and flies. The incidence of the disease is highest in the summer months. Concurrent infections may complicate the disease by such agents as *Mycoplasma bovoculi*, *Listeria monocytogenes*, *Chlamydia psittaci*, bovine herpesvirus 1, bovine adenoviruses and the nematodes *Thelazia* spp. Young animals under 2 years of age are most commonly affected and in a typical outbreak of disease about 60 per cent of young cattle, but only about 20 per cent of older animals are affected. The early signs of infectious bovine keratoconjunctivitis are lacrimation, blepharospasm and conjunctivitis. Later an ulcer develops on the cornea. Corneal opacity and oedema surround the ulcer and in severe cases vascularisation of the cornea occurs from the limbus to the ulcer. The corneal opacity then involves the entire cornea. In the healing stage, granulation tissue forms on the ulcer floor and a characteristic red cone of granulation tissue will project from the cornea (**349**). The granulation tissue and the ulcer itself will eventually regress leaving a white corneal scar. The scar may or may not be permanent. Mild cases are similar, but ulcers resolve without vascularisation occurring and most eyes become clinically normal in 2–3 weeks. **Table 94** summarises the sites of isolation and diseases of the moraxellae.

Laboratory Diagnosis

Specimens

A swab of lacrimal secretions is taken from deep in the inner canthus of the eye. Ideally the blood agar plates should be inoculated with the lacrimal secretions immediately after collection. If this is not possible, each swab should be placed in about 1–2 ml of sterile distilled water, to prevent desiccation, and the specimens taken to the laboratory within 2 hours of collection.

Direct microscopy

Gram-stained smears are of little practical use, but the fluorescent antibody technique on smears will demonstrate and identify *M. bovis* if sufficient bacterial cells are present.

Table 94. Summary of the diseases and sites of isolation of the *Moraxella* species.

Species	Host	Disease and/or site of isolation	
Moraxella bovis	Cattle	Infectious bovine keratoconjunctivitis ['pink eye' or 'New Forest disease' (UK)]	
	Horses	Rare isolates from horses with conjunctivitis	
M. lacunata	Many animal species	Isolated from guinea-pigs, aborted equine foetuses, goats with viral pneumonia and encephalitis, a goat with septicaemia, and from various pathological specimens from dogs and pigs. Its role in disease processes in animals is not known	
	Humans	May cause conjunctivitis	
M. phenylpyruvica	Sheep and cattle	Recovered from the urinogenital tract and brain	Pathogenicity for animals is unknown
	Pigs	Isolated from the urinogenital tract	
	Goats	Obtained from the intestinal tract	

Isolation

Lacrimal secretions should be inoculated as soon as possible after collection on blood agar and incubated at 35°C for 48–72 hours. The inoculation of a MacConkey plate is useful to gauge the degree of contamination by other Gram-negative bacteria. *M. bovis* is unable to grow on MacConkey agar.

Identification

Colonial appearance

After 48 hours' incubation the colonies of *M. bovis* are flat, round, small (1 mm diameter), greyish-white and friable, surrounded by a narrow zone of complete haemolysis (**350**). The appearance is not unlike that of a beta-haemolytic streptococcus. New isolates are often piliated and erode the agar, sinking into it. Colonial growth will autoagglutinate when suspended in saline. On subculture, colonial variation is common, pili are no longer formed and the colonies are butyrous and less likely to autoagglutinate. Some colonies can become non-haemolytic. The strains of *M. bovis* (*'M. equi'*) isolated from horses are non-haemolytic even on primary isolation. *M. lacunata* and *M. phenylpyruvica* are non-haemolytic on blood agar and some strains of *M. phenylpyruvica* will grow on MacConkey agar.

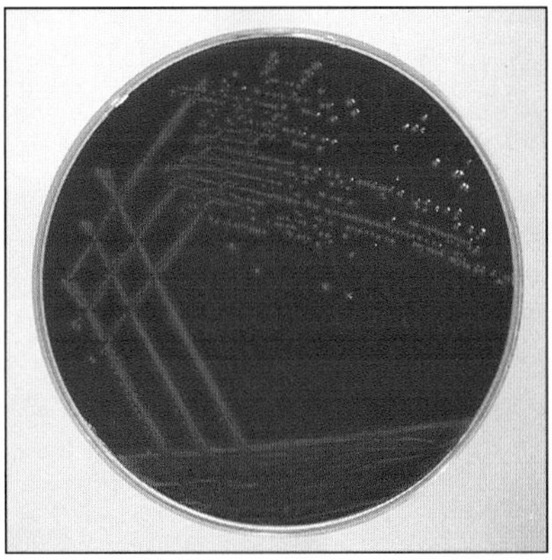

350

350 *Moraxella bovis* on sheep blood agar showing small haemolytic colonies. Some strains have larger zones of beta-haemolysis.

Table 95. Differentiation of the *Moraxella* species.

Characteristic	*Moraxella bovis*	*M. lacunata*	*M. phenylpyruvica*
Beta-haemolysis (blood)	(+)	–	–
Growth on MacConkey agar	–	–	(+)
Oxidase	+	+	+
Catalase	(+)	+	+
Nitrate reduction	–	+	(+)
Urease	–	–	+
Phenylalanine deaminase	–	–	+
Gelatinase	(+)	+	–

+ = positive reaction, (+) = most strains positive, – = negative reaction.

Microscopic appearance
Gram-stained smears from colonial growth reveal fat, Gram-negative rods or cocci, characteristically in pairs (**351**). Specific identification of *M. bovis* can be obtained by a fluorescent antibody technique on smears from colonies.

Biochemical reactions
None of the moraxellae attack carbohydrates with the formation of acid. They are non-motile, indole-negative and all are sensitive to penicillin. *M. bovis* will slowly pit a Loeffler serum slope, is relatively salt-tolerant and grows on media containing 5 per cent NaCl. Litmus milk medium inoculated with *M. bovis* becomes alkaline (blue) and develops three zones: a deep blue upper layer, a soft blue curd in the centre with the bottom white (reduced) and coagulated. **Table 95** gives the characteristics of the moraxellae isolated from animals. The *Moraxella* species have also been differentiated by the analysis of the cellular fatty acid.

Immunology
No serological tests are available.

Animal inoculation
The inoculation of virulent haemolytic and piliated strains of *M. bovis* intraperitoneally into mice or guinea-pigs results in a fatal infection.

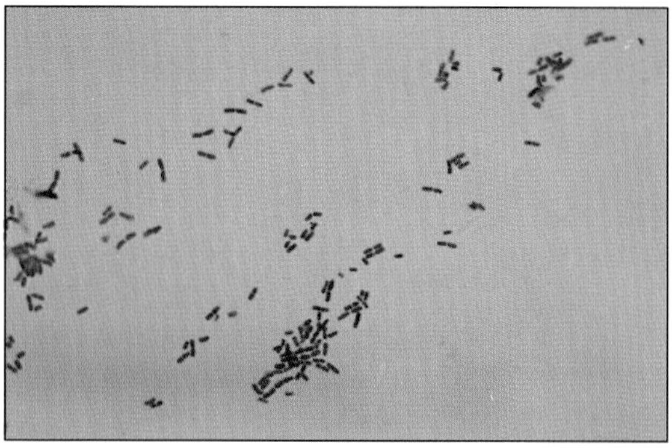

351

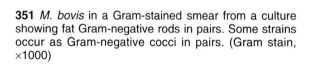

351 *M. bovis* in a Gram-stained smear from a culture showing fat Gram-negative rods in pairs. Some strains occur as Gram-negative cocci in pairs. (Gram stain, ×1000)

30 Glucose-Non-Fermenting, Gram-Negative Bacteria

The glucose-non-fermenting, Gram-negative bacteria use glucose oxidatively or not at all. A member of this group may be suspected when inoculation of triple sugar iron (TSI) agar results in a neutral or alkaline slant and a neutral butt without gas. The *Pseudomonas, Bordetella* and *Moraxella* species, which include significant veterinary pathogens, belong to this category and these are described in other chapters. Other genera belonging to glucose-non-fermenting bacteria, that are of minor veterinary importance, include *Alcaligenes, Acinetobacter, Branhamella, Flavobacterium, Neisseria, Weeksella* species and CDC Groups M5, M6 and EF-4. These bacteria are described in this chapter and their significance is summarised in **Table 96**.

Laboratory Diagnosis

Isolation

These glucose-non-fermenting bacteria are fairly commonly isolated from clinical specimens incubated under aerobic conditions. Most grow at 37°C although this may be above their optimum growth temperature. Pigment is produced by some of these bacteria, including *Flavobacterium* species (**352**).

Identification

A presumptive identification can be made on the basis of a few comparatively simple tests. A general comparison of the genera is given in **Table 97** and some characteristics of the species commonly isolated from clinical materials are shown in **Table 98.** The tests used are as follows:

- Gram-stained smears for morphology. The genera *Neisseria* (**353**), *Branhamella, Acinetobacter* and *Moraxella* can all be Gram-negative diplococci. However, if the smears are made from the culture near the edge of a zone of inhibition around a penicillin disc the following morphology is usually seen:

	Cocci	Rods
Neisseria	+	
Branhamella	+	
Acinetobacter		+
Moraxella		+

- Motility is shown by *Pseudomonas* (polar flagella). *Bordetella* and *Alcaligenes* are peritrichous. A flagellar stain (West *et al.*, 1977) can determine the flagellar arrangement (**Appendix 1**).
- The O-F and other tests are carried out as described in 'Identification of Bacteria' (Chapter 4). For additional tests for this group of non-fermentative bacteria see Clark *et al.* (1984) and Pickett *et al.* (1991).

Both automated and miniaturised commercial systems are available for the identification of the glucose-non-fermenting bacteria. Automated systems include Automicrobic System (Vitek Systems Inc.), Autobac IDX System (General Diagnostics) and AutoSCAN-4 (American Microscan). The range of miniaturised identification systems includes API 20E and API Rapid NFT (Analytab Products, Inc.), Oxi/Ferm (Roche, Inc.) and Flow NF (Flow Laboratories).

352

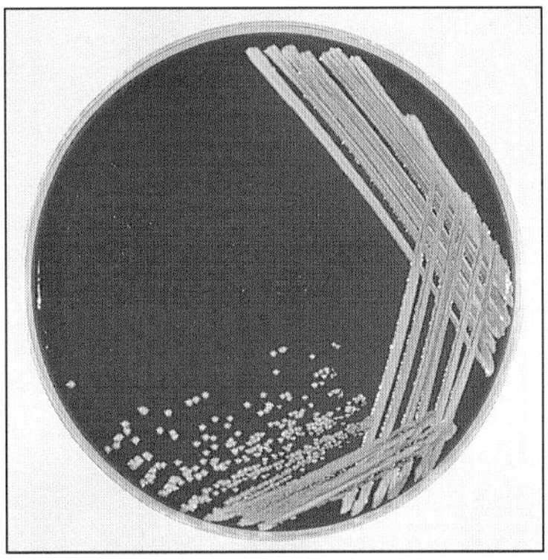

352 *Flavobacterium* species on sheep blood agar showing distinctive yellow pigmentation.

353

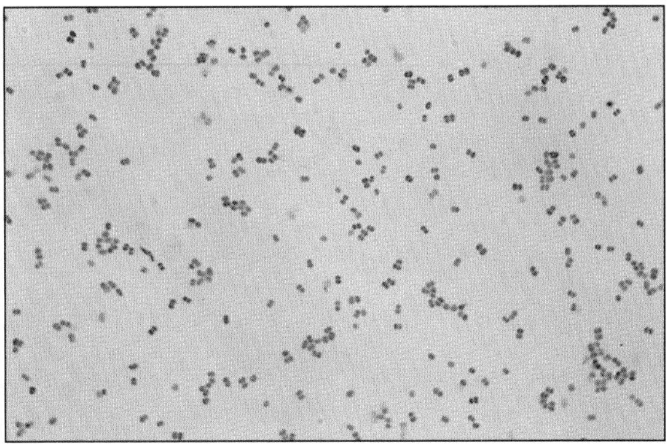

353 *Neisseria* species in a Gram-stained smear from a culture showing the typical fat cocci in pairs. (×1000)

Table 96. The significance of glucose-non-fermenting, Gram-negative bacteria.

Genus and species	Host(s)	Significance
Alcaligenes **species**	Many animals	Saprophytes that are occasionally isolated from the intestinal tract of vertebrates
	Turkeys	The *A. faecalis* strains isolated from turkeys with coryza are now classified as *Bordetella avium*
Acinetobacter **spp.**	Environment	Commonly found in soil, water, sewage, food and milk
	Animals and humans	Part of the normal flora. Can cause nosocomial infections, especially in immunocompromised patients
	Dogs	Has been isolated from the blood of sick animals
	Mink	Associated with bronchopneumonia
Branhamella caviae	Guinea-pigs and dogs	Isolated from the throats of guinea-pigs and from the conjunctiva of dogs
B. cuniculi	Rabbits	Isolated from the oral cavity of healthy rabbits
	Marine pinnipeds	Recovered from the nasopharynx
B. ovis	Sheep, cattle and goats	Considered a commensal of the conjunctiva and upper respiratory tract. Low pathogenicity
B. catarrhalis (*Neisseria catarrhalis*)	Humans	Part of normal nasopharyngeal flora but may cause bronchitis, pneumonia, sinusitis and otitis media
	Calves	Isolated from pneumonic lungs
	Dogs	Found in the upper respiratory tract, mouth and conjunctiva
		Pathogenic significance unclear in animals
Neisseria meningitidis	Humans	Meningococcal meningitis
N. gonorrhoeae	Humans	Gonorrhoea
N. animalis	Guinea-pigs	Isolated from the pharynx
N. denitrificans	Guinea-pigs	Isolated from the nasopharynx
N. canis	Cats and dogs	Isolated from the nasopharynx
N. flavescens and ***N. sicca***	Dogs	Oral cavity and pharynx of healthy dogs
N. lactamica	Dogs	Isolated from the conjunctival sac
N. mucosa	Dolphins	Respiratory tract of a healthy dolphin
Flavobacterium **spp.**	Environment	Soil and water: widespread
	Humans (animals)	Nosocomial infections
F. meningosepticum	Humans	Meningitis and septicaemia in hospitalised infants. Pneumonia and meningitis in compromised adults
Weeksella zoohelcum (previously *Flavobacterium* sp. group IIj)	Dogs and cats	Part of the normal flora of the oral cavity and paws. Not thought to be pathogenic for animals
	Humans	Infected animal bites and scratches
CDC group M5 (similar to *Moraxella* spp.)	Dogs	Commensal in canine oropharynx
	Humans	Infected dog bites
CDC group M6	Humans	Isolated from clinical specimens
CDC group EF-4	Dogs and cats	Oral cavity and nasopharynx. Of no pathological significance
	Humans	Infection of dog and cat bites

Table 97. Comparison of the characteristics of Gram-negative, glucose-non-fermenting bacteria.

	Morphology	Acid from oxidation of glucose*	Haemolysis	Growth on MacConkey agar	Motility	Catalase	Oxidase	Pigment	Other characteristics
Pseudo-monas	Medium -sized rods	+	+ β (*P. aeruginosa)*	+	(+) polar	+	+	(+)	*P. aeruginosa* has grape-like odour and bluish pigment
Moraxella	Fat rods or cocci in pairs	–	(+)β (*M. bovis*)	(–)	–	(+)	+	–	Pathogenic *M. bovis usually haemolytic*
Bordetella	Cocco-bacilli	–	(+)β (*B. bronchi-septica*)	(+)	(+) peri-trichous	+	+	–	*B. bronchiseptica*: urease + and *B. avium*: urease –. Both citrate-positive
Alcaligenes	Rods or cocco-bacilli	–	–	+	+ peri-trichous	+	+	–	*A. faecalis*: urease – and citrate +
Acineto-bacter	Rods or diplo-cocci	v	–	+	–	+	–	–	Blue colonies on EMB medium
Branha-mella	Cocci in pairs	–	(+)β	–	–	+	+	–	Remain coccal near a penicillin disc
Neisseria	Cocci in pairs	(+)	(–)	–	–	+	+	+ yellow	Growth enhanced by 10% carbon dioxide
Flavo-bacterium	Rods or cocco-bacilli	(+)	+α	(+)	–	+	+	+ yellow	*F. odoratum*: fruity odour
Weeksella	Rods	–	vα	–	–	+	+	–	Hydrolyse urease rapidly. Small, sticky colonies
CDC group M5	Cocco-bacilli	–	vβ	v	–	+	+	–	
CDC group M6	Cocco-bacilli	–	–	v	–	–	+	–	
CDC group EF-4	Short rods or cocco-bacilli	(+)	–	v	–	+	+	+ yellow to tan	EF-4a ferments glucose. EF-4b oxidises glucose. No other sugars utilized

* = O-F test, + = oxidative, and – = unreactive. R = resistant and S = susceptible.
+ = positive, (+) = most positive, (–) = most negative, – = negative, v = variable reactions.

Table 98. Reactions of some Gram-negative, glucose-non-fermenting bacteria.

	Acid from glucose	Haemolysis (blood agar)	Growth on MacConkey agar	Motility	Catalase	Oxidase	Nitrate reduction	Urease	Indole	Gelatinase	Penicilline susceptibility	Pigment
Alcaligenes faecalis	–	–	+	+	+	+	–	–	–	–	S	–
A. xylosidans* subsp. *xylosidans	v	–	+	+	+	+	+	–	–	–	S	–
subsp. *denitrificans*	–	–	+	+	+	+	+	v	–	–	S	–
A. piechaudii	–	–	+	+	+	+	+	–	–	–	S	–
Acinetobacter calcoaceticus	+	–	+	–	+	–	–	v	–	–	R	–
A. lwoffii	–	–	+	–	+	–	–	–	–	–	R	–
Branhamella catarrhalis	–	–	–	–	+	+	(+)	–	–	–	S	–
B. caviae	–	β weak	–	–	+	+	+	–	–	–	S	–
B. cuniculi	–	–	–	–	+	+	–	–	–	–	S	–
B. ovis	–	(β)	–	–	+	+	(+)	–	–	–	S	–
Neisseria canis	–	–	–	–	+	+	+	–	–	+	S	Yellow
N. flavescens	–	–	–	–	+	+	–	–	–	–	S	Yellow
N. sicca	O	v	–	–	+	+	–	–	–	–	S	v (yellow)
N. lactamica	O	–	–	–	+	+	–	–	–	–	S	Yellow
N. denitrificans	O	–	–	–	+	+	–	–	–	–	S	v (yellow)
N. mucosa	O	–	–	–	+	+	+	–	–	–	S	v (yellow)
Flavobacterium meningosepticum	O	(α)	+	–	+	+	–	–	+	+	R	v (yellow)
F. indologenes	O	v α	v	–	+	+	v	(–)	+	(+)	R	Yellow
F. odoratum	–	(α)	+	–	+	+	–	+	–	+	R	v (yellow)
F. multivorum	O	(α)	+	–	+	+	–	+	–	–	R	Yellow
Weeksella zoohelcum	–	v(α)	–	–	+	+	–	+	+	+	S	–
CDC group M5	–	v β	v	–	+	+	–	–	–	–	S	–
CDC group M6	–	–	v	–	–	+	+	–	–	–	(S)	–
CDC group EF-4	(O)	–	v	–	+	+	+	–	–	–	•	Yellow/tan

* O = oxidative, – = unreactive, (O) = most strains oxidative.
+ = positive reaction, (+) = most strains positive, (–) = most strains negative, – = negative reaction, v = variable, S = susceptible, R = resistant, • = data unavailable, β = beta-haemolysis, α = alpha haemolysis, ( ) = most strains.

References

Clark, W.A., Hollis, D.G., Weaver, R.E. and Riley, P. (1984). Identification of unusual pathogenic gram-negative aerobic and facultatively anaerobic bacteria. Centers for Disease Control, Atlanta, Georgia, USA.

Pickett, M.J., Hollis, D.G. and Bottone, E.J. (1991). Miscellaneous gram-negative bacteria, pp. 410–428. In Lennette, E.H., Balows, A., Hausler, W.J., Jr. and Shadomy, H.J. (ed.), *Manual of Clinical Microbiology*, 5th ed., American Society for Microbiology, Washington, DC, USA.

West, M., Burdash, N.M. and Freimuth, F. (1977). Simplified silver-plating for flagella, *Journal of Clinical Microbiology*, **6**: 414–419.

31 The Spirochaetes

The order Spirochaetales includes the families *Spirochaetaceae* and *Leptospiraceae*. The genera of significance in animals and humans are *Serpulina, Treponema* and *Borrelia* (*Spirochaetaceae*) and *Leptospira* (*Leptospiraceae*). Two of the *Treponema* species of veterinary interest have been placed recently in a new genus, *Serpulina*. The spirochaetes are slender, motile, flexuous, unicellular, helically coiled bacteria ranging from 0.1–3.0 µm in width. The outer sheath, the outermost layer of a spirochaete cell, is a multilayered membrane that completely surrounds the periplasmic flagella (axial filament) and the helical protoplasmic cylinder. The cylinder consists of the nuclear material, cytoplasm, cytoplasmic membrane and the peptidoglycan portion of the cell wall. The periplasmic flagella are wrapped around the cylinder and are in the periplasmic space of these Gram-negative bacteria. One end of each flagellum is inserted near a pole of the protoplasmic cylinder and attached by plate-like structures called insertion discs. The distal end of each flagellum is not inserted and extends to the centre of the cell and may overlap the flagellum from the opposite end (**Diagrams 35** and **36**). The periplasmic flagella facilitate the motility of the bacteria in viscid environments. **Table 99** and **Diagram 37** summarise the main differences between the genera *Leptospira, Serpulina (Treponema)* and *Borrelia*.

Changes in nomenclature

Previous name	**Present name**
Treponema hyodysenteriae	*Serpulina hyodysenteriae*
Treponema innocens	*Serpulina innocens*

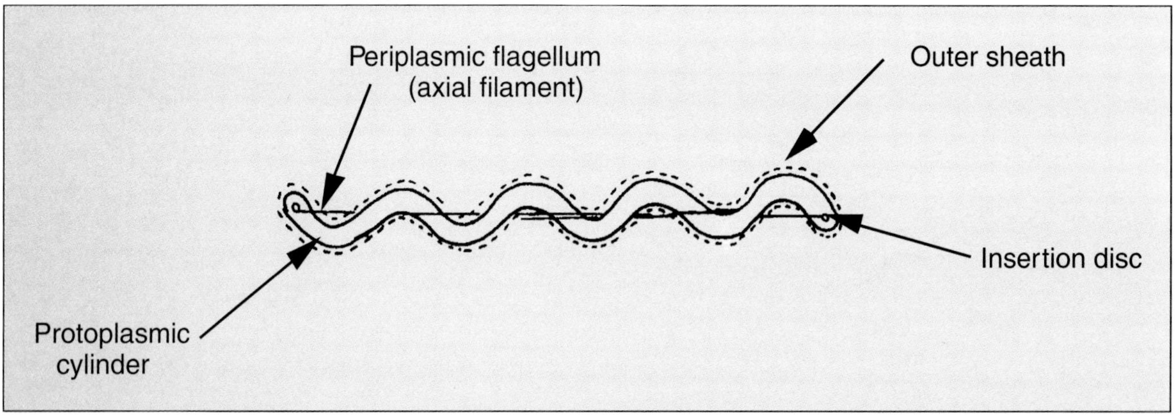

Diagram 35. A spirochaete.

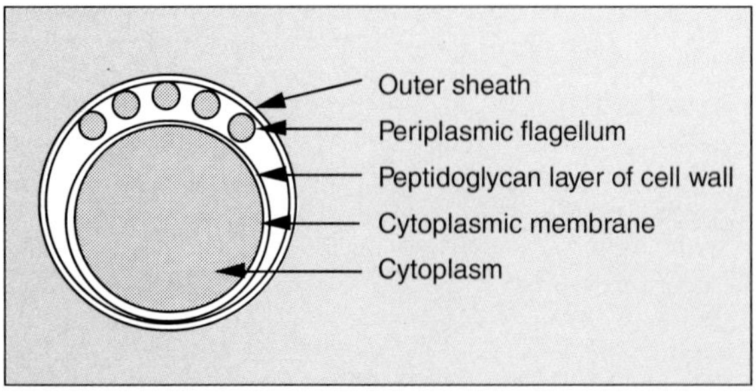

Diagram 36. Section through a spirochaete demonstrating structural features.

Leptospira interrogans serovars

The genus *Leptospira* is at present divided into two species, *L. interrogans* (parasitic) and *L. biflexa* (saprophytic). However, this classification is currently under review. It has been proposed that *L. interrogans* should be divided into six species: *L. borgpetersenii, L. interrogans, L. noguchii, L. santarosai, L. weilii* and *L. kirschneri* (Subcommittee on Taxonomy of *Leptospira*, 1992). As this species designation has not

Table 99. Main differences between the genera *Leptospira, Serpulina/Treponema* and *Borrelia*

	Leptospira	***Serpulina/Treponema***	***Borrelia***
Morphology	Many spirals, tight and fine	6–14 regular spirals with 1 μm amplitude.	4–8 loose spirals with 3μm amplitude
Length	6–20 μm	5–20 μm	3–20 μm
Width	0.1–0.2 μm	0.1–0.5 μm	0.2–0.5 μm
Periplasmic flagella	2	6–10	15–20
Insertion discs	3–5	1	2
Ends	One or both ends hooked	Pointed	Tapers to a fine filament
Atmosphere	Aerobic	Microaerophilic or anaerobic	Microaerophilic
Insect vectors required	No	No	Yes
DIRECT MICROSCOPY			
Dark-Field	+	+	+
DCF stain	–	+	+
Giemsa	–	Poor	+
Silver impregnation	+	+	+
FA technique	+	+	+
Movement	Rotation around central axis, flexing and translational, undulatory movement	Rotation and majestic translational movement and stiff flexion	Frequent reversal of translational movement, corkscrew-like and lashing

been validated by the International Committee on Systematic Bacteriology, the term *L. interrogans* will be used when referring to pathogenic leptospires. Some of the differences between *L. biflexa* and *L. interrogans* are shown in **Table 100**. The test for conversion of cells to spherical forms is conducted with 75 per cent washed cells in 1M NaCl between 20–30°C and occurs within 2 hours.

L. interrogans is divided into more than 180 serovars on the basis of antigenic composition. The serovars with antigens in common are placed in serogroups for diagnostic convenience. The term 'serogroup' has no taxonomic significance. Many serovars appear to have a certain animal species as a natural host, but animals and humans can be infected with a wide variety of serovars. The serovars causing disease in animals vary between countries and sometimes between regions in the same country. Usually the majority of infections in domestic animals, in a particular region, are caused by only a few serovars of *L. interrogans*. **Table 101** lists some of the serovars, and the serogroups to which they belong, that have been commonly isolated from domestic and wild animals.

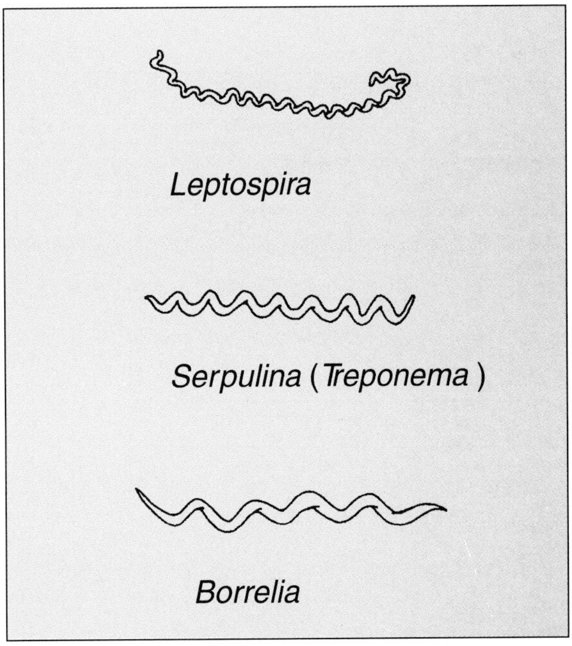

Diagram 37. A comparison of the general morphology of *Leptospira, Serpulina (Treponema)* and *Borrelia* (*see* also Table 99).

Table 100. Differential features of the *Leptospira* species.

	L. interrogans	***L. biflexa***
Pathogenicity	+	–
Growth at 13°C	–	+
Growth inhibited by 8-azaguanine (225 μg/ml)	+	–
Conversion of cells to spherical forms by 1M NaCl	+	–

+ = positive reaction, – = negative reaction.

Table 101. Common *Leptospira interrogans* serovars and their hosts.

		Occurrence in:						
Serogroup	**Serovar**	**Cattle**	**Pigs**	**Dogs**	**Horses**	**Sheep**	**Rodents**	**Other wildlife**
Australis	*australis*	+						+
	bratislava	+	+		+			
Autumnalis	*autumnalis*						+	+
Ballum	*ballum*						+	+
Bataviae	*bataviae*	+					+	
Canicola	*canicola*	+	+	+			+	+
Grippotyphosa	*grippotyphosa*	+	+	+	+		+	+
Hebdomadis	*hebdomadis*							+
	szwajizak	+						
Ictero-haemorrhagiae	*ictero-haemorrhagiae*	+	+	+	+		+	+
	copenhageni	+		+			+	
Pomona	*pomona*	+	+	+	+	+		+
Sejroe	*balcanica*	+						
	hardjo	+			+	+		
	saxkoebing	+						
	sejroe	+					+	+
Tarassovi	*tarassovi*		+					

Natural Habitat

Leptospires are present in the tubules of mammalian kidneys and are excreted in urine, often for several months. Characteristically, a reservoir host shows minimal or no clinical signs. A protracted shedder state can follow in animals that have recovered from clinical disease. Significant reservoirs of animal and human leptospirosis are shown in **Table 102.** Indirect exposure depends on wet environmental conditions and a mild climate that favour the survival of leptospires. Streams and ponds can be a source of infection as can aerosols of urine in cowsheds and milk from infected cows. Leptospirosis occurs worldwide.

Table 102. Significant reservoirs of pathogenic leptospires.

Leptospira interrogans serovar	Main reservoir(s)	Alternative reservoir(s)
autumnalis	wildlife	–
ballum	wildlife	–
bratislava	pigs, horses, cattle	–
canicola	dogs	cattle, pigs, rodents
grippotyphosa	wildlife	dogs, cattle, pigs
hardjo	cattle	sheep
icterohaemorrhagiae	rats	dogs, cattle, pigs
pomona	pigs, cattle	dogs, wildlife

Pathogenesis

Leptospires gain entry through mucous membranes or damaged skin from direct or indirect contact. After epithelial penetration there is haematogenous spread with localisation and proliferation in parenchymatous organs, particularly the liver, kidneys, spleen and sometimes meninges. In the kidneys the organisms reach and localise in the lumen of proximal convoluted tubules. Penetration and multiplication in the foetus can occur in pregnant animals leading to foetal death and resorption, abortion or weak offspring. The foetus, if infection occurs in the third trimester, can produce specific antibodies and may overcome the infection. Antibody production in infected animals begins a few days after the onset of leptospiraemia. The leptospires tend to persist in sites such as renal tubules, eyes and uterus where antibody activity is minimal.

Leptospires damage vascular endothelium resulting in haemorrhages. Serovars in the serogroups Autumnalis, Grippotyphosa, Icterohaemorrhagiae and Pomona produce a haemolysin that is probably responsible for the haemoglobinuria (redwater) in young calves infected with these serovars. Cytotoxic protein is produced by virulent strains but the role of the toxin is unknown. Virulence varies between the serovars and between two genotypes of *L. interrogans* serovar *hardjo* known as *hardjobovis* and *hardjoprajitno*. **Table 103** summarises the syndromes of leptospirosis in domestic animals. Serological evidence indicates that a variety of leptospiral serovars can infect cats but disease appears to be uncommon in this animal.

Laboratory Diagnosis

Specimens

- Whole blood (serum) for serological tests.
- Mid-stream urine for dark-field examination:
  a) If the urine cannot be examined within 20 minutes, it should be neutralised with N/10 HCl or N/10 NaOH.
  b)To preserve the morphology of the leptospires for several days, 20 ml of the mid-stream urine should be added immediately to 1.5 ml of 10 per cent formalin in a 30 ml bottle. Leptospires can disintegrate quite quickly in a urine sample, especially if it is acidic. By adding formalin, the leptospires will be killed but will retain their morphology for several days and can be examined by darkfield microscopy.
- Mid-stream urine for attempted culture should be diluted 1:10 with 1 per cent bovine serum albumin (BSA) for transport to the laboratory. Alternatively, a few drops of the urine can be aseptically added to culture media on a 'beside-the-cow' basis.
- Whole blood (5 ml) from an animal in the leptospiraemic phase can be collected in 0.5 ml of 1.0 per cent sodium oxalate or 0.1 ml of 1.0 per cent heparin for attempted culture of leptospires.
- Kidneys:
  a) A small plug of kidney, taken from inside the organ with a sterile glass Pasteur pipette, can be macerated with a little distilled water and examined under dark-field microscopy.
  b) Kidney tissue can be used for culture. Leptospires remain viable in unfrozen kidneys for several days after the death of the animal.

Table 103. Leptospirosis in domestic animals.

Host	*Disease syndrome*
Cattle	• Subclinical with or without leptospiruria
	• Milk-drop syndrome, with or without any other clinical signs (often *hardjo*)
	• Abortion and neonatal mortality: abortion 'storms' (*pomona*) and sporadic abortions (*hardjo*)
	• Infertility (often *hardjo*)
	• Haemoglobinuria, jaundice and fever in calves and, less commonly, in young adults. Serovars commonly involved are *pomona*, *grippotyphosa* and *icterohaemorrhagiae*. Occasionally, some animals show signs of meningitis
Pigs	• Subclinical, often with leptospiruria: especially with *pomona*. Pigs are considered to be the maintenance host for this serovar
	• Fever and focal non-suppurative mastitis and leptospiruria
	• Infertility, abortions and stillbirths: often *canicola*, *pomona* or *icterohaemorrhagiae*
	• Fever, anorexia, jaundice, haemoglobinuria and high mortality in young pigs: often *icterohaemorrhagiae*
Dogs	• Subclinical with leptospiruria: often *canicola*
	• Acute haemorrhagic disease: high fever, vomiting, prostration and often early death; usually *icterohaemorrhagiae*
	• Less acute icteric type: intense icterus, depression, fever, haemorrhages with blood in faeces and urine; *canicola* or *icterohaemorrhagiae*
	• Uraemic type: uraemia associated with extensive kidney damage, ulcerative stomatitis and uraemic breath. Death occurs in a high percentage of cases. These severe signs can occur 1–3 years after the initial infection: often *canicola*
	• Rarely a chronic, active hepatitis: seen in a *grippotyphosa* infection
Horses	• Recurrent iridocyclitis ('periodic ophthalmia' or 'moon blindness') which can result in blindness. Aetiology has not been conclusively determined
	• Occasionally abortion with foetuses of 6 months to term
	• Rarely fever, anorexia, depression and icterus
Sheep	• Mainly subclinical infections with leptospiruria: serovars such as *hardjo*
	• Occasionally, acute leptospirosis with depression, dyspnoea, haemoglobinuria, anaemia and high mortality in lambs: often *pomona*

c) Kidney and/or liver sections in 10 per cent formalin for histopathology (**354**) can be used.
d) Foetal kidney: cryostat sections or smears for FA technique. The morphological characteristics of the leptospires are usually retained better in porcine foetal kidneys than they are in those of foetal calves, which are frequently autolyzed.

- Foetal abomasal contents, cotyledons and uterine discharge should be collected for differential diagnosis in abortion cases.

In suspected leptospirosis in cattle and pigs, the infection should be investigated on a herd basis, with urine and serum samples taken from clinical and in-contact animals. Leptospires may be excreted in urine intermittently and leptospiral antibody titres can vary considerably between individual animals. **Diagram 38** summarises the specimens and laboratory tests for the diagnosis of leptospirosis.

354

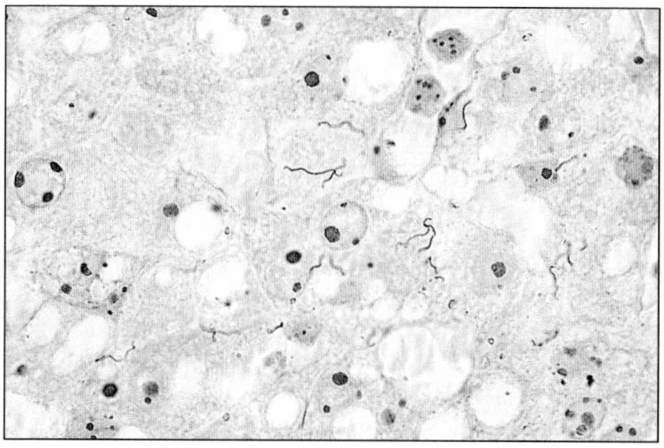

354 *Leptospira interrogans* serovar *icterohaemorrhagiae* in a section of canine liver. (Levaditi stain, ×1000)

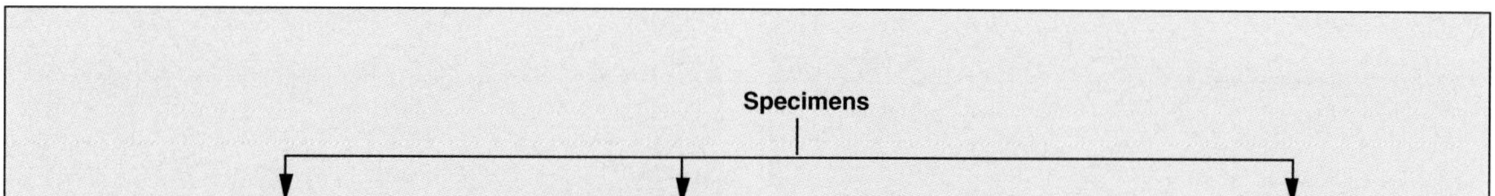

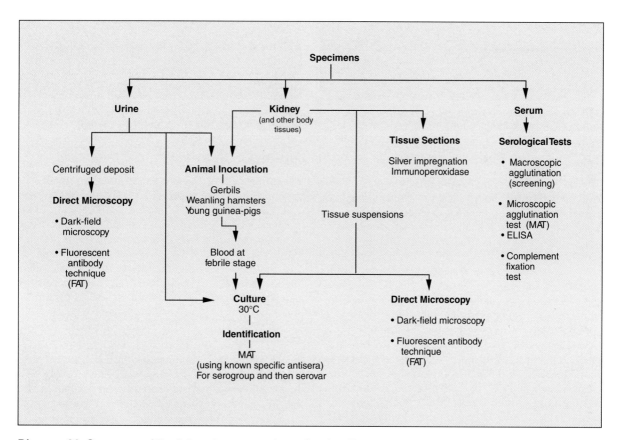

Diagram 38. Summary of the laboratory procedures for the diagnosis of leptospirosis.

Direct microscopy

L. interrogans serovars average only 0.1–0.2 µm in width and cannot be resolved under the light microscope. They stain weakly, or not at all, with both aniline and Romanowsky stains. But leptospires can be demonstrated in urine, other body fluids and tissues by darkfield microscopy (**355**) and fluorescent antibody (FA) technique (**356**). Urine is centrifuged at 9750 rcf for 10 minutes to concentrate the leptospires. Unclotted blood is centrifuged at low speed to sediment the red cells after which the plasma can be removed and centrifuged at 9750 rcf for 10 minutes. Tissue specimens are ground in a Ten Broeck tissue grinder and prepared as a 10 per cent suspension in 1 per cent bovine serum albumin (BSA, Fraction V powder). A drop of the preparation is placed on a microscope slide, under a cover slip, and examined by darkfield microscopy. The FA technique can be carried out on urinary deposits, tissue impression smears or tissue sections. Silver impregnation or immunoperoxidase staining techniques can be used on tissue sections of kidney, liver or other tissues.

Under darkfield microscopy leptospires can just be seen under the low power if large numbers are present. Under high-dry or oil-immersion objectives they appear as silver, beaded (due to the helical coils), slender rods with a distinct hook at each end (**355**). The *L. interrogans* serovars are obviously motile, if they are still viable, with flexuous movements and spinning around the long axis. The characteristic morphology is often less evident in FA or silver impregnation preparations.

355

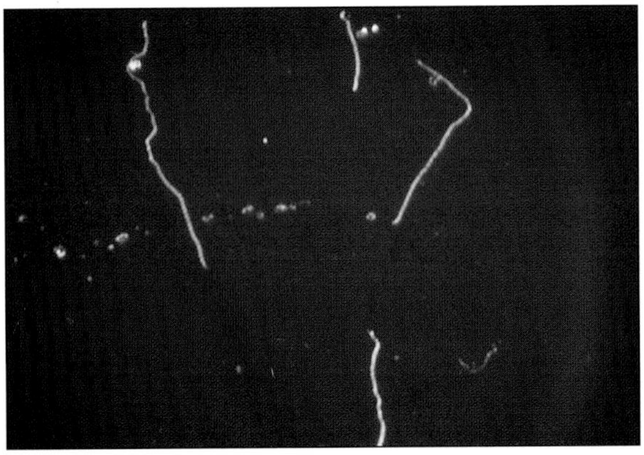

355 *Leptospira interrogans* serovar *canicola* from a young, actively dividing culture. (Darkfield, ×1000)

356

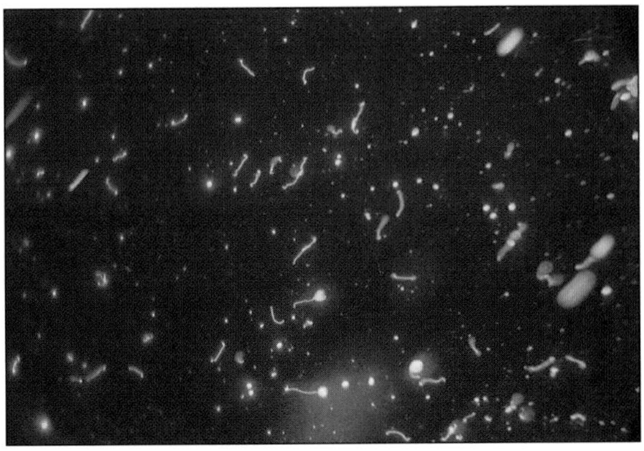

356 *Leptospira interrogans* serovar *hardjo* in a bovine urinary deposit. (Direct FA technique, ×400)

Culture

The media used for the culture of leptospires include Korthof and Stuart broths, Fletcher semisolid medium, EMJH and Tween 80-albumin medium (OAC) that are liquid but can be made semi-solid by the addition of 1.5 g agar/litre of medium. Contaminated samples (or cultures) can be filtered through a 0.45 µm bacteriological filter and inoculated into a medium containing 5-fluorouracil at 200 µg/ml. The media can be inoculated with 1–3 drops of carefully taken urine within a few minutes of collection, or urine diluted 1:10 with 1 per cent BSA, as soon as possible after collection.

A 10 per cent tissue suspension in 1 per cent BSA (1–2 drops) or a few drops of oxalated or heparinised blood can also be inoculated into media. The cultures are incubated at 30°C for up to 8 weeks. A drop of the culture is examined by darkfield microscopy once weekly. *L. hardjo* is one of the slowest growing serovars and *bratislava* is difficult to culture in laboratory media.

Animal inoculation

Weanling gerbils, hamsters and guinea-pigs can be inoculated intraperitoneally with 0.5–1.0 ml of neutralised urine, unclotted blood or a 10 per cent tissue suspension in EMJH or 1 per cent BSA. Cardiac blood is taken aseptically when a temperature rise is detected or at 5, 8, 10 and 14 days post-inoculation. Media are inoculated with 2–3 drops of the freshly collected blood. Animals surviving for 21 days post-inoculation can be euthanized and blood samples taken for serology.

Identification

FA techniques or immunoperoxidase staining are usually carried out on the original specimens. The cultures are typed by a microscopic agglutination test (MAT) using known specific antisera. An isolate used as antigen can be screened by a selected group of 12 or more serovar antisera to determine the serogroup after which the serovar is investigated. Isolates can be sent to a leptospiral reference laboratory for identification. Definitive culture typing is described in detail by Faine (1982).

Alternative methods for the classification of leptospires by the use of restriction endonuclease DNA analysis and monoclonal antibodies are being employed in some reference laboratories.

Serology

Macroscopic and microscopic agglutination tests, and to a lesser extent complement fixation tests and the ELISA technique are used for the detection of leptospiral antibodies in serum. The macroscopic agglutination test is a screening test and uses dead antigens but suffers from lack of specificity. The leptospiral microscopic agglutination test uses live leptospires as antigen and is highly sensitive and serovar-specific. Details of the macroscopic and microscopic agglutination tests are given by J. R. Cole, Jr. (1990) and the method for a microtiter procedure for the MAT can be obtained from the National Leptospirosis Reference Center, National Veterinary Services Laboratory, Box 844, Ames, Iowa 50010, USA. A titre of 1:100 is suspicious for infection in the MAT and those of 1:200 or more are regarded as positive. The titres are often 1:800 or more in an active infection.

Serpulina (Treponema) species

Members of the genus *Serpulina* (*Treponema*) are host-associated spirochaetes found in the oral cavity, intestinal tract and genital region of animals and humans. The cells are wider, are not as tightly coiled as the leptospires and can be stained by aniline dyes. The species of veterinary significance are *S. hyodysenteriae* (swine dysentery) and *T. paraluiscuniculi* (vent disease of rabbits). *S. innocens*, present in the faeces of pigs and dogs, is thought to be non-pathogenic, although some workers regard *S. innocens* as a strain of *S. hyodysenteriae* of lower virulence. Blaha *et al.* (1984) produced signs of swine dysentery in pigs with *S. innocens* after five passages. *Serpulina* species are anaerobic. *S. hyodysenteriae* and *S. innocens* are similar morphologically, culturally as well as biochemically and can be grown on laboratory media. *T. paraluiscuniculi* has not been cultured *in vitro*.

Normal Habitat

T. paraluiscuniculi produces a benign venereal disease of rabbits and is present in lesions in the genito-perineal area of rabbits. It causes latent infection in mice, guinea-pigs and hamsters. The treponemes can be found in the lymph nodes of these animals. The reservoir of *S. hyodysenteriae* is the intestinal tract of pigs, wild rats and mice. Recovered, asymptomatic pigs can excrete these organisms in faeces for 3 months or more. Survival of *S. hyodysenteriae* in soil or voided pig faeces is short, about 24–48 hours. Infection is by the faecal-oral route.

Pathogenesis

After *S. hyodysenteriae* enters a susceptible pig the spirochaete invades goblet cells of the colonic mucosa, multiplies in the crypts of Lieberkuhn and causes necrosis and erosion of the mucosal cells. The faecal material is watery and contains mucus, blood and necrotic debris. Only the large intestine is involved and colonic malabsorption occurs. The lipopolysaccharides of the bacterium are thought to play a part in the pathogenicity. *S. hyodysenteriae* is unable to produce dysentery in gnotobiotic pigs and it appears that the pathogen requires the interaction of other bacteria of the normal flora such as *Fusobacterium necrophorum, Bacteroides vulgatus, B. fragilis* and *Campylobacter coli*. The disease was once thought to be due to *C. coli* as these curved rods can often be seen in significant numbers in faecal smears from pigs with swine dysentery.

Laboratory Diagnosis

Specimens

Deep mucosal scrapings should be taken from a portion of the affected large intestine from a dead pig, or rectal swabs and faeces from several live affected pigs. The numbers of *S. hyodysenteriae* can sometimes be low in faeces and are difficult to see. Sections of affected colon in 10 per cent formalin should be taken for histopathology.

Direct microscopy

Various methods can be used to visualise *S. hyodysenteriae*:

- A portion of deep scrapings from the mucosa or faecal material is examined in a drop of water under darkfield microscopy.
- Fixed smears of mucosal scrapings or faeces are stained by dilute carbol fuchsin (DCF) for 4–6 minutes, by Victoria blue 4-R (**Appendix 1**) (**357**) or by a silver impregnation technique (**358**).
- Histological sections of colon can be stained by a silver impregnation stain or Victoria blue 4-R.
- Fluorescent antibody technique can be carried out on mucosal scrapings or faeces (**359**).

Three to 5 serpulinas per high power field is considered significant. *S. hyodysenteriae* is 0.3–0.4 μm in width, 6–8 μm in length, is loosely coiled with 2–4 coils and tapered ends. It is motile by flexing movements under the darkfield. Pigs normally have other non-pathogenic treponemes such as *S. innocens* and smaller, tightly coiled spirochaetes in the intestinal tract (**360**) that can complicate the interpretation of darkfield examination or stained smears.

357 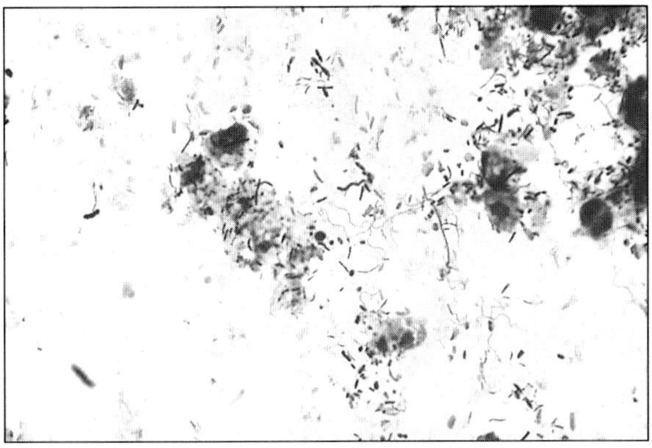

357 *Serpulina hyodysenteriae*: characteristic, delicate, spiral forms in a faecal smear from a pig with swine dysentery. (Victoria blue stain, ×1000)

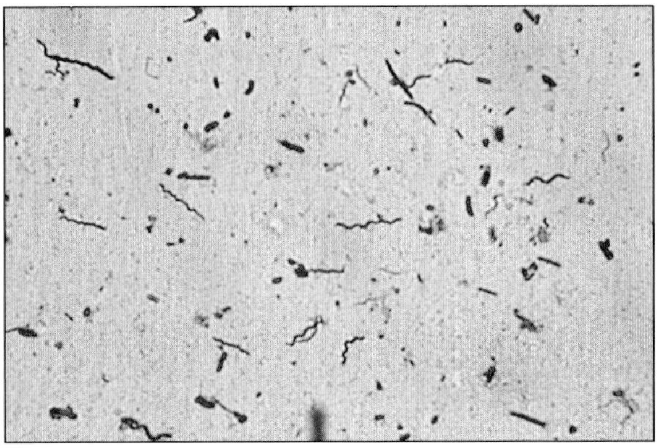 358

358 *S. hyodysenteriae* in porcine faeces. (Silver stain, ×1000)

359 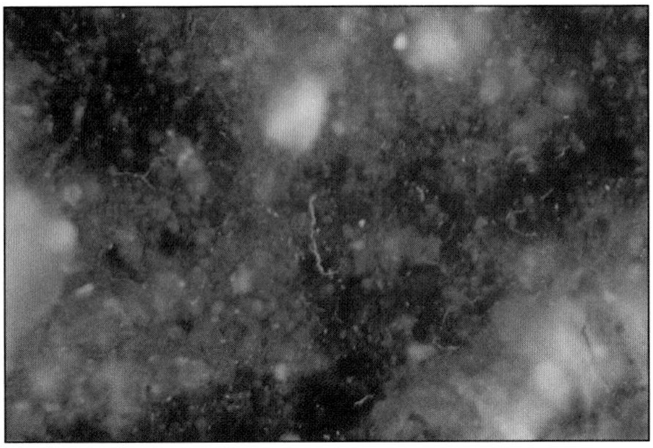

359 *S. hyodysenteriae* in a faecal smear from a pig with swine dysentery. (Direct FA technique, ×400)

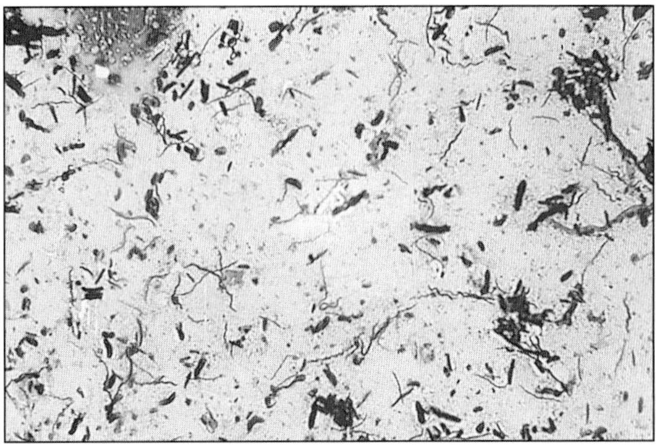 360

360 Spirochaetes in porcine faeces showing the characteristic morphology. (DCF stain, ×1000)

Isolation

S. hyodysenteriae is anaerobic but oxygen-tolerant and growth is enhanced by CO_2. It grows well on freshly poured, or pre-reduced, trypticase soy agar with 5 per cent blood and 400 μg/ml spectinomycin (Upjohn) at 42°C for 2–3 days under an atmosphere delivered by an $H_2 + CO_2$ envelope. Increased growth has been obtained using Fastidious Anaerobe Agar (Lab M) with 5 per cent sheep blood and also by the incorporation of 1 per cent sodium RNA (BHD Chemicals) into media. Sodium RNA can be sterilized by filtration or by autoclaving. It is a growth enhancer and also increases the difference in haemolysis between *S. hyodysenteriae* and *S. innocens*. Colonic mucosal scrapings or faeces can be prepared as a 1:10 suspension in saline and clarified by centrifugation at slow speed. The supernatant can be passed, serially, through 0.8, 0.65 and 0.45 μm filters and this material is used to streak the culture medium.

Identification

Colonial characteristics
After about 48 hours' incubation *S. hyodysenteriae* appears as small, translucent colonies with a zone of clear haemolysis. *S. innocens* is weakly beta-haemolytic.

Microscopic appearance
DCF-stained smears or darkfield microscopy from the colony reveals the typical helical serpulinas.

Biochemical characteristics
S. hyodysenteriae must be distinguished from *S. innocens*. Bélanger and Jacques (1991) suggest that a rapid and simple differentiation can be based on the combined results of the haemolysis, the haemolysis intensification test and an indole spot test.

- Haemolysis: a distinctive characteristic of *S. hyodysenteriae* is the production of a strong beta-haemolysis that is best demonstrated on sheep blood agar. *S. innocens* gives a weak reaction.

- Haemolysis intensification: this test can be carried out by cutting an agar block (0.5 cm^2) from a 4-day culture and by incubating the plate for another 4 days. The intensification of haemolysis is recognised by a lighter zone, 1–2mm wide, along the cut line. *S. hyodysenteriae* gives this reaction (ring phenomenon) but *S. innocens* does not (**361**).
- Indole spot test for anaerobes: the reagent is 1 per cent *p*-dimethyl-aminocinamaldehyde in 10 per cent conc. HCl (REMEL : Regional Media Laboratories Inc.). This is used to saturate a strip of filter paper in a Petri dish and growth from a culture of the test bacterium is smeared on the filter paper. *S. hyodysenteriae* gives a positive reaction indicated by a blue colour usually within 1–3 minutes. *S. innocens* is negative and the filter paper remains pink.

Neither *S. hyodysenteriae* nor *S. innocens* are very reactive biochemically and, with field strains particularly, they have many reactions in common. **Table 104** summarises the rapid presumptive differentiation between the two *Serpulina* species.

361

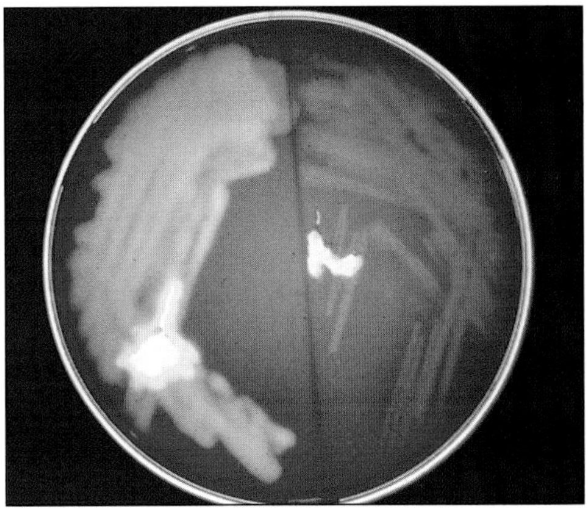

361 S. *hyodysenteriae* (left) showing strong beta-haemolysis and the 'ring phenomenon'. *S. innocens* (right) gives a weaker haemolysis and no 'ring phenomenon'.

Tests for *S. hyodysenteriae* antigens

- The fluorescent antibody test is useful as a screening test for identification of *S. hyodysenteriae* in specimens but both false-positive and false-negative results can occur (Lysons and Lemcke, 1983).
- A slide agglutination test (Burrows and Lemcke, 1981) and a microscopic agglutination test (Lysons, 1991) using absorbed polyclonal serum, have been described. The slide test requires a considerable amount of pure culture and non-specific clumping in the saline control can occur. The use of monoclonal antibodies would increase the specificity of both tests.

Tests for *S. hyodysenteriae* antibodies

Several serological procedures have been developed for the serodiagnosis of swine dysentery and for the detection of carrier animals. The ELISA (Joens *et al*., 1982) appears to be the most sensitive but should be used as a herd test, rather than for individual animals, as false-positive and false-negative results can occur.

Borrelia species

The *Borrelia* species are spirochaetes highly adapted to arthropod transmission. The infections affecting animals and humans have blood-borne phases and can become localised and generalised. The animal pathogens include *B. anserina* (avian spirochaetosis), *B. burgdorferi* (Lyme disease), *B. theileri* (tick spirochaetosis of cattle) and possibly *B. coriaceae* which may prove to be the agent of epizootic bovine abortion.

Table 104. Differentiation of *Serpulina hyodysenteriae* and *Serpulina innocens*.

	Serpulina hyodysenteriae	*Serpulina innocens*
Size (length)	6–8 μm	6–8 μm
Optimum temperature	42°C	37°C
Beta-haemolysis	strong	weak
Haemolysis intensification	+	–
Indole spot test	+	–

+ = positive reaction, – = negative reaction

Natural Habitat

Ticks are the main reservoir of borreliae affecting animals. The ticks can remain infected for years after feeding on affected animals. Other arthropods can be short-term vectors. **Table 105** indicates the vectors and geographical distribution of the *Borrelia* species of veterinary significance.

Pathogenesis and Pathogenicity

Endotoxin may be involved in the pathogenesis of the *Borrelia* spp. In Lyme disease, immune complexes and immunosuppression may also play a part in the pathogenesis. *B. burgdorferi* infects humans, dogs and other domestic animals. One of the first manifestations of Lyme disease in man is a skin rash around the site of the tick bite (**362**). A rash has not been reported in dogs where the characteristic clinical feature is a migratory arthritis.

Table 106 summarises the hosts and diseases caused by the borreliae of veterinary importance.

Table 105. Distribution and vectors of *Borrelia* species affecting domestic animals.

Borrelia species	Arthropod vector	Reservoir	Distribution	Disease
B. anserina	*Argas persica* *A. miniatus* *A. reflexus*	Birds	All continents	Avian spirochaetosis
B. burgdorferi	*Ixodes dammini*	Rodents (white-footed and deer mice)	East and midwest USA	Lyme disease
	I. pacificus	Rodents	Western USA	Lyme disease
	I. ricinus	Rodents	Europe	Lyme disease
	I. persulcatus	Rodents	Asia	Lyme disease
	?	?	Australia	Lyme disease
B. theileri	*Rhipicephalus* spp. *Margaropus* spp. *Boophilus* spp.	Cattle, horses, sheep	South Africa, Australia, North America, Europe	Tick spirochaetosis
B. coriaceae	*Ornithodoros coriaceus*	Deer (?) Cattle (?)	Western USA	Epizootic bovine abortion (?)

Table 106. Pathogenicity and hosts of *Borrelia* species affecting domestic animals

Borrelia species	*Hosts*	*Disease*
B. anserina	Chickens, ducks, geese, turkeys, pigeons, pheasants, canaries, and some wild birds	Avian spirochaetosis: affects all ages but higher mortality in young birds. There is fever, depression and anorexia. Birds become cyanotic and have a greenish diarrhoea. Later, paralysis and anaemia may develop. Mortality 10–99%. On postmortem the spleen is enlarged and mottled
B. burgdorferi (Lyme disease)	Humans	• Skin rash, stiffness of joints and neck, headache and lymphadenopathy. • Sequelae if not treated include chronic arthritis, cardiac signs and chronic neurological symptoms.
	Dogs	Sudden onset of lameness involving several joints. The affected joints are swollen and painful. The arthritis is characteristically migratory, affecting some joints and then shifting to others. The synovial fluid has a high neutrophil count. Some dogs can develop neural and renal disorders
	Cattle and horses	Arthritis, encephalitis, uveitis and laminitis
B. theileri	Cattle (sheep, horses)	Tick spirochaetosis: fever, weight loss, weakness and anaemia. There are one or more febrile attacks before recovery. The febrile attacks are of minor clinical significance in sheep and horses

362 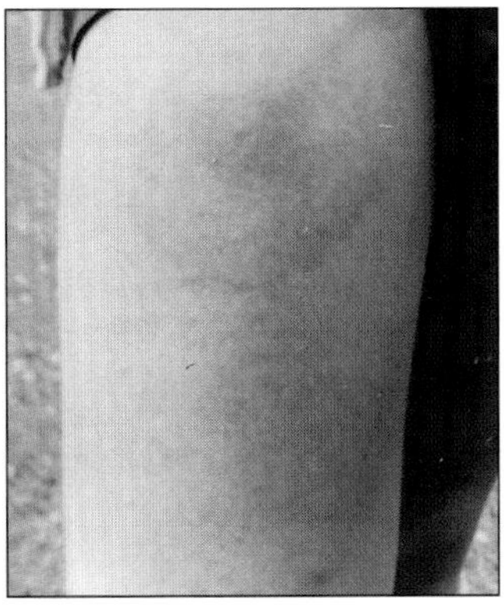

362 Typical skin rash of Lyme disease (*Borrelia burgdorferi*) in a human following a tick-bite (*Ixodes* species).

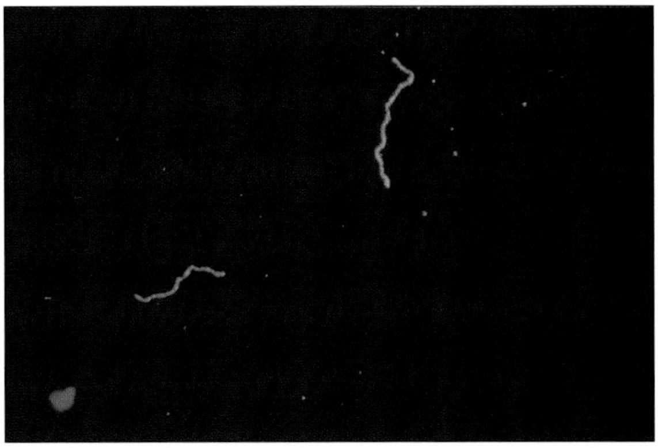 363

363 *Borrelia burgdorferi.* (Direct FA technique, ×1000)

Laboratory Diagnosis

***B. anserina*:**

- Demonstration of the borreliae (8–20 µm in length) in blood, spleen, and liver smears by darkfield microscopy, Giemsa stain or fluorescent antibody (FA) technique.
- Culture in 6–10-day-old fertile chicken or turkey eggs.
- Immunodiffusion tests can be used to demonstrate either antigen or antibody.

***B. theileri*:** demonstration of the organism in blood during the febrile stage. It is a large (20–30 µm in length), loosely coiled spirochaete that can be visualised using darkfield microscopy or Giemsa-stained smears.

***B. burgdorferi*:**

- Demonstration of *B. burgdorferi* (10–30µm in length) by darkfield microscopy, Giemsa-stained smears or FA technique (**363**) in tissues and joint fluids.
- Culture is difficult but the organism will grow in modified Kelley's medium (Barbour *et al.*, 1983). The cultures are examined weekly by darkfield microscopy for growth.
- Several molecular procedures have been developed to identify *B. burgdorferi* such as restriction endonuclease analysis, DNA hybridisation (Le Febvre *et al.*, 1989) and a polymerase chain reaction (Rosa and Schwan, 1989).
- Serology: indirect FA technique for antibodies or the more sensitive ELISA (Russell *et al.*, 1984) can be used.

References

Barbour, A.G., Burgdorfer, W., Hayes, S.F., Peter, O and Aeschlimann, T. (1983). Isolation of a cultivable spirochete from *Ixodes ricinus* ticks of Switzerland, *Current Microbiology*, **8**:123–126.

Bélanger, M. and Jacques, M. (1991). Evaluation of the An-Ident System and an Indole spot test for the rapid differentiation of porcine treponemes, *Journal of Clinical Microbiology*, **29**:1727–1729.

Blaha, T., Gunther, H., Flossmann, K.D. and Erler, E. (1984). Der epizootische Grundvorgang der Schweinedysenterie, *Zbl. Vet. Med.*,B **31**:451–465.

Burrows, M.R. and Lemcke, R.M. (1981). Identification of *Treponema hyodysenteriae* by a rapid slide agglutination test, *Veterinary Record*, **108**:187–189.

Cole, J.R., Jr. (1990). Spirochetes, pp. 41–60, In *Diagnostic Procedures in Veterinary Bacteriology and Mycology*. Carter, G.R. and Cole, J.R., Jr. (ed.), 5th ed. Academic Press Inc., New York.

Faine, S. (ed.) (1982). *Guidelines for the control of leptospirosis*, WHO offset publication no. 67. World Health Organization, Geneva.

Joens, L.A., Nord, N.A., Kinyon, J.M. and Egan, I.T. (1982). Enzyme-linked immunosorbent assay for detection of antibody to *Treponema hyodysenteriae* antigens, *Journal of Clinical Microbiology*, **15**:249–252.

LeFebvre, R.B., Perny, G.C. and Johnson, R.C. (1989). Characterisation of *Borrelia burgdorferi* isolates by restriction endonuclease analysis and DNA hybridization, *Journal of Clinical Microbiology*, **27**: 636–639.

Lysons, R.J. and Lemcke, R.M. (1983). Swine dysentery: to isolate or to fluoresce, *Veterinary Record*, **112**:203.

Lysons, R.J. (1991). Microscopic agglutination: a rapid test for identification of *Treponema hyodysenteriae*, *Veterinary Record*, **129**: 315–316.

Rosa, P.A. and Schwan, T.G. (1989). A specific and sensitive assay for the Lyme disease spirochete, *Borrelia burgdorferi*, using the polymerase chain reaction, *Journal of Infectious Diseases*, **160**: 1018–1029.

Russell, H., Sampson, J.S., Smith, G.P., Wilkinson, H.W. and Plikaytis, B. (1984). Enzyme-linked immunosorbent assay and indirect immunofluorescence assay for Lyme disease, *Journal of Infectious Diseases*, **149**: 465–470.

Subcommittee on Taxonomy of Leptospira. International Committee on Systematic Bacteriology (1992). *International Journal of Systematic Bacteriology*, **42**: 330–334.

32 Miscellaneous Gram-Negative Bacterial Pathogens

These bacteria, not included in any other pathogenic groups, are not commonly isolated but may have importance in either animal or human disease, or as zoonotic infections. A list of these bacteria, the diseases they cause, their hosts and natural habitats is given in **Table 107**.

Table 107. Hosts, habitat and diseases of the miscellaneous Gram-negative bacterial pathogens.

Bacterium	Natural habitat	Hosts	Disease(s)
Streptobacillus moniliformis	Nasopharynx of wild and laboratory rats, and small carnivores such as cats	Humans	Haverhill fever contracted by the ingestion of milk, water or food contaminated by the organism from rats
			Rat-bite fever follows the bite of a rat and occasionally bites from mice, pigs, squirrels, weasels, dogs or cats (*Spirillum minus* is another cause of rat-bite fever)
		Laboratory mice	Epizootics characterised by swelling of the extremities, arthritis, conjunctivitis, lymphadenitis and acute septicaemia. Abortion can occur
		Laboratory rats	Bronchopneumonia, middle-ear infections, and conjunctivitis
		Guinea-pigs	Abscesses in the cervical lymph nodes
		Turkeys	Tenosynovitis (associated with rat bites)
		Calves	Pneumonic lesions
Cat-scratch disease (fever) bacterium. (Unnamed Gram - negative, intracellular bacterium)	Commensal of buccal cavity of cats	Humans	Cat-scratch disease in man following the scratch, bite or lick of a cat or kitten
Chromo-bacterium violaceum	Soil and water of subtropical and tropical regions. (South-east Asia, South-east USA, North Australia and South America)	Humans	Skin abscesses or septicaemia with pulmonary, liver and subcutaneous abscesses
		Pigs, cattle, buffaloes and gibbons	Suppurative pneumonia and other infections
***Capnocytophaga ochraceae, C. gingivalis* or *C. sputigena* (CDC biogroup DF-1)**	Human buccal cavity	Humans	Juvenile periodontitis. Occasionally, a bacteraemia in immunosuppressed patients
***C. canimorsus* or *C. cynodegmi* (CDC biogroups DF-2 and DF-2-like, respectively)**	Nasopharynx, mouth and saliva of dogs and occasionally cats	Humans	Infections associated with dog bites. Local sepsis and occasionally septicaemia in compromised patients (immnunosuppressed, splenectomised patients and alcoholics).

Streptobacillus moniliformis

Streptobacillus moniliformis, the only species in the genus, is an extremely pleomorphic Gram-negative rod (0.1–0.7 × 1–5 μm) with rounded or pointed ends but can form unbranched filaments 10–150 μm long. The filaments bear knobby irregularities, hence the specific name, *moniliformis* (latin) meaning in the form of a 'necklace' or 'string of beads'. It is non-motile, catalase-negative, oxidase-negative and ferments a range of carbohydrates but is relatively unreactive in other biochemical tests. L-phase and transitional variants are readily formed. *S. moniliformis* is fastidious and requires media enriched with blood, serum or ascitic fluid. It may appeai to be an obligate anaerobe on primary isolation, but is a facultative anaerobe on subculture.

Natural Habitat and Pathogenicity

The reservoir is the nasopharynx of rats and possibly small carnivores. Rat-bite fever is most common in children from rat-infested accommodation, but humans can acquire the disease from the bite of a laboratory rat. The incubation period is usually 1–4 days and clinical signs include fever, headache, rash on hands and feet, and later polyarthralgia in about two-thirds of patients. The bite wound usually heals well but the disease can develop into a chronic relapsing illness. Transmission of the organism of Haverhill fever (Haverhill, Massachusetts) is by the ingestion of contaminated milk or water to which rats have had access. The clinical signs in Haverhill fever are similar to those of rat-bite fever but gastrointestinal and respiratory signs are more common. *Spirillum minus* (**364**) is another cause of rat-bite fever in man but this bacterium does not affect animals.

364

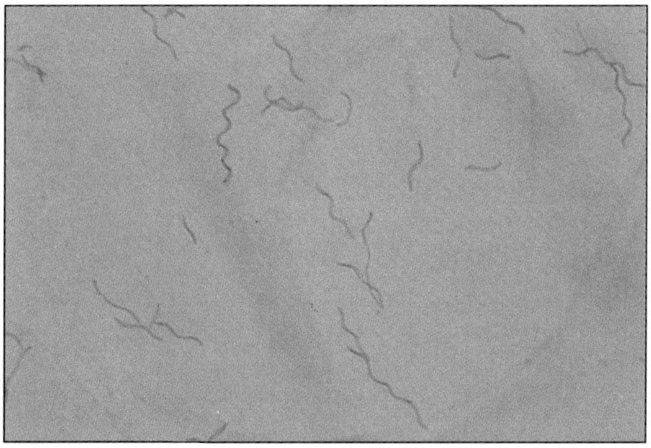

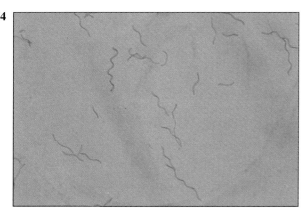

364 *Spirillum* species showing the spiral form of the bacterium. (DCF stain, ×1000)

Laboratory Diagnosis

Specimens

These include joint fluid, citrated blood (equal volume of blood and 2.5 per cent sodium citrate), pus, exudates and tissue lesions.

Direct microscopy

S. moniliformis may not stain strongly with the Gram stain so either the DCF or Giemsa stains should be used. It is less pleomorphic in animal and human tissues than in smears from cultures. An FA technique is also used in some laboratories.

Isolation

An enriched medium consisting of brain heart infusion base enriched with 15 per cent blood, 20 per cent horse or calf serum, or 5 per cent ascitic fluid is required. The L-forms are particularly fastidious. Rogosa (1985) suggested a heart infusion base enriched with 20 per cent horse serum and yeast extract. Thioglycollate and meat infusion broths, both with added serum, will support the growth of the bacterium. The sodium-polyanethol sulfonate (SPS) present in most blood culture media is inhibitory to this organism. Inoculated media are incubated in a humified 10 per cent CO_2 atmosphere at 35–37°C for up to 6 days.

Identification

Colonial morphology

Colonies on blood agar are non-haemolytic, round, greyish, smooth, glistening with a diameter of 1–2 mm in 2–3 days. Spontaneous L-form microcolonies may occur adjacent to the bacterial-type colonies. These L-form colonies have a 'fried-egg' appearance and can be stained by the Dienes stain (as for mycoplasmal microcolonies) and examined under a low-power objective of a microscope.

In thioglycollate and other semi-solid or liquid media, colonies of *S. moniliformis* grow as characteristic 'fluff-balls' near the bottom of the tube.

Microscopic appearance

Under the most favourable cultural conditions, *S. moniliformis* is rod-shaped and relatively uniform with an occasional short filament or knob-like irregularity (**365**). In less favourable conditions or in older cultures the bacterium is highly pleomorphic and consists of filaments and bulbous swellings. The L-form microcolonies yield small Gram-negative coccal forms.

Biochemical reactions

Fermentation tests on sugars are best conducted in cystine tryptic agar (CTA) base with added serum or ascitic fluid. The alkaline phosphatase test is usually positive. Some of the other biochemical reactions are given in **Table 108**.

Table 108. Summary of the biochemical reactions of *Streptobacillus moniliformis, Chromobacterium violaceum* and *Capnocytophaga* species.

			CDC biogroup DF-1				
	Streptobacillus moniliformis	*Chromobacterium violaceum*	*Capnocytophaga gingivalis*	*Capnocytophaga ochracea*	*Capnocytophaga sputigena*	*Capnocytophaga canimorsus* DF-2	*Capnocytophaga cynodegmi* DF-2-like
Catalase	–	+	–	–	–	+	+
Oxidase	–	+	–	–	–	+	+
Growth on MacConkey agar	–	+	–	–	–	–	–
Citrate	–	v	–	–	–	–	–
Urease	–	v	–	–	–	–	–
Lysine decarboxylase	–	–	–	–	–	–	–
Ornithine decarboxylase	–	–	–	–	–	–	–
Arginine dihydrolase	v	+	–	–	–	v	v
Indole	–	–*	–	–	–	–	–
Gelatin liquefaction	–	+	–	–	v	–	–
Nitrate reduction	–	+	–	–	+	–	–
Motility	–	+	–	–	–	–	–
Acid from:							
Galactose	+	–	–	+	–	+	v
Glucose	+	+	+	+	+	+	+
Fructose	+	+	•	•	•	•	•
Lactose	v	–	–	+	v	+	+
Maltose	+	v	+	+	+	+	+
Mannitol	–	–	–	–	–	–	–
Raffinose	–	–	v	v	–	–	+
Sucrose	v	v	+	+	+	–	+
Trehalose	–	+	•	•	•	•	•
Xylose	v	–	–	–	–	–	–
Pigment production	–	+	+	+	+	–	–

* = some non-pigmented strains are indole +; • = data not available.
+ = more than 80% of strains positive; v = 20–80% of strains positive; – = less than 20% of strains positive.

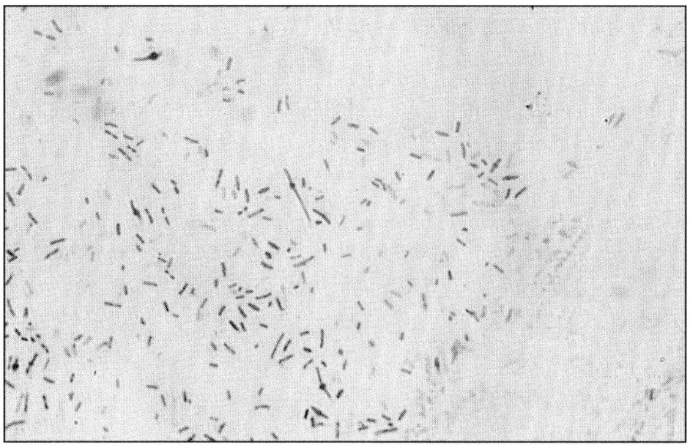

365 *Streptobacillus moniliformis*: a highly pleomorphic Gram-negative rod with filamentous forms bearing knobby irregularities. (Gram stain, ×1000)

A typical fatty acid profile can be identified for this bacterium using gas-liquid chromatography.

Serology
A tube agglutination test, using a formalin-treated cell suspension of antigen is used for the diagnosis of the disease in humans.

Animal inoculation
This is not used extensively and there is the possibility of the laboratory animals being already infected with *S. moniliformis*. Mice inoculated with a suspension of the bacterium, intravenously or intraperitoneally, develop either acute septicaemia or a chronic disease with arthritis. Adult rats are usually resistant but inoculated neonates may develop pneumonia.

Cat-scratch Disease Bacterium

Cat-scratch disease (CSD) or cat-scratch fever in man was first recognized in France during the 1930's but the aetiology remained obscure until Wear *et al.* (1983) observed a Gram-negative rod (0.3–0.5 × 0.5–1.0 μm) in lesions of CSD. The bacterium is non-acid-fast, appears to be intracellular and is stained best by the Warthin–Starry silver stain.

Natural Habitat

The bacterium is thought to be a commensal in the buccal cavity of cats. Cats associated with cases of CSD have been healthy and usually under one year of age.

Pathogenicity

Most cases of CSD in humans occur about 2 weeks after a bite, scratch or lick from a cat. There is a local lesion, regional lymphadenopathy, anorexia, fatigue, fever and headache. Occasionally a rash and general lymphadenopathy occurs with suppuration of an affected lymph node. Treatment of CSD is not recommended and one episode of the disease is thought to give lifelong immunity. Hadfield *et al.* (1985) have reviewed the disease.

Laboratory Diagnosis

A skin test, which is highly specific, has been used as a main diagnostic test for CSD in man. An antigen, prepared from heat-treated pus obtained from a suppurating lymph node, is used. A wheal or papule develops within 72 hours after injection with 0.1 ml of the antigen in affected patients.

Smears of a lymph node biopsy stained by the Warthin–Starry silver stain reveal pleomorphic, coccobacillary bacteria.

Antibodies in patients' serum when coupled with fluorescein or immunoperoxidase bind to the bacterium in lymph node biopsy material from a newly diagnosed patient.

Isolation methods for *in vitro* culture are being developed.

Chromobacterium violaceum

Chromobacterium violaceum is a Gram-negative rod, (0.6–0.9 × 1.5–3.5 μm) that is sometimes slightly curved with the cells occurring singly or in pairs. It often shows barred or bipolar staining in a Gram-stained smear. There is a characteristic type of flagellar arrangement with a short polar flagellum that stains poorly and has a relatively long wavelength and 1–4 longer lateral flagella that stain well and have a shorter wavelength, as shown in **Diagram 39**. Most strains are facultative anaerobes but a few are aerobic. It is catalase-positive, oxidase-positive and grows on nutrient and MacConkey agars at 37°C but not at 4°C. Most strains of *C. violaceum* produce a non-diffusible violet pigment, known as violacein, that requires the presence of oxygen for its formation.

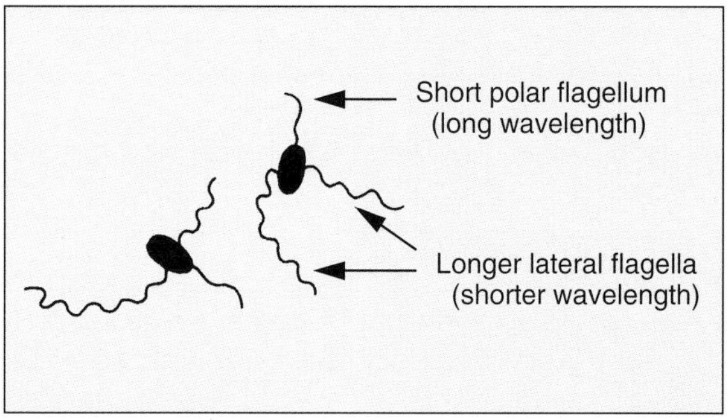

Diagram 39. Flagellar arrangement in *Chromobacterium violaceum*.

Natural Habitat

C. violaceum is a saprophyte in soil and water of subtropical and tropical regions, which occasionally causes disease in man and animals. Human cases have been reported from Southeast Asia, southeast USA, South America and northern Australia. Outbreaks of disease have been recorded in buffaloes, cattle, pigs and gibbons. Comparatively few infections occur, despite the bacterium being widespread within the regions where it occurs.

Pathogenicity

In humans, *C. violaceum* can cause skin abscesses or an overwhelming septicaemia with pulmonary, liver and subcutaneous abscesses, occasionally resembling meliodosis (*Pseudomonas pseudomallei*). Suppurative pneumonia and other infections occur in animals.

Laboratory Diagnosis

Isolation

Direct microscopy is not rewarding. Specimens such as pus or material from tissue lesions can be used to inoculate blood and MacConkey agars. Humidity may be important as the bacterium dies rapidly on over-dried plates or agar slopes on prolonged storage. Inoculated plates are incubated aerobically at 35–37°C for 24–48 hours.

Identification

Colonial morphology

After 24 hours on blood agar the colonies may appear black, 1–2 mm in diameter, round, smooth, with a variable degree of partial haemolysis and smell of cyanide. Violacein production is enhanced by tryptophan and aerobic incubation at room temperature (22°C). Non-pigmented strains occur and these are occasionally isolated from both soil and clinical specimens. Growth occurs on MacConkey agar.

Microscopic appearance

Gram-stained smears show Gram-negative rods that may be slightly curved and tend to show barred or bipolar staining. The unique flagellar arrangement is seen best in preparations from young cultures grown on solid media. The lateral flagella are rare in broth cultures.

Biochemical reactions

C. violaceum is included in the list of bacteria that can be identified by API 20E (Analytab Products). The catalase test is positive but may be weak. If the violet pigment interferes with the oxidase reaction, the test should be performed on a culture grown anaerobically, as oxygen is required for pigment production. The sugar fermentation test can be conducted in peptone water sugars. A biochemical profile of this bacterium is given in **Table 108**.

Capnocytophaga species

Capnocytophaga species belongs to the order Cytophagales, and lack flagella but exhibit a characteristic gliding motility. They are Gram-negative, slender or filamentous rods, are facultative anaerobes and capnophilic (carbon dioxide-loving). All ferment carbohydrates but never mannitol or xylose. The species in CDC biotype DF-1 group (DF for dysgonic fermenter) cannot always be differentiated by conventional tests but DNA-DNA hybridisation will distinguish the five species listed below.

Nomenclature

CDC biotype DF-1	*Capnocytophaga gingivalis*
	C. ochracea
	C. spurigena
CDC biotype DF-2	*C. canimorsus*
CDC biotype DF-2-like	*C. cynodegmi*

Natural Habitat and Pathogenicity

The *Capnocytophaga* species in the DF-1 group are commensals in the oral cavity of humans and are associated with juvenile periodontitis in man.

C. canimorsus and *C. cynodegmi* are commensals in the nasopharynx, mouth and saliva of dogs and probably cats. Infections in man are associated with dog bites (occasionally cat bites) and vary from local sepsis to severe septicaemia, with endocarditis and meningitis in immunosuppressed, splenectomised or alcoholic patients.

Laboratory Diagnosis

Isolation

Specimens are inoculated on a heart-infusion agar with 5–10 per cent blood and incubated under 5–10 per cent CO_2 at 35–37°C for 2–3 days. Gliding motility is seen best on heart-infusion agar with 5 per cent rabbit blood and agar content increased to 2–3 per cent.

Identification

Colonial morphology

The DF-1 group have colonies that are barely visible in 24 hours, but after 2–3 days' incubation they are 2–3 mm in diameter, flat, rough, yellow and have an irregular spreading margin forming finger-like projections. There may be slight haemolysis on blood agar and the growth has an odour of bitter almonds. The colonies of the DF-2 and DF-2-like bacteria are 1–2 mm in diameter after 48 hours' incubation and are convex, smooth and circular.

Microscopic appearance

The DF-1 bacteria appear as slender, fusiform or filamentous Gram-negative rods. The DF-2 and DF-2-like organisms are long, thin Gram-negative rods (1–3µm) with tapering ends. They can resemble the cells of *Fusobacterium* spp. in Gram-stained smears.

Biochemical characteristics

- In tests for acid-production from carbohydrates, particularly with the DF-2 and DF-2-like bacteria, positive results are more certain if relatively large inocula are used in small volumes of fluid. A broth base (3 ml) should contain 0.3 ml of filter-sterilised 10 per cent carbohydrate solution and 0.1 ml sterile rabbit serum.
- The indole test is performed in 4 ml heart infusion broth containing 0.1 ml sterile rabbit serum. The culture is incubated for 4 days at 35°C, indole is extracted with xylene and tested with Ehrlich reagent.
- Nitrate reduction is tested in 4 ml heart infusion broth containing 0.2 per cent potassium nitrate and 0.1 ml sterile rabbit serum. Tubes should be incubated for 4 days.

The biochemical results for the *Capnocytophaga* species are summarised in **Table 108**.

References

Hadfield, T.L., Schlagel, C. and Margileth, A. (1985). Stalking the cause of cat-scratch disease, *Diagnostic Medicine*, **8**:23.

Rogosa, M. (1985). *Streptobacillus moniliformis* and *Spirillum minus*, p. 400–406, In *Manual of Clinical Microbiology*, E.H. Lennette, A. Balows, W.J. Hausler, Jr. and H.J. Shadomy (ed.), 4th ed., American Society for Microbiology, Washington, DC., USA.

Wear, D.J., Margileth, A.M. and Hadfield T.L. (1983). Cat-scratch disease: a bacterial infection, *Science*, **221**: 1403.

33 The Chlamydiales (Order)

The order Chlamydiales is comprised of a single family *Chlamydiaceae* with only one genus *Chlamydia*. Chlamydiae possess a cell wall similar to other Gram-negative bacteria but lacking in muramic acid. They are obligate intracellular parasites with a requirement for high energy compounds such as ATP. This characteristic has led to the coining of the phrase 'energy parasites'. They have a unique developmental cycle with alternating, morphologically distinct, infectious and reproductive forms. The elementary body (EB) is small (200–300 nm in diameter), infectious and represents the extracellular form of the organism. The EB enters a cell by endocytosis and differentiates into the larger, non-infectious but metabolically active reticulate body (RB) inside an expanding vacuole. The RB multiplies by binary fission producing further RBs. At about 20 hours following infection, some of the RBs start to condense and mature within the inclusion to form EBs. In general, release of infectious EBs begins at about 40 hours post infection due to lysis of the cell (**Diagram 40**). At present only two species, *C. trachomatis* and *C. psittaci*, are officially recognised, although a third species, *C. pneumoniae*, has been proposed for those *C. psittaci* isolates referred to as TWAR strains. **Table 109** indicates the main differences between *C. trachomatis* and *C. psittaci*.

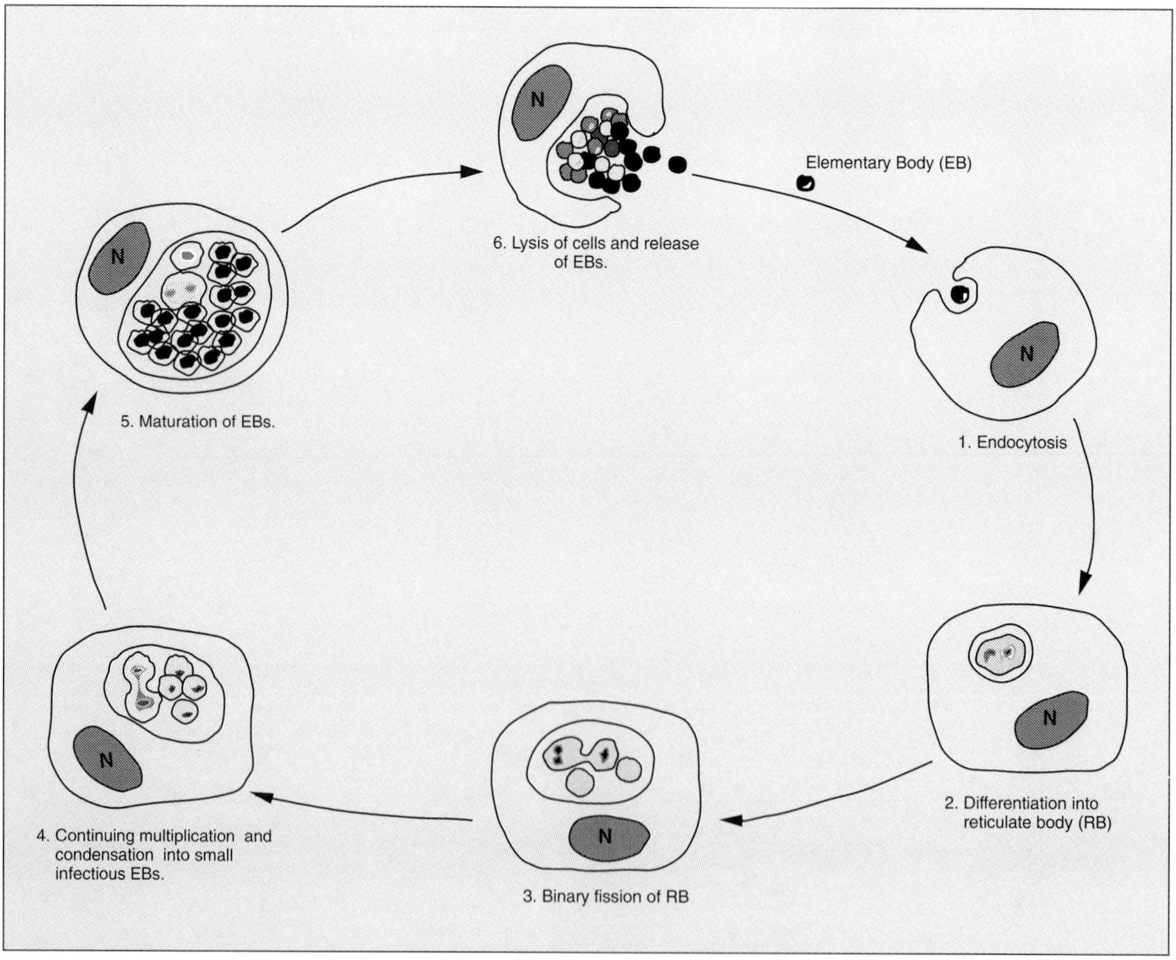

Diagram 40. Representation of the life cycle of chlamydiae (N = Nucleus).

Table 109. Differentiation of *Chlamydia trachomatis* and *C. psittaci.*

	C. trachomatis	*C. psittaci*
Inclusion morphology	Rigid, compact single inclusion per cell. Cell nucleus is displaced	Irregular, diffuse, multiple inclusions. No displacement of the cell nucleus
Presence of glycogen in the inclusions	+	–
Growth inhibited by sulphadiazine	+	–
Natural host range	Mice, humans	Birds, non-human mammals

Natural Habitat

Persistent infections are the rule rather than the exception in chlamydial disease. In both birds and mammals the gastrointestinal tract appears to be the natural habitat of *C. psittaci*. *C. psittaci* may give rise to an inapparent infection with prolonged faecal shedding. These infections are important for successfully perpetuating and maintaining the infection in animal populations. The elementary bodies are relatively resistant and remain viable for several days under suitable environmental conditions.

Pathogenicity

Chlamydiae do not show a strict host specificity. They infect over 150 avian species, a large number of mammalian species and an increasing number of isolations are being made from invertebrates. This enormous range of host species is matched by the diversity of recognised clinical conditions associated with chlamydial infections (**Table 110**). Despite the ability of *C. psittaci* to induce a variety of disease syndromes, one clinical condition usually predominates in an outbreak or series of related clinical cases. The presenting condition and its severity depend on:

a) Strain and virulence of the agent: the importance of the infecting strain is exemplified in sheep where *C. psittaci* has been associated with inapparent intestinal infection, abortion, conjunctivitis, polyarthritis, pneumonia and orchitis/epididymitis. Isolates from ovine abortion have been shown to be distinct from other ovine isolates by analysis of polypeptide and restriction endonuclease profiles (McClenaghan *et al.*, 1984; McClenaghan *et al.*, 1991).
b) Age, sex, physiological state and species of host.
c) Route of infection and degree of exposure to chlamydiae.
d) Environment and management practices.

Chlamydial abortion is described in cattle, sheep and goats. Enzootic abortion of ewes (EAE) is a major cause of economic loss in intensively managed sheep flocks and is regarded as a serious problem in many sheep-rearing areas of the world. Spread of infection occurs predominantly at the time of lambing. Ewes abort in the last 2–3 weeks of pregnancy and this is associated with a diffuse, necrotic placentitis. Despite serological and cultural evidence of *C. psittaci* infections in sheep, the disease appears to be extremely rare in countries such as Australia and Ireland.

Epizootic bovine abortion (EBA), or foothill abortion, occurs in range-cattle of western USA. Originally this disease was ascribed to *C. psittaci* infection, but doubt has been cast on the role of chlamydiae in this disease (Kimsey *et al.*, 1983). It is now thought that chlamydial abortion in cattle is a distinct entity while EBA may possibly be caused by a *Borrelia* species.

Outbreaks of chlamydial polyarthritis have been reported in feedlot lambs in both the USA and Australia. In other countries this condition tends to be sporadic.

Sporadic bovine encephalomyelitis has been reported from the USA, Japan and Israel. The disease is characterised by inflammation of vascular endothelium throughout the body. However, nervous signs tend to predominate. The prevalence of clinical disease in infected herds is usually low.

The first feline isolate of *C. psittaci* was obtained from a cat with pneumonia and the disease was named feline pneumonitis. The majority of subsequent isolations have been from cats with mild upper respiratory tract infections and conjunctivitis. Affected cats recover spontaneously but tend to suffer recurrent episodes of conjunctivitis at intervals of 10–14 days.

Psittacosis (ornithosis) is an acute or chronic infection of wild and domestic birds, characterised by respiratory, intestinal and systemic signs. It occurs worldwide and is transmissible to other animals, including man. Respiratory discharges and faeces are infective. A carrier state can exist after infection and stress may provoke recurrence of disease. The term 'psittacosis' refers to the infection in psittacine birds and 'ornitho-

Table 110. Clinical conditions associated with chlamydial infections.

Chlamydia species	Host species	Clinical condition
C. trachomatis	Human	• Ocular disease a) Trachoma b) Inclusion conjunctivitis of neonates • Genital tract infections a) Non-gonococcal urethritis b) Salpingitis c) Cervicitis d) Epididymitis • Respiratory disease of infants • Proctitis • Lymphogranuloma venereum • Arthritis
C. psittaci (TWAR strains) (*C. pneumoniae*)	Human	• Pneumonia
C. psittaci	Human (zoonosis)	• Psittacosis (ornithosis) • Abortion • Conjunctivitis
C. psittaci	Mammalian	• Intestinal infection and diarrhoea • Gastritis • Pneumonia • Abortion, e.g. enzootic abortion of ewes, chlamydial abortion of cows • Genital infections • Mastitis • Polyarthritis/polyserositis • Encephalomyelitis, e.g. sporadic bovine encephalomyelitis • Hepatitis • Conjunctivitis, e.g. feline pneumonitis
C. psittaci	Avian (psittacosis/ornithosis)	• Pneumonia and air sacculitis • Pericarditis • Encephalitis • Conjunctivitis • Intestinal infection and diarrhoea

sis' to the disease in non-psittacines. The clinical signs are similar in all birds, although chronic infections tend to occur more frequently in psittacine birds.

Laboratory Diagnosis

Specimens

These will vary with the clinical condition:

- **Abortion**: confirmation of a diagnosis of EAE is most easily accomplished by microscopic examination of stained smears from affected cotyledons or chorion for evidence of chlamydial EBs. If the placenta is unavailable, smears may be made from vaginal swabs taken within 1–2 days of abortion, or from the wet surface of an aborted foetus. Culture is more sensitive than direct microscopy on stained smears and may be used to diagnose EAE from uterine discharges or foetal tissues in the absence of foetal membranes (Johnson *et al.*, 1983).

Suitable samples for isolation of chlamydiae include a piece of affected cotyledon, vaginal swabs, foetal lung and foetal liver. These should be placed in

chlamydial transport medium (**Appendix 2**) and held at 4°C.
- **Polyarthritis**: aspirated synovial fluid should be collected.
- **Conjunctivitis**: a conjunctival swab is an acceptable specimen.
- **Systemic infection**: samples of lung, liver and spleen should be submitted.

Paired serum samples should be collected for serology, from both affected and clinically normal animals in the herd or flock.

Demonstration of Chlamydia psittaci

The direct demonstration of *C. psittaci* in clinical material (**366**, **367**) may be carried out by:
a) Direct microscopy: EBs in smears or tissue sections can be detected by the use of either chemical stains or immunological staining techniques such as immunofluorescence or immunoperoxidase (**91**). Suitable chemical stains include:
- Modified Ziehl–Neelsen (MZN) stain (**Appendix 1**). The EBs tend to occur in clumps and stain red against a blue background (methylene blue counter-stain) or a green background (malachite green counter-stain).
  If the MZN-stained smears are examined under darkfield microscopy, the EBs appear as bright green, coccal structures.
- Methylene blue stain (**Appendix 1**). If the stained smear is examined under darkfield illumination, the EBs are revealed as refractile, yellow-green bodies surrounded by a halo (Dagnall and Wilsmore, 1990).
- Method of Macchiavello (**Appendix 1**). The EBs stain red against a blue background.
- Method of Castaneda (**Appendix 1**). The EBs are stained blue with a reddish background.
- Giemsa stain (**Appendix 1**). This stain is useful for staining smears prepared from conjunctival scrapings, particularly those from cats with feline pneumonitis. Infected conjunctival epithelial cells contain basophilic intracytoplasmic aggregates of *C. psittaci*.

In stained smears, *Brucella* species and *Coxiella burnetii* may look very similar to *C. psittaci* but can be differentiated by immunological staining, serologically or following isolation of the organism.

b) Enzyme-linked immunosorbent assay (ELISA): very few ELISAs have been developed specifically for *C. psittaci* but many of those commercially available for *C. trachomatis* can be used because they detect the presence of the common genus-specific antigen (lipopolysaccharide) shared by both *Chlamydia* species.

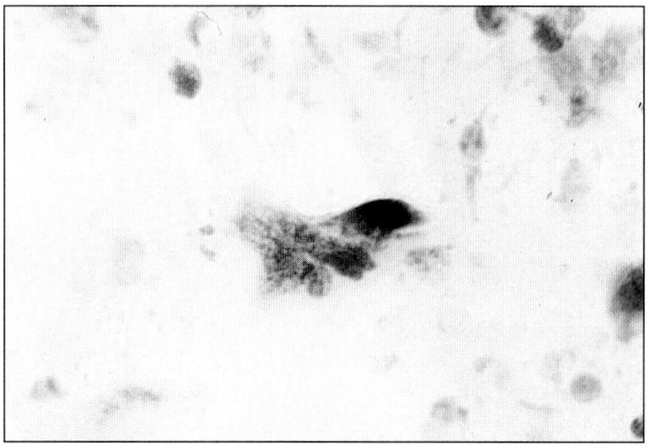
366

366 *Chlamydia psittaci* in an MZN-stained smear from a bird with psittacosis. (×1000)

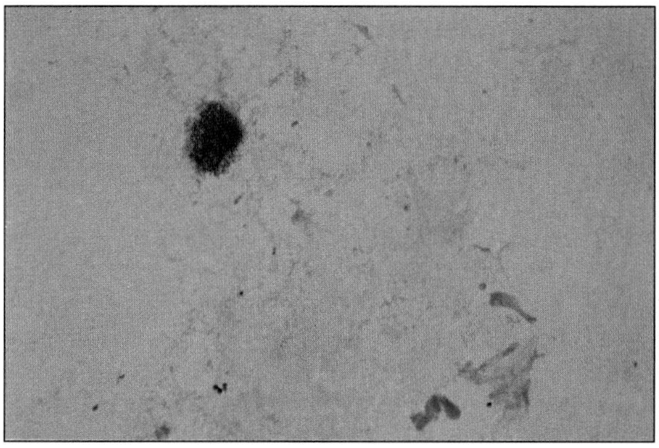
367

367 *C. psittaci* in an MZN-stained cotyledonary smear from an ewe with enzootic abortion (EAE). (×1000)

Isolation and cultivation

Three laboratory systems can be used:
a) Mouse inoculation: this technique was used in the past, but currently has little place in a routine diagnostic laboratory.
b) Inoculation of embryonated hens' eggs (6–7 days old) via the yolk sac route: this method will detect all chlamydiae but it is laborious and prone to bacterial contamination.
c) Cell culture (**368, 369**, **370**): a number of continuous cell lines are susceptible to chlamydial infection such as McCoy, baby hamster kidney-21, L929 and Vero. Chemical treatment of the cells with cycloheximide (1-2 μg/ml), 5-iodo-2-deoxyuridine (80 μg/ml), cytochalasin B (1 μg/ml) or diethylaminoethyl-dextran (30 μg/ml) and centrifugation of the sample onto the monolayer greatly enhances the sensitivity of the isolation procedure. The use of antibiotics in the tissue culture medium to which chlamydiae are sensitive (oxytetracycline, erythromycin, penicillin, tylosin) should be avoided.

368

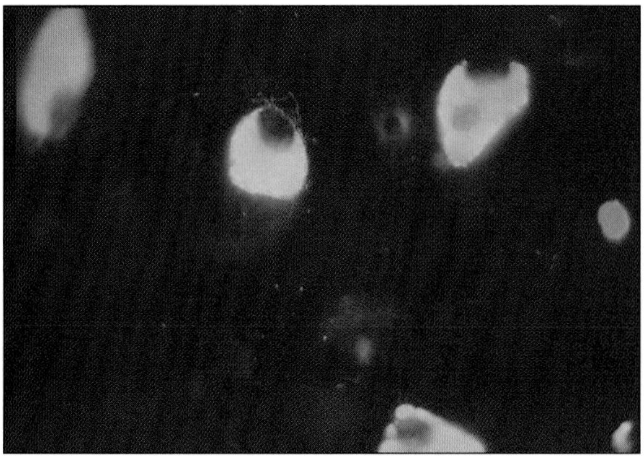

368 *C. psittaci* (EAE isolate) inclusions in McCoy cells. (Direct FA technique, ×400)

369

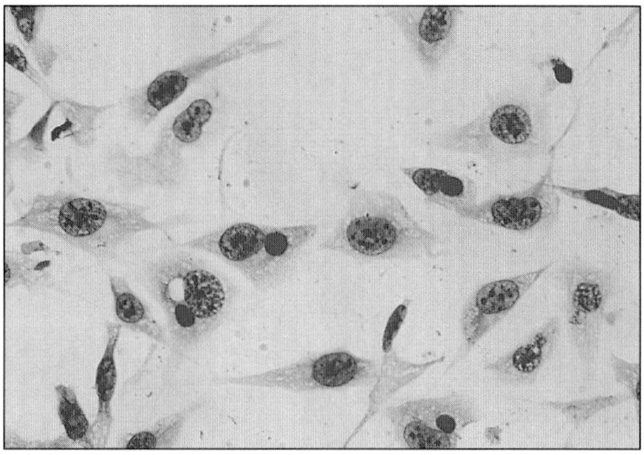

369 *C psittaci* (EAE) inclusions in McCoy cells. (Giemsa stain, ×400)

370

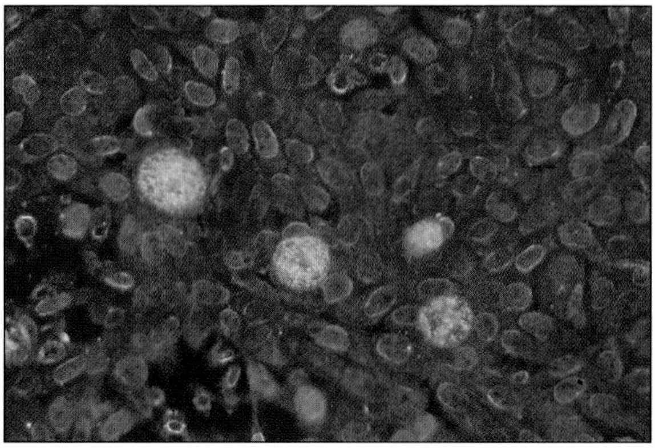

370 *C. trachomatis* in McCoy cells demonstrating autofluorescence of the inclusions (yellow) under dark-field illumination. *C. psittaci* inclusions show autofluorescence when stained with methylene blue but not with Giemsa stain. (Giemsa stain, ×400)

Serology

Because of the widespread distribution and frequently subclinical nature of chlamydial infections in animals, it is important to demonstrate an unusually high or rising antibody titre. A number of tests capable of detecting chlamydial antibody are available:

a) Complement fixation test (**84**): this is the standard serological test used. However, it is time consuming and not particularly sensitive.
b) Enzyme-linked immunosorbent assay (ELISA).
c) Indirect immunofluorescence antibody test.

These tests primarily detect antibodies directed against the common genus-specific antigen (LPS). Consequently, it is generally not possible to distinguish between infections caused by isolates of varying virulence, such as between abortion and non-abortion-inducing ruminant strains, by serological methods. In addition, cross-reactivity between *Chlamydia* spp. and other Gram-negative bacteria has been demonstrated, further complicating the interpretation of serological data.

Identification

The designation of all chlamydial isolates from mammals and birds as *C. psittaci* is unsatisfactory. Significant biological and antigenic differences have been found among *C. psittaci* isolates from different host species and even among isolates from a single host species (Spears and Storz, 1979; Perez-Martinez and Storz, 1985). Results from techniques such as polyacylamide gel electrophoresis, restriction endonuclease analysis and monoclonal antibody typing will inevitably lead to the formation of many new species of chlamydiae (Fukushi and Hirai, 1989; McClenaghan *et al.*, 1991; Andersen, 1991).

References

Andersen A.A. (1991). Comparison of avian *Chlamydia psittaci* isolates by restriction endonuclease analysis and serovar-specific monoclonal antibodies, *Journal of Clinical Microbiology*, **29**: 244–249.

Dagnall G.J.R. and Wilsmore A.J. (1990). A simple staining method for the identification of chlamydial elementary bodies in the foetal membranes of sheep affected by ovine enzootic abortion, *Veterinary Microbiology*, **21**: 232–239.

Fukushi H. and Hirai K. (1989). Genetic diversity of avian and mammalian *Chlamydia psittaci* strains and relation to host origin, *Journal of Bacteriology*, **171**: 2850–2855.

Johnson F.W.A., Clarkson M.J. and Spenser W.N. (1983). Direct isolation of the agent of enzootic abortion of ewes (*Chlamydia psittaci*) in cell cultures, *Veterinary Record*, **113**: 413–414.

Kimsey P.B., Kennedy P.C., Casaro A.P., BonDurant R.H., Oliver M.N. and Kendrick J.W. (1983). Studies on the pathogenesis of epizootic bovine abortion, *American Journal of Veterinary Research*, **44**: 1266–1271.

McClenaghan M., Herring A.J. and Aitken I.D. (1984). Comparison of *Chlamydia psittaci* isolates by DNA restriction endonuclease analysis, *Infection and Immunity*, **45**: 384–389.

McClenaghan M., Inglis N.F. and Herring A.J. (1991). Comparison of isolates of *Chlamydia psittaci* of ovine, avian and feline origin by analysis of polypeptide profiles from purified elementary bodies, *Veterinary Microbiology*, **26**: 269–278.

Perez-Martinez J.A. and Storz J. (1985). Antigenic diversity of *Chlamydia psittaci* of mammalian origin determined by micro-immunofluorescence, *Infection and Immunity*, **50**: 905–910.

Spears P. and Storz J. (1979). Biotyping of *Chlamydia psittaci* based on inclusion morphology and response to diethylaminoethyl-dextran and cycloheximide, *Infection and Immunity*, **24**: 224–232.

Bibliography

Barron A.L. (1988). *Microbiology of Chlamydia*, CRC Press Inc., Boca Raton, Florida, USA.

Storz J. (1971). *Chlamydia and Chlamydia-Induced Diseases*, Charles C. Thomas, Illinois, USA.

34 The Rickettsiales (Order)

The members of the Rickettsiales are minute obligate intracellular Gram-negative bacteria. They are rods or coccobacilli, visible under the light microscope and varying in size from about 0.3–0.6 μm in length. They are non-motile, aerobic and divide by binary fission. The members of the *Rickettsiaceae* (Family) have cell walls similar to those of other Gram-negative bacteria, but those belonging to the *Anaplasmataceae* have cells bounded by a two-layer membrane. They all stain poorly with basic aniline dyes, such as are used in the Gram stain, but they stain well with Giemsa and other Romanowsky stains. A few of the *Rickettsiaceae* will grow in conventional inert media, but most require living cells for propagation and are cultured in the yolk sac of fertile eggs or in tissue culture. Others, such as *Haemobartonella felis*, have not yet been propagated *in vitro*. The classification of the Rickettsiales of veterinary interest is indicated in **Table 111**.

Natural Habitat

The Rickettsiales are essentially parasites of arthropods, replicating in the cells of the gut. Some can be passed transovarially in ticks and mites but others, such as *Ehrlichia canis, Ehrlichia (Cytoecetes) phagocytophila* and *Cowdria ruminantium*, are passed transstadially but not transovarially. They are labile outside the cells of host or vector, with the exception of *Coxiella burnetii* which produces endospore-like forms that can survive in dust particles for 50 days or more. Complete inactivation of this bacterium may not occur even after exposure to pasteurisation temperatures of 63°C for 30 minutes or 72°C for 15 seconds. A number of the pathogenic Rickettsiales, such as *Ehrlichia canis, Anaplasma marginale* and *Haemobartonella felis* may persist in the host in a latent form.

Pathogenesis

Infection occurs from the bite of an affected arthropod or ingestion of infected flukes in salmon poisoning. The principal route is by aerosols in Q fever in man. Maternal transmission is thought to occur in feline infectious anaemia. Pathogenesis varies with the rickettsial species. Damage to the capillary endothelial cells may result in loss of plasma, shock and death. Toxins, haemolysins and endotoxin-like lipopolysaccharides are known to be produced by some members of the group. When mammalian erythrocytes are parasitised, as in anaplasmosis, anaemia probably results from the immune clearance of the altered erythrocytes. The principal diseases, hosts, means of transmission and cell parasitism of pathogens in the Rickettsiales are summarised in **Table 112**.

Table 111. Genera of the Rickettsiales of veterinary importance.

Family and tribe	Genus	Cells parasitised in host
Rickettsiaceae		
Tribe Rickettsieae	*Rickettsia*	Vascular endothelial cells
	Coxiella	In vacuoles of cells of reticuloendothelial system
Tribe Ehrlichieae	*Ehrlichia*	Circulating leukocytes
	Cowdria	Vascular endothelial cells
	Neorickettsia	Reticular cells of lymphoid tissue
Anaplasmataceae	*Anaplasma*	Within erythrocytes
	Aegyptianella	Within erythrocytes of birds
	Haemobartonella	On, or in, erythrocytes
	Eperythrozoon	On erythrocytes

Table 112. Significant rickettsial veterinary pathogens.

Agent	Main host(s) and distribution	Transmission (vector)	Disease and cells parasitised
Rickettsia rickettsii	Humans, dogs. Western hemisphere, especially Eastern USA	Endemic in ticks (*Dermacentor* spp.). Bites of infected ticks	Rocky Mountain spotted fever in man. Tick fever in dogs. Parasitises vascular endothelial cells
Coxiella burnetii	Humans, cattle and small ruminants. Other animals can be a reservoir. Worldwide	Agent in milk, birth fluids of ruminants and dust. Persists in ticks	Q fever in man: aerosol inhalation, ingestion, tick bites or laboratory accidents. Occasional abortions, weak offspring or infertility in ruminants; usually a tick-borne infection
Ehrlichia canis	Dogs. Americas, Asia, Africa, Caribbean and Mediterranean	Brown dog tick (*Rhipicephalus sanguineus*)	Canine ehrlichiosis (tropical canine pancytopenia). Agent in lymphocytes, monocytes and, rarely, in neutrophils
E. equi	Horses. USA only	Vector unknown, ticks suspected	Equine ehrlichiosis. Granulocytes and vascular endothelial cells affected
E. ondiri	Cattle. East Africa	*Amblyomma* ticks	Bovine petechial fever. Granulocytes and monocytes parasitised
E. (Cytoecetes) phagocytophila	Cattle, sheep and wild ruminants. UK, Ireland and Scandinavia	*Ixodes ricinus*	Tick-borne fever. Pregnant animals may abort. Present in granulocytes and monocytes
E. platys	Dogs. Similar distribution to *E. canis*	Vector not known	Canine infectious cyclic thrombocytopenia. Usually subclinical. Platelets affected
E. risticii	Horses. USA and Europe	Vector unknown. Seasonal, late spring to early autumn	Potomac horse fever (equine monocytic ehrlichiosis). Present in monocytes. Intestinal involvement
Cowdria ruminantium	Domestic and wild ruminants. Africa, Caribbean	*Amblyomma* ticks	Heartwater. Reticular cells, neutrophils, and vascular endothelial cells affected
Neorickettsia helminthoeca	Dogs, coyotes, foxes, bears and ferrets. Pacific coast of USA	Ingestion of salmon containing the agent	Salmon poisoning. Agent present in salmon within the metacercariae of the *Nanophyetus salmincola* fluke. Parasitises reticuloendothelial cells of the lymphoid system, including macrophages
Elokomin fluke fever agent	As above	As above	Salmon fever (milder form). This agent occurs with *N. helminthoeca* or separately. Reticuloendothelial cells of the lymphoid system
Anaplasma marginale	Cattle and other ruminants. Tropical and sub-tropical regions	Ticks and other arthropods and veterinary instruments	Anaplasmosis (gall sickness). Inside erythrocytes with a marginal distribution
A. centrale	As above	As above	Mild or subclinical disease. Can be used to protect animals against the more virulent *A. marginale*. Predominantly central in erythrocytes
A. ovis	Sheep, goats and deer	As above	Infrequent and mild disease. Predominantly marginal in erythrocytes
Aegyptianella pullorum	Poultry. South Africa, South-east Asia and Mediterranean	Ticks such as *Argas persicus*	Aegyptianellosis. Agent in erythrocytes, leukocytes and mononuclear cells. *A. persicus* can also transmit *Borrelia anserina*
Haemobartonella felis	Cats. Worldwide	Possibly via cat fights and biting arthropods. Intrauterine infection reported	Feline infectious anaemia. Present on or in erythrocytes
Eperythrozoon ovis	Sheep. Africa, USA, Australia and Europe	Biting arthropods	An infrequent disease in lambs with anaemia and failure to gain weight. Present on red cells
E.suis	Pigs. USA	Biting arthropods and instruments	Porcine eperythrozoonosis. Usually subclinical but occasionally icteroanaemia, embryonic death and abortion. Present on erythrocytes

Laboratory Diagnosis

Specimens

These will vary with the disease, but usually include unclotted blood for blood smears or affected tissue such as brain in suspected heartwater. Paired serum samples for serology are appropriate in some diseases.

Direct microscopy

As the Rickettsiales stain poorly with the Gram procedure, Giemsa and other Romanowsky stains, such as Gimenez, Macchiavello and Leishman (**371**, **372**, **373**), as well as fluorescent antibody (FA) staining (**374**), are used for both blood and tissue smears. A number of blood smears may have to be examined as the agents may appear in the blood only periodically.

371

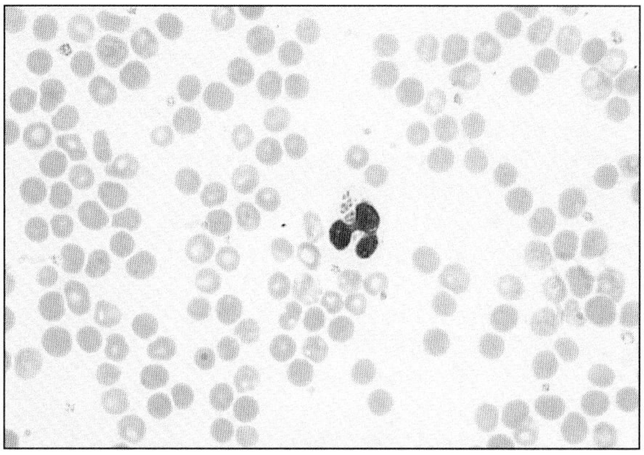

371 Leishman-stained blood smear from a sheep showing a neutrophil containing *Ehrlichia (Cytoecetes) phagocytophila.* (×1000)

372

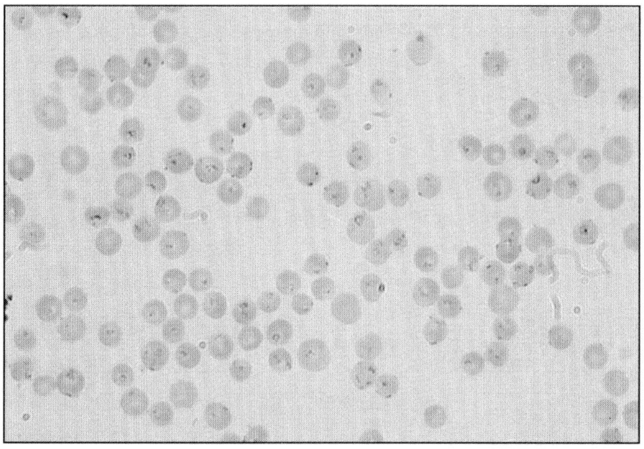

372 Leishman-stained blood smear from a sheep with *Eperythrozoon ovis* parasitising the red cells. (×1000)

373

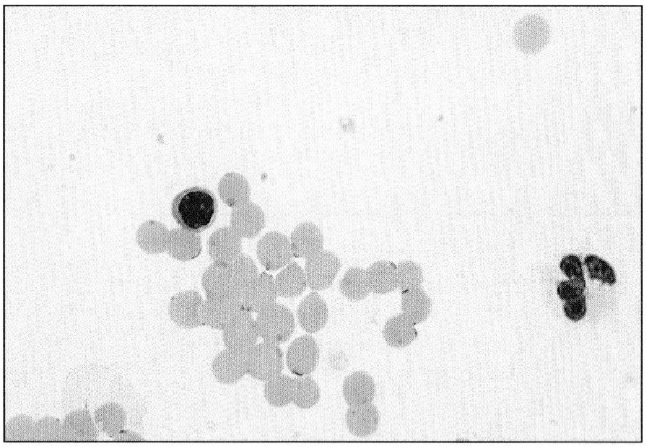

373 Leishman-stained blood smear from a cat with *Haemobartonella felis* on red blood cells. The rubricyte (top left) is indicative of red cell regeneration (×1000).

374

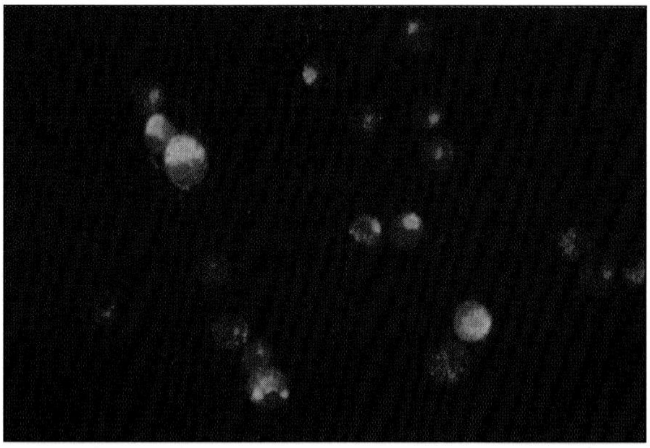

374 *Ehrlichia risticii* in mouse macrophages. (Direct FA technique, ×1000)

Isolation and cultivation

This is often difficult and is not usually necessary for a laboratory diagnosis. *Rochalimaea* spp. (human pathogens) will grow on inert media but the other Rickettsiales require living cells for replication; the yolk sac of chick embryos, cell cultures or laboratory animals (mice or guinea-pigs) are used.

Identification

This is usually based on the animal species, clinical signs and demonstration of the organisms in stained blood or tissue smears and by specific serological tests.

The laboratory diagnosis of the principal rickettsial diseases is summarised in **Table 113**.

Table 113. Laboratory diagnosis of important rickettsial diseases.

Disease	Laboratory diagnosis	Appearance of agent in Giemsa-stained smears
Q fever (*Coxiella burnetii*)	Fluorescent antibody (FA) or Giemsa-stained smears from ruminant placentas. Paired serum samples for serology (CFT, ELISA or micro-agglutination). Antibody rise 2–3 weeks post-infection	Small purple-red cocci (0.2–0.4 µm) or short rods within cells. Similar in appearance to *Chlamydia psittaci* when stained with the Giemsa stain
Canine ehrlichiosis (*Ehrlichia canis*)	Giemsa-stained blood smears, best at about the 13th day post-infection. Indirect FA test on serum for antibody	Purple-staining cells (0.5 µm diameter) or inclusions (morulae) up to 4.0 µm diameter in monocytes or lymphocytes
Equine ehrlichiosis (*Ehrlichia equi*)	Giemsa-stained blood or buffy coat smears. Inclusions can be seen 48 hours after onset of disease. Indirect FA test on serum for antibody	As for *E. canis* but cells or inclusions present in granulocytes, especially neutrophils
Potomac horse fever (*Ehrlichia risticii*)	FA or Giemsa-stained blood smears. ELISA or indirect FA for antibodies in serum	Purplish-staining agent in monocytes. Inclusions similar to other *Ehrlichia* spp.
Tick-borne fever (*Ehrlichia phagocytophila*)	FA or Giemsa-stained blood smears	Purplish inclusions varying from 0.7–3.0 µm in neutrophils, eosinophils, basophils and monocytes
Heartwater (*Cowdria ruminantium*)	FA or Giemsa-stained smears from brain tissue (cerebral cortex). Inoculation of mice or susceptible cattle	Purple-staining coccal (0.2–0.5 µm) or short bacillary forms in cytoplasm of vascular endothelial cells of capillaries in the brain
Salmon poisoning (*Neorickettsia helminthoeca*)	Clinical signs and the finding of fluke eggs (*Nanophyetus salmincola*) in faeces. Demonstration of the agent in lymph node aspirates	Purplish morulae in cytoplasm of macrophages with individual coccal (0.3–0.4 µm) forms scattered within the cells
Anaplasmosis (*Anaplasma marginale*)	Giemsa, acridine orange and FA staining of blood smears. Serology: indirect FA, CFT and card agglutination	Reddish-violet pleomorphic forms (0.2–0.4 µm diameter) within erythrocytes and near the periphery. Up to 50% of red cells may be parasitised
Avian aegyptianellosis (*Aegyptianella pullorum*)	Giemsa-stained blood smears. Inoculation of susceptible birds by parenteral routes or skin scarification with infected blood	Great variety of violet-reddish forms: oval, round and ring (0.3–3.9 µm diameter), and also larger inclusions in erythrocytes
Feline infectious anaemia (*Haemobartonella felis*)	Giemsa or FA-stained blood or tissue smears. Check Giemsa-stained smears daily for 1 week as the presence of the agent on red cells is inconsistent	Deep purple, small coccoid or rod-shaped (0.2 µm diameter) organisms on erythrocytes. A few ring-forms occasionally seen
Ovine eperythrozoonosis (*Eperythrozoon ovis*)	Giemsa-stained blood smears. With acridine orange staining there is bright orange fluorescence	Pale purple organisms in disc- or ring-forms (0.5–1.0 µm diameter). Rod-forms are most common at the margin of the erythrocytes
Porcine eperythrozoonosis (*Eperythrozoon suis*)	Giemsa or FA-stained blood smears. Serology: indirect FA or CFT	Bluish-violet cocci or ring-forms (up to 2.5 µm diameter) on erythrocytes. Largest species in the genus

Antibiotic susceptibility

Testing for antibiotic susceptibility is difficult in the Rickettsiales. Clinical trials indicate that tetracyclines or chloramphenicol are the drugs of choice. A comparatively long course of therapy is required to prevent relapses. Long-term medication of feed with tetracyclines has been used to eliminate the carrier state with *Anaplasma marginale*.

35 The Mycoplasmas (Class: Mollicutes)

The mollicutes are the smallest procaryotic cells capable of self-replication. The genomes of *Mycoplasma* and *Ureaplasma* species are 5×10^8 daltons compared to the average bacterial genome of 2.5×10^9 daltons. These organisms lack the genetic ability to form a cell wall and are enclosed in a plasma membrane composed of protein, glycoprotein, glycolipid and phospholipid. As there is no rigid cell wall, the mollicutes are plastic and pleomorphic. Cell forms include cocci, spirals, filaments and rings. Some divide by binary fission while others have a reproductive cycle involving the break-up of elongated forms into round forms. The minimal reproductive unit is about 0.3 µm in diameter, but because the cells are pliable they are able to pass through a 0.22 µm membrane filter. They characteristically form fried-egg-shaped microcolonies that grow into agar media. Most are facultative anaerobes or microaerophiles except for the *Anaeroplasma* species that are anaerobic.

The mollicutes have a Gram-negative-type cell, but stain poorly with the Gram-stain, producing better results with Giemsa and other Romanowsky stains. The six genera of the Class Mollicutes are often, collectively, referred to as the 'mycoplasmas'. The first organism in the class to be isolated was *Mycoplasma mycoides* subsp. *mycoides*, the cause of contagious bovine pleuropneumonia (CBPP). The collective term 'pleuropneumonia-like organisms' (PPLOs) has been used for mycoplasmas. The differential features of the genera are shown in **Table 114**. Species of the genera *Mycoplasma, Ureaplasma* and *Acholeplasma* are of significance in animals.

Natural Habitat

The mycoplasmas (mollicutes) occur worldwide as free-living saprophytes or as parasites of animals. Both pathogenic and non-pathogenic species are found as commensals on the mucous membranes of the upper respiratory, intestinal and genital tracts, on articular surfaces and in the bovine mammary gland. Outside the host, the pathogenic species can survive in microenvironments, protected from sunlight, for several days.

Table 114. Differential features of the genera of mycoplasmas (mollicutes).

Genus	Cholesterol requirement	Habitat	Other features
Mycoplasma	+	Animals	Many are animal pathogens Optimal pH 7.5 Microcolonies 0.1–0.6 mm in diameter
Ureaplasma	+	Animals	Some associated with disease Optimal pH 6.0. Produce urease Microcolonies 0.01–0.05 mm in diameter, called 'T-mycoplasmas' (T = tiny)
Acholeplasma	–	Animals, soil, sewage	Many saprophytic, a few associated with disease in animals Microcolonies 0.1–1.0 mm in diameter
Spiroplasma	+	Plants, insects	Some cause disease in plants and insects. Helical and motile forms
Anaeroplasma	v	Rumen of sheep and cattle	Commensals Anaerobic
Thermoplasma	–	Acid hot springs	Saprophyte. Optimal growth temperature 59°C and pH 1 to 2

+ = cholesterol required; – = cholesterol not required; v = strains vary in requirement for cholesterol.

Pathogenesis

The parasitic mycoplasmas tend to adhere firmly to the host's mucous membranes and some species have been shown to affix to cells by specific attachment structures. The organisms are extracellular and produce haemolysins, proteases, nucleases and other toxic factors that can lead to the death of host cells or to a chronic infection. One species, *Mycoplasma neurolyticum*, produces a neurotoxin. Some pathogenic species have a predilection for mesenchymal cells lining joints and serous cavities. The respiratory tract and lungs are frequent sites of infection. Mycoplasmas are capable of destroying the cilia of cells in the respiratory tract, thus predisposing to secondary bacterial invasion. Latency can occur and various stresses predispose to mycoplasmal diseases. The infections are frequently chronic or low grade. Infections are either endogenous or exogenous. Transmission is usually venereal, vertical or by aerosols and many important avian mycoplasmas are egg-transmitted.

The most important species of mycoplasmas and the diseases that they cause in poultry, other domestic animals and laboratory animals are listed in **Table 115**. The mycoplasmas tend to be fairly host specific. The distribution of the diseases is generally worldwide except where otherwise indicated in the table. In addition to the mycoplasmas shown in **Table 115,** there are many species isolated from birds and animals, whose disease status is at present uncertain. These are listed in **Table 116**.

Laboratory Diagnosis

Specimens

Mycoplasmas are fragile and specimens must be kept refrigerated and delivered to a laboratory within 24–48 hours of collection. The specimens should be obtained from animals at an early stage in the clinical disease. The samples may include mucosal scrapings, tracheal exudates, aspirates, pneumonic tissue from the edge of the lesion, cavity or joint fluids and mastitic milk. Swabs should be submitted in transport medium (see **Appendix 2**). A simple transport medium for mastitic milk samples is the sample itself with 5 mg/ml ampicillin. This is kept at room temperature which will allow replication of the mycoplasmas in the specimen during transportation. A milk sample without antibiotics should also be collected for the isolation of any bacterial pathogens present.

Direct microscopy

Fragility, pleomorphism and weak staining by various methods make direct examination of stained smears of little value for the diagnosis of mycoplasmal diseases. Some workers have developed a fluorescent antibody technique to identify *M. dispar* and ureaplasmas in bronchial epithelium of calves with pneumonia.

Isolation

Culture media

Mycoplasmas are fastidious organisms and most require specific growth factors, an isotonic medium and the absence of inhibitory substances for growth. The basic medium is a good quality beef infusion with supplements. *Mycoplasma* and *Ureaplasma* species require cholesterol and this is usually supplied by 20 per cent horse serum. Serum from individual animals can vary in growth-promoting properties, pooled serum from several animals is preferred. Other growth factors are catered for by the addition of yeast extract, DNA and possibly nucleotides. The water quality for the medium should be equal to that used for tissue culture. Penicillin can be added to discourage the growth of Gram-positive bacteria and thallium acetate to inhibit fungi and Gram-negative bacteria. The final pH of the medium is adjusted to between 7.2 and 7.8 for *Mycoplasma* species. Ureaplasmas require the addition of urea and a final pH of 6.0. Thallium acetate is omitted from ureaplasma media as it is harmful to them. Arginine is inhibitory to *M. bovoculi. M. dispar* may be sensitive to some penicillin G or benzyl penicillin preparations.

The basic medium and supplements can be bought commercially. Media usually take the form of broths or solid agar, although semi-solid agar and diphasic media (agar with broth overlay) are also used. The mycoplasmal transport medium (**Appendix 2**) can also make a suitable broth medium for isolation. This can be converted to an agar medium by substituting 28 g of PPLO agar (Difco) for the 16.8 g of PPLO broth. A modification of Hayflick's medium for ureaplasmas is given in **Appendix 2.** *Acholeplasma* species will usually grow on media suitable for *Mycoplasma* species. Special media may have to be prepared for the mycoplasmas that are particularly difficult to culture, such as *M. dispar* and *M. meleagridis*.

The agar media are poured into small Petri dishes of about 60 mm diameter, preferably made of glass, as poor growth has been reported when using certain types of plastic Petri dishes. Strict quality control measures are required for the media to obtain optimal growth of these fastidious organisms. Each new batch of basic medium and supplements should be pre-tested with a laboratory control strain to ensure that the medium sustains growth and does not contain any substances inhibitory to the mycoplasmas.

Table 115. Mycoplasmas causing significant disease in domestic and laboratory animals.

Species	Disease
POULTRY	
Mycoplasma gallisepticum	Chickens: chronic respiratory disease. Turkeys: infectious sinusitis. Infections in game birds and imported Amazon parrots
M. synoviae	Chickens and turkeys: infectious synovitis
M. meleagridis	Turkeys: *Mycoplasma meleagridis* disease (MM disease), an air sacculitis and bursitis in young birds
M. iowae	Turkey poults: air sacculitis, stunting and leg deformities. Mortality of turkey embryos can occur
M. anatis	Ducks: sinusitis
PIGS	
M. hyorhinis	Chronic progressive arthritis and polyserositis in 3–10-week-old pigs
M. hyosynoviae	Mycoplasmal polyarthritis in 12–24-week-old pigs
M. hyopneumoniae	Enzootic ('virus') pneumonia of pigs
CATTLE	
M. mycoides subsp. *mycoides* (small colony type)	Contagious bovine pleuropneumonia (CBPP) (Africa, Middle East, China)
M. bovis	Mastitis, arthritis, pneumonia, genital infections, abortion
M. bovigenitalium	Vaginitis, arthritis, mastitis, seminal vesiculitis
Ureaplasmas including *U. diversum*	Vulvovaginitis, pneumonia
M. dispar	Pneumonia (calves)
M. californicum	Mastitis
M. canadense	Mastitis
M. bovoculi	A predisposing cause of infectious bovine keratoconjunctivitis (*Moraxella bovis*)
GOATS	
M. mycoides subsp. *mycoides* (large colony type)	Septicaemia, polyarthritis, pneumonia, mastitis, conjunctivitis (North America)
M. mycoides subsp. *capri*	Contagious caprine pleuropneumonia (CCPP) (Africa, Mediterranean)
Mycoplasma strain F-38	Contagious caprine pleuropneumonia (CCPP) (Africa)
M. putrefaciens	Mastitis, arthritis
SHEEP	
M. ovipneumoniae	Pneumonia
SHEEP AND GOATS	
M. agalactiae	Contagious agalactia (USA, Mediterranean, Europe, Asia)
M. conjunctivae	Keratoconjunctivitis
M. capricolum	Polyarthritis, mastitis, pneumonia
Acholeplasma oculi	Keratoconjunctivitis
HORSES	
Mycoplasma felis	Pleuritis (a commensal that can enter the pleural cavity after severe exercise)
DOGS	
M. cynos	Pneumonia (part of 'kennel cough' complex)
CATS	
M. felis	Conjunctivitis
RATS AND MICE	
M. neurolyticum	Rolling disease
M. pulmonis	Pneumonia
M. arthritidis	Polyarthritis

Table 116. Mycoplasmas whose disease status is uncertain.

POULTRY	SHEEP AND GOATS	DOGS
M. cloacae	*M. arginini*	*M. canis*
M. gallinarum		*M. edwardii*
		M. maculosum
	HORSES	*M. molare*
PIGS		*M. opalescens*
	M. equigenitalium	*M. spumans*
M. flocculare	*M. equirhinis*	
M. hyopharyngitis	*M. fastidiosum*	**CATS**
M. sualvi	*M. salivarium*	
A. axanthum	*M. subdolum*	*M. feliminutum*
A. granularum	*A. equifetale*	*M. gateae*
	A. hippikon	
	A. laidlawii	
CATTLE		**GUINEA-PIG**
M. alvi		*M. caviae*
M. arginini		
M. bovirhinis		
M. verecundum		
A. laidlawii		
A. modicum		

Inoculation of culture media

For routine isolation of mycoplasmas, the specimen should be inoculated into two broths and onto two plates of agar (one suitable for mycoplasmas and the other for ureaplasmas). The inoculation technique will vary according to the nature of the specimen:

a) Fluid materials such as foetal fluids and exudates can be inoculated directly into broth medium and spread over the surface of the agar medium.
b) Some specimens such as semen, joint fluids and tissues may contain inhibitors for mycoplasmas. Both undiluted specimen and ten-fold dilutions in mycoplasmal broth (up to 10^{-6}) should be cultured.
c) With tissues, such as pieces of lung, a freshly cut surface on a block of the tissue can be moved across the surface of an agar plate for inoculation. Alternatively, the tissue can be homogenized in broth, ten-fold dilutions made and broth cultures inoculated.

The inoculated agar plates are incubated in a humid atmosphere at 37°C. It is advisable to inoculate and incubate duplicate plates, one aerobically and one under 5 per cent CO_2 and 95 per cent N_2. A candle-jar may be satisfactory. The plates should be examined after 48 and 96 hours' incubation. The plates are viewed under a stereoscopic microscope (transmitted light) or under the low-power objective of a light microscope, for the mycoplasmal 'fried-egg' microcolonies (**375**, **376**). The cultures can be regarded as negative if no microcolonies are seen after 14 days' incubation. The tubes of broth, incubated aerobically at 37°C, are checked daily for growth. When a slight turbidity is seen (or a colour change if the broth contains phenol red), loopfuls of the broth can be used to inoculate appropriate agar media.

Obtaining a pure culture

A pure culture of the isolated mycoplasma should be obtained before carrying out identification techniques. The microcolonies grow into the agar (**Diagram 41**) and it is difficult to obtain cells from the colonies on an inoculation loop, so different methods have to be used for subculture. An agar plate should be selected with well-separated colonies. Individual colonies are removed by cutting out a small block of agar, containing the colony, with a sterile scalpel. The block is transferred to a tube with 2–3 ml of broth and incubated at 37°C for 48 hours or longer. The broth culture, when showing turbidity, is taken into a sterile syringe and passed through a 0.45 µm membrane filter. The filtrate is diluted with fresh broth, 1:10 and 1:100, and a loopful of each dilution is inoculated onto agar medium and the plates incubated. This procedure should be repeated three times. At least four colonies should be used as not all microcolonies will grow after subculture.

An alternative method is to take a block of agar containing one colony and place it colony-side downwards on an agar plate. The agar block is pushed back and forth over the agar surface and the inoculated plate incubated for 48–72 hours. This procedure is repeated using three separate microcolonies.

375

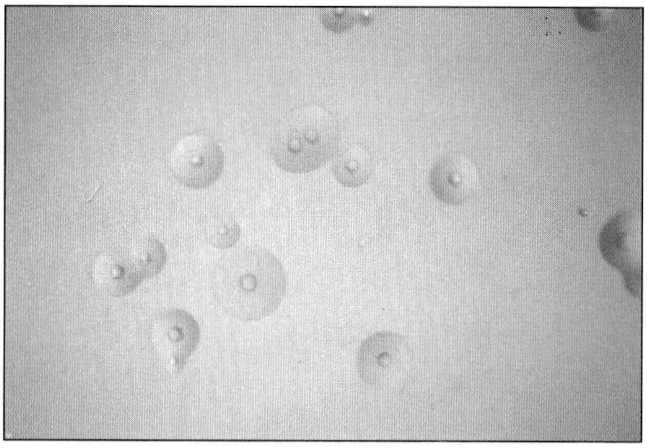

375 Unstained mycoplasmal microcolonies illustrating the characteristic appearance. (×10)

376

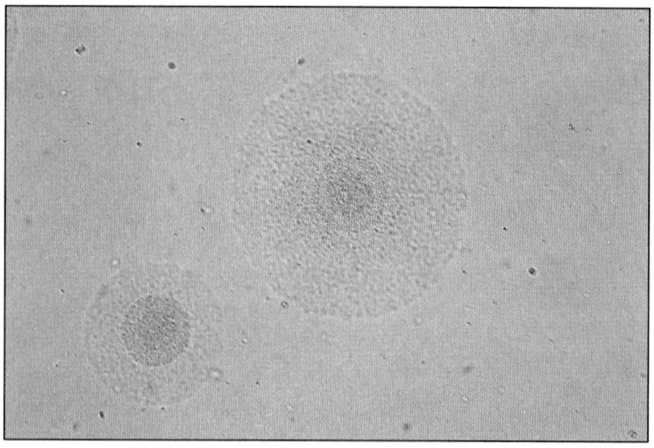

376 Unstained microcolonies of a *Mycoplasma* species showing the typical 'fried-egg' morphology. (×25)

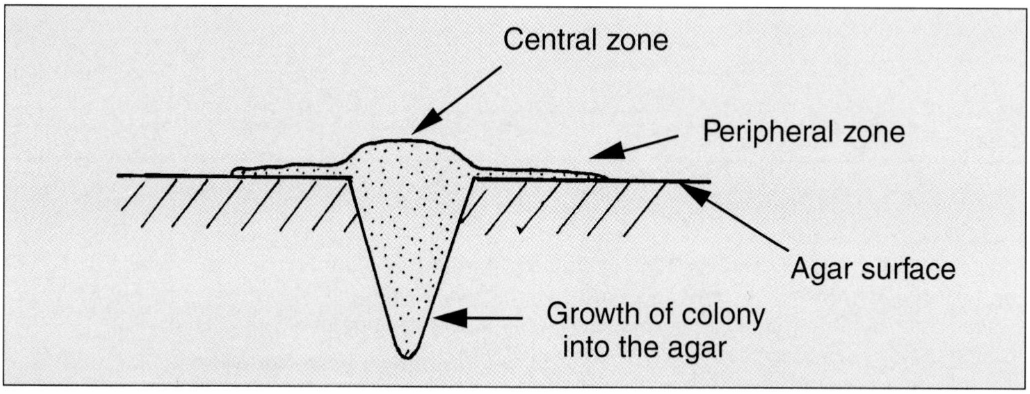

Diagram 41. Section through a mycoplasmal colony showing surface and subsurface growth.

Once a pure culture has been obtained, normal subculturing can be carried out using similar techniques, either from agar-to-agar or agar-broth-agar.

Identification

Differentiation from bacterial L-forms

Bacteria that have temporarily failed to form cell walls (L-forms) can produce microcolonies similar to those of the mycoplasmas. The failure to form cell walls is often due to the bacterium having been exposed to penicillin or other antibacterial agents that affect cell wall formation. Differentiation between mycoplasmal and bacterial L-form microcolonies may be necessary and can be carried out as follows:

a) Subculturing the suspected bacterial L-form on media without antibacterial substances should cause reversion of the L-form with the formation of normal-sized bacterial colonies. Up to five subcultures may be necessary.

b) Staining microcolonies with the Dienes' stain is useful for easier visualisation of microcolonies but this also aids in the differentiation between mycoplasmal and bacterial L-form microcolonies. Mycoplasmal colonies retain the Dienes' stain indefinitely, whereas bacterial L-form microcolonies tend to decolourise in about 15 minutes. Staining microcolonies with the Dienes' stain is carried out as follows:

- A block of agar containing microcolonies is placed, colony-side upwards, on a microscope slide. A light film of Dienes' stain is placed on a cover slip and allowed to dry. This is then put, stain-side downwards, on the microcolonies on the agar block. The preparation is examined under the low-power objective of a light microscope. The denser centre of the microcolonies, which grow down into the agar, stain dark blue. The less dense peripheral zone, representing surface growth, stains light blue (**377**). Dienes' stain can be bought commercially and the formula is given in **Appendix 1.**

Identification of the genus

Differentiation between the three genera of veterinary importance is shown in **Diagram 42**.

a) *Sensitivity to digitonin*: this test reflects the requirement of cholesterol for growth. *Mycoplasma* and *Ureaplasma* species are sensitive to digitonin, whereas *Acholeplasma* species are not. Sterile filter paper discs (6 mm diameter) are saturated with 0.025 ml of digitonin (1.5 per cent w/v in ethanol) and allowed to dry. An agar medium that gives good growth of the mycoplasma under test is flooded with a broth culture of the organism and the surplus liquid removed. The surface of the plate is allowed to dry and the digitonin disc placed on the agar surface. The plate is incubated and the test can be read when microcolonies are visible. The zone of inhibition around the disc should be 5 mm or more from the disc's edge to indicate sensitivity.

b) *Modified urease test*: the ureaplasmas are able to produce the enzyme urease. To test for this activity a mixture of equal parts of 10 per cent urea and 0.8 per cent manganese chloride is applied directly onto 40-hour-old microcolonies. A positive result is indicated by an immediate colour reaction on the surface of the ureaplasma colonies. The colour reaction goes from light to dark brown and finally to black due to the deposition of manganese on the surface of the colony.

c) *Size of the microcolonies*: this may also be an aid to the differentiation of the three genera isolated from animals (**Diagram 42**).

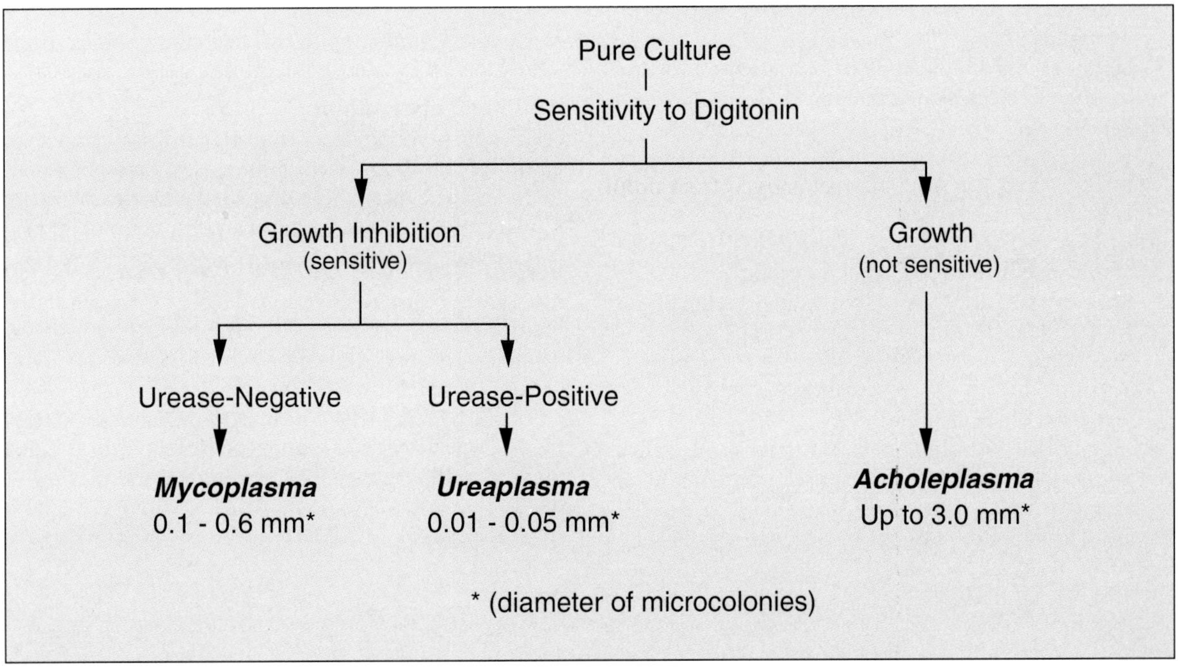

Diagram 42. Differentiation of the three mycoplasmal genera isolated from animals.

377

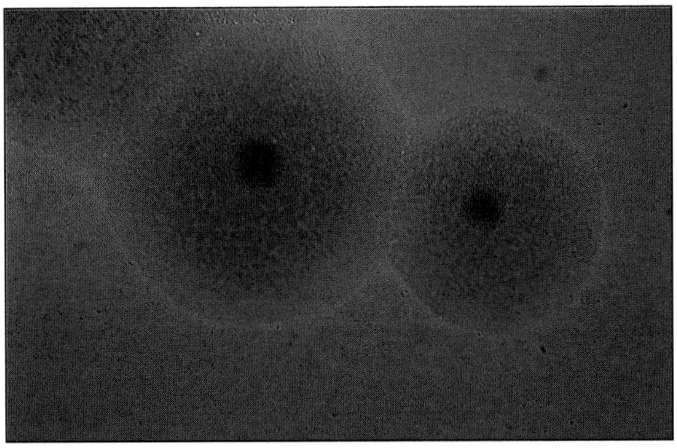

377 Mycoplasmal microcolonies. (Dienes' stain, ×25)

378

378 Microcolonies of *Mycoplasma bovis*. (Direct FA technique, ×25)

Identification of species

Knowledge of the animal species and disease process will often suggest a particular mycoplasmal species. Precise identification of the species often requires sophisticated techniques and cultures are usually submitted to reference laboratories for final identification. The following methods are examples of the techniques that can be used to identify mycoplasmal species:

- Fluorescent antibody (FA) staining to identify *M. dispar* and ureaplasmas in bronchial epithelium of calves with pneumonia.
- FA technique (direct and indirect) for staining mycoplasmal colonies (**378**). This is the best method of recognising mixed cultures and the method is used commonly with the avian mycoplasmas.
- Enzyme-linked immunoperoxidase used on sections of porcine bronchial epithelium to detect *M. hyopneumoniae.*
- Agar gel diffusion tests using known antisera to identify prepared mycoplasmal antigen from broth cultures.
- ELISA and complement fixation test (CFT) for antigen identification using known antisera.
- Species-specific DNA probes have been developed for the identification of some species.
- Biochemical tests such as glucose fermentation, arginine hydrolysis, phosphatase activity and reduction of tetrazolium.
- Metabolic inhibition tests: essentially neutralisation tests in which specific antisera prevent the use of a particular substrate.
- Growth inhibition tests using specific sera on filter paper discs.

Serological tests for mycoplasmal antibodies

A number of these tests are available and include:

- Rapid plate agglutination tests (coloured antigen) for screening poultry for the major mycoplasmal diseases.
- Haemagglutination-inhibition test for avian mycoplasmas.
- Indirect haemagglutination tests for *M. bovis* and *M. bovigenitalium.*
- Agar-gel diffusion tests for avian mycoplasmas using known antigen.
- Complement fixation test for the control of contagious bovine pleuropneumonia and for screening pigs for enzootic pneumonia.
- ELISA for the detection of antibodies in cattle due to an *M. bovis* infection.
- Latex agglutination test to detect antibodies to the mycoplasmas causing disease in goats.

Antibiotic susceptibility

Although mycoplasmas may develop resistance to antimicrobial drugs, susceptibility tests are not usually performed. Among the drugs used for treatment are tylosin, tetracyclines, tiamulin and fluoroquinolones.

Specific-Pathogen-Free (SPF) Programmes

Flocks of poultry (chickens and turkeys) free from the major avian mycoplasmas and pig herds free from *M. hyopneumoniae* have been established in some countries with specific-pathogen-free programmes. There are usually two phases in these programmes: detection of infections and culling or isolation of affected pigs or birds, followed by serological monitoring of the flocks or herds to demonstrate continued freedom from infection.

36 Mastitis

Mastitis, inflammation of the mammary gland, can be caused by physical or chemical agents but the majority of cases are infectious and usually caused by bacteria. Over 130 different microorganisms have been isolated from bovine mastitic milk samples but *Staphylococcus aureus*, streptococci and members of the *Enterobacteriaceae* are among the most common aetiological agents in cows and in other animal species. Invasion of the mammary gland by microorganisms is characterised by an increased leukocyte count in the milk, the majority of cells being neutrophils.

Epidemiology

Infection of the mammary gland is almost always via the teat canal. In cows this often occurs when the teat sphincter is slack, for a period of 20 minutes to 2 hours after milking. The pathogenic microorganisms generally come from one of two sources, the environment (*Escherichia coli* and other coliforms) or from inside the udder of other animals (*Staphylococcus aureus* and *Streptococcus agalactiae*) and are transmitted via the milking machine or milker's hands. Once the microorganisms have passed into the teat canal, they establish themselves there and multiply. Then the pathogenic microorganisms, with invasive ability, pass into the mammary gland where their capacity to adhere to the mammary epithelium is due to a virulence factor. This prevents the pathogens being swept out of the udder with the milk flow. Invasion of the mammary gland elicits an inflammatory reaction and signs of mastitis may be seen with changes in the appearance of the milk and an increased leukocyte count. Whether mastitis occurs depends on the interaction between microbial, host and environmental factors. These are summarised in **Table 117**.

Table 117. Interaction of bacterial, host and environmental factors leading to mastitis.

Microbial factors	Host factors	Environmental factors
• Ability to survive in the immediate environment of the animal.	• Genetic predisposition to mastitis. This is probably related to factors such as teat shape, sphincter tone, anatomy of the teat canal and susceptibility to weakening of the suspensory ligament ('pendulous udder')	• Presence of large numbers of potential pathogens in the immediate environment of the animal. This may indicate poor hygiene
• Ability to colonise the teat duct	• Age: older cows, especially after four lactations, are more prone to mastitis	• General management of animals. For example, 'coliform' mastitis is much more frequent in housed cattle and sheep
• Ability to adhere to mammary epithelium and not be flushed out with the milk flow	• Stage of lactation: cows are generally more susceptible just after calving and for the following 2 months	• Milking-machine malfunction or inadequate design
• The degree of invasiveness. For example, streptococci cause little pathological change to secretory cells but staphylococci initiate degenerative changes	• Presence of lesions on the teats that may predispose to inadequate milking or may harbour mastitis-producing bacteria	• Milking-shed environment including poor milking technique and hygiene
• Ability to resist phagocytosis and antibacterial substances in udder, including resistance to antibiotics	• Immunological factors such as the level of IgA, IgG1, lactoferrin and phagocytes in the mammary gland	• External trauma such as that arising from rough, muddy approaches to the milking shed or with ewes, the sucking of large vigorous lambs

Clinical Syndromes of Mastitis

The main clinical types of mastitis are:

- **Peracute**: swelling, pain, heat and abnormal secretion in the mammary gland are accompanied by signs of systemic disturbance such as fever, depression, anorexia, weakness and a rapid, weak pulse. The signs are those of a toxaemia or septicaemia. Gangrenous mastitis is included in this category.
- **Acute**: changes in the mammary gland are similar to those of peracute mastitis but the systemic signs are less severe.
- **Subacute**: no systemic reaction and the changes in the gland are less marked.
- **Chronic**: there are no systemic signs and very few external signs of change in the udder, but abnormal secretion in the gland occurs intermittently.
- **Subclinical**: the infection in the mammary gland is detectable only by bacterial culture or by tests to demonstrate a high leukocyte count in the milk. There is no obvious change in the appearance of the milk. *Staphylococcus aureus* is notorious for causing a high percentage (often 50 per cent of the herd) of subclinical infections in a dairy herd with a staphylococcal mastitis problem.

Mastitis in Domestic Animals other than Cattle

Mastitis has the greatest prevalence and economic importance in dairy cows but it can be significant in other domestic animals, particularly in sheep, goats and pigs. **Table 118** gives the most common aetiological agents in sheep and goats, their natural habitat and the type of clinical syndrome that is usually caused by each agent.

Sheep

Mastitis is not as common in sheep as it is in dairy cattle, but in ewes used for milk production it has been estimated that 7 per cent may have subclinical mastitis. Outbreaks of mastitis can occur in ewes housed for lambing, probably due to contamination of bedding from infected udder secretions. Suppurative lesions caused by *Corynebacterium pseudotuberculosis* (where the bacterium is endemic) can be found fairly frequently in ovine mammary glands. It is thought that most of these infections spread to the udder from the supramammary lymph nodes. The prevalence of mastitis increases with the age of the ewe and with the number and weight of the lambs. Trauma from large, vigorously sucking lambs may predispose the ewe to mastitis. The importance of mastitis in ewes lies not only in the deaths that occur from peracute cases, but also in the depressed weaning weights and lamb mortality from starvation. Maedi (lentivirus) leads to a progressive induration of the mammary tissue and the secretory tissue is gradually obliterated while the milk remains normal in appearance. While Maedi virus does not cause acute mastitis the effects on the lambs are similar, leading to poor growth or starvation. Contagious pustular dermatitis (orf) lesions can occur on udders and this could predispose to mastitis.

Goats

The aetiology of infectious mastitis in goats and cows is similar but goats are affected less frequently. *Staphylococcus aureus* and *Streptococcus agalactiae* are most commonly involved but coliform mastitis can occur sporadically or sometimes as a herd problem. It is important to remember that milk from goats can appear relatively normal even with severe inflammatory changes in the udder. The diagnosis, control and treatment of mastitis in goats are similar to those for dairy cows. Caprine-arthritis-encephalitis (lentivirus), like Maedi in sheep, causes a gradual destruction of secretory tissue.

Pigs

In pigs, mastitis is almost always a disease of recently farrowed gilts or sows, usually in the first 48 hours postpartum. Coliforms (*Escherichia coli*, *Klebsiella* and *Enterobacter* species) most commonly cause peracute mastitis. Coliform mastitis can occur as an entity in itself or the bacteria may play a part in the mastitis-metritis-agalactia syndrome, most frequently seen in gilts. Mortality in piglets, from starvation, can be high.

Subacute mastitis occurs in older sows and can involve either staphylococci or streptococci. One or more glands are affected and this may impair the ability of the sow to suckle and rear a large litter successfully.

Granulomatous lesions in the mammary glands are associated with chronic *Staphylococcus aureus* infections or with those caused by *Actinomyces bovis* or *Actinobacillus lignieresii.*

Other bacteria such as *Fusobacterium necrophorum* and *Actinomyces pyogenes* have also been reported as causing mastitis in sows. Clinically inapparent mammary abscesses occur commonly.

Horses

Mastitis in mares is comparatively rare, but it can be severe when it does occur. One or both mammary glands can be affected with a marked and painful swelling. There is fever, depression, and a stiff gait may be seen, or the mare may stand with hindlegs apart because of the discomfort. Gangrenous mastitis has also been reported in mares. *Staphylococcus aureus* and Lancefield group C streptococci are the main aetiological agents but *Pseudomonas aeruginosa* and *Corynebacterium pseudotuberculosis* have also been isolated from mastitic milk from mares. If a sick foal is not sucking, a distended and painful udder can mimic the signs of mastitis.

Table 118. Ovine and caprine mastitis: aetiological agents, natural habitat and clinical types.

Aetiological agent	***Natural habitat***	***Clinical type***
OVINE MASTITIS		
Staphylococcus aureus	Skin and mucous membrane	Acute or peracute (gangrenous)
Streptococcus agalactiae	Intramammary pathogen	Acute
S. dysgalactiae	Buccal cavity and other mucous membranes	Acute
S. uberis	Skin and faeces	Acute
Pasteurella multocida	Commensal on mucous membranes	Acute
P. haemolytica	Commensal on mucous membranes	Peracute (gangrenous) 'blue bag'
Escherichia coli	Faeces and bedding	Peracute
Actinomyces pyogenes	Skin and mucous membranes	Acute and suppurative
Pseudomonas aeruginosa	Soil or water	Acute or peracute (gangrenous)
Corynebacterium pseudotuberculosis	Gastrointestinal tract	Acute and suppurative. Spread may occur from supramammary lymph nodes
CAPRINE MASTITIS		
Staphylococcus aureus	Skin and mucous membranes	Subacute or peracute (gangrenous)
Streptococcus agalactiae	Intramammary pathogen	Subacute or acute
S. dysgalactiae	Buccal cavity and other mucous membranes	Acute
S. pyogenes	Human pathogen	Acute
Actinomyces pyogenes	Skin and mucous membranes	Acute
Escherichia coli and other 'coliforms'	Faeces and bedding	Acute: sporadic or as a herd problem
Yersinia pseudotuberculosis	Rodents are reservoirs	Acute or chronic
Mycoplasma putrefaciens *M. capricolum* *M. agalactiae*	Mucous membranes of urogenital, upper respiratory, upper intestinal tracts and external ear canal	Acute Acute Acute (contagious agalactia)

Dogs and cats

Mastitis is uncommon in dogs and cats. Cases occur within 6 weeks of parturition and may involve one or more mammary glands. Infectious agents gain access through the teat orifice, by penetrating wounds or via haematogenous spread. *Staphylococcus aureus*, *S. intermedius* and beta-haemolytic streptococci are the most common bacteria isolated. The clinical syndrome can be peracute, acute or subacute.

Rabbits

Staphylococcus aureus is the most common cause of mastitis in rabbits. The bacterium may be introduced via splinters from wooden nest boxes, teeth of nursing young or from an unhygienic environment.

Bovine Mastitis

Aetiology

Some of the many microorganisms that can cause bovine mastitis are summarised in **Table 119**. The natural habitat and type of mastitic syndrome are given for each pathogen. *Streptococcus agalactiae* is unusual in that it needs the mammary gland for its perpetuation in nature, although it can survive outside the gland for short periods. *Staphylococcus aureus* can be present in large numbers in both subclinical and clinical cases of mastitis. The subclinical cases (up to 50 per cent of the cows in some herds) can act as a source of infection for the rest of the animals.

Staphylococcus epidermidis (coagulase-negative) and *Corynebacterium bovis* are quite commonly isolated from milk samples. They are normal inhabitants of the teat canal and it is doubtful whether either of them ever causes clinical mastitis as their pathogenicity is low. *C. bovis* causes a persistent infection of the teat duct epithelium and a mild but significant rise in the leukocyte cell count. Because *C. bovis* is susceptible to disinfectants used as teat-dips, it has been suggested that the presence or absence of the bacterium could be used to monitor the efficiency of the teat-dipping procedure in a herd. *S. epidermidis* does not usually cause an increase in the leukocyte cell count of the milk. Its presence may be advantageous to the host as it tends to occupy attachment sites in the teat duct required by the coagulase-positive, pathogenic staphylococci.

Pathogenesis

Coliform mastitis

Escherichia coli and other coliforms can multiply rapidly in quarters with a low cell count. This elicits an inflammatory reaction that destroys a large proportion of the bacterial population. When Gram-negative bacteria die and lyse, endotoxin is released from the cell wall. This massive and sudden liberation of endotoxin results in a severe, life-threatening toxaemia. A unique feature of coliform infections is that, in the cows that recover, the udder tissue gradually returns to normal without fibrosis. In subsequent lactations the gland produces to its optimal capacity.

Staphylococcal mastitis

Affected quarters often have large numbers of *Staphylococcus aureus* in the milk. The staphylococci are easily transmitted during milking via the teat cups or milkers' hands. The type of mastitis produced by *S. aureus* ranges from subclinical to the peracute life-threatening form, one of which is gangrenous mastitis. Gangrenous mastitis is caused by the action of the alpha toxin that damages blood vessels, resulting in ischaemic coagulative necrosis of adjacent tissue. The affected quarter becomes purplish and cold and will eventually slough, if the animal survives the toxaemia. Both chronic and subclinical mastitis lead to a gradual replacement of secretory tissue with fibrous tissue and a subsequent loss of milk production by the affected quarter. The chronic and subclinical forms of mastitis respond poorly to antibiotics because of the development of a tissue barrier that prevents penetration of antibiotics to the site of infection. Dry-cow therapy is more likely to be successful but, even with this method, it is estimated that only 70 per cent of these persistent staphylococcal infections respond to therapy.

Streptococcal mastitis

Streptococcus agalactiae resides in the milk and on the surface of the milk channels but does not invade the tissue. There is rapid multiplication of the bacterium with a great outpouring of neutrophils into the ducts with damage to the ductal and acinar epithelium. Ducts are obstructed with cells and debris, causing involution of the acini in the affected lobules. Fibrosis of inter-alveolar tissue occurs, which also leads to a loss of secretory function.

'Summer mastitis'

This syndrome is seen most commonly in Europe. It is a dual infection of *Actinomyces pyogenes* and an anaerobe, often *Peptostreptococcus indolicus*. 'Summer mastitis' occurs in non-lactating heifers and cows at pasture in the summer months and tends to be more common during wet weather. It is thought to be fly-borne. A massive invasion of the mammary tissue via the teat canal results in a large proportion of the gland being affected at one time. There is a severe systemic reaction and loss of function of the entire quarter. The secretion from the affected quarter is foul-smelling, attributable to the activities of the anaerobic *P. indolicus*. If the cow survives, the quarter becomes extremely indurated and abscesses develop, later rupturing through the floor of the udder, often at the base of a teat.

Uncomplicated infections, with *A. pyogenes* alone, can also be serious with quarters so severely affected that there is a permanent loss of the quarter and sloughing can occur.

Pseudomonas aeruginosa mastitis

Infection by *Pseudomonas aeruginosa* can have a pathogenesis similar to coliform mastitis and a severe endotoxaemia can occur. The infection may result in a subclinical mastitis with the pathogen persisting in the mammary gland.

Table 119. Bovine mastitis: aetiological agents, natural habitat of agents and clinical types (aetiological agents are in approximate descending order of importance).

Aetiological agent	***Natural habitat***	***Clinical type of mastitis***
Staphylococcus aureus	Udder lesions, skin and mucous membranes	Subclinical, chronic, acute and peracute, including gangrenous mastitis. A high % of subclinical carriers can occur in a herd
Streptococcus agalactiae	Intramammary in the milk ducts	Acute or chronic with recurring clinical cases. Infection can occur in maiden heifers
S. dysgalactiae	Buccal cavity and genitalia of cattle	Acute
S. uberis	Skin, tonsils, vagina and faeces	Acute. Can occur in dry period
Escherichia coli *Klebsiella pneumoniae* *Enterobacter aerogenes*	Faeces, sawdust and other bedding. Disease of housed cows	'Coliform mastitis'. Peracute (toxaemia) and usually occurs just after calving in cows with low somatic cell counts. Life-threatening. Little or no fibrosis in udder of recovered animals
Actinomyces pyogenes	Skin and mucous membranes	Peracute, suppurative mastitis
A. pyogenes and *Peptostreptococcus indolicus* or occasionally some other anaerobe.	Both part of normal flora. Infection thought to be fly-borne	'Summer mastitis'. Most common in dry cows and heifers. Foul-smelling udder secretion. Loss of quarter or death can occur
LESS COMMON CAUSES OF BOVINE MASTITIS		
Streptococcus pyogenes	Human pathogen	Acute mastitis
S. pneumoniae	Human pathogen	Peracute. Fever present
S. equi subsp. *zooepidemicus*	Mucous membranes	Subacute or chronic
Enterococcus faecalis	Faeces and skin	Acute mastitis
Pseudomonas aeruginosa	Soil, water or faeces	Peracute (toxaemia) but can be chronic and persistent
Nocardia asteroides	Soil	Sporadic. Acute at first, becoming chronic. Granulomas in udder tissue
Serratia marcescens	Soil and faeces	Peracute (toxaemia) or chronic coliform mastitis
Pasteurella multocida	Mucous membranes (URT)	Acute mastitis
P. haemolytica	Mucous membranes (URT)	Peracute and severe or acute
Mycoplasma bovis *M. bovigenitalium* Other *Mycoplasma* spp.	Respiratory tract and mucous membranes	Acute with rapid onset. Most severe in recently calved animals. All quarters often affected. Dramatic drop in milk secretion, but rarely any systemic reaction
Mycobacterium bovis	Metastasis from existing tuberculous lesion	Induration and hypertrophy of tissue. Can often palpate lesions in udder after milking
M. fortuitum *M. smegmatis*	Soil, but also associated with oil-based intramammary preparations	Severe mastitis. Cows are either culled or die
Fusobacterium necrophorum	Part of normal anaerobic flora of animals	Acute, secretion viscid and stringy. No fibrosis in udder tissue
Bacillus cereus	Associated with feeding brewers' grains or via intramammary preparations	Peracute or acute
Leptospira interrogans serovars *hardjo* or *pomona*	Water, wet soil or urine of subclinical excretors	Agalactia, self-limiting
Candida albicans (yeast)	Mucocutaneous or environmental	Acute but often self-limiting
Cryptococcus neoformans (yeast)	Often introduced via intramammary tubes	Acute mastitis. Milk is mucoid. Severe swelling of udder
Aspergillus fumigatus (mould)	Often introduced via intramammary tubes	Acute (abscess formation) or chronic
Prototheca zopfii or *P. wickerhamii* (algae)	Mud, soil, faeces or water. Ubiquitous in environment	Chronic. Very difficult or impossible to treat

Mycoplasmal mastitis
Although mycoplasmal mastitis can be clinically severe, there is rarely any systemic involvement. Cows of all ages and at all stages of lactation can be affected, with those that have recently calved showing the most severe signs. There can be long-term persistence of the organisms in udders (up to 13 months) and some cows may become shedders of mycoplasmas without severe clinical signs being seen. The secretion from an affected quarter appears fairly normal in the early stages of infection, but if the milk is allowed to stand, a deposit of fine, flaky material settles out leaving a turbid, whey-like supernatant fluid. Leukocyte counts in the milk are usually very high, often over 20 million cells/ml.

***Nocardia asteroides* mastitis**
N. asteroides causes a destructive mastitis that can be acute initially, but is characterised by a granulomatous inflammation that leads to extensive fibrosis and formation of palpable nodules in the udder tissue. Once clinical changes are evident in the mammary gland there is little hope of successful treatment.

Infections by atypical mycobacteria and fungi
The atypical mycobacteria, such as *Mycobacterium fortuitum*, are thought to be accidentally introduced into the udder with oil-based antibiotics. Infections due to yeasts and moulds may also be introduced with intramammary antibiotics. Prolonged or repetitive use of antibiotics can aid their establishment in the udder. These infections tend to be chronic and refractory to treatment.

Infectious conditions of the skin of mammary glands
Traumatic or infectious conditions affecting the skin or subcutaneous tissue of the teats or udder may predispose the cow to mastitis. Painful lesions can lead to difficulty in applying the teat cups or result in the cow giving an incomplete let-down of milk. In the case of 'acne' or 'impetigo', caused by *Staphylococcus aureus*, the infection provides a source of organisms that could cause mastitis if they gained entry to the gland through the teat orifice. The infectious conditions affecting the external surface of teats and udder are summarised in **Table 120**.

Importance of Bovine Mastitis

Major economic loss can occur in a dairy herd with a mastitis problem. The losses due to mastitis are composed of some or all of the following:

- Deaths due to peracute forms of mastitis.
- Loss of cows through premature culling.
- Loss of milk production:
  a) Milk discarded during treatment and withholding periods.
  b) Permanent loss of production, in individual cows, due to secretory tissue being replaced by fibrous tissue.
- High leukocyte counts that lead to a loss of income if a bonus scheme is operating for milk with low cell counts.
- Cost of treatment and veterinary fees.

With all forms of mastitis, even subclinical, a cow will not reach her full production potential in that lactation. In most types of mastitis (except that caused by coliforms) some of the milk secretory tissue will have been permanently destroyed. In a herd with a mastitis problem due to *Staphylococcus aureus*, about 10 per cent of the cows may have clinical mastitis but another 50 per cent can have subclinical mastitis and act as a source of infection for further clinical cases.

Diagnosis of Bovine Mastitis

National mastitis schemes are operating in many countries where the bulk milk, from individual farms, is subjected to a total cell count, usually on a monthly basis. These regularly monitor the dairy herds and allow a herd, with a high cell count, to be investigated before the problem becomes too severe. The cell count levels taken to be significant will vary slightly with each scheme, but generally 200,000 cells/ml milk is considered to be an average somatic count, 500,000 cells/ml or more signifies a problem, and a herd with 400,000 cells/ml on three consecutive monthly tests requires investigation.

Essentially, bovine mastitis should be regarded as a herd problem and the methods of investigation and diagnosis should reflect this fact. The diagnostic methods include:
- Total and leukocyte cell counts for both herds and individual cows.
- Indirect chemical tests.
- Microbiological investigation to determine:
  a) The major pathogen(s) causing mastitis in a herd.
  b) The percentage of subclinical carrier cows.
  c) The antibiotic susceptibility patterns of the major pathogens so that effective treatment can be administered.
- Clinical examination and detection of abnormal milk using a strip cup.

Cell counts on milk
Many cell counting methods have been developed:
- Electronic somatic cell counts using equipment such as a Coulter counter. These are total cell counts as both exfoliated epithelial cells and leukocytes are counted.
- Direct microscopic count (modified Breed's smear): in this case leukocytes can be counted directly. A known volume of milk (0.01 ml) is

Table 120. Miscellaneous infectious conditions of the external surface of bovine udder and teats.

Infectious agents	Disease	Clinical findings
Bovine parapoxvirus (*Poxviridae*)	Pseudocowpox	Starts as small inflammatory papule, usually on teats, that develops into a dark red scab. The central area desquamates leaving a ring- or horse-shoe-shaped scab. Heals in 4–6 weeks
Bovine herpesvirus 2 (*Herpesviridae*)	Bovine ulcerative mammillitis	The lesions are usually on teats but can spread to the udder. Lesion starts as a local thickening of the skin, followed by severe oedema and erythema. Detachment of the epithelium occurs leaving a raw ulcerated surface. Uncomplicated cases heal in 3–4 weeks
Bovine papillomaviruses (*Papovaviridae*)	Bovine papillomas types 1 and 2	Cutaneous fibropapilloma that can be large and cauliflower-like. Usually occurs on head and neck but can be present on the skin of the udder
	type 5	Teat fibropapilloma with a 'rice-grain' appearance, being flat and white
	type 6	Teat papilloma or 'frond warts' that are thin and pedunculated
Bovine orthopoxvirus (*Poxviridae*)	Cowpox (Vaccinia virus infection)	Now a rare condition in cattle. In Britain the infection is probably more common in cats. A pustule occurs and ruptures. The exudate forms a scab covering an ulcerated area. The lesion may take several weeks to heal. The vaccinia virus (smallpox vaccination) can cause similar lesions
Fusobacterium necrophorum (±***Staphylococcus aureus***)	'Black pox' or 'black spot'	Deep crater-shaped ulcers with raised edges and a central black spot. Almost always at the tip of the teat and usually invades the sphincter
S. aureus	'Udder impetigo'	Small pustules (2–4 mm diameter) that involve the subcutaneous tissue. Lesions are often situated at the base of the teats
Pithomyces chartarum	Facial eczema (mycotoxicosis)	Photosensitisation with reddening of the teats and udder with eventual sloughing of the skin
Aphthovirus (*Picornaviridae*)	Foot-and-mouth disease	Vesicles on teats of milking cow
Vesiculovirus (*Rhabdoviridae*)	Vesicular stomatitis	Vesicles may occur on the teats of milking cows
Orbivirus (*Reoviridae*)	Blue tongue	Lesions may occur on teats of milking cows

spread over 1 cm^2 on a microscope slide, defatted and stained by a methylene-blue-based stain. The microscope is calibrated and from an average number of leukocytes per field (counting 50 fields) the number of leukocytes/ml of milk can be calculated. If comparatively large numbers of pathogenic bacteria are present in the milk sample, these may also be seen in the stained smear (**379**).

- Individual indirect cell counts using the Californian Mastitis Test (CMT): this test can be used in the field or in the laboratory and is based on the quantity of DNA in the milk and hence the number of leukocytes and other cells present. A squirt of milk from each quarter of the udder is placed in each of four shallow cups in the CMT paddle. An equal amount of commercial CMT reagent or 14 per cent Teepol (Shell) is added to each cup. A gentle circular motion is applied to the mixtures, in a horizontal plane and a positive gelling reaction occurs in a few seconds with positive samples (**380**). **Table 121** gives the interpretation of the CMT scores (0–3), the visible reaction, the approximate total cell count and expected percentage of neutrophils in the total count (Schalm *et al.*, 1971). The CMT gives a good indication of the leukocyte count of the milk.
- Other tests that depend on the development of gels are the Brabant, Wisconsin and NAGase mastitis tests. The NAGase mastitis test is easily automated and it is based on a cell-associated enzyme in milk, N-acetyl-D-glucosaminidase. High levels of the enzyme indicate a high cell count.

Leukocytes in milk samples disintegrate quite rapidly

Table 121. Correlation between the Californian mastitis test result and the somatic cell count.

CMT score	Interpretation	Visible reaction	Total cell count (/ml)
0	Negative	Milk fluid and normal	0–200,000 0–25% neutrophils
T	Trace	Slight precipitation	150,000–500,000 30–40% neutrophils
1	Weak positive	Distinct precipitation but no gel formation	400,000–1,500,000 40–60% neutrophils
2	Distinct positive	Mixture thickens with a gel formation	800,000–5,000,000 60–70% neutrophils
3	Strong positive	Viscosity greatly increased. Strong gel that is cohesive with a convex surface	≥ 5,000,000 70–80% neutrophils

379

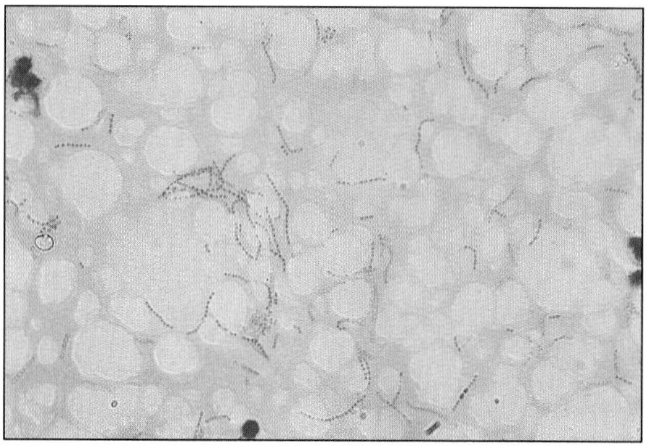

379 Newman stain showing chains of streptococci in a bovine mastitic milk sample. (×1000)

380

380 Californian Mastitis Test (CMT) designed as a 'beside-the-cow' test to detect subclinical mastitis. Interpretation: normal 0 (top left); positive 1+ (top right); positive 2+ (bottom right) and positive 3+ (bottom left).

on storage so the cell counts should be conducted within 2 hours of milk collection. Alternatively formalin could be added to the milk sample as a preservative.

There are normal variations in cell counts on milk:

a) Cows in early and late lactation have higher counts than those expected in mid-lactation. However, the milk from all four quarters will be equally affected.
b) Individual variation exists between cows and, normally, a cow will maintain a certain cell count level throughout life.
c) The presence of *Corynebacterium bovis*, regarded as non-pathogenic, in the teat duct will cause a rise in the cell count.

In cases of chronic mastitis, the highest cell count is obtained from strippings at the end of milking. The counts vary, too, depending on the pathogen present. Infections associated with *Streptococcus agalactiae* cause higher cell counts than those in staphylococcal mastitis.

Indirect chemical tests to detect mastitis

- Some tests are based on the increase in sodium and chloride ions in mastitic milk and consequently an increase in electrical conductivity of the milk.
- The serum albumin concentration in milk increases if epithelial damage is present. A radial immunodiffusion test has been developed based on this fact.
- An anti-trypsin test measures the trypsin-inhibitor capacity of milk. Anti-trypsin activity tends to be naturally high at the beginning of a lactation due to the colostrum levels, but later in a lactation the values are high only if serum anti-trypsin has leaked through damaged mammary epithelium. This method lends itself to automation.

Microbial investigation of mastitis

Milk sample collection

It is vital that a milk sample for microbiology is taken so as to ensure that the potential pathogen(s) in the sample came from the inside of the mammary gland and not from dust or faecal particles on the udder surface. It is not always possible for the veterinarian to collect the sample, but if the farmer is instructed in the correct procedure, good quality samples can be obtained. It is essential to obtain a milk sample before the cow has been treated with either intramammary or systemic antimicrobial agents. The main points in a good collection technique are to:

- Wipe the teat thoroughly with 70 per cent ethyl alcohol, paying particular attention to the teat orifice.
- Carry out the collection as swiftly as possible.
- Hold the sterile collection bottle nearly **horizontal** and keep the lid in the crook of the little finger so that the lid does not become contaminated.

A composite milk sample is satisfactory unless it is necessary to investigate the quarters separately. The first stream of milk, from the teat canal, usually has a higher cell count and bacterial population than that in the mammary gland. As the results from the examination of this 'fore-milk' are more a reflection of the condition existing within the teat than in the mammary gland itself, it is usually recommended that the first few squirts of milk from each quarter be discarded. The milk sample should, ideally, be kept refrigerated from the time of collection to the time of bacteriological examination. If mycoplasmal mastitis is suspected a simple transport medium is the milk sample itself with ampicillin added at 5 mg/ml. The milk sample, in this case, is held at the ambient temperature to allow mycoplasmal growth during transportation. A second milk sample should be submitted, without the ampicillin, as a check for other mastitis-producing pathogens.

Direct microscopy

The milk sample can be centrifuged and a stained smear made from the deposit. A Gram-stain is used routinely to detect Gram-positive pathogens such as staphylococci (**381**), streptococci (**382**) and this will also reveal yeasts, such as *Candida albicans*, that stain deeply by crystal violet. An MZN-stained smear can be made if *Nocardia asteroides* is suspected and a ZN-stained smear for the rare cases when bacteria such as *Mycobacterium fortuitum* or *M. bovis* (**383**) are present.

Culture

Most of the bacterial pathogens causing mastitis grow on ox or sheep blood agar. A MacConkey agar plate is streaked in parallel to detect *Enterococcus faecalis* and any Gram-negative bacteria that are able to grow on the medium. Edwards medium is highly selective

381

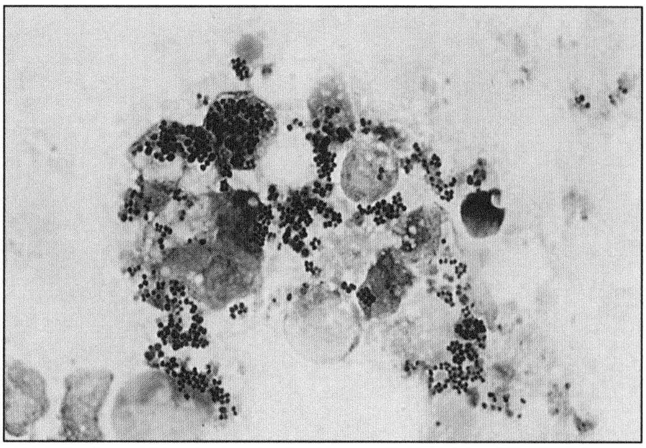

381 *Staphylococcus aureus* in a bovine mastitic milk sample. (Gram stain, ×1000)

382

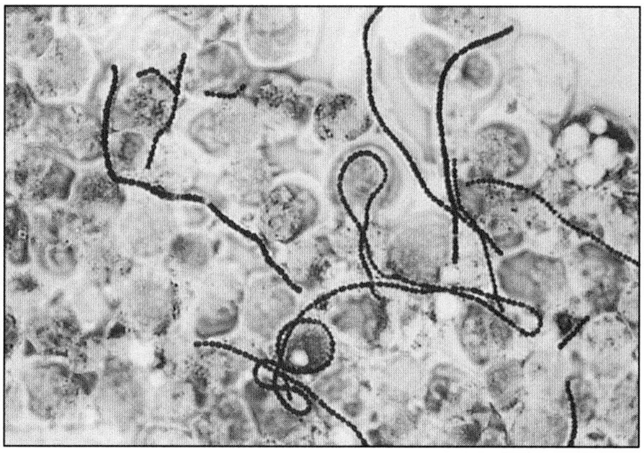

382 Streptococcal chains in a bovine mastitic milk sample. (Gram stain, ×1000)

383

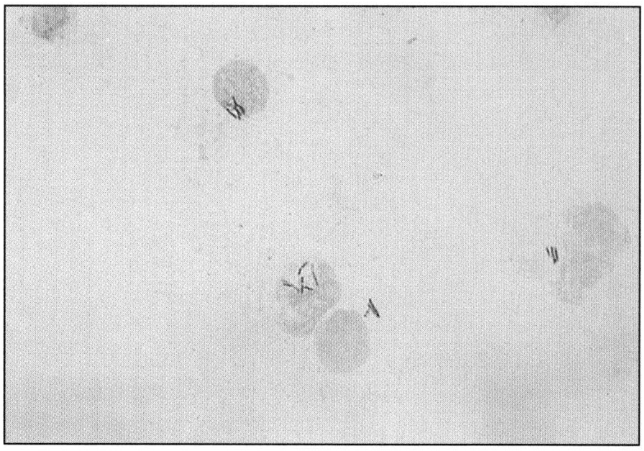

383 *Mycobacterium bovis* in a bovine milk sample from a case of tuberculous mastitis. (ZN stain, ×1000)

for streptococci and also acts as an indicator medium for haemolysis and for the hydrolysis of aesculin. A Sabouraud dextrose agar plate can be inoculated if a fungal pathogen is suspected. However, pathogens such as *Candida albicans* and *Aspergillus fumigatus* form colonies on blood agar at 37°C in 2–3 days, if there is little or no competition from faster-growing bacteria.

If a large number of milk samples are to be cultured on a herd basis, quarter-plating (Chapter 4) the samples on aesculin blood agar, alone, is satisfactory. This medium is not selective and will support the growth of the majority of the bacterial pathogens. Aesculin blood agar consists of blood agar with 0.05–0.1 per cent aesculin added as a sterile solution when the blood agar base has cooled to 50°C.

The inoculated plates are incubated aerobically, unless specific anaerobes such as *Fusobacterium necrophorum* or *Peptostreptococcus indolicus* are being sought. All the mastitis-producing microorganisms grow at 37°C. It is advisable to incubate the plates for up to 5 days to accommodate the slow-growing fungi and bacteria such as *Nocardia asteroides*.

Identification

The main characteristics and tests for the presumptive identification of the mastitis-causing pathogens are summarised in **Table 122**. For more detailed information the chapter on the relevant bacterium or fungus should be consulted.

Staphylococcus aureus

a) Colonial appearance: round, shiny, golden-yellow colonies surrounded by a zone of double-haemolysis on blood agar (**384**). There is no growth on most formulations of MacConkey agar, especially those with added crystal violet. No growth occurs on Edwards medium.
b) Gram-stained smear: Gram-positive cocci.
c) Coagulase test (**116**): this test is necessary to ensure that the isolate is coagulase-positive and therefore a pathogenic strain.
d) Purple agar with 1 per cent maltose: as a presumptive check that the isolate is *S. aureus* (**118**).
e) Antibiotic susceptibility test: many *S. aureus* strains are resistant to penicillin and to other commonly used antibiotics, so a susceptibility test is necessary.
f) Commercial identification systems such as Staph-Trac© (Analytab Products) and Minitek© Gram-Positive set (BBL) are available. However, these increase the cost of identification and their use is not usually necessary.

Mastitis-producing Streptococci

a) Colonial appearance: small, translucent colonies at 24 hours' incubation on blood agar with alpha-haemolysis, beta-haemolysis or gamma-haemolysis (**119**, **120**).
b) Growth on Edwards medium: all the streptococci are able to grow on this selective medium. *Streptococcus uberis* (**385**) and *Enterococcus faecalis* (**386**) hydrolyze aesculin but *S. agalactiae* (**387**) and *S. dysgalactiae* (**390**) do not. Aesculin hydrolysis on Edwards medium is indicated by a darkening of the medium and colonies (**388**). This can be seen more clearly under ultra-violet (UV) light (Wood's lamp) (**389**). The aesculin in the medium glows a dull-blue colour under UV light.
c) Gram-stained smear: scattered Gram-positive cocci. Streptococci are not usually seen in chains from colonies on a solid medium.
d) Catalase test: streptococci are catalase-negative whereas the staphylococci are catalase-positive.
e) Growth on MacConkey agar: *E. faecalis* and some of the other Lancefield Group D streptococci tolerate the bile salts in MacConkey agar and grow as red, pin-point colonies (**391**).
f) CAMP test: only *S. agalactiae* gives a sharp arrowhead enhancement of haemolysis caused by the beta-haemolysin of *Staphylococcus aureus* (**392**).
g) Lancefield Grouping: this is not usually necessary but can be carried out by a latex agglutination kit (**122**) such as Streptex© (Wellcome) which covers Groups A, B, C, D and G or Phadebact Streptococcus Test© (Remel Inc.) for Groups A, B, C and G. *S. uberis* does not belong to a Lancefield Group, but if the isolate is in a pure culture from a carefully taken milk sample, and corresponds to the characteristics in **Table 123**, then a presumptive identification can be made.
h) Optochin and bacitracin susceptibility: if the mainly human pathogens, *S. pyogenes* and *S. pneumoniae* are suspected, bacitracin or optochin susceptibility tests could be carried out. Group A streptococci (*S. pyogenes*) are susceptible to bacitracin (**133**) but the other beta-haemolytic streptococci are resistant. *S. pneumoniae* is susceptible to optochin (**134**) but the other alpha-haemolytic streptococci are not.
i) Commercial systems are available for the identification of streptococci, such as API 20Strep© and Rapid Strep© (Analytab Products).
j) Antibiotic Susceptibility Test: some of the mastitis-producing streptococci have become resistant to penicillin, so an antibiotic susceptibility test on the isolate is advisable. This has to be carried out on a blood agar plate as most streptococci fail to grow on media without blood or serum.

Table 122. Bacteria capable of causing bovine mastitis showing some of their main characteristics leading to a presumptive identification.

Mastitis-causing bacteria	Gram reaction	Shape	Catalase	Oxidase	Haemolysis	Growth on MacConkey agar	Aesculin hydrolysis (Edwards medium)	CAMP test	Lancefield group	Other characteristics and confirmatory tests
Streptococcus agalactiae	+	C	–	–	β, γ, α	–	–	+	B	CAMP test positive
S. dysgalactiae	+	C	–	–	α	–	–	–	C	Alpha-haemolytic, CAMP-negative
S. uberis	+	C	–	–	α, γ	–	+	–	–	Aesculin-splitter, no growth on MacConkey agar
Enterococcus faecalis	+	C	–	–	α, γ	+	+	–	D	Pin-point red colonies on MacConkey agar. Aesculin hydrolysis
S. pyogenes	+	C	–	–	β	–	–	–	A	Susceptible to bacitracin (0.04 unit disc)
S. pneumoniae	+	C	–	–	α	–	±	–	–	Susceptible to optochin. Often mucoid
S. equi* subsp. *zooepidemicus	+	C	–	–	β	–	–	–	C	Group C: trehalose –, sorbitol +, lactose+ and maltose+(–)
Staphylococcus aureus	+	C	+	–	+	–				Golden-yellow pigment; double-zoned haemolysis; coagulase+ and ferments maltose on purple agar base plus 1% maltose
Escherichia coli	–	R	+	–	±	+				'IMViC' test +/+/–/–. Metallic sheen on EMB agar. Occasionally mucoid, quite often haemolytic
Klebsiella pneumoniae	–	R	+	–	–	+				Mucoid colonies, non-motile
Enterobacter aerogenes	–	R	+	–	–	+				Mucoid colonies, motile ('IMViC' test –/–/+/+)
Serratia marcescens	–	R	+	–	–	+				Red pigment at 25°C, some strains at 37°C
Pseudomonas aeruginosa	–	R	+	+	±	+				Greenish-blue pigment, fruity smell
Actinomyces pyogenes	+	R	–	–	+	–				Small colonies, hazy haemolysis. Pits Loeffler serum slope
Nocardia asteroides	+	F	+	–	±	–				Powdery white colonies, adherent to medium. MZN +. Requires 3–4 days' incubation. Growth on Sabouraud agar
Pasteurella multocida	–	R	+	+	–	–				Colonies with sweetish smell. Non-haemolytic, indole+ and no growth on MacConkey agar
P. haemolytica	–	R	v	+	+	+				No smell. Haemolytic, indole – and red, pin-point colonies on MacConkey agar
Bacillus cereus	+	R	+	–	+	–				Forms endospores. Wide zone of haemolysis. Large, flat, dry and granular colonies

C = coccus, R = rod, F = filamentous, + = positive reaction, ± = most strains positive, v = strains vary, – = negative reaction. 'IMViC' = indole, methyl red, Voges–Proskauer and citrate tests. MZN = modified Ziehl–Neelsen stain.

384

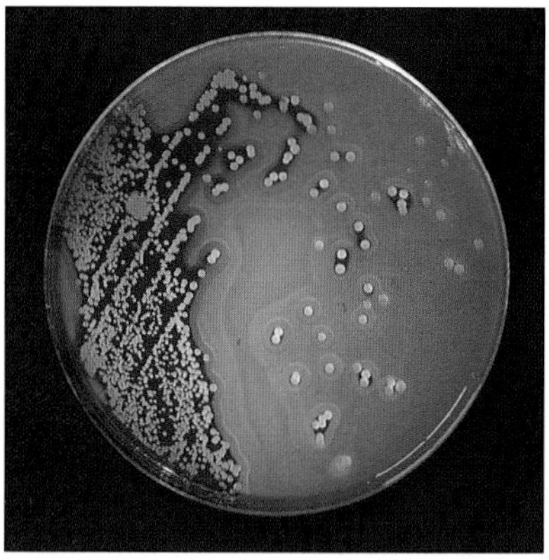

385

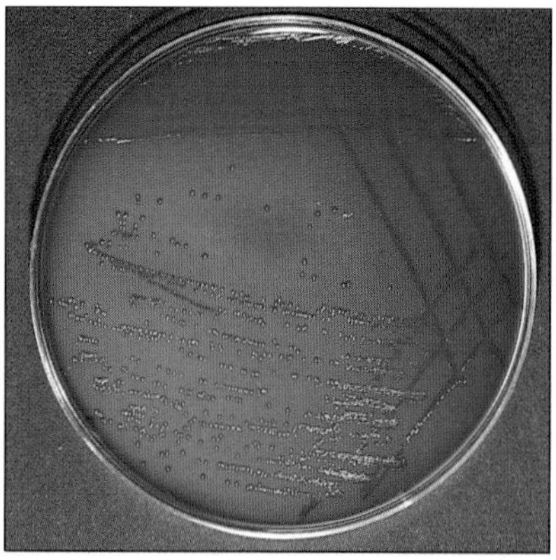

384 Sheep blood agar inoculated with a bovine mastitic milk sample giving an almost pure culture of *S. aureus*. Note the characteristic 'target' haemolysis.

385 *Streptococcus uberis* on Edwards medium: alpha-haemolytic and aesculin splitting.

386

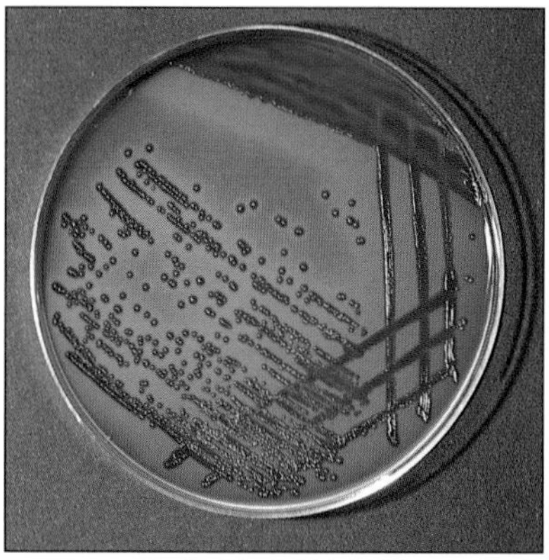

387

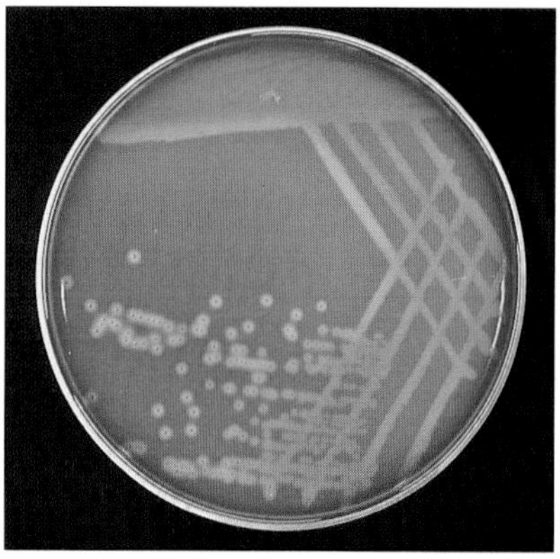

386 *Enterococcus faecalis* on Edwards medium: alpha-haemolytic and aesculin splitting.

387 *S. agalactiae* on Edwards medium: beta-haemolytic but non-aesculin splitting.

388

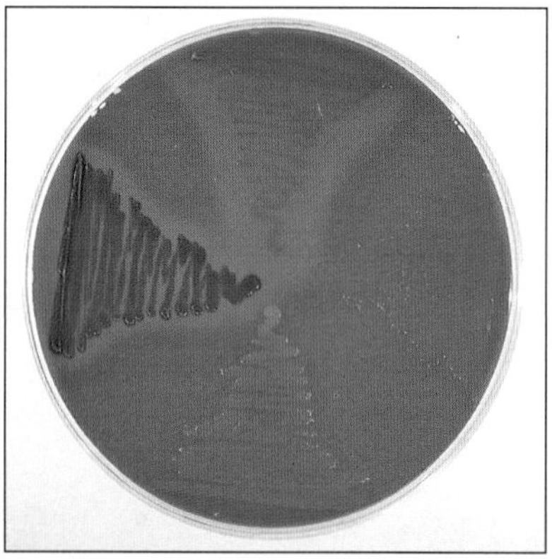

389

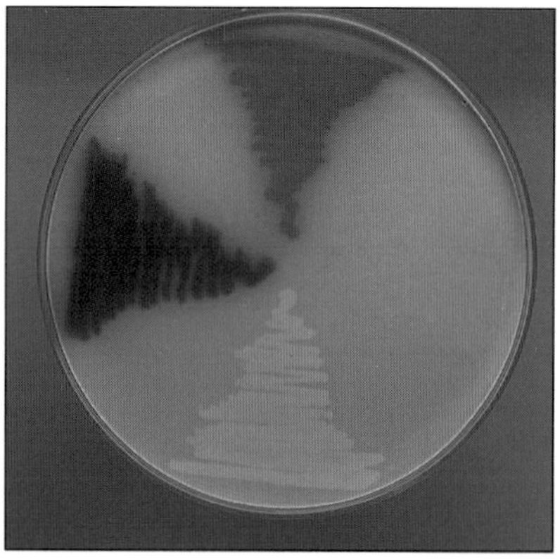

388 Aesculin hydrolysis on Edwards medium (under ordinary light, compare with **389**): *E. faecalis* (left) and *S. uberis* (top) split aesculin, whereas *S. dysgalactiae* (right) and *S. agalactiae* (bottom) fail to do so.

389 Wood's lamp (UV light) aids the detection of aesculin hydrolysis in Edwards medium: *E. faecalis* (left) and *S. uberis* (top) split aesculin, whereas *S. dysgalactiae* (right) and *S. agalactiae* (bottom) fail to do so (compare with **388** showing the same plate under ordinary light).

390

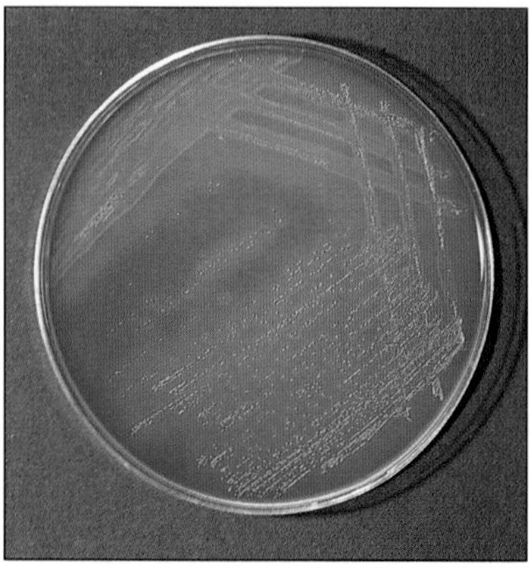

391

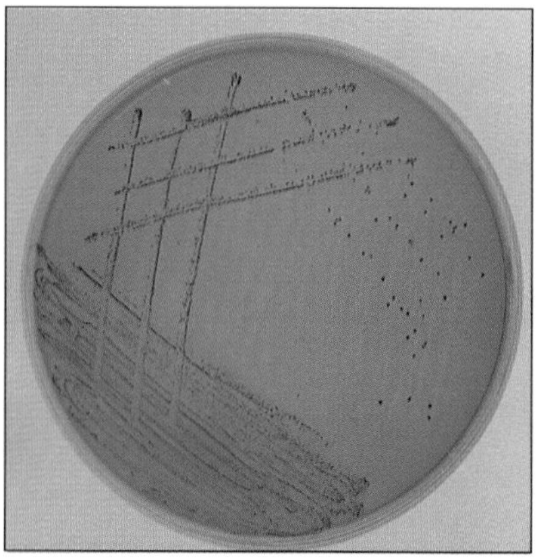

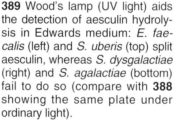

390 *S. dysgalactiae* on Edwards medium: alpha-haemolytic and non-aesculin splitting.

391 *E. faecalis* on MacConkey agar showing pin-point red colonies. This is one of the few Gram-positive bacteria able to tolerate the inhibitors in this medium.

392

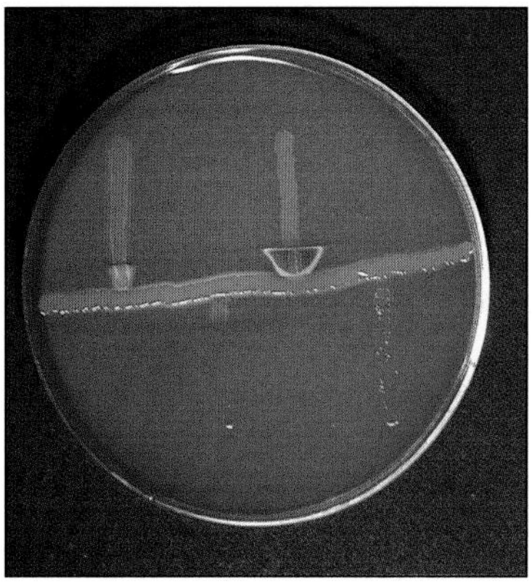

392 CAMP test on sheep blood agar with *Staphylococcus aureus* as the horizontal streak. *S. agalactiae* (top right) gives a positive CAMP reaction with 'arrow-head' enhancement of the partial haemolysis caused by the staphylococcal beta-haemolysin. *S. dysgalactiae* (bottom right) gives no reaction, while the weak reactions of *S. uberis* (bottom left) and *E. faecalis* (top left) can easily be distinguished from the positive reaction of *S. agalactiae*.

393

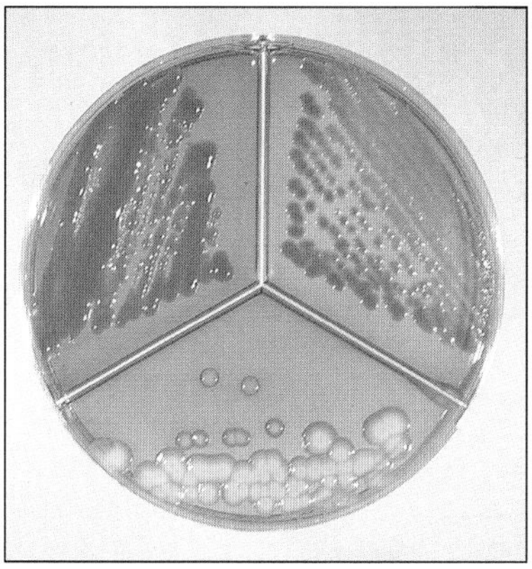

393 A mucoid isolate of *E. coli* (left), *Enterobacter aerogenes* (right) and *Klebsiella pneumoniae* (bottom) on MacConkey agar. All are lactose fermenters but colonies of *E. coli* are invariably a more vivid pink colour.

394

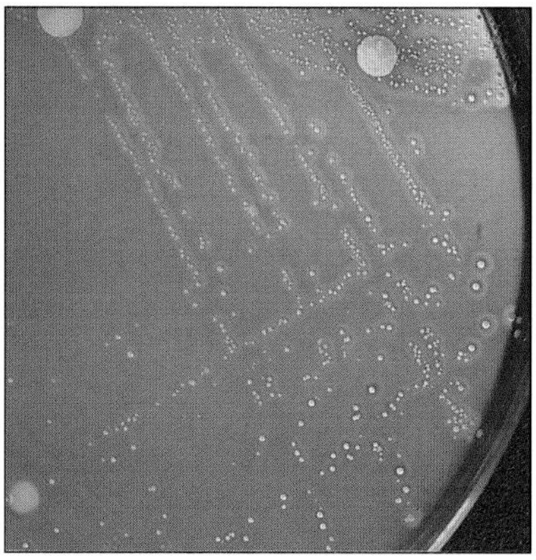

394 Sheep blood agar inoculated with a mastitic milk sample yielding an almost pure growth of *Actinomyces pyogenes:* tiny colonies even after 72 hours' incubation, with a hazy type of beta-haemolysis. The large colonies are those of contaminating bacteria.

Table 123. Summary of the characteristics of the streptococci that cause bovine mastitis.

	Haemolysis	Aesculin hydrolysis	Growth on MacConkey agar	CAMP test	Lancefield group	Optochin susceptibility	Bacitracin susceptibility
Streptococcus agalactiae	β, γ or α	–	–	+	B	•	R
S. dysgalactiae	α	–	–	–	C	R	•
S. uberis	α or γ	+	–	–	–	R	•
Enterococcus faecalis	α or γ	+	+	–	D	R	•
S. equi* subsp. *zooepidemicus	β	–	–	–	C	•	R
S. pyogenes	β	–	–	–	A	•	S
S. pneumoniae	α	–	–	–	–	S	•

+ = positive reaction, – = negative reaction, • = not applicable, R = resistant, S = susceptible.

Coliform bacteria

Escherichia coli, *Klebsiella pneumoniae* and *Enterobacter aerogenes* are the members of the *Enterobacteriaceae* most commonly involved in bovine mastitis and less commonly *Serratia marcescens*.

a) Colonial appearance: *K. pneumoniae, E. aerogenes* and, very occasionally, strains of *E. coli* have mucoid colonies (**393**). The colonies of *E. coli* are round, discrete and bright-pink (lactose-fermenting) on MacConkey agar (**281**). Within the *Enterobacteriaceae, E. coli* is the member most commonly haemolytic on blood agar (**280**), although this is a variable characteristic. The coliforms do not grow on Edwards medium.

b) Gram-stained smear: medium-sized Gram-negative rods.

c) Oxidase test: many of the Gram-negative bacteria are oxidase-positive but members of the *Enterobacteriaceae* are exceptional in being oxidase-negative.

d) Motility test in semi-solid medium (Chapter 4): to distinguish between *K. pneumoniae* and *E. aerogenes* which both, characteristically, have mucoid colonies. *K. pneumoniae* is non-motile but *E. aerogenes* is motile.

e) 'IMViC' test: used for the presumptive identification of *E. coli* which is invariably indole+/MR+/VP-/citrate- (**283**). **Table 124** shows the corresponding reactions for *K. pneumoniae* and *E. aerogenes*.

f) Pigment production: *Serratia marcescens* produces the distinctive red pigment, prodigiosin, at 25°C and less reliably at 37°C (**277**). The pigment can be seen in the colonies on MacConkey agar as the bacterium would otherwise have pale colonies of a non-lactose fermenter.

g) Commercial identification systems, such as API 20E© (Analytab Products), are available for the *Enterobacteriaceae* (Chapter 4).

h) Antibiotic susceptibility test: this is extremely important for members of the *Enterobacteriaceae* because of the presence of R-factor plasmids that can code for multiple drug resistance.

Actinomyces pyogenes

a) Colonial appearance: after 24 hours' incubation the colonies on blood agar are so small that they are difficult to see but there will be a hazy haemolysis along the streak lines. After longer incubation, the small colonies are visible surrounded by a hazy type of haemolysis (**394**). No growth occurs on MacConkey agar or Edwards medium.

b) Gram-stained smear: pleomorphic Gram-positive rods (**160**).

c) Catalase test: *A. pyogenes* is catalase-negative.

d) Loeffler serum slope: pitting of the slope by *A. pyogenes* in 24–48 hours (**169**) gives a presumptive identification.

e) Antibiotic susceptibility test: the test is difficult to read with this slow-growing bacterium; however, it is usually sensitive to penicillin.

Pseudomonas aeruginosa

a) Colonial appearance: on blood agar *P. aeruginosa* produces large, flat colonies, usually haemolytic, with the green-blue pigment, pyocyanin, most obvious on the areas of heavy growth (**298**). The colonies have a characteristic fruity odour. On MacConkey agar the bacterium is a non-lactose fermenter, but the green-blue pigment is often superimposed on what would otherwise be pale colonies (**300**). There is no growth on Edwards medium.

b) Gram-stained smear: medium-sized Gram-negative rods.

c) Oxidase test: *P. aeruginosa* is strongly oxidase-positive. This helps to distinguish it from the oxidase-negative coliforms.

d) Pyocyanin-enhancing medium [Pseudomonas agar P (Difco)]: as pyocyanin is unique to *P. aeruginosa*, demonstration of this pigment gives a good presumptive identification of this bacterium. The pyocyanin-enhancing medium (**395**) is useful for strains of the organism that are poor producers of the pigment.

e) Antibiotic susceptibility test: this is essential for *P. aeruginosa* as the organism is notoriously drug-resistant (**306**).

Nocardia asteroides

In chronic cases palpable granulomas are often present in the udder tissue.

a) Direct microscopy: *N. asteroides* is slow growing and difficult to culture, especially if the animal has been treated with antibiotics. An MZN-stained smear on the deposit from a centrifuged milk sample may be the best way to reach a diagnosis. The appearance is similar to that seen in **171**.

b) Colonial appearance: after 3–4 days' incubation, white, powdery colonies, embedded in the agar are seen (**396**). Some strains are haemolytic. The colonies of *N. asteroides* have no smell, which distinguishes them from *Streptomyces* species, which have similar colonies (**175**) but a pungent, earthy odour. *Streptomyces* spp. are common contaminants on plates that have been left at room temperature for a few days.

c) Growth on Sabouraud dextrose agar: *N. asteroides* and some *Streptomyces* species have the unusual ability of being able to grow on Sabouraud agar (**174**), which is a selective medium for fungi.

395

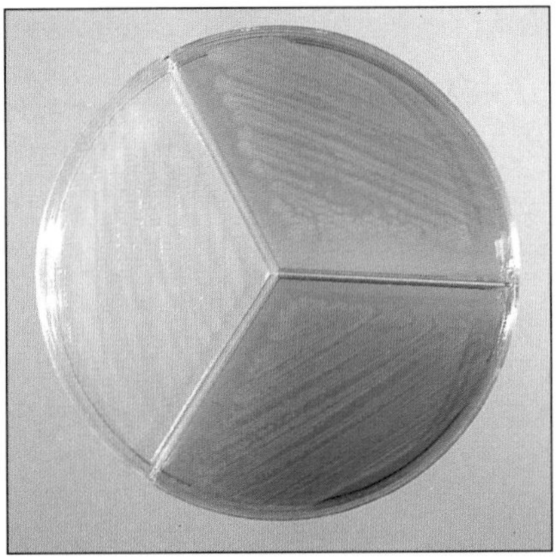

396

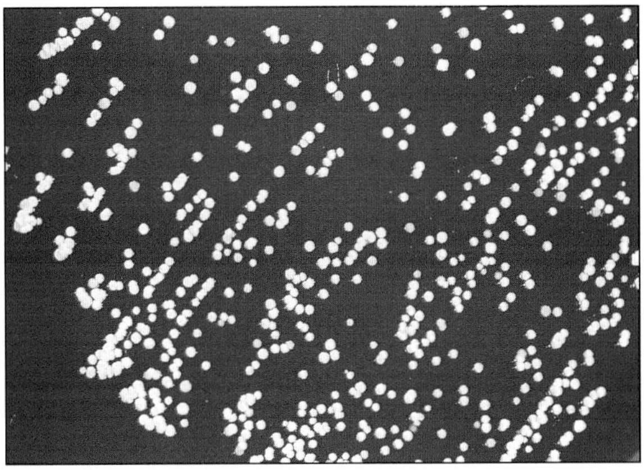

395 *Pseudomonas* strains on 'Pseudomonas agar P' which enhances pyocyanin production. No pyocyanin is detectable from the strain on the left, while the strains on the right are good pyocyanin producers.

396 White powdery colonies of *Nocardia asteroides* on sheep blood agar after 96 hours' incubation.

Table 124. Summary of the characteristics of the mastitis-producing coliform bacteria.

	Escherichia coli	*Klebsiella pneumoniae*	*Enterobacter aerogenes*
Motility	+ (–)	–	+
Indole production	+	–	–
Methyl red (MR) test	+	– (+)	–
Voges–Proskauer (VP) test	–	+	+
Citrate utilisation	–	+	+
Mucoid colonies	– (+)	+	+
Haemolysis	v	–	–

+ = positive reaction, (+) = some strains positive, – = negative reaction, (–) = some strains negative, v = variable reaction.

Pasteurella species

a) Colonial appearance: *P. multocida* has medium-sized, shiny (sometimes mucoid), non-haemolytic colonies that have a pinkish tinge on blood agar (**320**). The colonies have a delicate, but highly characteristic, sweetish odour. *P. haemolytica* is haemolytic and the colonies on blood agar are similar to those of a beta-haemolytic streptococcus (**323**). *P. multocida* will not grow on MacConkey agar but *P. haemolytica* grows as small, red, pinpoint colonies (**324**) not unlike those of *Enterococcus faecalis*. No growth occurs on Edwards medium.
b) Gram-stained smear: Gram-negative rods that tend to be coccobacillary.
c) Catalase test: the pasteurellae are catalase-positive and this distinguishes them from streptococci.
d) Oxidase test: both *P. multocida* and *P. haemolytica* are oxidase-positive, whereas members of the *Enterobacteriaceae* are oxidase-negative.
e) Indole production: *P. multocida* is a good indole producer. This and other characteristics are summarised in **Table 125**. For further details of biochemical reactions, *see* Chapter 22.

Bacillus cereus

a) Colonial appearance: large, flat, granular, haemolytic colonies on blood agar (**217**). It will not grow on MacConkey agar or Edwards medium.
b) Gram-stained smear: large Gram-positive rods, some of which may be sporing. The endospores are seen as clear, unstained, oval areas within the mother cells.
c) *See* Chapter 15 for confirmatory tests.

Table 125. Summary of the main distinguishing factors of the pasteurellae capable of causing bovine and ovine mastitis.

	Pasteurella multocida	*Pasteurella haemolytica*
Catalase	+	+
Oxidase	+	+
Haemolysis	–	+
Growth on MacConkey agar	–	+ (pin-point red colonies)
Indole production	+	–
Characteristic sweetish smell	+	–

+ = positive reaction, – = negative reaction.

Mycoplasma species

Special media are required for the isolation of *Mycoplasma* species (Chapter 35). Indications that mycoplasmas are involved in the outbreak of mastitis include the fact that no major pathogens could be isolated by routine cultural methods. There are severe changes to the mammary glands but usually no systemic involvement.

Mycobacteria

Infection of the udder by mycobacteria is rare except in countries where bovine tuberculosis is still relatively common. If a mycobacterial infection is suspected, possibly due to finding palpable lesions in the udder, about 50 ml of milk should be carefully collected. There are usually relatively few mycobacteria present, so the milk sample is centrifuged to concentrate the bacteria in the deposit. This deposit is used to prepare a ZN-stained smear (**383**) and for culture (Chapter 12).

Leptospiral agalactia

This is not a typical mastitis, although *Leptospira interrogans* serovars *hardjo* or *pomona* are present in the mammary glands. The presumptive diagnosis of this condition is based on the clinical signs and later the demonstration of leptospires in the urine by dark-field microscopy (Chapter 31).

Fungal pathogens

These potential pathogens are ubiquitous and usually introduced accidentally into the udder via the nozzle of intramammary antibiotic tubes. Because fungi are ubiquitous, a repeat milk sample might be advisable. A large number of fungi have been isolated from mastitic milk samples and these include *Candida albicans* (**397**), *Cryptococcus neoformans, Aspergillus fumigatus* (**398**), *Trichosporon* and *Saccharomyces* species. For identification methods, consult the section on Mycology.

397

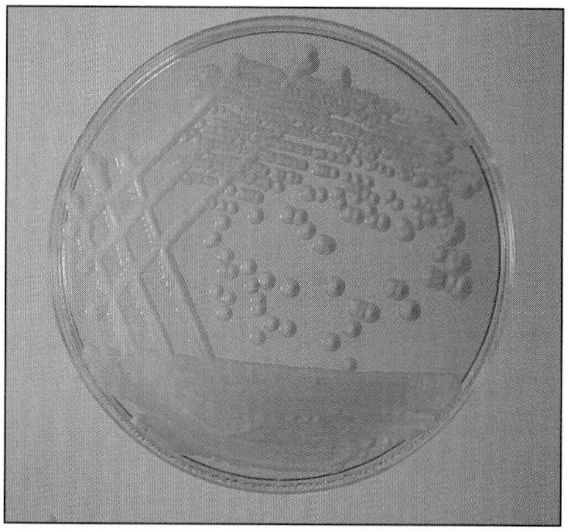

397 *Candida albicans* on Sabouraud agar growing as white, high convex, shiny colonies with a pleasant beer-like smell.

398

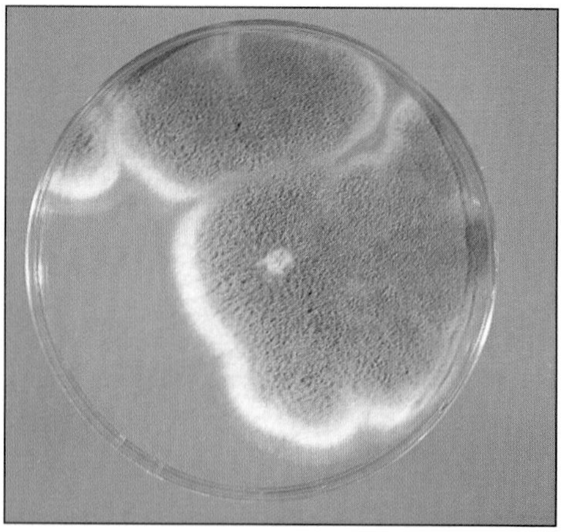

398 *Aspergillus fumigatus* on Sabouraud agar: blue-green powdery colony.

***Prototheca* species**

P. zopfii and *P. wickerhamii* are achlorophyllic algae but they form bacterial-like colonies on blood agar at 25–37°C, after 48–72 hours incubation. The colonies are small and greyish-white. Microscopic examination reveals large sporangia (8-25 μm) containing 4–8 daughter cells. They are best demonstrated in wet preparations or in smears stained by the Wright or Giemsa method. These algae are widespread in water, soil, mud and cattle faeces. A repeat milk sample might be warranted to check that the algae did in fact come from inside the mammary gland. The prognosis is poor for this type of mastitis, as there is little response to treatment and the algae can persist in the udder tissue for over 100 days. Affected animals are a danger to the other cows in the herd and immediate culling is advised.

Investigation of Mastitis Problem Herds

Key Factors of Mastitis Control

• **Milking machine**	
• **Herd management:**	General Milking shed (parlour)
• **Teat dipping routine**	
• **Treatment:**	Clinical cases Dry-cow therapy
• **Culling:**	Cows with chronic or persistent mastitis

An indication that a dairy herd has a mastitis problem may come either from the owner or from milk-company records that show cell counts, on the bulk milk, of over 400,000/ml on three consecutive occasions. The following points are a guide to the investigation of the problem so that successful control measures can be suggested, based on the key factors of mastitis control:

- The milking machine must be checked thoroughly to ensure that it is functioning correctly. This could be carried out, and a report obtained, before visiting the farm for further investigations.
- It is advantageous to arrive at the farm about an hour before milking time. The herd records of mastitis cases, cell counts and herd production can then be examined. The design and general hygiene in the milking shed should be noted. The owner could be questioned about the general herd management, treatment of mastitis cases, and culling rate for cows with persistent mastitis.
- During milking the following points should be observed: the milking technique, whether teat-dipping is being carried out efficiently, and the general level of hygiene in the shed.
- Milk samples should be taken from any new cases of mastitis and random samples from about 10–20 per cent of the herd for bacteriological examination. The samples should be obtained **before** the milker has washed the udder or carried out any other pre-milking procedure. While taking the milk samples, the udder and teats can be examined for lesions. Evidence of hyperkeratosis of the epithelium at the teat orifice, or ecchymosis at the end of the teat, which might indicate excessively high vacuum levels in the past, should be noted.

Laboratory investigation of the milk samples should provide the following information:

a) The major pathogen(s) causing the mastitis.
b) The percentage of subclinical cases of mastitis in the herd.
c) The antibiotic susceptibility/resistance patterns of the pathogens. These will provide a guide to effective treatment of clinical cases and dry-cow therapy.

Treatment

The drug of choice in the treatment of mastitis is one to which the bacteria are sensitive and which achieves high concentrations in the mammary gland without provoking tissue changes. The intramammary route for the administration of antimicrobial drugs to cows with mastitis is usually practical and convenient.

The antimicrobial agent should be distributed well throughout the mammary gland and maintain sufficient concentrations to clear the bacteria from the tissue. The formulation of the intramammary preparation, including the amount of drug present and the physical properties of the carrier influences the distribution and persistence in the mammary tissue. The distribution of drugs, following intramammary administration, occurs principally by passive diffusion. It is also a suitable route for administration of long-acting antimicrobial preparations at 'drying off' as part of a mastitis control programme. Dry-cow therapy is designed to eliminate infections present in the udder and prevent new infections during the dry period. Control of mastitis, therefore, is based on prevention of new infections through teat dipping and hygiene standards, that minimize transfer of infection via the milking machine, and elimination of old infections.

Successful treatment of clinical mastitis often requires a history of the herd, isolation, identification and susceptibility pattern of the bacteria involved, and relevant information on the milking machine and on the milking routine. The quality of milk samples submitted for susceptibility testing is central to the information provided to the diagnostic laboratory. Although treatment may have to proceed before laboratory reports are at hand, it should, if necessary, be immediately revised as soon as susceptibility results are available.

An antimicrobial preparation for intramammary use should not be selected on the basis of the broadest spectrum as it is the **susceptibility of the bacterium being treated that should determine selection**. Penicillin G is frequently the most effective antibiotic against *Streptococcus* species, *Actinomyces pyogenes* and susceptible *Staphylococcus aureus*. Treatment of *S. agalactiae* and *S. dysgalactiae* with penicillin G or a semisynthetic penicillin is generally more than 95 per cent effective. *S. uberis* and enterococcal infections respond less favourably to this treatment.

Results of treatment of staphylococcal mastitis are highly variable and depend to a large extent on the duration of infection, the susceptibility pattern of the bacteria, the degree of inflammation at the time of treatment, the extent of tissue damage and the stage of lactation. Many *S. aureus* isolates produce penicillinase (beta-lactamase) which render penicillin G and similar antibiotics ineffective. Cloxacillin and nafcillin are effective against such bacteria and clavulanic acid acts as an inhibitor of beta-lactamase. Cephalosporins and erythromycin are likely to have good activity against beta-lactamase-producing *S. aureus*. Although *S. aureus* isolates may be susceptible *in vitro* to a selected antibiotic, poor penetration into chronic lesions can allow survival of bacteria. Therapeutic success may be achieved by continuing treatment for longer than 72 hours and by infusing intramammary preparations at more closely spaced intervals.

Coliform mastitis usually requires treatment with an aminoglycoside such as framycetin, streptomycin or neomycin. This form of mastitis often proceeds to severe toxaemia, hence prompt treatment with antibiotics and supportive fluid therapy may be indicated. Both local and systemic therapy may be beneficial in coliform mastitis as aminoglycosides are often poorly distributed following intramammary administration. When an infection is peracute and life-threatening, systemic treatment and stripping out the quarters every 2 hours can aid in the elimination of both bacteria and endotoxin.

Mastitis due to *Nocardia* spp., fungi or algae is generally unresponsive to treatment. Early recognition of the aetiological agent and immediate culling is the most appropriate approach.

Selection of intramammary preparations requires consideration of the spectrum of activity, distribution throughout the mammary gland, frequency of administration, cost and milk withholding-time. **Table 14** should be consulted for the range of antimicrobial drugs suitable for treatment of bacteria causing mastitis. A narrow-spectrum drug, such as penicillin, is most appropriate for streptococcal infections and penicillin-susceptible staphylococci. Susceptibility testing is an essential step in mastitis control programmes, especially when dealing with long-standing infections in well-managed dairy herds. Antimicrobial therapy is only one aspect of mastitis control and, apart from dry-cow therapy, probably the least rewarding of the measures used in mastitis control programmes.

Reference

Schalm, O.W., Carroll, E.J. and Jain, N.C. (1971). *Bovine Mastitis*. Lea and Febiger, Philadelphia, USA.

Bibliography

McKellar, Q.A. (1991). Intramammary treatment of mastitis in cows, *In Practice*, **13**: 244–249.

37 Bacterial Food Poisoning

The term 'food poisoning' embraces a group of acute illnesses caused by the ingestion of foods that contain substances or agents injurious to humans. These substances may be chemical or biological in origin and can be summarised as follows:

- Foods that are themselves inherently poisonous such as some toadstools; undercooked red kidney beans (*Phaseolus vulgaris*) that contain a haemagglutinin; apricot kernels containing cyanide and shellfish or fish (paralytic shellfish poisoning and ciguatera poisoning, respectively) that have fed on toxic marine dinoflagellates.
- Foods that cause hypersensitivity reactions in some individuals. This may or may not be due to antibiotic residues in the foods.
- Chemicals such as heavy metals, insecticides, herbicides or fungicides that accidentally contaminate foodstuffs.
- Parasites including *Trichinella spiralis, Entamoeba histolytica* and *Giardia lamblia* that may be present in foods.
- Viruses that are commonly food-borne such as the Norwalk agent, human rotaviruses and the Hawaii and Montgomery County agents.
- Mycotoxins produced in foods as the result of fungal activity. Important examples of these are the aflatoxins (*Aspergillus flavus*) and the alkaloids responsible for ergotism in cereals contaminated by the sclerotia of *Claviceps purpurea*.
- Scombrotoxin, believed to consist of histamine and saurine, produced by the bacterium *Providencia morganii* growing in fish such as tuna, pilchards and mackerel. *P. morganii* is able to produce histamine from histidine present in these scombroid fish.
- Bacterial food poisoning, manifested as an acute gastroenteritis, due to the ingestion of the bacteria themselves and/or their toxins present in food. The pathogenesis of these bacterial pathogens can be divided into:
  a) Intoxications in which there is ingestion of preformed toxins produced by bacteria during their growth in the food. The bacteria responsible include *Staphylococcus aureus*, *Bacillus cereus*, *Clostridium botulinum*, enterotoxigenic *Escherichia coli* producing the heat stable toxin and some streptococci that are capable of elaborating an enterotoxin in foods.
  b) Infections where bacteria have multiplied in the food and are ingested in relatively large numbers. They may also multiply *in vivo* producing enterotoxins in the intestinal tract. Bacteria causing these acute food-borne infections include salmonellae, *Vibrio parahaemolyticus, Campylobacter jejuni, Yersinia enterocolitica* and *Aeromonas hydrophila*.
  c) An intermediate type of pathogenesis involves *Clostridium perfringens*. This bacterium does not readily form toxin when growing in food, but if ingested in comparatively large numbers, it liberates an enterotoxin when sporulating in the intestine.

Various Gram-negative bacteria, such as *Proteus, Providencia, Citrobacter* and *Pseudomonas* species, when present in large numbers in foods, have occasionally been suspected of causing outbreaks of food poisoning.

Other bacteria that are commonly food-borne but are not usually included among the food-poisoning bacteria comprise *Salmonella typhi* (typhoid fever), *S. paratyphi* A, B and C, *Shigella* species, *Vibrio cholerae* (cholera) and *Listeria monocytogenes*. All except *L. monocytogenes* cause enteric disease and some strains of *S. paratyphi* B can give rise to gastroenteritis with clinical signs similar to those of food poisoning. These bacteria and the other substances such as chemicals, viruses, parasites and mycotoxins, previously cited, that cause food poisoning have been mentioned as they are important in the differential diagnosis. Most of the cases of food poisoning caused by chemicals are characterised by vomiting within a few minutes to half-an-hour after ingestion of the contaminated food. This contrasts with the incubation period for bacterial food poisoning in which clinical signs are rarely seen until 2 hours or more post-ingestion.

This section will deal with the bacteria regarded as causing classical food poisoning but also with those that can cause a gastroenteritis or enteritis with signs similar to those of food poisoning. Bacteria such as *Yersinia enterocolitica, Campylobacter jejuni* and *Escherichia coli* can be food-borne and animals play a part in the epidemiology. Two recently recognised food poisoning syndromes are caused by *Aeromonas hydrophila* and certain streptococci, respectively. **Table 126** summarises the distinguishing features of these bacteria together with their epidemiology, pathogenesis and clinical signs. Listeriosis is referred to at the end of this Chapter. *Listeria monocytogenes* does not cause the clinical signs of food poisoning but it is a zoonotic bacterium and has recently been recognised as 'an emerging food-borne pathogen'.

Table 126. Food-poisoning bacteria: epidemiology, pathogenesis and clinical signs.

Food-poisoning pathogen	*Reservoir* Animal	Human	Environment	Main food sources	Infective dose (bacteria)	Pathogenesis	Incubation period (hours)	*Clinical signs* Vomiting	Diarrhoea	Fever		Duration of illness
***Salmonella* species** **Gram-negative rods**	•	(•)		Cooked meat and poultry. Raw egg dishes and seafoods	10^3–10^8	Enteric infection	6–36 (usually 12–24)	(+)	++	+	Acute gastro-enteritis. Abdominal pain and fever almost always present. Dehydration is severe in infants	1–7 days
Clostridium perfringens **Gram-positive rods. Anaerobe. Endospores formed**	•	•		Cooked meat and poultry, especially bulk-meat or re-warmed dishes	10^8–10^9 or $\geq 10^5$/g food	Enterotoxin formed in intestines when cells sporulate	8–20	–	+++	–	Diarrhoea and abdominal pain. Vomiting uncommon. Rarely fatal and course is short	12–24 hours
Staphylococcus aureus **Gram-positive cocci**		•		Baked ham, poultry, meat pies, cream, custard and ice-cream	10^6/g food (1.0µg toxin)	Enterotoxin produced in food (intoxication)	2–6	+++	+	–	Severe vomiting that can mimic seasickness. Rapid recovery	6–24 hours
Bacillus cereus EMETIC SYNDROME			•	Cooked rice, occasionally spaghetti	Over 10^5/g food	Enterotoxin produced in food (intoxication)	1–5	+++	(+)	–	Sudden onset of vomiting and stomach cramps. Some patients may develop diarrhoea	6–24 hours
DIARRHOEAL SYNDROME Gram-positive rods, Endospores formed			•	Casseroles, sausages, cooked meat, poultry, fish, soups, and dairy produce	Over 10^5/g food	Enterotoxin mainly formed in intestines	8–16	(+)	++	–	Profuse watery diarrhoea and abdominal pain. Vomiting can also occur	12–24 hours
Vibrio parahaemolyticus **Gram-negative rods, halophilic**			•	Raw fish- and shell fish or recontaminated cooked seafoods	10^3–10^7 or over 10^5/g food	Enteric infection	2–48 (usually 12–18)	+	++	(+)	Profuse diarrhoea often leading to severe dehydration. Fever and vomiting can occur	2–5 days

Table 126. Food-poisoning bacteria: epidemiology, pathogenesis and clinical signs *(continued).*

Food-poisoning pathogen	*Reservoir*			Main food sources	Infective dose (bacteria)	Pathogenesis	Incubation period (hours)	*Clinical signs*				Duration of illness
	Animal	Human	Environment					Vomiting	Diarrhoea	Fever		
Clostridium botulinum **Gram-positive rods, anaerobic. Endospores formed**			•	Smoked, pickled or canned foods: meat, fish, non-acid fruit and vegetables	1–2 μg toxin	Neuro-toxin pre-formed in foods (intoxication)	12–96 (usually 18–36)	(+)	–	–	Sometimes early vomiting. Main signs: blurred or double-vision, pharyngeal paralysis, aphonia and death from respiratory paralysis	Death 1–8 days or slow recovery over 6–8 months
Escherichia coli **Gram-negative rods**	(•)	•		Many raw and cooked foods and contaminated water	10^6–10^{10}	Enterotoxins ST or LT, or both, produced in intestine (infection + toxin).	12–72	+	++	–	'Travellers' diarrhoea' with sudden onset of diarrhoea, abdominal cramps and vomiting	1–7 days
Yersinia enterocolitica **Gram-negative rods. Growth at 4°C**	•			Contaminated water, dairy products, meat, especially pork	10^9	Enteric infection	24–36 (range 3–5 days)	–	+++	+	Diarrhoea, fever and abdominal pain. Vomiting is rare. Occurs mainly in children	3–5 days
Campylobacter jejuni **(*C. coli*, 5–10% of infections) Gram-negative curved rods**	•	(•)		Undercooked chicken, raw fish and shellfish, milk and water	5 x 10^2 –10^6	Enteric infection	3–5 days	–	+++	+	Abdominal pain can precede diarrhoea. Occurs mainly in school children and young adults. Can be another form of 'travellers' diarrhoea'	Few days with excretion for up to 3 months
Aeromonas hydrophila **Gram-negative rods. Slow growth at 4°C**			•	Contaminated water, milk and seafoods	Over 10^6 per g of food	Enterotoxin produced in food and intestine (?)	2–48	(+)	++	+	Recently recognised in India, Ethiopia, USA and Australia	2–7 days
Streptococci (Group A and *Enterococcus faecalis*) Gram-positive cocci	(•)	•		Raw milk, custard, chocolate pudding, ham and shrimp cocktails	10^9–10^{10}	Enterotoxin probably produced in food (intoxication)	3–22	++	+	–	Gastroenteritis resembling that of staphylococcal food poisoning	24–48 hours

• = main reservoir, (•) = less important reservoir
+++ = most prominent clinical sign, ++ = prominent sign, + = sign present, – = sign absent.

Food Spoilage Versus Food Poisoning

Many bacteria and fungi can cause food spoilage that is recognised by off-odours, unpleasant tastes and changes in the appearance of the food, such as in colour, consistency or gas production. Food-poisoning bacteria usually give no indication of their presence in the food. Cans of meat or fish containing *Clostridium botulinum* and its toxin may or may not be 'blown' (gas production) or the food changed in appearance. If spoilage has occurred it is more likely to be due to harmless spoilage organisms also present in the cans. *Clostridium perfringens* has usually achieved numbers sufficient to cause food poisoning before any manifestations of its presence, such as gas production, occur. Counts of *Bacillus cereus* in cooked rice can reach 10^6–10^9/g, which are in excess of those needed to cause food poisoning, without any change in the flavour or appearance of the rice. However, *B. cereus* can sometimes cause spoilage in milk or cream when the numbers are thought to be too low to initiate food poisoning.

Thus, food that is normal in taste and appearance, as well as spoilt food, may or may not contain levels of pathogenic bacteria sufficient to cause food poisoning. However, if a food has obviously suffered spoilage it is less likely to be consumed.

Factors Associated with Outbreaks of Food Poisoning

The occurrence of food poisoning depends on a specific set of circumstances. Some or all of the following factors may be present:

- The food is contaminated with bacteria capable of causing food poisoning from a human, animal or environmental source. Cross-contamination from raw to cooked food can occur.
- The food must be suitable for bacterial multiplication, so that the levels of bacteria or bacterial toxins in the food are sufficient to cause food poisoning.
- The bacterium, itself, must produce toxins or have other virulence factors that cause clinical signs of food poisoning. It must also have the ability to survive in the food during manufacture, storage, distribution and preparation. This survival is aided by poor techniques such as under-cooking.
- Susceptible humans must eat food containing critical levels of bacteria and/or bacterial toxins.

Diagram 43 shows some of the factors, and the interaction of these factors, that could lead to an outbreak of food poisoning.

The Incidence of Bacterial Food Poisoning

The prevalence of food poisoning varies from country to country and depends on such factors as farming practices, types of foods commonly consumed, food hygiene standards and methods of food preparation. In England and Wales the number of cases of food poisoning reported by laboratories in 1984 were 15,312 (Hobbs and Roberts, 1989). Of these 87 per cent were due to *Salmonella* species, 11 per cent involved *Clostridium perfringens*, 1 per cent were caused by *Staphylococcus aureus* and 1 per cent by *Bacillus cereus*. Enteritis caused by *Yersinia enterocolitica* and *Campylobacter jejuni* is common in most countries but the cases are not usually included in the food poisoning statistics. *C. jejuni* is one of the most common causes of acute bacterial diarrhoea in school children and young adults. Enteritis due to *Y. enterocolitica* is important in some countries; in Germany the incidence is ranked second to that of salmonellae. In western Europe, America and Australasia, *Salmonella typhimurium* tends to be the most common *Salmonella* species involved in food poisoning but in Britain, since 1988, the reported cases of *S. enteritidis* have exceeded those of *S. typhimurium*. Almost all of the increase has been due to *S. enteritidis* phage type (PT) 4 (Gilbert and Roberts, 1990).

Clinical Signs of Food Poisoning in Humans

Food poisoning caused by bacteria, such as salmonellae, that provoke infections and multiply in the body, tend to have relatively long incubation periods of several hours to several days. The disease is characterised by fever, abdominal pain, diarrhoea and to a lesser extent by nausea and vomiting. The duration of illness is about 2–5 days. Food poisoning in which preformed bacterial toxins are ingested has a shorter incubation period, usually 2–6 hours. The patient is afebrile and signs include nausea, vomiting and to a lesser extent diarrhoea. Recovery is relatively rapid and is usually complete in 24–48 hours. In botulism, caused by the ingestion of a pre-formed neurotoxin, the incubation period is usually 12–36 hours but can vary from 8 hours to 8 days. The prominent signs are vomiting, constipation, blurred vision, ocular paresis, pharyngeal paralysis and sometimes aphonia. General consciousness remains until near death. Death is caused by asphyxia due to paralysis of the respiratory muscles. In patients that survive, complete recovery may take several months.

The case-fatality rate in bacterial gastroenteritis is generally low, most of the deaths occurring at the extremes of age, in the very young and the old. In bot-

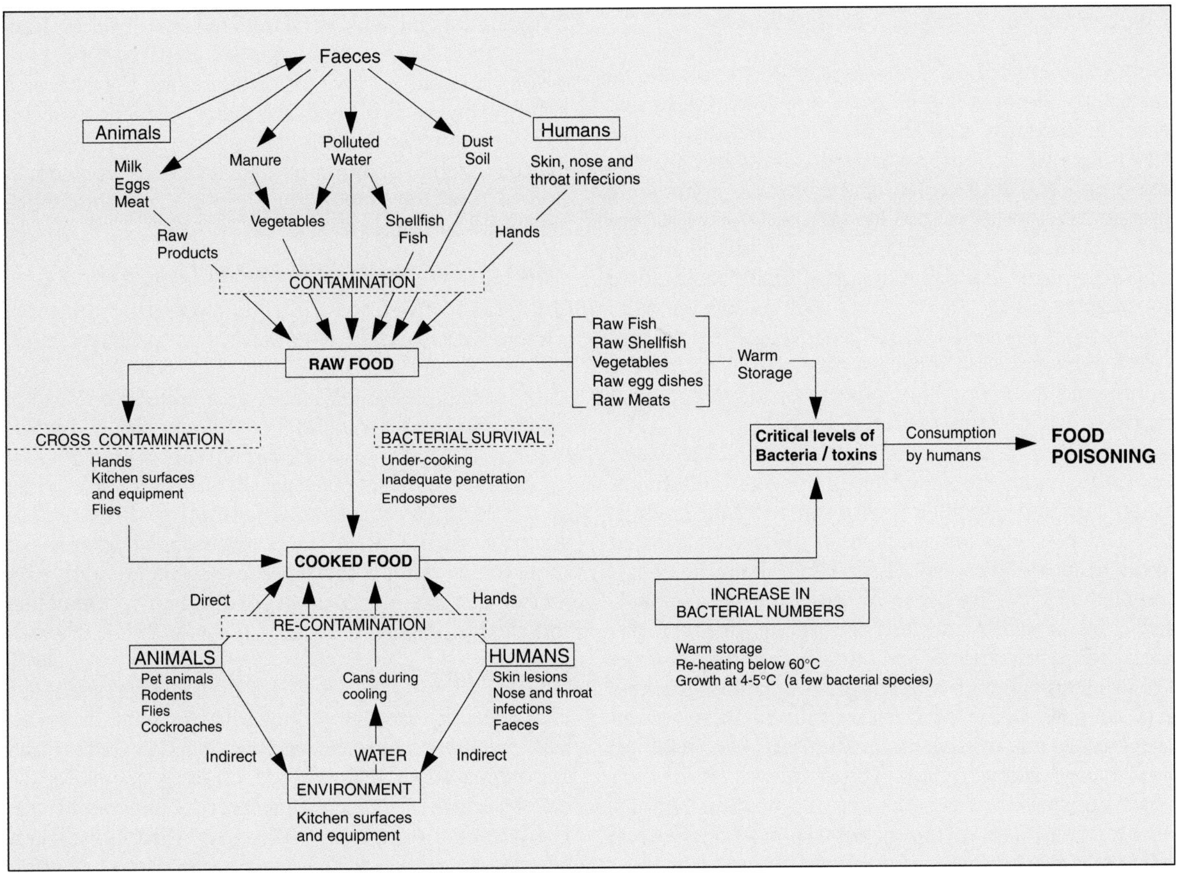

Diagram 43. Sources and methods of contamination of food, for human consumption, by food-poisoning bacteria. It shows the conditions that favour the attainment of critical levels of bacteria, or their toxins, required to initiate food poisoning.

ulism, however, high fatality rates can occur particularly with type A and B outbreaks and occasionally in those involving *C. botulinum* type E.

Bacteria Responsible for Food Poisoning

Salmonella food poisoning

*Salmonella*e are medium-sized, Gram-negative, non-spore-forming rods (**399**). They are facultative anaerobes with an optimum growth temperature of 37°C and have the ability to grow well at 43°C, but are killed at 60°C within 10 minutes. *Salmonella*e can survive for several months in microenvironments such as faecal particles, moist soil and stream sediments. Over 2,000 *Salmonella* species can cause food poisoning and they are worldwide in distribution, although the predominant species vary between countries (*see* also Chapter 18).

399

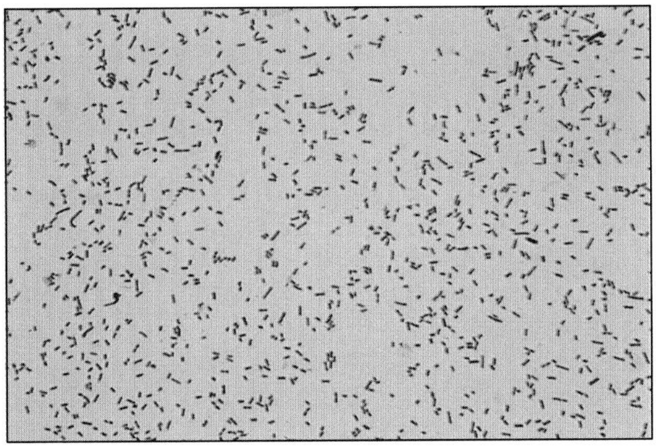

399 *Salmonella dublin* in a Gram-stained smear from culture. Salmonellae, *E. coli*, *Aeromonas hydrophila* and *Yersinia enterocolitica* are indistinguishable in Gram-stained smears. (×1000)

Epidemiology of Salmonella food poisoning

Both man and animals are susceptible to salmonella infections. While some of these infections cause disease, the majority probably lead to subclinical cases resulting in a healthy carrier state with intermittent excretion of salmonellae in the faeces. Whether a human develops disease following ingestion of salmonellae depends on the dose of organisms, the species of *Salmonella* and upon specific and non-specific immunological factors. Species such as *S. typhimurium* and *S. enteritidis* usually cause gastroenteritis (food poisoning) but other species such as *S. choleraesuis* and *S. dublin* have greater invasive potential and may cause a more serious disease.

The majority of food-poisoning outbreaks caused by salmonellae follow the consumption of food directly or indirectly associated with infection in animals. The chain of transmission is often from contaminated animal foodstuffs to animals and then from the contaminated carcasses to man. Transmission of salmonellae from animal to animal can occur on the farm. However, when animals and birds are crowded together during transportation or while being held in markets and slaughter-houses, the rate of salmonella excretion is increased due to stresses imposed by overcrowding, anxiety and lack of water. This in turn increases the cross-infection between the animals or birds. At the slaughter-house salmonellae can be transferred between carcasses by human agencies.

Although food-animals, such as intensively reared pigs, poultry and calves are probably the major reservoir of infection for man, they are not the only source of salmonellae as infection occurs quite frequently in wildlife, especially in rodents and in pet animals such as dogs, cats and terrapins. Shellfish and fish may become infected in polluted waters. Poultry eggs become contaminated with salmonellae either via the infected oviduct of a bird, through cracks in the shell, from faeces on the shell surface or through the pores in the shell. This may occur when warm, newly-laid eggs contact cool contaminated fluid. Not many *Salmonella* species cause oviduct infections but it is thought that *S. enteritidis* PT4 is able to do so, with the result that the eggs may be contaminated **before** being laid.

Raw food has usually been contaminated, via the animal food chain, prior to reaching the kitchen. Before cooking, this raw food may contaminate cooked food via contaminated surfaces, kitchen utensils or human hands. Food may be exposed to food-poisoning bacteria after cooking also, directly or indirectly, by rodents or pet animals that are excreting salmonellae and by flies. The human excretor of food-poisoning salmonellae, is thought to play a minor part in contaminating foods. Food-handlers tend to be victims not sources, as they can become infected through frequent contact with contaminated raw food or from tasting food during preparation. A prepared food, containing a small number of salmonellae, may become highly dangerous if kept at a warm ambient temperature for a few hours before being served, thus allowing the salmonellae to multiply. The sources of contamination of food for human consumption by salmonellae are shown in **Diagram 44**.

Foods incriminated in Salmonella food poisoning

Outbreaks of salmonella food poisoning are most commonly associated with meat, meat products, poultry meat or raw-egg dishes and, less commonly, with milk, cream, seafoods, salad vegetables and canned meats. Foods with a high fat content, or a good buffering capacity, may protect a relatively small dose of salmonellae during their passage through the acidic region of the stomach, thus permitting a lower dose than normal to initiate food poisoning. Examples of such foods, which have been associated with outbreaks of salmonella food poisoning, are chocolate and cheddar cheese.

The infecting dose

The infecting dose of salmonellae for humans varies with *Salmonella* species, the type of food substrate and the immunological status of the human, including the colonisation resistance of the normal intestinal flora. The intestinal flora is not fully established in neonates. In older individuals it may have been altered by antibiotic therapy, thus rendering them more vulnerable to the salmonellae. From human volunteer experiments, the dose required to produce signs of food poisoning varies from about 10^3–10^8 salmonellae, but minimal infective doses can be considerably lower than this for young infants and adults in poor health.

Pathogenesis

The pathogenesis of salmonella food poisoning (enteritis) follows three distinct phases:

- Colonisation of the intestine;
- Invasion of the intestinal epithelium;
- Stimulation of fluid secretion into the intestinal lumen with the consequent watery diarrhoea.

Colonisation of the distal small intestine and colon is opposed by the resident normal flora of the intestine. These block access to attachment sites required by the salmonellae. Indigenous fusiform bacteria produce volatile organic acids that inhibit the growth of salmonellae. Any factors disrupting the normal flora, such as antibiotic therapy or water deprivation increase the host's susceptibility to both enteric and septicaemic salmonellosis. The invasion phase involves the villous tips of the ileum and colon. The salmonellae penetrate the brush borders and enter the cells provoking an inflammatory response. The intestinal mucosa releases prostaglandins which activate adenylate cyclase with the resultant net secretion of water, bicarbonate and chloride into the intestinal lumen. The inflammatory response also triggers the release of vasoactive substances that increase the permeability of the mucosal vasculature also leading to fluid secretion into the lumen. Enterotoxins similar to the heat-labile (LT) and heat-stable (ST) toxins of *Escherichia coli* are produced by some *Salmonella* species. However, workers have failed to find enterotoxins in patients with enteric salmonellosis and watery diarrhoea, infected with a variety of *Salmonella* species. It is thought that the salmonella enterotoxins may not play a major role in the diarrhoea seen in salmonella food poisoning.

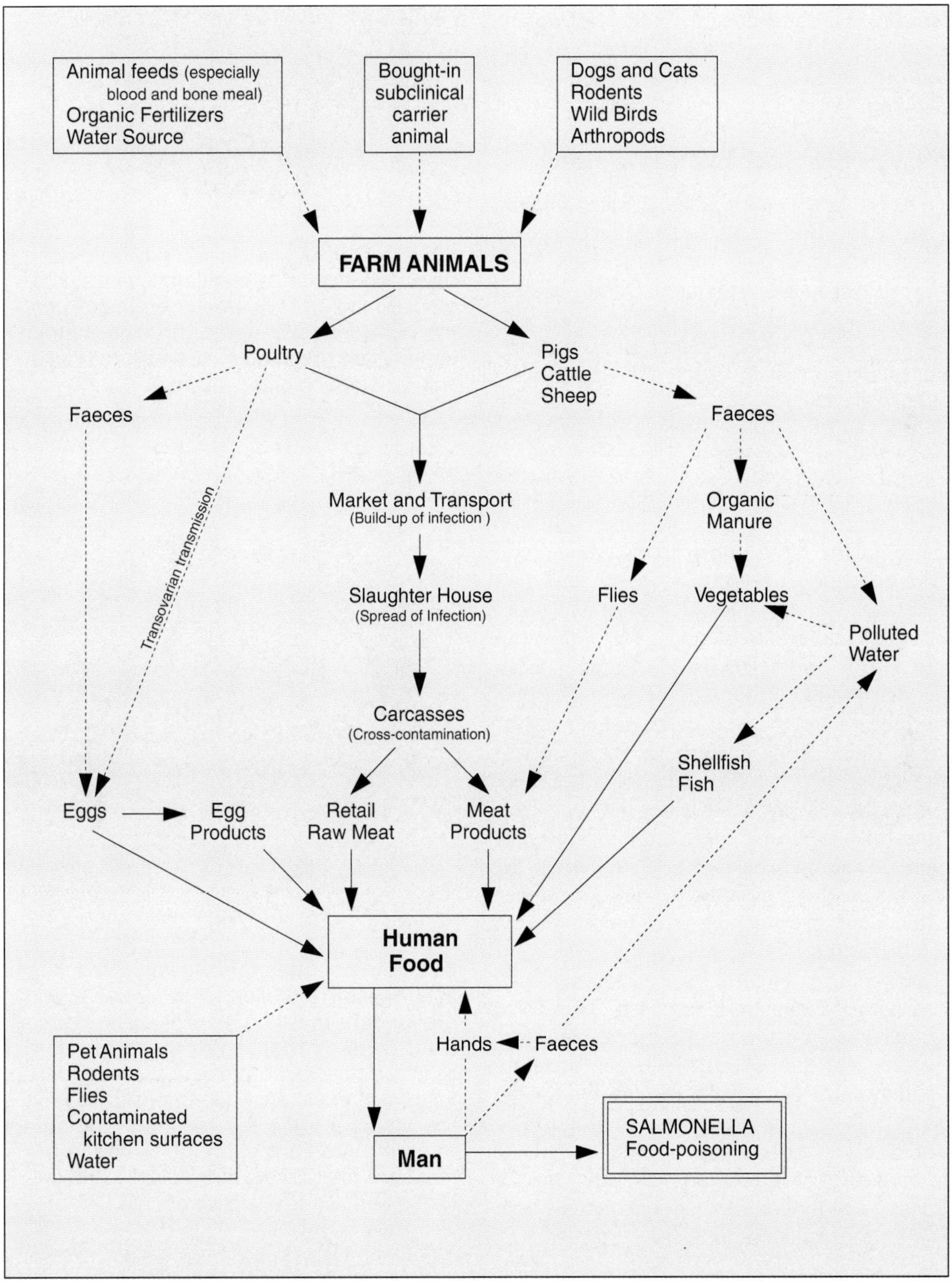

Diagram 44. Spread of salmonellae and sources of contamination for human food.
- - - - - - - - indirect contamination
——————— salmonellae carried directly

Diagnosis of Salmonella food poisoning

The Public Health laboratory investigating an outbreak of food poisoning will endeavour to isolate the *Salmonella* species both from the patient's faeces and from the suspect food. Phage typing is available for some salmonellae including *S. typhimurium* and *S. enteritidis*. The methods and media used are similar to those described in Chapter 18 (*Enterobacteriaceae*). Selective media for salmonellae include MacConkey, brilliant green and XLD agars (**400**). Bismuth sulphite agar (**401**) is often used for the isolation of salmonellae from foods.

400

400 *S. enteritidis* on MacConkey agar (left): pale, non-lactose fermenting colonies; XLD medium (right): H_2S production (black centre) surrounded by a red (alkaline) border and brilliant green agar (bottom): red (alkaline) colouration of the medium.

Clostridium perfringens Food Poisoning

C. perfringens is a large, Gram-positive, anaerobic rod that produces endospores (**402**). These endospores allow survival outside the body for long periods. They are highly resistant to many chemical and physical factors including desiccation and heat. The most heat-resistant spores can survive a temperature of 100°C for 60 minutes and so remain viable after many types of cooking (*see* also Chapter 17).

Epidemiology of C. perfringens

C. perfringens type A is the type commonly implicated in food poisoning. It is widely distributed in soil and is present in the large intestine of animals and humans. The endospores present in faeces pass into the soil or sewage systems and water. Dust and vegetation return the bacterium to animals. The spores can contaminate a wide variety of foods including meat, poultry and their products. Some of the spores survive the cooking process and the heat shock or heat activation, provided by cooking, encourages the spores to germinate. When slowly cooling meat, or other food, reaches a temperature of 50°C or lower the vegetative cells can multiply attaining their optimal and very rapid replication rate between 43°C–47°C, when the generation time is 10–12 minutes. Cooking provides many suitable anaerobic microenvironments within a food for this bacterium.

In food poisoning, a large dose of vegetative cells are ingested with the contaminated food. This large number of bacteria alter the normal intestinal flora, multiply briefly and then sporulate in the intestine. During sporulation the enterotoxin responsible for the signs of food poisoning is produced. Production of endospores rarely occurs when *C. perfringens* is growing in foodstuffs or in ordinary laboratory media, although special sporulation medium encourages both spore and entertoxin production.

401

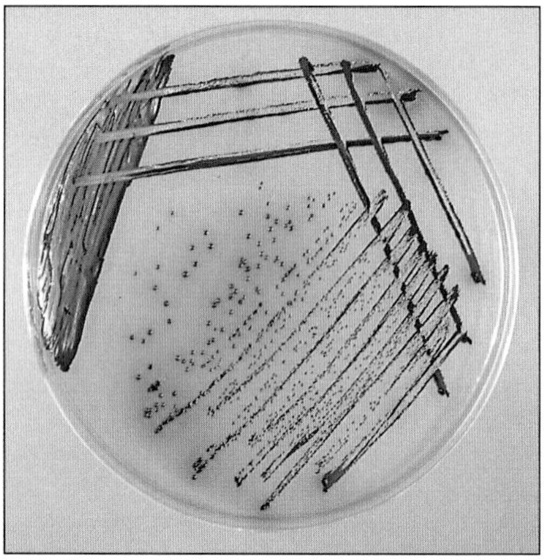

401 *S. typhimurium* on bismuth sulphite agar (Wilson and Blair) for the isolation and preliminary identification of salmonellae from water, foods and other products. Sulphur compounds produce a substrate for H_2S production, while metallic salts in the medium stain the colonies and surrounding medium black or brown in the presence of H_2S. Bismuth sulphite and brilliant green act as selective agents.

402 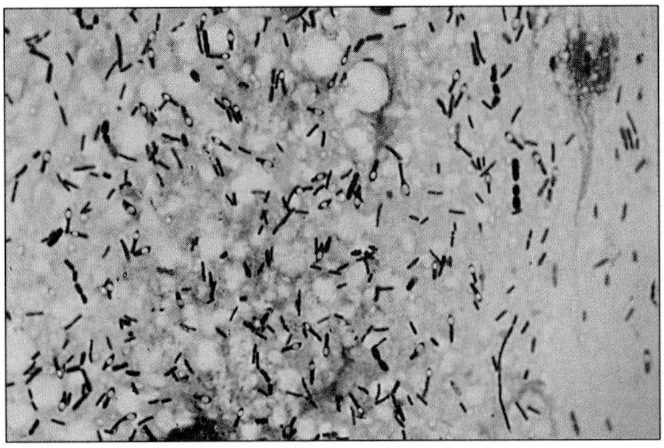

402 Spores of *Clostridium perfringens*. These are produced in the intestines of patients and are associated with the elaboration of the toxin responsible for food poisoning. (Gram stain, ×1000)

Foods incriminated in C. perfringens Type A food poisoning

Meat and poultry dishes cooked slowly, in advance of serving, or allowed to stand at room temperature for several hours, encourages growth of *C. perfringens*, which can lead to food poisoning. Large joints or bulk cooking of meat allows the survival of endospores, especially at the centre of the meat mass. It also provides favourable anaerobic conditions for germination and growth when the large mass of meat is cooling. Re-heating cooked meat dishes at temperatures below 50°C also encourages rapid multiplication of the bacterium, high viable counts of 10^6/g of *C. perfringens* being reached in a few hours.

The infecting dose

It has been estimated that *C. perfringens* type A present at the level of 10^5/g of food or a dose of 10^8–10^9 viable vegetative cells can initiate food poisoning.

Pathogenesis

The enterotoxin causes vasodilation, increases capillary permeability and intestinal motility leading to diarrhoea. *C. perfringens* food poisoning has an incubation period of 8–20 hours and is characterised by diarrhoea and abdominal pain. Vomiting is infrequent and fever uncommon. Recovery occurs in 10–24 hours. Fatalities are rare but have been recorded in debilitated and elderly patients.

Enterotoxin is also produced by *C. perfringens* type C and D strains. Disease in man caused by type C is rare but has occurred in natives of New Guinea with a staple diet of bananas and sweet potatoes. These foods have a trypsin-inhibitor and the type C enterotoxin is not destroyed because of the low levels of digestive proteases in the intestine. The disease 'pig-bel', or enteritis necroticans, is precipitated at feasts when large amounts of slow spit-roasted pork are consumed, often contaminated with the spores of *C. perfringens* type C. **'**Pig-bel' is frequently a fatal intoxication.

Diagnosis of C. perfringens food poisoning

Viable counts are carried out by Public Health laboratories for *C. perfringens* on faeces of patients and on samples of suspected food. *C. perfringens* occurs mainly in the vegetative state in foods while spores predominate in faeces. Selective media such as 'Perfringens agar' (Oxoid) (**403**) are available commercially. At least one of the following criteria should be satisfied for laboratory confirmation of a *C. perfringens* food poisoning outbreak:

- The number of *C. perfringens* in the incriminated food is over 10^6/g.
- The median faecal spore count of *C. perfringens* is over 10^6/g (counts are about 10^3–10^4/g in faeces of normal humans).
- Serological typing of the isolates should be carried out. This is based on type-specific capsular polysaccharides. The isolates from the food and from the patients' faeces should belong to the same serotype.

Serological methods for the detection of enterotoxin in the faeces of affected people have been developed.

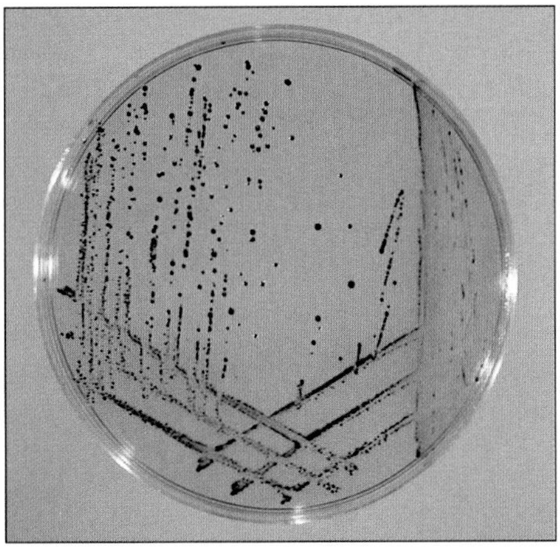 403

403 *C. perfringens* on 'Perfringens agar' developed for enumerating this bacterium in foods. The addition of sulphadiazine, oleandomycin and polymyxin renders the medium selective. *C. perfringens* produces H_2S on the medium, yielding black, shiny colonies.

Staphylococcal Food Poisoning

Staphylococcus aureus is a Gram-positive coccus (**404**) that is non-spore-forming, salt-tolerant and facultatively anaerobic. The majority of food poisoning strains are of human origin. These are usually coagulase-positive and produce enterotoxin A that is heat tolerant (*see* also Chapter 8).

Epidemiology of S. aureus food-poisoning strains

In many outbreaks of staphylococcal food poisoning, a human food-handler is implicated who contaminates the food and then, under favourable conditions, *S. aureus* will multiply and produce enterotoxin. It is estimated that between 20 per cent and 50 per cent of the population are carriers of *S. aureus*, on the hands or in the nasal cavity, and of these 15 per cent carry food-poisoning strains. Approximately 30 per cent of the human population have small numbers of staphylococci in the intestine. If the normal flora is disturbed, as can happen after antibiotic therapy, staphylococci may become a dominant organism for short periods in the intestines and be excreted in very large numbers.

Staphylococci are present on the mucous membranes of the nose and throat and in the pores and hair follicles of normal skin, particularly in damp areas such as axillae and perineum. Non-septic lesions in the skin can harbour more than usual numbers of the organisms while septic lesions, such as boils or carbuncles, can release very large numbers of staphylococci. Washing the skin with soap and water usually eliminates many of the surface Gram-negative bacteria but Gram-positive cocci tend to rise to the surface of the skin from pores and can be present in even larger numbers on the surface after washing. Scrubbing disturbs the superficial layers of the skin and may further spread staphylococci. The salt-tolerance of the staphylococci give them a selective advantage on skin, as sweat has a high salt content.

Very occasionally a *S. aureus* strain from cows is implicated in an outbreak of food poisoning. The suspected food in this case is usually raw milk or a raw milk product. Milk can also be contaminated by a human carrier. Five serologically distinct enterotoxins (A–E) are recognised, with enterotoxin A most frequently involved in food-poisoning outbreaks. The enterotoxins are proteins with a disulphide bridge and are resistant to most proteolytic enzymes and are relatively heat-tolerant. *S. aureus* itself is destroyed by pasteurisation and cooking but enterotoxin A is destroyed only gradually at 100°C for 30 minutes and can survive short or light cooking.

Incriminated foods

S. aureus grows readily in non-acid cooked foods such as baked ham, meat pies, poultry, raw milk, cream, custards and ice cream. The amount of enterotoxin produced is determined by factors such as the composition of the food, competition from other microorganisms and by temperature and time. Beef broth may contain a dialysable factor that increases the thermal stability of the enterotoxins. The presence of other bacteria affects the production of enterotoxin apparently by limiting the multiplication of the staphylococci. The optimum temperature for the production of enterotoxin is 35–40°C. Canned meat has occasionally been implicated in food poisoning. The staphylococci might enter the can through minute defects in the seam during the cooling process or, more probably, the meat could be contaminated after the can has been opened.

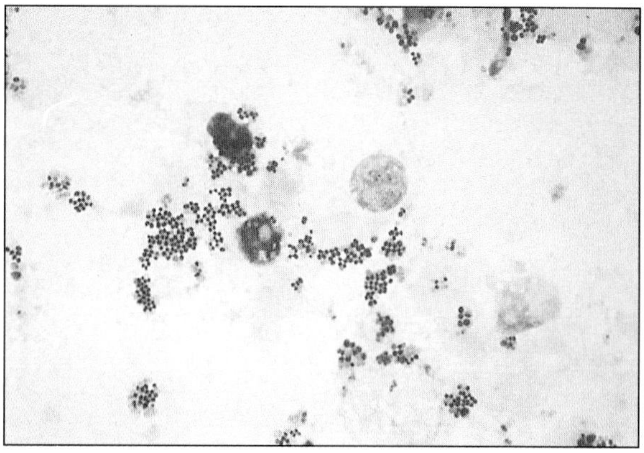

404

404 *Staphylococcus aureus* in a milk sample showing characteristic grape-like clusters. (Gram stain, ×1000)

The infecting dose

As little as 1. 0 μg of enterotoxin A can be sufficient to produce food poisoning. *S. aureus* present in a food at levels of 10^6/g or more could be dangerous if conditions for enterotoxin production are favourable.

Pathogenicity

As staphylococcal food poisoning is an intoxication, the onset is rapid and the incubation period can be as short as 2 hours, but is usually 4–6 hours. The enterotoxin may be regarded as a neurotoxin, as vomiting is initiated by its action on the emetic centre. The signs include severe vomiting that can mimic sea-sickness. There can be abdominal pain and diarrhoea, sometimes followed by dehydration and collapse. Recovery is rapid, usually within 24 hours.

Diagnosis of Staphylococcal food poisoning

It should be established that the strain or phage-type of the *S. aureus* isolated from specimens of vomit and faeces is identical to those from the suspected food and, if possible, from the hand and nose of one or more of the food-handlers. The numbers of *S. aureus* in an incriminated food should be between 10^6–10^8/g, as a count of 10^5/g or less may not be significant. The isolated strain of *S. aureus* must be shown to be enterotoxigenic. Selective-indicator media for the isolation of *S. aureus* include mannitol salt agar (**405**) and Baird–Parker medium (**406**). Phage typing (**407** and **408**) is used extensively for tracing the sources of outbreaks of staphylococcal food poisoning. ELISA and reversed passive latex agglutination kits for the detection of staphylococcal enterotoxins in foods are commercially available. The ELISA has been found to be the more sensitive of the two tests.

405

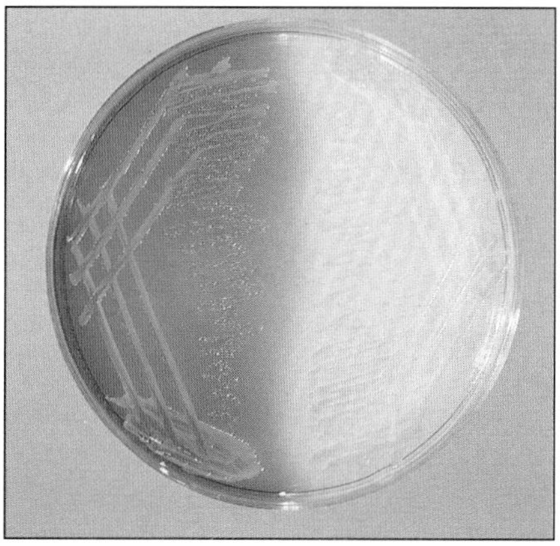

405 *S. aureus* (left) and *S. epidermidis* (right) on mannitol salt agar. Most pathogenic bacteria, other than the staphylococci, are inhibited by the 7.5 per cent NaCl. Coagulase-positive strains usually ferment the mannitol producing a pH change from pink to yellow (phenol red indicator).

406

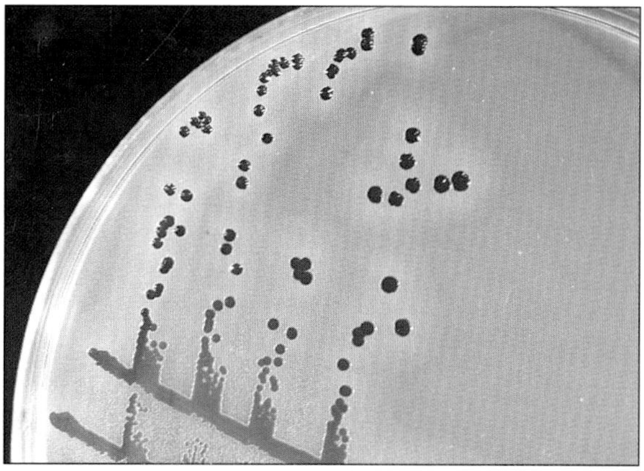

406 Close-up of *S. aureus* colonies on selective Baird–Parker medium. *S. aureus* reduces tellurite to form black shiny colonies. Opaque halos surrounding them are probably due to the action of a lipase. Some strains also produce a smaller, clear zone around the colonies due to proteolytic activity.

407

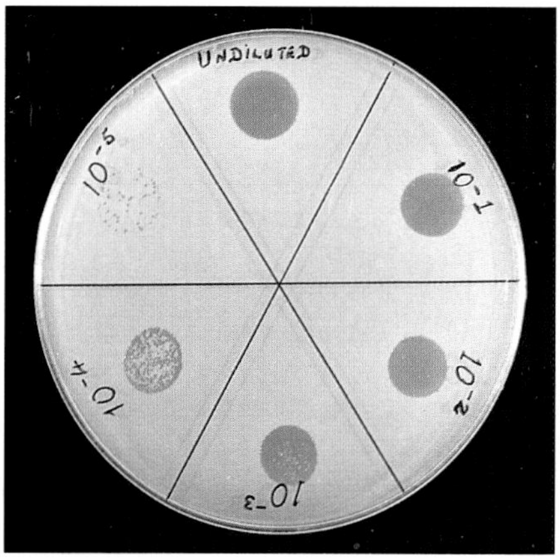

407 Titration of a *S. aureus* bacteriophage to determine the RTD (routine test dilution) for phage typing. The RTD in this case is 10^{-3} (confluent lysis). Dilutions of the phage (20µl amounts) were added to a lawn of the host bacterium on a nutrient agar plate.

408

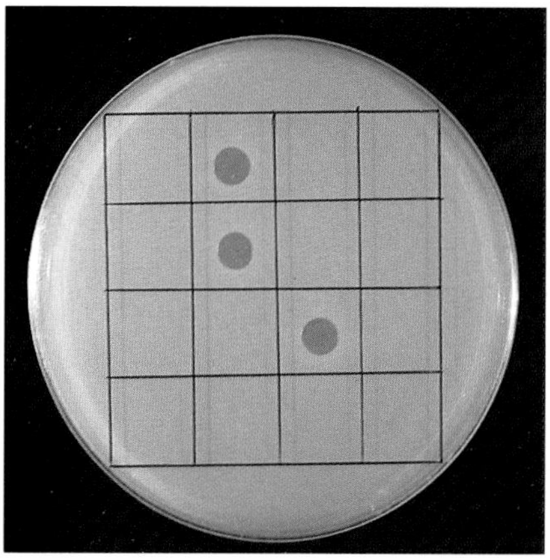

408 Phage typing of a *S. aureus* strain using 16 different staphylococcal phages. *S. aureus,* grown as a lawn on a nutrient agar plate, has been lysed by the second phage across in rows 1 and 2 and by the third phage across in row 3 of the grid.

Bacillus cereus Food Poisoning

Bacillus cereus is a large Gram-positive, spore-forming rod that is facultatively anaerobic. It grows within a temperature range of 10–48°C with an optimum between 28–35°C. The endospores (**409**) are formed freely, in almost every cell, under conditions favourable for growth. The endospores of some strains, found in raw rice, can survive boiling and stir-fry procedures. *B. cereus* can produce two distinct forms of food poisoning, the diarrhoeal syndrome caused by a heat-labile enterotoxin and an emetic syndrome involving a very heat-stable enterotoxin that is not destroyed after 90 minutes at 121°C (*see* also Chapter 15).

Epidemiology

The endospores of *B. cereus* are widespread in the environment and are commonly found in dehydrated cereals. The bacterium can be responsible for spoilage of milk and dairy products.

Incriminated foods

A great diversity of foods has been implicated in the diarrhoeal syndrome including casseroles, sausages, other cooked meats and poultry, cooked vegetables, fish, soups, sauces, dessert dishes, pasta, milk and other dairy products. The heat used in cooking may have been insufficient to kill all the spores, or the cooked dishes could have been contaminated after cooking. Products such as spices and cereals are often heavily contaminated with *B. cereus* endospores and this could contribute to the levels in the foods. *B. cereus* can attain numbers of 10^5–10^9/g in contaminated cooked food.

In contrast to the diarrhoeal syndrome, the majority of outbreaks of the emetic syndrome have been associated with boiled or cooked rice served in Chinese restaurants and 'take-aways' or 'take-outs'. The practices precipitating food poisoning include:

- The preparation of large amounts of boiled rice in advance of requirement, in which some *B. cereus* spores survive.
- Holding the cooked rice at warm ambient temperatures for long periods when the *B. cereus* spores germinate and the vegetative cells multiply and form enterotoxin. The enterotoxin can survive the stir-fry procedure which is often used before serving the rice dish.

Other foods that have occasionally been associated with the emetic syndrome include cooked spaghetti, dairy products and infant foods.

409

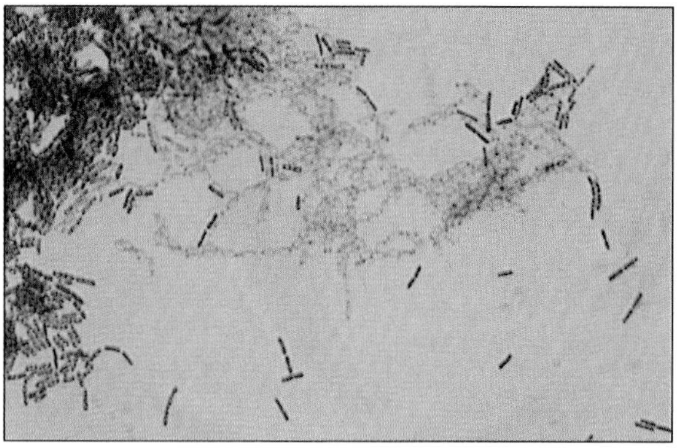

409 Spores of *Bacillus cereus* from culture: the endospores are stained by malachite green while the cytoplasm of the mother cells is counterstained by safranin (red). (×1000)

Pathogenicity

Enterotoxin, as well as large numbers of *B. cereus* vegetative cells and spores, is ingested with the contaminated food. The emetic syndrome is an intoxication and is similar to staphylococcal food poisoning. The diarrhoeal syndrome is similar, clinically, to *C. perfringens* food poisoning. Some pre-formed heat-labile toxin may be ingested but more may be formed in the intestines. This thermolabile enterotoxin, of the diarrhoeal syndrome, is an antigenic protein of medium molecular weight and is produced to some extent by all the *B. cereus* food-poisoning strains. It is capable of causing fluid accumulation in rabbit ileal loops and altering the vascular permeability in rabbit skin. The emetic syndrome is associated with the thermostable toxin, of low molecular weight, which is formed more readily, and almost exclusively, when *B. cereus* is growing on a cooked rice substrate.

The diarrhoeal syndrome is characterised by an incubation period of 8-16 hours, profuse watery diarrhoea and abdominal pain. Nausea and vomiting are occasionally seen. The symptoms persist for 12–24 hours. The incubation period in the emetic syndrome is short (1–5 hours) and there is a sudden onset of vomiting and stomach cramps. Later some of the patients may develop diarrhoea. Recovery is usually rapid.

Diagnosis of B. cereus food poisoning

Association of a suspected food with *B. cereus* food poisoning requires the demonstration of large numbers (over 10^5/g) of the bacterium in the food, acute phase vomitus and faeces of patients. Selective-indicator media for *B. cereus* usually contain polymyxin B to suppress the Gram-negative bacteria, egg yolk as *B. cereus* produces a lecithinase and mannitol as a fermentable sugar. Such a medium, '*Bacillus cereus* selective agar' (Oxoid) is illustrated (**410**). *B. cereus* yields lecithinase+/mannitol- colonies.

A serological typing scheme has been developed for *B. cereus*, based on the type-specificity of the flagellar (H) antigens. *B. cereus* strains of the same serotype should be present in the suspected food, vomitus and faecal samples.

410

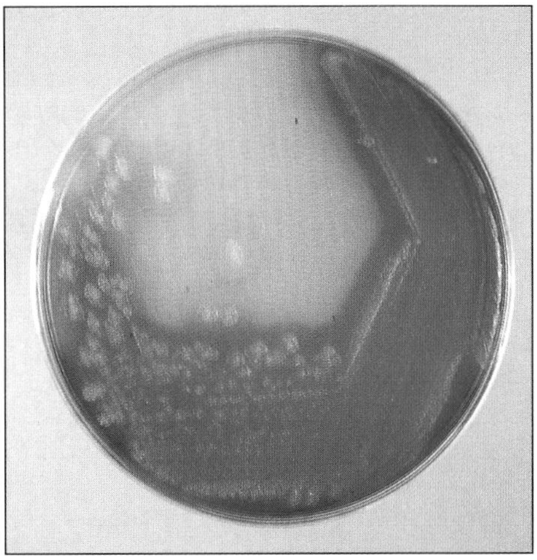

410 *B. cereus* on '*Bacillus cereus* selective agar' developed for the isolation and enumeration of this bacterium from foods. The typical colonies are crenated and have a distinctive peacock-blue colour surrounded by precipitation of the egg yolk in the medium giving a hazy turquoise zone. The medium is rendered selective by the addition of polymyxin B. The pH indicator is bromothymol blue.

Food Poisoning caused by other *Bacillus* species

B. subtilis and *B. licheniformis* have occasionally been implicated in outbreaks of food poisoning.

Vibrio parahaemolyticus Food Poisoning

V. parahaemolyticus is a halophilic vibrio and is a Gram-negative, straight rod (**411**). It is oxidase-positive, motile, lysine-decarboxylase-positive and argenine-dihydrolase-negative (*see* also Chapter 20). This bacterium produces at least one enterotoxin and is implicated in 70 per cent or more of the food-poisoning incidents in Japan.

Epidemiology

Warm weather allows growth and multiplication of *V. parahaemolyticus* and the organism can be isolated from water, fish, shellfish and other seafoods in coastal waters off Japan and other Far-Eastern countries and off the Pacific coast of the Americas. *V. parahaemolyticus* has been isolated from British and German coastal waters and sea creatures, during the summer months, but the numbers are small. Food poisoning outbreaks in the UK and other western countries, caused by *V. parahaemolyticus*, are often traceable to imports of prawns, crab and other seafoods from the Far East.

Incriminated Foods

Raw fish and shellfish are the most important sources in Japan but cooked seafoods, re-contaminated after cooking, are implicated in outbreaks of food poisoning in countries where fish is not usually eaten raw. Over 10^5/g of *V. parahaemolyticus* are present in suspected foods during outbreaks.

411

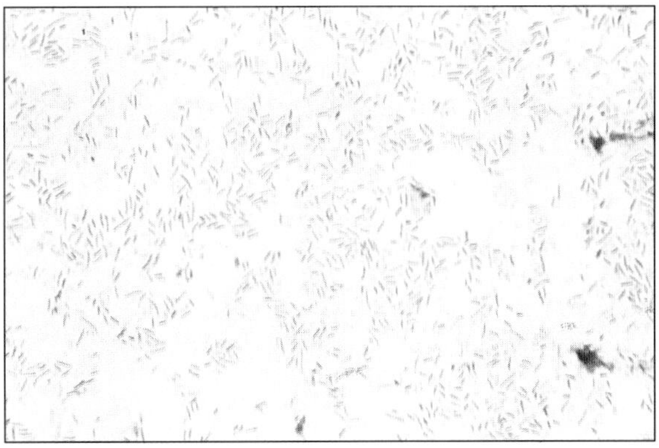

411 *Vibrio parahaemolyticus* in a Gram-stained smear from culture showing the Gram-negative straight rods. (×1000)

Pathogenicity

The illness is similar to a mild attack of cholera. There is an average incubation time of about 15 hours followed by a sudden onset of profuse diarrhoea, often leading to dehydration. The diarrhoea is accompanied by acute abdominal pain and to a lesser extent by vomiting and fever. The duration is usually 2–5 days but ill-effects may last longer.

Diagnosis

V. parahaemolyticus is non-haemolytic on sheep blood agar (**412**) and will tolerate the 7·5 per cent salt in mannitol salt agar (**413**), fermenting the mannitol. A selective medium such as 'Cholera medium TCBS' (Oxoid) is suggested for the isolation of *V. parahaemolyticus* from faecal samples. It produces blue-green colonies, 3–5 mm in diameter in 24 hours on this medium. Oxidase is a useful screening test as all *Vibrio* species (other than *V. metschnikovii)* are oxidase-positive. Most are urease-negative but a few urease-positive strains of *V. parahaemolyticus* are emerging. Other biochemical reactions are given in Chapter 20.

412

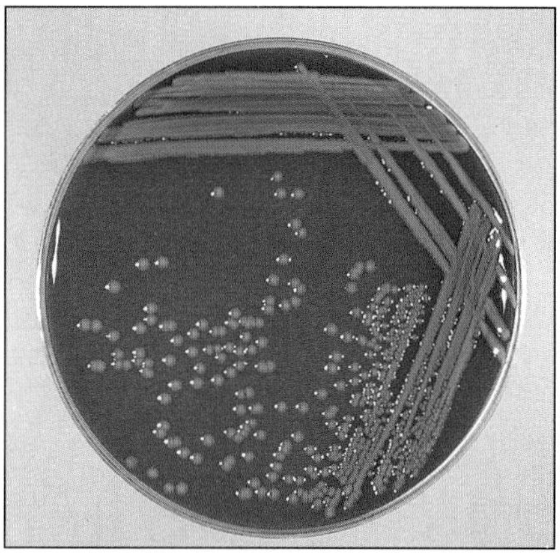

412 *V. parahaemolyticus* on sheep blood agar after 24 hours' incubation at 37°C.

413

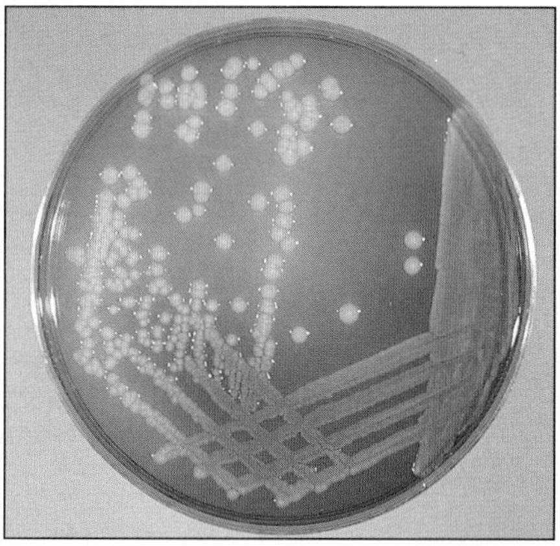

413 *V. parahaemolyticus* on mannitol salt demonstrating its innate halophilic character and ability to ferment mannitol.

Botulism: *Clostridium botulinum* Food Poisoning

Clostridium botulinum is a Gram-positive, anaerobic rod that produces resistant endospores (*see* also Chapter 17). The endospores are killed at 121°C in 15 minutes. There are seven types, A–G (type G has been renamed *C. argentinense*) which produce potent neurotoxins that are identical in action but differ in antigenicity and potency. The toxins are inactivated in 20 minutes at 100°C. *C. botulinum* has an optimum pH of 7. 0–7. 6 and the optimal temperature range for growth and toxin production is 35–40°C. There is no growth at 45°C or above and little or no growth below 10°C. Type E is the exception, as it has a lower optimum temperature for growth and is capable of slow growth and toxin production down to 5°C. The endospores of type E are more heat-labile than those of the other types.

Epidemiology

C. botulinum occurs worldwide but the types tend to be regional in distribution. Botulism is less common in humans than it is in animals. Human botulism is due mainly to types A, B and E. The occurrence of the disease tends to reflect the number, distribution and type of *C. botulinum* spores in the environment (soil, marine and freshwater sediments) and the dietary preferences of the population. The food-poisoning form of botulism is due to the consumption of food in which *C. botulinum* has been growing and producing neurotoxin. The disease is essentially an intoxication, but it is suspected that in some cases, *C. botulinum* organisms ingested with the food, can multiply and produce toxin in the intestine. It is thought that this may account for the relapses and delayed deaths sometimes observed.

Incriminated foods

The majority of reported outbreaks of botulism have been caused by food that has been smoked, pickled or canned, allowed to stand at ambient temperatures for a period and then eaten without cooking. In Europe, sausages, ham, preserved meats, pâtés and brawn are often involved. In North America the contaminated foods are commonly home-canned or bottled vegetables and fruits such as olives, string beans and spinach. Raw, salted, smoked and canned fish have also been associated with outbreaks of botulism. In many cases no spoilage of the *C. botulinum* contaminated food is obvious. With canned foods, *C. botulinum* spores, from heavily contaminated water, can enter small defects in the seams of the can during cooling. Canned foods have an anaerobic environment suitable for the growth of *C. botulinum*. Toxin can also be produced within microenvironments in loose wet mincemeat.

Many factors influence the germination, multiplication and toxigenesis of *C. botulinum* in foods. A heavy initial contamination by spores, insufficient heating, a neutral or slightly alkaline pH and anaerobic conditions favour the growth of *C. botulinum*. A pH below 4.6 inhibits germination of spores, 10 per cent salt inhibits growth, and boiling the preserved food (100°C for 20 minutes) will destroy the toxin. Type E botulism is associated with raw fish; the rarity of type E in cooked foods is attributed to the heat-lability of type E endospores.

Pathogenesis

Much of the ingested toxin is destroyed, or otherwise inactivated, in the intestine. For botulism to occur the initial dose and potency of the toxin must be sufficient for residual toxin to be absorbed by the mucosa of the upper small intestine. The lethal dose of type A toxin in the mouse by the oral route is $5–25 \times 10^4$ times that of the intraperitoneal dose. The action of neurotoxin is peripheral and affects both the efferent autonomic nervous system and somatic nerves to skeletal muscles. Only the cholinergic nerves are sensitive to the toxin as it interferes with the release of acetylcholine, leading to a flaccid paralysis. The early signs are vomiting, blurred or double-vision, ocular paresis, pharyngeal paralysis and aphonia. Later, in fatal cases, there is coma and asphyxia due to paralysis of the respiratory muscles. Incubation periods vary from 8 hours to 8 days (usually 12–36 hours). Generally, the earlier the symptoms appear, the higher the fatality rate.

Diagnosis

An attempt can be made to demonstrate toxin in patient's serum or in an extract of the suspect food by mouse inoculation (intravenous or intraperitoneal). The classical 'wasp-waist' sign (**245**) develops in an affected mouse. The toxin can be typed in a mouse neutralisation test if antitoxins A–F are available. Mouse inoculation is the most sensitive method of detecting *C. botulinum* toxin but a recently developed ELISA, using monoclonal antibodies, is a promising alternative.

Examination of faeces, mud and soil can be made for the presence of *C. botulinum* by methods discussed by Smith (1990). Extracts of the samples are inoculated into a suitable anaerobic broth (with and without preliminary heating at 60°C for 1 hour) and incubated at 30°C for 6–8 days. Filtrates are inoculated into mice or tested by ELISA for toxin. Once the toxicity of a culture is established an attempt can be made to isolate *C. botulinum* from the broth. Heating at 80°C for 30 minutes will kill any non-sporing contaminants and subcultures can then be made onto blood agar (**414**), lactose egg-yolk agar or a selective medium. Part of the broth should be plated without heating, or at a temperature below 80°C, for the isolation of type E. Pure cultures of *C. botulinum* can be examined for toxin production. However, types A, B and E not uncommonly give rise to non-toxic variants and types C and D may lose the bacteriophage that governs their toxigenesis.

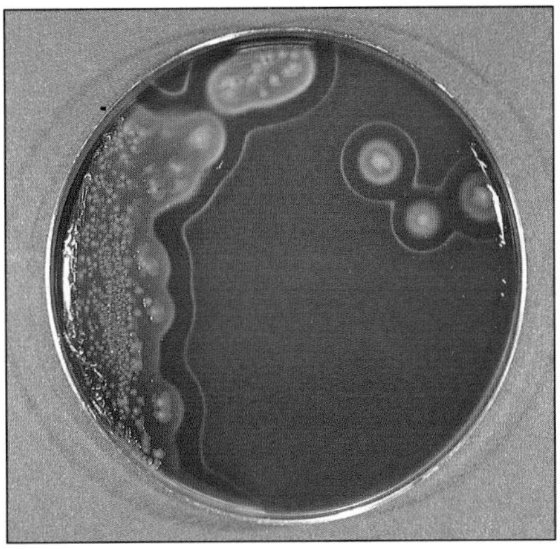

414

414 *Clostridium botulinum* on sheep blood agar.

Escherichia coli: Travellers' Diarrhoea

E. coli is a non-sporing, medium-sized, Gram-negative rod that is motile and facultatively anaerobic (*see* also Chapter 18). The organism is a commensal in the intestinal tract of man and animals. Enterotoxigenic strains of the bacterium produce enterotoxin, either a heat-labile toxin (LT) or a heat-stable toxin (ST), or both. Almost 50 per cent of the human enterotoxigenic strains of *E. coli* form only ST. LT is closely related to the toxin (CT) of *Vibrio cholerae* in structure, antigenicity and activity.

Travellers' diarrhoea, caused by enterotoxigenic strains, is a worldwide illness of brief duration and occurs most frequently among people travelling from areas of good hygiene and a temperate climate to areas with a lower standard of hygiene and often a tropical climate. Outbreaks of *E. coli* diarrhoea have also occurred on cruise liners. The disease can be both water-borne and food-borne. The incubation period is 12–72 hours followed by the rapid onset of diarrhoea and other signs such as nausea, vomiting and abdominal cramps.

Symptoms in adults are initiated by large doses (10^6–10^{10} organisms) of enterotoxigenic *E. coli*. The organisms enter the kitchen in raw foods, via contaminated water or from a food-handler who is a faecal carrier. *E. coli* is killed by heat used in cooking but the organisms can readily pass to cooked foods via hands, surfaces, containers and other kitchen equipment. *E. coli* can also be ingested with salad vegetables that are eaten raw and have been washed with contaminated water. During epidemics of travellers' diarrhoea large numbers of the enterotoxigenic *E. coli* will be excreted in the faeces of affected individuals.

Selective-indicator media for isolation of *E. coli* are available such as MacConkey agar (**415**), eosin methylene blue (EMB) agar (**416**) and 'Fluorocult' (Merck) culture media (**417**). The isolated *E. coli* can be serotyped for epidemiological studies as it has been found that many of the human enterotoxigenic strains fall into a restricted range of O serogroups. Detection of the enterotoxins, LT and ST, should be carried out. LT activity can be detected in tissue cultures of Yl mouse-adrenal cells and Chinese-hamster ovary cells. The detection of ST requires the inoculation of enterotoxin preparations into ligated ileal loops of rabbits or into the milk-filled stomachs of infant mice. The ileal loops and mouse stomachs are examined for dilatation due to the accumulation of fluid. A recently developed ELISA, with monoclonal antibodies, has been used for enterotoxin detection. Another recent development is the production of DNA probes for the genes coding for ST or LT and these probes can be used on faecal, food or water samples suspected of containing enterotoxigenic *E. coli*.

Yersinia enterocolitica Enteritis

Y. enterocolitica is a medium-sized, non-sporing, Gram-negative rod. It is a member of the *Enterobacteriaceae*, motile and facultatively anaerobic. The optimum temperature is 25°C but it is capable of growth at 4°C and 37°C. Up to 5 per cent NaCl can be tolerated by this organism (*see* also Chapter 18). Many strains produce an enterotoxin that closely resembles the ST enterotoxin of *E. coli*. The role of the enterotoxin in the pathogenesis of yersinial enteritis is unknown as the toxin is not formed at 37°C nor under the anaerobic conditions of the intestinal tract.

Epidemiology

Symptomless faecal carriers occur in many species of animals and birds. Strains frequently associated with disease in man have been isolated from the tongue, tonsils and intestinal contents of pigs in many countries. Animals, particularly pigs, are an important reservoir of infection for humans.

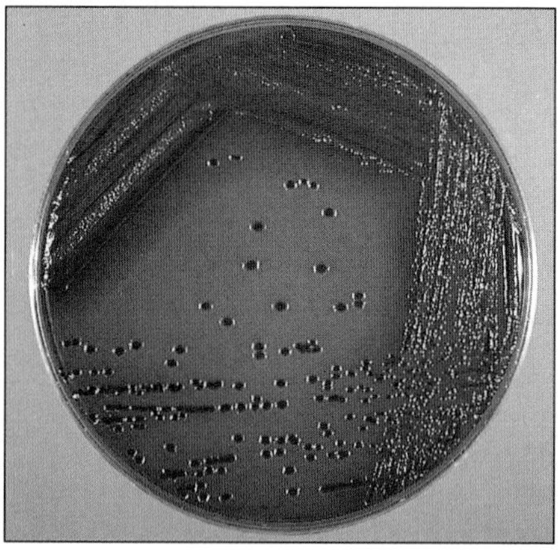
415

415 *Escherichia coli* on MacConkey agar: bright pink, discrete colonies.

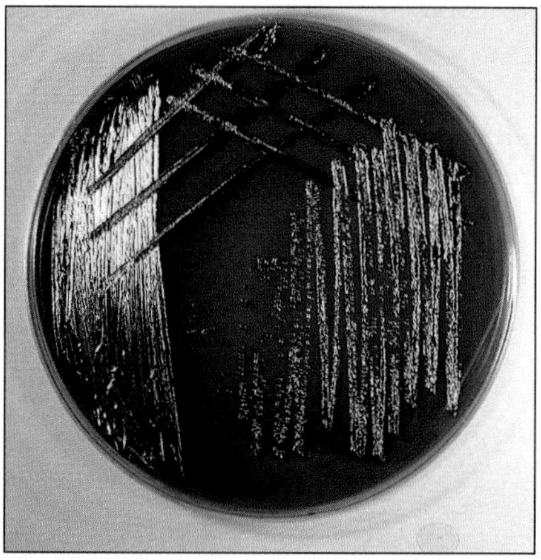
416

416 *E. coli* colonies on EMB medium showing the characteristic metallic sheen.

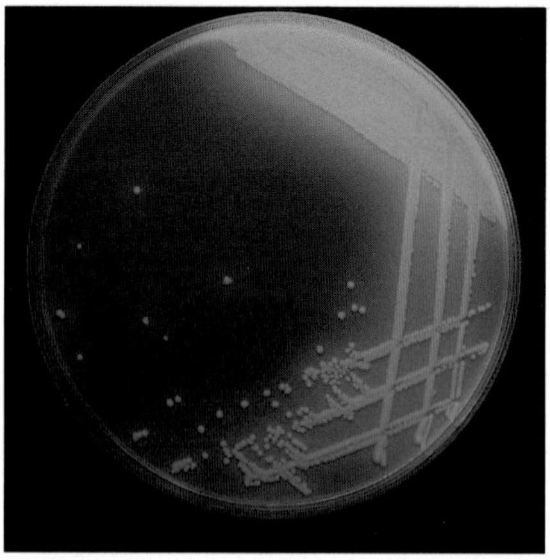
417

417 *E. coli* on 'Fluorocult' medium giving blue fluorescence under long-wave UV light. This indicates the presence of methylumbelliferone due to the action of beta-D-glucuronidase on the substrate in the medium.

Pathogenesis

In man, *Y. enterocolitica* can cause a terminal ileitis and a pseudo-appendicitis syndrome. Much more common is enteritis or gastroenteritis in children, mainly under 7 years of age, although adults can also be affected. The importance of *Y. enterocolitica* as a cause of human gastroenteritis varies from country to country. In Germany it is ranked just below the salmonellae in significance. The disease is seen mainly in the autumn and winter and the incubation period is usually 24–36 hours but can be 3–5 days. Replication of the organism probably occurs mainly in the lower ileum or upper colon and is largely confined to the epithelial cells of the intestinal mucosa. The disease is characterised by diarrhoea, fever and abdominal pain. Nausea and vomiting are rare. Symptoms may persist for 2 weeks or more with some serogroups, with relapses several months after the initial infection, but fatalities are rare. The asymptomatic carrier state in humans is not common. It has been estimated that the infective dose for man is 10^9 organisms.

Incriminated food

Cooked food can be contaminated by animal faeces, via flies, and occasionally from a human source. Water, milk, cream, pork, tongue, beef, lamb and blood sausage have been incriminated in outbreaks of illness.

Diagnosis

Samples of faeces, food, milk and water can be cultured. Usually a cold, enrichment broth is first inoculated with the specimens, incubated at 4°C or 22°C for several days and then subcultures are made onto selective-indicator media such as MacConkey agar (**418**), or 'Yersinia selective agar' (Oxoid) (**419**) containing an antibiotic supplement of cefsulodin 15 mg/l, irgasan 4 mg/l and novobiocin 2·5 mg/l. The inoculated cultures are incubated aerobically at 22–28°C. Both the biogroup and O serogroup of each isolate should be identified.

Because of the difficulty of isolating *Y. enterocolitica* from grossly contaminated samples, DNA probes, based on the virulence plasmid, have been developed. These can be used to detect the organism in samples of food or faeces. Serological diagnosis in man can be used in areas where a limited number of O serogroups predominate. Paired serum samples are taken, at an interval of l0 days, in order to demonstrate a rising antibody titre. ELISA and agglutination tests are used.

418

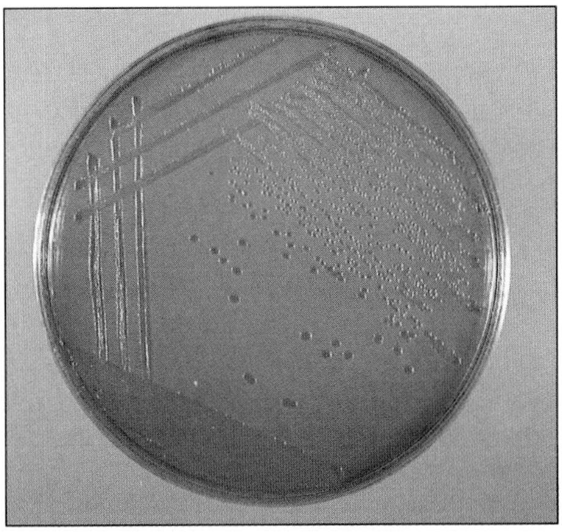

418 *Yersinia enterocolitica* on MacConkey agar showing pale pink colonies (weak fermentation of lactose).

419

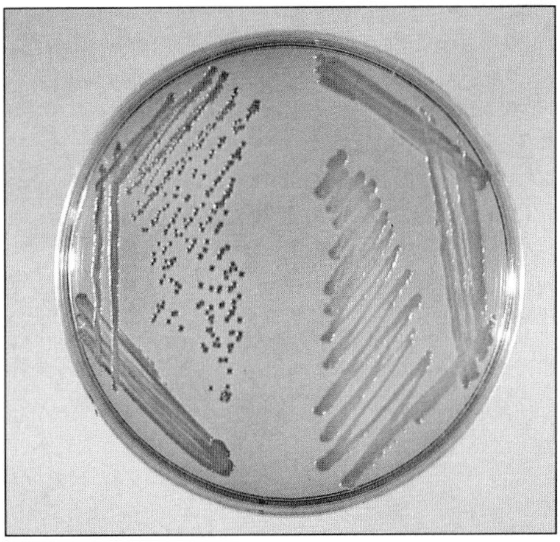

419 *Y. enterocolitica* (left) and *Y. pseudotuberculosis* (right) on 'Yersinia selective agar' formulated for the isolation and enumeration of *Y. enterocolitica* from clinical specimens and foods. Typical colonies of the bacterium develop dark red 'bullseyes' surrounded by a transparent border (neutral red indicator).

Campylobacter jejuni : Campylobacter Enteritis

C. jejuni is a microaerophilic, Gram-negative, curved rod that grows at 37°C and 43°C but not at 25°C (*see* also Chapter 25). Some strains of *C. coli* may also cause campylobacter enteritis, accounting for 5–10 per cent of infections. *C. jejuni* is found regularly in the intestines of a wide range of wild and domesticated animals, especially birds. It is thought that this may account for its optimum temperature of 42°C. *C. coli* is commonly found in the intestines of pigs. Both *C. coli* and *C. jejuni* are only transiently resident in the human intestine and man is a minor reservoir of infection. Natural bodies of water are probably an important source of infection for domestic stock. Animals restricted to a mains water supply have low infection rates. Surveys have shown faecal carrier rates of up to 70 per cent in sheep, cattle and pigs and up to 100 per cent in chickens.

Pathogenicity

In most developed countries, campylobacter enteritis is now the commonest form of acute bacterial diarrhoea. All ages are susceptible but infection falls gradually with age, the highest incidence being in school children and young adults. Infection in pre-school children is often subclinical. In developing countries, transmission of campylobacters is high and many children suffer several episodes of infection during the first 2 years of life and thereby gain an immunity. Campylobacter enteritis is a common form of travellers' diarrhoea in people normally resident in developed countries. The incubation period ranges from 3–5 days.

Clinically, campylobacter enteritis is not reliably distinguishable from other bacterial infections of the intestines. About one-third of patients seeking medical aid suffer a prodromal influenza-like illness, which may last for 2–3 days before the onset of diarrhoea. This may represent an invasive stage of the infection. Abdominal pain may also precede the onset of diarrhoea. If this pain is severe (usually in young adults) it can be misdiagnosed as appendicitis. The onset of diarrhoea is abrupt and the faeces profuse and watery. Spontaneous resolution occurs in a few days and the campylobacters are shed in the faeces for a variable period, but usually for less than 3 months.

Incriminated foods

Most human infections are acquired from the consumption of contaminated food, milk or water. Occupational exposure and direct transmission from animals to humans occur in a few cases. Drinking raw milk, non-mains water, raw or rare-cooked fish, shellfish and mushrooms carries a high risk of infection. The greatest danger is in the consumption of undercooked chicken. Normal cooking will kill the campylobacters but barbecue, fondue or rotisserie methods may not be adequate. Campylobacters are unable to multiply in food at normal ambient temperatures (25°C) but cross-contamination from raw to cooked food can occur by human hands, flies and cockroaches, and the estimated infective dose (5×10^2-10^6 organisms) is relatively small.

Diagnosis

Campylobacters can be recognised in faeces either in wet-preparations, under phase-contrast or darkfield microscopy, or in fixed smears stained by dilute carbol fuchsin (DCF) stain. In wet preparations of freshly voided faeces the characteristic rapid, jerking motility can be seen. Culture is a more sensitive technique and isolates are needed for speciation. Selective media such as charcoal-cefoperazone-deoxycholate agar (**420**) can be used. The inoculated plates are incubated under an atmosphere of 6 per cent O_2, 10 per cent CO_2 and 84 per cent N_2 at 42°C for 2–3 days (*see* Chapter 25 for identification). A serological test is sometimes used. The antigen, common to *C. jejuni, C. coli* and *C. laridis,* is present in bacterial extracts prepared by ultrasonic disintegration. A mixture of several strains is made and this is satisfactory for use in a CFT or the more sensitive ELISA.

Aeromonas hydrophila Diarrhoea

A. hydrophila is a medium-sized, motile, Gram-negative rod and is facultatively anaerobic (*see* also Chapter 20). Most strains will grow at 30–37°C but the optimum range is 22-28°C. There may be slow growth at 5°C. It grows well on blood and MacConkey agars (**421**).

Outbreaks of food-borne diarrhoea, associated with this bacterium, have been reported recently in India and Ethiopia. The strains isolated from faecal samples of affected humans were shown to produce enterotoxin. Surveys of diarrhoea in children and adults, in both the USA and Australia, have suggested that *A. hydrophila* may be a significant cause of enteritis in humans.

Water, milk and seafoods are considered to be the main sources of the organism for humans, but refrigerated meat and poultry may be important as strains of *A. hydrophila* have the ability to grow and multiply slowly at 5°C.

Streptococcal Food Poisoning

Streptococci are Gram-positive cocci that tend to occur in chains (**422**) (*see* also Chapter 9). There is some evidence that beta-haemolytic Group A and some alpha-haemolytic streptococci, such as *Enterococcus faecalis*, can cause outbreaks of gastroenteritis if present in large numbers in a food. The infective dose has been estimated at 10^9–10^{10} organisms. Raw milk, custard, chocolate pudding, boiled eggs in salad, ham and shrimp cocktails have all been incriminated. In most cases similar strains of streptococci have been isolated from both affected patients and food-handlers. It is thought that a food-handler with a streptococcal throat infection may be the reservoir of infection. Enterotoxins that give clinical signs similar to those seen in staphylococcal food poisoning are thought to be involved. Vomiting, abdominal pain and/or diarrhoea are the signs seen.

Investigation of Food-Poisoning Outbreaks

A suspected outbreak of food poisoning should be reported immediately to the local health authority. As soon as it has been established that an outbreak of illness has occurred and that it was probably food-borne, the manager of the kitchen or factory that appears to be the source of the suspect food

420 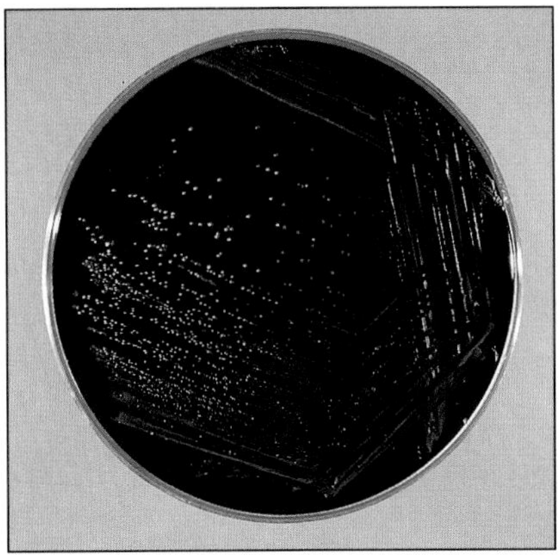

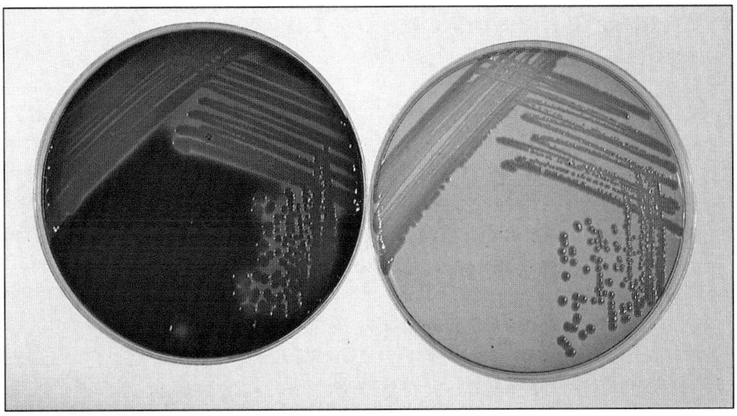 421

420 *Campylobacter jejuni* on selective charcoal-cefoperazone-deoxycholate agar.

421 *Aeromonas hydrophila* on sheep blood agar (left) and MacConkey agar (right).

422 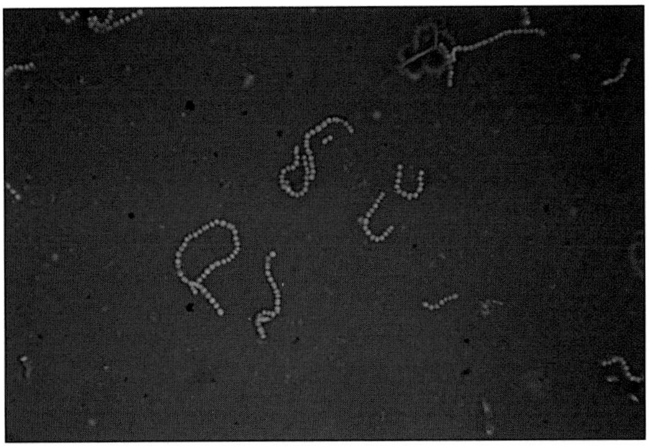

422 Characteristic chains of streptococci. (Nigrosin stain, ×1000)

should be warned not to discard any foodstuffs. Epidemiological investigations are carried out to:

- Determine the nature of the food-poisoning agent.
- Discover what food or foods were contaminated and that they were in fact the source of the outbreak.
- Ascertain the manner in which the food was contaminated.
- Ensure that all clinical cases and carriers of the agent are identified.

Control measures are enforced to stop the outbreak of illness if it is still continuing. The procedures for investigating the outbreak include:

- Obtaining a full list of the affected people, their clinical histories and the foods that they have consumed over the past 2–3 days. The clinical signs may suggest the pathogen involved, for example, the short incubation periods in both staphylococcal and *Bacillus cereus* (emetic syndrome) food poisoning. The histories may indicate that all the affected people went to a particular restaurant or attended the same banquet. If so, it should be ascertained which food or foods were consumed by all the affected people.
- Full details of the origin, method of preparation and storage of the suspected foods should be obtained.
- Specimens for laboratory examination should include:
  a) Samples of the suspected food.
  b) Vomitus and faeces from patients.
  c) Blood, spleen, liver and intestines from any fatal cases.

Once sufficient evidence has been obtained from the preliminary clinical and bacteriological examinations to indicate the pathogen and the probable source, further investigations may be necessary to:

a) Discover how the food was contaminated
b) The reservoir of the causative pathogen
c) Any precipitating factors such as inadequate cooking or poor storage methods. This entails:

- Inspection of the food-handlers for skin lesions on their hands or arms. If necessary samples should be obtained for laboratory examination from skin lesions, nasal and throat swabs and faeces from all the food-handlers involved with the incriminated food.
- Examination of the food preparation area for cleanliness, rodent and fly exclusion and general methods of cooking, food preparation and storage.
- Washings and swabs from surfaces in the kitchen, such as food preparation areas, kitchen containers and utensils, could be taken for laboratory examination.
- If a commercial canned food or other food on sale to the general public is involved, the places of sale must be identified and the product promptly removed from the shelves.

Pinegar and Suffield (1982) have summarised the steps to be taken in food-poisoning outbreaks.

Preventative Measures against Bacterial Food Poisoning

Essentially bacterial food poisoning is caused by the ingestion of large numbers of pathogenic bacteria, or pre-formed toxin, with food. The presence of these numerous bacteria may be due to:

- A high initial population of the pathogenic bacterium contaminating the raw food or individual ingredients of the food.
- Conditions in the food conducive to growth, multiplication and toxin production (where applicable). Contributing factors are often inadequate cooking, storage at ambient temperatures for several hours or re-heating food at temperatures that allow growth.

The presence of a high initial population of food-poisoning bacteria in the raw food creates several hazards:

- As the death curve for bacterial cells is exponential, the higher the initial population the greater the risk of some survivors after cooking or pasteurisation.
- Food-handlers may become infected from the heavily contaminated raw food, as can happen with salmonellae.
- The contaminated raw food may in turn contaminate kitchen surfaces, containers and utensils. There is also the risk of direct cross-contamination between raw and cooked foods.
- Bacteria that are able to grow slowly at refrigeration temperatures, such as *Listeria monocytogenes*, may reach critical levels while apparently in safe storage conditions.
- Foods eaten raw, such as shellfish and salad vegetables, could initiate food poisoning if the initial contamination rate was high.

Methods for controlling the initial population of potential pathogens or excluding them from raw food products include:

- Pasteurisation of raw milk for liquid consumption or for the manufacture of milk products.
- Salad vegetables, to be eaten raw, should be thoroughly washed in clean, uncontaminated water.
- Shellfish should be taken from non-polluted areas and kept for 2 days in clear chlorinated water. Shrimps and prawns should be boiled and shelled.
- The heat treatment used in canning of foods must be sufficient to kill all the spores of *Clostridium botulinum*. The water in which the cans are cooled must be of a potable standard otherwise *C. botulinum* spores or staphylococci may enter the cans through minute defects in the seams, which reseal themselves when the can has completely cooled.
- Food-poisoning bacteria, such as *Staphylococcus aureus*, where humans are the main reservoir, depend for their exclusion from food on the strict personal hygiene of food-handlers. Hands should be washed after nose-blowing, visits to the toilet and handling raw food. Soap and water as well as a disinfectant hand rinse (such as chlorhexidine) should be used. The handling of cooked food must be kept to a minimum. Personnel with nose, throat, eye or skin infections, or with diarrhoea, should be temporarily excluded from food-handling areas.
- Rodents, domestic pets, flies and other insects should be excluded from food preparation areas.
- Contamination of raw foods by salmonellae is the most difficult to control as the infection chain starts on the farm. An increase in infected animals and excretion rates occurs at market and during transportation. There is spread of contamination at the slaughter-house and during processing of raw meat (*see* **Diagram 44**). Salmonellae are killed by adequate cooking but care must be taken to prevent recontamination of cooked meat and poultry by raw meat. Raw meat and cooked meat and other foods must be kept completely separated. Hands should be washed thoroughly after handling raw meats and preparation surfaces washed down and disinfected on a routine basis.
- Dishes prepared from raw eggs and raw egg products require particular care, because the eggs could contain small numbers of salmonellae. It has been recommended that duck eggs should be boiled for 15 minutes and hen eggs for 7 minutes to kill all salmonellae. Dishes containing raw eggs must be kept at ambient temperatures for a minimal time during preparation, storage and serving. Salmonellae do not grow at refrigeration temperature (4–5°C).
- The water supply to kitchens and food preparation areas should be of good quality, preferably chlorinated, and free of potential pathogenic bacteria.

Preventative measures should also be taken to ensure that there is no growth and multiplication of food-poisoning bacteria in cooked food:

- Food must be cooked adequately with the aim of killing all vegetative bacteria present. Some cooking techniques such as barbecue or rotisserie methods may leave some viable salmonellae or *Campylobacter jejuni* cells. Large quantities of meat or other foods should be divided into smaller portions for cooking. This ensures that the central part of the food will be adequately heated and that the food will cool more rapidly after cooking.
- Food should be left at room temperature for as short a time as possible after cooking. The food should be covered and held at either 4°C or at over 60°C, if it is being reheated.

In summary, the general strategy should be to:

- Limit the initial contamination of raw foods.
- Use cooking procedures that are adequate to kill all vegetative cells.
- Prevent recontamination of cooked foods by being particularly careful to keep raw and cooked foods totally separate.
- Keep cooked food refrigerated to prevent growth and multiplication of survivors or germination of endospores.
- Use a temperature of 60°C or above, if reheating.

Listeria monocytogenes as a Food-borne Pathogen

L. monocytogenes has long been known as a pathogen affecting animals. Although it does not cause clinical signs of food poisoning, it has recently been recognised as an emerging food-borne pathogen causing severe outbreaks of disease and deaths in humans.

L. monocytogenes is a motile, Gram-positive rod with a tendency to be coccobacillary in shape (*see* also Chapter 13). It has unusually wide pH and temperature growth ranges that facilitate its growth in foods under a variety of conditions. *L. monocytogenes* grows best at neutral to a slightly alkaline pH but will tolerate a range from pH 5·6–9·6. The optimum growth temperature is 30–37°C but growth and multiplication occur in a temperature range of 3–45°C. The bacterium can also multiply in high salt concentrations of up to 10 per cent sodium chloride. *L. monocytogenes* is one of the few pathogenic bacteria able to grow under refrigeration temperatures (4–5°C); the others include *Yersinia enterocolitica*, *Aeromonas hydrophila* and *Clostridium botulinum* type E. At the other end of the temperature scale there is evidence that there can be survival of *L. monocytogenes* after HTST pasteurisation (72·2°C for 16·4 seconds). This survival in milk may be aided by the intracellular location of the bacterium.

Epidemiology and pathogenicity

L. monocytogenes is worldwide in distribution and widespread in the farm environment, being found in animal faeces, soil, water, silage and on plants. It can also be excreted in bovine milk. Animals and humans are infected mainly by the oral route, although infection can occur through breaks in the mucous membranes, skin, and by the respiratory and urogenital routes. Visceral or septicaemic listeriosis occurs in many species of young animals and birds, whereas abortion and neural listeriosis ('circling disease') most commonly occur in sheep, cattle and goats. In the majority of healthy adult humans the infection is subclinical. The incidence of listeriosis is highest in neonates, pregnant women, the elderly and in immunosuppressed individuals. Immunosuppression can occur in patients with conditions such as AIDS, malignancies and cirrhosis of the liver, or in people receiving corticosteroids or cancer therapy. The septicaemic and neural form of the disease tends to occur in neonates and the immunosuppressed, while pregnant women may suffer an influenza-like syndrome with subsequent abortion or premature birth. Listeriosis in humans is usually sporadic but epidemics have occurred, several of food-borne origin. The mortality rate in human listeriosis is estimated to be 30 per cent.

Foods involved

The foods incriminated in outbreaks of listeriosis include coleslaw (cabbages contaminated by manure), milk, chocolate milk and soft cheeses. *L. monocytogenes* can be isolated from meats, but poultry has been the only meat so far linked with an outbreak of listeriosis. Several surveys have been carried out to determine the incidence of *L. monocytogenes* in raw milk. It was found to be about 4 per cent in California, but ranged during the year from 14–75 per cent in a milk factory in Madrid. Some foods support the growth of *L. monocytogenes* better than others. In chocolate milk it was shown that both the cane sugar and cocoa powder, present in the product, enhanced the growth of the pathogen. The bacterium grows well in liquid milk products, even at refrigeration temperatures.

Cheeses made with unpasteurised milk are particularly dangerous. In cheeses, the growth of *L. monocytogenes* may be retarded, but not stopped, by lactic starter cultures. Listeriae are entrapped in the curd with only a small percentage of the organisms escaping in the whey. Once in the curd, the multiplication or survival of *L. monocytogenes* varies in the different types of cheeses and with the ripening processes. There is growth in feta cheese, but death of most bacterial cells occurs in cottage cheese which tends to have a low pH. During ripening, the numbers of listeriae decrease gradually in cheeses such as Cheddar and Colby but increase markedly in Camembert.

Diagnosis

L. monocytogenes is often difficult to isolate from foods due to the other microflora that are present. This is particularly so in the case of mould-ripened cheese and cabbage. Cassiday and Brackett (1989) have reviewed the methods and media employed for the isolation and enumeration of the pathogen from foods. Enrichment procedures are often required for this organism. This involves inoculation of either non-selective or selective broths that are incubated at 4°C for up to 8 weeks. Subcultures are made weekly onto agar media. Selective indicator media include those containing aesculin (that is hydrolyzed by *L. monocytogenes*) with the addition of antimicrobial agents (**423**).

423 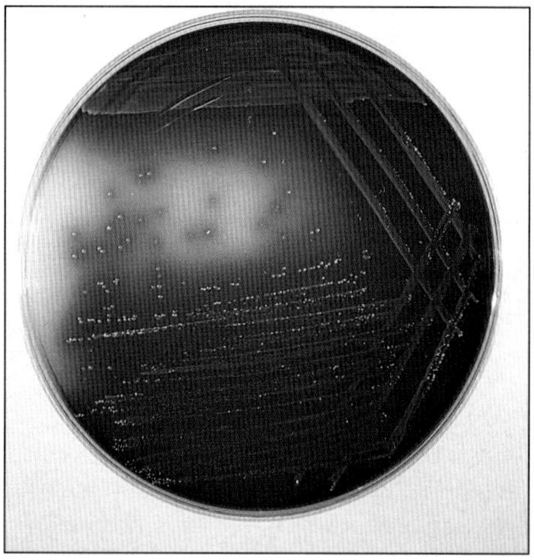

An ELISA, using monoclonal antibodies, has been developed to identify listeriae in foods, and also a DNA probe for the detection of the bacterium in dairy products. Early fluorescent antibody techniques resulted in false-positives due to cross-reactivity with other microorganisms. Automated fluorescent antibody procedures using flow cytometry are currently under investigation.

423 *Listeria monocytogenes* on 'Listeria selective agar'. This medium is used for the detection of the bacterium in clinical specimens and foods. Hydrolysis of aesculin results in black zones around the colonies due to the formation of black, iron phenolic compounds.

References

Cassiday, P. K. and Brackett, R. E. (1989). Methods and media to isolate and enumerate *Listeria monocytogenes* : A review, *Journal of Food Protection*, **52**: 207–214.

Gilbert, R. J. and Roberts, D. (1990). Foodborne gastroenteritis, pp. 490–512. In Topley and Wilson's *Principles of Bacteriology, Virology and Immunology*. Vol 3, 8th ed. , Smith, G. R. and Easman, C. S. F. (eds.). Edward Arnold, London.

Hobbs, B. C. and Roberts, D (1989). *Food poisoning and food hygiene*. 5th ed., Edward Arnold, London.

Pinegar, J. A. and Suffield, A. (1982). In Corry, J. E. L. , Roberts, D. and Skinner, F. A. (eds.), *Isolation and identification methods for food poisoning organisms*. Academic Press, London.

Smith, G. R. (1990). Botulism, pp. 514–529, In Topley and Wilson's *Principles of Bacteriology, Virology and Immunology*. Vol. 3, 8th ed. , Smith, G. R. and Easman, C. S. F. (eds.). Edward Arnold, London.

Bibliography

International Commission on Microbiological Specifications for Foods (ICMSF), (1978). *Micro-organisms in foods. 1. Their significance and methods of enumeration*, 2nd ed. University of Toronto Press, Canada.

International Organization for Standardization (1975). ISO (international standard) 3565, *Meat and meat products: detection of salmonellae* (reference method).

Pearson, L. J. and Marth, E. H. (1990). '*Listeria monocytogenes* – Threat to a safe food supply': A review. *Journal of Dairy Science*, **73**: 912–928.

Riemann, H. and Bryan, F. L. (1979). *Food-borne infections and intoxications*, 2nd ed., Academic Press, London.

Section 3: Mycology

38 Introduction to the Pathogenic Fungi

General Characteristics of the Fungi

The fungi have an eucaryotic cell type and are therefore insensitive to most bacterial antibiotics. They are non-photosynthetic and usually non-motile. Although their optimum pH is about 6, they can tolerate more acidic conditions. They are strict aerobes with an optimum temperature for growth of 20–30°C, but the pathogenic fungi causing systemic mycoses can tolerate 37°C. Fungi are comparatively slow-growing on laboratory media, the Zygomycetes and *Aspergillus* species may show growth in 2–3 days but the incubation time for some of the dermatophytes may be as long as 3–5 weeks. The fungi can be divided into moulds and yeasts. The moulds are filamentous with branching filaments or hyphae 2–10 µm in diameter. The hyphae in most moulds have cross-walls or septa but the Zygomycetes rarely produce septa and are termed non-septate. The branching hyphae grow to form a tangled mass known as the mycelium. Moulds usually form large fluffy colonies on laboratory media and produce aerial fruiting hyphae that bear asexual spores.

Yeasts are oval, spherical or elongated cells, about 3–5 µm in diameter, and form moist colonies that are usually larger, but not unlike bacterial colonies. The yeasts reproduce by budding or by both budding and spore formation. The terms 'mould' and 'yeast' have no taxonomic significance and are not mutually exclusive. Some of the fungal pathogens are dimorphic, being yeasts or yeast-like in animal tissues and when grown on enriched media at 37°C, but are moulds in their natural environment and when grown on media at 25°C. Yeasts such as *Candida albicans* can grow in animal tissue as elongated cells, joined together, that resemble septate hyphae, known as pseudohyphae.

Classification of the Fungi

The four traditional classes of the fungi, based on the sexual spores, are shown in **Table 127**. The majority of the animal and human pathogens are in the Deuteromycetes with a few in the Zygomycetes. The perfect or sexual stage has not been found for members of the Deuteromycetes and they are known as the Fungi Imperfecti. If the sexual state is discovered, the fungus is given another generic name and placed in the appropriate Class depending on the type of sexual spores that are produced. Most of the pathogenic fungi, previously in the Fungi Imperfecti, have been found to be Ascomycetes when the sexual state is found. The name for the asexual state of a fungus is known as the anamorph and that for the sexual state the teleomorph. For example, the teleomorph of the yeast *Cryptococcus neoformans* is *Filobasidiella neoformans*, one of the few pathogenic fungi in the Basidiomycetes.

General Features of Fungal Infections

A few of the dermatophytes are considered to be obligate parasites but the majority of the pathogenic fungi are widespread in the environment as saprophytes or present as commensals associated with animals and humans. Most are therefore opportunistic pathogens and predisposing factors often contribute to the establishment of fungal infections. These include an alteration in the normal flora of the host by prolonged administration of antibiotics, immunosuppression, concurrent infections, breaks in the skin or mucous membranes, constantly moist areas of skin or the exposure to a large infective dose, as with *Aspergillus fumigatus* spores in brooder pneumonia of chicks. Fungal diseases do not usually assume epidemic proportions except in certain instances such as explosive outbreaks of ringworm. No exotoxins or endotoxins have conclusively been shown to be produced, but mycotoxicoses occur due to animals ingesting preformed toxic metabolic products produced during fungal growth in animal feedstuffs.

Chronic fungal infections lead to a granulomatous reaction that resembles the host's response to a foreign body or to an actinomycete infection. Immunity to fungal infections is considered to be more cell-mediated than antibody-mediated. Antibodies are produced in most mycoses, except in ringworm, but they do not appear to be protective. Hypersensitivity may develop to a particular fungus in the infected or exposed host. This may lead to skin rashes in humans and can be useful in various diagnostic skin tests such as the use of histoplasmin for the diagnosis of histoplasmosis. The mycoses are sometimes divided into deep (systemic) and superficial mycoses, however several pathogenic fungi such as *Candida albicans* and *Aspergillus fumigatus* are capable of causing both superficial and deep infections.

Table 127. The traditional classes of fungi.

	Deuteromycetes (Fungi Imperfecti)	*Ascomycetes*	*Basidiomycetes*	*Zygomycetes*
Septate hyphae	+	+	+	–
Sexual state present	–	+	+	+
Asexual spores	Conidia Arthrospores Chlamydospores Blastospores	Conidia mainly Arthrospores Chlamydospores	Conidia Odia Arthrospores	Sporangiospores
Sexual spores	–	Ascospores	Basidiospores	Zygospores

General Methods for the Diagnosis of the Mycoses

History, clinical signs, gross pathology, and in a few cases intradermal skin tests, are all of value in the diagnosis of fungal infections, backed by a laboratory investigation of clinical specimens. These investigations include direct microscopy, isolation and identification of the pathogen. The identification is based on colonial characteristics, examination of the fruiting heads and on biochemical reactions in the case of the yeasts and certain of the dermatophytes.

Direct Microscopic Examination of Clinical Specimens

Table 128 indicates methods used for the examination of fungal elements in clinical specimens and **Table 129** gives a brief summary of the diagnostic features of some of the fungi found in veterinary diagnostic samples. The specimens suspected of containing fungal pathogens range from hairs and skin scrapings for dermatophytes to exudates, biopsies and tissues. Because many of the potentially pathogenic fungi are ubiquitous it is important to take tissue for histopathology whenever possible, to demonstrate fungal hyphae or yeast cells actually invading the tissue, often with a tissue reaction. If there is a correlation between the fungus that is isolated and the histopathological findings, there can be greater confidence in the diagnosis of the disease or condition.

Histopathological sections can be made from biopsies or from tissues. Frozen sections are prepared from fresh tissue while more permanent tissue sections are made from material fixed in 10 per cent formalin. These sections can be stained by methods such as the periodic acid-Schiff (PAS), which will differentially stain the fungal elements pink, or by silver impregnation stains such as Gomori's methenamine silver stain, where the fungal elements stain dark brown or black. Calcofluor white powder (Fluorescent Brightener 20, Sigma Chemical Co.) is a cotton brightener which binds to chitin in the fungal cell walls. On excitation with light of wavelength 500 nm the bound calcofluor white fluoresces blue-green. As a working solution (0.1 per cent w/v) it can be used mixed with exudates, incorporated into 10 per cent KOH solutions or employed to stain histological sections (*see* **Appendix 1** for details). Wright or Giemsa stains can be used on impression smears from biopsies or bone marrow for *Histoplasma capsulatum*. A Gram stain or simple methylene blue stain is useful for many of the yeasts. India ink or nigrosin wet preparations are used to demonstrate the characteristically large capsule of *Cryptococcus neoformans*. This and other comparatively rapid methods involving wet preparations to visualise fungi in diagnostic specimens are described below. To clear and clarify the specimens, so that the fungal elements can be seen, 10–20 per cent potassium hydroxide (KOH) and other chemicals are used.

Table 128. Summary of the methods employed for the direct microscopic examination of fungi.

Technique	Use	Fungi
10–20% KOH wet preparations	Clears specimens to make fungi more visible. Examine under low- and high-dry objectives or phase contrast	Fungal elements of most moulds and yeasts. Dimorphic fungi as yeast-like forms in tissue. Arthrospores on affected hairs for dermatophytes
Calcofluor white (0.1%)	Fluorescence of fungal elements under ultra violet microscope. Visualisation of fungi made easier	Detection of most fungal elements in wet preparations and in tissue sections
India ink or nigrosin	Wet preparation with cerebral spinal fluid or clear exudates	*Cryptococcus neoformans* to demonstrate the characteristically large capsule
Gram or methylene blue stain	Fixed smears of tissues or exudates	Yeast cells such as *Candida albicans* as well as any bacteria that are present. *C. neoformans* stains poorly by these methods
Fluorescent antibody technique	Frozen sections or fixed smears	Available in specialized laboratories for some of the dimorphic fungi such as *Blastomyces dermatitidis*
Periodic acid-Schiff (PAS) + counterstain (haematoxylin)	Frozen or paraffin-embedded histological sections from biopsies or tissues	Most fungal elements can be demonstrated in tissues by this method. The fungi stain pink. Any tissue reaction caused by the fungal invasion can also be observed
Methenamine silver stain + counterstain	Frozen or paraffin-embedded histological sections from biopsies or tissues	Most fungal elements in tissues will be stained a dark brown by this method and are easy to see. Visualisation of any internal structures may be harder than with the PAS-haematoxylin stain
Wright or Giemsa stain	Fixed bone-marrow smears or impression smears from biopsies	Demonstration is limited to *Histoplasma capsulatum*

KOH Wet Mount

- Place 1–2 drops of 10–20 per cent KOH on a microscope slide and add a small amount of the specimen to the drop of KOH and mix well.
- Gently pass the slide through a low flame of a Bunsen burner to hasten clearing (do not boil or over-heat). This is an optional stage.
- Place a coverslip on top of the preparation and press down gently.
- Allow to stand for 1–2 hours, or even overnight in a moist chamber. The time required will depend on the density of the specimen. The KOH will partially clear the proteinaceous debris.
- Examine under phase contrast or under the low and high-dry objectives of a light microscope.

Table 129. Morphological features of pathogenic fungi in diagnostic specimens.

Fungus	Techniques	Summary of diagnostic features
Aspergillus fumigatus	KOH, calcofluor white, periodic acid-Schiff (PAS) or silver-impregnation stains	Septate hyphae, dichotomous branching at a 45° angle. Hyphae 3–6 μm and rarely up to 12 μm in diameter. Tissue reaction is granulomatous or necrotising, but may not occur in an immuno-suppressed host. May see distorted fruiting heads if fungus spreads into an air-space in the body
Zygomycetes: *Rhizopus, Mucor, Rhizomucor, Absidia* and *Mortierella* spp.	KOH, calcofluor white, PAS or silver-impregnation stains	Large, bulging, non-septate hyphae that can be twisted and fragmented. About 10–20 μm in diameter (range 3–25 μm) with irregular branching. The invading hyphae of *Mortierella wolfii* tend to be finer (2–12 μm diameter) than the other zygomycetes
Candida albicans	Gram stain, KOH, PAS or silver-impregnation stains	Budding cells, oval or round, 3–4 μm diameter. Pseudohyphae may be present in tissue; these have regular points of constriction between individual elongated yeast cells. They must be distinguished from moulds with septate hyphae
Malassezia pachydermatis (*Pityrosporum canis*)	Gram stain, methylene blue, KOH or calcofluor white	Bottle-shaped, small yeast (1–2 x 2–4 μm). Unipolar budding and reproduction is by bud-fission in which the bud detaches from the mother cell by a septum
Cryptococcus neoformans	India ink, KOH, PAS, or Mayer's mucicarmine stain	Spherical budding yeast cells, 2–15 μm diameter, usually surrounded by a large capsule. Produces pinched-off buds, sometimes multiple. Cells vary greatly in size in a single preparation. Encapsulated pseudohyphae are very occasionally seen
Blastomyces dermatitidis	KOH, calcofluor white, FA technique, PAS or silver-impregnation stains	Large, budding yeast 8–15 μm (range 2–30 μm) in diameter with very thick walls. Buds are connected by a broad base. Intracytoplasmic contents are usually evident
Histoplasma capsulatum	Wright, Giemsa, PAS or silver-impregnation stains	Small budding yeast, spherical to oval, 2–5 μm, intracellular in monocytic cells. A clear halo can be seen around the darker-staining cell. Buds are single with narrow bases. The fungus is difficult to detect in unstained preparations
Coccidioides immitis	PAS and silver-impregnation stains, KOH + calcofluor white	Large spherules present in tissue. When mature, up to 200 μm in diameter and contain numerous non-budding endospores (2–5 μm). Immature spherules vary in size and do not contain endospores
Sporothrix schenckii	Gram stain or KOH on exudates. PAS or silver-impregnation stains on biopsies	Small, cigar-shaped yeasts, 2–6 μm. May exhibit multiple budding. Only a small number are usually present in exudates and they may be hard to see
Dermatophytes *Microsporum* and *Trichophyton* spp.	KOH, KOH + calcofluor white, DMSO + KOH, blue-black ink + KOH	Septate hyphae (2–3 μm diameter) surround affected hairs and fragment into arthrospores. Some hyphae may still be present but more usually a sheath of refractile round arthrospores (2–8 μm diameter) is present. These arthrospores must not be confused with fat globules or hair-pigment granules (melanosomes)

Table 129. Morphological features of pathogenic fungi in diagnostic specimens. *(continued)*

Fungus	Techniques	Summary of diagnostic features
Fungi in mycetomas	KOH, calcofluor white, PAS and silver-impregnation stains	Irregular granules, 0.5–3.0 mm and variously coloured, are present in biopsies or scrapings. Within crushed granules are intertwined hyphae (2–5 μm) with swollen cells (15 μm or more) at the periphery
Fungi in chromo-blastomycoses	KOH, calcofluor white, PAS and silver-impregnation stains	Single-celled or clustered, spherical (4–12 μm), thick-walled bodies and darkly pigmented (sclerotic) bodies. Hyphae may be present (2–6 μm) and are seen in skin scrapings and aspirates
Rhinosporidium seeberi	KOH, calcofluor white, and PAS stains	Large sporangia (up to 200–300 μm diameter) in wet mounts from nasal polyps or discharges and in tissue sections from biopsies. The sporangia contain numerous endospores. Must be distinguished from the spherules of *Coccidioides immitis*
Chrysosporidium (Emmonsia) parvum* var. *parvum ***C. parvum* var. *crescens***	PAS stain on tissue sections; can also be seen in a haematoxylin-eosin-stained section	Thick-walled, spherical adiaspores up to 400 μm in diameter at the centre of a granulomatous lesion. The adiaspores contain multiple minute nuclei in the periphery of the cytoplasm and have triple-layered walls, each layer of variable thickness. They do not reproduce in the host. Must be distinguished from the spherules of *C. immitis*

Modifications to the KOH Wet Mount Method

a) *DMSO plus KOH*: penetration and clarity of specimens can be improved by the addition of dimethyl sulphoxide (DMSO) to the KOH. A formulation of 20 per cent KOH and 36 per cent DMSO is used.

b) *Blue-black ink to stain fungal elements*: the fungi take up the dye (ink) selectively and the specimen is satisfactorily cleared by the 10 per cent KOH. The working solution is given in **Appendix 1**.

c) *Calcofluor white and KOH*: 1 drop of 0.1 per cent calcofluor white and 1 drop of 10 per cent KOH are placed on the microscope slide. A small amount of the specimen is added to it and a coverslip placed on top. After clearing, the preparation is examined under a microscope with excitation light of 500 nm. The fungal elements fluoresce blue-green making them more obvious.

India Ink or Nigrosin Preparations

(*See* **Appendix 1** for preparation of the solutions)

- Place a drop of India ink or nigrosin on a microscope slide.
- Add a loopful of cerebrospinal fluid or clear exudate suspected of containing *Cryptococcus neoformans* to the India ink or nigrosin and mix well.
- Place a coverslip on the mixture and press down gently. Blot-up any excess fluid.
- Examine under the low and high-dry objectives of the light microscope.

Isolation and Subculture of Fungi

Media for fungi

The media designed for isolating pathogenic fungi from clinical specimens need to be selective against the faster-growing bacteria and often more rapidly-growing contaminating fungi. Sabouraud dextrose agar is the medium most commonly used. It has a pH of 5·6 and is therefore inhibitory to bacteria while supporting the growth of fungi that are acid tolerant. The medium may be made more selective by the addition of chloramphenicol (antibacterial). Cycloheximide (actidione) has an antifungal action against some of the contaminant fungi. Cycloheximide can also be inhibitory for some of the pathogenic fungi such as the Zygomycetes, *Aspergillus* spp., *Cryptococcus neoformans* and some of the dimorphic fungi such as *Blastomyces dermatitidis*. It is advisable to use media with and without the selective agent cycloheximide, especially when attempting to isolate the dimorphic fungi. Enriched medium, such as brain-heart infusion agar with 5 per cent blood, is used for the dimorphic fungi. When incubated at 37°C, this aids conversion from the mycelial to the yeast phase of growth.

Emmons' modification of Sabouraud dextrose agar with a pH of 6·9 has been suggested for the dermatophytes as although they tolerate pH 5·6 they prefer a pH nearer neutrality. Yeast extract is added to the medium to provide growth factors needed by some of the *Trichophyton* spp. and also chloramphenicol (0·05 g/l) and cycloheximide (0·4 g/l) (see page 387 for full formula). The selective agents can be obtained commercially as an antibiotic supplement such as Oxoid SR75. Both chloramphenicol and cycloheximide have the advantage of being stable at autoclaving temperatures. Sabouraud dextrose agar and Emmons' modification can be obtained from commercial firms such as Difco, BBL and Oxoid. **Table 130** summarises the media, incubation temperatures and times for some of the pathogenic fungi.

Using an incubation temperature of 37°C is in itself a selective procedure. Many of the potentially pathogenic fungi, such as *Aspergillus* spp., prefer lower incubation temperatures but fungi that invade animal tissue can tolerate 37°C. This incubation temperature will deter many non-pathogenic fungi.

Table 130. Summary of the most appropriate media for some of the pathogenic fungi.

Fungal pathogen	Sabouraud dextrose agar	Sabouraud dextrose agar + chloramphenicol	Sabouraud dextrose agar + C + C	Emmon's Sabouraud agar + yeast extract and C + C*	Brain heart infusion agar + 5% blood	Brain heart infusion agar + 5% blood and C + C*	Incubation temperature	Incubation time
Aspergillus spp.	✓	✓					37°C	1–4 days
Zygomycetes	✓	✓					37°C	1–4 days
Candida albicans	✓		✓				37°C	1–4 days
Cryptococcus neoformans	✓	✓					37°C	1–2 weeks
Dermatophytes	✓		✓	✓			25°C	2–6 weeks
Dimorphic fungi (yeast phase)					✓	✓	37°C	1–4 weeks
Dimorphic fungi (mould phase)	✓	✓	✓				25°C	1–4 weeks
Fungi causing subcutaneous mycoses	✓	✓					25°C	2–3 weeks

* C + C = chloramphenicol (0.05 g/l) and cycloheximide (0.4 g/l).

The dermatophytes are slow-growing and need a lengthy incubation period so there is the danger of the agar media drying up in conventional plastic Petri dishes. This can be overcome by several methods:

- By taping the plastic Petri dishes. However, as fungi are strict aerobes the tape should be removed and replaced daily to supply the fungus with the required oxygen.
- By pouring the medium, in 20–25 ml volumes, into sterile glass Petri dishes. The greater depth of agar helps to retain the moisture. These can also be taped if necessary.
- The medium can be poured as slopes in 30 ml universal bottles. These are inoculated and incubated with loose caps. The disadvantage is that any suspect fungi growing on the slopes are more difficult to examine or to subculture.

Inoculation of media

Culture media for the isolation of yeasts can be streaked with an inoculum from the specimen as for bacterial cultures. When attempting to isolate a mould, the surface of the agar should be lightly cross-hatched in about five sites with a sterile scalpel, the cuts being at right angles to each other. The specimens, such as small bits of tissue, scabs or hairs, can then be gently pushed into the agar at these cross-hatched areas. If the specimens are merely pushed into the agar surface there is the danger of the agar splitting during incubation.

Subculturing fungal colonies

Yeasts can be subcultured, and pure cultures obtained, in the same manner as for bacteria. With moulds different techniques are needed:

- *If the mould colony is sporing*: a mould colony usually starts to produce spores from the centre outwards and it is often the spores that give a colony its characteristic colour. An inoculating loop can be made slightly moist and sticky by pushing it into a portion of sterile agar, and can then be used to collect spores from the colony. The inoculum is introduced just under the surface of the agar in the centre of the new plate if the fungus is a rapid-grower, or in four segments of the plate if the fungus is a slow-grower or produces small colonies.
- *If a mould colony is not sporing* : the staling phase and then death of a fungal colony starts with the hyphae in the centre of the colony. It is therefore important to subculture hyphae from the edge of a colony. A small block of agar (about 5 mm^2) is cut out of the centre of the subculture plate with a sterile scalpel and the agar block discarded. Using the same scalpel, a similar-sized block of agar is cut, including fungal hyphae, from the edge of the colony to be subcultured. The agar block containing hyphae is transferred to the subculture plate and placed snugly, mould-side up, into the previously cut hole. The cut hyphae will regenerate and grow out from the surface of the block.

Methods for the Examination of the Microscopic Aspects of Fungal Colonies

- **Use of the dissecting microscope**: the culture plate with the mould colony to be examined is placed under the microscope with its lid removed. The pattern of mycelial growth and characteristics of the fruiting heads can be seen. As an example, a greenish, velvety colony is often either an *Aspergillus* or a *Penicillium* species, and the differentiation can usually be made on the morphology of the fruiting head under this low magnification. However, the results should be checked by wet mount preparations that are examined under the light microscope.
- **Wet mount method**: mycology needles are used to remove a small portion of the colony to be studied, including a little of the underlying agar. The sample is taken half-way between the centre and the periphery of the colony. With a yeast colony a little of the growth can be taken with an inoculating loop. The fungal growth is transferred to a drop of lactophenol cotton blue (LPCB) stain (*see* **Appendix 1** for the formula) on a microscope slide. A coverslip is applied and pressure used directly over the colony fragment (an eraser on the end of a pencil is useful for this) to spread out the hyphae and other structures. The fungal structures are stained by the LPCB dye. The preparation is examined under the low and high-dry objectives of the light microscope. The disadvantage of this method is that it is difficult to preserve the continuity between the spores, fruiting structures and hyphae. Due to the rigorous treatment these delicate structures receive during the preparation of the wet mount disintegration may occur. This can be partially overcome by the 'sticky-tape' method and to a greater extent by the use of the slide culture technique.
- **Sticky tape preparation**: clear tape must be used such as Sellotape©, Scotch© brand tape No. 600 or an equivalent. Frosted tape is unsuitable. A 6 cm length of 2 cm wide tape is taken between the thumb and middle finger, with the index finger in the centre of the loop that is held sticky side downwards. The adhesive side is then pressed firmly down, with the index finger, on the centre of the colony to be examined. The fruiting heads and spores stick to the tape and can be gently pulled

from the mat of mycelium. The inoculated tape is placed over a drop of LPCB on a microscope slide. The tape is pulled taut and the free sticky ends folded over each end of the slide. The tape acts as a cover slip and the preparation can be examined under the light microscope.

- **Slide culture technique** : although this method is not suitable for making a rapid diagnosis, it is excellent for demonstrating the fruiting heads and attachment of spores in a relatively undisturbed state. This may be necessary for the definitive identification of some fungal pathogens.

Slide Culture Technique

- A bent glass rod is placed in a glass Petri dish with a circular piece of filter paper at the bottom. A microscope slide and a coverslip (22 mm^2) are also placed in the Petri dish. This is then wrapped in paper and autoclaved.
- A small block of agar is cut from an Emmons' Sabouraud dextrose agar plate previously poured to a depth of about 4 mm. The block (18 mm^2) can be cut with a sterile scalpel or with the open end of a sterile test tube (18 x 150 mm) pushed into the agar. The block can be expelled from the test tube by gently heating the tube, causing expansion of the enclosed air. The block of agar is placed in the centre of the microscope slide in the Petri dish. The slide itself is raised from the bottom of the dish by the glass rod. All procedures must be carried out aseptically.
- Spores or a small portion of the fungal colony to be studied are inoculated into four points in the side of the agar block. The sterile cover slip is then placed on top of the agar block with sterile forceps (**424**).
- The circular filter paper at the bottom of the dish is moistened with sterile distilled water and the lid replaced. The preparation is incubated at the optimum temperature for the fungus. The fungus should grow on the inoculated sides of the agar block and under the edges of the coverslip. Examine regularly to make sure the preparation does not become dry and to judge when sufficient growth has occurred and fruiting bodies have formed at the edge of the cover slip.
- The coverslip is removed from the agar block and placed fungal side down, on a drop of LPCB on a microscope slide. Gentle pressure is applied to the coverslip and excess stain removed. For a semi-permanent mount the coverslip can be generously 'ringed' with a clear nail polish.
- If the fungus has grown down from the agar block onto the microscope slide, this can serve as a second mount. The agar block is removed and discarded, a drop of LPCB placed on the area of fungal growth and a coverslip applied. It is examined under the light microscope.

424

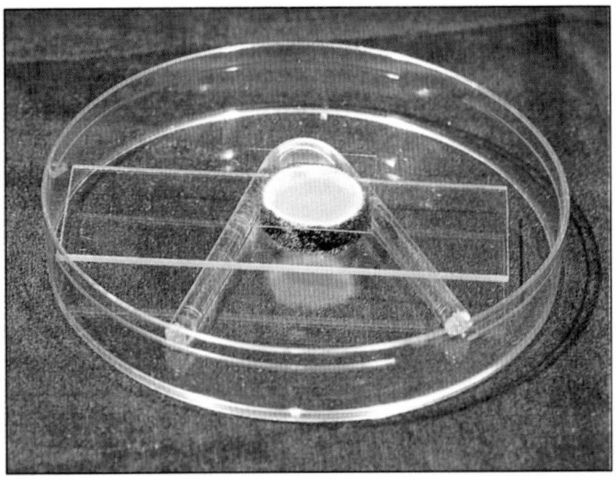

424 Slide culture technique: agar plug inoculated with *Aspergillus niger.*

425

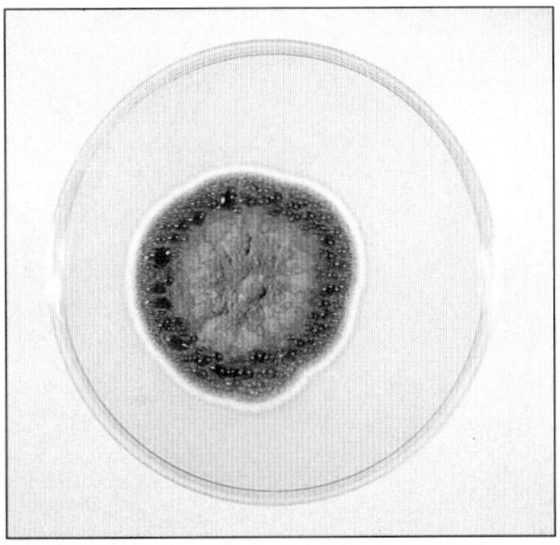

425 *Penicillium* sp. on Sabouraud agar, 6 days.

Identification of the Pathogenic Fungi

The specific methods for the identification of individual pathogenic fungi are given in the relevant mycology chapters. In general, more emphasis is placed on morphological structures than is the case when identifying bacteria. The procedures include:

- The direct microscopic appearance of the fungus in the clinical specimen.
- The morphology of the colony and the type of pigmentation.
- The microscopic appearance of the fruiting heads and spores from mould colonies and the morphology of the yeasts and the type of budding.
- Biochemical tests can be used for yeasts and, to a more limited extent, to identify some moulds.
- Other tests specific to the particular fungus such as the germ-tube test for *Candida albicans*.

426

426 *Penicillium* sp. showing characteristic arrangement of conidiophores, phialides and conidia. Slide culture preparation. (×400).

Serological Tests for Fungal Diseases

This aspect of the diagnosis of the animal mycoses has not received much attention. Serological tests are carried out for some of the mycotic diseases in humans and a few diagnostic reagents are available commercially. The difficulties are the comparative rarity of the diseases themselves, the expense of the serological reagents and the interpretation of the results in animals. Serological tests are mentioned in the relevant mycology sections on specific fungal pathogens, where they are used.

Fungi Commonly Encountered on Laboratory Media

The main significance of these saprophytic contaminants is that they must be recognized and distinguished from pathogenic fungi. Some of these common contaminants are also potential pathogens. A presumptive identification, to a generic level, can often be made on the basis of the colonial appearance and the microscopic characteristics of the fruiting heads and spores (**425**, **426**). **Diagram 45** shows the microscopic aspects of some of the commonly encountered fungi. **Table 131** relates to the diagram and briefly describes the colonial and microscopic characteristics of the fungi that are illustrated. **Table 132** explains some of the commonly used mycological terms.

Safety Aspects in Mycology

Many of the fungi causing disease in animals are also pathogenic for humans. Great care should be exercised when handling material and cultures suspected of containing pathogenic fungi, particularly *Coccidioides immitis* which produces highly infective arthrospores at 25°C and 37°C that easily form an aerosol. *Cryptococcus neoformans* and the dimorphic fungi, such as *Blastomyces dermatitidis*, can cause serious disease in man. Ideally an approved biological safety hood should be used for all mycological procedures.

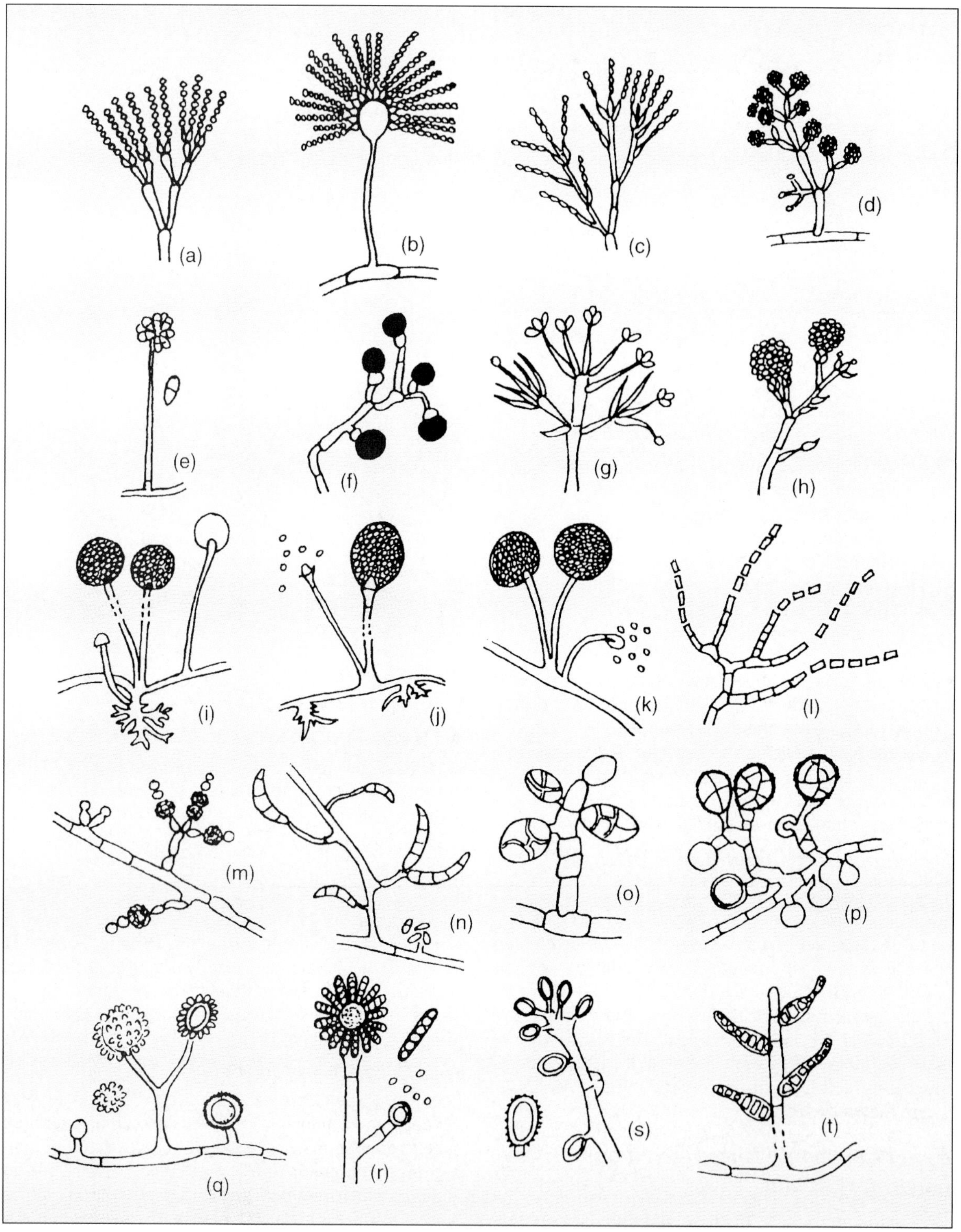

Diagram 45. Microscopic aspects of some genera of the common contaminating fungi.
a) Penicillium ***b) Aspergillus*** ***c) Paeciliomyces*** ***d) Trichoderma*** ***e) Trichothecium***
f) Nigrospora ***g) Verticillium*** ***h) Gliocladium*** ***i) Rhizopus*** ***j) Absidia***
k) Mucor ***l) Geotrichum*** ***m) Scopulariopsis*** ***n) Fusarium*** ***o) Stemphylium***
p) Epicoccum ***q) Sepedonium*** ***r) Syncephalastrum*** ***s) Sporotrichum*** ***t) Helminthosporium***

Table 131. Colonial and microscopic characteristics of the common contaminating fungi (illustrated in **Diagram 45**).

(a) *Penicillium* spp.
Colony: usually blue-green and velvety, occasionally species have other colours and textures
Microscopic: brush-like arrangement of fruiting head. Conidiophores have secondary branches (metulae) bearing whorls of phialides from which the smooth or rough and round conidia (2.5–5.0 μm) are borne

(b) *Aspergillus* spp.
Colony: often bluish-green and velvety but shades of yellow, brown or black occur depending on the species
Microscopic: conidiophore unbranched and rising from a foot cell. A swollen vesicle is produced at the tip of the conidiophore and from this arise the phialides, or metulae and then phialides. The latter produce chains of round conidia (2–5 μm)

(c) *Paeciliomyces* spp.
Colony: flat surface, powdery or velvety, yellowish-brown or light pastel shades of pink, violet or grey-green
Microscopic: resembles *Penicillium* but the phialides are more elongated and taper into a long slender tube. The conidia are elliptical or oblong and occur in chains

(d) *Trichoderma* spp.
Colony: greenish and cottony, becoming powdery with a lawn-like growth
Microscopic: conidiophores are short and often branched at wide angles. Phialides are flask-shaped and the conidia are clustered at the tips of the phialides

(e) *Trichothecium* spp.
Colony: pink or orange-pink and woolly
Microscopic: conidiophores are long and unbranched. Conidia (8–10 x 12–18 μm) are smooth, two-celled and pear or club-shaped. They are produced in alternating directions and remain side by side. The conidia are similar to those produced by *Microsporum nanum*

(f) *Nigrospora* spp.
Colony: compact and woolly, white at first but black areas appear due to the production of black, globose conidia. The reverse is black
Microscopic: short conidiophores that swell and then taper to the point of conidia formation. Conidia are large (10–14 μm), black, round but slightly flattened

(g) *Verticillium* spp.
Colony: powdery or velvety, white becoming brown, pink, yellow, red or green, and spreading.
Microscopic: conidiophores are simple or branched and in whorls. Phialides are very elongated with a pointed apex; while conidia are oval to cylindrical, single-celled, and appear in clusters at the ends of phialides but are easily disturbed.

(h) *Gliocladium* spp.
Colony: white at first, centre becoming green. Fluffy to granular and lawn-like
Microscopic: conidiophores and phialides are similar to *Penicillium* but the conidia from several phialides clump together to form large green balls

(i) *Rhizopus* spp.
Colony: fills Petri dish in 5 days with dense grey, woolly mycelium. Sporangia can often be seen as small black dots
Microscopic: rhizoids are nodal. Sporangiophores are long, usually branched, and terminate in dark round sporangia (60–350 μm). Stolons connect the groups of sporangiophores

(j) *Absidia* spp.
Colony: rapid growth of woolly, white-to-grey mycelium filling the Petri dish
Microscopic: rhizoids are internodal and not very obvious. Sporangiophores are often branched and widen (forming an apophysis) just below the columella. Sporangia are pear-shaped (20–120 μm). A short collarette often remains when the sporangial wall disintegrates

(k) *Mucor* spp.
Colony: rapid spread of colony over the surface of the agar but growth is low. White turning grey or yellowish
Microscopic: no rhizoids. Sporangiophores often branched and bear round sporangia (60–300 μm)

(continued)

Table 131. Colonial and microscopic characteristics of the common contaminating fungi (illustrated in **Diagram 45**). *(continued)*

(l) *Geotrichum* spp.

Colony: whitish, flat, moist and yeast-like with a granular surface. Some strains produce short, white, cottony aerial hyphae

Microscopic: the septate mycelium fragments into arthrospores (arthroconidia), which are formed consecutively and become round. No blastoconidia are produced

(m) *Scopulariopsis* spp.

Colony: white and glabrous, becoming powdery and some shade of yellow, buff or brown

Microscopic: fairly short conidiophores and the conidia-bearing cells can be cylindrical. Conidia are comparatively large (4–9 μm), thick-walled and spiny when mature

(n) *Fusarium* spp.

Colony: tendency to produce delicate rose-pink, lavender or purple pigments seen on obverse and reverse of the colony

Microscopic: produces large (3–8 × 11–70 μm), banana-shaped macroconidia from phialides, and simple conidiophores can bear small (2–4 × 4–8 μm), one- to two-celled conidia singly or in clusters resembling those of *Acremonium* spp.

(o) *Stemphylium* spp.

Colony: black, almost yeast-like colony with white aerial hyphae

Microscopic: conidiophores are simple, occasionally branched, and bearing muriform conidia

(p) *Epicoccum* spp.

Colony: yellow to orange at first, becoming black with age. A diffusible pigment (yellowish or red) may colour the agar

Microscopic: clusters of conidiophores bear round to pear-shaped muriform conidia (15–30 μm), which are dark brown and warty

(q) *Sepedonium* spp.

Colony: white and waxy, becoming fluffy and yellow with age

Microscopic: simple or branched conidiophores bearing large (7–17 μm), round, rough, knobby conidia. Rather similar to *Histoplasma capsulatum* but no microconidia are produced and *Sepedonium* spp. do not convert to a yeast form

(r) *Syncephalastrum* spp.

Colony: rapidly fills the Petri dish with a white cottony growth that turns dark grey with age

Microscopic: an occasional septum is seen. Short branched sporangiophores have swollen tips bearing chains of spores enclosed in tubular sporangia (4–9 × 9–60 μm). Rhizoids can be present

(s) *Sporotrichum* spp.

Colony: velvety to granular surface, white becoming tan, pinkish, yellow or orange. Reverse is tan

Microscopic: short, simple conidiophores that bear single-celled, ovoid, yellow conidia. The conidia tend to retain a portion of conidiophore after separation. Clamp connections present at the septa

(t) *Helminthosporium* spp.

Colony: cottony and dark grey to black. Reverse is black

Microscopic: unbranched conidiophores that are brown, slightly curved, with conidia forming along the sides. The latter are large, dark, multi-celled and club-shaped

Table 132. A short glossary of mycological terms.

Adiaspores: spores that increase in size without replication.

Aerial hyphae: hyphae above the agar surface. They often produce fruiting structures.

Aleuriospore: a non-deciduous spore that fractures the wall of the hypha, to which it was attached, as it breaks off.

Anthropophilic: describes fungi that usually infect humans only.

Apical: located at the tip of a pointed extremity.

Apophysis: the swelling of a sporangiophore immediately below the columella.

Arthrospore (arthoconidium): an asexual spore formed by the fragmentation of a hypha. The resulting spores can be rectangular, barrel-shaped, or can become rounded.

Ascocarp: a mycelial sac within which are formed asci and ascospores of the Ascomycetes.

Ascospore: the sexual spore of the Ascomycetes.

Ascus: a sac-like structure usually containing eight ascospores that are produced during the sexual reproduction of an Ascomycete.

Basidiospore: the sexual spore of the Basidiomycetes

Blastospore: an asexual spore produced by a budding process along hyphae.

Chlamydospore: a thick-walled, resistant spore formed by the direct differentiation of the mycelium.

Clavate: club-shaped.

Columella: the sterile, inflated end of a sporangium extending into a sporangium.

Conidiophore: specialised aerial hypha bearing conidia.

Conidium (pl. conidia): asexual fungal spores that are abstricted in various ways from a conidiophore.

Dematiaceous: referring to fungi that have dark brown or black hyphae.

Dermatophytes: a fungus belonging to the genera *Microsporum, Trichophyton* and *Epidermophyton* (human), which has the ability to infect skin, hair, nails or claws of animals and man. 'The ringworm fungi'.

Dimorphic: having two morphological forms, such as the dimorphic fungi which have a mould and yeast phase.

Ectothrix: outside the hair shaft.

Endothrix: inside the hair shaft.

Fusiform: spindle-shaped.

Geophilic: fungi whose natural habitat is the soil (literally 'soil-loving').

Germ tube: the tube-like process from a germinating spore that develops into a hypha. In the case of *Candida albicans* it is a tube extruded from the yeast cell that evolves into a pseudohypha.

Glabrous: smooth and usually bare (without aerial hyphae). Term that is used when describing a colony.

Hyaline: colourless or transparent.

Hyphae (sing. hypha): filaments that collectively make up the mycelium of a fungus.

Imperfect state: asexual state. The phase of a life cycle in which there is no sexual reproduction.

Internodes: areas on a stolon between the points where sporangiophores are borne.

Macroconidium (pl. macroconidia): a large conidium – as opposed to a smaller conidium, which is called a microconidium.

Metulae (sing. metula): secondary branches of a conidiophore that support the phialides (*Penicillium* and some *Aspergillus* species).

Microconidium: the smaller of the two types of conidia (see above).

Muriform: the term applied to multicelled conidia divided by both vertical and horizontal septa.

Mycelium: a mat of intertwined and branching hyphae.

(continued)

Table 132. A short glossary of mycological terms. (*continued*)

Mycosis (pl. mycoses): a disease produced by a fungus.

Mycotoxicosis (pl. mycotoxicoses): a disease caused by the ingestion of pre-formed toxic fungal metabolites, produced while the fungus is growing on foodstuffs.

Nodal: denoting the area on a stolon where sporangiophores are borne.

Perfect state: sexual stage.

Phialide: special portion of the conidiophore from which the conidia are borne.

Pseudohyphae (sing. pseudohypha): filaments formed by elongated yeast cells that have failed to separate from each other.

Pyriform: pear-shaped.

Rhizoid: root-like branched hyphae extending into the substrate.

Rugose: wrinkled or folded.

Saprophyte: an organism that obtains nutrients from dead organic matter.

Sclerotium (pl. sclerotia): a hard, compact mass of mycelium. It represents the resting stage of fungi such as *Claviceps purpurea*.

Septate: describes hyphae that are divided by cross-walls.

Sessile: attached directly to the base, without a stalk.

Simple: unbranched.

Spherule: a thick-walled, closed, usually spherical structure enclosing asexual spores.

Sporangiophore: a specialised hyphal branch bearing a sporangium.

Sporangiospore: a spore borne within a sporangium.

Sporangium: a closed structure enclosing asexual sporangiospores produced by cleavage.

Sterigmata: specialised projections borne on a vesicle and producing conidia.

Stolon: a horizontal hypha that sprouts where it touches the substrate and often produces rhizoids and sporangiophores at that point.

Stroma: a cushion-like mat of fungal elements.

Sympodial: describes the growth of a conidiophore after the main axis has produced a terminal conidium. The growth appears twisted and the other conidia are produced below, and to one side, of the previous conidium.

Truncate: cut off sharply or ending abruptly.

Tuberculate: covered with knob-like structures.

Vegetative: refers to hyphae involved in food-absorption as opposed to producing spores.

Verrucose: covered with wart-like projections.

Vesicle: a bladder-like structure.

Zoophilic: fungi that infect animals.

Zygospore: a thick-walled, sexual spore produced through the fusion of two gametangia. Produced by the Zygomycetes.

39 The Dermatophytes

The dermatophytes are a group of closely related fungi that utilize keratin for growth. They tend to be confined to the superficial integument including the outer stratum corneum of the skin, nails, claws and hair of animals and man. The classical lesions are circular and known as ringworm. Traditionally the dermatophytes are placed in the Fungi Imperfecti but the perfect state has been described for some and they are classified as Ascomycetes. *Nannizzia* sp. is the teleomorph for *Microsporum* sp. and *Arthroderma* sp. for any *Trichophyton* species that are found to have a sexual state.

Over 38 species of dermatophytes are known. Those affecting animals are placed in one of two genera, *Microsporum* or *Trichophyton*. The dermatophyte species affecting animals are known as ectothrix as the septate hyphae invading the skin and hairs fragment into arthrospores and these form a sheath around the infected structures. Macroconidia and microconidia are produced in laboratory cultures. The *Microsporum* spp. tend to produce spindle (**427**) or boat-shaped (**428**) macroconidia whereas those of the *Trichophyton* spp. are usually elongated, cigar-shaped with almost parallel sides (**429**). The macroconidia of *M. nanum* are unique in being pear-shaped and usually two-celled (**430**). **Diagram 46** gives the main points of differentiation between the two genera. The colonies of many of the dermatophytes are pigmented and accordingly both the obverse and reverse of the colonies should be examined.

427

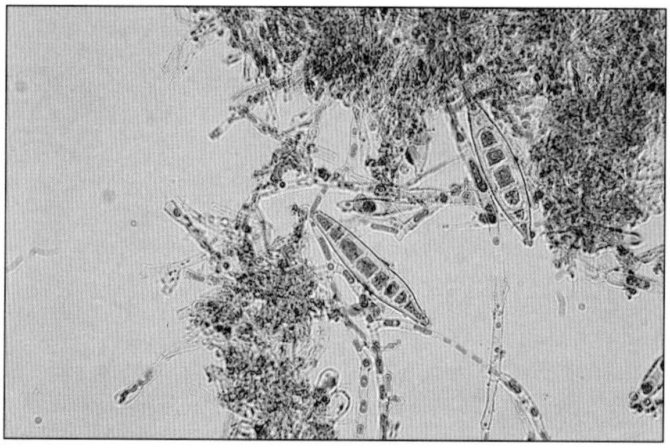

427 *Microsporum canis* : spindle-shaped macroconidia. [Lactophenol Cotton Blue (LPCB), ×400]

428

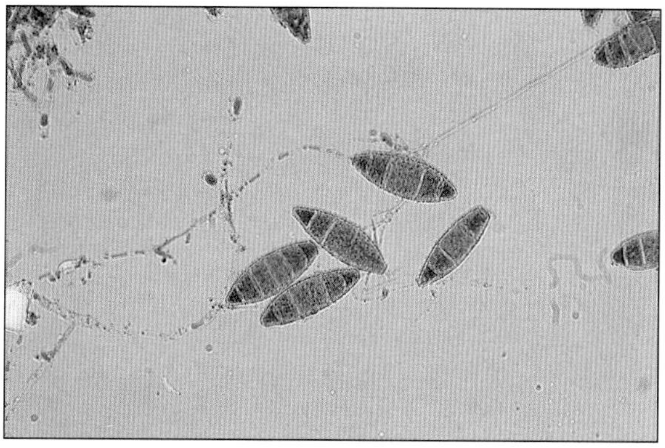

428 *M. gypseum* : numerous, boat-shaped macroconidia. (LPCB, ×400)

429

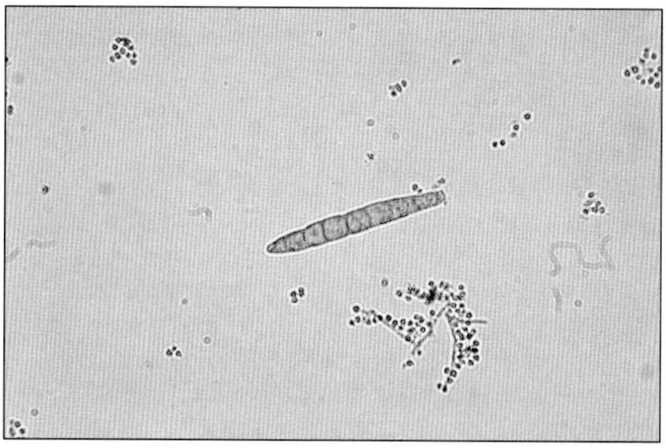

429 *Trichophyton mentagrophytes* : numerous microconidia and a cigar-shaped macroconidium. (LPCB, ×400)

430

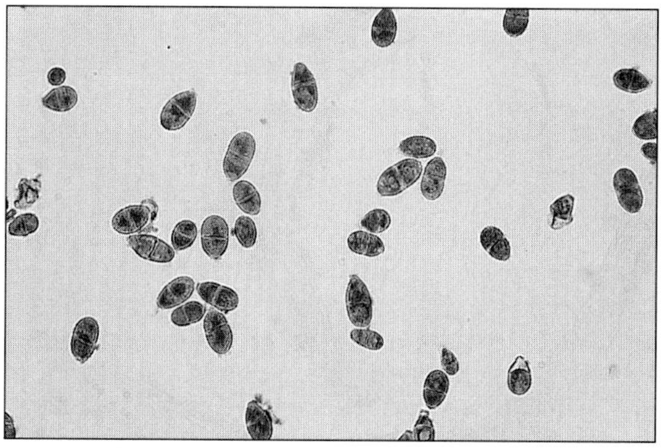

430 *M. nanum:* macroconidia that are characteristically pear-shaped and two-celled. (LPCB, ×400)

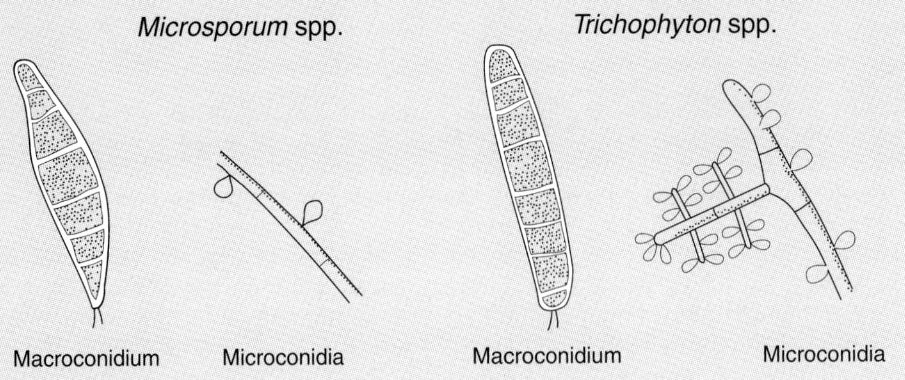

	Microsporum spp.	*Trichophyton* spp.
Macroconidia	Large thick-walled and divided into many cells by transverse septa. Tend to be spindle or boat-shaped.	Few or absent in some species. If present they are elongated and cigar or pencil-shaped. The walls are thin and smooth. Divided by septa into 3-8 cells.
Microconidia	Relatively few or absent. If present they are tear-shaped and borne singly on the hyphae.	Usually numerous and borne singly along the hyphae or in grape-like clusters.

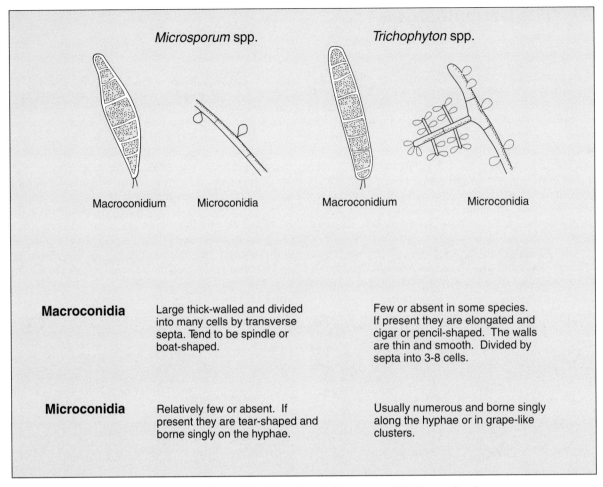

Diagram 46. Microscopic differentiation of the dermatophyte genera affecting animals.

Natural Habitat

The geophilic (soil-loving) dermatophytes inhabit the soil and can exist there as free-living saprophytes. *Microsporum gypseum* and *M. nanum* are examples of geophilic dermatophytes that can also cause lesions in animals and in man. The zoophilic dermatophytes are primarily parasites of animals, although they constitute zoonoses for humans. Humans are the main host for the anthropophilic dermatophytes and these very rarely cause ringworm in animals and are not considered in this chapter. Some dermatophytes have become adapted for survival in the skin of specific host animals, for example:

- *Microsporum canis* : cats
- *Microsporum persicolor* : voles
- *Trichophyton erinacei :* European hedgehogs
- *Trichophyton mentagrophytes* : rodents
- *Trichophyton verrucosum* : cattle.

These dermatophytes are able to cause subclinical or inapparent infections in the host animal, although they can also produce clinical lesions. Most of the dermatophytes causing lesions in animals are also capable of producing ringworm in humans.

Pathogenesis

The ability of the dermatophytes to hydrolyse keratin may cause some damage to the epidermis and hair follicles. The mechanism by which they cause lesions is through a hypersensitivity reaction, in the host, to the fungal metabolic products. The host mounts an inflammatory response that is harmful to the fungus, so the dermatophyte moves away peripherally towards normal skin. The result is the commonly seen circular lesions of alopecia with healing at the centre and inflammation at the edge (**431**). There appears to be a balanced host-parasite relationship where a dermatophyte has become adapted to a specific host animal. While these animals may not show actual lesions they

431

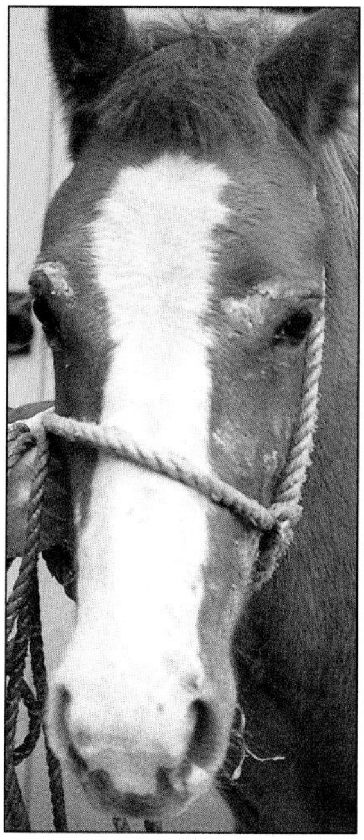

431 *T. equinum*: typical ringworm lesions in a horse (bridle area).

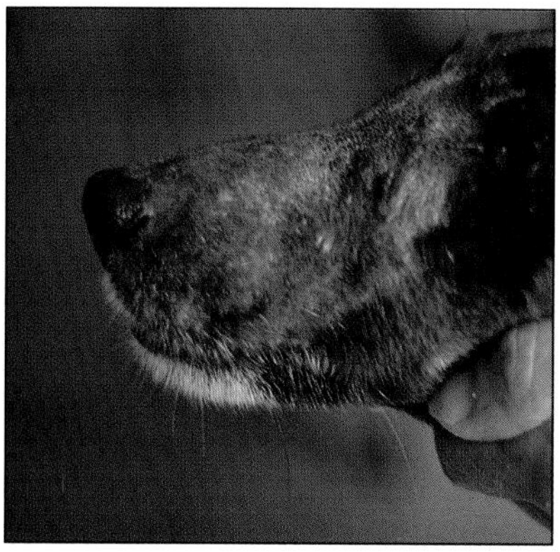

432

432 *T. erinacei* : muzzle alopecia in a terrier known to worry hedgehogs.

can act as a reservoir of infection. The manifestations of dermatophyte infections can vary and may be summarised as:

- Subclinical or inapparent infections.
- Classical round ringworm lesions.
- Serious generalised lesions that may be complicated by mange mites or by secondary bacterial infection, often by *Staphylococcus aureus* or *S. intermedius*.
- Nodular or tumourous lesions called kerions, seen most commonly in dogs.

Epidemiology

The important dermatophytoses of animals are perpetuated by infected animals of the same species with direct contact being the common mode of transmission. However, infection can also occur from reservoirs such as rodents (*T. mentagrophytes*), hedgehogs (*T. erinacei*) (**432**) soil (*M. gypseum*) and fomites such as bedding, grooming gear and harness containing hairs with infective arthrospores. These arthrospores can remain viable on shed hairs and skin particles for at least 6–12 months. The reservoir, transmission and site of the lesions can often be related to the actual dermatophyte involved. This is especially evident in the horse and the dog as indicated in **Table 133** and is an argument in favour of attempting to isolate and identify the dermatophyte in an infection. **Diagram 47** lists the main dermatophytes affecting animals, the main hosts of each and their geographical distribution.

Laboratory Diagnosis

Preliminary examination : Wood's Lamp.

M. canis, *M. distortum*, *M. audouinii* (human) and *M. ferrugineum* (human) produce certain metabolites when growing on hairs and skin that will fluoresce a vivid apple-green under ultraviolet light (366 nm) of a Wood's Lamp. The animal itself can be examined with the lamp in a dark room and the site of the lesions will fluoresce. The technique is particularly useful for suspected inapparent infections in kittens that almost always involve *M. canis*. The infected areas are often the face, front paws and abdominal area of these kittens. Alternatively, the lamp can be used to examine plucked hairs or skin scrapings taken from lesions. It is estimated that about 50 per cent of *M. canis* infections give this fluorescence, so negative cases should always be submitted for further laboratory examinations. If the owner has applied a topical ointment to the lesion, this can sometimes lead to spurious fluorescence.

Specimens

The following points should be noted:

- Hairs should be plucked from the lesions, never cut with scissors, as the basal portion of the hairs often contain the most useful diagnostic material. Any stubby or damaged-looking hairs that may be present should be collected.
- Scab material should be obtained from the edge of the lesion as this is the site where the dermatophyte is most likely to be viable. A blunt scalpel blade is used to scrape until blood is just drawn. The scrapings and the scalpel blade should be submitted with adherent material. This specimen will also be useful for detecting mange mites, if present.

Dermatophyte, hosts and distribution	Colonial and Microscopic Appearance	Macroconidium
Microsporum canis Worldwide Dogs, cats (humans). Less common in other animals.	Growth rapid. Surface white and silky at centre with bright yellow periphery. Reverse side bright yellow or orange. Some strains are poor producers of macroconidia but they are numerous in others. They are spindle-shaped and mature spores end in a distinct knob. Cells 6 - 15, Size 8 - 20 µm x 40 - 150 µm.	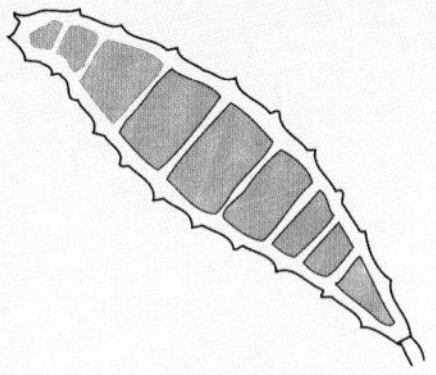
Microsporum distortum Australia, New Zealand and North America. Dogs and less commonly in cats.	Growth fairly rapid. Surface white to tan and reverse white or yellowish tan. Colony is velvety to fluffy with a tendency to form radial grooves. Usually abundant macroconidia, they are distorted, thick-walled and multicellular. Size 12 - 27 x 30 - 60 µm.	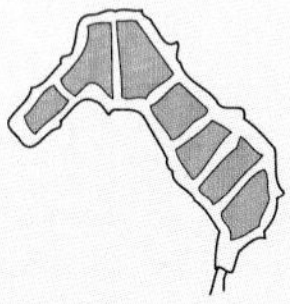
Microsporum gallinae Europe, North and South America. Chickens and turkeys.	Colonial type can vary: 1. Velvety with irregular folds over entire surface. Obverse pink and reverse strawberry-pink with diffusion into the agar. 2. White and fluffy with fewer folds and reverse orange-pink. Abundant macroconidia, fusiform with blunt spatulate tips, often curved. Walls smooth and thick, 2 - 10 celled. Size 6 - 8 x 15 - 50 µm.	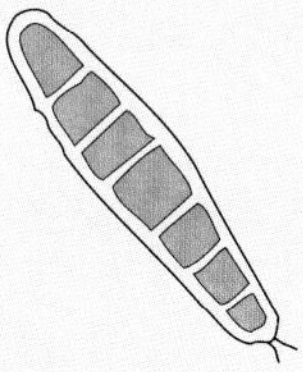
Microsporum gypseum Worldwide in soil. Rodents, horses and dogs. Less common in other animals and in humans.	Fairly rapid growth. Colony is flat, powdery with a fringed border. Obverse is buff to cinnamon-brown and reverse pale yellow to tan (occasionally reddish in a few strains). Has an odour similar to that of a mouse colony. Macroconidia very abundant, boat-shaped with rounded ends and walls thick and rough, 4 - 6 celled. Size 8 - 12 x 30 - 50 µm.	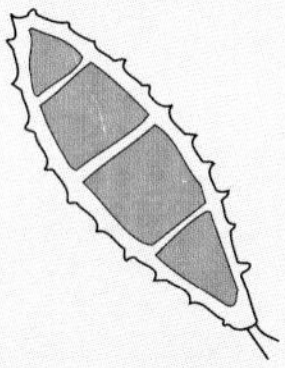
Microsporum nanum Has been found in soil. Europe, North and South America, Australia and New Zealand. Pigs, rarely in other animals and humans.	Colony is flat, white and cottony at first, later granular and buff-coloured. Reverse is orange, becoming reddish-brown. Abundant and characteristic macroconidia, pear-shaped with thick spiny walls. Usually 2 celled but occasionally 1 or 3 celled. Size 4 - 8 x 12 - 18 µm.	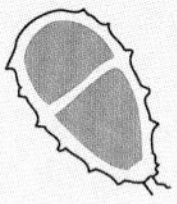
Trichophyton equinum var. autotrophicum Australia and New Zealand. Horses.	Colony at first fluffy, white with a raised centre, later white to buff with folded centre. Reverse is yellow becoming dark rose-red. Differs from *T. equinum* in not requiring nicotinic acid for growth. Macroconidia not reported.	

Diagram 47. Dermatophytes of animals with their cultural and microscopic characteristics.

Dermatophyte, hosts and distribution	Colonial and Microscopic Appearance	Macroconidium
Trichophyton equinum Worldwide. Horses, occasionally other animals and humans.	Growth comparatively rapid, colony initially flat, white and fluffy but later velvety with central folding. Cream to tan in colour, reverse is yellow to reddish-brown. Macroconidia are rare. If present they are slightly club-shaped, smooth, thin-walled with 3 - 5 cells. Variable size and shape. Chlamydospores are abundant in old cultures.	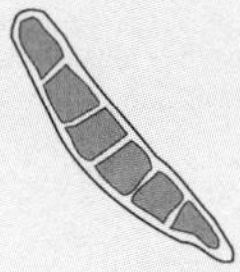
Trichophyton erinacei England, France and New Zealand. European hedgehogs and dogs.	Colony flat with raised centre and finely granular. Fringed sub-surface border, obverse white to cream and the reverse is a brilliant yellow. Macroconidia sparse. If present they have an irregular shape and size, smooth-walled and 2 - 6 celled. Intermediate forms between macro- and microconidia are seen. Thought to be a variant of *T. mentagrophytes*.	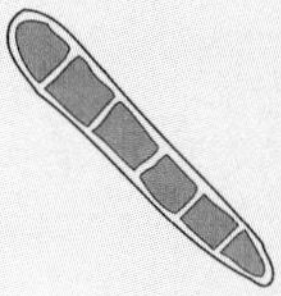
Trichophyton mentagrophytes Worldwide. Rodents, dogs, horses and many other animals and humans.	Two colony forms occur, the granular (most common from animals) and the downy type. 1. Finely or coarsely granular, cream to light buff obverse and reverse varying from buff-tan to dark-brown. 2. The downy type is woolly and white with older colonies becoming cream-tan. Reverse varies from white through yellow to reddish-brown. Macroconidia comparatively abundant from the granular type colony. Cigar-shaped, thin-walled, 3 - 7 celled and size 4 - 8 x 20 - 50 µm.	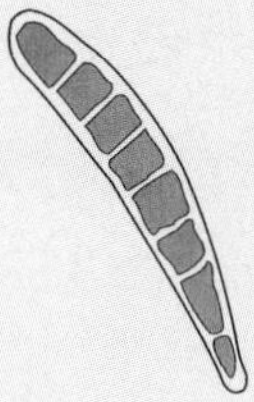
Trichophyton mentagrophytes var. quinckeanum Unknown distribution. Mice.	Colony initially white and fluffy, becoming downy and deeply folded. The reverse is deep yellow becoming orange-brown. Macroconidia are rare. Smooth, thin-walled, cigar-shaped to club-shaped and 4 - 6 celled.	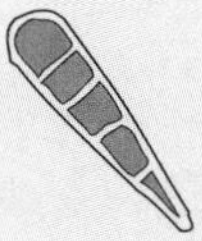
Trichophyton simii Soil in India. Endemic in Brazil, Guinea and India. Monkeys and poultry.	Finely granular colony with diffuse margin, white to pale or rose-buff with folding in some strains. Reverse white and later reddish-brown. Macroconidia abundant, cylindrical to fusiform, thin-walled, smooth, 3 - 10 celled and size 6 - 11 x 35 - 85 µm. Formation of endochlamydospores in older macroconidia.	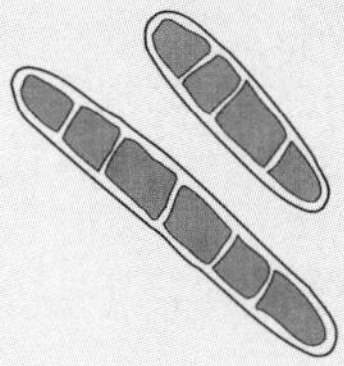
Trichophyton verrucosum Worldwide. Cattle, occasionally sheep, horses, humans.	Slow growing, small, white, velvety, heaped and folded colony. Colour varies from white or whitish-grey and occasionally yellowish-ochre. Reverse is white. Macroconidia extremely rare but chlamydospores forming chains are characteristic of this species.	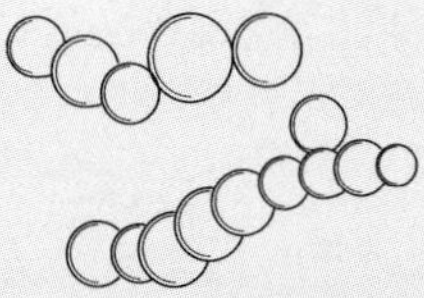 Chlamydospores

Diagram 47. Dermatophytes of animals with their cultural and microscopic characteristics (*continued*).

Table 133. Dermatophytes affecting dogs and horses.

Dermatophyte	*Reservoir/activity*	*Site of lesions*
DOG		
Microsporum canis	Other infected dogs or cats	Anywhere on the body
M. gypseum	Compulsive burying of objects in soil	Muzzle above the nose
Trichophyton erinacei	Avid "hedgehog-worriers"	Muzzle, face, front paws and legs
T. mentagrophytes	Good rat catchers	Muzzle, face, front paws and legs
HORSE		
M. canis	Stable cats (inapparent infection?)	Anywhere on the body
M. gypseum	Horses at pasture: rolling in soil	Back-line and sides of body
T. equinum	Harness, saddle blankets, etc.	In harness areas
	Grooming gear	Anywhere on the body

Specimens *(continued)*

- A paper envelope can be held under the lesion when scrapings are being taken to catch the scab material, scurf and damaged hairs. The specimens should be submitted to the laboratory in the envelope (inside additional wrapping) as they tend to remain drier and less contaminated in paper, compared to a glass or plastic container.
- Scrapings and clipping from claws should be taken from as near the base as possible.
- In suspected inapparent infections, if the site of the dermatophyte infection has not been disclosed by a Wood's Lamp examination, the animal should be brushed, using a brush that can be sterilised or discarded after use. The hairs and scurf are collected in a container held under the animal.
- Where the specimens tend to be very contaminated by bacteria and saprophytic fungi, particularly in the case of pigs, consideration might be given to wiping the lesions with 70 per cent alcohol and then allowing the area to dry thoroughly before collecting the specimens.

433

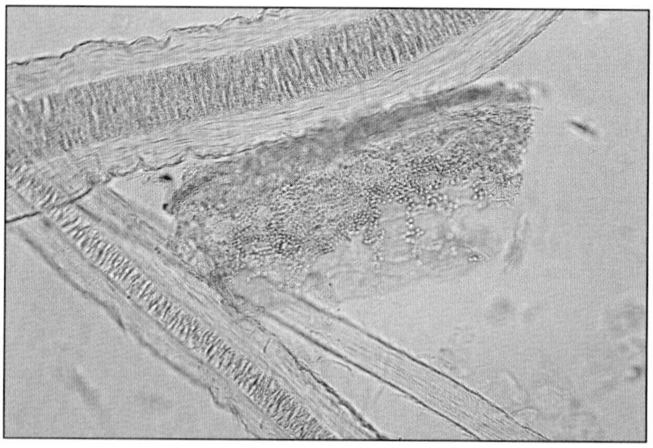

433 *T. verrucosum*: infected bovine hair with arthrospores. (10% KOH wet preparation, ×400)

Direct Microscopy

The KOH wet preparation method or modifications (see Introductory Mycology section) is used for hairs, scabs or claw scrapings. The preparation is examined under the low power objective, with the condenser lowered slightly, focusing on any abnormal-looking hairs. The high-dry objective is used to visualise the round, refractile arthrospores surrounding the hair or on pieces of scab material (**433**). Occasionally the septate hyphae of the dermatophyte can be seen forming chains of arthrospores. Care must be taken not to mistake normal skin or hair structures such as fat globules or hair pigment granules (melanosomes) for arthrospores. Arthrospores vary slightly in size depending on the dermatophyte involved; those of *T. verrucosum* are particularly large (about 5–6 µm in diameter) and easy to see. Mange mites, if present, will be also visible in the cleared KOH preparations.

If skin lesions from cattle, sheep or horses are being examined and scab material or scurfy pieces of skin are available, a Gram-stained or Giemsa-stained smear should be made from them and examined for *Dermatophilus congolensis*.

Isolation

Even if either the Wood's Lamp examination or the direct microscopy for arthrospores has proved to be positive, it is still useful to attempt to isolate and identify the dermatophyte for epidemiology and control aspects. A culture medium for dermatophytes must cater for the few species that require specific growth factors. These can be satisfied by the use of commercial media such as those developed for the *Trichophyton* species (Difco):

- *T. verrucosum* : requires thiamine or thiamine and inositol (Bacto-Trichophyton agar 3).
- *T. equinum* : requires nicotinic acid (Bacto-Trichophyton agar 5).
- *M. gallinae* : thiamine stimulates growth (Bacto-Trichophyton agar 3).

A more practical alternative is to prepare a medium on which all the dermatophytes, likely to affect animals, will grow. Such a medium is prepared as follows:

- Emmons' Sabouraud dextrose agar (pH 6.9)
- Yeast extract 2–4 per cent (to supply specific growth factors)
- Chloramphenicol 0.05 g/litre (antibacterial)
- Cyclohexamide 0.4 g/litre (to inhibit some faster-growing fungi).

A light inoculum of hairs and skin scrapings can be scattered over the surface of the agar and gently pressed down on the medium with a swab or sterile forceps. The dermatophyte cultures are incubated aerobically at 25°C. The plates should be examined twice weekly and not discarded as negative for 3 weeks, for most of the dermatophytes. The plates should be kept for 5 weeks in the case of *T. verrucosum*. Some of the more rapidly-growing dermatophytes, such as *M. canis*, may be recognised after 4–6 days' incubation. *T. verrucosum* is an exception among the dermatophytes in tolerating 37°C. When attempting to isolate this dermatophyte duplicate cultures can be made, one plate being incubated at 25°C and the other at 37°C. The plates may be taped to prevent the agar drying, but as the dermatophytes are strict aerobes the tape should be removed and replaced once daily.

Dermatophyte test medium (DTM) that can be obtained commercially (Pitman–Moore, Pfizer, Difco) is a selective and differential medium for dermatophytes containing the pH indicator phenol red. The dermatophytes change the medium from yellow to red. The medium is also changed, but usually more slowly, by some saprophytic fungi, yeasts and bacteria. It is not advisable to use this medium alone for primary isolation as its colour obscures the characteristic pigmentation of dermatophytes that is useful for species identification.

Identification

Usually a dermatophyte can be identified by knowledge of the animal host from which it was isolated, the colonial appearance and the microscopic characteristics of the colonies. If there is any doubt about a particular isolate, a subculture on an agar slant should be submitted to a mycology reference laboratory.

Colonial appearance

Considerations such as rate of growth, texture and pigmentation of the obverse and reverse sides of the colony should be noted. The characteristics of the colonies of dermatophytes commonly affecting animals are given in **Diagram 47** and illustrated (**434–447**).

434

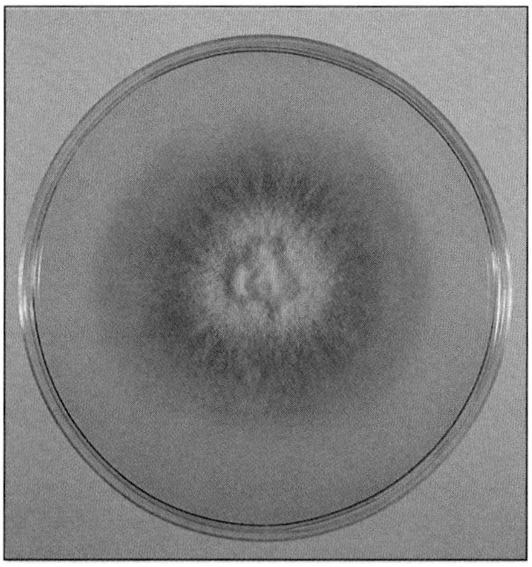

434 *M. canis* on Sabouraud agar, 10 days.

435

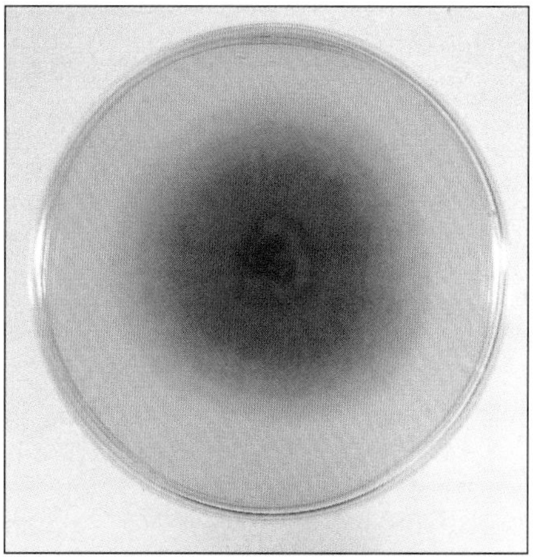

435 *M. canis* on Sabouraud agar, 10 days. Reverse.

436

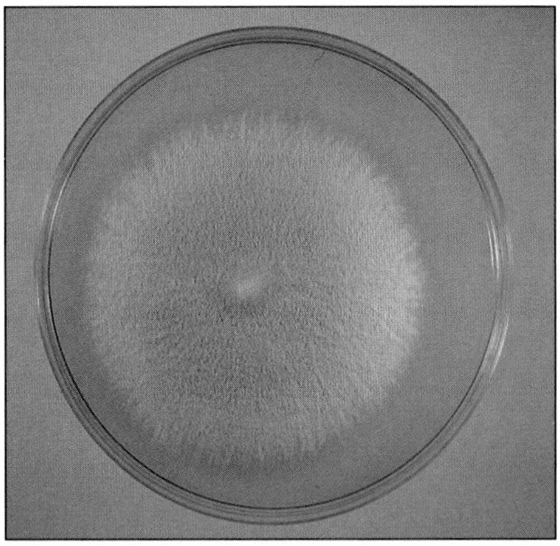

436 *M. gypseum* on Sabouraud agar, 12 days.

437

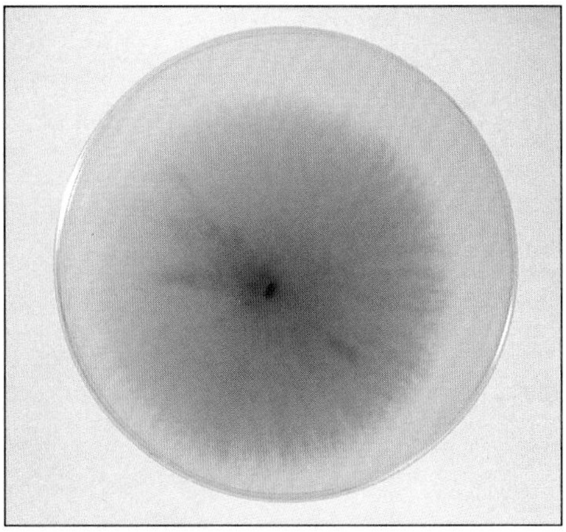

437 *M. gypseum* on Sabouraud agar, 12 days. Reverse.

438

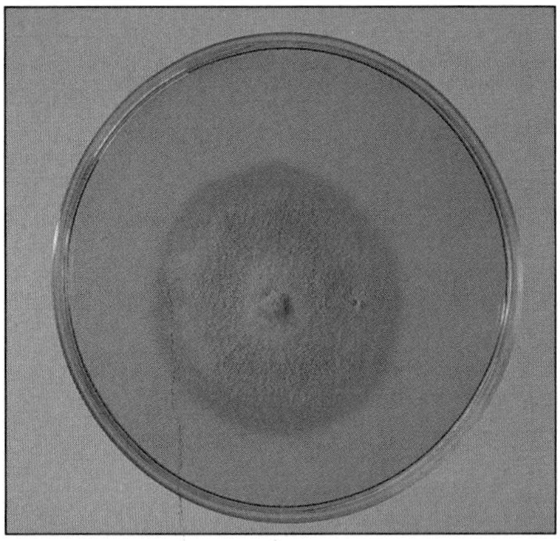

438 *M. nanum* on Sabouraud agar, 13 days.

439

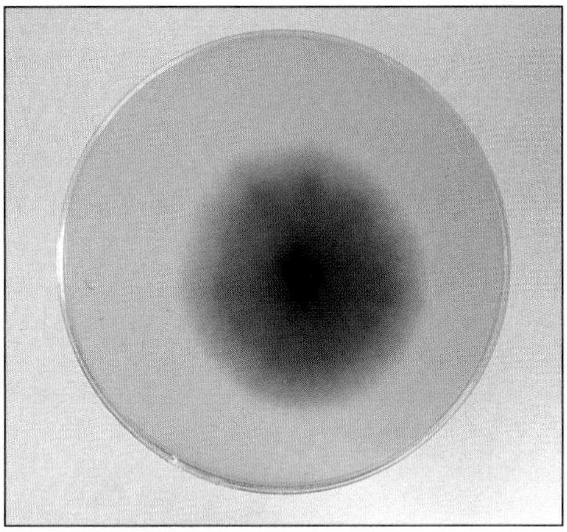

439 *M. nanum* on Sabouraud agar, 13 days. Reverse.

440

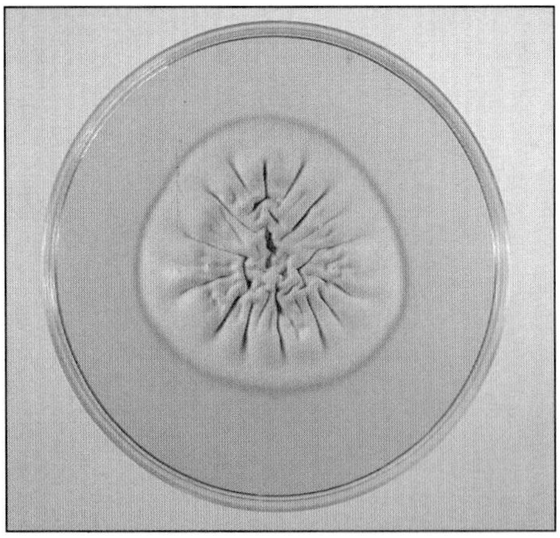

440 *T. equinum* on Sabouraud agar, 15 days.

441

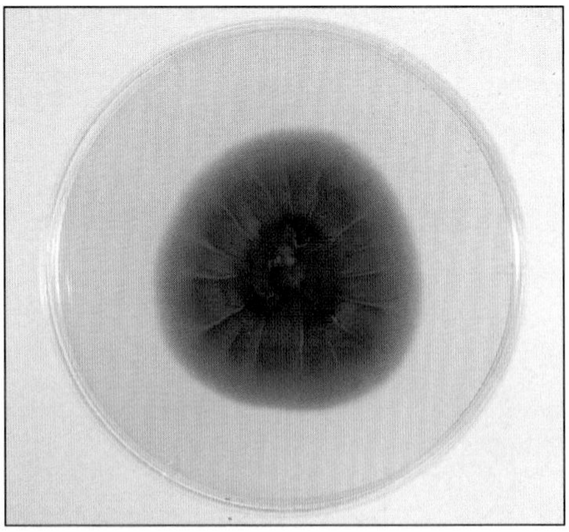

441 *T. equinum* on Sabouraud agar, 15 days. Reverse.

442

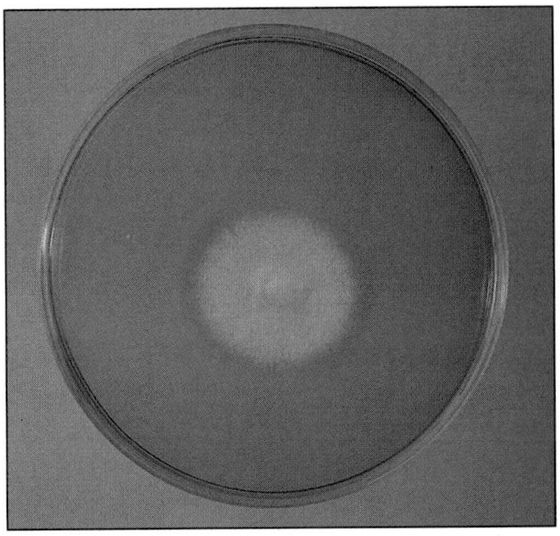

442 *T. erinacei* on Sabouraud agar, 12 days.

443

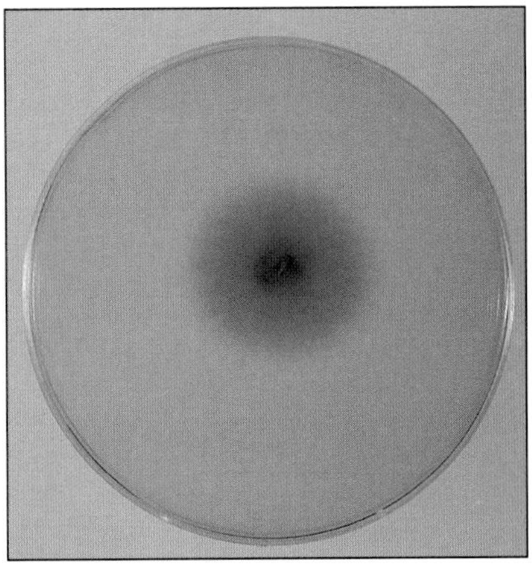

443 *T. erinacei* on Sabouraud agar, 12 days. Reverse.

444

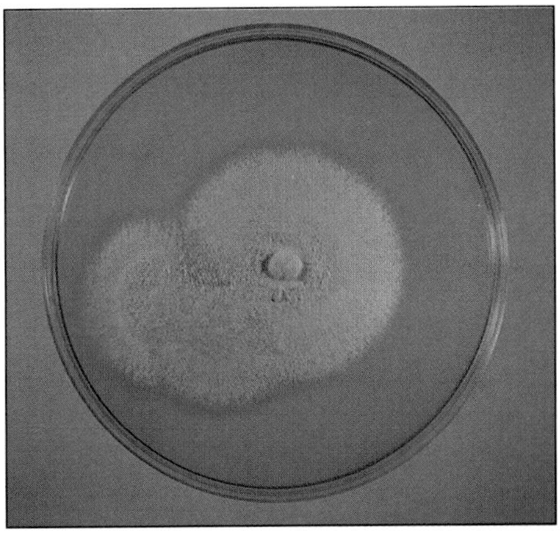

444 *T. mentagrophytes* on Sabouraud agar, 12 days. Granular-type colony.

445

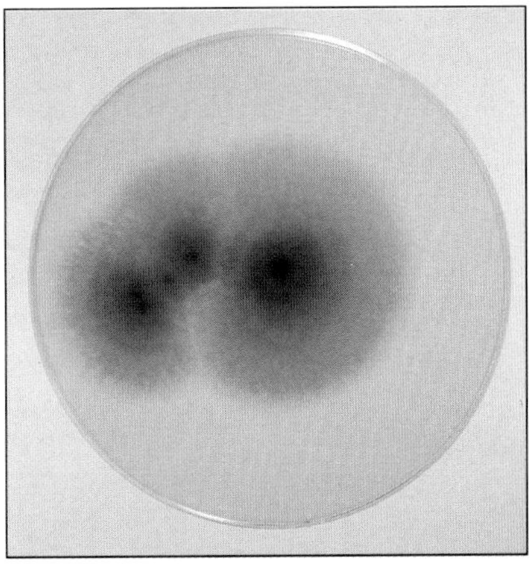

445 *T. mentagrophytes* on Sabouraud agar, 12 days. Reverse.

446

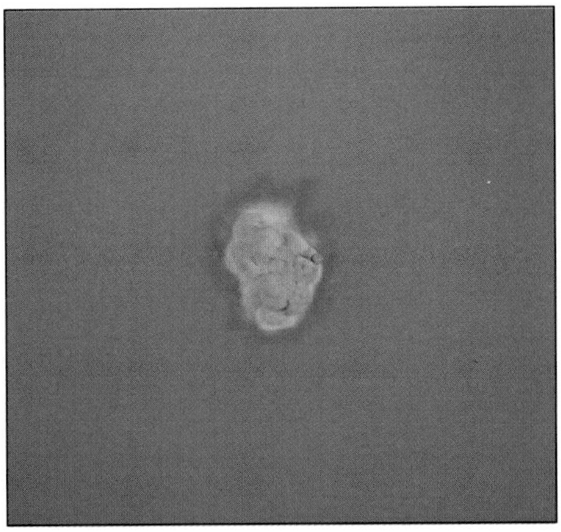

446 *T. verrucosum* on Sabouraud agar, 21 days.

447

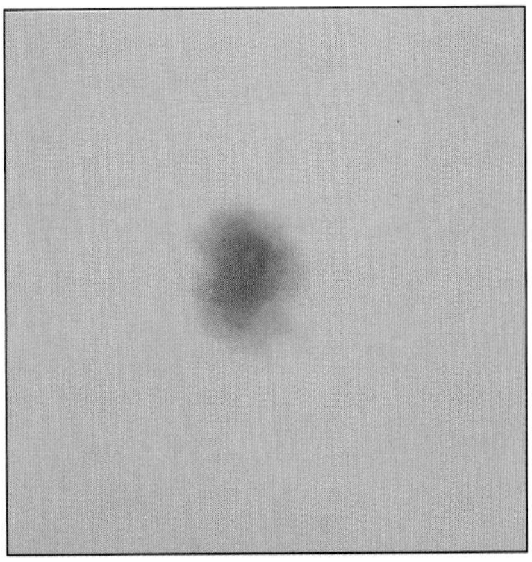

447 *T. verrucosum* on Sabouraud agar, 21 days. Reverse.

Microscopic appearance of the colony

The method for preparing lactophenol cotton blue (LPCB) wet preparations for examination of fungal colonies is given in the Introductory Mycology Section. **Diagram 47** indicates the microscopic appearance of the macroconidia for the dermatophytes commonly affecting animals. The genera *Microsporum* and *Trichophyton* can be differentiated on the appearance of their macroconidia and on other characteristics (**Diagram 46**).

Hair perforation test

This test is used mainly in medical mycology as an aid to distinguish *T. mentagrophytes* from *T. rubrum*. *T. mentagrophytes* has the ability to invade the hair shaft and produce conical perforations of the hair, seen in LPCB preparations as wedge-shaped areas (**448**). *T. rubrum* does not penetrate the hair but grows on the surface.

The conventional method, given in many of the standard medical textbooks, involves adding sterile hairs to a moist piece of filter paper in a Petri dish and adding a portion of the fungal colony to the hairs. However, the authors have found that the method outlined below was more reproducible and convenient:

Hair Perforation Test

- Collect hairs from a young child with fair hair.
- Sterilize the hairs by autoclaving at 121°C for 15 minutes.
- Layer the sterile hairs on a 3–5-day-old subculture of the dermatophyte under test and incubate at 25°C.
- Examine the hairs daily from the seventh day of incubation onwards by mounting a few hairs in lactophenol cotton blue and examining them microscopically, using the low and then the high-dry objective.

448

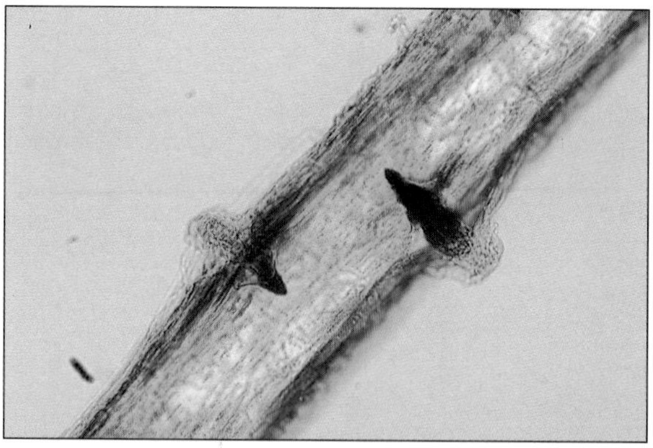

448 *T. mentagrophytes*: illustrating *in vitro* hair-penetration by this dermatophyte seen as wedge-shaped, dark-blue areas. (LPCB, ×400)

40 *Aspergillus* species

Several hundred species of *Aspergillus* have been described and most are harmless saprophytes. It is estimated that *Aspergillus fumigatus* is responsible for 90–95 per cent of aspergillosis infections in animals. Other *Aspergillus* species that occasionally cause infections include *A. niger*, *A. flavus*, *A. terreus* and possibly *A. nidulans*. *A. flavus* is more commonly involved in aflatoxicosis. Most of the aspergilli are classified as Fungi Imperfecti but the perfect state has been found in a few such as *A. nidulans* that can produce ascospores. The *Aspergillus* spp. are rapidly growing moulds with septate hyphae. Many have highly coloured colonies ranging from bluish-green through yellow to black, due to the profuse production of pigmented spores (conidia). The chains of small (2-3 µm) oval or spherical conidia are borne on the tips of phialides radially positioned over the surface of the swollen tip (vesicle) of the aerial hypha (conidiophore) (**Diagram 48**). *Aspergillus* species can cause disease in several ways. They can be invasive, cause mycotoxicoses and are involved in allergic reactions in humans.

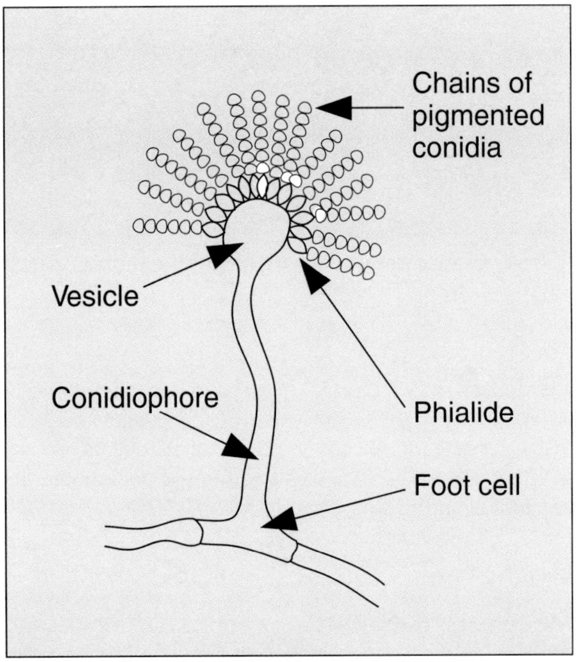

Diagram 48. Fruiting head of an *Aspergillus* species.

Natural Habitat

The aspergilli are ubiquitous, can be isolated from soil, air and vegetation and are worldwide in distribution. They are common laboratory contaminants.

Pathogenesis

A. fumigatus produces haemolysins, proteolytic enzymes and other toxic factors but their role in the pathogenesis of aspergillosis is not known. Infection is acquired from environmental sources, generally by inhalation or ingestion. It is an opportunistic pathogen depending on impaired, overwhelmed or by-passed host defences. In pulmonary infections, suppurative exudates accumulate in the bronchioles. Mycelial growth may extend into blood vessels and dissemination occurs. Granulomas can develop in many body organs and are visible as yellowish-grey nodules. If *Aspergillus fumigatus* breaks out into an air space in the body, such as the air sacs in chicken, distorted fruiting heads can be formed (**449**). **Table 134** indicates the diseases that can be caused in animals and man by *A. fumigatus*.

Laboratory Diagnosis

Specimens

Specimens should include pneumonic lung, granulomatous nodules, centrifuged mastitic milk, foetal lesions, foetal stomach contents, cotyledons, ear swabs, skin scrapings and biopsies from nasal granulomas and fungal plaques in the guttural pouch. Because the aspergilli are ubiquitous, where possible,

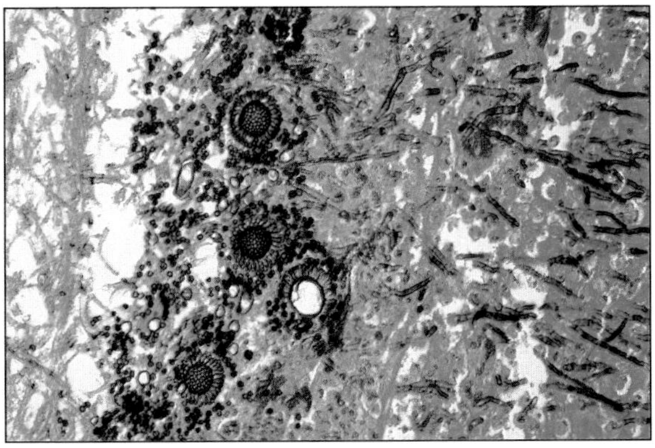

449 *Aspergillus fumigatus:* infected air sac of a swan showing septate hyphae and fruiting heads. (Methenamine silver stain, ×400)

Table 134. Diseases and main hosts of *Aspergillus fumigatus*.

Host(s)	*Disease(s)*
Horses	Guttural pouch mycosis Nasal granulomas Corneal infections Intestinal aspergillosis in foals
Dogs	Otitis externa (often as part of a mixed infection) Chronic rhinitis Occasionally, generalised aspergillosis with thrombosis of vessels
Cattle	Mycotic abortion Mycotic pneumonia Mastitis (abscesses and granulomas in mammary gland) Intestinal aspergillosis in calves
Birds: poultry, wild birds, water fowl and penguins in captivity	Brooder pneumonia (newly-hatched chicks) Pneumonia and air-sac infections (acute and chronic) Generalised aspergillosis (yellowish nodules in body organs)
Many animals	Mycotic pneumonia Superficial infections of skin and cornea Mycotoxicosis (some strains may produce tremorgens)
Humans	Immunocompromised humans and those on prolonged courses of antibiotics are most at risk. Invasive disease can involve lungs, skin, nasal sinuses, external ear, bronchi, bones and meninges. Inhalation of spores can lead to a hypersensitive state

tissue for histopathology should be obtained. For example, in the case of a nasal granuloma, the significance of culturing *A. fumigatus* from nasal discharge would be difficult to interpret, whereas a more confident diagnosis could be made if invading, septate hyphae were seen in a histological section from a biopsy, together with the isolation of *A. fumigatus*. If the biopsied tissue is very small, it can be wrapped in a piece of paper before placing it in the 10 per cent formalin so that it can be more easily found again.

Direct microscopy

Tissue scrapings and other material can be examined after clearing in 10 per cent KOH or by modifications of this method (*see* Introductory Mycology section). Histopathological sections should be prepared and examined wherever possible. These are stained by the PAS stain (**450**) or methenamine silver stain (**449**) and septate hyphae invading the tissue should be noted.

Isolation

Sabouraud dextrose agar, with and without 0.05 g/l chloramphenicol, is used. The aspergilli are sensitive to cycloheximide. The surface of the agar should be cross-hatched to a depth of about 2 mm in 4–5 well-separated areas on the plate. Small pieces of tissue (about one-quarter the size of a finger nail) are placed on the cross-hatched areas and gently pushed into the agar. The inoculated plates for *A. fumigatus*, or other aspergilli causing a systemic mycosis, are incubated aerobically at 37°C for up to 5 days. The colonies usually appear in 2–3 days. *A. fumigatus* can tolerate temperatures up to 45°C but some of the other aspergilli, that are occasionally involved in superficial mycoses, may not grow at 37°C, so an additional plate could be inoculated and incubated at 25°C.

450

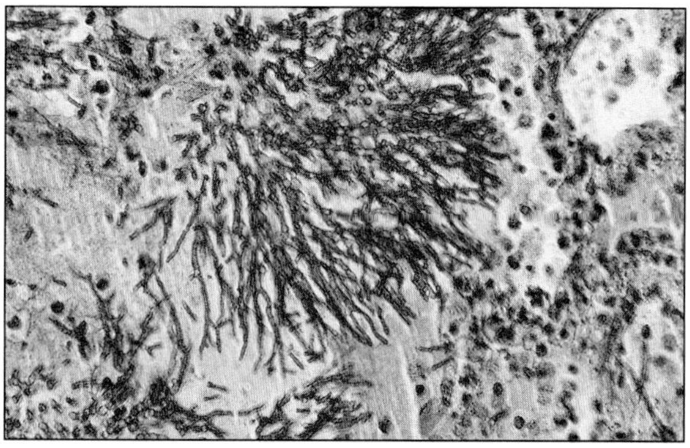

450 *A. fumigatus:* equine mycotic pneumonia. (PAS stain, ×400)

Identification

Colonial morphology

A. fumigatus : white fluffy colony when it first appears, rapidly becoming velvety or granular and a bright bluish-green in colour (**451**). Older colonies can assume a smokey battle-ship grey colouration.

451

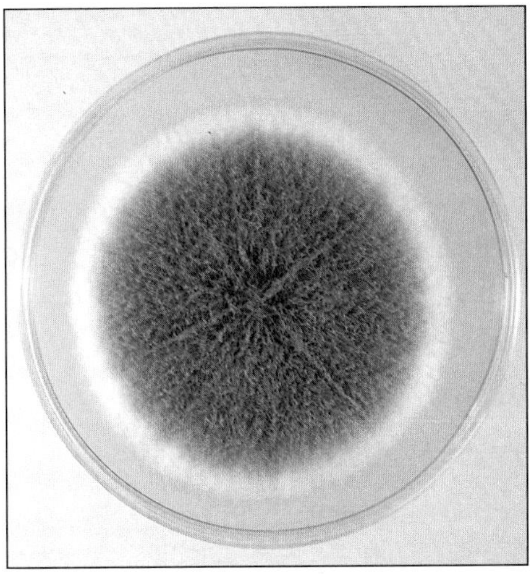

451 *A. fumigatus* on Sabouraud agar, 5 days.

452

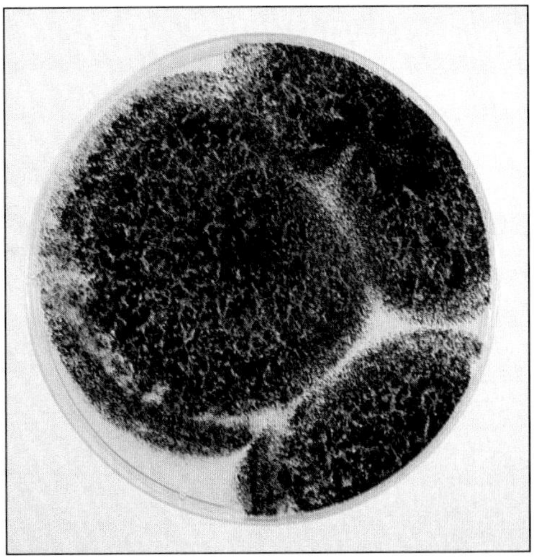

452 *A. niger* on Sabouraud agar, 5 days.

453

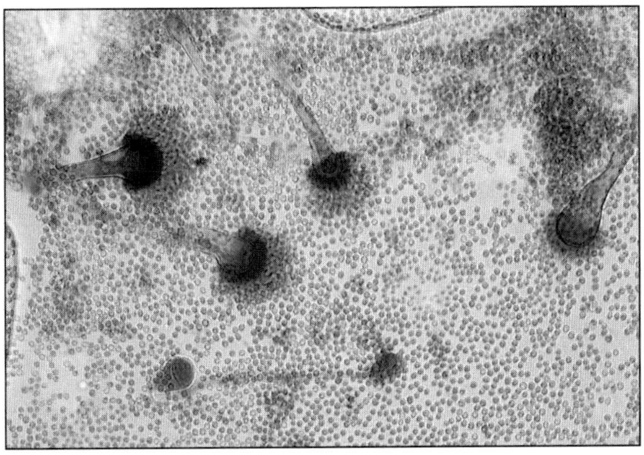

453 *A. fumigatus*: conidiophores and conidia. Transparent adhesive tape preparation. (LPCB, ×400)

A. niger : white at first when very young but soon developing a black pepper effect as the black conidia are produced (**452**). The reverse remains buff or cream-coloured. This distinguishes *A. niger* from the dematiaceous moulds.

A. flavus : cottony aerial mycelium when young but soon becomes a yellow-green with a sugary texture (**481**).

A. terreus : white becoming a cinnamon-buff colour and sugary in texture as the profuse sporulation occurs.

Microscopic appearance

Mounts are made in lactophenol cotton blue (LPCB) from the colony. The introductory mycology section should be consulted for the methods involved. The characteristic fruiting heads indicate the genus but rather more experience is required for speciation.

A. fumigatus : conidiophores are moderate in length and have a characteristic 'foot cell' at their bases. The vesicles are dome-shaped and the upper one-half to two-thirds bear phialides from which long chains of globose, spiny, green conidia (2–3 μm) are borne (**453**). These chains tend to sweep inwards.

A. niger : this has very large fruiting heads that look like small black balls under the dissecting microscope. The spherical vesicle bears large metulae that support the smaller phialides. The conidia are black and rough (**Diagram 49**).

A. flavus : the vesicles are round with sporulation over the entire surface. Phialides alone or phialides and metulae are present. The conidia are 3–5 μm in diameter, yellowish, elliptical or spherical and become spiny with age.

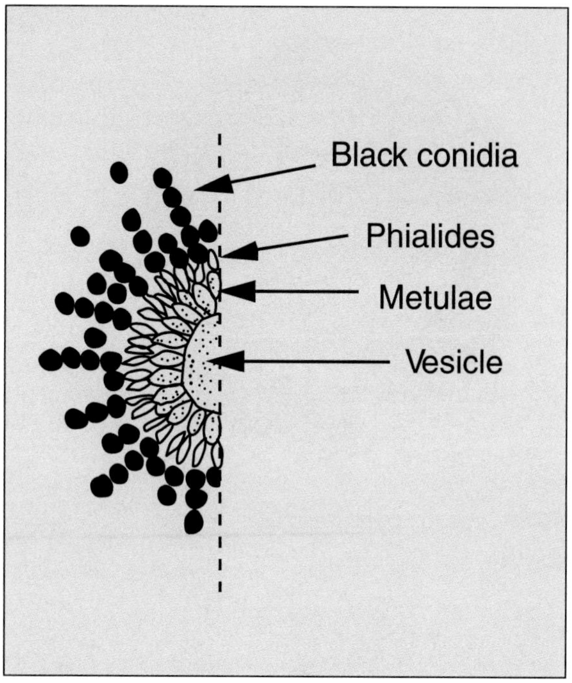

Diagram 49. Section through the fruiting head of *Aspergillus niger.*

A. terreus : the vesicles are small and dome-shaped and both phialides and metulae are present. The conidia (2–3 μm diameter) are elliptical. Aleuriospores are produced on submerged hyphae.

Serology

Commercial immunodiffusion kits for the serological diagnosis of aspergillosis are available. The interpretation of positive results may be equivocal because animals are frequently exposed to aspergilli that are widespread in the environment.

General Diagnosis

Mycologists base the speciation of the aspergilli on colony pigmentation, the size and length of the conidiophores, the shape and size of the vesicles, presence or absence of metulae; position of phialides, size, shape and appearance of the conidia, the length of the chains of spores and other criteria. This requires considerable expertise. However, if the culture is pure and correlates with the histopathological findings, then a presumptive identification can be made of the aspergilli known to be pathogenic based on their colonial and microscopic appearance. If there is doubt, the culture could be referred to a mycology reference laboratory.

41 The Pathogenic Yeasts

1. *Candida albicans*

There are more than 150 species of *Candida* but only *C. albicans* is commonly associated with disease in animals. *C. albicans* grows as a budding yeast cell, oval and 3.5–6.0 × 6.0–10.0 μm in size, on agar cultures and in animal tissues. Pseudohyphae are also produced in animal tissue by elongation of individual yeast cells that fail to separate. These can be mistaken for septate hyphae of moulds. Thick-walled resting cells, known as chlamydospores, are produced *in vitro* on certain media such as cornmeal-tween 80 agar. *C. albicans* will grow on ordinary media over a wide range of pH and temperatures. At both 25°C and 37°C it produces white, shiny, high-convex colonies in 24–48 hours.

Natural Habitat

C. albicans is a commensal of mucocutaneous areas, particularly of the intestinal and genital tracts of many animal species and humans. Most infections are endogenous in origin and predisposing causes, such as immunosuppression, prolonged antibiotic therapy and malnutrition, can initiate infections. In cattle, the yeast may be introduced into the udder on the nozzle of tubes of intramammary antibiotics. *C. albicans* is worldwide in distribution.

Pathogenesis

Neuraminidase and proteases may play a part in virulence and cell-wall glycoproteins have an endotoxin-like activity. Infections caused by *C. albicans* frequently involve mucous membranes. Inflammatory responses are predominantly neutrophilic and granulomatous lesions are rare. In severe, chronic infections, such as infection of the crop in chickens, the wall of the crop is thickened and covered by a corrugated pseudomembrane of yellowish-grey necrotic material, giving it the characteristic 'terry-towelling' effect. Infections due to *C. albicans* have been given several names including moniliasis, candidosis and candidiasis. **Table 135** lists the diseases caused by *C. albicans* in animals and in humans. There have been reports of mastitis in cattle being caused by other *Candida* spp. including *C. tropicalis, C. pseudotropicalis, C. parapsilosis, C. guilliermondii, C. krusei* and *C. rugosa*.

Table 135. Diseases and main hosts of *Candida albicans*.

Host(s)	*Disease(s)*
Chickens, turkeys, pigeons and other birds	'Thrush' of the mouth, oesophagus or crop. It can be a stunting condition with high mortality in young birds
Foals	Isolated from ulcerative lesions in the stomach
Mares and stallions	Genital infections
Calves	Pneumonic, enteric and generalised candidiasis. Often seen in animals following prolonged antibiotic therapy
Cows	Mastitis: mild and self-limiting. Spontaneous recovery may occur within a week
Kittens and pups	Mycotic stomatitis
Kittens	Enteritis
Bitches	Genital tract infections
Canines	Generalised infection with lesions in muscles, bones and skin (rare)
Felines	Pyothorax
Lower primates and marine mammals	Mucocutaneous candidiasis
Humans	Mycotic stomatitis in infants
	Nail infections
	Infections of genital tract, skin, lungs and other organs

Laboratory Diagnosis

Specimens

These may include scrapings from lesions, centrifuged milk samples, and biopsy or tissue samples in 10 per cent formalin for histopathology.

Direct microscopy

C. albicans can be demonstrated in specimens by Gram-stained smears (**454**), 10 per cent KOH preparations, or in tissue sections stained by PAS-haematoxylin or methenamine silver stains. *C. albicans* stains purple-blue with the Gram-stain. In tissue sections it appears as thin-walled, oval, budding yeast cells and/or in the form of pseudohyphae (**Diagram 50**).

Isolation

C. albicans grows well on blood agar or on Sabouraud dextrose agar with and without inhibitors. However, some of the other *Candida* spp. may be inhibited by cycloheximide. The plates are streaked with a small inoculum as for bacteria. The cultures are incubated at 37°C, aerobically, for up to 5 days.

Identification

Colonial appearance

Colonies of *C. albicans* usually appear in 1–3 days. They are white to cream, shiny, high-convex and have a pleasant 'beery' smell. They can attain a diameter of 4–5 mm (**455**).

Microscopic appearance

A small amount of growth can be placed in lactophenol cotton blue as a wet preparation, or a Gram or methylene blue stain can be used on fixed smears from the colonies. *C. albicans* produces thin-walled, budding yeast cells on Sabouraud dextrose agar or blood agar.

Demonstration of germ tubes

A small inoculum from an isolated colony is suspended in 0.5 ml of sheep, bovine, rabbit or human serum and is incubated at 37°C for 2–3 hours. A drop of the preparation is examined under phase contrast or the high-dry objective of the light microscope (with the condenser slightly lowered). Small tubes will be seen projecting from some of the yeast cells (**456**). This is characteristic of *C. albicans* (**Diagram 51**). Some strains of *C. tropicalis* can occasionally produce pseudo-germ tubes but require at least 3 hours' incubation or more.

454

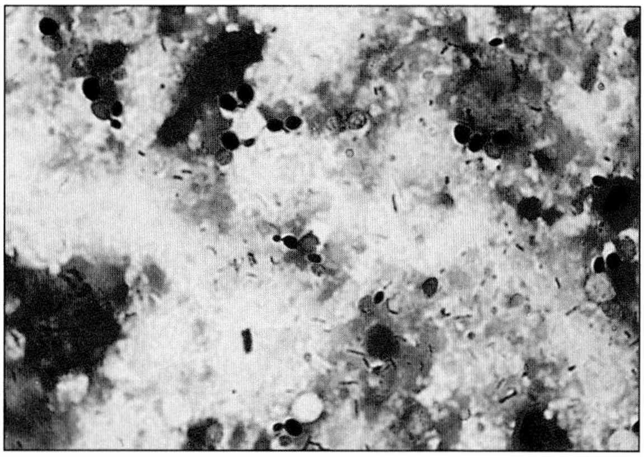

454 *Candida albicans* in a faecal smear from a calf. (Gram stain, ×1000)

455

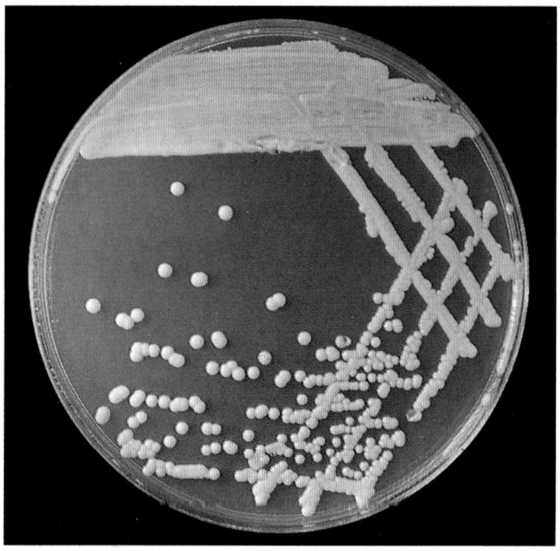

455 *C. albicans* on Sabouraud agar, 5 days.

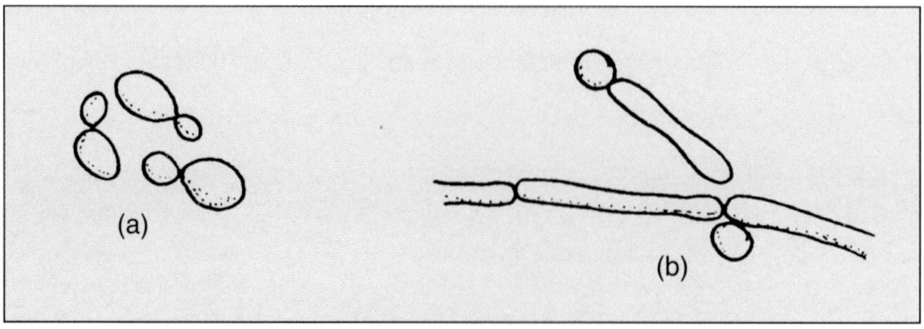

Diagram 50. *Candida albicans*: a) Yeast cell; b) Pseudohyphae.

456 

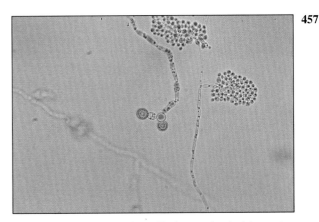

456 *C. albicans* forming germ tubes following 2 hours' incubation at 37°C in serum. (Unstained, ×400)

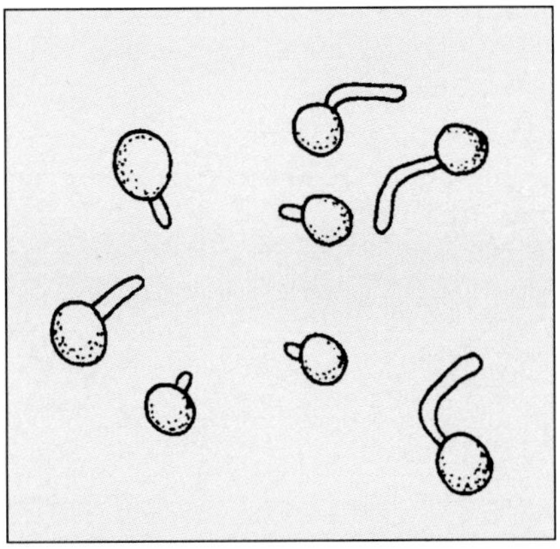

Diagram 51. *Candida albicans* producing germ tubes.

Chlamydospore production

A plate of cornmeal-tween 80 or chlamydospore agar is inoculated by making three parallel cuts in the medium 1.0 cm apart. The cuts are made at 45° to the surface to facilitate later microscopic examination. Subsurface inoculation is made, as chlamydospore production is enhanced by lowered oxygen tension. The inoculated plates are incubated at 30°C for 2–4 days. A thin coverslip is placed on the surface of the agar and the preparation examined under the low and high-dry objectives for the thick-walled chlamydospores (8–12 μm) borne on the tips of pseudohyphae (**457**). Clusters of smaller blastospores may also be present (**Diagram 52**).

Biochemical tests

Positive tests with the production of germ tubes and chlamydospores are sufficient for a presumptive identification of *C. albicans*. For a definite identification of *C. albicans* and some of the other *Candida* spp. either conventional biochemical tests can be used or commercial systems such as API 20C, API-yeast-Ident (Analytab Products) and Uni-Yeast-Tek system (Flow Labs.).

BiGGY agar (Oxoid)

Bismuth-sulphite-glucose-glycine-yeast agar can be used for the isolation and/or presumptive identification of *Candida* spp. Most bacterial contaminants are inhibited by the bismuth sulphite. *C. albicans*, *C. krusei* and *C. tropicalis* strongly reduce the bismuth sulphite to bismuth sulphide. *C. albicans* gives smooth, circular, brownish colonies with a slight white fringe and no colour diffusion into the surrounding medium (**458**). The colonies of *C. tropicalis* are similar but there is diffuse blackening of the medium after 72 hours. *C. krusei* gives large, flat, wrinkled, silvery, brown-black colonies with a brown periphery and yellow diffusion in the surrounding medium.

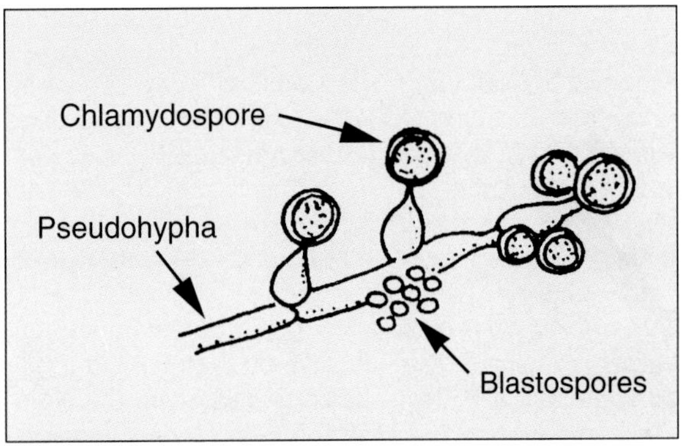

Diagram 52. Chlamydospores of *Candida albicans* (blastospores are also present).

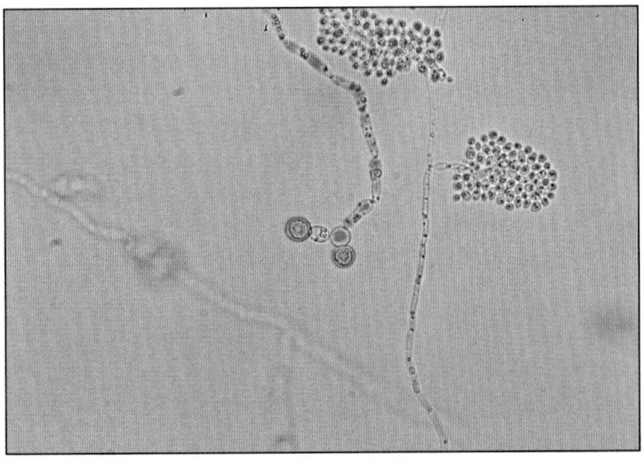 457

457 *C. albicans*: thick-walled, terminal chlamydospores, pseudohyphae and two clusters of smaller blastospores. (Unstained, ×400)

458

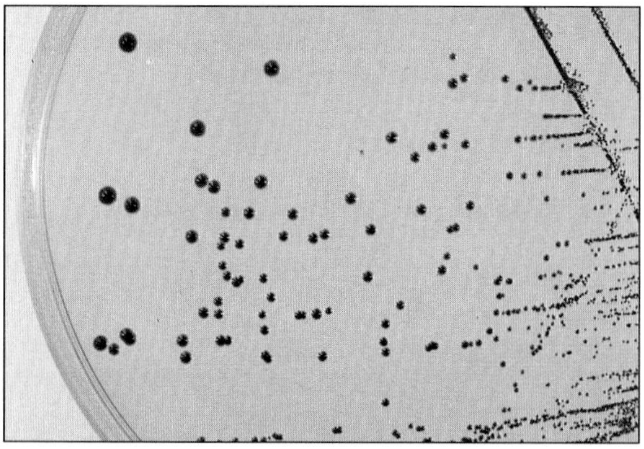

458 *C. albicans* on 'BiGGY' agar, 7 days.

2. *Cryptococcus neoformans*

Of the 19 species of *Cryptococcus*, only *C. neoformans* is pathogenic for animals and humans. It is a spherical to oval, thin-walled, budding yeast that varies greatly in diameter (2.5–20.0 µm). The cells are surrounded by a mucoid polysaccharide capsule that varies in thickness, but in animal tissues it is usually very large, the width of the capsule exceeding the diameter of the parent cell. Daughter cells are usually single and bud from the parent cell by a thin neck. *C. neoformans* is a member of the Fungi Imperfecti, but experimentally, strains have been converted to the perfect state of a Basidiomycete and given the teleomorphic name *Filobasidiella neoformans*. Cryptococcosis (European blastomycosis, torulosis) is a subacute or chronic infection involving the central nervous system, the respiratory system and the eye.

Natural Habitat

C. neoformans is present in dust and has been isolated from the skin, mucous membranes and intestinal tract of normal animals and birds. It is concentrated in pigeon faeces due to their high content of creatinine. The creatinine inhibits many other microorganisms but can be utilized by *C. neoformans*. It can survive in pigeon droppings for more than a year. *C. neoformans* has a worldwide distribution.

Pathogenesis

The virulence of *C. neoformans* is largely associated with the antiphagocytic and immunosuppressive capsule. The cryptococcal lesions, on gross examination, resemble myxomatous neoplasms. They consist of capsular slime, yeast cells, some inflammatory cells and later histiocytes, epithelioid and giant cells. The histiocytes and giant cells often contain *C. neoformans*. The route of infection is usually respiratory, often with localisation in the nasal cavity or paranasal sinuses and later extension to the brain and meninges. Infection of the meninges can resemble tubercular meningitis. Extension to the optic nerve may result in blindness. Subcutaneous granulomas occasionally occur, often in the cervical or pedal regions. *C. neoformans* can probably affect any mammal but cryptococcosis is most commonly seen in cats, dogs, cattle, horses and humans.

Table 136 summarises the diseases and conditions caused by *C. neoformans* in animals and humans.

Laboratory Diagnosis

Great care should be exercised when handling material suspected of containing *Cryptococcus neoformans* as it can cause serious disease in humans. An approved biological safety hood should be used when carrying out procedures with materials or cultures thought to contain this yeast.

Specimens

Cerebrospinal fluid, lesions or exudates, mastitic milk, biopsies and tissues should be collected.

Direct microscopy

India ink or nigrosin preparations can be made with cerebrospinal fluid or clear exudates *(see* Introductory Mycology Section). These stains will demonstrate the characteristic capsule (**459**) (**Diagram 53**). Histological sections on biopsies of tissue from lesions can be stained by the PAS-haematoxylin stain. This will stain, or outline, the yeast cell but not the capsule, which appears as a clear area surrounding the cell. In Mayer's mucicarmine stain, the wall of the yeast and the capsule are stained red, which is diagnostic for *C. neoformans*.

Isolation

C. neoformans will grow well on blood agar or on Sabouraud dextrose agar (without cycloheximide). The plates are streaked out as for bacteria and incubated aerobically at 37°C for up to 2 weeks. Capsular growth can be enhanced by culture on chocolate agar under 5 per cent CO_2 at 37°C. The majority of the saprophytic *Cryptococcus* species are unable to grow at 37°C, whereas *C. neoformans* can grow at temperatures up to 40°C.

Identification

Colonial morphology

Colonial growth is often not apparent until after nearly 2 weeks' incubation. The colonies are smooth, moist, shiny and become very mucoid with age. They are initially white but develop a yellowish shade. Mucoid yeast colonies are produced at 25°C and at 37°C indicating that it is a yeast and not a dimorphic fungus (**460**).

Table 136. Diseases and main hosts of *Cryptococcus neoformans*.

Host(s)	***Disease(s)***
Dogs and cats	Subcutaneous and nasal granulomas, central nervous system lesions and blindness
Horses	Nasal-passage granulomas with nasal discharge. Less commonly, lesions in lungs and skin
Cattle	Mastitis with severe swelling and firmness of the mammary glands. Milk yield is reduced and the milk is mucoid in texture. Very rarely, metastasis to the lungs occurs
Other animals	Very uncommon in other animals
Humans	Cryptococcosis is associated with either immunodepression or extensive exposure. Infections usually involve the lungs and central nervous system (cryptococcal meningitis)

459

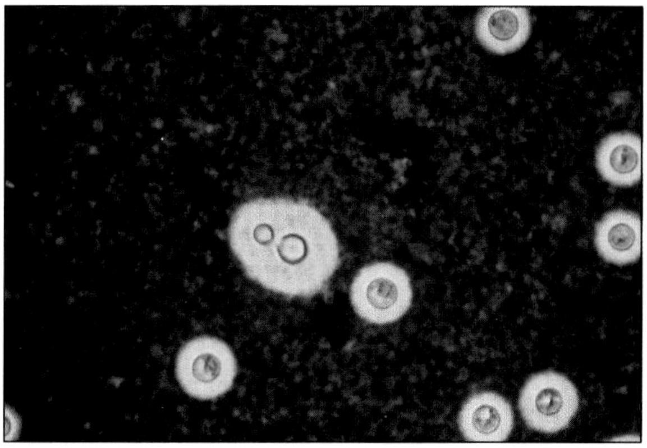

459 *Cryptococcus neoformans* in exudate. (Nigrosin stain, ×1000)

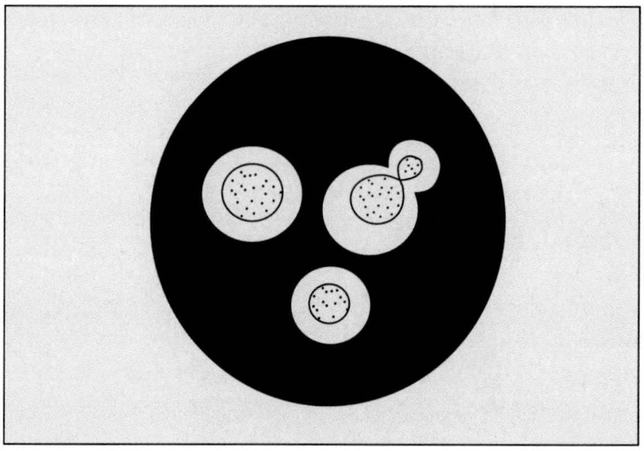

Diagram 53. *Cryptococcus neoformans* in a nigrosin preparation illustrating the large capsules.

460

460 *C. neoformans* on Sabouraud agar incubated at 25°C (left) and 37°C (right), confirming that it is in the yeast-form at both temperatures.

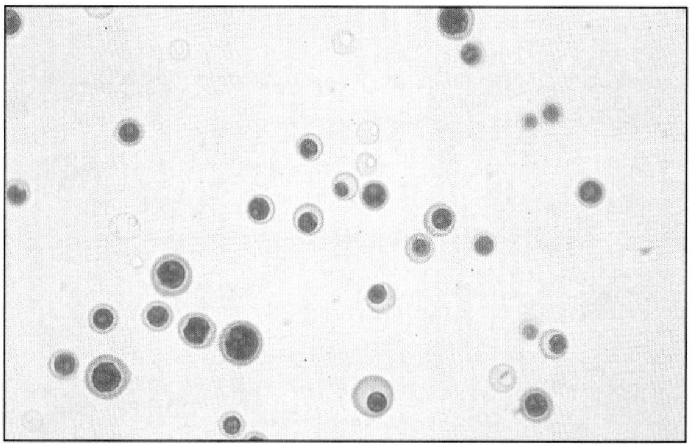

461

461 *C. neoformans* from a culture. The capsules are less prominent than they are in direct microscopic smears from tissues. (LPCB, ×1000)

Microscopic appearance

LPCB or nigrosin wet preparations should reveal a spherical, budding yeast surrounded by a capsule. The capsules are often not as large when growth is taken from laboratory media as those present around *C. neoformans* cells from animal tissue (**461**).

Ability to grow at 37°C

This differentiates *C. neoformans* from the majority of the *Cryptococcus* species.

Biochemical tests

a) Urease production: *Cryptococcus* spp. will produce urease on a heavily inoculated Christensen's urea agar slope.

b) Melanin production on Niger or birdseed agar: *C. neoformans* is one of the few *Cryptococcus* spp. that can use creatinine and produce melanin-pigmented (brown) colonies on media containing di- and polyphenolic compounds. The formula for the medium is given in **Appendix 2** but it can be obtained commercially. The plates are heavily inoculated and then incubated aerobically at 37°C for at least a week. The dark brown pigment develops first on the area of heaviest growth and later spreads to the remainder of the inoculated plate.

c) Biochemical profile: the API 20C (Analytab Products) and Uni-Yeast-Tek (Flow Laboratories) systems can be used to obtain a biochemical profile of the isolate for full identification.

Mouse inoculation.

Mice can be inoculated intraperitoneally. If they do not die, they are euthanised after 2 weeks when gelatinous lesions are found in the abdominal cavity and possibly in the lungs. *C. neoformans* is the only *Cryptococcus* sp. that is pathogenic for mice.

Immunology

Slide latex agglutination test kits have been designed to demonstrate antigen in serum and cerebrospinal fluid. Indirect FA tests have been used to detect antibody. Antibody may not be constantly demonstrated due to its combination with circulating antigen.

Summary of the characteristics for the presumptive identification of C. neoformans

- Demonstration of a budding yeast with a large capsule
- Growth at 37°C. Smooth, shiny colonies becoming mucoid.
- Production of a brown pigment on birdseed agar.
- Urease production.

These characteristics give a good presumptive identification of the yeast. Confirmation requires a larger range of biochemical tests.

3. *Malassezia pachydermatis (Pityrosporum canis)*

Malassezia pachydermatis is a lipophilic yeast that reproduces by unipolar budding. It occurs as a commensal on the oily areas of skin and ears of dogs, cats and probably other animals. In some cases of otitis externa in dogs the yeast appears to be present in larger numbers than usual and it may play a part in the pathological process. It is a small, bottle-shaped (1–2 × 2–4 µm) yeast reproducing by a process known as bud fission in which the bud detaches from the parent cell by the production of a septum. While the bud and parent cell are joined there is a wide base between them.

Direct examination

Gram-stained and methylene blue-stained smears or 10 per cent KOH preparations of exudate from dogs with otitis externa will often reveal this yeast with its characteristic morphology. A presumptive identification can be made on the microscopic appearance alone but confirmatory culture is advisable.

Isolation

Discharge from ears or skin scrapings can be streaked out on Sabouraud dextrose agar and the culture incubated at 25°C for up to 2 weeks. Colonies may be confined to the well of the plate but growth can be improved by placing a film of sterile olive or coconut oil on the surface of the agar before inoculation. The colonies are small, smooth, and often have an odour not unlike a 'wet-dog' (**462**). The typical bottle-shaped yeast cells with a wide septum between mother and daughter cells can be seen in stained smears from cultures (**463**).

4. Other yeasts that may be occasionally pathogenic

Torulopsis glabrata

This yeast occurs as a commensal in animals and is also found in the soil. There have been reports of it causing bovine mastitis and systemic infections in dogs and monkeys.

Trichosporon beigelii (cutaneum)

T. beigelii is a saprophytic yeast but it has been recovered from a nasal granuloma and a bladder infection in cats, skin infections in horses and monkeys and mastitis in cattle and sheep. *Trichosporon capitum* has been reported as a cause of mastitis in cattle.

These three yeasts can be identified by the API 20C (Analytab Products) system.

462

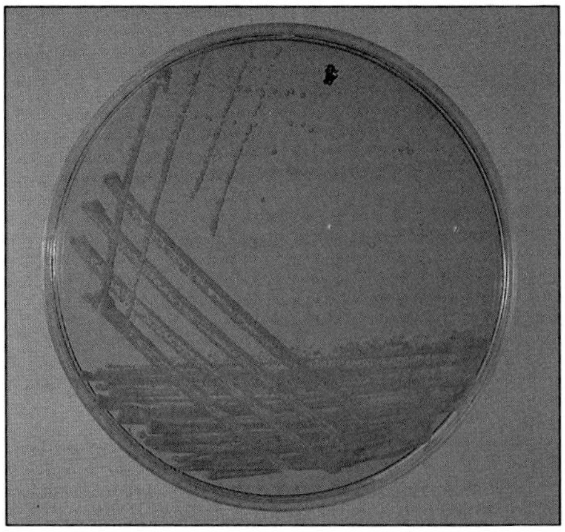

462 *Malassezia pachydermatis* on Sabouraud agar, 10 days.

463

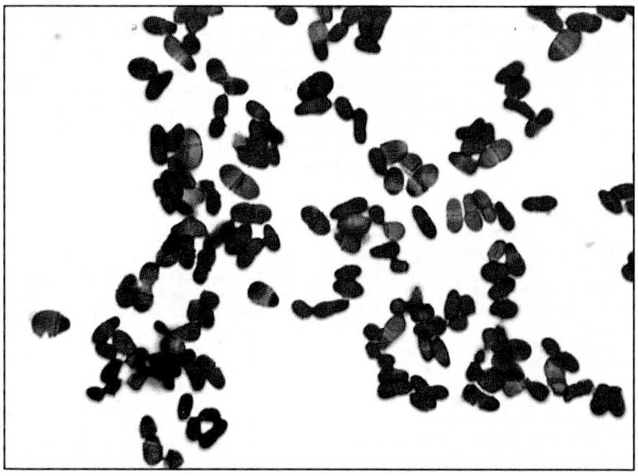

463 *M. pachydermatis* from a culture. Parent cells and buds are characteristically joined by a broad base. (Methylene blue stain, ×1000)

5. *Geotrichum candidum*

Geotrichum candidum is a mould but the colonial growth and odour are very yeast-like (**464**). It is widespread in nature and its isolation is not necessarily significant. There is doubt regarding its ability to invade animal tissues, however, it may be associated with bovine mastitis or intestinal infections especially if the animals have been on a prolonged course of antibiotic therapy. The fungus does not produce conidiophores but the hyaline, septate hyphae fragment into rectangular, one-celled arthrospores that tend to remain in chains (*see* **Diagram 45**, in the Introductory Mycology Section).

464

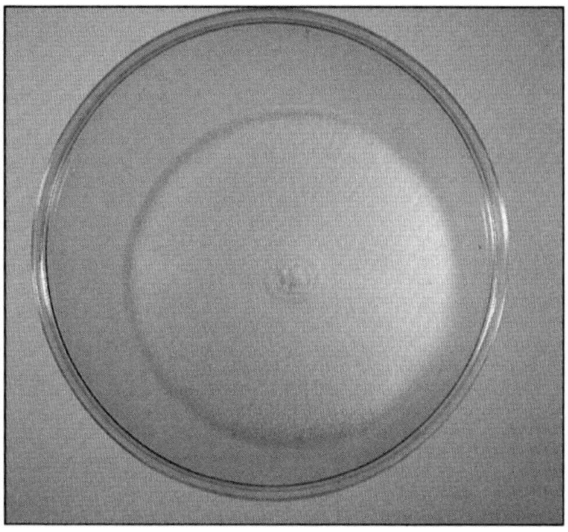

464 *Geotrichum* sp. on Sabouraud agar, 5 days.

42 The Dimorphic Fungi

The dimorphic fungi occur in two growth forms: a mould when growing saprophytically in the environment or when on culture media at 25–30°C, and a yeast or yeast-like form in animal tissues or when cultured on enriched media at 37°C. The mould or mycelial phase tends to be the more stable of the two. These fungi cause deep or systemic mycoses in animals and humans. **Table 137** indicates the fungi included in this group and gives the main hosts and diseases, the reservoir of infection and their geographical distribution.

Laboratory Diagnosis

Safety aspects

All of the dimorphic fungi can cause disease in humans and should be treated with respect. Cultures of *Coccidioides immitis,* particularly, represent a major biohazard for laboratory personnel because the arthrospores, produced on media at 25°C–37°C, can easily form an infective aerosol. Culture of this dimorphic fungus should either be avoided or appropriate precautions must be taken. These include the use of a biological safety cabinet when handling any material suspected of containing the organism, especially cultures. Culturing on slopes in screw-capped bottles is recommended and if culture plates are used these must be taped. The cultures of *C. immitis* should be covered with sterile water or saline before introducing an inoculation needle to prevent dispersion of the arthrospores. All microscopic preparations must be done in a biohazard cabinet. Cultures should be autoclaved as soon as the final diagnosis of *C. immitis* has been made.

Direct microscopy

Table 129 *(see* Introductory Mycology Section) and **Table 138** indicate the microscopic morphology of the dimorphic fungi in animal tissue and in cultures at 25°C and 37°C. Histopathological sections are most useful for demonstrating the yeast-forms in animal tissues (**465, 466**).

Yeast conversion of the dimorphic fungi

For full identification of these fungi an attempt to convert them to the yeast phase should be made on enriched media at 37°C. This is possible with all of them, with varying degrees of difficulty. The exception is *C. immitis* which remains in the mould-form in cultures at 25°C and 37°C. The mould or mycelial phase is the more stable one. For the mould phase, inoculated plates of Sabouraud dextrose agar, with and without chloramphenicol (0.05 g/l) and cycloheximide (0.4 g/l), are incubated at 25°C. When suspicious colonies have grown (3–4 days for *Sporothrix schenckii* and 2–4 weeks for the other fungi) a heavy subculture is made on brain-heart infusion agar plus 5–10 per cent sheep blood on slopes in 30 ml screw-capped bottles. A few drops of sterile water can be placed in the bottle to provide moisture during incubation. The caps of the bottles are slightly loosened to allow oxygen to reach the cultures. The plates are incubated at 37°C and *S. schenckii* should show growth in 3–5 days but *Blastomyces dermatitidis* and *Histoplasma capsulatum* may require 2–4 weeks. The colonies are examined in lactophenol cotton blue preparations for yeast cells. *C. immitis* has experimentally been forced to produce the spherule phase *in vitro* using a liquid medium at 40°C, but it is a difficult technique and not carried out routinely. *Histoplasma farciminosum* needs slightly different techniques and is considered separately at the end of this section.

Colonial morphology

- ***Sporothrix schenckii***

  At 25°C growth is visible in 3–5 days. Colonies are white to cream at first, becoming wrinkled with delicate aerial hyphae, and then later become dark and leathery (**467**, **468**).

  At 37°C colonies are yeast-like, smooth, soft, and cream to tan in colour. Growth occurs in about 3–5 days.

- ***Blastomyces dermatitidis***

  At 25°C growth occurs in about 2–4 weeks. The colonies are small and produce white, cottony aerial hyphae, becoming greyish or dark brown with age (**469**).

  At 37°C the waxy, yeast-like colonies are wrinkled and cream to tan in colour. They can have radiating 'prickles' from the surface.

- ***Histoplasma capsulatum***

  At 25°C white to cream colonies with cottony aerial hyphae are seen. They turn grey to brown with age and require 2–4 weeks' incubation. The colonies are similar to those of *B. dermatitidis.*

  At 37°C the colonies are smooth, yeast-like and white in colour. The fungus has a tendency to revert to the more stable mycelial form.

Table 137. Diseases and distribution of the dimorphic fungi.

Dimorphic fungus	*Main host(s)*	*Disease(s)*	*Site of lesion*	*Reservoir*	*Geographical distribution*
Sporothrix schenckii	Horses, dogs, cats and humans	Sporotrichosis (lymphangitis of limbs in horse)	Subcutaneous nodules, rarely systemic	Old wooden posts, soil, vegetation, rose thorns	Worldwide
Blastomyces dermatitidis	Dogs and humans. Rare in other species	North American blastomycosis	Primary in lungs of dog, metastases in skin and other organs	Probably soil associated with low pH, manure and decaying vegetation	Endemic in eastern third of USA. Cases reported in Africa, Asia and Europe
Histoplasma capsulatum	Dogs, cats and humans. Clinical disease rare in other animals	Histoplasmosis	Primary in lungs and secondary in intestines	Soil enriched with bat or bird faeces	Endemic areas in Mississippi and Ohio river valleys. Sporadic cases in many other parts of the world
H. farciminosum	Equidae	Epizootic lymphangitis. (African Farcy)	Lymphatics, lymph nodes and systemic	Reservoir not known	Africa and Asia, with cases reported in France, Italy, Russia and Egypt
Coccidioides immitis	Dogs and humans. Subclinical in other animals	Coccidioido-mycosis	Primary in lungs and secondary in bones and other organs	Soil in low-elevation deserts	Endemic in south-western states of USA, Mexico, South and Central America
Paracoccidioides brasiliensis	Humans. Infection not reported in animals	Paracoccidioido-mycosis	Chronic mycosis of skin, mucous membranes and internal organs	Soil	Mexico, Central and South America

465

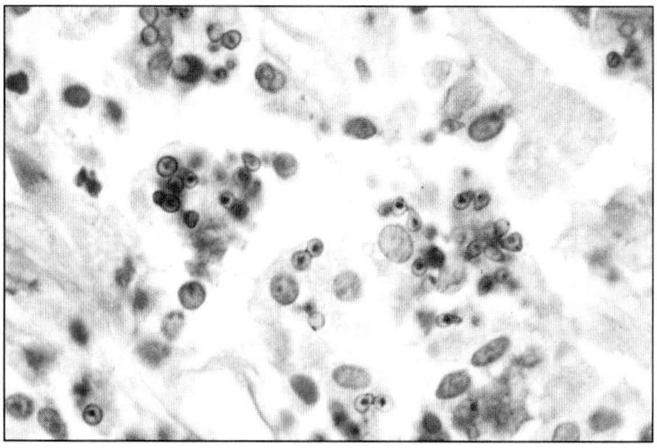

465 *Blastomyces dermatitidis:* yeast form in tissue. (PAS-haematoxylin stain, ×1000)

466

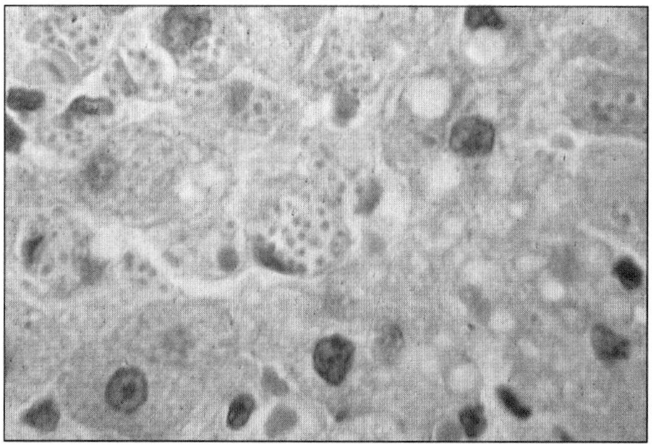

466 *Histoplasma capsulatum:* yeast form in Kupffer's cells (dog's liver). (Silver stain, ×1000)

Table 138. Microscopic morphology of the dimorphic fungi.

Dimorphic Fungus	Animal Tissue (37°C)	Culture (37°C) Brain-heart + 5% blood agar	Culture or Environment (25°C) Sabouraud agar
Sporothrix schenckii	 Cigar-shaped, budding yeast cells, that may occur within neutrophils. Usually very few present. (2 - 4 μm diameter). Asteroid bodies occur.	 Single or multiple-budding yeast cells, 2 - 4 μm in diameter.	 Fine branching hyphae with 2 - 3 μm pyriform conidia in flowerettes from short conidiophores. Conidia connected by thread-like process.
Blastomyces dermatitidis	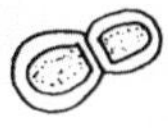 Large (8 -10 μm) round or oval, thick-walled yeast cells. Buds on a broad base, single buds. Cytoplasmic granulation is often obvious.	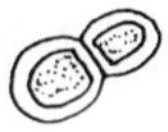 Large (8 - 10 μm), round or oval thick-walled yeast cells budding on a broad base.	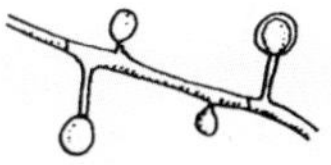 Small (2 - 3 μm) oval or pear-shaped conidia borne on tips of short conidiophores from septate hyphae.
Histoplasma capsulatum	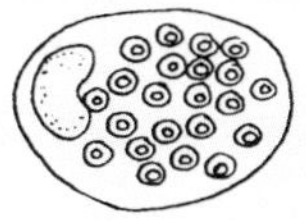 Small (2 - 5 μm) budding yeast cells intracellular in phagocytic cells. The yeast cells are usually surrounded by a halo.	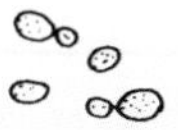 Oval, budding yeast cells (3 - 4 μm diameter) with a narrow neck between mother and daughter cells.	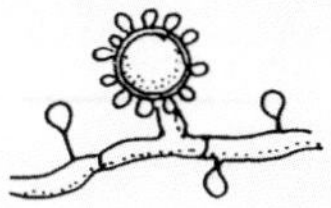 Two types of conidia: large (8 - 14 μm) tuberculate macroconidia that are sunflower-like and small tear-drop microconidia.
Coccidioides immitis	 Spherules (15 - 60 μm), the mature forms filled with endospores. No endospores in immature spherules.	 Septate hyphae branching at right-angles. With age the hyphae dissociate into barrel-shaped arthrospores. These are separated by clear, non-viable cells. Arthrospores are wider than the hyphae. Cannot be converted easily to spherule form *in vitro*.	

467

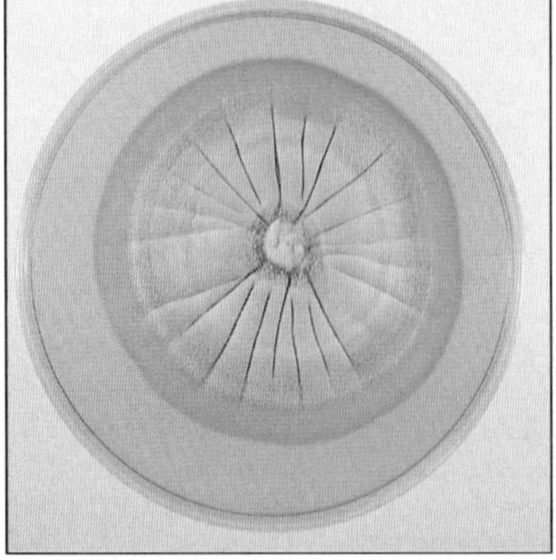

467 *Sporothrix schenckii* on Sabouraud agar at 25°C, 13 days.

468

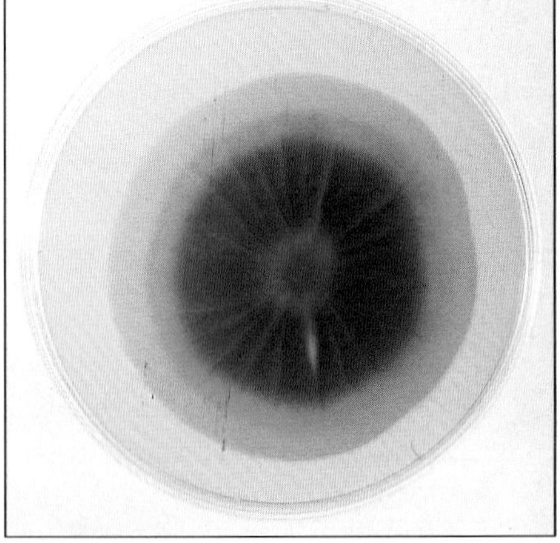

468 *S. schenckii* on Sabouraud agar, 13 days. Reverse.

469

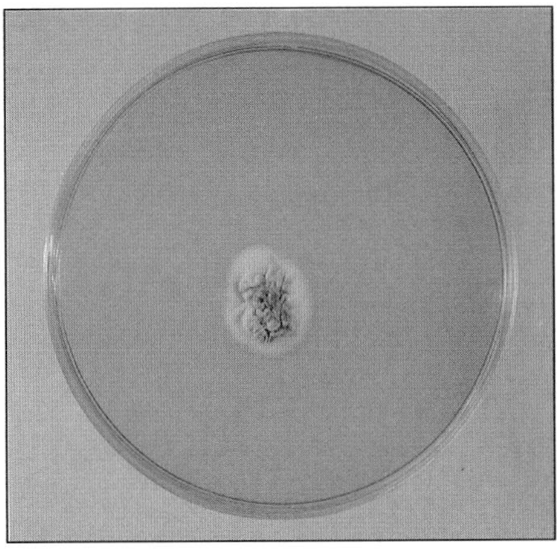

469 *B. dermatitidis* on Sabouraud agar. (25°C, 16 days)

470

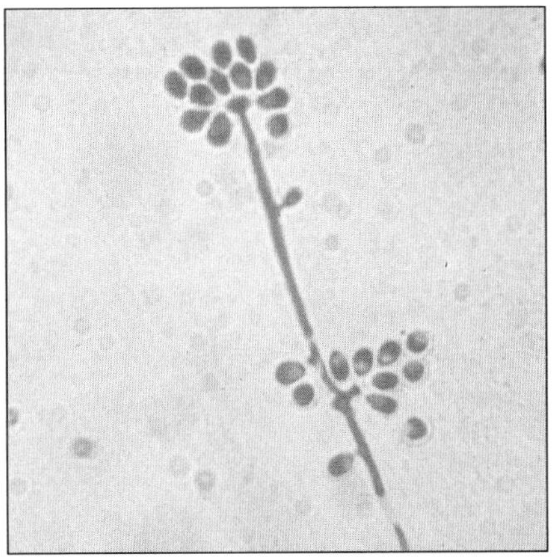

470 *S. schenckii* conidiophore and conidia. Culture incubated at 25°C. (LPCB, ×400)

- ***Coccidioides immitis***
  At 25°C and 37°C delicate cobweb growth in 3–21 days occurs. It causes a greenish discolouration on blood agar. Colonies have fluffy areas alternating with areas adherent to the agar surface.

Microscopic appearance

Table 138 summarises the microscopic appearance of the dimorphic fungi in animal tissue and from cultures at 25°C and 37°C. The mould-form (**470**) and yeast-form (**471**) of *S. schenckii* are illustrated.

471

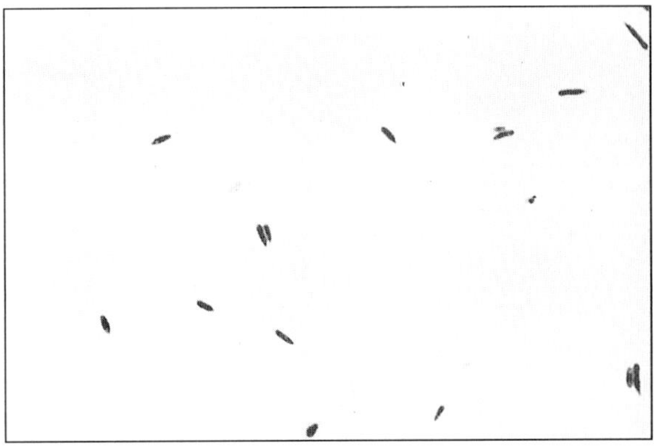

471 *S. schenckii* yeast cells. Culture incubated at 37°C. (LPCB, ×1000)

Exoantigen test

This test is used in some laboratories for *B. dermatitidis*, *H. capsulatum* and *C. immitis*. It is a relatively simple and rapid method for identification and, if it is positive, obviates the necessity to convert the fungus to the yeast phase. The method is an immunodiffusion test that, with reference antisera, detects cell-free antigens (exoantigens) extracted and concentrated from a mycelial colony. Kaufman and Standard (1987) have described the technique. Positive control antisera can be obtained commercially (Immuno-Mycologics, Meridian Diagnostics and Scott-Nolan Laboratories).

B. dermatitidis has the exoantigen **A**; *H. capsulatum* has **h** and **m**; and *C. immitis* **HS**, **F** or **HL**.

Immunological tests

C. immitis infection gives a strong immunological response and the serological tests are more reliable than for the other mycoses. With *B. dermatitidis* many of the serological tests cross-react with the other dimorphic fungi and are of limited value. **Table 139** summarises the immunological tests, and their interpretation, used to diagnose mycoses caused by the main dimorphic fungi. Immunodiffusion kits are available commercially from American MicroScan; Whittaker M.A. Bioproducts; Meridian Diagnostics; Immuno-Mycologics; and Nolan Biological Laboratories.

Mouse inoculation tests

All the dimorphic fungi cause lesions in mice. There is little need for animal inoculation now because of the availability of other specific confirmatory tests. However, mouse inoculation might be the only method of recovering the fungi from heavily contaminated specimens.

- ***Sporothrix schenckii***: mice are inoculated intratesticularly. Orchitis develops in 2–4 weeks.

Table 139. Immunological tests for the main dimorphic fungi.

Test	*Interpretation*
Blastomyces dematitidis	
Skin test	Lacks sensitivity and specificity: not useful
Immunodiffusion	Useful in humans, data for animals currently unavailable
ELISA	Useful test in humans, no data for animals available yet
Counterimmunoelectrophoresis	Reliable in dogs (Barta *et al.*,1983)
FA technique	For identification of yeast-form in tissue
Complement fixation test	Reliable for dogs
Histoplasma capsulatum	
Skin test (histoplasmin)	Becomes positive following exposure and remains positive indefinitely. Lack of reaction may indicate anergy
Complement fixation test	Antibodies present soon after infection and disappear after about 9 months
Immunodiffusion	Detection of **h** antibodies indicates active infection and antibodies to **m** antigen a past infection
Latex agglutination (latex particles coated with histoplasmin)	Useful screening test. Detects antibodies early in disease
FA technique	Identification of yeast-form in tissue
Coccidioides immitis	
Skin test (coccidioidin)	A negative test in the presence of signs of disease indicates dissemination
Immunodiffusion	Multiple bands tend to be associated with active infection and a single band with a chronic condition
Complement fixation test	Antibodies rise in disseminated disease and tend to remain high
Latex agglutination test	Antibodies detected early in disease (IgM) and the test later becomes negative
Sporothrix schenckii	
Immunodiffusion and CFT	Antibodies are demonstrable only in the rare systemic infections
FA technique	For identification of yeast-forms in tissue or exudates

- ***Blastomyces dermatitidis***: mice or guinea-pigs are inoculated intraperitoneally. Lesions may be found in liver, spleen, lungs and lymph nodes in 3 weeks.
- ***Histoplasma capsulatum***: mice are inoculated intraperitoneally and the yeast-form can be recovered from the liver and spleen in 2-4 weeks.
- ***Coccidioides immitis***: intraperitoneal inoculation into mice. The mice are euthanised 7–10 days post-inoculation and nodules are found in the peritoneal cavity, lungs and spleen. These nodules are examined for the characteristic spherules produced by the fungus.

Summary of the Diagnostic Tests for the Identification of the Dimorphic Fungi

- Direct microscopy: wet mounts of exudates and tissues.
  : histopathology on tissue sections.
- Culture and demonstration of the mould phase at 25–30°C and the yeast phase on enriched medium at 37°C.
- Microscopic appearance from cultures to observe the fruiting structures and spores (colonies at 25–30°C) and yeast-forms (colonies at 37°C).
- Exoantigen tests.
- Immunological and serological tests.
- Mouse inoculation.

Histoplasma farciminosum

About 90 per cent of the cases of epizootic lymphangitis (African farcy) have been reported in horses and the remainder in mules and donkeys. The legs and neck are most commonly involved. Nodular, granulomatous and ulcerative lesions of skin, subcutaneous tissue and lymphatic vessels occur. The disease can become disseminated. The natural habitat, other than infected animals, remains unknown. Transmission is thought to occur mainly through breaks in the skin or via biting insects. The morphology and life cycle of the fungus are similar to those of *H. capsulatum*.

Direct microscopy

Wet mounts of pus, or exudates or biopsies can be examined for the intracellular, pear-shaped, double-contoured yeast cells (2–4 µm). They are usually present inside mononuclear cells or neutrophils. Budding normally occurs from the pointed end of the yeast cell.

Culture

The isolation of *H. farciminosum* is a difficult and slow process. Two Sabouraud dextrose agar plates, with and without the antimicrobial agents, are inoculated with material taken aseptically from unruptured nodules and incubated at 25–30°C for 2–8 weeks. For the conversion to the yeast phase, Hartley digest agar with 10 per cent horse serum is inoculated and incubated at 37°C, under 20 per cent CO_2, for 2–8 weeks.

Identification

Colonial and microscopic appearance

At 25°C the colonies appear as minute grey flakes, later becoming dry and very wrinkled. The colonies are often composed of sterile hyphae although very rarely chlamydospores (5–10 µm), arthrospores and blastospores are present. At 37°C the small, grey, flaky colonies are composed of yeast cells and some hyphae.

Immunology

The immunodiffusion test on mycelial extracts detects the **h** and **m** genus-specific exoantigens.

Skin sensitivity develops after exposure to the fungus and a skin test may be positive in the absence of clinical signs.

An ELISA and an indirect fluorescent antibody test have been developed for use in the detection of antibodies to this fungus.

Mouse inoculation

Mice can be inoculated intraperitoneally with a suspension of material suspected of containing the fungus. Impression smears from the liver and spleen, 2–4 weeks post-inoculation, should reveal the yeast-form of the fungus.

Adiaspiromycosis

Adiaspiromycosis is a self-limiting respiratory infection seen in humans, rodents, dogs and a variety of wild mammals. The infection results from the inhalation of the conidia from the soil fungi *Chrysosporium parvum* var. *parvum* (*Emmonsia parva*) or *C. parvum* var. *crescens* (*E. crescens*). The inhaled conidia do not replicate in the host's lungs but simply enlarge to form thick-walled adiaspores. In var. *parvum* these are up to 40 µm in diameter and up to 400 µm in var. *crescens*. These adiaspores could be confused with the spherules of *Coccidioides immitis*. Infections are usually discovered incidentally in the course of histopathological examination of lung tissue for other infections. The condition has been reported in Africa, Asia, Europe, New Zealand, North America and South America. On direct microscopy the two varieties may be distinguished by the relative sizes of the adiaspores and by the fact that the adiaspores remain uninucleate in var. *parvum* but become multinucleate in var. *crescens*.

Culture

Both varieties grow well on Sabouraud dextrose agar to produce mycelial colonies at 25°C. Strains may vary considerably in appearance. Young cultures are colourless and smooth but become white or brown with white aerial hyphae. The reverse ranges from white to yellow or brown. Microscopically these mycelial cultures show septate hyphae and ovoid, one-celled, smooth-walled or rough-walled conidia with a broadly truncated base and basal scar, often with a remnant of hyphal wall left attached (**Diagram 54**).

On brain heart infusion blood agar at 37°C, var. *crescens* produces adiaspores that are 25–400 μm in diameter with walls 70 μm thick. The incubation temperature must be raised to 40°C before var. *parvum* will produce adiaspores. These are 10–25 μm in diameter with walls about 2 μm thick.

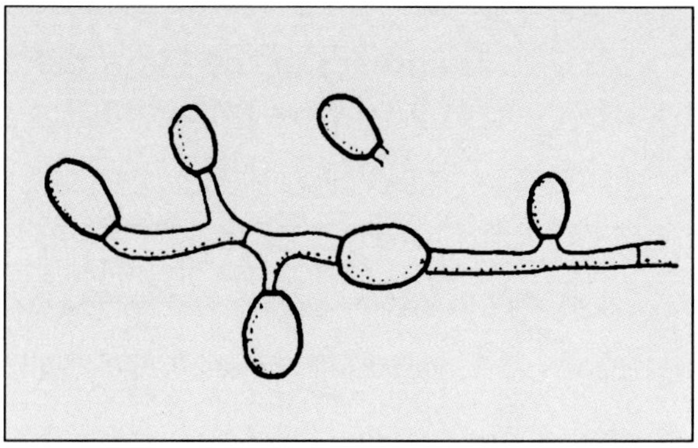

Diagram 54. *Chrysosporium* sp. from a mycelial cuture (25°C) illustrating one-celled conidia (2-10 μm) with truncated base and broad basal scar.

References

Barta, O., Hubbert, N.L., Pier, A.C. and Pourciau, S.S. (1983). Counterimmunoelectrophoresis (immunoelectroosmosis) and serum electrophoretic patterns in serologic diagnosis of canine blastomycosis, *American Journal of Veterinary Res*earch, **44**: 218–222.

Kaufman, L. and Standard, P.G. (1987). Specific and rapid identification of medically important fungi by exoantigen detection, *Annual Review of Microbiol*ogy, **41**: 209–225.

43 The Pathogenic Zygomycetes

Zygomycosis (older term phycomycosis) is an all-embracing term for an infection due to a fungus in the taxonomic Class Zygomycetes. Most of these fungi are saprophytes and widespread in the environment but some of them can be opportunistic pathogens. Clinical disease usually occurs if the host's defences are lowered or if large numbers of spores are ingested or inhaled. The hyphae, in cultures and in animal tissues, are wide (5–15 μm diameter), irregular and ballooning (**472**). They lack septa except near fruiting structures or if there has been damage to hyphae in older cultures. The sexual spores are thick-walled zygospores (**473**, **Diagram 55**) produced through fusion of gametangia often from two different strains of the species.

The Class contains two Orders—Mucorales and Entomophthorales. Pathogenic species within the Order Mucorales are found within the genera *Absidia*, *Mucor*, *Rhizopus*, *Rhizomucor* and *Mortierella*. The Order Entomophthorales includes *Basidiobolus ranarum* and *Conidiobolus coronatus*, both of which can cause disease in animals and humans.

472

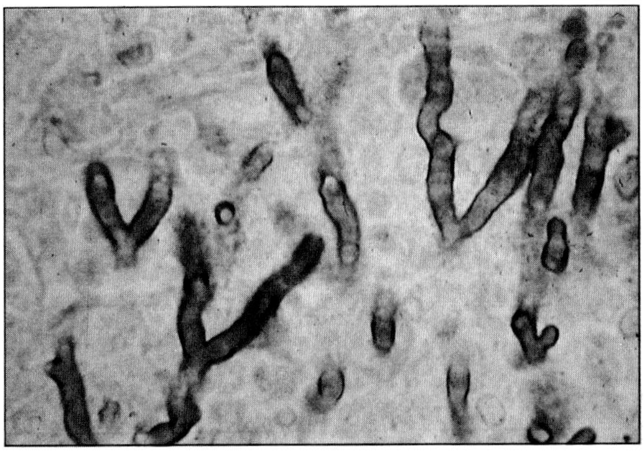

472 *Rhizopus* sp. non-septate hyphae in wall of bovine rumen. (Silver stain, ×1000)

473

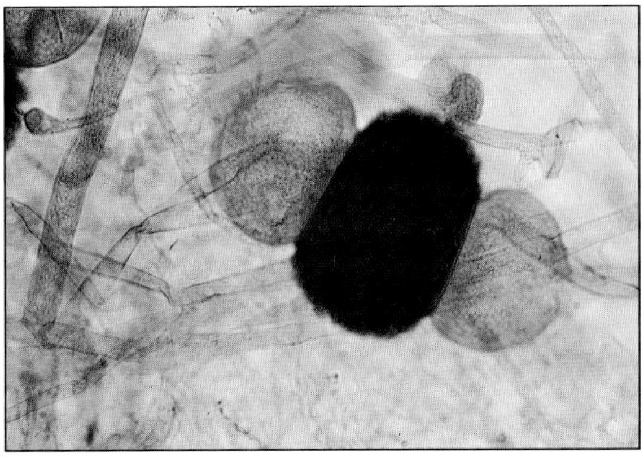

473 *Rhizopus* species: zygospore. (×400)

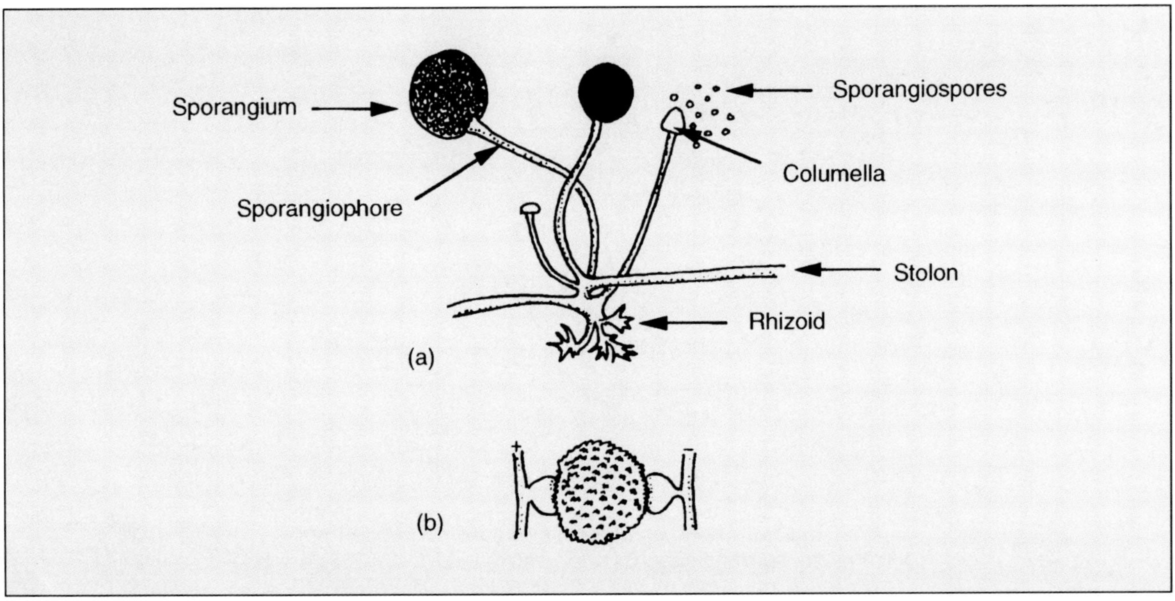

Diagram 55. A mucoraceous Zygomycete showing: (a) asexual fruiting structures, (b) sexual zygospore.

The Mucoraceous Zygomycetes (Order Mucorales)

Members of this Order characteristically have rapidly growing colonies with greyish, woolly mycelium which fills the Petri dish in 3–5 days and often reaches and raises the lid of the plate (**474**). An exception to this is *Mortierella wolfii* which has a white, low, velvety growth (**475**). The aerial hyphae (sporangiophores) arise from stolons and usually end in a columella, which is enclosed in the sac-like sporangium. Within the sporangium the asexual spores (sporangiospores) are formed. These spores may be colourless, yellowish or brown and the sporangia, packed with spores, often appear as dark, pin-head-sized dots within the woolly grey mycelium. As the spores mature, the wall of the sporangium becomes fragile and ruptures, releasing myriads of spores (**Diagram 55**). The root-like rhizoids are produced by several genera and promote anchorage to the substrate (**476**).

474

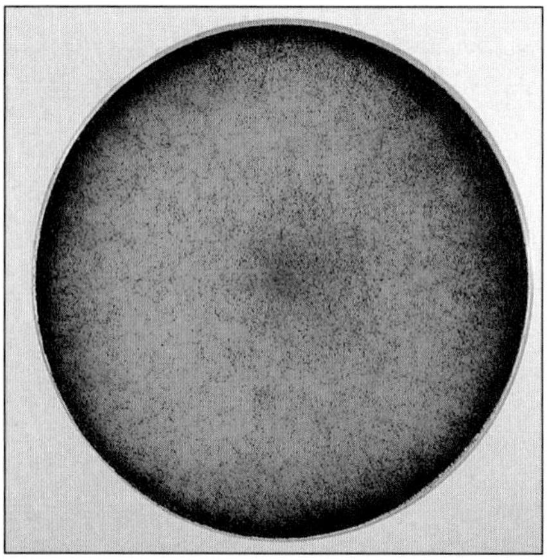

474 *Rhizopus* species on Sabouraud agar, 7 days.

475

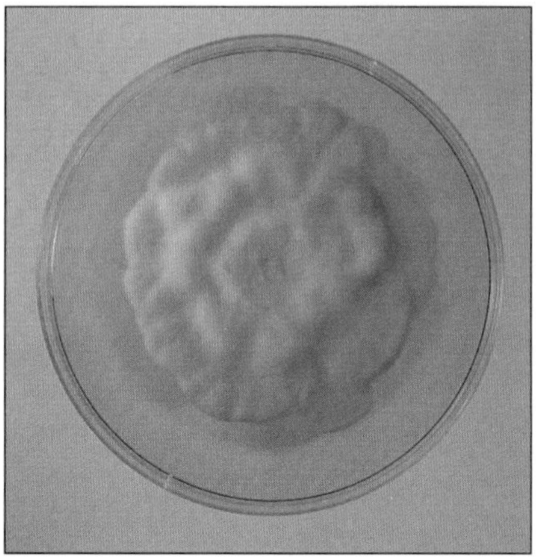

475 *Mortierella wolfii* on Sabouraud agar, 4 days.

476

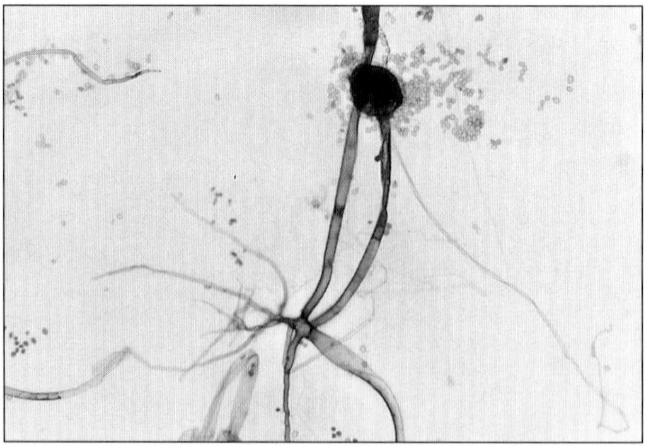

476 *Rhizopus* species: sporangiophore, collapsed sporangium, sporangiospores and a rhizoid. (LPCB, ×100)

Natural Habitat

The mucoraceous Zygomycetes are usually saprophytes and widespread in soil, vegetation and air. Specimens are often contaminated by them. Cycloheximide, added to Sabouraud dextrose agar, will inhibit these fungi. *Mortierella wolfii* is not easy to find in the environment although it has been recovered, with difficulty, from soil associated with a silage stack and from debris in cattle yards. Most of the Zygomycetes occur worldwide. Infection in cattle due to *M. wolfii* has been reported in New Zealand, Australia, UK and North America.

Pathogenesis

The most common form of zygomycosis (sometimes referred to as mucormycosis) affects the lymph nodes of the respiratory and intestinal tracts. The lymph nodes may enlarge and show caseous necrosis. The internal organs can also be involved. This type of infection is most common in cattle, pigs and dogs. Young animals are more prone to intestinal tract infections clinically manifested by diarrhoea. A granulomatous reaction, ulceration and caseous necrosis occur in the intestinal tract. In acute and severe infections the fungi invade the blood vessels and cause a necrotising vasculitis with thrombosis and haemorrhage.

Mortierella wolfii causes abortion in cattle, and in about 5 per cent of these animals an acute, fulminating pneumonia; death occurs within 48 hours of the abortion. Lesions are characteristic of mycotic abortion and resemble those seen in *Aspergillus fumigatus* infections. The placenta is thickened and 'wooden' in appearance (**477**) and the foetus may have ringworm-like fungal skin plaques (**478**). In fatal pneumonia the lungs are heavy, red and wet with most lobes affected (**479**). A variable amount of fluid is present in the thoracic cavity. Other Zygomycetes have been isolated from cases of abortion in cattle but, as these other fungi are more widely distributed than *M. wolfii*, the isolation should be confirmed by histological examination of cotyledons and foetal lesions. **Table 140** summarises the main hosts and diseases caused by the mucoraceous Zygomycetes. Zygomycosis has also been reported in horses, mink, guinea-pigs, mice, poultry and exotic birds.

Laboratory Diagnosis

Specimens

Lesion biopsies or tissues from dead animals, cotyledons, foetal abomasal contents and uterine discharges may be collected. Thin sections, in 10 per cent formalin, for histopathology should be taken when possible.

477

477 Bovine mycotic placentitis in this instance caused by *M. wolfii.*

478

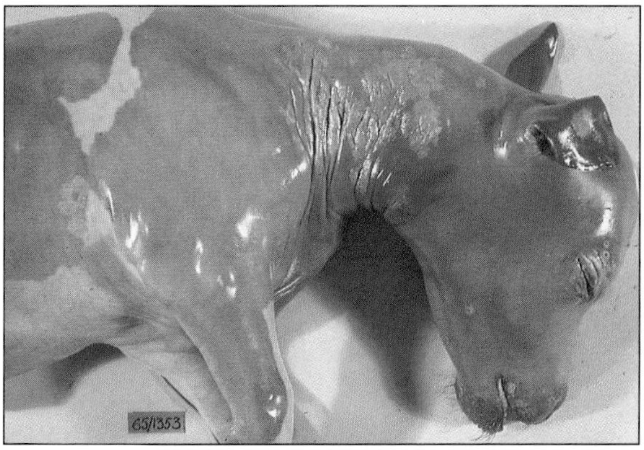

478 Fungal plaques on a bovine foetus characteristic of mycotic abortion *(M. wolfii).*

479

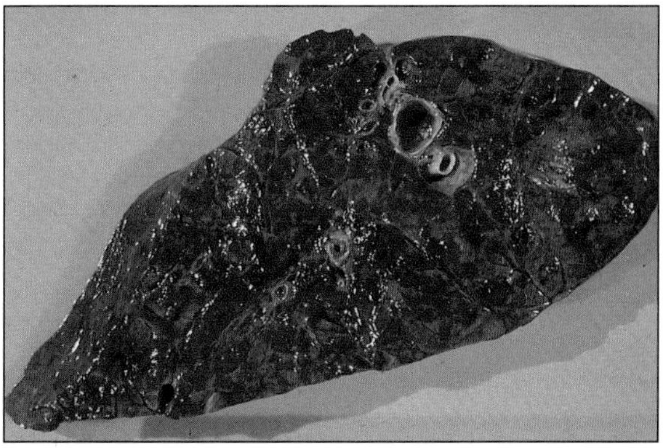

479 Bovine lung. Acute, diffuse, mycotic pneumonia caused by *M. wolfii.*

Table 140. Diseases caused by the mucoraceous Zygomycetes.

Host	*Disease or condition*
Pigs	Association of Zygomycetes with gastric ulcers
	Lesions in mediastinal and submandibular lymph nodes
	Embolic lesions in liver and lungs
Cattle	Abortions (sporadic) due to *M. wolfii* and possibly other zygomycetes
	Fatal pneumonia due to *M. wolfii* in about 5% of cows that abort
	Lesions in bronchial, mesenteric and mediastinal lymph nodes
	Nasal and abomasal ulcers
Dogs	Gastrointestinal infections marked by ulcerative lesions and mesenteric lymphadenitis

Direct microscopy

Histological examination of tissue sections is needed for these ubiquitous fungi. They are PAS-positive (pink staining) and this or a methenamine silver stain can be used. The hyphae are broad (10–15 µm diameter), irregular, non-septate and branching. The hyphae of *M. wolfii* are somewhat finer (2–12 µm) but irregular. Fungal hyphae may also be seen in KOH wet preparations.

Culture

Sabouraud dextrose agar without cycloheximide or Emmons' modification can be used. The inoculated plates are incubated aerobically for up to 10 days although some colonial growth is usually seen in 48 hours. The species capable of causing systemic infections can tolerate 37°C. This temperature has the advantage of discouraging any of the purely saprophytic fungi. If the lesions are of a superficial nature a duplicate plate could be inoculated and incubated at 25°C. The Zygomycetes will grow well on blood agar plates. They may be growing as contaminants or as significant pathogens in cases of unsuspected mycotic infections. *M. wolfii* does not compete well with the faster-growing bacteria on blood agar or with fungal contaminants on mycology plates.

Identification

At present there are no identification kits, nucleic acid probes or commercially available immunological or serological means for identifying the Zygomycetes. Accordingly there must be heavy reliance on histological examination of affected tissues and the colonial and microscopic appearance of the isolated Zygomycete. **Table 141** indicates the morphology of some of the species capable of causing disease in animals. The cultures may lend themselves to identification to a generic level but subcultures could be submitted to a mycology reference laboratory for species confirmation.

The Entomophthoraceous Zygomycetes (Order Entomophthorales)

The Order contains *Basidiobolus ranarum* and *Conidiobolus coranatus* which are recognised pathogens in humans and have also been recorded as causing rare infections in animals. In humans they are often responsible for subcutaneous mycoses (zygomycoses). In direct microscopy from specimens, broad, irregularly branching hyphae (5–18 µm), sparsely septate and enclosed by eosinophilic material, will be seen. They exhibit an entirely different colonial appearance to that of the mucoraceous Zygomycetes. They grow in 3–5 days as flat, waxy colonies that often develop radial grooves but become fuzzy with age and the hyphae are often septate in older cultures. Colonies are difficult to remove from the agar and the colour varies from tan to grey-brown. A brief description is given of the colonial and microscopic appearance of these two fungi but the identification should be confirmed by a mycology reference laboratory.

Laboratory Diagnosis

Colonial appearance

- ***Basidiobolus ranarum:*** the colonies are thin, waxy and buff to grey. They become radially folded, greyish-brown, and covered with white aerial hyphae on longer incubation. The reverse side is white in colour. Some strains have an earthy odour similar to that of *Streptomyces* spp.

Table 141. Colonial and microscopic morphology of some clinically significant mucoraceous Zygomycetes.

Genus and species	*Colonial morphology*	*Microscopic appearance*
***Absidia corymbifera* (*A. ramosa*)**	Rapid growth. Woolly and white, becoming olive-grey. Fills Petri dish like *Rhizopus*. It is thermotolerant and grows at 45°C	Rhizoids are internodal (arising from stolons between the sporangiophores). Sporangiophores are long (450 µm) and branch repeatedly. Sporangia are pear-shaped and contain globose to oval spores (2–4 µm). Columella is round and merges with an apophysis (swelling) below the sporangium
Mortierella wolfii	Colony white and velvety with a lobulated outline giving a rosette appearance. Spreads rapidly over plate	Colonies on Sabouraud agar are sterile or produce very few sporangia. On hay or potato-carrot agar sporangia are produced more freely in 2–10 days. Sporangiophores are 80–250 µm long and tapered, each bearing up to 5 branches. The lack of a columella is characteristic of the genus. The sporangia are globose, smooth and fragile. Spores are colourless and ovoid or kidney-shaped (2–5 x 6–12 µm). Rhizoids are nodal
Mucor circinelloides	Colonies spread right across Petri dish but growth is low. Pale grey or yellowish-brown at 37°C	No rhizoids. Branching or simple sporangiophores that are short. Sporangia are round (25–30 µm) and columella is variable in shape but (unlike *Absidia*) is entirely within the sporangium. Spores are hyaline, ellipsoidal (4–7 µm) and smooth-walled
Rhizomucor pusillus ***R. miehei*** (species were once in *Mucor* genus)	Rapidly growing, cottony colonies. White but surmounted by brown sporangia. Thermophilic with growth at 50–60°C	Rhizoids poorly developed and internodal. Sporangiophores from surface hyphae and usually branched. Sporangia are spherical (40–60 µm) and brown or grey. Columella shape is variable. Spores are hyaline, 3–4 µm, and smooth-walled. *R. miehei* has sporangia with spiny walls
Rhizopus arrhizus ***R. microsporus*** ***R. rhizopodoformis***	Coarse, rampant growth. Petri dish is filled in 5 days with dense, woolly mycelium. White at first, becoming greyish and surmounted by black pin-head-sized sporangia. Growth reaches lid of the plate and may raise it	Rhizoids are nodal (immediately under the sporangiophores), well-developed, and usually obvious in wet mounts or even under the dissecting microscope. The long sporangiophores arise singly or in groups from nodes on the stolon. Sporangia (60–350 µm) are black, round, and contain globose spores. The columella is hemispherical

- ***Conidiobolus coronatus:*** flat, waxy, buff or grey colonies becoming covered with white aerial hyphae. The colonies become tan to brown with age and the reverse is white. The Petri dish lid becomes covered with spores that have been forcibly ejected from the conidiophores.

Microscopic Appearance

The main microscopic features are shown in **Diagrams 56** and **57**.

- ***Basidiobolus ranarum:*** hyphae are wide (8–20 µm) with a few septa that can become more numerous during sporulation. Sporangiophores, each bearing a single-celled sporangium, arise from the hyphae. Cleavage of the sporangial contents leads to the production of sporangiospores within the sporangium. When mature, the sporangium is forcibly ejected together with fragments of the sporangiophore. Zygospores, formed by the conjugation of two adjacent hyphal cells, are common. They are thick, smooth walled, 20–50 μm in diameter, and have a prominent beak-like appendage on one side that is the remnant of a copulatory tube (**Diagram 56**).

- ***Conidiobolus coronatus:*** the hyphae have few septa and give rise to sporophores which at their tips produce spherical spores 10–30 μm in diameter. At maturity, the spores are forcibly ejected and bear a broad tapering projection at the site of previous attachment. Spores may develop short hair-like appendages or germinate to produce single or multiple hyphal tubes. These in turn can become sporophores each bearing secondary spores. A spore may also replicate by producing a number of short extensions that give rise to a corona of secondary spores (**Diagram 57**).

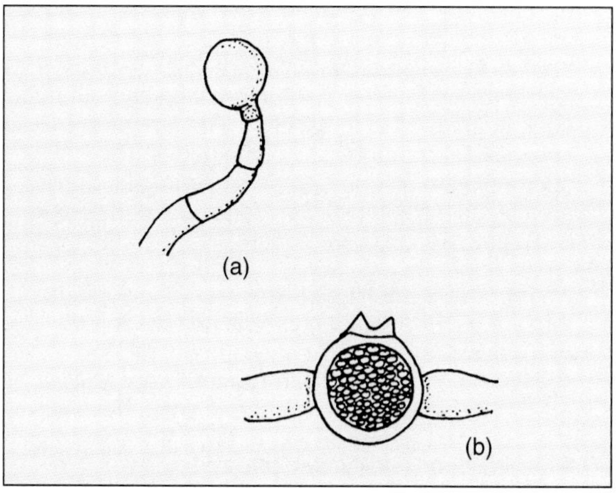

Diagram 56. *Basidiobolus* sp. (an entomophthoraceous Zygomycete): a) single-celled sporangium produced on a hypha; b) zygospore with beak-like appendage.

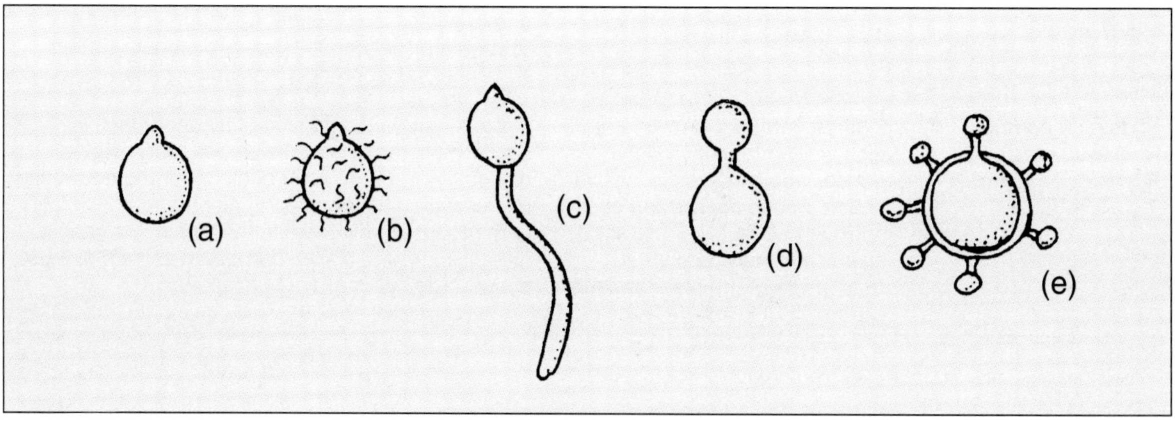

Diagram 57. *Conidiobolus coronatus* (an entomophthoraceous zygomycete: a) conidium; b) conidium with hair-like appendages; c) germination of a conidium to form a hyphal tube; d) conidium bearing a secondary conidium; e) conidium bearing a corona of secondary conidia.

44 The Subcutaneous Mycoses

The subcutaneous mycoses are all transmitted in a similar manner. Fungi that are normally saprophytes in soil, vegetable debris, water or on plants become implanted in the skin due to trauma, with the subsequent development of a subcutaneous infection that is usually chronic. Occasionally there is spread of the infection with involvement of other organs. With the exception of sporotrichosis, which occurs worldwide, the subcutaneous mycoses are most common in tropical and subtropical regions. **Table 142** summarises the subcutaneous mycoses and gives the causative fungi, distribution, main hosts, type of lesions produced, and the appearance of the fungi on direct microscopy of the specimen.

The fungi involved in chromoblastomycosis and phaeohyphomycosis are dematiaceous or darkly pigmented (**480**). The actual fungal species isolated from a lesion often reflects its relative abundance in the particular geographical area. For example, *Phialophora verrucosa* is the most common cause of chromoblastomycosis in humans when the condition occurs in temperate regions.

- Dark sclerotic bodies (5–12 µm diameter) are characteristic of chromoblastomycosis. These thick-walled, muriform cells are thought to represent an intermediate vegetative form, phenotypically arrested between a yeast and a mould (**Diagram 58**).
- The lesions of phaeohyphomycosis, once included with chromoblastomycosis, contain dematiaceous hyphae and yeast-like cells but no sclerotic bodies (**Diagram 59**).
- Mycetomas ('fungal tumours') can be caused by a range of fungi or by the procaryotic actinomycetes. One of the characteristics of mycetomas is the granules (0.5–3.0 mm) that occur in the pus from the lesions. In mycetomas caused by fungi, the granules are composed of intertwined fungal hyphae between 2 and 5 µm in width and inflammatory components. The filaments present in the actinomycotic granules are only 0.5–1.0 µm in width.
- *Rhinosporidium seeberi*, causing rhinosporidiosis, has only recently been cultured *in vitro* using a human rectal tumour cell line. The normal habitat of the fungus is thought to be stagnant water. A diagram of the large sporangia seen on microscopic examination of polyps or other lesions is shown in **Diagram 60**.
- *Pythium insidiosum* is an aquatic microorganism and the cause of cutaneous pythiosis. Some taxonomists now place this organism in the Kingdom Protista, although it is usually discussed in the mycology section for convenience.
- *Loboa loboi*, the aetiological agent of lobomycosis in humans and dolphins, is thought to be a yeast but has not yet been grown *in vitro*. The organism, from lesions in dolphins, has been successfully passaged in mice by foot-pad inoculation.

480

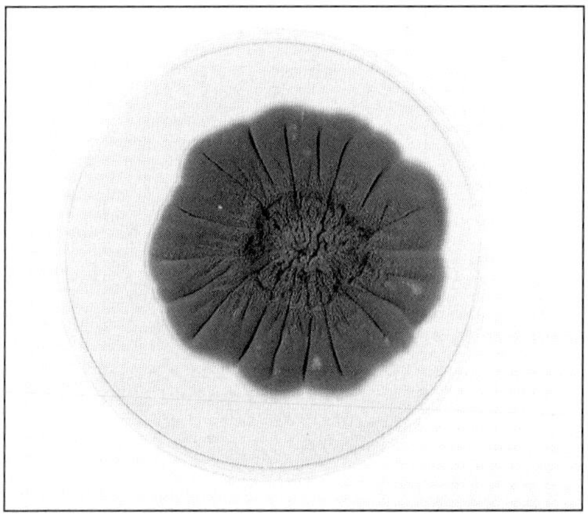

480 *Cladosporium* sp. (a dematiaceous mould) on Sabouraud agar, 5 days.

Laboratory Diagnosis

Direct microscopy

Table 142 summarises the subcutaneous mycoses and indicates the morphological appearance of the fungi on direct microscopy. Wet mounts of exudates or tissues and histopathological sections of biopsies or tissues, stained by the PAS or silver stains, should yield a great deal of information. For practical purposes, the clinical lesions together with direct microscopy can give a presumptive diagnosis of most of the superficial mycoses.

Culture

Sabouraud dextrose agar, with and without antimicrobial agents, will support the growth of the normally saprophytic fungi involved in chromoblastomycosis, phaeohyphomycosis and mycetomas. A few are difficult to isolate and others are very slow-growing, requiring an incubation time of up to 6 weeks. An incubation temperature of 25–30°C is suitable.

Table 142. The subcutaneous mycoses.

Disease	*Fungal agent(s)*	*Distribution*	*Host(s)*	*Lesions*	*Direct microscopy*
Sporotrichosis (*see* Chapter 42)	*Sporothrix schenckii*	Worldwide	Horses, dogs, cats and humans	Ulcerating cutaneous nodes that follow lymphatic vessels, often on the hind legs of horses. Ulcers heal but re-erupt. Lymphatics become thickened. In dogs and cats dissemination to other organs occurs	Single or multiple-budding, cigar-shaped yeast cells (2–4 μm). Rather low numbers in clinical specimens
Epizootic lymphangitis (African Farcy) (*see* Chapter 42)	*Histoplasma farciminosum*	Africa, Asia with cases reported in France, Italy, Russia and Egypt	Equidae	Chronic pyogranulomatous infection of equine skin and lymphatics. Regional lymph nodes are involved and dissemination can occur	Intracellular, pear-shaped, double-contoured yeast cells (2–4 μm). Usually inside mononuclear cells or neutrophils. Budding occurs most commonly at the pointed end of the cells
Subcutaneous zygomycosis (*see* Chapter 43)	*Basidiobolus ranarum* *Conidiobolus coronatus*	Africa, Asia, Caribbean, South America, and a few cases in USA	Humans and rarely in animals	In humans, *C. coronatus* tends to cause a rhinofacial infection and *B. ranarum* is associated with lesions on arms and body. Nodular subcutaneous lesions are produced. If untreated the lesions are slowly progressive	Broad, irregularly branching, hyphae (5–18 μm), sparsely septate and enclosed in eosinophilic material
Chromoblastomycosis	*Fonsecaea pedrosoi* *F. compacta* *Phialophora verrucosa* *Cladosporium carrionii* *Rhinocladiella aquaspersa*	Africa, Australia, South America and Japan, and *P. verrucosa* in temperate regions	Humans and rarely in animals	Legs and feet are most commonly affected. Lesion begins as a nodule but growth becomes large and cauliflower- or wart-like. The initial and satellite growths remain localised and persist for many years	Dark sclerotic bodies (5–12 μm) are present. They are thick-walled, muriform (divided by vertical and horizontal septa), and are usually chestnut brown. In the crusts, dematiaceous (dark), septate, branched hyphae may be seen
Phaeohyphomycosis	*Wangiella dermatitidis* *Exophiala jeanselmei* *Phialophora verrucosa* *Cladosporium trichoides* *Bipolaris specifera* *Aureobasidium pullulans* *Alternaria* spp.	Tropical regions mainly	Humans, cats, dogs, horses, cattle and goats	The superficial lesions are similar to those in chromoblastomycosis but systemic infections can occur involving a wide range of tissues	Dematiaceous hyphae and yeast-like cells, or both, may be present. The hyphae are regular or irregular with swollen ends. No sclerotic bodies are seen

(*continued*)

Table 142. The subcutaneous mycoses. (*continued*)

Disease	*Fungal agent(s)*	*Distribution*	*Host(s)*	*Lesions*	*Direct microscopy*
Mycetomas	*Pseudallescheria boydii* *Exophiala jeanselmei* *Curvularia geniculata* *Madurella mycetomatis*	Africa, Asia, South and Central America, and less commonly in some temperate regions	Humans and more rarely in cattle, horses, dogs and cats	Mycetomas ('fungal tumours') can be caused by both fungi and actinomycetes. Characterised by granulomatous swellings with sinus tracts discharging pus and granules. Slowly progressive and can involve adjacent tissue	Granules in the pus are small (0.5–3.0 mm), irregularly shaped, and of various colours. Microscopically, the granules consist of broad (2–5 μm), intertwined hyphae and swollen cells (15 μm or more) at periphery. Actinomycete filaments are only 0.5–1.0 μm wide
Cutaneous pythiosis	*Pythium insidiosum* (*'Hyphomyces destruens'*)	Tropical and subtropical regions mainly	Humans, horses, cattle and dogs	The skin lesions are pyogranulomatous or fibrogranulomatous. The disease is chronic and progressive. In the horse the lesions are large (up to 45 cm), discharging swellings, usually on extremities, ventral trunk or head. Nasal mucosa can be involved	Masses of hyphae (4 μm diameter) are seen in histological sections. These are mixed with a variety of inflammatory cells such as necrotic macrophages, epithelioid cells and giant cells, forming necrotic masses
Rhinosporidiosis	*Rhinosporidium seeberi*	Tropics and subtropics	Humans, horses, cattle, mules, dogs, goats and some wild water fowl	Granulomatous, mucocutaneous infection. Large polyps, tumours or wart-like lesions on nasal and ocular mucous membranes. The growths are highly vascularised, sessile or pedunculated. About 90% of infections involve the nasal cavity of male animals	Large sporangia (up to 200–300 μm diameter) that contain thousands of endospores
Lobomycosis (Keloidal blastomycosis)	*Loboa loboi*	Dolphins off Florida coast and in Surinam River, Surinam	Humans and dolphins	Lesions are nodular and keloidal in appearance. Early in infection they are freely moving, subcutaneous nodules, but later verrucose, nodular plaques are formed. The lesions spread by peripheral extension	Spherical, thick-walled yeast cells (5–12 μm diameter) are present. Buds are single but sequential budding can lead to chains of cells linked by tubular isthmuses

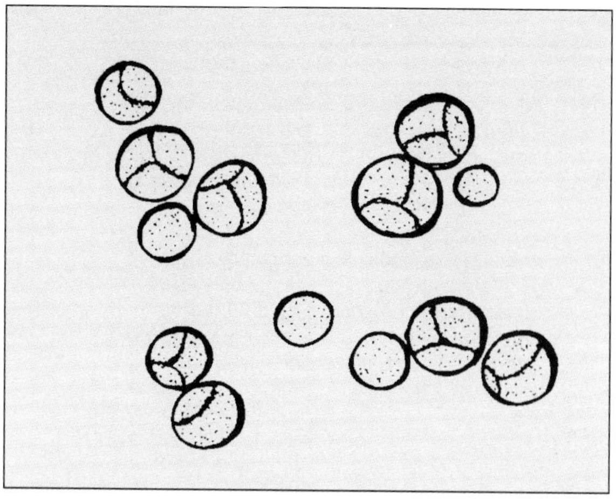

Diagram 58. Muriform sclerotic bodies (5–12 μm in diameter) seen in chromoblastomycosis.

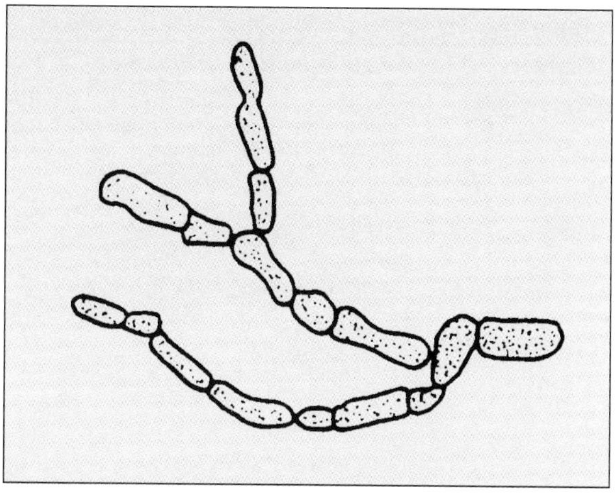

Diagram 59. Dematiaceous hyphae (5–10 μm in diameter) in an aspirate from a case of phaeohyphomycosis.

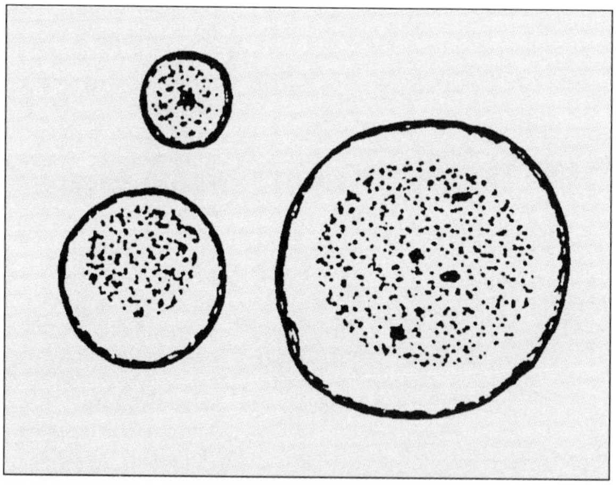

Diagram 60. Sporangia of various sizes (6–300 μm in diameter) that occur in nasal polyps in rhinosporidiosis.

Identification

For the fungi associated with chromoblastomycosis, phaeohyphomycosis and mycetomas, the identification is based on the macroscopic and microscopic morphology of the isolated fungus. With the dematiaceous fungi, particularly, the identification requires a considerable amount of experience. Slide cultures will often be necessary to see clearly the continuity between the conidia, conidia-bearing structures, conidiophores and vegetative hyphae.

The nomenclature is complicated. Fungi should be classified by sexual structures if they are produced and where possible the teleomorphic name (that of the sexual state) is used. However, a teleomorph may produce more than one asexual form (anamorph). For example, *Pseudallescheria boydii* is the teleomorph of both the anamorphs *Scedosporium apiospermum* and *Graphium* species. But *S. apiospermum* is an anamorph that can be produced by several species of both *Pseudallescheria* and *Petriella*. The aid of a reference mycology laboratory should be sought for the identification of the cultures. Schematic drawings of the asexual fruitings heads of some of the more common fungi involved in chromoblastomycosis, phaeohyphomycosis and fungal mycetomas are shown in **Diagram 61**. A summary of the colonial and microscopic features of each is given in **Table 143**.

Table 143. Colonial and microscopic characteristics of some fungi associated with the subcutaneous mycoses (illustrated in **Diagram 61**).

(a) *Alternaria spp.*
Colony: dark greenish-black to grey-brown with a light border. Reverse is black.
Microscopic: hyphae dark and septate. Conidia are large (7–10 × 23–34 μm), brown, muriform, club-shaped and occur singly or in chains

(b) *Curvularia* spp.
Colony: olive-green to brown or black with a pinkish-grey woolly surface. Reverse is black.
Microscopic: conidiophores simple or branched and bent where conidia are formed. Conidia are large (8–14 × 21–35 μm), often four-celled, and appear curved due to the swelling of a central cell

(c) *Bipolaris (Drechslera)* spp.
Colony: greyish-brown forming a matted centre and raised lighter periphery.
Microscopic: dark septate conidiophores with knobby, bent appearance. Conidia are brown, thick-walled, oblong to cylindrical (6–12 × 16–35 μm) and contain four or more cells

(d) *Cladosporium carrionii*
Colony: dark, velvety and olive-green, with a dark reverse.
Microscopic: conidiophores of varying lengths that produce long branching chains of brown, smooth-walled, oval, pointed conidia. The conidia are easily dispersed

(e) *Aureobasidium (Pullularia) pullulans*
Colony: mature colony is black, shiny, leathery and yeast-like, with a lighter fringe. The reverse is black.
Microscopic: two types of hyphae are produced:
1 thick, dark-walled and closely septate, some forming tubes that produce oval conidia
2 hyaline hyphae producing conidia directly from the walls

(f) *Acremonium (Cephalosporium)* spp.
Colony: at first folded and felt-like but becoming overgrown with loose, white, cottony aerial hyphae. Colour varies from white to grey or rose. The reverse is white to pinkish.
Microscopic: unbranched, tapering phialides rise directly from the septate hyphae. The conidia are oblong (2–3 × 4–8 μm), one- or occasionally two-celled, and borne in easily disrupted clusters at the tips of the phialides

(g) *Fonsecaea (Phialophora) compacta*
Colony: slow-growing, velvety to woolly, olive to black. The reverse is dark.
Microscopic: dark, erect conidiophores bearing compact masses of conidial chains that are not easily dissociated. The conidia are one-celled, cask-shaped to round, and usually three per chain

(h) *Phialophora verrucosa*
Colony: mat-like to heaped and granular. The obverse is olive-grey and the reverse is black.
Microscopic: conidiophores, when present, are short and the phialides flask-shaped. Distinct collarettes are present at the apices of the phialides. The conidia (1–3 × 2–4 μm) are hyaline and occur in balls which may slip down the phialides

(i) *Rhinocladiella* spp.
Colony: rapid-growing, velvety, slightly elevated and olive-black.
Microscopic: the conidiophores are sympodial, unbranched, erect and usually darker than the vegetative hyphae. Conidia are one-celled (rarely two-celled), fusiform, elliptical, obovate, smooth, light-brown with a dark basal scar

(j) *Wangiella* spp.
Colony: yeast-like at first, then developing some aerial hyphae with age. Oiive to black.
Microscopic: conidiophores indistinguishable from the hyphae. The conidia-bearing cells are phialides without collarettes. Conidia are one-celled, dark and smooth, forming balls at the apices of the phialides, these balls then slide down the sides of the phialides

(k) *Scedosporium* spp.
Colony: rapidly growing, cottony, and smokey grey to dark-brown.
Microscopic: one-celled conidia may occur singly along the hyphae or in clusters at the apices of the annellides (conidia-bearing structures). The latter are often characterised by swollen rings. The conidia are obovate, truncate, subhyaline to light black

(l) *Exophiala* spp.
Colony: appearance varies; slow to rapid growth, moist and yeast-like at first, but becomes woolly with age and grey to black. Some colonies remain black and yeast-like. The reverse is black.
Microscopic: conidiophores are dark, simple or hypha-like. The conidia-bearing cells are tapered to a narrow apex. The conidia are oval, one-celled, and accumulate in balls at the apices of the conidia-bearing cells

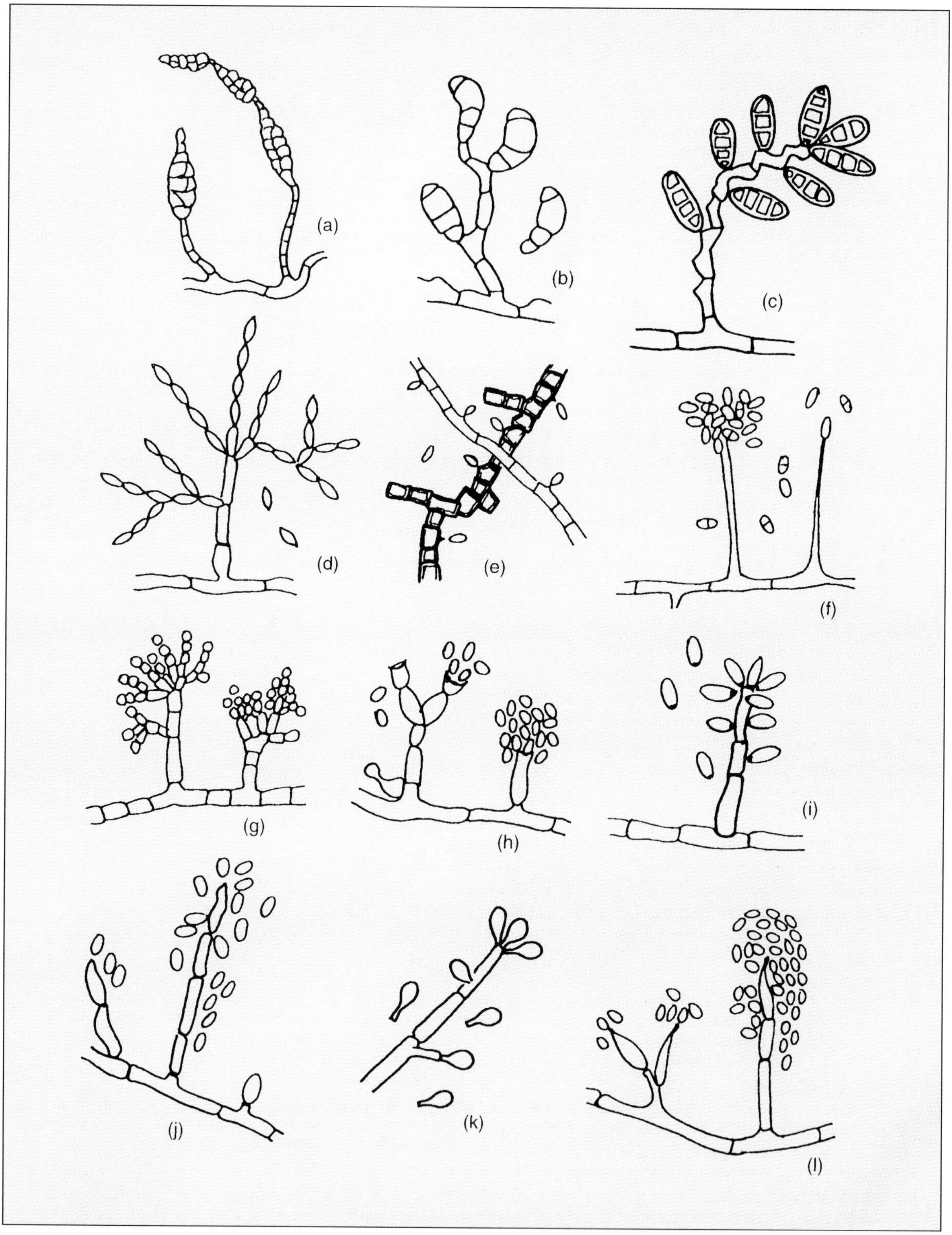

Diagram 61. Fruiting heads of the genera of some fungi associated with subcutaneous mycoses.
a) *Alternaria;* b) *Curvularia;* c) *Bipolaris* (*Drechslera*)*;* d) *Cladosporium;* e) *Aureobasidium* (*Pullularia*)*;* f) *Acremonium* (*Cephalosporium*); g) *Fonsecaea compacta* (*Phialophora compacta*); h) *Phialophora verrucosa;* i) *Rhinocladiella;* j) *Wangiella;* k) *Scedosporium;* l)*Exophiala.*

45 Mycotoxins and Mycotoxicoses

General Features of Mycotoxin Formation

Moulds can grow on a wide range of organic matter including growing crops and stored feed. Those of greatest significance in veterinary medicine produce secondary metabolites which are toxic for many animal species including man. Fungal toxins are referred to as mycotoxins and the diseases they produce are termed mycotoxicoses. Production of mycotoxins occurs as a result of normal fungal metabolism. No specific role for these metabolites in the life cycle of the fungus has been demonstrated. Some mycotoxicoses such as ergotism have been known since the Middle Ages, while many other diseases arising from the ingestion of mycotoxin-contaminated pasture or feed have been recognised only in recent decades.

Mycotoxicoses are not infections but are acute or chronic intoxications produced by toxic metabolites of fungal origin. Many of the toxigenic fungi are widespread throughout the world and over 100 known species are capable of elaborating mycotoxins. More than one fungal species may produce the same mycotoxin while individual moulds may produce two or more different mycotoxins. Because taxonomic classification of fungi is based almost exclusively on morphological rather than physiological considerations, there is frequently little correlation between the pattern of toxin production by particular fungi and their phylogenetic classification. Toxin production occurs only under specific conditions of moisture, temperature, suitability of substrate and appropriate oxygen tension. The optimum conditions for toxin production are relatively specific for each fungus. *Fusarium sporotrichoides* elaborates its toxin at freezing temperatures while *Aspergillus flavus* requires a temperature of 25°C. Only some strains of a single species, such as *Aspergillus flavus,* have the ability to produce toxins, even under favourable conditions. Some strains of fungi have a distinct preference for certain substrates and consequently there may be a regional prevalence of particular mycotoxicoses depending on the types of crops or pasture cultivated in the region.

The susceptibility of different crops to mould infection is governed by the presence of suitable substrates. The seed or kernel may be the preferred target of some fungi because of the ready availability of carbohydrates, while the fibrous part of the plant with its high cellulose content may be invaded by other fungi capable of using this substrate. Damage to the seed coat by insects, mechanical harvesting, severe frost or other factors may predispose crops to fungal attack. Insects may also serve as carriers of fungal spores.

Mycotoxicoses are diseases in which many factors interact. Animals vary widely in their susceptibility to mycotoxins. Younger animals tend to be more susceptible than adults and there is also considerable variation in species and individual susceptibility. The conditions which favour mycotoxin production and the factors which influence the severity of mycotoxicoses are summarised in **Diagram 62**. Since mycotoxins are most likely to be concentrated in highest amounts in stored feeds, groups of animals at risk include poultry, pigs, dairy and feedlot cattle fed on contaminated feed. In some countries, particularly New Zealand, mycotoxicoses such as 'facial eczema' are associated with standing pasture and accordingly prevailing climatic conditions determine the occurrence of disease.

Characteristics of mycotoxins

As mycotoxins are low molecular weight, non-antigenic substances in their naturally occurring forms, acquired immunity does not occur following exposure. Many of the major mycotoxins are heat-stable and consequently can retain their toxicity after processing temperatures used for pelleting or other milling procedures. Each fungal toxin, if present in the diet at a sufficient concentration, usually affects specific target organs or tissues. Secondary mycotoxic disease may be more difficult to recognise because low levels of toxin intake may not result in a specific mycotoxicosis but in a heightened susceptibility to intercurrent infections due to immunosuppression. Some of the more important characteristics of mycotoxins are presented in **Table 144**.

Mycotoxicoses

The severity of mycotoxicoses in animals and their clinical recognition is determined by many factors including the species of toxigenic fungus, the concentration of mycotoxin in the food, the age, sex and health status of the exposed animal, the target organ or tissue affected, and the duration of exposure to contaminated feed. The features which characterise mycotoxic diseases include sporadic and seasonal occurrence, lack of transmissibility, association with certain batches of stored food or particular types of pasture, and disappointing response to drug treatment effective against infectious diseases (**Table 145**). Clinical and

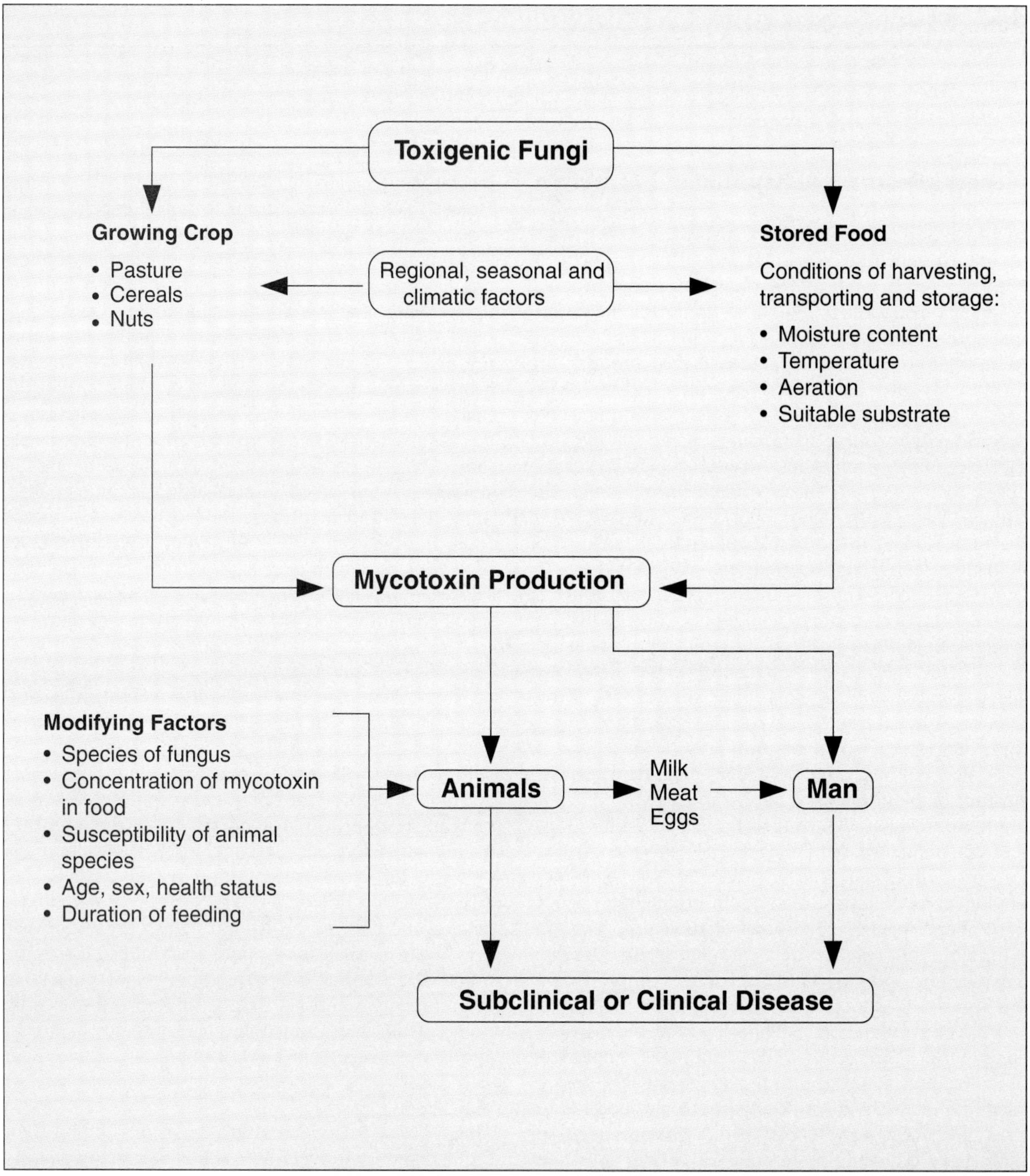

Diagram 62. Factors influencing the production of mycotoxins on growing crops or stored foods. Mycotoxicoses in animals depend on many factors, some relating to regional distribution of toxigenic fungi and their preference for certain substrates, others relating to the species and age of animals exposed to contaminated food. Human mycotoxicoses can result from consumption of contaminated food, especially stored food, such as cereals or nuts, and occasionally from food of animal origin, particularly milk or dairy products.

Table 144. Characteristics of mycotoxins.

- Mycotoxins are secondary fungal metabolites
- They include a diverse group of compounds with a wide spectrum of toxic effects produced by different toxigenic fungi
- They have a low molecular weight, are non-antigenic, heat-stable, and are active at low concentrations
- The absence of acquired immunity means that exposure does not confer protection
- Many fungal toxins affect target organs or tissues
- They may induce a range of clinical effects; some have potent carcinogenic, mutagenic and teratogenic properties, others are immunosuppressive

Table 145. Clinical features of mycotoxicoses.

- Outbreaks are often seasonal and sporadic, and may be associated with certain batches of stored food or particular types of pasture
- The diseases produced are not transmissible to in-contact animals
- Initially, the cause may be obscure, with vague signs of illness such as decreased growth rate or immunosuppression
- Exposure does not confer protection; rather, continued exposure leads to progressive deterioration
- Treatments such as antibiotics, employed for infectious diseases, are usually ineffective
- Recovery generally depends on the type and amount of mycotoxin ingested and the duration of exposure to contaminated food
- The only acceptable evidence for the presence of mycotoxicoses in animals is the laboratory demonstration of mycotoxins in suspect food, or in the tissues, secretions or excretions of affected animals
- Characteristic lesions in target organs of affected animals is important supporting diagnostic evidence

laboratory procedures may demonstrate pathological changes in the animal, characteristic of a particular intoxication. Diagnosis of a mycotoxicosis, however, requires the demonstration of biologically effective concentrations of the fungal toxin in the feed available to the animal, or in the animal's tissues, secretions or excretions. **Table 146** summarises the principal features of those mycotoxicoses which can be recognised clinically. Subclinical and chronic disease associated with the consumption of fungal toxins in feed is an aspect of mycotoxicology which, although difficult to quantify, is of likely economic and public health significance.

Mycotoxicoses

Aflatoxicosis

Aflatoxins are a group of approximately 20 related toxic compounds produced by some strains of *Aspergillus flavus* (**481**) and *Aspergillus parasiticus* during growth on natural substrates including growing crops and stored food. These fungi are ubiquitous, saphrophytic moulds which grow on a variety of cereal grains and foodstuffs such as maize, cottonseed and groundnuts. About half of the strains of *A. flavus* and *A. parasiticus* are toxigenic under optimal environmental conditions. High humidity and high temperatures

Table 146. Mycotoxicoses of domestic animals, including poultry and fish.

Disease	*Fungus*	*Crop or substrate*	*Myco-toxin(s)*	*Animals affected*	*Pathogenesis/clinical syndrome*
Aflatoxicosis	*Aspergillus flavus* *Aspergillus parasiticus*	Stored grain, groundnuts, maize, most nutcrops	Aflatoxins B_1, B_2, G_1, G_2	Cattle, pigs, poultry, trout, dogs, rats	Aflatoxins are hepatotoxic, teratogenic, mutagenic and carcinogenic. Clinical effects include a drop in milk production in dairy cows, nervous signs in young animals, decreased food conversions in most species and immunosuppression
Diplodiosis	*Diplodia zeae*	Maize	Unidentified neurotoxin	Cattle, sheep	Neurotoxic disease. Clinical signs include ataxia, paralysis, lacrimation and salivation
Ergotism	*Claviceps purpurea*	Seedheads of many grasses and grain	Ergot alkaloids: ergotamine, ergometrine	Cattle, sheep, pigs, horses, poultry	Vascular spasm leads to the gangrenous form of the disease; convulsions caused by stimulation followed by depression of the central nervous system
Facial eczema	*Pithomyces chartarum*	Pasture litter	Sporidesmin	Sheep, cattle	Hepatotoxicosis, followed by photosensitisation
Fescue toxicity	*Acremonium coenophialum*	Tall fescue grass	Thought to be a vasco-constrictive mycotoxin	Cattle	Ischaemic necrosis of the extremities, unthriftiness, heat intolerance and hyperthermia
***Fusarium* toxicoses: Oestrogenism**	*Fusarium graminearum*	Maize, barley and cereals	Zearalenone	Pigs and sometimes other species	Mycotoxin has oestrogenic activity: hyperaemia and oedema of the vulva, enlargement of the mammary glands, and infertility in sows and cows
Leukoencephalomalacia	*Fusarium monilforme*	Maize	Fumonisins B_1 (A_1, A_2, B_2)	Horses and donkeys	Liquefactive necrosis of areas in the cerebral hemispheres leading to ataxia, tremors, recumbency and death
Food refusal/ emetic syndrome	*F. graminearum*	Cereals	Deoxy-nivalenol	Pigs and other species	Local and central stimulation that leads to emesis
Trichothecene toxicosis	Many *Fusarium* species	Cereals	T-2 toxin, diacetoxy-scirpenol	Many species	Epithelial necrotising agents, immunosuppressive agents, haemorrhagic syndromes and immunosuppression
Lupinosis	*Phomopsis leptostromiformis*	Growing lupins	Phomopsins A and B	Cattle, sheep	The mycotoxins are hepatotoxic: inappetence, ruminal stasis, jaundice and photosensitisation

(*continued*)

Table 146. Mycotoxicoses of domestic animals, including poultry and fish. (*continued*)

Disease	*Fungus*	*Crop or substrate*	*Myco-toxin(s)*	*Animals affected*	*Pathogenesis/clinical syndrome*
Myrothecio-toxicosis	*Myrothecium roridium* and other *Myrothecium* species	Ryegrass, white clover	Macrocyclic tricho-thecenes	Cattle, sheep	Abdominal pain, depression, salivation, dehydration: the syndrome resembles a chemical rumenitis
Ochra-toxicosis	*Aspergillus ochraceus* *Penicillium viridicatum*	Barley, wheat, maize	Dihydroiso-coumarin derivatives, ochratoxin A	Pigs, poultry	Toxins interfere with protein synthesis, and lead to degenerative kidney changes, weight loss, polydipsia and polyuria
Slaframine toxicosis	*Rhizoctonia leguminicola*	Red clover, pasture or hay	Slaframine	Cattle, sheep, horses	Cholinergic mycotoxin: excessive salivation, lacrimation, urination and diarrhoea
Stachy-botryo-toxicosis	*Stachybotrys atra*	Hay or straw	Trichothe-cene toxins	Horses, cattle, sheep and other species	Toxins have a radiomimetic effect: atrophy and necrosis of lymphoid tissue, aplastic anaemia, degenerative changes in liver and kidneys, stomatitis, necrosis of oral mucosa, severe depression, impaired blood clotting, secondary bacterial infection and death
Tremorgen intoxication	Many fungi including: *Penicillium cyclopium*	Mouldy feed	Penitrem A and related substances	Many species, particularly ruminants	Many tremorgens act as neurotoxins. Clinical signs include tremors, ataxia, and convulsive episodes
	Acremonium loliae	Perennial ryegrass	Lolitrems		
	Claviceps paspali and other *Claviceps* species	Paspalum grass	Paspaline, paspalicine, paspalinine and paspalitrems		
	Aspergillus clavatus	Sprouting grain or millers' culms	Not identified		

481

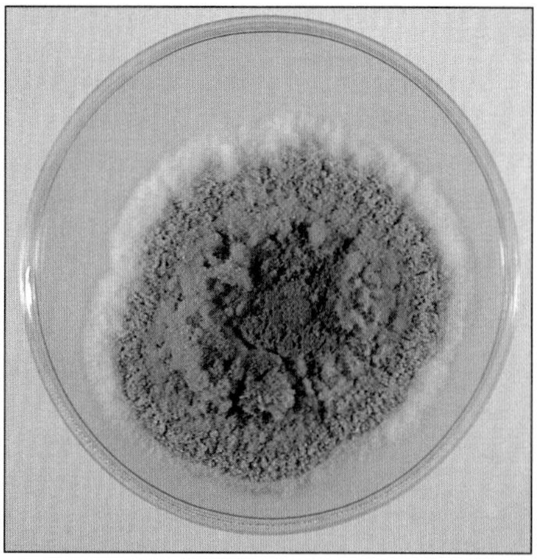

481 *Aspergillus flavus* (toxigenic strain) on Sabouraud agar, 5 days.

during pre-harvesting, harvesting, transportation and storage, as well as damage to field crops by insects, drought and mechanical injury during harvesting, favour the growth of *A. flavus* and toxin production (**Diagram 62**). Although other fungi such as *Penicillium* spp. and *Rhizopus* spp. are capable of producing aflatoxins, their relevance to livestock production has not yet been established. The name 'aflatoxin' derives from *Aspergillus*(a-), *flavus* (fla-) and toxin.

One of the first, well-documented outbreaks of aflatoxicosis occurred in East Anglia, England, in 1960 when more than 100,000 turkey poults died of an unknown disease ('turkey X disease'). Subsequently, it was demonstrated that these birds died from a toxin present in pelleted feed which formed a major part of their diet. A shipment of groundnut meal containing aflatoxin had been used as a protein supplement in the turkey rations. Examination of the incriminated groundnut meal revealed the presence of mould mycelia and thin layer chromatography showed the presence of several compounds which fluoresced under ultra-violet light. The fungus was identified as *Aspergillus flavus* and the toxic metabolites were called aflatoxins. Since that time, numerous outbreaks of aflatoxicosis have been described worldwide.

Aflatoxins

Aflatoxins are a group of related bisfuranocoumarin compounds with toxic, carcinogenic, teratogenic and mutagenic activity. The four major aflatoxins are B_1, B_2, G_1, and G_2. Aflatoxins M_1 and M_2 are hydroxylated metabolites of B_1 and B_2 that are excreted in the milk of lactating animals such as dairy cows. Aflatoxin B_1 (AFB_1) is the most commonly occurring and also the most toxic and carcinogenic member of the group (**Diagram 63**). These mycotoxins are named according to their position and fluorescent colour on thin layer chromatograms, when viewed under ultra-violet light. AFB_1 and AFB_2 produce a blue and AFG_1 and AFG_2 a green fluorescence. Most of the other aflatoxins are metabolites formed endogenously in animals after ingestion or administration of aflatoxins. These toxins are stable compounds in food and feed products and are relatively resistant to heat. They retain much of their activity after exposure to dry heat at 250°C and moist heat at 120°C but may be degraded by sunlight. They have a low molecular weight (AFB_1: 312; AFM_1: 328) and are non-antigenic in their native state.

When growing in maize, *A. flavus* usually produces B_1 and B_2 aflatoxins, while *A. parasiticus* produces all four of the major aflatoxins. On soybeans, only low concentrations of AFB_1 are produced by both species. Mould growth and toxin formation require a moisture content of the substrate greater than 15 per cent, a temperature of at least 25°C and adequate aeration. Toxin formation can occur in a matter of hours when favourable conditions exist.

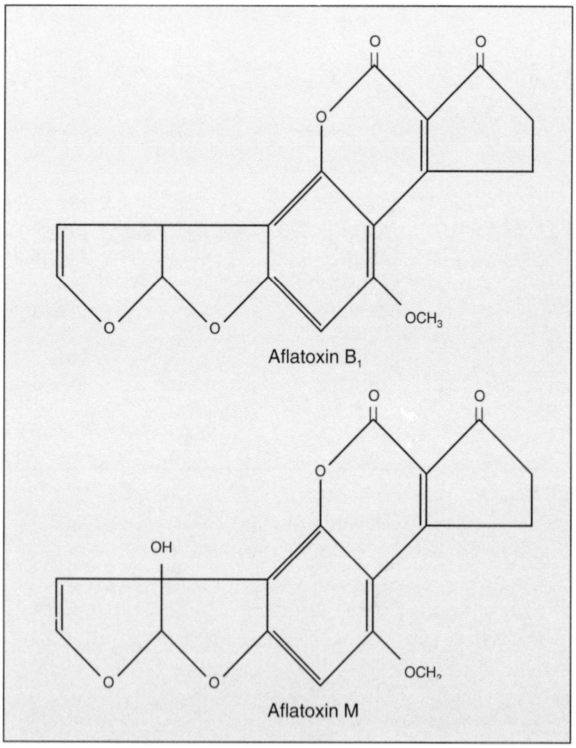

Diagram 63. Structure of aflatoxin B_1 and its hydroxylated metabolite M_1, which may be secreted in milk.

Biological effects of aflatoxins

The toxic effects of aflatoxins are dependent on dose, time and species. The toxins are absorbed from the stomach and metabolised in the liver to a range of toxic and non-toxic metabolites which are then excreted in urine and milk. The major biological effects of aflatoxins include inhibition of RNA and protein synthesis, impairment of hepatic function, carcinogenesis and immunosuppression. AFB_1 is bioactivated in the liver to a highly reactive intermediate compound which reacts with various nucleophiles in the cell and binds covalently with DNA, RNA and protein. After deliberate administration of AFB_1 there is marked interference with protein synthesis at the translational level, which seems to correlate with disaggregation of polyribosomes in the endoplasmic reticulum. Many of the toxic responses observed in animals, resulting from AFB_1 activity, can be attributed to alterations in carbohydrate and lipid metabolism and interference with mitochondrial respiration.

The biological effects of aflatoxins, observed clinically, can be divided into two categories, short-term effects and long-term effects, depending on the dosage level and frequency of exposure to the toxin. Short-term effects include acute toxicity with clinical evidence of hepatic injury and nervous signs such as ataxia and convulsions. In acutely affected animals death may occur suddenly. Long-term consumption of low levels of aflatoxins probably consititutes a much more serious veterinary problem than acute, fulminating outbreaks of aflatoxicosis. With chronic aflatoxicosis there is reduction in efficiency of food conversion, depressed daily weight gain, decreased milk production in dairy cattle and enhanced susceptibility to intercurrent infections in most species due to immunosuppression. AFB_1 is also an extremely potent hepatocarcinogen in many species of animals.

The principal target organ of these mycotoxins in all species is the liver. Depending on the severity and duration of the intoxication, lesions in the liver may vary from acute swelling with hepatocellular necrosis and bile retention to cirrhosis and marked bile duct hyperplasia. Reduction of hepatic function and increased serum enzyme activities indicative of hepatic cell necrosis are common sequelae to aflatoxin-induced hepatic injury. Although not related to pyrrolizidine alkaloids, aflatoxins produce similar liver changes. As liver function is progressively altered, other effects such as coagulopathy, icterus, serosal and mucosal haemorrhages may occur. Acute hepatic failure and massive haemorrhage due to impaired blood clotting and increased capillary fragility leading to death, may occur with higher doses.

In addition to liver damage, higher doses may cause degenerative changes in the proximal tubules of the kidney. The thymus is also affected and AFB_1 induces thymic cortical aplasia leading to depressed cell-mediated responses. Humoral immunity appears to be affected minimally, but reduced complement levels and decreased phagocytic activity have been reported in aflatoxin-treated animals. Young animals are reported to be notably more susceptible to aflatoxin poisoning than mature animals of the same species. Young pigs, calves, turkey poults and ducklings are particularly susceptible to these fungal toxins.

Aflatoxins are extremely potent hepatocarcinogens in many animal species. AFB_1 is one of the most carcinogenic compounds known for the rat and for the rainbow trout. Experimentally, the carcinogenic activity of pure AFB_1 has been confirmed in rats, duck, pigs, trout and monkeys. Epidemiological studies of primary hepatocellular carcinoma in man indicate that aflatoxins are aetiologicaly involved. Hepatitis B virus and aflatoxins are believed to act synergistically as hepatocarcinogens in man. Teratogenic and embryotoxic effects of aflatoxins have been reported in chickens, Syrian golden hamsters, mice and pigs.

Diagnosis of aflatoxicosis : clinical aspects

Sporadic outbreaks of disease associated with a particular consignment of feed, accompanied by unthriftiness, inappetance and vague signs of illness may suggest the presence of a hepatotoxin in the rations. Clinical signs vary with the susceptibility of individual species to aflatoxins. Prominent signs in calves include blindness, circling, grinding of the teeth, diarrhoea and tenesmus. Elevated plasma aspartate aminotransferase, gamma-glutamyl transferase and alkaline phosphatase are likely findings.

Terminally, convulsions may occur. Postmortem findings include a pale, firm fibrosed liver usually with centrilobular necrosis and bile duct proliferation. The kidneys of affected cattle may be yellow. Ascites and oedema of the mesentery may be present. Blood coagulation defects are likely to occur when extensive liver damage is present. In dairy cattle, aflatoxins M_1 and M_2, hydroxylated metabolites of B_1 and B_2, are excreted in the milk and are of public health significance, if present in appreciable amounts. Milk containing AFB_1 proved hepatotoxic when fed to ducklings. AFM_1 is stable in cheese made from naturally contaminated milk for at least 3 months.

Ruminants vary in their susceptibility to these toxins. Aflatoxicosis has been described in goats, but sheep appear to be very resistant and most of the aflatoxin administered to sheep appears to be degraded in the body.

Aflatoxicosis in pigs produces a range of clinical signs including drowsiness, inappetance, jaundice, weight loss and yellow urine. Changes in the activities of liver-specific enzymes usually parallel the extent of hepatic lesions. A depression of acquired immune responses has been observed in pigs with aflatoxicosis.

Ducklings are considered to be the avian species most susceptible to aflatoxins. Signs of acute disease include anorexia, poor growth rate, ataxia and opisthotonos, followed by death. In birds over three weeks of age, subcutaneous haemorrhages of legs and feet may be evident. In both acute and chronic aflatoxicosis, liver lesions are a common finding. Prolonged exposure to low levels of aflatoxins leads to marked nodular hyperplasia of the liver, bile duct proliferation, fibrosis and hepatocellular carcinoma. Because of their rapid response to aflatoxins, ducklings have been used for biological assays.

Acute toxicity of aflatoxins in chickens and turkeys may be characterised by haemorrhage in many tissues, liver necrosis and jaundice. Aflatoxicosis increases the susceptibility of turkeys to pasteurellosis and salmonellosis. Chickens become more susceptible to coccidiosis and Marek's disease, presumably because of the immunosuppressive effect of aflatoxins. When layers are fed contaminated feed, aflatoxins are present in eggs, principally in the yolk.

Outbreaks of aflatoxicosis have been described in dogs scavenging garbage, and they appear to be particularly susceptible to AFB_1. Chronic aflatoxicosis in dogs is associated with loss of weight, jaundice and ascites. Lesions include subserosal and submucosal haemorrhages in the thoracic and peritoneal cavities and a yellow, mottled liver.

Laboratory Investigation of Outbreaks

Diagnosis of aflatoxin poisoning in animals requires careful consideration of epidemiological factors, clinical signs in affected animals and postmortem findings. Laboratory examination of suspect material may assist in identifying potentially toxigenic fungi growing in food. Chemical identification of mycotoxins in food samples submitted and biological assays for toxicity are important confirmatory steps in field investigations. Cultural examination of food may show the presence of potentially toxigenic fungi (*A. flavus* and *A. parasiticus*) but this is not diagnostic of aflatoxicosis. Laboratory confirmation of mycotoxicoses requires the demonstration of toxigenic strains of *A. flavus* or *A. parasiticus* and of potentially toxic levels of mycotoxins in the food or tissues in conjunction with appropriate clinical or pathological findings. Concentrations of AFB_1 in excess of 100 µg/kg of feed are considered toxic for cattle; the lowest detectable level of this aflatoxin is still toxic for trout.

The analytical procedures for detecting mycotoxins generally follow a standard pattern: sampling, extraction, clean-up, separation, detection, quantitation and confirmation. Mycotoxins are rarely uniformly distributed in natural products such as cereal grains. Aflatoxins are generally found in high concentrations at sites where toxigenic fungi have invaded the crop or stored feed. Accordingly, when investigating suspected field outbreaks of aflatoxicosis, a representative sample of the entire batch is necessary in addition to a sample from contaminated areas. A 5 kg sample, taken into a clean dry container should be labelled with relevant information and stored at −20°C. Mycotoxin formation is continuous when temperature, moisture, aeration and substrate are favourable; therefore, it is necessary to stop aflatoxin formation in the sample at the time of collection by freezing.

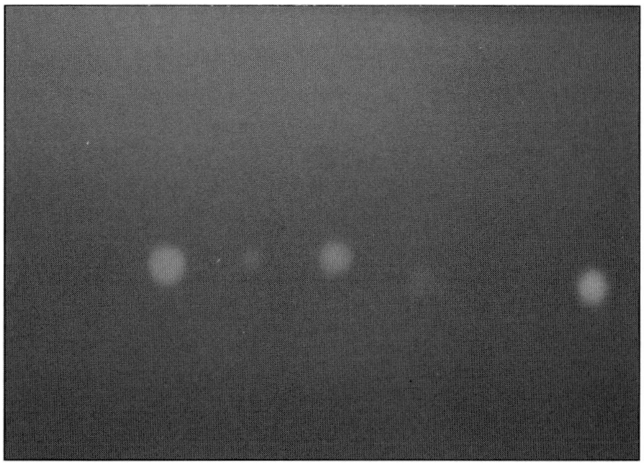

482 Thin layer chromatography plate of aflatoxins under ultra-violet light showing blue, B_2 (extreme left) and green, G_2 (extreme right) fluorescence.

A number of rapid, economical analytical methods are available for determining aflatoxin levels in a wide range of agricultural food products. Thin layer chromatography (TLC) is relatively inexpensive and a number of samples can be analysed simultaneously. The chromatogram is viewed under ultra-violet light for blue or green fluorescent spots that agree in colour and location with internal and external standards (**482**). Many different TLC methods have been reported, including two-dimensional chromatography for differentiating co-extracted substances. In recent years minicolumn detection methods have been employed for rapid screening procedures. High-performance liquid chromatography (HPLC) has been introduced for aflatoxin detection and it has the advantage of being easier, faster and giving more reproducible results than TLC. HPLC is a more sensitive but also a more expensive procedure.

Although aflatoxins are low molecular weight substances, they can be conjugated to protein or polypeptide carriers and subsequently used for antiserum production. A number of immunoassay methods are currently available including radioimmunassay (RIA) and enzyme-linked immunosorbent assay (ELISA). A particularly promising method for detecting aflatoxins is the development of monoclonal antibody assays.

Biological assays for aflatoxins include bile duct proliferation in one-day-old ducklings, chick-embryo bioassay for AFB_1, brine shrimp larvae tests, mutagenicity tests with different bacteria and trout-embryo bioassay for carcinogenicity.

Control and Prevention of Aflatoxicosis

Prevention of contamination at all stages of food production, storage and use is the preferred method of preventing aflatoxicosis. This can be accomplished by reducing fungal infections in growing crops, by rapid drying of harvested crops or by using effective antifungal preservatives. Effective methods have been devised for minimising fungal damage of many food crops, for detection of aflatoxins at low levels and for detoxification of contaminated products. Quality control measures can be only applied, however, where the necessary level of technological services are available during harvesting, storage, food processing and distribution. The risk of aflatoxin contamination is often greatest in regions where detection measures are unavailable, particularly where climatic conditions favour mould damage to crops and stored food.

Decontamination strategies proposed include physical removal, thermal inactivation, irradiation, microbial degradation and chemical treatment of aflatoxin-contaminated feeds. Physical removal and chemical inactivation of toxins in contaminated feed have been used commercially in some countries. Acids, alkalis, aldehydes, oxidising agents and selected gases have been used for degrading aflatoxins. Ammonia gas at elevated temperatures and pressures cleaves aflatoxin molecules and it is claimed that ammoniated feeds can be safely fed to animals. Suspect feed or aflatoxin-contaminated material that has been decontaminated should not be fed to dairy cattle because of the danger of AFM_1 transfer into milk; such feed is also unsuitable for young pigs, calves or turkey poults because of their high susceptibility to aflatoxin. High affinity inorganic adsorptive compounds such as hydrated sodium calcium aluminosilicate added to the diet have been reported to bind aflatoxins and reduce their bioavailability and toxicity. In many countries legislation has been introduced to regulate the maximum acceptable levels of aflatoxins in animal feeds and human food.

After harvesting and during shipment, storage or compounding of agricultural crops, growth of *A. flavus* and *A. parasiticus* is influenced by moisture levels, temperature, aeration, mould spore density, conditions of storage particularly leakage of water into storage containers or condensation, and by biological heat and the chemical nature of the crop. The most critical environmental factors for aflatoxin production are moisture content, temperature and time. Where physical methods of storage are unsatisfactory, chemical preservatives should be considered. The addition of selected chemicals at harvesting can reduce fungal growth where other methods cannot be used. Various organic acids such as benzoic and propionic acid have been widely used as preservatives for stored agricultural products. It should be noted, however, that sub-inhibitory levels of propionic acid may actually stimulate aflatoxin biosynthesis.

Ergotism

Ergotism in animals results from the ingestion of grasses and cereals, particularly rye, infected with fungal species of the genus *Claviceps*, notably *Claviceps purpurea*. The fungus colonises the seed head and the ovarian tissue is destroyed and replaced by a soft mycelial mat which enlarges, hardens and darkens to form sclerotia. The word ergot, which is the French term for a rooster's spur, accurately describes the compacted mass of hyphae that projects from the ear as a dark, purplish-black, misshapen replica of the original seed. Although ergotism is one of the oldest mycotoxic conditions known, with descriptions of the condition dating to biblical times, it was not until the middle of the last century that an association between the ingestion of rye infected with *Claviceps purpurea* and clinical ergotism was established. Major epidemics of human ergotism were recorded in central Europe from the 12th to the 18th century. Few epidemics of ergotism in the human population have been recorded in recent years. Periodic outbreaks in India and Ethiopia have been recorded. However, ergot poisoning of livestock from consumption of infected grain or grazing pastures containing grasses with ergotized seed heads is relatively common.

Claviceps purpurea has a complex life cycle which takes a year to complete. The sclerotia overwinter on the soil and during the following spring they produce several long-stalked, mushroom-like stromata with globose heads containing perithecia that open to the surface. Within each perithecium, several cylindrical asci are formed, each containing thread-like ascospores. When susceptible grasses are flowering in the spring, the ascospores are forcibly discharged from the perithecia. The wind-borne ascospores germinate when they reach cereal rye, rye-grasses or other suitable grass flowers. The heaviest infestation occurs when germination of the overwintered sclerotia coincides with the flowering of the grasses. The tissue of the ovary is replaced by a soft mycelial mat that becomes covered by layers of short conidiophores bearing one-celled conidia mixed with sticky exudate. At this stage insects may spread the conidia and cause secondary infections of other plants. The mycelial mats in infected ovaries enlarge, harden and darken to form sclerotia that completely replace the grain or grass seed and repeat the life cycle (**Diagram 64**).

Ergot alkaloids

The sclerotia or ergots contain the toxic alkaloids (**483**). Ergot alkaloids are derivatives of lysergic and isolysergic acids. Although more than 40 alkaloids

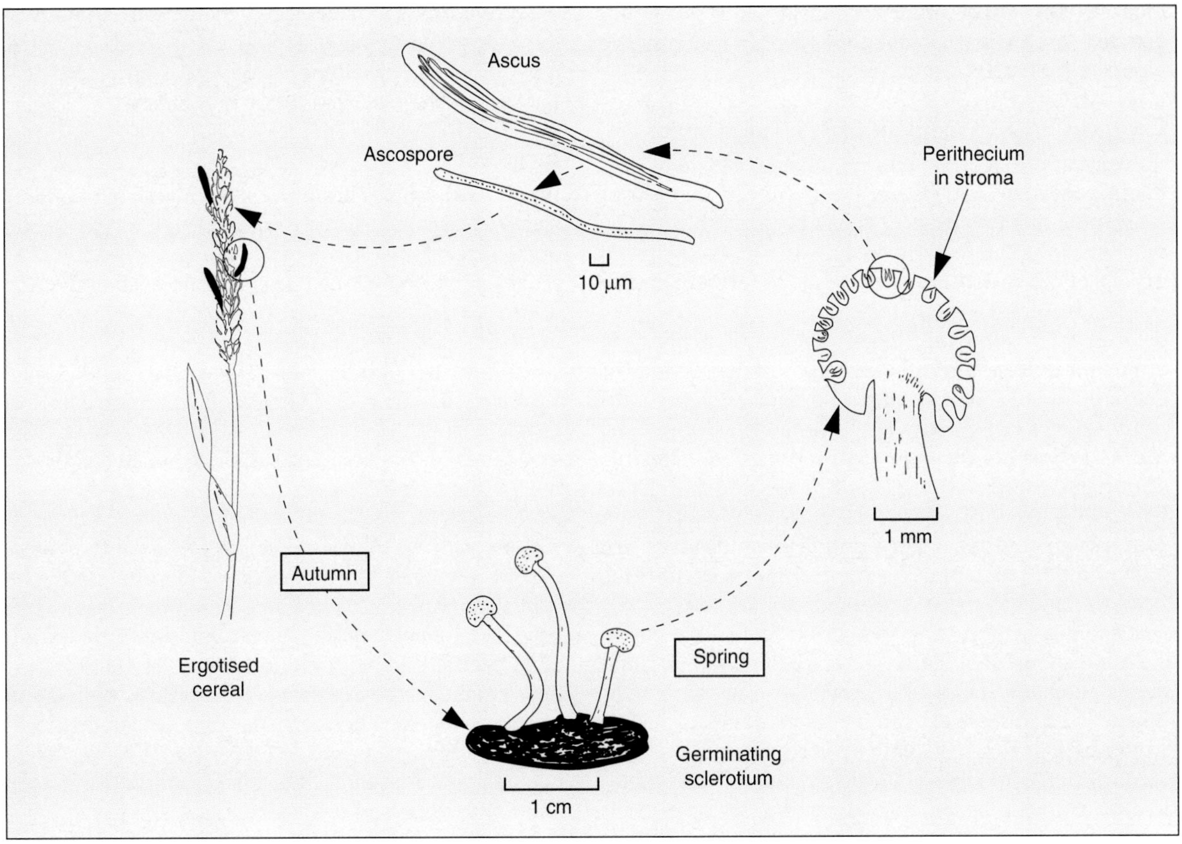

Diagram 64. Life cycle of *Claviceps purpurea*.

have been isolated from the sclerotia of *C. purpurea*, the major toxic alkaloids are ergotamine and ergometrine. In addition, the sclerotia contain a large number of amines and other compounds with physiological activity.

Two forms of ergotism are observed in animals, gangrenous and convulsive ergotism. The ergot alkaloids, particularly ergotamine, stimulate and then depress the central nervous system when taken in large amounts. They also exert a direct stimulatory effect on constrictor adrenergic nerves supplying arteriolar smooth musculature. When consumed in small amounts over long periods, they produce arteriolar spasm, capillary and endothelial damage resulting in vascular stasis, ischaemia and gangrene of the affected part. Convulsive ergotism, characterised by neurotoxicity, is an acute form of the disease, in which the intake of alkaloids is presumed to be substantially higher than in the chronic, gangrenous form of the disease. Towards the end of pregnancy, ergot alkaloids exert an oxytocin-like effect on the pregnant uterus, but this response occurs only late in gestation. While abortions have been described in cattle consuming ergotised grass, it is doubtful if the mycotoxins produced by *C. purpurea* are responsible for abortion in this species. Premature births, low litter size and

483

483 Ryegrass with sclerotia of *Claviceps purpurea*. Arrow indicates a free sclerotium.

mummified foetuses have been attributed to chronic poisoning in pigs, and experimental exposure to ergotamine has caused abortions and foetal deaths in sheep.

Clinical findings

Gangrenous ergotism affects most species of domestic animals. Gangrenous necrosis of the extremities – nose, ears, tail, teats and limbs – following arteriospasm, congestion and endothelial cell degeneration of the capillary bed is a common clinical finding in many species of animals. In poultry, the tongue, comb and wattles are frequently affected.

Cattle grazing ergotised pasture or fed contaminated grain or silage develop lameness and gangrene as a major clinical sign of ergot toxicity. Lameness may appear about 2 weeks after initial ingestion, depending on the concentration of alkaloids in the ergot and the quantity of ergot in the feed. Hind limbs are usually affected before forelimbs. An elevated body temperature and increased pulse and respiration rates may accompany the lameness. Affected limbs are often swollen with a defined line separating the normal tissue from the dry, gangrenous extremity (**484**). A cold environment predisposes the extremities to gangrene. The affected part, which gradually loses sensation, may eventually slough. The tips of the ears or tail may become necrotic and the teats and udder may appear unusually pale. The nervous form of ergotism, attributed to *C. purpurea,* is uncommon in cattle, and is accompanied by muscular incoordination, tremors, blindness and convulsions. Dairy cattle on ergotised feed show a sharp drop in milk production.

Sheep with the acute form of ergotism may show nervous signs. Gangrenous ergotism seems to be uncommon in this species. Ulceration and necrosis of the tongue and the mucosa of the digestive tract may be observed in chronic forms of the disease. Classical signs of convulsive and gangrenous ergotism are not described in pigs. Chronic ergotism results in lack of udder development and agalactia in sows and the birth of small pigs. Low litter sizes, premature births and lack of vitality in both sows and piglets have been associated with the chronic form of this disease.

Diagnosis

Diagnosis of ergotism is based on the demonstration of sclerotia on pasture, in grains or in silage and on the signs of disease in affected animals. Extraction of ergot alkaloids and detection by chromatography, or biological testing may be necessary in suspect ground grain meals or in processed feeds.

Prevention of ergotism

The occurrence of ergotism on growing pasture or in grain is related to the susceptibility of the crop, climatic factors and agricultural practices. Sclerotia are stimulated to germinate by low winter soil temperatures and wet springs. Ergot infestation of grain fields can be minimised by using clean seed, crop rotation and deep cultivation. Climatic factors such as cold wet springs, which prolong the flowering period, tend to increase the level of primary infection. Ascospore dispersal is favoured by windy conditions and maturation of the sclerotia is enhanced by warm summers. Frequent grazing or topping of pastures prone to ergot infection during the summer months will reduce flower head production and help prevent the disease, especially following wet springs.

Ergotised grain or pastures containing ergotised grasses should not be used for animal feeding. Separation of ergots from sound grain can be achieved by mechanical or flotation techniques when small batches of grain are contaminated. Growing ergot-resistant grains such as wheat or barley, rather than rye, may be advisable in areas where ergotism is a recurring problem.

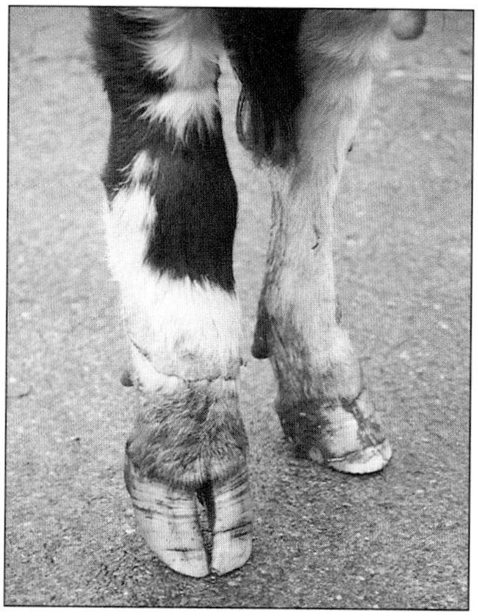

484

484 Ergotism in a cow: a swollen right hind leg showing a line of separation and terminal gangrene. The left hind limb is unaffected.

Fescue Toxicity

Tall fescue grass (*Festuca arundinacea)* is widely cultivated in the USA and in parts of Australia and New Zealand. Cattle and sheep grazing pastures consisting mainly of this grass may develop sporadic intoxication. Two diseases are associated with tall fescue grass, namely fescue foot and fescue summer toxicosis.

Fescue foot is a disease of cattle grazing fescue pastures. Affected cattle develop ischaemic necrosis of the skin of the extremities including the fetlock, ears and tail. The hooves and portions of the tail may slough off and the disease is clinically indistinguishable from gangrenous ergotism. The herd incidence of disease may be as high as 10 per cent. The poisonous effects of tall fescue grass are associated with the presence in the grass of the endophytic fungus, *Acremonium coenophialum*, but other fungi may be involved also. Although the lesions are thought to be caused by vasoconstrictive agents, probably mycotoxins, the toxic substances responsible for fescue foot has not yet been identified.

Summer fescue toxicosis is associated with high environmental temperatures. Clinical signs include reduced weight gain, a drop in milk production, a dull rough hair coat and elevated body temperature. The disease is related to the presence of toxic substances in the pasture, presumably mycotoxins. The disease has also been observed in animals fed on hay made from affected pastures.

Facial Eczema

Facial eczema is a hepatogenous photosensitivity of sheep and cattle caused by the ingestion of toxic conidia of *Pithomyces chartarum*. The mycotoxin, sporidesmin, is contained in the conidia which become dispersed by wind and water and adhere to pasture which is then ingested by grazing animals. The fungus is widely distributed and the disease has been reported in New Zealand, Australia, South Africa, several South American countries and southern France.

Sporidesmin is produced in the fungal spores as *P. chartarum* grows in dead leaves at the base of the pasture (**Diagram 65**), and the spores are then carried up through the dead material by new grass growth, which is then consumed by grazing animals. Field outbreaks of facial eczema typically occur after a period of warm, rainy weather, following a drought during the summer. The drought results in the death of grass leaves, which act as a substrate for the saprophytic growth of *P. chartarum*. Although the disease is commonly associated with ryegrass pastures the fungus is capable of growing on many kinds of dead leaf material. Climatic conditions which favour sporulation of the fungus have been accurately defined in New Zealand.

Sporidesmin produces liver damage and biliary occlusion which result in secondary photosensitisation. Phylloerythrin, a photodynamic pigment that is a degradation product of chlorophyll produced in the digestive tract of ruminants, is normally excreted in bile. As a result of the progressive liver damage, there is inadequate excretion of absorbed phylloerythrin and it accumulates in the blood. When the accumulated phylloerythrin reaches the peripheral circulation and becomes activated by absorbing energy from ultraviolet radiation when the skin is exposed to sunlight, photosensitisation occurs. The liver damage also results in the accumulation in the blood of the haemoglobin degradation product bilirubin which leads to jaundice. There is a latent period of 10–14 days between ingestion of sporidesmin and manifestation of photosensitivity. This interval corresponds to the time taken for the development of obliterative cholangitis and the subsequent retention of phylloerythrin.

Clinical signs of facial eczema occur suddenly when animals are exposed to sunlight. In sheep the first observable signs are swelling and reddening of the exposed parts of the skin on the head, ears and lips. The animals stop grazing when their lips become irritated. Exudation, with scab formation follows and eventually skin necrosis (**485**). Jaundice is invariably present.

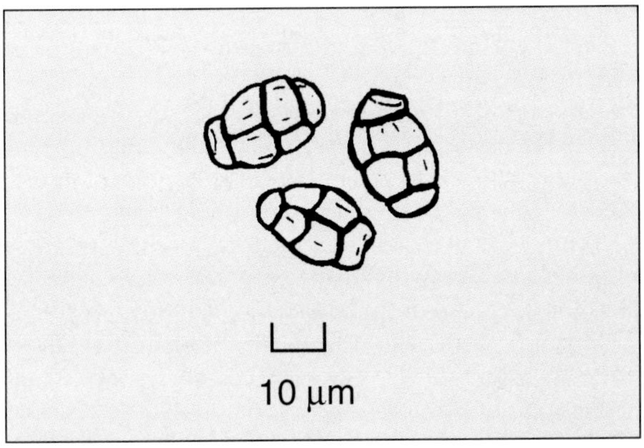

Diagram 65. Macroconidia of *Pithomyces chartarum*.

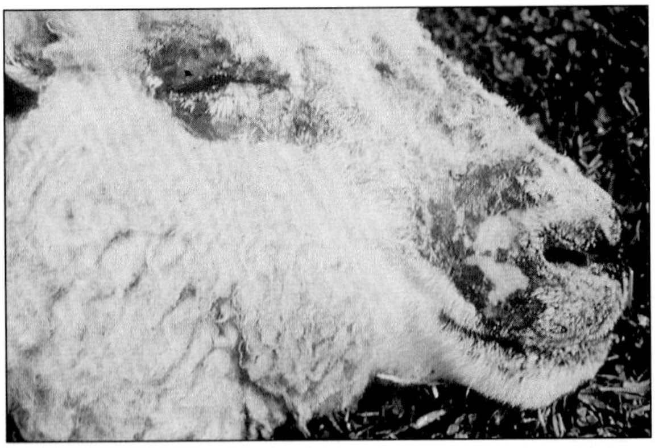

485 Sheep with facial eczema (*Pithomyces chartarum*). Lesions of photosensitization on non-wool areas.

Affected animals have photophobia and seek out any shade available. Mortality can be high if photosensitive animals are not removed from sunlight or death can occur at a later stage due to liver damage. In cattle, apart from lesions on the head and non-pigmented parts of the body, the teats and udder may be severely affected.

Postmortem findings include jaundice and a swollen mottled liver with thickened bile ducts. Moderate to severe portal fibroplasia and bile duct proliferation are constantly present. Other findings are degenerative and haemorrhagic lesions in the gall and urinary bladders.

Two important control measures have been used with success in New Zealand:

- Recognition and avoidance of toxic pastures;
- Reduction in the numbers of toxic conidia in pastures by applying fungicides.

Daily oral administration of zinc salts to dairy cattle has been reported to reduce the toxic effects of sporidesmin. Vaccination against sporidesmin has been tried without success.

Fusarium Toxicoses

The genus *Fusarium* is the largest single group of fungi with known toxigenic capability. Fungi of this genus and related organisms with sexual (perfect) reproduction produce a number of mycotoxins which are capable of causing a variety of diseases in animals. Because of their close association with plants and their relatively high water activity requirements for growth, fusaria are usually well established in a crop before harvesting and may cause many problems in cereals following a late harvest after a wet summer. Although some species such as *Fusarium moniliforme* are associated particularly with tropical and subtropical climates and others such as *F. sporotrichorides* with cold climates, many species occur in temperate parts of the world. Two general categories of toxins are produced by *Fusarium* species: the oestrogenic metabolites such as zearalenone (also referred to as F-2 toxin) and the trichothecene toxins. The fusaria generally produce mycotoxins at temperatures below those supporting optimal mycelial growth and some species have been reported to form several toxins. These fungi can germinate, grow and produce toxins in a wide variety of plants and feedstuffs. *Fusarium* species tend to produce highly coloured colonies: both the obverse (**486**) and reverse (**487**), with banana-shaped macroconidia (**488**).

Oestrogenism produced by zearalenone

This oestrogenic mycotoxicosis was first described in the USA more than 60 years ago. The disease, then termed vulvovaginitis, was associated with the consumption of mouldy maize by gilts. *Fusarium graminearum* (perfect state *Gibberella zeae*) and other

486

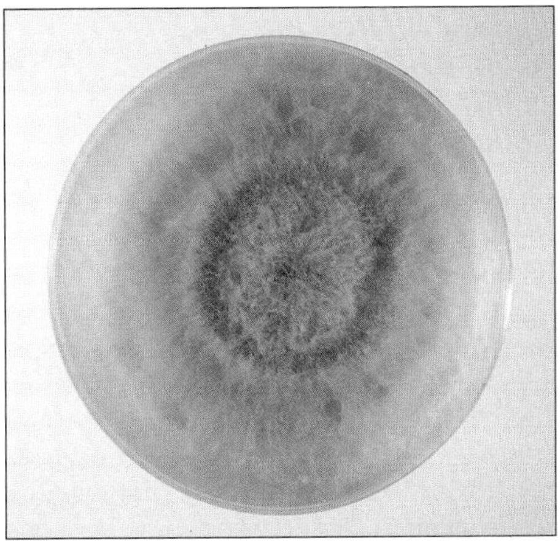

486 *Fusarium* sp. on Sabouraud agar, 7 days.

487

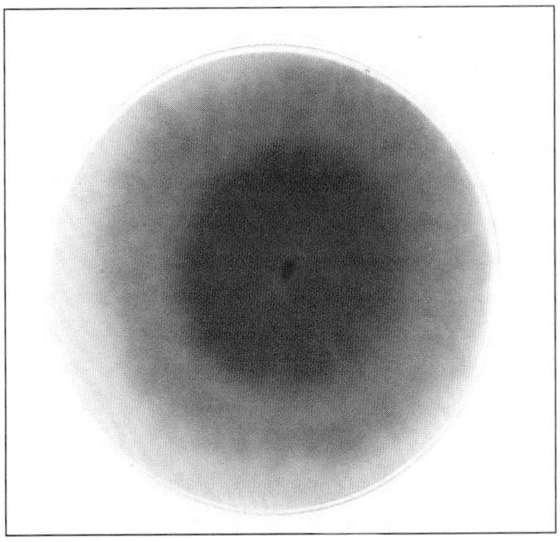

487 *Fusarium* sp. on Sabouraud agar, 7 days. Reverse.

488

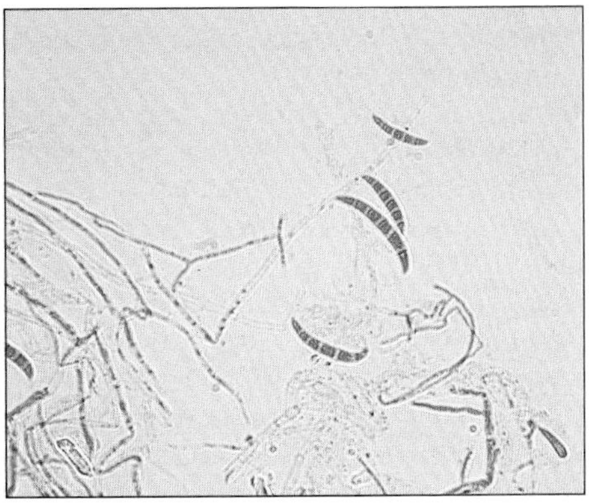

488 *Fusarium* sp. showing typical banana-shaped macroconidia. (LPCB, ×400)

Fusarium species growing on maize, barley and other grains produce zearalenone, a phenolic resocyclic acid lactone with oestrogenic activity. The target organ system is the reproductive tract and pigs are the animals most commonly affected by this mycotoxin, although other species may also be affected. Oestrogenic syndromes in pigs fed mouldy grain have been recognised in Europe, North America, the former Soviet Union, Japan, Australia and South Africa.

Infection of susceptible maize by *F. graminearum*, a phytopathogen, is favoured by cool, wet weather during the final stages of growth of the crop. The production of zearalenone in infected maize is enhanced by conditions of high moisture and alternating moderate and low temperatures. Sometimes *Fusarium* fungi may be absent from the sample of feed but zearalenone may still be present as it is a stable compound.

Pigs are the most sensitive animals, especially sexually immature gilts. Clinical signs of oestrogenism occur about one week after consumption of contaminated feed and include hyperaemia and oedema of the vulva, enlargement of the mammary glands and, in severe cases, prolapse of the vagina and rectum due to mucosal irritation and straining. Although the morbidity rate is high and may approach 100 per cent, the mortality rate is usually low. Affected sows may have anoestrus, pseudopregnancies, infertility or decreased litter size with small, weak piglets that may have splayed limbs. In boars, swelling of the prepuce and atrophy of the testicles have been reported.

Porcine oestrogenism is sometimes erroneously referred to as 'vulvovaginitis'. The latter term is an incorrect description as there are usually no inflammatory changes in the genital tract and the alterations are principally physiological in nature due to interstitial oedema and proliferation of epithelium in the vagina and cervix. The thickened endometrium is oedematous and the submucosal glands are hyperplastic. Clinical signs usually disappear about 7 days after withdrawal of contaminated feed.

Cattle appear to be much more resistant than pigs to the oestrogenic effects of cereals infected with *F. graminearum*, although zearalenone can lower the conception rate in heifers. Chickens appear to be affected minimally by this mycotoxin.

Zearalenone can be demonstrated in feeds by thin-layer or gas chromatography. Administration of extracts of zearalenone-contaminated feed to sexually immature female mice causes uterine engorgement and hypertrophy. This mycotoxin is secreted into milk, if dairy cattle are fed *F. graminearum*-infected cereals and may be of public health concern.

Equine leukoencephalomalacia

Leukoencephalomalacia, also called mouldy corn poisoning, is a highly fatal neuromycotoxicosis of Equidae caused by the ingestion of maize infected with *Fusarium moniliforme*. The disease is characterised by focal malacic lesions of the cerebral white matter.

The fungus infects maize in the field and produces mycotoxins called fumonisins. Four of these have been identified, A_1, A_2, B_1, B_2. The major toxin appears to be B_1. Two forms of *F. moniliforme* toxicosis have been reported in Equidae. The most common form of the disease is the neurotoxic form which is characterised by acute and severe neurological disease and is usually fatal. This form of the disease is called equine leukoencephalomalacia. A hepatotoxic form of the disease manifested by icterus, haemorrhages and oedema has also been reported. It seems probable that the neurological and hepatic forms of the disease are caused by the same toxin and that the size of dose determines the outcome. Equine leukoencephalomalacia has been recognised in the USA since the latter part of the 19th century and subsequently in many other countries, including South Africa, Egypt, China, Brazil and Argentina. Field outbreaks of disease are associated with the feeding of mouldy maize, particularly in wet seasons preceded by a drought.

The first signs of disease appear abruptly and consist of unthriftiness, reluctance to move backwards, muscular tremors, locomotory disturbance and ataxia. As the disease progresses, nervous signs become more pronounced and the animal walks into objects and later may become frenzied and run wildly into objects. Inability to swallow and paralysis of the lower lip may be evident. Death after 2 or 3 days may be preceded by recumbency and paddling limb movements. The pathognomonic lesions of this disease are one or more areas of liquefactive necrosis in the white matter of the cerebral hemispheres. Histopathologically, the encephalomalacic areas appear as large irregular empty spaces where the white matter of the brain has disintegrated. Perivascular haemorrhage and oedema may also be present. Pathological changes in the liver may be seen in some horses with jaundice and scattered haemorrhages evident. Liver damage consists of fibrosis, fatty changes and bile-duct proliferation.

Trichothecene toxins produced by Fusarium species

Many *Fusarium* species produce a number of sesquiterpene metabolites called trichothecenes and some of these mycotoxins, especially deoxynivalenol (DON), T-2 toxin and diacetoxyscirpenol (DAS) are implicated in mycotoxicoses in different species of animals. Although trichothecene toxicoses involve a broad spectrum of clinical syndromes, those that are most frequently recognised include vomiting, food refusal, ulceration and inflammation of the alimentary mucosa, haematological changes and nervous disturbances. The degree to which these effects are manifested varies greatly among different species. Fusaria may produce several different mycotoxins depending on the substrate and growth conditions. Accordingly, multiple toxins may occur in the same batch of mouldy food. The pharmacological effects of trichothecenes depend on impairment of the chemoreceptor area of the medulla oblongata, increased vascular permeability, and on their activity on mucosal epithelium and skin.

Food refusal and emetic syndromes

One of the most commonly occurring mycotoxins in animal feeds is DON (also know as vomitoxin). Pigs refuse to consume *F. graminearum*-infected maize, barley or mixed feeds containing these cereals. Emesis may occur in animals consuming small quantities of infected cereals. DON appears to be the principal mycotoxin responsible for the feed-refusal and emetic syndromes produced by this fungus, although other tricothecenes such as T-2 toxin and DAS may also be involved. *F. graminearum* also produces the oestrogenic mycotoxin, zearalenone. Emesis is apparently not due to local irritation of the gastric mucosa as parenteral administration of emetic extracts induces vomition, suggesting that the response is centrally mediated.

Other biological effects of trichothecene toxins

T-2 toxin and DAS are potent epithelial necrotizing agents causing necrosis and ulceration of epithelial surfaces of the skin, mouth and upper gastrointestinal tract following contact with contaminated feed. T-2 toxin has an immunosuppressive effect on growing pigs, affecting both T and B lymphocytes. Lymphocytopenia follows exposure to T-2 contaminated feed. Serum protein levels decline, haemorrhagic lesions occur in the digestive tract and degenerative lesions are seen in the liver and kidneys. There is some evidence that T-2 toxin is both mutagenic and teratogenic.

Trichothecene intoxication can result in haemorrhagic syndromes in cattle, pigs and poultry. Toxic effects on the marrow, thrombocytopenia and other induced blood changes result in defective blood coagulation. Many of these changes are attributed to the inhibitory effects of trichothecenes on protein synthesis.

Ochratoxicosis

Several toxigenic strains of *Aspergillus ochraceus* and *Penicillium viridicatum* produce ochratoxins, a group of related isocoumarin derivatives. Ochratoxin A is the principal nephrotoxic mycotoxin in this group. It impairs kidney function in many species, particularly pigs and poultry. A condition known as 'mycotoxic nephropathy' affects a significant percentage of pigs in Denmark. This disease has been recognized for more than 60 years and the fungus in question produces oxalic acid, citrinin and ochratoxin which act as nephrotoxins. The lesions of Balkan nephropathy, a human disease widely recognised in many parts of eastern Europe, are similar to the renal lesions seen in ochratoxin A-induced porcine nephropathy.

Natural production of ochratoxin occurs primarily in barley, wheat and maize in the northern hemisphere. Some *Penicillium* species may produce these toxins at temperatures as low as 5°C. The saprophytic fungi grow on harvested crops during storage and produce mycotoxins on suitable substrates. Ochratoxin A is a stable compound which is only partially destroyed by heat processing and autoclaving. The primary target organ of ochratoxin is the kidney, but high doses may produce liver damage also.

In pigs, clinical signs of chronic ochratoxicosis include reduced food intake, loss of body weight, depression, polydipsia and polyuria. The mortality rate is usually low. Macroscopically, the kidneys may be enlarged and pale with a mottled surface. Histologically, the renal lesions are characterised by degeneration of the proximal tubules and interstitial fibrosis of the cortex. Poultry affected by ochratoxicosis show a depressed growth rate, coagulopathy and poor quality eggshells. Postmortem findings include pale kidneys and liver, intestinal congestion and sometimes ruptured large intestines. Ruminants, especially adult animals, appear to be less susceptible to ochratoxicosis than monogastric animals. The flora of the adult rumen has been shown to degrade ochratoxin A.

Ochratoxins can cross the placental barrier in some species and exert a teratogenic effect. They are also immunosuppressive, particularly in relation to antibody production. Many of the biological effects of ochratoxin A relate to interference with protein synthesis at the translational level. Ochratoxin formation is primarily a grain storage problem and detection of these mycotoxins requires solvent extraction followed by thin layer chromatography of separated fractions. Ochratoxins fluoresce yellow-green under ultraviolet light.

Stachybotryotoxicosis

Stachybotryotoxicosis is a trichothecene-induced disease produced by *Stachybotrys atra* growing on hay or straw. Most reports of this intoxication come from the former Soviet Union or eastern European countries. The saprophytic, cellulose-degrading fungus, *S. atra*, growing on harvested crops produces mycotoxins during storage. Contaminated material is toxic not only when ingested but also when used as bedding, as the conidia are toxic on contact or when inhaled. Horses, cattle, sheep and other species may be affected. At least five trichothecenes are produced by *S. atra*. These toxins are cytotoxic and have a marked radiomimetic effect especially on lymphoid tissues, bone marrow and epithelial tissue. Damage to these tissues results in leukopenia, anaemia, thrombocytopenia and immunosuppression. Under field conditions, poisoning is usually chronic following low level intake of toxins over a prolonged period.

The disease, which is best characterised in the horse, progresses in stages. After consumption of mould-infested hay or straw over several weeks, stomatitis, hyperaemia and necrosis of the oral mucosa and the tongue may be evident. The second stage may be asymptomatic with depression of leukocytes and thrombocytes. This is followed by a severe and progressive depression of leukocytes and thrombocytes, presumably due to toxic effects on haematopoietic tissues. There is impaired clotting with consequential haemorrhage.

Secondary bacterial infection results in an elevated temperature and death frequently results from haemorrhages and septicaemia. Cattle appear to be less susceptible to stachybotryotoxicosis than horses but otherwise the course of the disease is similar. Postmortem findings include haemorrhages and necrosis in many tissues. Necrosis is particularly evident along the gastrointestinal tract and ulcers may be present in the large intestine. Histopathological lesions include degenerative changes in the liver, kidney and myocardium. Atrophy and necrosis of lymphoid tissue and aplastic anaemia are also present.

Disease outbreaks can be minimised by ensuring that straw and hay are kept dry during storage and by prompt removal of contaminated feed and bedding.

Tremorgens

A group of compounds called tremorgens produced by toxigenic fungi produce nervous signs when ingested or experimentally administered to animals. Many of these mycotoxins induce similar clinical signs and some are chemically related in that they possess an indole moiety.

Penitrems

Penicillium cyclopium and other *Penicillium* species produce a range of toxins associated with outbreaks of nervous diseases in domestic animals. Penitrem A, verruculogen and related substances produce tremors, ataxia and convulsive episodes in animals. Penitrem A has been incriminated in nervous diseases in sheep, horses and dogs, chickens and rabbits fed mouldy feed. After experimental administration, fine tremors may be evident within minutes. Although initially involving the limbs, they may progress to whole body tremors with periods of extensor rigidity, opisthotonos and finally recumbency with paddling. In mild cases, recovery can occur in a few days when the toxic food is removed from the diet. Large doses of penitrems may result in convulsive responses and death. Pathological changes are difficult to demonstrate in animals that have been dosed with penitrems.

Ryegrass staggers

Tremorgenic toxins are now generally believed to be responsible for a number of diseases referred to as ryegrass staggers, Bermuda grass staggers, phalaris staggers and related neurological syndromes characterised by tremors, incoordination and locomotor disturbance.

Tremors and ataxia in cattle, sheep and horses grazing perennial ryegrass (*Lolium perenne*) have been recorded in many countries, particularly Australia, New Zealand, the USA and parts of Europe. Toxicity is associated with the presence of an endophyte *Acremonium loliae* growing within the plant, especially in the leaves. The active tremorgens have been isolated from endophyte-infected ryegrass plants and chemically characterised as lolitrems, which are probably secondary metabolites of the fungus. Perennial ryegrass staggers is usually associated with the grazing of ryegrass pastures but conserved forage may also induce the disease. Clinical signs include locomotor incoordination or abnormal staggering gait, stumbling and collapse followed by severe muscular spasms. Morbidity may be high with up to 80 per cent of a flock or herd affected after a few days on infected pasture but mortality is usually low. The disease often occurs when the pasture is short and animals graze close to the roots where, presumably, more mycotoxin is present. Deaths are uncommon if animals are promptly moved to safe pasture. No specific pathological tissue changes have been attributed to the mycotoxins causing perennial ryegrass staggers.

Cattle grazing Bermuda grass pastures in the southern USA late in the year sometimes develop muscular tremors and posterior paralysis. Clinical signs resemble convulsive ergotism but sclerotia of *C. purpurea* are absent from toxic pasture. Although a *Claviceps* species isolated from toxic Bermuda grass was reported to produce ergot alkaloids in culture the aetiology of the natural disease remains obscure.

Annual ryegrass staggers in cattle and sheep results from the ingestion of seedheads of annual ryegrass (*Lolium* spp.) containing nematode galls infected with the bacterium *Corynebacterium rathayi*. Accordingly, this disease is not caused by a mycotoxin. It is referred to here, however, because of the similarities of these tremorgenic syndromes involving grasses, regardless of their aetiology. The toxicity of the ryegrass is associated with the bacterial galls, the active principle being a toxin, termed corynetoxin. The toxin is believed to interfere with glycosylation enzymes, leading to depletion of essential glycoproteins. Clinical signs are similar to paspalum staggers (referred to later) but convulsions and muscular twitching is more severe. During seizures affected animals tend to lie on their sides with opisthotonos and limbs rigidly extended. Lesions include perivascular oedema and occasional haemorrhages in the brain and meninges. Degenerative liver changes have been recorded in experimentally induced annual ryegrass toxicosis in sheep.

Claviceps paspali: Paspalum staggers

A number of mycotoxicoses including fescue toxicosis and paspalum staggers share some common characteristics with classical ergotism. Paspalum staggers is a neurotoxic syndrome of animals, particularly cattle and occasionally sheep and horses, grazing pastures or eating hay that contains dallis grass or other grasses of the genus *Paspalum* parasitized by *Claviceps paspali*. The fungus colonises the ovaries of susceptible grasses in a manner similar to the colonisation of rye grass by *Claviceps purpurea*.

Infestation of the seed head is favoured by wet, humid summers and the disease occurs sporadically in the southern USA, South Africa, Australia, New Zealand and occasionally in Europe. Although the sclerotia of *C. paspali* contain alkaloids, these are not tremorgenic and the neurotoxicity is attributed to paspaline, paspalicine, paspalinine and paspalitrems. Unlike the gangrenous syndrome produced by the alkaloids of *C. purpurea, C. paspali* mycotoxins do not produce gangrene of the limbs or extremities. Affected animals show hyperexcitability, muscle tremors and incoordinated movements. Signs become more pronounced with exercise. Severely affected animals show ataxia and may become recumbent with paddling of the limbs. The appetite is unaffected and although occasional animals may die from sustained seizures resulting in interference with respiration, or from misadventure, the majority of animals recover when removed from affected pasture. No specific lesions have been described for paspalum staggers. High mowing of toxic seed heads followed by raking or heavy grazing during spring and summer to prevent formation of seed heads are measures aimed at preventing this mycotoxicosis.

Aspergillus clavatus tremors

Aspergillus clavatus has been associated with hypersensitivity, muscle tremors and ataxia in cattle grazing sprouted wheat or fed contaminated millers' culms. This fungus is known to produce a number of mycotoxins such as patulin and cytochalasin E. However, the mycotoxins produced by *A. clavatus* responsible for nervous signs have not been identified. The disease has been reported in China, South Africa and in several European countries. Clinical signs are not evident when animals are at rest. On exercise they may show ataxia, frothing from the mouth and knuckling of the hindlimbs. Changes described in the brain and spinal cord include degenerative changes in neurons and focal gliosis.

Lupinosis

Lupins (*Lupinus* species) are capable of causing two distinct forms of poisoning in livestock; one, a nervous syndrome caused by alkaloids in the seeds and a second disease referred to as lupinosis caused by fungi growing on the plant. *Phomopsis leptostromiformis* is a phytopathogen of certain *Lupinus* species which induces a hepatic syndrome in sheep and sometimes other species, called lupinosis. This disease has been reported in Australia, New Zealand, South Africa and in a number of European countries.

The fungus invades the growing plant in the field and after death of the diseased plant, *P. leptostromiformis* continues to grow saprophytically on the dead plant tissue. Mycotoxins produced by the fungus, particularly during the saprophytic phase of growth, are potent hepatotoxins. Clinical signs of acute disease include inappetance, ruminal stasis and jaundice. Photosensitisation may also occur. The most obvious postmortem findings in acute lupinosis are discoloration of the liver, jaundice and serosal haemorrhages. Bilary duct proliferation, fibroplasia of portal ducts and cirrhosis, are common histological findings. At least two mycotoxins referred to as phomopsins are produced by *P. leptostromiformis*. A mixture of phomopsins A and B has been reported to cause typical lupinosis in sheep. Phomopsin A, the principal mycotoxin, has been characterised as a cyclic hexapeptide.

Myrotheciotoxicosis

Myrothecium roridum and other *Myrothecium* species are noted plant pathogens and produce toxic metabolites belonging to the trichothecene group. A high prevalence of these fungi on ryegrass and white clover plants has been reported in New Zealand. Clinical signs appear within 48 hours of introduction of animals to affected pastures and resemble those of non-specific chemical rumenitis. Depression, salivation, abdominal pain, haemoconcentration and dehydration are features of this disease. Postmortem findings include ruminal and abomasal distension with hyperaemia and desquamation of the mucosa and patchy ulceration. Microscopically there is widespread necrosis of the mucosa of the abomasum, reticulum and rumen. A syndrome similar to myrotheciotoxicosis is seen in cattle grazing Kikuyu grass.

Slaframine Toxicosis

Slaframine is a cholinergic mycotoxin produced by the growth of *Rhizoctonia leguminicola* on red clover pasture or hay. Clinical manifestations in cattle and horses include excessive salivation ('slobbers'), urination, lacrimation and diarrhoea. Decreased milk production, weight loss and abortion may occur in dairy cattle.

Diplodiosis

Diplodiosis is a neuromycotoxicosis of cattle and sheep grazing on harvested maize lands that contain mouldy cobs of maize. The fungus involved, *Diplodia zeae*, which is responsible for stem and ear rot of maize, produces characteristic fruiting bodies or pycnidia on affected parts towards the end of the growing season. Ingestion of maize infested with this fungus produces a characteristic neurotoxic disease. Affected animals show ataxia, paresis, paralysis, lacrimation and salivation. The mortality rate may be high if grazing animals are not removed promptly from contaminated lands. Gross pathological changes are not usually present in diplodiosis and the mycotoxin responsible has not yet been identified.

Bibliography

Berry, C.L. (1988). The pathology of mycotoxins, *Journal of Pathology*, **154**: 301–311.

Cheeke, P.R. and Shull, L.R. (1985). *Natural toxicants in feeds and poisonous plants*, AVI Publishing Company, Inc., Westport, Connecticut, USA.

Humphreys, D.J. (1988). *Veterinary Toxicology*, Third Edition, Bailliere Tindall, London.

Kellerman, T.S., Coetzer, J.A.W. and Naude, T.W. (1988). *Plant poisoning and mycotoxicoses of livestock in Southern Africa*, Oxford University Press, Cape Town, South Africa.

Marasas, W.F.O. and Nelson, P.E. (1987). *Mycotoxicology*, The Pennsylvania State University Press, University Park, Penn., USA.

Moss, M.O. (1989). Mycotoxins of *Aspergillus* and other filamentous fungi, *Journal of Applied Bacteriology* Symposium Supplement, 69S–81S.

Pier, A.C. (1981). Mycotoxins and animal health, *Advances in Veterinary Science and Comparative Medicine*, **25**: 185–243.

Richard, J.L. and Thurston, J.R. (1986). *Diagnosis of mycotoxicoses*, Martinus Nijhoff Publishers, Dordrecht, Holland.

Seawright, A.A. (1982). *Animal Health in Australia,* Volume 2. Chemical and Plant Poisons. Australian Government Publishing Service, Canberra.

Section 4: Virology (including Prions)

46 Classification and Characterisation of Viruses

Viruses are divided into families on the basis of size, symmetry (icosahedral, helical or complex), substructure, type of nucleic acid genome (RNA or DNA), form of nucleic acid genome (single or double stranded, segmented, cyclic or linear, haploid or diploid) and mode of replication. The classification of genera and species within a family are based on criteria which include host species, pathogenesis, nucleic acid homology and antigenicity. Within some families, subfamilies have also been established. Families are named with the suffix *-viridae*. Subfamilies have the suffix *-virinae* and genera the suffix *-virus*. The classification and nomenclature of viruses used in this chapter follows that of the International Committee on Taxonomy of Viruses (Francki *et al.*, 1991).

There are approximately 200 viral species in some 20 viral families (**Tables 147** and **148**) known to infect the eight main domestic animal species (cattle, sheep, goats, pigs, horses, dogs, cats and poultry). The important viruses affecting domestic animal species are summarised in **Tables 149** and **150**. A brief description of the viral diseases is presented in **Section 6**.

Families of DNA Viruses of Veterinary Importance

Parvoviridae

Parvoviruses (from the Latin word *parvus* for small) have a genome of single-stranded DNA. They are remarkably resistant, surviving in a wide pH range and remaining active at room temperature for a long period. Most species have a narrow host range. There are three genera. The genus *Parvovirus* contains most of the important veterinary parvoviruses. Defective adeno-associated parvoviruses are assigned to the genus *Dependovirus*, while members of the genus *Densovirus* infect insects.

Papovaviridae

The papovaviruses (**pa**pilloma virus, **po**lyoma virus, and **va**cuolating agent - original members of the family) are, with few exceptions, oncogenic. There are two genera, *Papillomavirus* and *Polyomavirus*. Papillomaviruses induce warts on the host species of origin. Polyomavirus infections are silent in the natural population but oncogenic in laboratory animals following experimental inoculation.

Adenoviridae

Mammalian adenoviruses (from the Greek word *adenos* meaning gland) belong to the genus *Mastadenovirus*, while avian adenoviruses belong to the genus *Aviadenovirus*.

Hepadnaviridae

The name *Hepadnaviridae* has been proposed for a group of related viruses including human hepatitis B virus and similar viruses found in woodchucks, squirrels and Pekin ducks. There are two genera *Orthohepadnavirus* and *Avihepadnavirus*.

Herpesviridae

The name of the family is derived from the Greek word *herpes*, meaning creeping, and refers to the multiple recurrence of cold-sore eruptions in humans infected with herpes simplex virus. The family contains a large number of important veterinary and human pathogens. Three subfamilies are recognised: *Alphaherpesvirinae* (rapidly growing, cytolytic viruses), *Betaherpesvirinae* (slowly growing viruses which often produce greatly enlarged cells, giving rise to the name cytomegalovirus), and *Gammaherpesvirinae* (viruses which replicate in lymphocytes and may cause transformation of infected cells). There are six genera: *Simplexvirus* and *Varicellovirus* (*Alphaherpesvirinae*); *Cytomegalovirus* and *Muromegalovirus* (*Betaherpesvirinae*); *Lymphocryptovirus* and *Rhadinovirus* (*Gammaherpesvirinae*). Lifelong infection, usually in latent form, is a feature of all herpesvirus infections.

Poxviridae

Pox is an Old English word meaning vesicular skin lesions. Poxviruses are the largest of all viruses. The family is divided into two subfamilies: *Chordopoxvirinae* (containing the poxviruses of vertebrates) and *Entomopoxvirinae* (consisting of three genera of insect viruses). The *Chordopoxvirinae* subfamily is made up of eight genera: *Orthopoxvirus, Avipoxvirus, Capripoxvirus, Leporipoxvirus, Suipoxvirus, Parapoxvirus, Molluscipoxvirus* and *Yatapoxvirus*. Poxviruses replicate in the cytoplasm.

Table 147. DNA viruses: characteristics of families and domestic animals affected.

Animal species		*Parvoviridae* (a)	*Papovaviridae*	*Adenoviridae*	Unclassified (b)	*Hepadnaviridae*	*Herpesviridae*	*Poxviridae*
	SYMMETRY	Icosahedral	Icosahedral	Icosahedral	Icosahedral	Icosahedral	Icosahedral	Complex
	ENVELOPE	–	–	–	+	–	+	+
	DIAMETER (nm)	18–26	45–55	70–90	175–215	42–47	120-200	115-260 x 200-350
Cattle			Bovine papillomavirus (1–6)				Bovine herpesvirus (BHV): BHV 1 (IBR/IPV) BHV 2 (ulcerative mammillitis/pseudo-lumpy skin disease) Alcelaphine herpesvirus 1 (malignant catarrhal fever) Ovine herpesvirus 2 (malignant catarrhal fever) Porcine herpesvirus 1 (pseudorabies)	Bovine papular stomatitis virus Pseudocowpox virus Cowpox virus Lumpy skin disease virus
Sheep/ goats							Porcine herpesvirus 1 (pseudorabies) Ovine herpesvirus 2	Orf virus Sheeppox virus Goatpox virus
Pigs		Porcine parvovirus (SMEDI)			African swine fever virus		Porcine herpesvirus 1 (Aujeszky's disease) Porcine herpesvirus 2 (inclusion body rhinitis)	Swinepox virus
Horses			Equine papillomaviruses	Equine adenovirus 1			Equine herpesvirus (EHV): EHV 1 (abortion) EHV 2 cytomegalovirus EHV 3 (coital exanthema) EHV 4 (rhinopneumonitis)	

(*continued*)

Table 147. DNA viruses: characteristics of families and domestic animals affected. (*continued*)

Animal species		*Parvoviridae*[a]	*Papovaviridae*	*Adenoviridae*	Unclassified[b]	*Hepadnaviridae*	*Herpesviridae*	*Poxviridae*
	SYMMETRY	Icosahedral	Icosahedral	Icosahedral	Icosahedral	Icosahedral	Icosahedral	Complex
	ENVELOPE	–	–	–	+	–	+	+
	DIAMETER (nm)	18–26	45–55	70–90	175–215	42–47	120-200	115-260 x 200-350
Dogs		Canine parvovirus	Canine oral papilloma virus	Canine adenovirus 1 (infectious canine hepatitis) Canine adenovirus 2 (infectious canine tracheo-bronchitis)			Canine herpesvirus Porcine herpesvirus 1 (pseudorabies)	
Cats		Feline parvovirus (panleuko-penia)					Feline herpesvirus 1 (rhinotracheitis) Porcine herpesvirus 1 (pseudorabies)	Cowpox virus
Avian species				Egg drop syndrome virus		Duck hepatitis B virus	Avian herpesvirus (infectious laryngo-tracheitis) Marek's disease virus Duck herpesvirus (Duck plague)	Fowlpox virus Pigeonpox virus Canarypox virus

SMEDI = Stillbirth, mummification, embryonic death, infertility
IBR/IPV = Infectious bovine rhinotracheitis/infectious pustular vulvovaginitis

(a) All DNA viruses have double stranded DNA except *Parvoviridae*
(b) DNA viruses replicate in the nucleus with the exception of *Poxviridae* and African swine fever virus
+ = envelope present, – = nonenveloped

Table 148. RNA viruses: characteristics of families and domestic animals affected*.

Animal species		*Picornaviridae*	*Caliciviridae*	*Reoviridae (a)*	*Birnaviridae (a)*	*Flaviviridae*	*Togaviridae*	*Orthomyxoviridae (b)*	*Paramyxoviridae*	*Coronaviridae*	*Rhabdoviridae*	*Bunyaviridae*	*Retroviridae (b)*	*Torovirus*	*Unclassified*	
	SYMMETRY	**Icosa-hedral**	**Icosa-hedral**	**Icosa-hedral**	**Icosa-hedral**	**Icosa-hedral**	**Icosa-hedral**	**Helical**	**Helical**	**Helical**	**Helical**	**Helical**	**Icosa-hedral**	**Helical**	**Icosa-hedral**	**Unknown**
	ENVELOPE	–	–	–	–	+	+	+	+	+	+	+	+	+	–	+
	DIAMETER (nm)	**24-30**	**30-37**	**60-80**	**55-65**	**40-60**	**60-70**	**80-120**	**100-500**	**60-220**	**75 x 180**	**90-120**	**80-100**	**120-140**	**28-30**	**85-125**
Cattle		Foot-and-mouth disease virus (1–7)** Bovine enterovirus (1–7) Bovine rhinovirus (1–3)		Blue tongue virus (1–24) Bovine rotavirus		Bovine viral diarrhoea virus			Rinder-pest virus Bovine resp. syncytial virus Bovine para-influenza virus 3	Bovine corona-virus	Rabies virus Vesicular stomatitis virus Bovine ephemeral fever virus	Rift Valley fever virus Akabane virus	Bovine leukaemia virus	Breda virus	Astro-virus	
Sheep/ Goats		Foot-and-mouth disease virus (1–7)		Blue tongue virus (1–24) Ovine rotavirus		Louping ill virus Wessels-bron virus Border disease virus			Bovine resp. syncytial virus Ovine para-influenza virus 3 Peste-des-petits-ruminants virus		Rabies virus	Rift Valley fever virus Akabane virus Nairobi sheep disease virus	Maedi-Visna virus Ovine pulmonary adenomatosis virus (Jaagsiekte) Caprine arthritis-encephalitis virus		Astro-virus	Borna disease virus

(*continued*)

Table 148. RNA viruses: characteristics of families and domestic animals affected*. (*continued*)

Animal species	*Picornaviridae*	*Caliciviridae*	*Reoviridae (a)*	*Birnaviridae (a)*	*Flaviviridae*	*Togaviridae*	*Orthomyxoviridae (b)*	*Paramyxoviridae*	*Coronaviridae*	*Rhabdoviridae*	*Bunyaviridae*	*Retroviridae (b)*	*Torovirus*	*Unclassified*	
SYMMETRY	**Icosa - hedral**	**Icosa - hedral**	**Icosa - hedral**	**Icosa- hedral**	**Icosa - hedral**	**Icosa- hedral**	**Helical**	**Helical**	**Helical**	**Helical**	**Helical**	**Icosa- hedral**	**Helical**	**Icosa- hedral**	**Unknown**
ENVELOPE	–	–	–	–	+	+	+	+	+	+	+	+	+	–	+
DIAMETER (nm)	24-30	30-37	60-80	55-65	40-60	60-70	80-120	100-500	60-220	75 x 180	90-120	80-100	120-140	28-30	85-125
Pigs	Foot-and-mouth disease virus (1–7) Swine vesicular disease virus Porcine entero-viruses (1–11) (Teschen-Talfan/ SMEDI)	Vesicular exanthema virus	Porcine rotavirus		Japanese enceph-alitis virus Swine fever virus (Hog cholera)		Swine influenza virus		Trans-missible gastro-enteritis virus Porcine haemag-glutinating encephalo-myelitis virus Porcine epidemic diarrhoea virus	Rabies virus Vesicular stomatitis virus				Astro-virus	
Horses	Equine rhinovirus (1–3)		African horse sickness virus (1–9) Equine rotavirus			Equine arteritis virus Equine enceph-alomyelitis viruses (Western, Eastern and Venezuelan)	Equine influenza virus (sub-types 1 and 2)			Rabies virus Vesicular stomatitis virus		Equine infectious anaemia virus	Berne virus		Borna disease virus

(*continued*)

Table 148. RNA viruses: characteristics of families and domestic animals affected*. (*continued*)

Animal species		*Picornaviridae*	*Caliciviridae*	*Reoviridae* (a)	*Birnaviridae* (a)	*Flaviviridae*	*Togaviridae*	*Orthomyxoviridae* (b)	*Paramyxoviridae*	*Coronaviridae*	*Rhabdoviridae*	*Bunyaviridae*	*Retroviridae* (b)	*Torovirus*	*Unclassified*	
	SYMMETRY	Icosa-hedral	Icosa-hedral	Icosa-hedral	Icosa-hedral	Icosa-hedral	Icosa-hedral	Helical	Helical	Helical	Helical	Helical	Icosa-hedral	Helical	Icosa-hedral	Unknown
	ENVELOPE	–	–	–	–	+	+	+	+	+	+	+	+	+	–	+
	DIAMETER (nm)	24-30	30-37	60-80	55-65	40-60	60-70	80-120	100-500	60-220	75 x 180	90-120	80-100	120-140	28-30	85-125
Dogs				Canine rotavirus					Canine distemper virus Canine para-influenza virus 2	Canine corona-virus	Rabies virus					
Cats			Feline calici-virus	Feline rotavirus						Feline infectious peritonitis virus	Rabies virus		Feline leukaemia virus Feline sarcoma virus Feline immuno-deficiency virus		Astro-virus	
Avian species		Avian enceph-alomyelitis virus (Epidemic tremor)		Avian reo-virus (1–11)	Infec-tious bursal disease virus (Gum-boro)			Avian influ-enza virus (Fowl plague)	Avian paramyxo-virus 1 (New-castle disease)	Infectious bronchitis virus Transmis-sible turkey enteritis virus (Blue-comb)			Avian leukosis/ sarcoma viruses			

* The *Filoviridae* and the *Arenaviridae* have not been included in this table.
* * indicates the number of types, subtypes or serotypes.
(a) All RNA viruses have single stranded RNA, except *Reoviridae* and *Birnaviridae*.
(b) RNA viruses replicate in the cytoplasm, with the exception of *Orthomyxoviridae* and *Retroviridae*.
SMEDI = stillbirth, mummification, embryonic death, infertility.
+ = envelope present, – = nonenveloped.

Table 149. DNA viruses of domestic animals.

Family	Sub-family	Genus	Virus (Disease)
Parvoviridae		*Parvovirus*	Porcine parvovirus (SMEDI) Canine parvovirus Feline parvovirus (feline panleukopenia) Aleutian mink disease virus
Papovaviridae		*Papillomavirus*	Bovine papillomaviruses 1 and 2 (cutaneous papillomas) Bovine papillomavirus 3 (cutaneous papilloma) Bovine papillomavirus 4 (alimentary tract and bladder papillomas) Bovine papillomaviruses 5 and 6 (teat papillomas) Ovine papillomavirus Porcine genital papillomavirus Equine papillomavirus (warts) Equine papillomavirus ? (equine sarcoids) Canine oral papillomavirus Rabbit oral papillomavirus
Adenoviridae		*Mastadenovirus*	Bovine adenoviruses (1–9) Ovine adenoviruses (1–6) Caprine adenoviruses (1–2) Porcine adenoviruses (1–4) Equine adenoviruses (1–2) Canine adenovirus 1 (infectious canine hepatitis) Canine adenovirus 2 (infectious canine tracheobronchitis)
		Aviadenovirus	Avian adenoviruses Chicken: egg drop syndrome, inclusion body hepatitis Turkey: bronchitis, marble spleen disease, enteritis Quail: bronchitis
Hepadnaviridae		*Avihepadnavirus*	Duck hepatitis B virus
Herpesviridae	*Alphaherpes-virinae*	*Simplexvirus*	Bovine herpesvirus 2 (Allerton virus) (ulcerative mammillitis/pseudolumpy skin disease)
		Varicellovirus	Bovine herpesvirus 1 (infectious bovine rhinotracheitis, infectious pustular vulvovaginitis) Porcine herpesvirus 1 (Aujeszky's disease/pseudorabies) Equine herpesvirus 1 and 4 (abortion/rhinopneumonitis) Equine herpesvirus 3 (coital exanthema) Canine herpesvirus ('fading puppy' syndrome) Feline herpesvirus 1 (feline viral rhinotracheitis) Avian herpesvirus (infectious laryngotracheitis of chicken) Duck herpesvirus (duck plague)

(continued)

Table 149. DNA viruses of domestic animals. (*continued*)

Family	Sub-family	Genus	Virus (Disease)
Herpesviridae (continued)	*Betaherpes-virinae*	*Muromegalovirus*	Bovine herpesvirus 4 (bovine cytomegalovirus infection) Equine herpesvirus 2 (equine cytomegalovirus infection) Porcine herpesvirus 2 (inclusion body rhinitis)
	Gammaherpes-virinae	*Lymphocrypto-virus*	Ovine herpesvirus 2 (sheep-associated malignant catarrhal fever) Alcelaphine herpesvirus 1 (wildebeest-associated malignant catarrhal fever) Marek's disease virus
Poxviridae	*Chordo-poxvirinae*	*Orthopoxvirus*	Cowpox virus Horsepox virus
		Avipoxvirus	Fowlpox virus Pigeonpox virus Canarypox virus
		Parapoxvirus	Pseudocowpox virus Bovine papular stomatitis virus Orf virus
		Capripoxvirus	Sheeppox virus Goatpox virus Lumpy skin disease virus (Neethling virus)
		Suipoxvirus	Swinepox virus
		Leporipoxvirus	Myxoma virus (myxomatosis of rabbits)
Unclassified			African swine fever virus Chicken anaemia agent (chicken anaemia virus)

Unclassified DNA viruses

In 1984 the International Committee on Taxonomy of Viruses removed African swine fever virus from the family *Iridoviridae*. To date this virus has not been allocated to any other family or given full family status. African swine fever virus replicates in the cytoplasm.

Chicken anaemia agent is a single-stranded, circular, unclassified DNA virus, similar in size to parvoviruses.

Families of RNA Viruses of Veterinary Importance

Picornaviridae

The name is derived from *pico*, an Old Spanish word meaning very small and from the type of nucleic acid, RNA. The family consists of five genera: *Enterovirus, Rhinovirus, Aphthovirus, Cardiovirus* and *Hepatovirus.*

Caliciviridae

Calix is the Latin word for cup or goblet and refers to the unique morphology of these viruses. Dark cup-shaped depressions are seen in negatively stained preparations of the virus by electron microscopy. Originally the caliciviruses were classified as a genus of the family *Picornaviridae*. There is a single genus, *Calicivirus*.

Reoviridae

The name 'reovirus' (**r**espiratory **e**nteric **o**rphan viruses) was originally proposed for a group of viruses, isolated mainly from the respiratory and intestinal tracts, that were not associated with disease. The genome consists of 10–12 segments of linear double stranded RNA and the family now contains nine genera: *Orthoreovirus, Orbivirus, Rotavirus, Coltivirus, Aquareovirus, Phytoreovirus, Fijivirus, Cypovirus* and Plant reovirus 3. The first four genera listed contain many important animal pathogens, while the others contain viruses of fish, invertebrates and plants.

Table 150. RNA viruses of domestic animals.

Family	Sub-family	Genus	Virus (Disease)
Picornaviridae		*Enterovirus*	Bovine enteroviruses (1–7) Porcine enteroviruses (1–11) (Teschen–Talfan disease, SMEDI) Swine vesicular disease virus Avian encephalomyelitis virus (epidemic tremor) Duck hepatitis virus
		Rhinovirus	Bovine rhinoviruses (1–3) Equine rhinoviruses (1–3)
		Aphthovirus	Foot-and-mouth disease virus (A, O, C, Asia 1, SAT 1,2,3)
		Cardiovirus	Encephalomyocarditis virus
Caliciviridae		*Calicivirus*	Vesicular exanthema virus Feline calicivirus
Reoviridae		*Orthoreovirus*	Mammalian reoviruses 1–3 Avian reoviruses 1–11
		Orbivirus	Bluetongue virus (1–24) African horse sickness virus (1–9)
		Rotavirus	Bovine rotavirus Ovine rotavirus Porcine rotavirus Equine rotavirus Canine rotavirus Feline rotavirus Avian rotavirus
		Coltivirus	Colorado tick fever virus
Birnaviridae		*Birnavirus*	Infectious bursal disease virus (Gumboro disease)
Flaviviridae		*Flavivirus*	Louping ill virus Wesselsbron disease virus Japanese encephalitis virus
		Pestivirus	Bovine viral diarrhoea virus/Mucosal disease Border disease virus ('Hairy-shaker' disease) Swine fever virus (Hog cholera)
Togaviridae		*Alphavirus*	Equine encephalomyelitis viruses (Western, Eastern, Venezuelan)
		Arterivirus	Equine arteritis virus Lelystad virus (Porcine reproductive and respiratory syndrome)
Orthomyxoviridae		*Influenzavirus* A and B	Swine influenzavirus Equine influenzavirus (subtypes 1 and 2) Avian influenzavirus (Fowl plague)
Paramyxoviridae	*Paramyxovirinae*	*Paramyxovirus*	Bovine parainfluenzavirus 3 Ovine parainfluenzavirus 3 Canine parainfluenzavirus 2 Avian paramyxovirus 1 (Newcastle disease) Avian paramyxoviruses (2–9)
		Morbillivirus	Rinderpest virus Peste-des-petits-ruminants virus Canine distemper virus
	Pneumovirinae	*Pneumovirus*	Bovine respiratory syncytial virus

(*continued*)

Table 150. RNA viruses of domestic animals. (*continued*)

Family	Sub-family	Genus	Virus (Disease)
Coronaviridae		*Coronavirus*	Bovine coronavirus Porcine transmissible gastroenteritis virus Haemagglutinating encephalomyelitis virus (vomiting and wasting disease) Porcine epidemic diarrhoea virus Canine coronavirus Feline infectious peritonitis virus Feline enteric coronavirus Infectious bronchitis virus (1–10) Transmissible turkey enteritis virus (Bluecomb disease)
Rhabdoviridae		*Vesiculovirus*	Vesicular stomatitis virus
		Lyssavirus	Rabies virus
		Unclassified	Bovine ephemeral fever virus
Bunyaviridae		*Bunyavirus*	Akabane virus
		Phlebovirus	Rift Valley fever virus
		Nairovirus	Nairobi sheep disease virus
Retroviridae		HTLV-BLV group	Bovine leukaemia virus
		Avian type C retrovirus group	Avian leukosis-sarcoma viruses
		Mammalian type C retrovirus group	Feline leukaemia and sarcoma viruses Avian reticuloendotheliosis viruses
		Lentivirus	Maedi–Visna virus (ovine progressive pneumonia) Caprine arthritis-encephalitis virus Equine infectious anaemia virus Feline immunodeficiency virus Bovine immunodeficiency virus
		Unclassified	Ovine pulmonary adenomatosis virus (Jaagsiekte)
Not given family status		***Torovirus***	Breda virus Porcine torovirus Berne virus
Unclassified			Astrovirus Borna disease virus Rabbit haemorrhagic disease virus

Birnaviridae

The name is derived from the fact that the viruses in this family contain linear double-stranded RNA (bi=two, RNA). There is one genus, *Birnavirus*.

Flaviviridae

Formerly these viruses were classified with the togaviruses. *Flavus* is the Latin word for yellow and the name of the family is derived from the type species, yellow fever virus. There are three genera, *Flavivirus* (group B arboviruses), *Pestivirus* and Hepatitis C virus. Members of the genus *Flavivirus* are transmitted by arthropods.

Togaviridae

The Latin word *toga* means a gown or cloak and refers to the fact that these viruses are enveloped. There are three genera: *Alphavirus, Rubivirus* and *Arterivirus*. All species of the genus *Alphavirus* (group A arboviruses) replicate in arthropod vectors.

Orthomyxoviridae

The name is derived from the Greek words *orthos* and *myxa,* meaning straight or correct and mucus, respectively. There are two genera, *Influenzavirus* A and B, and *Influenzavirus* C. Replication takes place in the nucleus and cytoplasm of host cells.

Paramyxoviridae

The Greek word *para* means by the side of (similar to) while *myxa* means mucus. The family contains two subfamilies, *Paramyxovirinae* and *Pneumovirinae* containing three genera, *Paramyxovirus and Morbillivirus* (*Paramyxovirinae*) and *Pneumovirus* (*Pneumovirinae*).

Coronaviridae

The Latin word *corona* means a crown and refers to the large club-shaped, glycoprotein peplomers in the lipoprotein envelope of these viruses. There is a single genus, *Coronavirus*.

Rhabdoviridae

The name for this large family of bullet-shaped, enveloped viruses comes from the Greek word for a rod, rhabdos. The natural host range of rhabdoviruses is very wide and includes vertebrates, invertebrates, arthropods and plants. Two genera affecting animals, *Vesiculovirus* and *Lyssavirus*, and one affecting plants, Plant rhabdovirus, have been defined within the family.

Bunyaviridae

Bunyamwera is the locality in Uganda where the type species of the family was first isolated. There are five genera, *Bunyavirus, Phlebovirus, Nairovirus, Hantavirus* and *Tospovirus*. Nearly all of these viruses are arthropod-borne.

Retroviridae

The Latin word *retro* means backwards and refers to the reverse transcription of virus RNA into DNA during replication. The DNA is integrated into the chromasomal DNA of the host as provirus. Retroviruses are diploid, possessing a genome consisting of two identical molecules of single-stranded RNA. The family comprises seven genera, Mammalian type B oncovirus, Mammalian type C retrovirus, Avian type C retrovirus, Type D retrovirus, *Spumavirus*, Human T-cell lymphotropic virus-Bovine leukaemia virus (HTLV-BLV) and *Lentivirus*.

Torovirus

The genus *Torovirus* has been proposed for a group of antigenically related viruses which have an elongated, tubular nucleocapsid that may be bent into an open torus, thereby conferring a kidney-shaped mophology to the virion.

Unclassified RNA viruses

Astrovirus is the unofficial name accorded to viruses visualised by electron microscopy with a star-like appearance that are distinguishable from calicivirus particles. It has been suggested that rabbit haemorrhagic disease virus is a calicivirus. Borna disease virus is largely uncharacterised and accordingly unclassified.

Reference

Francki R.I.B., Fauquet C.M., Knudson D.L. and Brown F. (1991). *Classification and Nomenclature of Viruses, Fifth Report of the International Committee on Taxonomy of Viruses*. Springer-Verlag, Wien.

47 Kit-set Tests Available for Veterinary Virology

There has been a steady increase in the number of kit-sets available for use in veterinary diagnostic virology in the last decade. These commercially available tests are designed to detect the presence of either antigen or antibody in the specimen.

The enzyme-linked immunosorbent assay (ELISA) has been successfully adapted for the detection and quantitation of antigen and antibodies of many important veterinary viruses. Sandwich ELISA, competitive ELISA or direct ELISA may be used to detect antigen, while indirect or competitive ELISA is generally used for the detection of antibody. The ELISA is frequently carried out on a solid microwell format but membranes, combs and sticks may also be used as the solid phase.

Other less commonly applied test systems include latex agglutination, complement fixation, haemagglutination, agar gel immunodiffusion and immunofluorescence. **Tables 151–155** list a representative sample of currently available diagnostic kit-sets for veterinary viruses of domestic animals.

Table 151. Kit-set tests for the diagnosis of viral infections in ruminants.

Virus	Detection of antigen or antibody	Test system	Sample	Company
Bluetongue virus	Antibody	Agar gel immunodiffusion using cell culture, soluble viral antigen	Serum	Veterinary Diagnostic Technology Inc., Wheat Ridge, CO 80033, USA
	Antibody	Modified direct CFT using cell culture, soluble viral antigen	Serum	Veterinary Diagnostic Technology Inc., Wheat Ridge, CO 80033, USA
Bovine leukaemia virus (BLV)	Antibody	Agar gel immunodiffusion using viral glycoprotein as antigen	Serum	Pitman-Moore Inc., Washington Crossing, NJ 08560, USA
	Antibody	Agar gel immunodiffusion using viral glycoprotein as antigen	Serum	Rhone Merieux Inc., Athens, GA 30601, USA
	Antibody	Indirect ELISA using viral antigen-coated wells	Serum, plasma, milk	Svanova Biotech, Uppsala, Sweden
	Antibody	Indirect ELISA using monoclonal antibody (MAb) captured glycoprotein 51 (gp 51)	Serum, plasma, milk	Svanova Biotech, Uppsala, Sweden
Maedi–Visna (MV) virus	Antibody	Agar gel immunodiffusion using the envelope glycoprotein 135 (gp135) of the virus	Serum	Central Veterinary Laboratory, Weybridge, Surrey, England

(continued)

Table 151. Kit-set tests for the diagnosis of viral infections in ruminants. (*continued*)

Virus	Detection of antigen or antibody	Test system	Sample	Company
Caprine arthritis-encephalitis (CAE) virus	Antibody	Agar gel immunodiffusion using a core protein (p28) of the virus	Serum	Central Veterinary Laboratory, Weybridge, Surrey, England
MV/CAE viruses	Antibody	Agar gel immunodiffusion using gp135 of MV virus and p28 of CAE virus as antigens	Serum	Central Veterinary Laboratory, Weybridge, Surrey, England
	Antibody	Agar gel immunodiffusion using gp 135 of MV virus and p28 of CAE virus as antigen	Serum	Veterinary Diagnostic Technology Inc., Wheat Ridge, CO 80033
	Antibody	Agar gel immunodiffusion using p28 of CAE virus as antigen	Serum	Institut Pourquier, Montpelier, France
Respiratory syncytial virus	Antibody	Indirect ELISA using viral antigen-coated wells	Serum, plasma, milk	Svanova Biotech, Uppsala, Sweden
Parainfluenza-virus 3	Antibody	Indirect ELISA using viral antigen-coated wells	Serum, plasma, milk	Svanova Biotech, Uppsala, Sweden
Bovine herpesvirus 1 (infectious bovine rhino-tracheitis/ infectious pustular vulvo-vaginitis)	Antibody	Indirect ELISA using viral antigen-coated wells	Serum, plasma, milk	Svanova Biotech, Uppsala, Sweden
Bovine viral diarrhoea virus	Antibody	Indirect ELISA using viral antigen-coated wells	Serum, plasma, milk	Svanova Biotech, Uppsala, Sweden
Bovine coronavirus	Antibody	Indirect ELISA using viral antigen-coated wells	Serum, plasma, milk	Svanova Biotech, Uppsala, Sweden
Calf diarrhoea (*E. coli*, K99, group A rotaviruses and bovine coronaviruses)	Antigen	Direct ELISA using antibody-coated wells and conjugated MAb detection system	Faeces	Cambridge Veterinary Sciences, Littleport, Cambridgeshire, England

Table 152. Kit-set tests for the diagnosis of viral infections in pigs.

Virus	Detection of antigen or antibody	Test system	Sample	Company
Aujeszky's disease virus	Antibody directed against glyco-protein I (gI)	Blocking ELISA using gI antigen and anti-gI monoclonal antibody conjugate	Serum, plasma	Svanova Biotech, Uppsala, Sweden
	Antibody directed against glycoprotein II (gII)	Blocking ELISA using gII antigen and anti-gII monoclonal antibody conjugate	Serum, plasma	Svanova Biotech, Uppsala, Sweden
	Antibody	Competitive ELISA using viral antigen and conjugated anti-viral antibody	Serum	Fermenta Animal Health Co., Kansas City, MO 64190-1350, USA
	Antibody	Latex agglutination using viral antigen bound to latex particles	Serum, plasma	Viral Antigens Inc., Memphis, TN 38134-5611, USA
	Antibody	Indirect ELISA (screening) using chemically treated viral antigen	Serum	IDEXX Corp., Portland, ME 04101, USA
	Antibody	Indirect ELISA (verification) using chemically treated viral antigen and normal host cell antigen	Serum	IDEXX Corp., Portland, ME 04101, USA
	Antibody directed against glycoprotein I	Indirect ELISA using gI as antigen	Serum	SmithKline Beecham Animal Health, Exton, PA 19341, USA
	Antibody directed against glyco-protein X (gpX)	Competitive ELISA using chemically treated viral antigen and conjugated anti-gpX MAb	Serum	IDEXX Corp., Portland, ME 04101, USA
	Antibody directed against gpX	Competitive ELISA using gpX recombinant protein antigen produced in *E. coli* and conjugated anti-gpX antibody	Serum	AGDIA, Inc., Elkhart, IN 45614, USA
Porcine parvovirus (PPV)	Antibody	Indirect ELISA using viral antigen	Serum	Central Veterinary Laboratory, Weybridge, Surrey, England
	Antibody	Competitve ELISA using viral antigen-coated wells and anti-PPV conjugate	Serum, plasma	Svanova Biotech, Uppsala, Sweden
Trans-missible gastro-enteritis (TGE) virus	Antibodies specific for TGE virus	Blocking ELISA using viral antigen and monoclonal antibodies directed against both TGE virus and porcine respiratory coronavirus	Serum, plasma	Svanova Biotech, Uppsala, Sweden

Table 153. Kit-set tests for the diagnosis of viral infections in dogs and cats.

Virus	Detection of antigen or antibody	Test system	Sample	Company
Canine parvovirus	Antigen	Direct ELISA using MAb capture and conjugated MAb detection system	Faeces	Fermenta Animal Health Co., Kansas City, MO 64190-1350, USA
	Antigen	Haemagglutination	Faeces	Greenbriar Vet. Services Inc., Delaware, OH 43015, USA
	Antigen	Direct ELISA (modified dot blot) using MAb capture and conjugated MAb detection system	Faeces	IDEXX Corp., Portland, ME 04101, USA
Canine distemper virus	Antibody	Indirect immunofluorescence using polymer beads coated with viral antigen in a colloid film	Serum, plasma	Microbiological Assoc. Inc., Bethesda, MD 20816, USA
Feline immuno-deficiency virus	Antibody	ELISA (modified dot blot) using inactivated viral antigen binding to conjugated viral antigen/antibody complex	Serum, plasma, whole blood	IDEXX Corp., Portland, ME 04101, USA
	Antibody	Indirect ELISA using inactivated viral antigen	Plasma, serum	IDEXX Corp., Portland, ME 04101, USA
Feline infectious peritonitis virus	Antibody	Competitive ELISA using competition for inactivated viral antigen between the sample and conjugated MAb	Serum	Fermenta Animal Health Co., Kansas City, MO 64190-1350, USA
Feline leukaemia virus	Group specific antigen (p27)	Direct ELISA (modified dot blot) using membrane-bound MAb capture of conjugated MAb/viral antigen complex	Serum, plasma, whole blood, tears	IDEXX Corp., Portland, ME 04101, USA
	Antigen (p27)	Direct ELISA using MAb capture and conjugated MAb detection system	Serum, plasma	IDEXX Corp, Portland, ME 04101, USA
	Antigen (p27)	Direct ELISA using MAb capture and conjugated MAb detection system	Blood, serum, plasma, saliva, tears	Fermenta Animal Health Co., Kansas City, MO 64190-1350, USA
	Antigen (p27)	Direct ELISA using MAb capture and conjugated MAb detection system	Blood, plasma, serum	Cambridge Bioscience Corp., Worcester, MA 01605, USA
	Antigen (p27)	Direct ELISA using MAb capture and conjugated MAb detection system	Blood, plasma, serum, saliva, tears	Synbiotics Corp., San Diego, CA 92127, USA

Table 154. Kit-set tests for the diagnosis of viral infections in avian species.

Virus	Detection of antigen or antibody	Test system	Sample	Company
Avian encephalo-myelitis virus	Antibody	Indirect ELISA using chemically treated viral antigen	Serum	IDEXX Corp., Portland, ME 04101, USA
Avian reovirus	Antibody	Indirect ELISA using inactivated viral antigen (strains 1733 and S1133)	Serum	IDEXX Corp., Portland, ME 04101, USA
	Antibody	Indirect ELISA using purified and inactivated viral antigen	Serum	Kirkegaerd and Perry Laboratories, Inc., Gaithersburg, MD 20879, USA
Avian leukosis virus	Antigen	Direct ELISA using antibody capture and conjugated antibody detection system	Serum, egg albumen, cloacal swab	University Laboratories Inc., Highland Park, NJ 08904, USA
	Antigen	Direct ELISA using antibody capture and conjugated antibody detection system	Serum, egg albumen	IDEXX Corp., Portland, ME 04101, USA
	Antibody	Indirect ELISA using viral antigen	Serum	IDEXX Corp., Portland, ME 04101, USA
Infectious bronchitis virus	Antibody	Indirect ELISA using inactivated viral antigen	Serum	IDEXX Corp., Portland, ME 04101, USA
	Antibody	Indirect ELISA using purified viral antigen	Serum	Kirkegaerd and Perry Laboratories, Inc., Gaithersburg, MD 20879, USA
Infectious bursal disease virus	Antibody	Indirect ELISA using viral antigen	Serum	IDEXX Corp., Portland, ME 04101, USA
	Antibody	Indirect ELISA using purified and inactivated viral antigen	Serum	Kirkegaerd and Perry Laboratories, Inc., Gaithersburg, MD 20879, USA
Newcastle disease virus	Antibody	Indirect ELISA using inactivated viral antigen	Serum	IDEXX Corp., Portland, ME 04101, USA
	Antibody	Indirect ELISA using purified and inactivated viral antigen	Serum	Kirkegaerd and Perry Laboratories, Inc., Gaithersburg, MD 20879, USA
Reticulo-endotheliosis virus	Antibody	Indirect ELISA using viral antigen	Serum	IDEXX Corp., Portland, ME 04101, USA

Table 155. Kit-set tests for the diagnosis of viral infections in horses.

Virus	Detection of antigen or antibody	Test system	Sample	Company
Equine infectious anaemia virus	Antibody	Agar gel immunodiffusion using viral antigen	Serum	Pitman-Moore Inc., Washington Crossing, NJ 08560, USA
	Antibody	Agar gel immunodiffusion using purified viral antigen	Serum	Fermenta Animal Health Co., Kansas City, MO 64190-1350, USA
	Antibody to the viral antigen p26	Competitive/direct ELISA using competition for a bound MAb between the sample and conjugated p26 antigen	Serum	Fermenta Animal Health Co., Kansas City, MO 64190-1350, USA

Abbreviations

BLV	bovine leukaemia virus
CAE	caprine arthritis-encephalitis
CFT	complement fixation test
ELISA	enzyme-linked immunosorbent assay
g	glycoprotein
gp	glycoprotein
MAb	monoclonal antibody
MV	maedi-visna
p	protein
PPV	porcine parvovirus
TGE	transmissible gastroenteritis

Bibliography

Bunn, T. and Hoffmann, L. (1990). *Proceedings of Commercial Diagnostic Technology in Animal Health Monitoring*. Iowa State University, Ames, Iowa, USA.

48 Prions (Proteinaceous Infectious Particles)

A prion is a small, proteinaceous, infectious particle. It is highly resistant to heat, formalin, phenol and chloroform. Prions contain little or no nucleic acid, thus resisting inactivation by procedures which modify nucleic acids. The infectivity of the most resistant strains of scrapie-agent is lost only after autoclaving (moist heat) at 136°C for 4 minutes or after 24 hours at 160°C using dry heat. Prion protein is encoded by a cellular gene which appears to be most active in nerve cells. Prions are associated with a number of diseases (transmissible encephalopathies) in both man and animals (**Table 156**). All the prion-associated diseases share a number of common features:

- The diseases are confined to the central nervous system (CNS).
- There is a prolonged incubation period ranging from 2 months to more than 20 years before the onset of clinical signs.
- The clinical course is progressive, frequently prolonged (weeks to years) and invariably leads to death of the animal.
- Characteristic histopathological changes in the brain, including a reactive astrocytosis and frequently vacuolation of neurons are associated with these diseases.

Pathogenesis

Three different manifestations of CNS degeneration may occur:

- ***Slow infection.*** Kuru has been confined to the people, particularly the Fore people, inhabiting the Eastern Highlands of Papua New Guinea. It was the result of a slowly progressive infection transmitted by ritualistic cannibalism.
- ***Sporadic disease.*** Although some cases of Creutzfeldt–Jakob disease (CJD) can be attributed to the use of human cadaver tissues such as corneal transplants or to the injection of human growth hormone, the majority of cases appear to be sporadic with the transmissible agent of CJD being generated *de novo* in each affected individual.
- ***Genetic disorder.*** Familial cases of CJD, in which at least two family members have been strongly suspected or proven to have the disease, account for 5–10 per cent of all CJD cases. The majority of Gerstmann–Straussler syndrome (GSS) cases are familial. Pedigree studies of familial GSS cases suggest that the phenotype is inherited as an autosomal dominant disorder. A point mutation in the prion protein gene greatly increases susceptibility to the disease and the probability of spontaneous occurrence of infectious prion protein. At present, GSS and familial CJD are the only human diseases believed to be both genetic and infectious.

Table 156. Prion-associated diseases (transmissible encephalopathies).

Disease	Natural host
• Scrapie	Sheep and goats
• Transmissible mink encephalopathy	Mink
• Chronic wasting disease	Mule deer and elk
• Bovine spongiform encephalopathy	Cattle
• Feline spongiform encephalopathy	Domestic cats
• Kuru	Humans
• Creutzfeldt–Jakob disease	Humans
• Gerstmann–Straussler syndrome	Humans

Prion protein (PrP^c) is a glycoprotein found in the cells of normal mammals. It is generally found situated in the cell membrane. The concentration of PrP^c increases when neuroblasts differentiate into neurons indicating a possible role in cell-cell recognition. Mutations in the gene encoding for the naturally produced PrP^c can modulate the development of a prion disease or even lead to an encephalopathy in the absence of any infective process. PrP^{sc} is also a glycoprotein and is thought to have the same amino-acid sequence as PrP^c (provided both prion proteins were derived from the same animal species). PrP^{sc} accumulates within the body of the cell and is thought to have a half-life many times that of PrP^c. PrP^{sc} is believed to alter the way in which PrP^c is post-translationally modified and thus direct some of the normally produced native protein (PrP^c) to the infectious type (PrP^{sc}). The newly synthesised prions (sometimes designated $PrP^{c(x)}$) reflect the influence of the host cell's PrP gene in their amino-acid sequence and do not necessarily have the same sequence as the infecting PrP^{sc} molecules.

For the infection to proceed, the newly synthesised prions must leave the cell of origin and transfer to other cells in the same animal to create an ever increasing number of modified cells and pathological change. The tropism of prions for nervous tissue and their mechanism of transfer from cell to cell, or animal to animal, are not yet fully understood. Experimentally, prions have been shown to infect across mammalian species barriers. The occurrence of transmissible mink encephalopathy in ranch-reared mink and of bovine spongiform encephalopathy in cattle has been attributed to the incorporation of sheep offal, from scrapie-affected carcases, in the diet of these animals. Whether or not cattle are a dead-end host, as is the case with mink, is currently under investigation. Although epidemiological studies have so far failed to link CJD cases in man with the consumption of scrapie-infected sheep tissues, a number of countries have already taken steps to prevent ruminant offal and meat from affected animals entering the human food chain.

Prion proteins

PrP^{SC} = scrapie prion protein isoform
(Molecular weight = 33,000–35,000)

PrP^{C}= cellular prion protein isoform
(Molecular weight = 33,000–35,000)

Scrapie

Scrapie has been recognised in sheep for over 200 years in parts of Europe and the disease has also been recorded in Asia, Africa and the Americas. It is an afebrile, chronic, progressive, degenerative disease of the central nervous system with an incubation period of about 1–4 years. The majority of cases occur in sheep, and occasionally goats, of between 2.5 and 4.5 years of age.

The signs include an intense pruritis, fine tremors of the head and neck, ataxia with a weaving gait, paralysis and eventually death. Loss of fleece is marked as the affected sheep rub against posts and constantly bite at their legs and flanks. A strong nibbling reflex can easily be elicited. Affected animals may be seen standing with staring eyes and a fixed gaze. Death occurs from 6 weeks to 6 months after the first clinical signs become obvious.

Offspring of affected ewes often contract the infection. Ingestion of placental material from scrapie-infected ewes has been shown to be a means of transmission. The scrapie-agent can remain viable on pasture for up to 3 years. Studies have indicated that scrapie can be transmitted horizontally under close-penning conditions. There are high titres of the scrapie-agent in brain, spleen, spinal cord and lymph nodes, making sheep offal a potential danger if included in animal feeds.

The susceptibility of sheep to scrapie is genetically controlled and a genetic predisposition to the disease occurs in Suffolk sheep and some other breeds. Experimental transmission of the scrapie-agent has been successful from sheep to goats, mice, hamsters, guinea-pigs, cats, mink and monkeys but not to apes.

Bovine Spongiform Encephalopathy (BSE)

The first cases of BSE occurred in the UK in April, 1985. Epidemiological findings indicate that the disease is an extended common source epidemic. Infected cattle are thought to have been exposed as calves to the scrapie agent in cattle feedstuffs containing ruminant-derived protein, in the form of meat and bone meal. A high degree of homology between sheep and bovine PrP (98%) has been demonstrated (Prusiner *et al.*, 1993). In 1981–82 there was a reduction in the use of hydrocarbon solvent extraction of fat from meat and bone meal. Changes in the rendering process occurred coincidentally with the estimated time of exposure of cattle to the prion and suggested that the rendering process was no longer inactivating the scrapie agent. This exposure to the prion was constant until 1988, when inclusion of ruminant-derived protein in cattle rations was banned.

There is no evidence of a breed or sex predisposition to the disease and usually only one, or a few clinical cases occur in any one herd. The number of affected dairy cattle has been far greater than that for beef animals. This might be explained by concentrate rations being fed more commonly to dairy calves.

The incubation period for BSE is estimated to be between 2 and 15 years. The peak incidence occurs in 4-year-old and 5-year-old cattle and about 96 per cent of clinical cases have been recorded in animals between 3 years and 6 years of age. Presenting signs of BSE, initially reported by farmers, include nervousness, kicking in the milking parlour, locomotory difficulty, reduction in milk yield and loss of weight.

Generally there are abnormalities in mental status, posture, movement and sensation. Altered behaviour in affected animals includes, apprehension, low head carriage, excessive ear movements and a reluctance to pass through doorways or from one type of floor surface to another. Pruritis is not a predominant sign, but has been reported in a few cases. Weight loss occurs but usually appetite remains normal. Milk yield is reduced, but this may reflect the difficulty in milking the affected cows. There can be hyperaesthesia to touch and sound but 'mad cow disease' is a misnomer as frenzy and aggressiveness occur only in a few cases. The aggression tends to be directed at other cattle rather than at humans. There are tremors and hindlimb ataxia that leads to difficulty in turning. Affected animals fall and remain recumbent for long periods. Day-to-day variations in the presence and intensity of nervous signs occur. Maintaining the affected animal in quiet, familiar surroundings often results in a temporary reduction in the severity of the signs. The course of BSE is about 1–3 months, terminating in death of a few animals but most animals are slaughtered because of unmanageable behaviour or trauma due to falling.

If the food-borne route was the sole means of transmission, it has been estimated that the incidence of disease will gradually decrease and cease by the turn of the century. There is no evidence, so far, that cattle-to-cattle transmission occurs. However, if horizontal transmission is possible, then the disease will be perpetuated.

Feline Spongiform Encephalopathy

The first case of feline spongiform encephalopathy (FSE) was reported in 1990 and more than twenty other cases have since been described (Gruffydd-Jones *et al.*, 1991). Although cats have been fed sheep-derived food for many years, the changes in rendering of ovine materials may have allowed survival of the scrapie agent and led to increased exposure to this agent in their diets.

The disease has been reported in cats 5 to 12-years-old. The main clinical signs include ataxia, paticularly of the hind limbs, that becomes progressively worse and leads to a crouching stance, changes in temperament to either timidity or aggression, hyperaesthesia, head-nodding or fine continuous tremors of head and ears, hypersalivation unrelated to feeding and abnormal grooming activity.

Laboratory Diagnosis

In naturally occurring infections in cattle and sheep, prions do not appear to elicit a demonstrable immune response. Consequently, no blood-based tests exist, at present, to indicate exposure to the scrapie or BSE agents. The laboratory diagnosis is generally based solely on histopathology with other confirmatory techniques being conducted at research establishments.

- **Histopathology**: Specimens include the whole head if it can be transported to the laboratory within a reasonable time after death, or the brain, particularly the medulla, fixed in 10 per cent formalin. In the ovine brain, neuronal vacuolation (**489**) and evidence of diffuse astrocytic proliferation are vir-

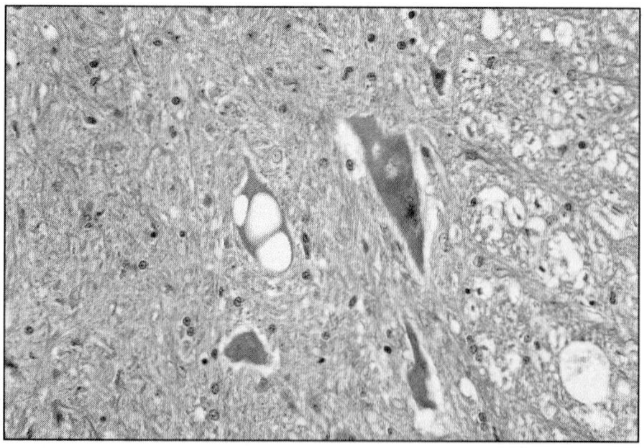

489 Multilocular vacuolation of a neuron characteristic of the spongiform encephalopathies. Section of the brain stem from a sheep with scrapie. (H&E stain, × 400)

tually diagnostic for scrapie. Lesions in the bovine brain are similar, but vacuolation of the neuropil is an additional feature. The nature and distribution of the vacuolation in FSE is quite distinctive and combined with other pathological features is characteristic for this disease.

- **Immunoblotting**: Although antibodies have not been detected in naturally occurring cases of prion-associated disease, it has been possible to produce antibodies to PrP27–30 (the proteinase-K resistant portion of PrP^{Sc}). These antibodies are used to detect the presence of PrP27–30 in detergent extracts of affected brain tissue.
- **Electron microscopy**: Prion protein may be visualised by negative stain transmission electron microscopy as 'prion rods' (Prusiner *et al.*, 1983) or as 'scrapie-associated fibrils' (Merz *et al.*, 1981) depending on the extraction treatment used on the affected brain tissue.

References

Gruffydd-Jones, T.J., Galloway, P.E. and Pearson, G.R. (1991). Feline spongiform encephalopathy, *Journal of Small Animal Practice*, **33**: 471–476.

Merz, P.A., Sommerville, R.A., Wisniewski, H.M. and Igbal, K. (1981). Abnormal fibrils from scrapie-infected brain, *Acta Neuropathol*, **54**: 63–74.

Prusiner, S.B., McKinley, M.P., Bowman, K.A., Bolton, D.C., Bendheim, P.E., Groth, D.F. and Glenner, G.G. (1983). Scrapie prions aggregate to form amyloid-like birefringent rods, *Cell*, **35**: 349–358.

Prusiner, S.B., Füzi, M., Scott, M., Serban, D., Serban, H., Taraboulos, A., Gabriel, J.M., Wells, G.A.H., Wilesmith, J.W., Bradley, R., DeArmond, S.J. and Kristensson, K. (1993). Immunologic and molecular biologic studies of prion proteins in bovine spongiform encephalopathy, *Journal of Infectious Diseases*, **167**: 602–613.

Biblography

Chesebro, B.W. (Editor) (1991). *Current Topics in Microbiology and Immunology*, number 172. *Transmissible Spongiform Encephalopathies: Scrapie, BSE and Related Human Disorders*, Springer–Verlag. Berlin, Germany.

Prusiner, S.B. and McKinley, M.P. (Editors) (1987). *Prions: Novel Infectious Pathogens Causing Scrapie and Creutzfeldt–Jakob Disease*. Academic Press Inc., San Diego, California, USA.

Prusiner, S.B., Torchia, M. and Westaway, D. (1991). Molecular biology and genetics of prions – implications for sheep scrapie, 'mad cows' and the BSE epidemic, *Cornell Veterinarian*, **81**: 85–101.

Section 5: Zoonoses and Control of Infectious Diseases

49 Zoonoses

Human populations encounter animals with varying frequency depending on their occupation, geographical location and the prevailing culture of the country. Whether living in an urban or rural environment, animals are constantly present and humans may have close contact with them on their farms (food-producing animals), in their homes (dogs, cats, caged birds), through leisure activities (horses, wildlife) or by virtue of their occupation as veterinarians or animal nurses.

Apart from their obvious benefits as a source of food, draught power, transportation and companionship, animals may occasionally have a negative impact on the human population through pollution of the environment, as a cause of traffic accidents, injury to humans through bite wounds and attacks on other susceptible species (dogs attacking sheep). Health hazards associated with animals are related to communicable diseases. The term zoonoses is applied to those diseases and infections which are naturally transmitted between vertebrate animals and man.

Transmission of disease may be direct, simply by contact with an animal, or indirect, through food, non-edible products, secretions or excretions (**Diagram 66**). Apart from food-borne zoonoses (Chapter 37) the importance of particular zoonotic diseases often varies with a person's occupation, the nature and type of animals present, and the diseases prevalent in the animal population in a particular geographical region. The more frequent and direct the contact with animals, the greater the risk of acquiring a zoonotic infection. Farmers, owners of companion animals, workers in slaughterhouses or by-product processing plants, veterinarians and laboratory staff dealing with infectious material, workers in zoos and circuses and personnel engaged in servicing sanitary services are more likely to acquire zoonotic diseases than workers who have infrequent contact with animals.

Zoonoses may be classified according to their aetiology as being of bacterial, fungal, viral or parasitic origin. Selected zoonotic diseases in each category are briefly reviewed.

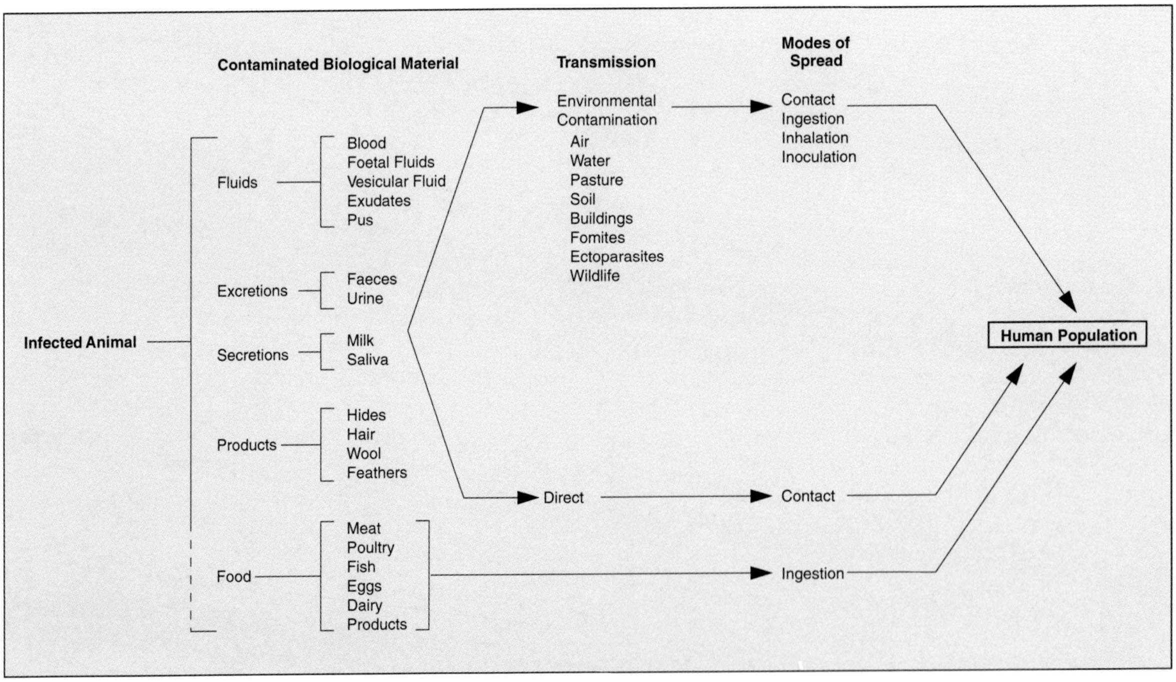

Diagram 66. Direct and indirect transmission of infectious agents from animals to the human population. Human diseases acquired in this manner are referred to as zoonoses.

Taenia saginata infection (beef tapeworm)

Animals vary in their ability to act as reservoirs of zoonotic diseases and also in their ability to transmit infectious agents to humans. Rarely, humans act as definitive hosts of parasites which infect animals. The beef tapeworm *Taenia saginata*, which is found in the human small intestine and occurs worldwide, is an example of such a parasite. The adult tapeworm which may be up to 10 metres in length, is composed of segments or proglottides each of which may contain up to 100,000 eggs. Gravid segments may contaminate pasture through sewage disposal, indiscriminate defaecation by infected humans in the vicinity of cattle, or by flooding of pasture land with contaminated water. Embryonated eggs which are immediately infective must be eaten by the only intermediate host, cattle, to develop further. Following ingestion by a susceptible bovine animal, the oncosphere, after hatching in the small intestine, travels via the blood to striated muscle and various organs including the heart, tongue and liver. Within three months the cysticerci which develop are infective for humans (**490**).

The cysticerci may survive from months to years and when they die, they frequently degenerate into a caseous mass which may become calcified. Human infection follows the ingestion of raw or inadequately cooked beef, which has not been frozen, containing viable cysticerci. The immature tapeworm evaginates its scolex, attaches to the mucosa of the jejunum, and develops into a mature tapeworm in about 10 weeks. Eggs can remain viable for several months in the environment and the adult worm may remain in the small intestine for many years. In humans, the presence of an adult tapeworm may produce mild discomfort such as light abdominal pain and diarrhoea. Infection with this parasite is also of concern for public health and aesthetic reasons. An economic aspect of this disease is carcase condemnation arising from heavy infestation with the cysticerci of *T. saginata* as well as the cost of inspecting meat for the parasite.

Control measures include education of the public on measures relating to the prevention of infection in cattle. As the adult *T. saginata* has a long life span, identification and treatment of infected human carriers is an essential step in prevention of infection. High standards of human sanitation, thorough cooking of beef, inspection of beef carcases for cysticerci followed by condemnation of those heavily infected, and freezing of lightly infected carcases at –10°C for 10 days, are essential preventive measures.

490

490 *Taenia saginata*: viable cysticerci dissected from the carcase of a heifer experimentally infected with eggs of *T. saginata* seven months earlier.

Toxocariasis (*Toxocara canis*)

Human infection, particularly in young children, caused by the accidental ingestion of eggs of *Toxocara canis*, the common dog roundworm, results in a disease syndrome termed visceral larva migrans or toxocariasis. The worm's life cycle is completed only in its canine host. Most dogs are infected *in utero* by larvae that have been reactivated in the pregnant bitch. Larvae which have migrated across the placenta, mature in pups after birth. Toxocariasis is acquired primarily by exposure to soil contaminated with dog faeces. Eggs, which require about 2 weeks to become infectious, can survive in soil for many months. When eggs are ingested, larvae penetrate the wall of the intestine but are unable to complete their life cycle in the human host and migrate through the host's tissues, producing eosinophilic granulomas. The life cycle of *T. canis*, together with the environmental contamination which may result from faecal dispersal, transmission of infection and its possible sequelae, are shown in **Diagram 67**.

Systemic toxocariasis, which is termed visceral larva migrans, results in hepatosplenomegaly, pulmonary infiltrates, seizures and, sometimes, behavioural disorders. Eosinophilia is a prominent sign and usually persists for several months. Ocular larva migrans, an infrequent but serious disease, usually presents as unilateral reduced vision without systemic symptoms or eosinophilia. Involvement of the central nervous system, or myocardium, the most serious form of disease produced by *T. canis*, is rare but may be fatal.

Prevention and control of visceral larva migrans requires the cooperation of the veterinary and medical professions. Education of pet owners on matters relating to *T. canis* infections and their prevention, combined with advice to parents of young children on the hazards of poor hygiene and geophagia in gardens frequented by dogs, would greatly reduce the probability of infection. Strict enforcement of legislation relating to fouling of public places by dogs would decrease the risk of exposure in parks and playgrounds. The largest reservoir of environmental contamination, pups and

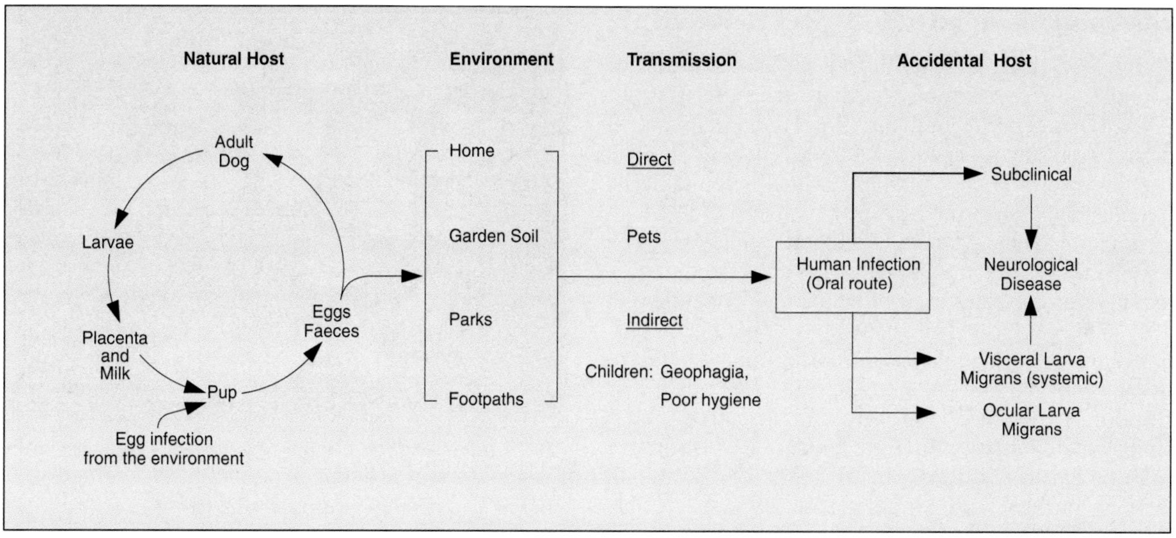

Diagram 67. The life cycle of *Toxocara canis* showing how the dispersal of eggs can lead to environmental contamination. Young children in close contact with young dogs shedding eggs are particularly at risk. Children may also acquire infection from contaminated soil in gardens or parks, from sand boxes or from fouled pavements.

nursing bitches, should be routinely treated with effective anthelmintics at 2, 4, 6 and 8 weeks after the pups are born and every 6 months thereafter.

The prevalence of intestinal parasites in dogs is high and the common association of children with pups tends to increase the possibility of human infection. Young children should be instructed routinely to wash their hands after playing with dogs (and other animals). Pet shops offering young dogs for sale should be required to treat them with anthelmintics and maintain a good standard of hygiene to prevent reinfection. Elimination of stray dogs, especially in urban areas, is an important step in the prevention of faecal environmental contamination by parasitic ova including the eggs of *T. canis*.

Cryptosporidiosis *(Cryptosporidium species)*

Cryptosporidium species have a wide host range and appear to exhibit little species specificity. These coccidian parasites are found in close association with epithelial cells of many species of animals including man. The gastrointestinal tract is most commonly affected in young ruminants and this parasite is thought to be of considerable importance in the calf diarrhoea complex. Major outbreaks of cryptosporidiosis have been reported in calves, lambs, pigs and other species. Oocysts, present in the faeces of animals (**491**), are presumed to be the infectious form for humans. Cryptosporidiosis may be acquired from infected animals and from faeces, contaminated food or water. Mild cases of disease have been reported in farm workers. Immunosuppressed persons, the very young and the very old, should avoid contact with this parasite as it may cause severe, and in some instances, intractable diarrhoea.

491

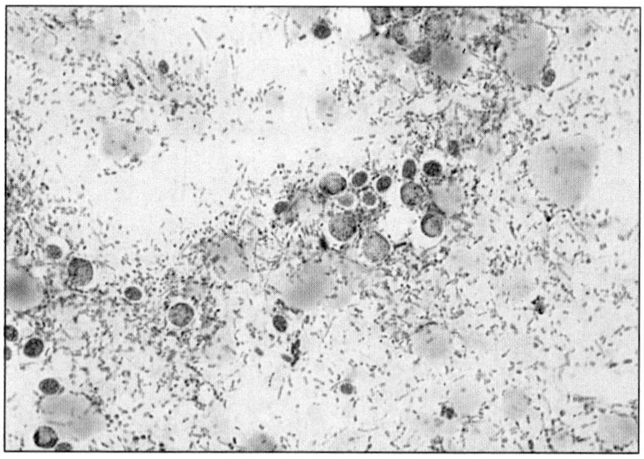

491 *Cryptosporidium* sp. in calf faeces appearing as round, reddish oocysts, stained with safranin-methylene blue stain. (×1000)

Orf *(Parapoxvirus)*

A number of infectious diseases including ringworm, orf (a parapoxvirus) and leptospirosis can be readily acquired when handling infected animals. Lesions of orf are frequently seen in sheep and goat populations as scabby lesions, often involving the muzzle and lips, which bleed after mild trauma (**492**). Occasionally, the teats of ewes and the gums and tongues of lambs may be affected. Human infection can occur among persons occupationally exposed, particularly sheep handlers. Infection of humans occurs usually as a single lesion on a finger, hand, forearm or face. Rarely, generalised disease has been reported.

492

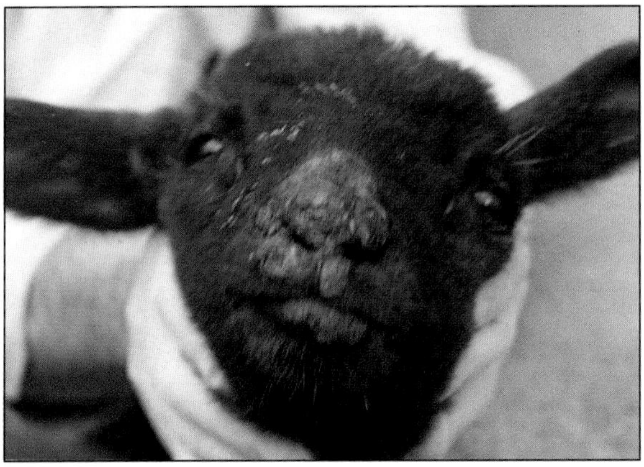

492 Lamb with lesions of orf (contagious pustular dermatitis) caused by a parapoxvirus. Orf is transmitted to humans by direct contact with an infected animal, or the vaccine which contains live virus.

493

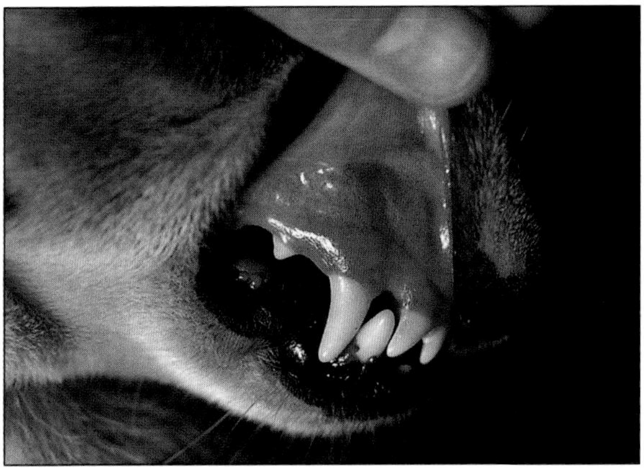

493 Leptospiral jaundice in a young dog.

Leptospirosis *(Leptospira interrogans* serovars*)*

Dogs, cattle, pigs, rats and a number of other animal species shed leptospires in their urine. Direct contact with urine or tissues of infected animals such as dogs (**493**) can lead to human infection. Leptospires may penetrate the skin, especially if abraded, or mucous membranes, sometimes leading to severe systemic disease. Common features of leptospirosis are fever with sudden onset, headache, severe myalgia, meningitis, haemolytic anaemia and jaundice, depending on the serovar involved. Diagnosis can be confirmed by isolation of leptospires from the blood (first week of illness), by serology using the microscopic agglutination test or immunofluorescence, and by isolation of leptospires from the urine after the second week of infection.

Toxoplasmosis *(Toxoplasma gondii)*

Toxoplasmosis is of major importance in many countries as a cause of perinatal mortality in sheep. It also causes sporadic disease in a wide range of animals including humans. *Toxoplasma gondii* is an intracellular protozoan parasite which occurs worldwide. Domestic cats and occasionally other members of the cat family (Felidae) are believed to be responsible for the widespread transmission of this parasite through faecal contamination of pasture, soil (including garden soil), feed concentrates and occasionally water. Although infection with *T. gondii* is relatively common in the human population, disease arising from infection with this parasite is uncommon. In immunosuppressed patients, such as those suffering from cancer or AIDS, toxoplasmosis is a serious disease which may be fatal. Congenital infection, which occurs only after a woman acquires a primary infection while pregnant, may produce serious defects in the developing foetus. Abnormalities may include bilateral retinochoroiditis, hydrocephalus and other neurological changes. Infections postnatally are usually less severe and many human infections are asymptomatic.

The definitive host of *T. gondii* is the domestic cat or feral cats. Feline infection usually begins with cats eating raw meat or prey with tissue cysts containing bradyzoites. Occasionally, faecal oocysts may be responsible for feline infections, or tachyzoites in an acutely infected prey animal. Following the development of sexual forms in the intestinal epithelium of the cat (the only species in which this occurs), oocysts are shed in the faeces but are not infective for about 72 hours. Faecal oocysts are highly resistant to environmental conditions. They are shed in large numbers especially by young cats, and are capable of infecting a wide range of animals including man. Tissue cysts containing bradyzoites are the forms most frequently seen in tissues of animals, but in acute toxoplasmosis, tachyzoites may be present. Tachyzoites can be cultured in monolayers (**494**).

Prevention of infection in the human population is concerned with awareness of the role of the cat in the cycle of infection and implementation of suitable hygienic measures. Pregnant women should avoid contact with cats unless the animals are fed canned or cooked food and kept indoors. Cat litter boxes should be emptied daily before oocysts sporulate and the litter disposed of in a safe manner. Pregnant women and other people in high-risk categories should not assist at lambing time, particularly if abortions have occurred in the flock. As meat, especially mutton or pork, may occasionally contain cysts of *T. gondii*, it should be adequately cooked. The faecal oocysts are resistant to many household disinfectants. Boiling water or iodophors should be used where contamination with this parasite is suspected.

494

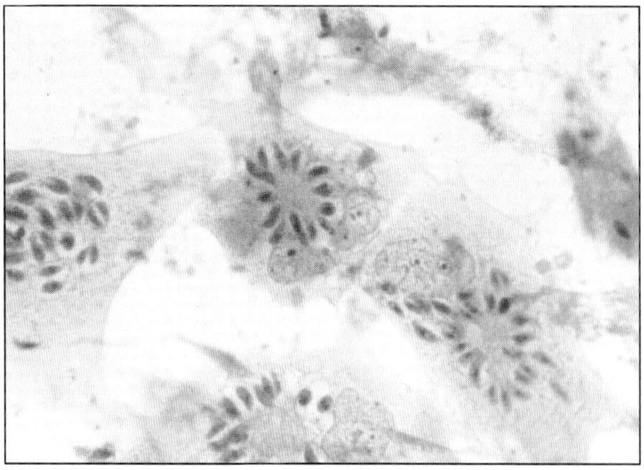

494 *Toxoplasma gondii:* tachyzoites growing in a monolayer. In some heavily parasitised cells, the tachyzoites are arranged in a rosette manner. Monolayers usually disintegrate within a few days of exposure to this parasite. (H&E stain, x 1000)

495

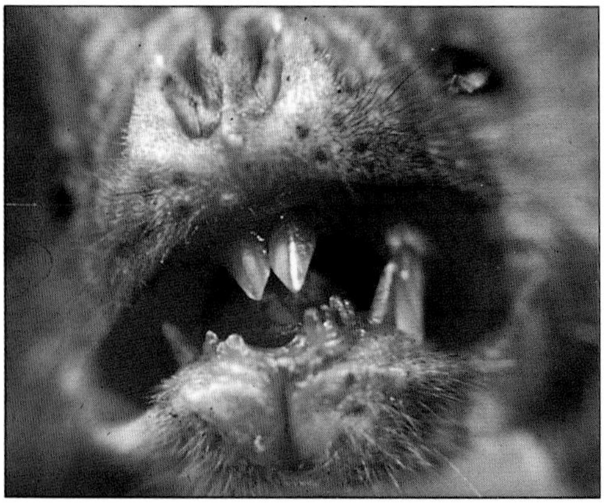

495 Close-up of the razor-sharp incisor teeth of a vampire bat. The bat punctures the skin of its victim with the incisor teeth and laps the blood from the puncture wound with its tongue (*see* **496**).

Rabies *(Lyssavirus, Rhabdoviridae)*

Several species of animals may cause injuries to humans through biting, scratching or kicking. Biting is the most common injury caused by domestic animals, mainly by dogs. A high proportion of cat-bite wounds become infected by *Pasteurella* species and other pathogenic bacteria. In countries where rabies is endemic, bite wounds from dogs and cats may assume life-threatening proportions until it has been definitively established that the virus was not present in the animal's tissues. Rabies virus enters the body through skin abrasions or bite wounds contaminated by virus-laden saliva from rabid animals. Occasionally, scratches by infected animals, or aerosols in caves frequented by bats harbouring the virus, may transmit infection. Rabies is widely distributed throughout the world with the exception of Australia, New Zealand, Japan, a number of European countries and some Caribbean islands. Wild mammals serve as a large and mainly uncontrollable reservoir of **sylvatic** rabies, which is an increasing threat to the human population and to domestic animals in many countries (**Diagram 68**).

In western Europe, which was formerly relatively free of rabies, foxes have reintroduced the disease. A noticeable increase in bat infections with a virus similar to that of the rabies virus has also been observed. Wildlife reservoirs of disease in North America include skunks, racoons and foxes (the Arctic fox in polar regions). In several countries of Central and South America vampire-bat rabies is a threat to livestock and to the human population. The mouth-parts, particularly the incisor teeth of a vampire bat (**495**), allow it to collect blood from the puncture wounds it inflicts on its victims, often cattle, horses (**496**) and

496

496 Horse bitten by a vampire bat. These bats are associated with the transmission of rabies virus.

other species of animals, thus facilitating the spread of the rabies virus. Canine rabies is still of major importance in many countries (**urban** rabies) and infected dogs are responsible for most of the human rabies cases recorded annually.

The control of rabies in a given country depends on whether it is free of the disease, has land frontiers or is an island. Strict quarantine of animals imported from rabies regions for at least 6 months has maintained the disease-free status of a number of European countries, Australia and New Zealand. However, once rabies is introduced into a region, especially if wildlife become infected, eradication is difficult and protracted. Immunization of dogs and cats is an effective measure where disease is endemic and attenuated, live virus vaccines are usually efficacious and safe. Control of urban rabies requires the elimination of stray dogs and cats combined with mass vaccination of all mammalian pets. Vaccination of wildlife against rabies was first attempted in the USA 30 years ago, after other methods of control had failed. Subsequently, it was demonstrated that an attenuated rabies virus strain given orally protected foxes against rabies. Currently, oral immunisation of wildlife in fat or fish meal bait, in Switzerland, Germany and neighbouring countries has succeeded in halting the advance of rabies in Europe. Field trials in Canada, with a similar oral vaccine distributed from aircraft, appear to have been less successful.

Individual zoonotic diseases may be acquired from a limited range of animal species. However, some zoonoses such as those caused by parasites may be acquired only from a single species. Conversely, diseases such as rabies, leptospirosis and brucellosis may be acquired from many different species of animals. **Tables 157–159** list the major zoonotic diseases of animals.

Control and prevention of zoonoses requires consideration of the infectious agent, its epidemiology, the nature and severity of the disease it produces in humans, and the frequency of its occurrence in animal and human populations. **Tables 160** and **161** outline measures appropriate for the prevention of zoonoses from companion and food-producing animals.

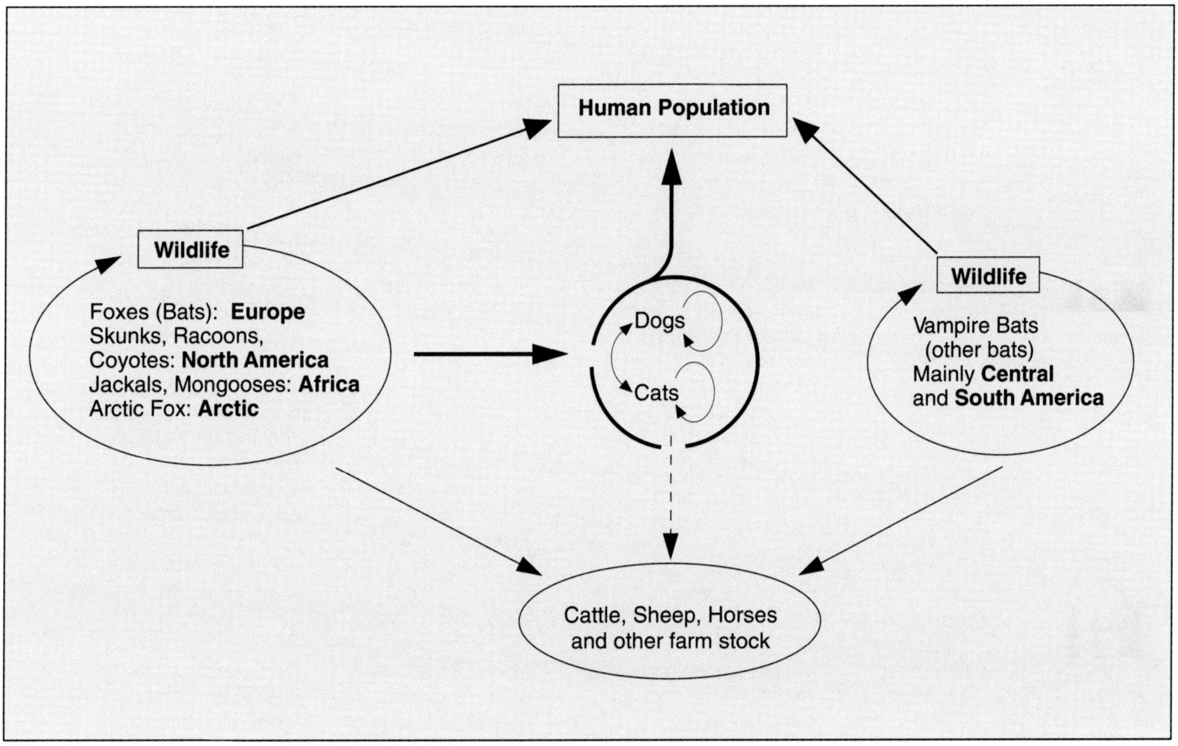

Diagram 68. Transmission of rabies from wildlife reservoirs to humans and to domestic animals.

Table 157. Viral zoonoses.

Disease	Infectious agent	Reservoir/vectors or mode of transmission	Distribution	Disease in humans
	Herpesviridae			
B-Virus disease of monkeys	*Cercopithecine herpesvirus 1* (*Herpesvirus simiae*)	Old World monkeys or monkey-cell cultures/monkey bites (infected saliva) or scratches; contact with infected material or by inhalation	Wherever infected monkeys are present or where their tissues are used in laboratories	Usually fatal ascending encephalomyelitis; occasional recoveries with considerable residual CNS damage
	Poxviridae			
Cowpox	*Orthopoxvirus*	Dairy cattle, domestic cats, rodents/direct contact during milking or handling infected cats or rodents	Many countries	Local erythema, followed by vesicle, pustule and scab formation. Self-limiting occupational disease
	Filoviridae			
Ebola-Marburg virus diseases	Two antigenically distinct viruses	Presumed animal reservoir (rodents?)/ contact with infected monkeys or their tissues. (Person-to-person transmission occurs by direct contact, infected blood or secretions.)	Eastern and Southern Africa	Severe haemorrhagic fever, CNS involvement, profuse diarrhoea, vomition, uraemia and haemorrhages. Fatal cases show necrotic foci in many organs including the brain. Case-fatality rates for Ebola infections range from 50–90%. **Highly contagious**
	Picornaviridae			
Foot-and-mouth disease	*Aphthovirus*	Cattle, sheep, pigs and many wildlife species/ contact	Africa, Asia, South America, some European countries	Vesicles on mucous membranes and on the skin. Mild disease, uncommon in humans
	Orthomyxoviridae			
Swine influenza	*Influenzavirus*	Suspected animal reservoirs, especially pigs/ airborne spread	Worldwide	Self-limiting acute upper respiratory tract infection
	Paramyxoviridae			
Newcastle disease	*Avian paramyxovirus 1*	Domestic and wild birds/ direct contact	Many countries	Occasional human infection; conjunctivitis, laryngitis, pharyngitis and tracheitis. Uncommon in humans

(*continued*)

Table 157. Viral zoonoses. (*continued*)

Disease	Infectious agent	Reservoir/vectors or mode of transmission	Distribution	Disease in humans
	Poxviridae			
Orf (contagious pustular dermatitis)	*Parapoxvirus*	Sheep and goats/direct contact with an infected animal or orf vaccine (live)	Worldwide	Usually a solitary lesion up to 3 cm in diameter on hands, arms or face; lesion regresses after 4–6 weeks
Pseudo-cowpox (milkers' nodule)	*Parapoxvirus*	Cattle/direct contact	Worldwide	Lesion on hands or arms; rarely becomes pustular. Mild self-limiting disease
	Rhabdoviridae			
Rabies	*Lyssavirus*	Wildlife including foxes, skunks, racoons, jackals, bats, domestic animals – particularly dogs and cats/ bite wounds, scratches (rarely aerosols)	North, South and Central America, Africa, Asia, many European countries	An invariably fatal acute encephalomyelitis. The disease progresses to paresis or paralysis; the fear of water (hydro-phobia) arises because of the difficulty in swallowing
	ARTHROPOD-BORNE AND RODENT-BORNE VIRAL DISEASES			
	Arenaviridae			
Lassa fever	*Arenavirus*	Wild rodents (in Africa, a house rat *Mastomys natalensis*)/direct or indirect contact	West and Central Africa	Highly contagious; produces a serious febrile illness; case fatality rate about 15%
Lympho-cytic chorio-meningitis	*Arenavirus*	House mouse (*Mus musculus*)/probably oral or respiratory route	Europe, Americas	Influenza-like symptoms, occasionally meningoencephalitis
	Bunyaviridae			
Nairobi sheep disease	*Nairovirus*	Sheep and goats/the tick *Rhipicephalus appendiculatus*	Eastern Africa, India	Fever
Rift valley fever	*Phlebovirus*	Sheep, goats and cattle/ mosquitoes (*Culex* and *Aedes*)	Africa	Fever, haemorrhages, encephalitis; case fatality rate is usually low but occasionally may be high
	Reoviridae			
Colorado tick fever	*Coltivirus*	Small mammals such as ground squirrels and chipmunks/tick (*Dermacentor andersoni*)	High altitude (above 2,000 metres) Western USA and Western Canada	Acute, febrile diphasic disease. Symptoms include headache, muscle and joint pains, usually of short duration. Deaths uncommon

(*continued*)

Table 157. Viral zoonoses. (*continued*)

Disease	Infectious agent	Reservoir/vectors or mode of transmission	Distribution	Disease in humans
	Togaviridae			
Eastern equine encephalo-myelitis (EEE)	*Alphavirus*	Birds/mosquito (*Culiseta melanura*)	Canada, Eastern USA, Central America	Encephalitis, often severe infection; case fatality rate may be high
Western equine encephalo-myelitis (WEE)	*Alphavirus*	Birds/mosquito (*Culex pipens*; *C. tarsalis*)	Western and Central USA, Canada, South America	Encephalitis, less severe than EEE and most patients recover
Venezuelan equine encephalo-myelitis (VEE)	*Alphavirus*	Rodents, horses/ mosquito (*Aedes, Culex*)	South America, Central America, Southern USA	Influenza-like disease with a diphasic course; some patients may show encephalitis. Mortality rate is low
	Flaviviridae			
Japanese encephalitis	*Flavivirus*	Birds, pigs/mosquito (*Culex*)	Western Pacific Islands from the Philippines to Japan, China, Korea, India, Indonesia	Encephalitis; case fatality rate is up to 10% (higher in older age groups). Residual neurological damage may occur in 30% of patients
Louping ill	*Flavivirus*	Sheep, occasionally cattle, horses, deer and grouse/ticks (*Ixodes ricinus*)	Scotland, Northern England, Ireland, parts of Western Europe	Biphasic disease, first phase influenza-like, followed by a meningoencephalitic syndrome (usually mild)
Tick-borne arboviral encepha-litides	*Flavivirus*	Mammals including a range of wildlife species such as hedgehogs and bats/ ticks (*Ixodes persulcatus*, *I. ricinus*)	Former Soviet Union (Russian spring-summer encephalitis), North America, Scandinavia, Asia	Often biphasic diseases, initial febrile stage, fever and meningoencepha-litis. Far Eastern Tick-borne disease may result in residual damage
St. Louis encephalitis	*Flavivirus*	Birds/mosquito (*Culex*)	USA, Canada, Central and parts of South America	Encephalitis, occa-sionally hepatitis; case fatality rate is approximately 10% but higher in older age groups
Yellow fever	*Flavivirus*	Mainly monkeys (possibly marsupials)/ mosquito (*Aedes*)	Africa, Central and South America	Acute febrile disease of varying severity; cell destruction in liver, spleen, kidney, bone marrow and lymph nodes. Jaundice and haemorrhages occur. Case fatality rate may approach 50% in epidemics

Table 158. Bacterial, chlamydial, rickettsial and fungal zoonoses.

Disease	Infectious agent	Reservoir/vector or mode of transmission	Distribution	Disease in humans
Anthrax	*Bacillus anthracis*	Animals, especially cattle, sheep and goats, also wildlife, animal products including hides and wool (contaminated soil may be a source of infection)/ contact, inhalation, ingestion	Parts of Europe, Asia, Africa, North, Central and South America; sporadic occurrence in many other parts of the world	Acute bacterial disease following inhalation or ingestion, producing septicaemia and death if untreated. Cutaneous anthrax progresses more slowly but may prove fatal if untreated
Boutonneuse fever	*Rickettsia conorii*	Dogs, rodents and other animals/ticks, *Rhipicephalus sanguineus* in Mediterranean countries; *Amblyomma, Rhipicephalus, Boophilus* and *Hyalomma* species in South Africa	Countries adjacent to Mediterranean, Caspian and Black seas; Africa and India	Febrile illness; lesion at site of tick bite, rash on palms and soles may persist for 7 days. Case fatality rate is low
Brucellosis (undulant fever)	*Brucella abortus* *Brucella melitensis* *Brucella suis* *Brucella canis*	Cattle (*B. abortus*), goats, sheep (*B. melitensis*), pigs (*B. suis*), dogs (*B. canis*)/ Contact with infected biological material, ingestion of raw milk and dairy products from infected animals, airborne infection occasionally	Worldwide, especially Mediterranean countries, North and East Africa, India, Asia, Central and South America	Disease spectrum is influenced by the species; often insidious with malaise, chills, fever and profuse sweating, headaches, arthralgia; fever may be intermittent; may develop into a chronic disease with many non-specific manifestations. Case fatality rate, even without treatment, is low
Campylobacter enteritis	*Campylobacter jejuni* (*Campylobacter coli* occasionally)	Cattle, poultry and other domestic animals, especially dogs; unpasteurised or recontaminated milk, foods recontaminated by contact with raw meat (especially poultry)/ ingestion of contaminated food, milk or water; contact with infected pets, often involving young children	Worldwide, especially in temperate zones	Acute enteric disease of variable severity which usually lasts for about one week. Diarrhoea (sometimes with blood), abdominal pain, fever, nausea and vomition are common manifestations

(*continued*)

Table 158. Bacterial, chlamydial, rickettsial and fungal zoonoses. (*continued*)

Disease	Infectious agent	Reservoir/vector or mode of transmission	Distribution	Disease in humans
Cat scratch disease	Pleomorphic, Gram-negative bacterium	Reservoir uncertain, cats appear to be carriers of the bacterium/scratch or bite wounds, also licking of the skin	Worldwide but uncommon; seasonal pattern may occur (late summer, autumn or winter)	A benign, self-limiting illness. A primary skin lesion may occur at site of a scratch or bite, followed by development of regional lymphadenopathy. Course of disease is variable and may last for several weeks
Chlamydial infections	*Chlamydia psittaci* (ovine strain)	Sheep, particularly flocks with enzootic (chlamydial) abortion/inhalation, ingestion	Many countries	Conjunctivitis, pneumonia; also, abortion in pregnant women has occurred following contact with flocks with chlamydial abortion. Serious disease with meningitis and disseminated intravascular coagulation has also been recorded
Cryptococcosis	*Cryptococcus neoformans*	Debris of pigeon roosts and pigeon droppings (birds are not infected); also, from soil/inhalation	Worldwide	Pneumonia; lesions in the skin, kidneys, bone and liver may occur; meningitis is frequently seen, especially in immunosuppressed individuals, and may terminate fatally if not treated
Dermatophilosis	*Dermatophiilus congolensis*	Many animals, including cattle, sheep, horses and wildlife/ contact	Worldwide	Usually occurs as discrete, cutaneous pustules on the hands, forearms or legs
Erysipelothrix infection (Erysipeloid)	*Erysipelothrix rhusiopathiae*	Pigs, fish, meat or poultry/ contact (skin abrasions)	Worldwide	Infection is usually limited to the skin at the site of trauma, most commonly on the fingers or hands. The lesion is raised and red or purple with a burning sensation (erysipeloid). Rarely, endocarditis occurs. Butchers, farmers, fish handlers, poultry workers and veterinarians are at greatest risk of infection

(*continued*)

Table 158. Bacterial, chlamydial, rickettsial and fungal zoonoses. (*continued*)

Disease	Infectious agent	Reservoir/vector or mode of transmission	Distribution	Disease in humans
Glanders	*Pseudomonas mallei*	Horses, mules and donkeys; also laboratory accidents/ contact and inhalation	Asia, eastern Europe	Short incubation period. Usually ulcers of the skin or mucous membranes. Inhalation may lead to fatal pneumonia. Infection by any route may be fatal unless treated promptly
Leptospirosis	*Leptospira interrogans* serovars	Many wild or domestic animals, especially rats, pigs, cattle and dogs/ contact with infected material, particularly urine of infected animals and contaminated water. Splashing of urine on mucous membranes or ingestion of contaminated food	Worldwide; some serovars may occur in defined geographical regions	Subclinical disease often occurs. Common features of infection with most serovars include fever, headache, chills, myalgia. The fever is frequently diphasic. Meningitis, haemolytic anaemia and jaundice may also occur. Acute illness lasts up to 10 days. Case fatality rate may be up to 20% in older patients without treatment
Listeriosis	*Listeria monocytogenes*	Soil, pasture, water, silage, infected domestic and wild animals/ingestion of unpasteurised milk, dairy products or contaminated uncooked vegetables	Worldwide	Mild, febrile, influenza-like syndrome, but may cause congenital infection in pregnant women leading to abortion or meningitis in the baby. In older age groups meningitis may occur, especially in immunosuppressed individuals. Rarely, endocarditis may follow a primary bacteraemia
Lyme disease	*Borrelia burgdorferi*	Wild rodents, deer and other animals/ticks (*Ixodes* spp.)	Most countries	Clinical illness includes distinctive skin lesions and systemic symptoms (early); and arthritis, neurological and, occasionally, cardiac involvement (later). Skin lesions occur in a high percentage of patients. Other findings include fatigue, fever, chills, musculoskeletal pain and lymphadenopathy. In untreated patients, complications may persist for many months or even years

(*continued*)

Table 158. Bacterial, chlamydial, rickettsial and fungal zoonoses. (*continued*)

Disease	Infectious agent	Reservoir/vector or mode of transmission	Distribution	Disease in humans
Melioidosis	*Pseudomonas pseudomallei*	The bacterium is a saprophyte in soil and a variety of domestic and wild animals can become infected/ contact with infected material or by inhalation of contaminated dust or soil	South-east Asia, Australia, Central and South America	A wide range of clinical manifestations are reported, including pulmonary disease, fatal septicaemia and chronic abscess formation. Mortality rate is high in untreated cases
Murine typhus	*Rickettsia typhi*	Rodents and possibly other small mammals and domestic animals may transport infected fleas, thus facilitating contact with humans/ rat flea (*Xenopsylla cheopis*) contamination of bite wound by flea faeces or following inhalation of dried, infective flea faeces	Worldwide	Sudden onset with headaches, chills, fever and general pains. After the sixth day macular eruptions appear on chest and abdomen. The course of the disease is about 3 weeks. Case fatality rate is usually less than 1% but increases with age
Pasteurellosis	*Pasteurella canis* and *P. dagmatis*	Cats, dogs/bite or scratch wounds	Worldwide	Local infections resulting from animal bites; cellulitis, swelling and local pain frequently occur at site of wound. Lymphadenitis and abscess formation occur in occasional cases
Plague (Urban and Sylvatic)	*Yersinia pestis*	Urban plague: rats/ fleas (*Xenopsylla cheopis*). Sylvatic plague: squirrels, prairie dogs, rabbits, rats/fleas	Areas of USA, former Soviet Union, Africa, Indonesia, Vietnam and parts of South America	In previous centuries *Y. pestis* produced pandemics of 'black death' with millions of fatalities. In **bubonic plague**, the bacteria reach the regional lymph nodes and infected nodes become inflammed, tender and may suppurate (buboes). Dissemination via the bloodstream may lead to pneumonia and meningitis. Secondary **pneumonic plague** is of particular importance as aerosolised droplets may allow person-to-person transfer to occur and this form of disease is invariably fatal. Case fatality rate in untreated bubonic plague is about 50%

(*continued*)

Table 158. Bacterial, chlamydial, rickettsial and fungal zoonoses. (*continued*)

Disease	Infectious agent	Reservoir/vector or mode of transmission	Distribution	Disease in humans
Psittacosis ornithosis	*Chlamydia psittaci*	Avian species (healthy birds may be carriers)/ inhalation from desiccated droppings or directly from infected birds	Worldwide	Initially mild, influenza-like disease; later, headache, fever, myalgia and arthralgia may occur. Pulmonary changes include consolidation and an unproductive cough. Encephalitis and myocarditis are occasional complications, especially in older patients
Q fever	*Coxiella burnetii*	Cattle, sheep, goats and cats and perhaps feral rodents/often airborne infection but can also be acquired by direct contact with contaminated material and by ingestion of raw milk	Worldwide	There is considerable variation in the duration and severity of this disease. Usually acute, self-limiting, febrile illness resembling influenza. Pneumonitis and pericarditis occur infrequently. Rarely, subacute or chronic endocarditis may present years after the initial infection
Rat-bite fever	*Streptobacillus moniliformis*	Rats, squirrels and weasels/bite wounds or secretions of animals carrying the bacterium in their upper respiratory tracts. Contaminated milk or water may transmit infection	Worldwide, but uncommon	Abrupt onset with fever, petechial rash, headache and polyarthritis; endocarditis may occur. In untreated cases, mortality rate may be up to 10%. (When not associated with a rat bite, the disease is often referred to as **Haverhill fever**)
Rat-bite fever	*Spirillum minus*	Rats/bite wounds	Worldwide, but uncommon	The bacterium invades the local lymph node causing lymphadenitis, rash and a relapsing fever

(*continued*)

Table 158. Bacterial, chlamydial, rickettsial and fungal zoonoses (*continued*).

Disease	Infectious agent	Reservoir/vector or mode of transmission	Distribution	Disease in humans
Relapsing fever (tick-borne)	*Borrelia recurrentis*	Wild rodents/ticks (*Ornithodoros* spp.)	North, Central and South America, Africa, Middle East	After an abrupt onset with fever, muscle aches and headache, large numbers of borreliae may be present in the blood and urine. Fever then declines and the borreliae disappear from the blood for about one week. Fever returns and the cycle is repeated up to ten or more times, usually with decreasing severity. The case fatality rate is usually less than 5% in untreated cases
Ringworm (Dermatomycoses)	*Microsporum canis* *Microsporum gallinae* *Microsporum gypseum* *Microsporum nanum* *Trichophyton equinum* *Trichophyton mentagrophytes* *Trichophyton verrucosum* *Trichophyton simii*	Cat, dog Poultry Soil, dog, horse and other species Soil, pig Horse Rodents, dog, horse and other species Cattle Monkeys	Worldwide (with a few exceptions)	The dermatophytes infect only epidermis, hair and occasionally nails. Lesions tend to expand equally in all directions with raised borders. Irritation and itching may be a feature of the disease. Redness, oedema, scaling and vesicle formation may occur. If hair is invaded, hair shafts become fragile and break off a short distance above the skin, leaving short stubs, usually in a balding circular patch
Rocky Mountain spotted fever	*Rickettsia rickettsii*	Rodents, dogs/ticks *Dermacentor* and *Amblyomma* spp.)	Mexico, North and South America	Characterised by a sudden onset with fever which may persist up to 3 weeks; muscle pain, severe headache and a maculopapular rash which appears about the third day. Damage to small blood vessels leads to a vasculitis. Myocarditis and intravascular coagulation may occur at a later stage. The case fatality rate is up to 20% in the absence of specific therapy

(*continued*)

Table 158. Bacterial, chlamydial, rickettsial and fungal zoonoses. (*continued*)

Disease	Infectious agent	Reservoir/vector or mode of transmission	Distribution	Disease in humans
Salmonellosis	Numerous *Salmonella* serotypes such as *Salmonella typhimurium* and *Salmonella enteritidis*	Domestic and wild animals including poultry/ by ingestion of contaminated food (raw or undercooked) of animal origin, especially processed meats, poultry, eggs and raw milk. Contaminated water may also be a source of human infection	Worldwide	Gastroenteritis usually follows ingestion of contaminated food or drinking water. Signs of illness begin 12–48 hours after consumption of the food. Fever, nausea, vomiting and abdominal cramps occur, with profuse diarrhoea which may persist up to 1 week; there may be septicaemia and even death in older patients
Streptococcal infection	*Streptococcus suis* Type 2	Pigs/cuts and abrasions	Western Europe	Meningitis, arthritis, septicaemia and occasional deaths
Tetanus	*Clostridium tetani*	Soils or fomites contaminated with faeces, especially horse faeces/puncture or other wounds contaminated with soil, faeces or dust	Worldwide	The incubation period may range from days to weeks; convulsive tonic contractions of voluntary muscles, especially of the jaw ('lockjaw') and opisthotonos; external stimuli may precipitate a tetanic seizure. The case fatality rate may approach 90% in babies and older patients
Tuberculosis	*Mycobacterium bovis* (*Mycobacterium avium* in immnunosuppressed individuals)	Cattle (*M. bovis*), poultry (*M. avium*), occasionally other species/inhalation or ingestion (raw milk or unpasteurised dairy products)	Countries where *Mycobacterium bovis* infection in cattle is endemic. Uncommon in countries with advanced eradication schemes	A chronic progressive disease with early lesions in cervical or mesenteric lymph nodes if ingested; dissemination to bones and joints occurs at a later stage. If inhaled, pulmonary tuberculosis results, indistinguishable from that caused by *M. tuberculosis*. Ultimately the disease may lead to emaciation and death if not treated

(*continued*)

Table 158. Bacterial, chlamydial, rickettsial and fungal zoonoses. (*continued*)

Disease	Infectious agent	Reservoir/vector or mode of transmission	Distribution	Disease in humans
Tularaemia	*Francisella tularensis*	Numerous wild animals, especially rabbits, hares, muskrats, beavers, rodents and some domestic animals; water polluted by carcasses of infected rodents/bites of infected arthropods – deer fly (*Chrysops discalis*), mosquito (*Aedes* species), ticks (*Dermacentor* and *Amblyomma* species); contact with infected animals (trappers); insufficiently cooked rabbit or hare meat; drinking contaminated water; laboratory infections have also been reported	North America, Scandinavia, former Soviet Union, China, Japan	Often present as an ulcer in the skin at the site of introduction. The organisms are carried to the regional lymph node which enlarges, becomes painful and may suppurate. Inhalation of infectious material may be followed by pneumonic disease. Typhoidal tularaemia follows ingestion of the organisms. This condition resembles typhoid fever with diarrhoea, vomiting and fever. The case fatality rate for typhoidal or pulmonary disease is about 10%
Yersiniosis	*Yersinia pseudotuberculosis* *Yersinia enterocolitica*	Rodents and domestic animals including birds (*Y. pseudotuberculosis*); pigs and other species (*Y. enterocolitica*) /ingestion of food or water that has been faecally contaminated; Raw pork has often been associated with *Y. enterocolitica* infection	Worldwide	*Y. pseudotuberculosis:* abdominal pain in the right lower quadrant suggestive of appendicitis, due to mesenteric lymphadenitis *Y. enterocolitica*: an acute but self-limiting gastroenteritis indistinguishable from a salmonella infection

Table 159. Parasitic zoonoses.

Disease	Infectious agent	Host/vector or mode of transmission	Distribution	Disease in humans
Angiostrongyliasis	*Angiostrongylus cantonensis, Angiostrongylus costaricensis*	Rats (*Rattus* species)/ ingestion of shrimps, crabs or large edible snails containing infective larvae	Hawaii, many Pacific islands, Vietnam, Malaysia, China, Indonesia, Phillipines, parts of Australia, east Africa, southern USA and Central America	Larvae of *A. cantonensis* migrate to the brain, spinal cord or eye where they become immature adults. Severe headache, neck stiffness and facial paralysis may occur. The small intestine is the usual site of infection of *A. costaricensis* where it produces a granulomatous inflammation
Anisakiasis	*Anisakis simplex, Pseudoterranova decipiens* and a number of other larval nematodes	Saltwater fish, squid or octopus/ consumption of uncooked, smoked or marinated fish or fish products	Japan, Netherlands, Scandinavia, USA, Central America	The larval stage of *P. decipiens* rarely penetrates the stomach or intestinal wall, but the larval stage of *A. simplex* is capable of penetrating the wall of the stomach or intestine, leading to abdominal pain, nausea, vomiting, diarrhoea and eosinophilic granuloma. The motile larvae may migrate to the oropharynx, causing a cough
Babesiosis	*Babesia divergens* (Europe) *Babesia microti* and other *Babesia* species	Cattle (*B. divergens*) Probably rodents (*B. microti*)/ticks (*Ixodes dammini* – USA and *Ixodes ricinus* – Europe)	USA, France, Ireland, Scotland, former Soviet Union, former Yugoslavia. Relatively uncommon	Often a severe and potentially fatal disease (especially in patients who have been splenectomised). Fever, fatigue and haemolytic anaemia are features of the disease
Balantidiasis	*Balantidium coli*	Pigs, guinea-pigs and perhaps other animals/ ingestion of faecally contaminated water or food	Worldwide in areas of poor environmental sanitation	Abdominal pain, nausea, vomiting, and diarrhoea and dysentery, are frequently observed, but asymptomatic carriage of *B. coli* is also reported

(*continued*)

Table 159. Parasitic zoonoses. (*continued*)

Disease	Infectious agent	Host/vector or mode of transmission	Distribution	Disease in humans
Capillariasis	*Capillaria philippensis*	Fish-eating birds/ ingestion of raw or inadequately cooked fish containing infective larvae	Philippines, Thailand, Japan, Egypt, Iran	Diarrhoea, malabsorption syndrome, dehydration and progressive emaciation are features of the disease. Large numbers of worms may accumulate in the small intestine and produce serious disease. Case fatality rate may be up to 10%.
Clonorchiasis	*Clonorchis sinensis*	Cats, dogs , pigs as well as humans/ ingestion of undercooked or raw freshwater fish or crayfish containing larvae	South-east Asia, USA (immigrants from Asia)	Chronic disease. Loss of appetite, diarrhoea, liver enlargement and progressive ascites may occur. Sometimes asymptomatic
Cheyletiella infection	*Cheyletiella parasitivoraz, Cheyletiella yasguri Cheyletiella blakei*	Rabbit (*C. parasitivorax*), dog (*C. yasguri*), cat (*C. blakei*)/contact	Worldwide	Usually a mild disease. Pruritic skin lesions on arms and body with papular eruptions and occasionally severe irritation
Cryptosporidiosis	*Cryptosporidium* species	Cattle, sheep, pigs, also infective humans/ingestion of food or water containing sporulated oocysts	Worldwide	Immunologically competent humans develop a profuse diarrhoea, nausea and abdominal pain which may last 2 weeks. Immunosuppressed people develop a severe, long-lasting infection which may persist for many months with life-threatening consequences
Dermanyssus infection	*Dermanyssus gallinae* (red mite of poultry)	Birds, both wild and domestic/contact, direct attack by mites	Worldwide	Dermatitis with intense pruritis. The northern fowl mite *Ornithonyssus sylviarum* and the tropical mite *Ornithonyssus bursa* may leave the nests of birds in buildings and also attack humans
Dirofilariasis	*Dirofilaria immitis* (dog heartworm)	Dogs and other canidae/mosquito bite	USA, Japan, Asia, Australia	These worms may lodge in and obstruct small pulmonary arteries, producing infarcts. Common symptoms include chest pain, cough and haemoptysis

(*continued*)

Table 159. Parasitic zoonoses. (*continued*)

Disease	Infectious agent	Host/vector or mode of transmission	Distribution	Disease in humans
Diphyllo-bothriasis (fish tapeworm infection)	*Diphyllo-bothrium latum*	Humans, also dogs, bears and other fish-eating mammals/ ingestion of raw or undercooked fresh water fish containing the larval (plerocercoid) stage	Scandinavia, former Soviet Union, northern Europe, Japan, Canada, Alaska	May produce diarrhoea and mild abdominal pain. A vitamin B_{12} deficiency is said to occur in some people infected with this tapeworm
Dipylidiasis (dog tapeworm infection)	*Dipylidium caninum* (the most common and wide-spread adult tapeworm of dogs and cats)	Dogs and cats/ accidental ingestion of the dog flea (*Ctenocephalides canis*) or cat flea (*Ctenocephalides felis*) containing the cysticercoid of *D. caninum*	Worldwide	Usually occurs in young children. A self-limiting infection of little conseqence apart from aesthetic considerations
Echino-coccosis	*Echinococcus granulosus*	Dogs, wolves, dingos and other canidae/ accidental ingestions (hand-to-mouth) of tapeworm eggs	Middle East, Africa, Australia, New Zealand, Asia including China. Parts of Europe, Canada, Alaska, South America	Hyatid cysts may develop in many organs and tissues of the body. They continue to grow until of sufficient size to cause clinical symptoms. Cysts in vital organs such as the brain may cause severe symptoms and even death
	E. multi-locularis	Dogs, foxes and other Canidae/ accidental ingestion (hand-to-mouth)	Arctic regions, parts of temperate Europe	Multilocular hydatid disease. Growth in man simulates neoplasia. Because of the metastases and infiltrative process in organs, surgery is rarely successful
Fleas	*Ctenocephalides canis* (dog flea) *Ctenocephalides felis* (cat flea) *Xenopsylla cheopis* (rat flea) *Echidnophaga gallinacea* (sticktight flea of poultry)	The rat flea (*X. cheopis*) may be a vector of murine typhus (*Rickettsia typhi*) and plague (*Yersinia pestis*). The dog flea (*C. canis*) is an intermediate host for the dog tapeworm (*Dipylidium caninum*)	Worldwide	Fleas may cause irritation when they attack humans if their preferred host is no longer available. They may also transmit serious infections to the human population such as plague and murine typhus. Many humans become sensitised to flea bites (salivary antigen) and develop an immediate type hypersensitivity - reaction at the bite site
Fascioliasis	*Fasciola hepatica* (liver fluke)	Sheep and cattle/ human infection is acquired by ingestion of aquatic vegetation such as watercress in salads on which metacercariae have encysted	Sheep and cattle-raising areas of Europe, South America, Caribbean, Australasia and Middle East	Metacercariae migrate from the intestine to the bile ducts by passing through the wall of the intestine into the dominal cavity and entering the liver through its outer surface. Liver damage and enlargement occur, particularly if large numbers of flukes are present. Biliary colic and obstructive jaundice may occur. Eosinophilia is a feature of the disease

(*continued*)

Table 159. Parasitic zoonoses. (*continued*)

Disease	Infectious agent	Host/vector or mode of transmission	Distribution	Disease in humans
Fasciolopsiasis	*Fasciolopsis buski*	Pigs, dogs and also humans/human infection results from ingestion of aquatic plants on which metacercariae have encysted	South-east Asia, especially China, Thailand, and parts of India	Local inflammation and occasionally ulceration of the wall of the small intestine, particularly the duodenum. Severe infections may produce abdominal pain; the severity of disease relates to the number of flukes present
Giardiasis	*Giardia lamblia*	Beavers, dogs and also humans/ingestion of food or water contaminated with faeces containing *Giardia* cysts	Worldwide	Intestinal symptoms include diarrhoea, abdominal cramps, fatigue and weight loss. Immunosuppressed patients are especially liable to massive infection. Conversely some infections with this parasite may be asymptomatic
Gnathostomiasis	*Gnathostoma spinigerum*	Cats and dogs/ ingestion of undercooked fish or poultry containing third-stage larvae (fish and poultry may act as second intermediate hosts)	Southern Europe, Africa, Asia and Australia	Larvae migrate through the tissues (visceral larva migrans) forming transient inflammatory lesions, including abscesses, in various parts of the body. Larvae may invade the brain producing focal cerebral lesions
Heterophyiasis	*Heterophyes heterophyes*	Dogs and cats/ ingestion of infected raw or undercooked fish containing the metacercariae	Middle East, Asia	Generally a mild infection. In occasional severe cases haemorrhagic diarrhoea may develop
Hookworm disease (cutaneous larva migrans)	*Ancylostoma caninum, Ancylostoma braziliense Uncinaria stenocephala*	Dogs and cats/contact with damp, sandy soil contaminated with dog and cat faeces	Worldwide	Larvae enter the skin and and migrate intracutaneously producing dermatitis ('creeping eruptions'). Intense itching is a feature of the disease, which is usually self-limiting after several weeks.
Hymenolepiasis	*Hymenolepis nana*	Rodents (intermediate hosts, infected insects) /ingestion in food or water of larvae-bearing insects or eggs	Southern USA, Latin America, Australia, Mediterranean countries, Near East, India	Abdominal pain, enteritis, vague symptoms such as weight loss and weakness
Leishmaniasis (cutaneous)	*Leishmania mexicana, Leishmania tropica* and many other species	Humans, wild rodents, carnivores including dogs, marsupials and other unknown hosts/ bites of infected sandflies of the genus *Phlebotomus*	Pakistan, Middle East, Africa, Mexico, South America	The disease, which may present in many forms, usually starts as an ulcer on the skin. Lesions may be single or multiple and may last from weeks to months. Recurrence after apparent cure may occur

(*continued*)

Table 159. Parasitic zoonoses. (continued)

Disease	Infectious agent	Host/vector or mode of transmission	Distribution	Disease in humans
Leishmaniasis (visceral)	*Leishmania donovani* and other species and subspecies	Believed to be wild canidae, domestic dogs, rodents and humans/bites of infected sandflies	Pakistan, China, former Soviet Union, Middle East, parts of Africa, Central and South America	A chronic systemic disease characterised by fever, hepatosplenomegaly, lymphadenopathy, anaemia and progressive emaciation. Untreated the disease may be fatal
Metagonimus infection	*Metagonimus yokogawai*	Dogs, cats, pigs and humans/ingestion of infected raw or undercooked freshwater fish containing the metacercariae	Asia, Egypt, Turkey, Balkan States	Mild diarrhoea and abdominal pain, often asymptomatic
Paragonimiasis (lung fluke disease)	*Paragonimus westermani* and other species	Humans, dogs, cats, pigs and wild carnivores/ingestion of infected raw or undercooked freshwater crabs or crayfish containing the metacercariae	Asia, particularly Korea and Japan; also, China, Phillippines, Africa, Central and South America; occasionally in USA and Canada	The immature fluke penetrates the small intestine, migrates to the pleural cavity and penetrates the lung tissue. Symptoms include a cough, chest pains and haemoptysis. The parasite may migrate to other sites including the CNS
Sarcocystis infection	*Sarcocystis hominis, Sarcocystis suihominis* and perhaps other species	Cattle, sheep, pigs/ingestion of raw or undercooked beef, lamb or pork containing cysts of the parasite	Worldwide	Subcutaneous swellings and eosinophilia have been attributed to infection with this parasite
Scabies (acariasis)	*Sarcoptes scabiei*	Dogs, pigs, cattle/contact	Worldwide	Skin lesions on the hands, arms and occasionally on the body. Lesions usually consist of papules or vesicles and may be intensely pruritic
Schistosomiasis	*Schistosoma japonicum*	Many species of animals including dogs, cats, pigs, primates, wild rodents and humans. Contact with water containing free-swimming, fork-tailed cercariae	Africa, Asia, parts of South America	Symptoms include abdominal pain, diarrhoea and hepatosplenomegaly. Chronic infections give rise to liver fibrosis and portal hypertension. The larvae of some schistosomes of birds and mammals may penetrate the human skin and cause a dermatitis sometimes referred to as 'swimmers' itch'
Sparganosis	*Spirometra* species	Carnivores (adult worms), frogs, snakes and other animals (larval forms)/man infected by eating intermediate hosts or by applying native poultices such as raw frog tissue	North and South America, Orient	Disease is characterised by the presence of large larvae in muscles and subcutaneous tissues, causing oedema and inflammation

(*continued*)

Table 159. Parasitic zoonoses. (*continued*)

Disease	Infectious agent	Host/vector or mode of transmission	Distribution	Disease in humans
Strongyloidiasis	*Strongyloides stercoralis, Strongyloides fuelleborni*	Many species of animals including dogs, cats, non-human primates/contact with moist soil contaminated with faeces	Tropical and temperate regions of the world	Transient dermatitis, pneumonitis and abdominal symptoms, depending on the stage and severity of disease
Taeniasis	*Taenia saginata* (beef tapeworm)	Humans (cattle are intermediate hosts)/ingestion of raw or undercooked beef containing viable cysticerci	Worldwide	Mild abdominal pain and diarrhoea may occur occasionally
Taeniasis	*Taenia solium* (pork tapeworm)	Humans (pigs are the intermediate hosts)/ ingestion of infected raw or undercooked pork containing viable cysticerci ('measly pork')	Central America, Africa, South-east Asia and Eastern Europe	Intestinal infection in man follows ingestion of cysticerci in raw or undercooked pork and leads to the development of adult tapeworms in the intestine. **Human cysticercosis** occurs following ingestion of the eggs of *T. solium* and is a more serious form of infection. The cysticerci may develop in any organ or tissue in the body. Serious disease may develop from localisation in the CNS or the eye
Ticks	Many species including *Ixodes, Dermacentor, Rhipicephalus, Amblyomma*	Ticks may transmit bacterial disease such as Lyme disease (*Borrelia burgdorferi*), protozoan diseases such as babesiosis (*Babesia divergens*), rickettsial diseases such as Rocky Mountain spotted fever (*Rickettsia rickettsii*) and viral diseases such as louping ill (flavivirus)	Worldwide	Apart from the diseases they transmit, ticks cause skin irritation while feeding on their hosts. Some are reported to cause flaccid paralysis due to injection of toxin

(*continued*)

Table 159. Parasitic zoonoses. (*continued*)

Disease	Infectious agent	Host/vector or mode of transmission	Distribution	Disease in humans
Toxocariasis (visceral larva migrans)	*Toxocara canis*, *Toxocara cati*	Dog (*T. canis*), Cat (*T. cati*)/ingestion of *Toxocara* eggs from contaminated soil (geophagia), contaminated uncooked vegetables, or from the coat of the animal shedding eggs	Probably worldwide	Primarily a disease of young children. Larvae penetrate the wall of the intestine but are unable to complete their life cycle. They migrate through their host's tissues, producing eosinophilic granulomas. Systemic toxocariasis (visceral larva migrans) results in hepatosplenomegaly, pulmonary infiltrates and sometimes seizures. Larvae migrating to the eye (ocular larva migrans) or CNS may produce serious disease
Toxoplasmosis	*Toxoplasma gondii*	Cats or other feline animals, various intermediate hosts/ ingestion of: • Tissue cysts in raw or undercooked meat, particularly mutton or pork • Faecal oocysts from the environment (sand boxes, soil, water) • Bradyzoites while assisting at lambing if toxoplasma abortions are occurring in the flock • Tachyzoites present in raw goat's milk (rare)	Worldwide	Only a small proportion of infected people show evidence of infection. The disease may be severe, even fatal, in immunosuppressed individuals, and primary infection during pregnancy can lead to congenital infection with serious consequences. Lymphadenopathy, myalgia and many non-specific symptoms may occur postnatally. If cysts form in the eye, CNS or myocardium, the outcome may be serious
Trichinosis	*Trichinella spiralis*	Pigs, and many other species of meat-eating animals including wild animals and marine mammals. *T. spiralis* is not host-specific/ ingestion of raw or insufficiently cooked meat, usually pork, containing encysted larvae	Worldwide	Initially there may be fever, non-specific gastroenteritis, myositis and circumorbital oedema. If the infection is severe, cardiac and neurological complications may occur

(*continued*)

Table 159. Parasitic zoonoses. (*continued*)

Disease	Infectious agent	Host/vector or mode of transmission	Distribution	Disease in humans
Trypano-somiasis	*Trypanosoma cruzi* (Chagas' disease)	Many species of domestic and wild animals, including dogs, cats, rodents and humans/ infection occurs when infected blood-sucking vectors (triatomid insects) contaminate wounds with infected faeces after feeding	Western Hemisphere: Mexico, South America and Southern USA	Acute disease occurs usually in children. An inflammatory response at the site of infection (chagoma) may last for 2 months. Unilateral bipalpebral oedema presents in a high proportion of acute cases. Fever, malaise, lymphadenopathy and hepatosplenomegaly are features of the disease. Occasionally, myocarditis and meningoencephalitis may occur
	Trypanosoma brucei gambiense *T. brucei rhodesiense*	Wild and domestic animals are reservoirs, especially cattle, bush buck and impala/bite of an infected tsetse fly (*Glossina* species)	Tropical Africa	Fever, intense headache and generalised lymphadenopathy may occur. CNS signs may be evident a few weeks after infection. Without treatment, the disease may be fatal within weeks

Table 160. Prevention of zoonoses acquired from companion animals.

- Education of pet owners and farmers so that they are aware of zoonotic diseases
- Practical hygiene, especially as it relates to children and those immediately at risk of infection
- Control of infectious diseases through strategic use of anthelmintics, chemotherapy and vaccination
- Isolation of sick animals, especially childrens' pets
- Control of stray animals, particularly in urban areas, by appointment of dog wardens
- Exclusion of pets from food shops and restaurants (except guide dogs)
- Prevention of indiscriminate fouling of pavements, gardens and parks by dogs (and cats)
- Mandatory identification of all pets by name tags or other appropriate systems
- Quarantine measures should be strictly enforced to exclude exotic zoonoses such as rabies
- Continuing education of the public through the mass media on matters relating to the care of animals, the diseases they acquire and disease prevention

Table 161. Prevention of zoonoses acquired from food-producing animals.

- Education of farmers and producers on the measures appropriate for the prevention of zoonotic diseases in farm stock, irrespective of the species of animal involved – cattle, sheep, goats, pigs or poultry
- Buildings should be rodent-proof and well maintained
- Hygiene standards in meat plants should minimise the possibility of cross-contamination
- Carcases should be inspected for evidence of zoonotic disease and where necessary detained for confirmatory tests
- Pasteurisation plants should operate to high efficiency and retain their records for periodic inspection
- In the event of an outbreak of a zoonotic disease in dairy cows, unpasteurised milk consumed on the farm should be heated to 100°C
- Vigilance is required to ensure that wholesome food, which is produced hygienically, is also stored properly, preferably at 4°C, and not re-contaminated before it reaches the consumer
- Utensils, cutlery and other items used for preparing raw meat dishes or eggs should **not** be used for cooked meat unless thoroughly washed
- Adequate cooking is an essential step in the prevention of meat-borne zoonoses such as trichinosis and toxoplasmosis
- Slurry generated on dairy, beef and pig farms should be stored before dispersal. If zoonotic diseases have occurred on the farm of origin, longer storage and dispersal on land for tillage should be considered

Bibliography

Balows, A. (1991). *Manual of Clinical Microbiology*, Fifth Edition, American Society for Microbiology, Washington, D.C., USA.

Benenson, A.S. (1990). *Control of Communicable Diseases in Man*, American Public Health Association, Washington, D.C., USA.

Brooks, G.F., Butel, J.S. and Ornston, L.N. (1991). *Jawetz, Melnick and Adelberg's Medical Microbiology,* Prentice-Hall International Inc., Englewood Cliffs, New Jersey, USA.

Davis, B.D., Dulbecco, R., Eisen, H.N. and Ginsberg, H.S. (1990). *Microbiology*, Fourth Edition. J.B. Lippincott Co., Philadelphia, USA.

Despommier, D.D. and Karapelou, J.W. (1987). Parasite Life Cycles. Springer-Verlag, New York, USA.

Murray, P.R., Drew, W.L., Kobayashi, G.S. and Thompson, J.H. (1990). *Medical Microbiology*, Wolfe Medical Publications Ltd., London.

Soulsby, E.J.L. (1982). *Helminths, Arthropods and Protozoa of Domesticated Animals,* Baillière Tindall, London.

Urquhart, G.M., Armour, J., Duncan, J.L., Dunn, A.M. and Jennings, F.W. (1987). *Veterinary Parasitology,* Longman Scientific and Technical, Essex, UK.

50 Control of Infectious Diseases

Infectious agents are still responsible for substantial, recurring losses in animal production despite advances in diagnostic procedures for infectious diseases, improved chemotherapy, chemoprophylaxis and the availability of a wide range of vaccines, some arising from recent developments in molecular biology. The selection of antimicrobial substances for the treatment of bacterial and fungal diseases has been reviewed in Chapter 7. Other aspects of disease control will be considered briefly in this chapter.

Transmission of Infectious Agents

Infected animals frequently shed pathogenic organisms, often in large numbers, and the resulting environmental contamination is an important method of transmitting infection to healthy animals. Salmonellosis, paratuberculosis, leptospirosis, and rotavirus infections are examples of disease where substantial environmental contamination may occur. Survival of pathogenic microorganisms in the environment is influenced by the method of shedding (faeces, urine, aerosols, saliva, exudates), the number shed and their ability to withstand adverse conditions such as dehydration, ultra-violet light, temperature and pH changes and the application of physical or chemical disinfectants. Pathogenic mycobacteria, a number of animal viruses such as parvoviruses, bacterial endospores, coccidial oocysts and parasitic ova are capable of prolonged survival in animal products, in faeces, soil, water, animal dwellings and on pasture. **Diagram 69** illustrates direct and indirect methods of transmission of infectious agents from infected to susceptible animals and also methods of interrupting the transfer of infection through control of animal movement, disinfection and vaccination.

Principles of Disease Control

Maintaining animals free of infectious diseases relies on well-defined procedures which may be used at farm level, regionally, nationally or internationally.

Procedures for the prevention and control of infectious diseases which are of international importance or public health significance include:

- Control of animal movement (quarantine and isolation).
- Testing of suspect groups of animals, slaughtering those infected and disposing of the carcases by burial or burning.
- Depopulation of herd or animal species within a specified area.
- Disinfection of premises, transport vehicles, equipment and other contaminated sources of infection.
- Vaccination of susceptible animals.

Diseases which are endemic in a country may be dealt with in different ways depending on their impact on animal or human health and the financial resources available. Chemotherapy, chemoprophylaxis, improved management systems, control of insect vectors or wildlife reservoirs are additional measures which may be used. Long incubation periods and the development of a carrier state, with the likelihood of intermittent shedding, render effective control and prevention of many infectious diseases somewhat uncertain. Land frontiers, unstable governments and wildlife reservoirs of infection diminish the chances of eradicating endemic or exotic animal diseases from a country.

Methods used for the prevention, treatment or control of an infectious agent depend on its status within the country, its public health significance and its international importance. **Table 162** compares the measures appropriate for five contrasting infectious agents. It is evident that while vector control is of little consequence for strangles, anthrax and foot-and-mouth disease, it is crucial for the control of African swine fever. Likewise, while vaccination is widely used for the prevention of anthrax and foot-and-mouth disease, it cannot at present be used for the control of African swine fever, histoplasmosis or ringworm infections. Accordingly, the development of a disease control programme requires careful consideration of the infectious agent, its mode of spread, survival in the environment, susceptibility to chemotherapy and disinfection, the availability of an effective vaccine, reservoirs of the organism in nature (soil, water, wildlife reservoirs) and the cost of eradication. A comparison of four substances, each with a defined role in disease control, is presented in **Table 163**. The role of two of these, disinfectants and vaccines, in disease control programmes will be considered in more detail.

Disinfection

Infectious agents vary widely in their susceptibility to physical and chemical disinfection procedures. The physical methods available for inactivating infectious agents include heat (moist or dry), ultra-violet light and gamma irradiation. Thermal inactivation is one of the most widely used physical methods. Heat-treatment of milk at 72°C for 15 seconds (pasteurisation) is widely used for destroying potential pathogens of

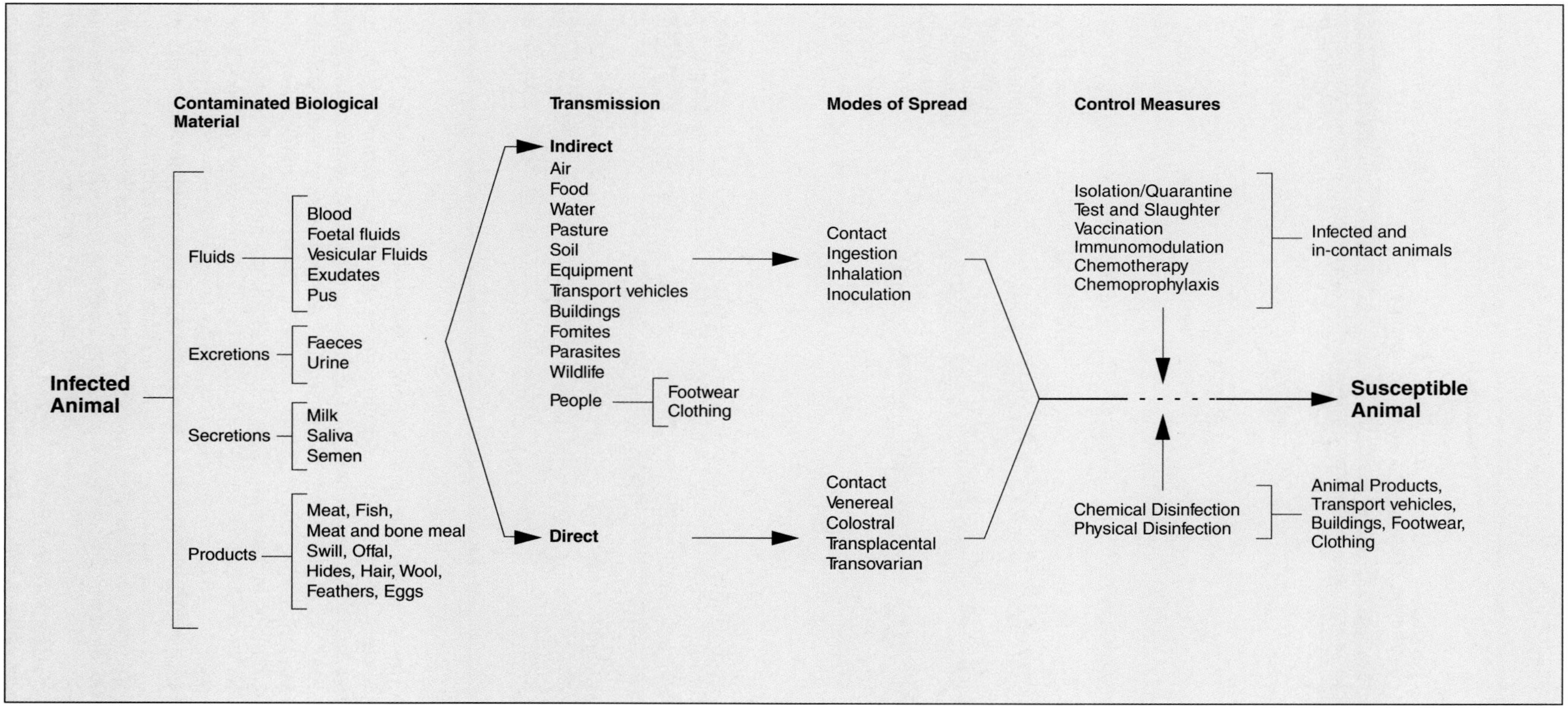

Diagram 69. Direct and indirect transmission of infectious agents from infected to susceptible animals. Control measures vary with the nature and significance of the infectious agent and include methods aimed at interrupting the transfer of infection by preventing contact between host and agent, treating infected animals, raising the immune status of susceptible animals or by eliminating infectious agents from the animals' immediate environment with appropriate disinfection procedures.

Table 162. Measures appropriate for the prevention, treatment and control of selected infectious agents.

Infectious agent	Host/ disease	Measures: Isolation/ quarantine	Vector control	Disinfection	Chemotherapy	Vaccination	Comments
Streptococcus equi subspecies *equi*	Horse/ strangles	++	–	++	+	±	Protection from vaccination questionable
Bacillus anthracis	Many species/ anthrax	++	–	++	+	++	Vaccination permitted where disease is endemic
Foot-and-mouth disease virus	Many species/ foot-and-mouth disease	++	–	++	–	++	When vaccination is permitted, vaccinal strain must match field strain for protection in endemic areas
African swine fever virus	Pig/African swine fever	++	++	++	–	–	Soft ticks of the genus *Ornithodoros* are the vectors of the virus
Histoplasma capsulatum	Many species/ histoplasmosis	–	–	+	++	–	Soil-borne fungus; opportunistic infection
Microsporum canis	Many species/ ringworm	++	–	+	++	–	Parasitic fungus transmitted by direct and indirect contact

++: effective measure; +: of limited benefit; ±: doubtful measure; —: not applicable.

public health significance. However, there is wide variation in the susceptibility of infectious agents to thermal inactivation. Most vegetative bacteria and many viruses are rapidly inactivated at 100°C. Foot-and-mouth disease virus can survive a temperature of 95°C for 15 seconds but is reliably inactivated at 148°C in 3 seconds. A temperature of 121°C for 15 minutes is required for the destruction of bacterial endospores. A group of infectious agents, referred to as prions, are highly resistant to chemical disinfectants and heat. These agents can withstand a temperature of 160°C (dry heat) for 24 hours. Autoclaving (moist heat) at 121°C for up to 5 hours is recommended for their inactivation.

Chemical disinfection is more versatile and accordingly is more widely used for inactivation of infectious agents than physical disinfection. Compounds with antimicrobial activity range from mineral acids, alkalis and alcohols to quaternary ammonium compounds. The spectrum of activity of each chemical disinfectant has a direct bearing on its usefulness in disinfection programmes. Selection and use of chemical disinfectants requires a detailed knowledge of their range of activity, safety, ability to act in the presence of organic matter, stability, susceptibility to neutralisation by pH changes, soaps or detergents and their cost. Some chemicals used in disinfection procedures are corrosive, toxic, irritant or potentially carcinogenic. All disinfectants should be handled with care especialy alkylating agents such as aldehydes. Acids and alkalis should not be mixed; rubber gloves and a face mask should be worn when working

Table 163. Comparison of chemotherapeutic and other substances used for the control and prevention of infectious diseases.

Charac - teristics of substance	Immunomodulators	Vaccines	Antimicrobial substances	Disinfectants
Nature/ source	Classified according to origin: physiological products, substances of microbial origin and synthetic compounds	Microorganisms or their products; synthetic antigens also used	Antibiotics, sulphonamides, coccidiostats, other antimicrobial compounds produced naturally or synthesised	Most are synthetic; some are naturally-occurring compounds
***In vivo* activity**	Usually acts on particular cell types; may induce (or inhibit) formation of substances active in immune reactions	Induces changes in T and/or B cells which result in protective immunity	Selectively toxic to microorganisms in host's tissues	Not suitable for *in vivo* use
Range of activity	Broad	Confined to the antigenic material present in vaccine	Varies with compound; usually active against a range of microorganisms	Broad
Specificity	Non-specific	Specific	Non-specific	Non-specific
Mode of action	Stimulate (or suppress) elements of specific or non-specific immunity	Stimulates specific humoral or cell-mediated immunity for the antigens present in vaccine	Interferes with metabolism of organism	Precipitate proteins, denature lipids, react with DNA and RNA
Practical application	May be used therapeutically or prophylactically	Used prophylactically	Used chemo-prophylactically and therapeutically	Used to destroy infectious agents on inanimate objects
Duration of activity	Varies with compound, usually weeks to months	Months	Days	Limited to time of application
Safety for food industry	Usually safe, varies with compound	Safe	Risk of residues in milk, meat and eggs	Usually safe, if selected carefully for specific purposes
Environmental considerations	Unlikely to present problems; may depend on compound and route of excretion	Live vaccines may present some potential problems	May promote drug resistance in some microorganisms	Occasional risk of environmental pollution from improper use

with corrosive chemicals. Formalin (formaldehyde gas dissovled in water) and chlorine-containing compounds should never be used together or after each other as a potent carcinogen may be formed if the two chemicals interact.

Table 164 lists the antimicrobial activity of some chemical disinfectants used in veterinary medicine. It should be noted that infectious agents vary widely in their susceptibility to chemical disinfectants. Mycoplasmas, enveloped viruses, Gram-positive bacteria and Gram-negative bacteria are susceptible to most chemical disinfectants; non-enveloped viruses and acid-fast bacteria are moderately resistant, while bacterial endospores and coccidial oocysts are highly resistant to chemical compounds. Prions are extremely resistant to chemical inactivation. Disinfection recommendations for this group of infectious agents is based more on precedent than factual scientific data. Organic solvents, long treatment with alkylating agents, halogen disinfectants or oxidising agents may progressively inactivate these resistant entities.

Despite advances in vaccine production techniques, many economically important diseases still cannot be controlled by immunisation. Staphylococcal mastitis, African swine fever and equine infectious anaemia are examples of diseases for which vaccines are not yet available. Accordingly, chemical disinfection, alongside other appropriate measures, plays a central role in the control of both exotic and endemic diseases. Disinfection programmes, however, must be carefully designed to ensure the destruction of diverse groups of infectious agents. The spectrum of activity of a disinfectant is the first consideration when embarking on a disinfection procedure. The concentration used, the presence of organic matter or other interfering substances, the time allowed for inactivation of the infectious agents and the temperature, are other variables which influence the outcome of a disinfection programme. As the concentration of disinfectant increases the time required for inactivation of susceptible bacteria is reduced (**497**, **498**). Halogen disinfectants are adversely affected by organic matter (**499**, **500**), thus they are not effective in a dirty environment.

Thorough cleaning is the first step in a disinfection programme and this measure alone will remove most of the infectious agents from equipment or from a building. Chemical compounds are used to destroy residual infectivity, especially in inaccessible locations and also to render a building safe before cleaning when dealing with zoonotic diseases such as anthrax.

The design and implementation of an efficient disinfection programme requires an evaluation of the disease status of the farm, an accurate record of disease problems, review of isolation facilities for replacement animals, the type of building being disinfected and its suitability for a particular disinfection routine. Factors which may contribute to the failure of a disinfection programme are presented in **Table 165**.

497

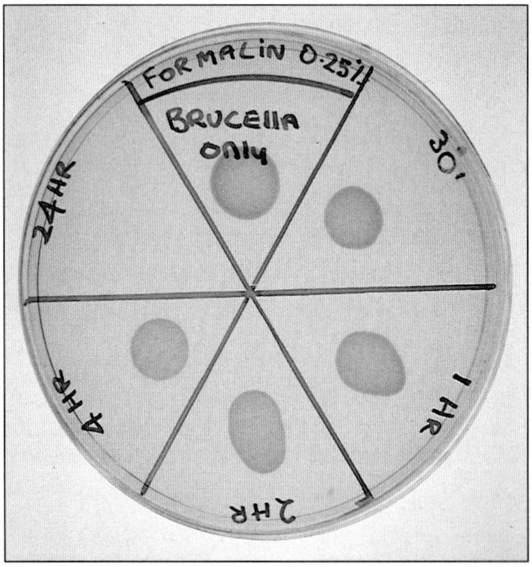

497 Disinfectant activity of 0.25 per cent formalin against *Brucella abortus* at different time intervals. At this concentration *B. abortus* can survive for more than 4 hours, but is completely inactivated within 24 hours.

498

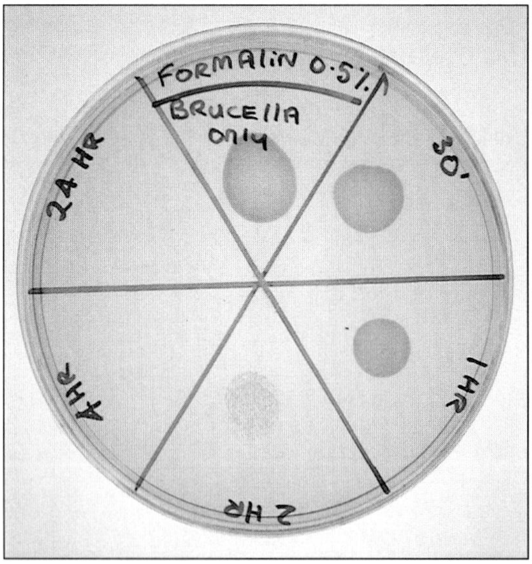

498 Disinfectant activity of 0.5 per cent formalin against *Brucella abortus* at different time intervals. At 2 hours' exposure there was evidence of survival but complete inactivation was achieved from 4 hours onwards.

Table 164. The antimicrobial activity of chemical disinfectants [(a)].

Microorganisms (in order of increasing resistance to disinfectants)	Chemical disinfectants									
						Halogens				
	Acids (mineral)	Alcohols	Aldehydes	Alkalis	Biguanides	Chlorine	Iodine	Oxidising agents	Phenolic compound	QACs[(b)]
Susceptible										
Mycoplasmas	++	++	++	++	++	++	++	++	++	++
Gram-positive bacteria	++	++	++	+	++	++	++	++	++	++
Gram-negative bacteria	++	++	++	+	+	++	+	+	++	+
Rickettsiae	+	+	+	+	±	+	+	+	+	±
Enveloped viruses	+	+	++	+	±	++	+	+	±[(c)]	±
Resistant										
Chlamydiae	±	±	+	+	±	+	+	+	±	–
Fungal spores	±	±	+	+	±	+	+	±	±	±
Non-enveloped viruses	±	–	+	±	–	+	±	±	–	–
Acid-fast bacteria	–	+	+	±	–	+	+	±	±	–
Highly resistant										
Bacterial endospores	±	–	+	±	–	+	+	+[(d)]	–	–
Coccidial oocysts [(e)]	–	–	–	±[(f)]	–	–	±	–	±[(g)]	–
Extremely resistant										
Prions [(h)]	–	–	±	±	–	±	±	±	–	–

++, highly effective; +, effective; ±, limited activity; –, no activity; (a), individual members may vary from the activity listed for that category; (b), quaternary ammonium compounds; (c), varies with the composition of disinfectant; (d), peracetic acid, a strong oxidising agent, is sporicidal; (e), methyl bromide (a halogen) is coccidiocidal; (f), ammonium hydroxide is coccidiocidal; (g), some have activity against coccidia; (h), limited data available on chemical disinfection – more than 4 hours at 121°C is required for inactivation of prions.

499

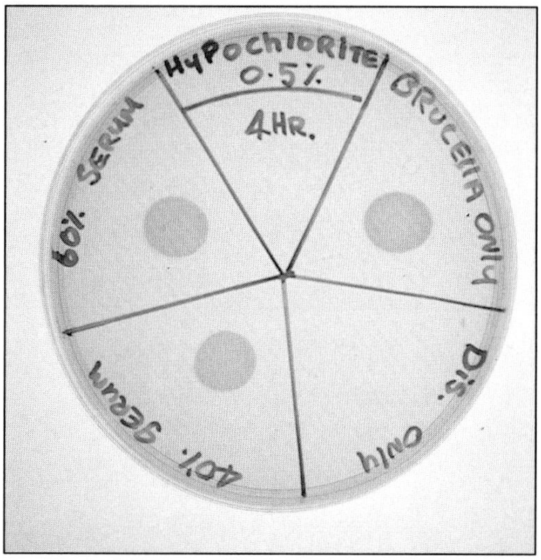

500

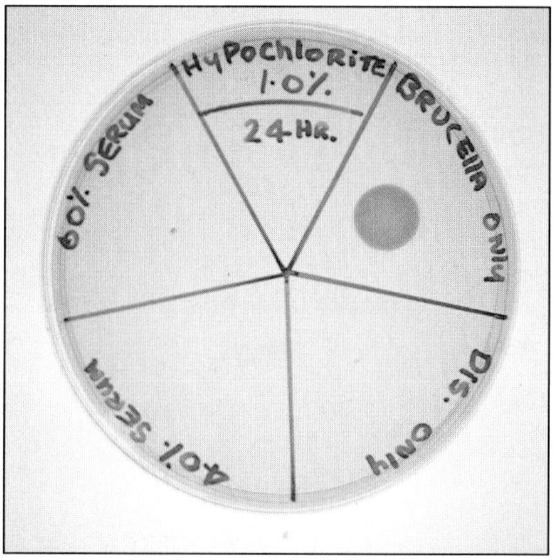

499 The effect of organic matter on the activity of 0.5 per cent sodium hypochlorite. In the absence of serum (organic matter) the brucellae (*Brucella abortus*) are inactivated within 4 hours by the disinfectant (bottom right), whereas in the presence of 40 per cent or 60 per cent serum the brucellae survive.

500 The effect of 1.0 per cent sodium hypochlorite against *Brucella abortus* over 24 hours in the presence or absence of organic matter (serum). This concentration of the disinfectant inactivated the bacterium in the presence of both 40 per cent and 60 per cent serum.

Table 165. Factors which may contribute to the failure of a disinfection programme.

Disinfectant	Environmental and other factors	Apparent failure
Compound selected incapable of inactivating the infectious agent	Inadequate cleaning with high levels of organic matter still present on surfaces	Reintroduction of infectious agents via carrier animals, rodents, insects, food or fomites
Diluted past its optimal concentration	Lack of penetration of porous materials by gaseous disinfectants	
Insufficient time allowed for disinfectant to inactivate the infectious agent	Inactivation of quarternary ammonium compounds by detergents or partial inactivation of some disinfectants by the pH or other characteristic of the material being treated	
Temperature too low for optimal activity		
Relative humidity too low for a gaseous disinfectant		

Vaccination

The body's ability to maintain itself free of infectious diseases derives from natural barriers in the structure and function of tissues (non-specific immunity) and from highly specialised and adaptable cells capable of responding specifically to invading pathogens and their products.

Diagram 70 illustrates the principal elements of non-specific and specific (acquired) immunity. In most animals, the combination of natural resistance and stimulation of adaptive immune responses to infectious agents is usually adequate for disease prevention. Vaccines, if effective, are the preferred method of control of specific infectious diseases. There are, however, a number of important diseases (some of which are listed in **Table 162**) which cannot be controlled by vaccination. Specific immunity is divided into two types, active and passive. Transfer of antibodies to newborn animals in colostrum or artificially via injection of antiserum confers passive immunity of limited duration (weeks). Active immune responses following an encounter with an infectious agent or through vaccination result in the development of cell-mediated and/or humoral (antibody-mediated) immunity which persists for months or years.

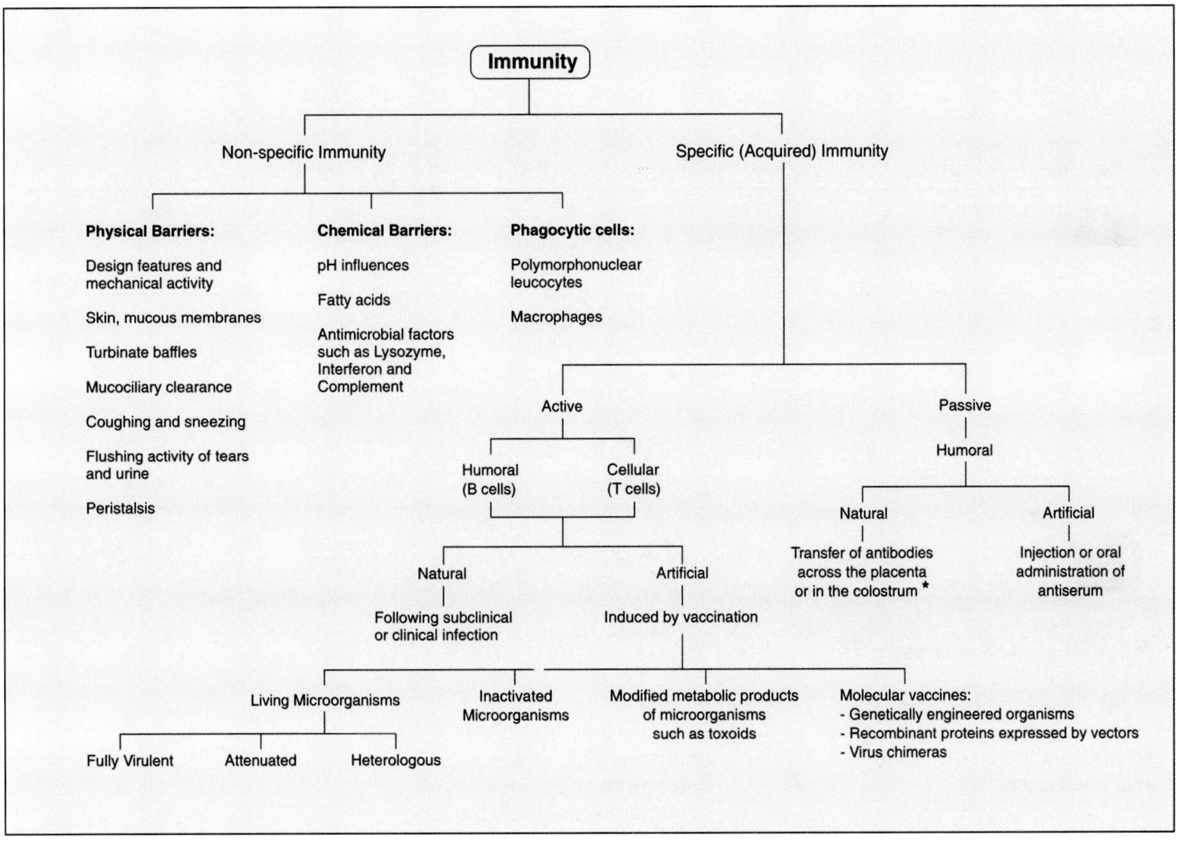

Diagram 70. A comparison of the principal elements of non-specific immunity and specific (acquired) immunity. Unlike non-specific immunity which is innate and lacks specificity, specific immunity is both inducible and specific, and results in the development of immunological memory.

*The colostrum of some species such as cattle, is rich in T and B lymphocytes and macrophages. These transferred cells may be able to participate in cell-mediated reactions until rejected by the neonatal animal.

Vaccines used in veterinary medicine include living and inactivated microorganisms, modified metabolic products such as toxoids and molecular vaccines produced by genetic engineering or recombinant technology (**Diagram 70**). The relative merits of living and inactivated vaccines are presented in **Table 166**. A point of particular importance relating to live viral vaccines is the adverse effect of maternal antibodies in young animals on such vaccines. Accordingly, vaccination of pups and kittens with live viral vaccines should be deferred until maternally acquired antibodies have dropped to a low level, usually after 12 weeks of age. If vaccination is carried out before 12 weeks of age, a heterologous vaccine, if available, should be used (measles vaccine will protect puppies against canine distemper). As a general rule, live vaccines should not be administered to pregnant animals because of the risk of foetal infection.

The immune response to vaccination typically follows a normal distribution with a small group of animals responding poorly to immunisation and the remainder producing a good or strong response (**Diagram 71**). Congenital or acquired immunosuppression may interfere with an animal's ability to respond to vaccination. Congenital deficiencies in cell-mediated or humoral immunity are uncommon in small animals but are reported in large animals, especially horses. Acquired causes of immunosuppression occur more frequently due to the administration of immunosuppressive drugs such as corticosteroids, or due to infections with viruses, such as canine parvovirus or feline leukaemia virus.

Table 166. A comparison of the relative merits of live and inactivated vaccines.

Comparative feature	Live vaccine (May be attenuated, avirulent or occasionally fully virulent)	Inactivated vaccine (Killed microorganisms or toxoids)
Safety	Some risks (reversion to virulence, foetal infection, contaminating microorganisms)	Usually safe (unlikely to contain contaminating microorganisms)
Stability	Affected by temperature and sunlight. Short shelf life	Stable on storage
Duration of immunity	Up to one year or longer	Usually only months
Onset of immunity	Rapid	Slow
Antibody response	(IgM), IgG, IgA	Predominantly IgG
Cell-mediated response	Good	Poor
Local secretory immunity	Usually present	Absent
Antigenic dose	Low	High
Route of administration	Natural route or by injection	By injection
Number of doses administered	Single (usually)	Multiple (boosters needed)
Requirement for adjuvants	Not required	Required
Influence of maternal antibody	Very significant (adverse effect)	Minimal effect
Cost	Low	High

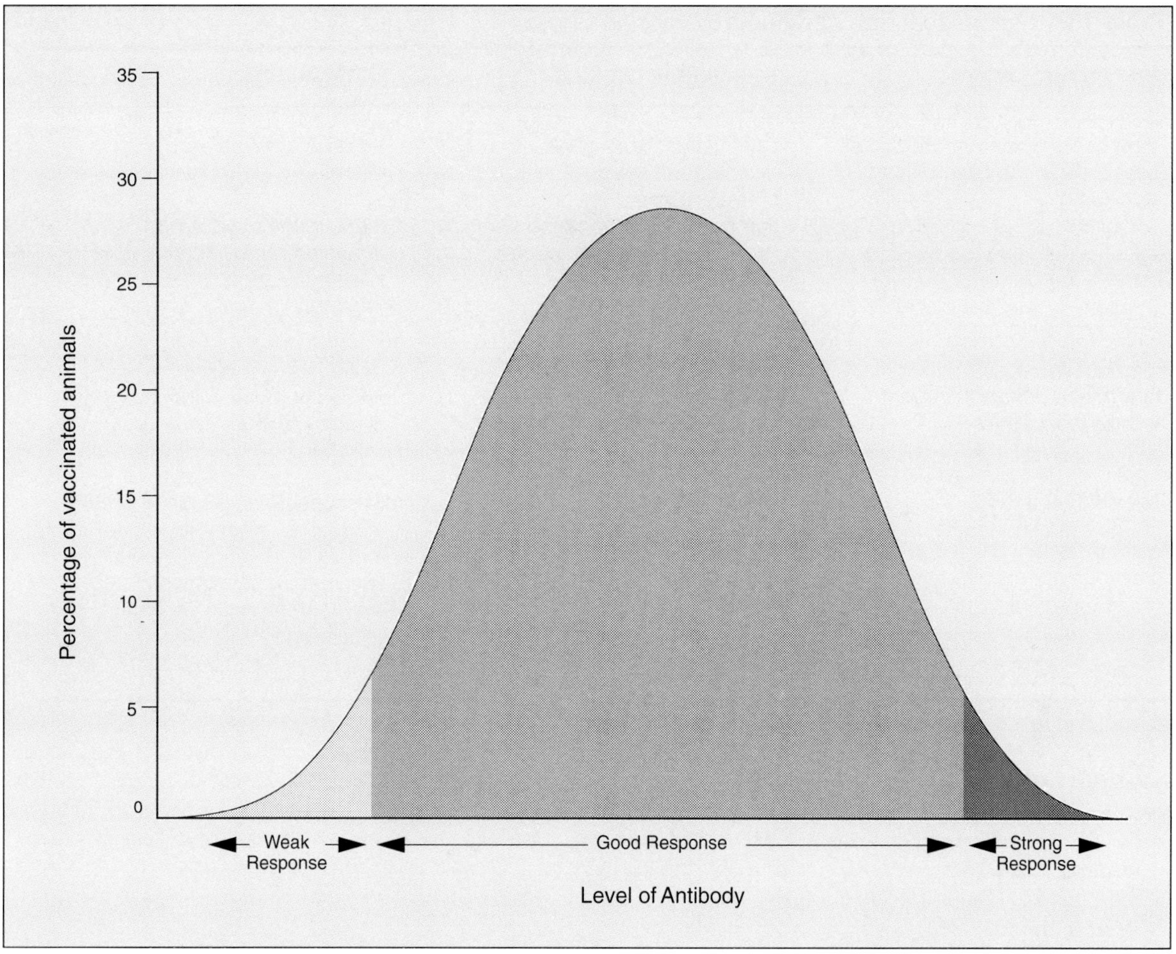

Diagram 71. The antibody responses of a randomly selected population of normal animals to vaccination. Cell-mediated responses to vaccines capable of inducing such reactions tend to follow a similar pattern.

Inactivated vaccines, especially for large animals frequently contain adjuvants, substances which enhance the immune response to the vaccine. Substances with adjuvant activity include aluminium salts, surface active agents, oil emulsions and immune-stimulating complexes (ISCOMS). Aluminium hydroxide and Freund's adjuvant (complete and incomplete) are two adjuvants widely used in veterinary medicine. In recent years ISCOMS have been gaining in popularity and ISCOM-based commercial vaccines are now available.

The duration of protection conferred by vaccination is variable and depends on the disease, the level and type of immunity required to protect the animal, the type of vaccine administered and the degree of challenge experienced by the vaccinated animal. The route of entry of the infectious agent into the animal's body may require a local immune response for protection (as in the upper respiratory tract for protection against respiratory viruses). Inactivated vaccines produce a weaker immune response of shorter duration than their live counterparts and revaccination at 6-month intervals may be required. Attenuated vaccines produce a more sustained immune response of longer duration and revaccination at one-year or two-year intervals is usually satisfactory. Under field conditions, animals may be challenged by the infectious agent against which they have been vaccinated and this natural exposure will reinforce their level of protection. Vaccinated animals reared in relative isolation may become susceptible to infection earlier than animals in frequent contact with others shedding the homologous infectious agent.

Vaccination failure may arise because of host-related or vaccine-related factors, or because of improper reconstitution or administration of a vaccine. The factors which may contribute to failure of a vaccination procedure are presented in **Table 167**.

Table 167. Factors that may contribute to vaccine failure.

Host-related factors	Vaccine-related factors	Other factors
Presence of maternal antibodies (especially in the case of live vaccines) Animal incubating the disease	Unsatisfactory vaccine: incapable of stimulating protective immunity because of composition, or because strain (serotype) prevalent in area was not included in vaccine	Vaccine administered incorrectly (by the wrong route) Inactivation of vaccine administered by scarification by a topically applied disinfectant (alcohol)
Immunosuppression: genetically determined, infectious agents or drug-induced	Inadequately passaged vaccine and consequently still virulent for some susceptible species	Vaccine (lyophilized) reconstituted with inappropriate diluent (such as water containing chlorine)
Age, nutrition, poor management and housing with resultant heavy challenge of infectious agents	Vaccine potency lost: inactivation of live vaccine due to sunlight, improper storage or contamination of multi-dose container Vaccine past expiry date	Susceptible agents in vaccine affected by drugs administered to host animal (irradiated larval vaccine: anthelmintics; avirulent bacterial vaccine: antibiotics)

Bibliography

Bittle, J.L. and Murphy, F.A. (1989). Vaccine Biotechnology, *Advances in Veterinary Science and Comparative Medicine*, **33**: 1-444.

Dhein, C.R. and Gorham, J.R. (1986). Host response to vaccination, *Veterinary Clinics of North America: Small Animal Practice*, **16**: 1227–1245.

Linton, A.H., Hugo, W.B. and Russell, A.D. (1987). *Disinfection in Veterinary and Farm Animal Practice,* Blackwell Scientific Publications, Oxford.

Mowat, A. McI and Donachie, A.M. (1991). ISCOMS– a novel strategy for mucosal immunization? *Immunology Today*, **12**: 383–385.

Quinn, P.J. (1990). Mechanisms of action of some immunomodulators used in veterinary medicine, in *Advances in Veterinary Science and Comparative Medicine*, **35**: 43–99.

Quinn, P.J. (1991). Disinfection and disease prevention in veterinary medicine, in *Disinfection, Sterilization and Preservation,*. Edited by S.S. Block. pp. 846–868, Lea and Febiger, Philadelphia.

Section 6: A Systems Approach to Infectious Diseases on a Species Basis

A microbiological approach to infectious diseases affecting domestic animals, including birds, is presented in Tables 168-174. These tables are organised on a species and systems basis. Species are dealt with in the following order: cattle (**Table 168**), sheep/goats (**Table 169**), pigs (**Table 170**), horses (**Table 171**), dogs (**Table 172**), cats (**Table 173**) and domestic birds (**Table 174**). The diseases listed under poultry may affect several avian species. Those conditions mentioned under turkeys, ducks, geese and pigeons are more common in, or specific to, these particular birds.

These tables are intended as a brief review of the major infectious diseases, including clinical aspects and diagnostic tests appropriate for each disease. Some of the diseases listed may occur infrequently in some geographical areas, whereas they may be endemic in other specific regions or countries.

Diseases where sudden death occurs do not lend themselves to description on a systems basis. Such diseases include anthrax and the clostridial enterotoxaemias. Lesions on the skin of the mammary glands and teats are covered in Chapter 36. Relevant information on chemotherapy relating to the bacterial and fungal diseases is presented in Chapter 7.

ABBREVIATIONS

+ve	positive
-ve	negative
Ab	antibody
Ag	antigen
AGID	agar gel immunodiffusion test
BA	blood agar
CAM	chorioallantoic membrane
CFT	complement fixation test
CNS	central nervous system
CO_2	carbon dioxide
CPE	cytopathic effect
CSF	cerebrospinal fluid
DCF	dilute carbol fuchsin stain
EM	electron microscopy
ELISA	enzyme-linked immunosorbent assay
FA	fluorescent antibody technique
HA	haemagglutination
HAI	haemagglutination inhibition test
IB	inclusion body
i/c	intracerebral
IFA	indirect fluorescent antibody test
IHA	indirect haemagglutination test (passive)
IN	intranuclear
i/p	intraperitoneal
i/v	intravenous
KOH	potassium hydroxide
Lab.	laboratory
MAT	microscopic agglutination test
MZN	modified Ziehl-Neelsen stain
PM	postmortem
RIA	radioimmunoassay
SMEDI	stillbirth, mummification, embryonic death, infertility
TC	tissue culture
URT	upper respiratory tract
VN	virus neutralization test
ZN	Ziehl-Neelsen stain

Table 168. Infectious diseases of cattle.

BUCCAL CAVITY: cattle

Disease	Agent(s)	Comments	Diagnosis
ACTINOBACILLOSIS **(Wooden or timber tongue)**	*Actinobacillus lignieresii*	Wooden tongue is a characteristic form of the disease. A hard, tumorous mass develops in the substance of the tongue. There is anorexia, excess salivation and the tongue is hard and 'wooden'	• Clinical signs • Microscopy and culture if exudate is present
ACTINOMYCOSIS	*Actinomyces bovis*	Actinomycosis of the jaw can predispose to suppurative alveolar periostitis, often involving the fourth or third molars. The condition should be considered when loose cheek teeth occur in cattle	• Microscopy on sulphur granules in pus or exudate • Culture, if necessary
BLUETONGUE (BT)	*Orbivirus* (*Reoviridae*)	Infections are often subclinical or mild in cattle. Occasionally well-developed lesions in mouth, encrusted muzzle ('burnt' appearance), nasal discharge, laminitis (severe) and a patchy dermatitis are present	• History of endemic area • Virus isolation: i/v inoculation of 10-12 day old embryonated eggs • Serology: VN, AGID, ELISA, modified CFT
BOVINE PAPULAR STOMATITIS	*Parapoxvirus* (*Poxviridae*)	Papular and erosive lesions occur in buccal mucosa and on muzzle of animals under 6 months of age. No systemic involvement	• History of milkers' nodules in humans feeding calves • Lesions and no illness • EM on biopsy of lesions
BOVINE VIRAL DIARRHOEA (BVD) AND MUCOSAL DISEASE	*Pestivirus* (*Flaviviridae*)	BVD: Oral lesions are seen in about 75 per cent of animals with the diarrhoeal syndrome. There is diffuse reddening of mucosa, followed by focal lesions that develop into discrete, rounded, shallow erosions (1-2 cm) Mucosal disease: Erosive stomatitis is one characteristic sign. The oral lesions are discrete, rounded, sharply - defined depressions. Disease occurs in 12-36 month-old animals. There is also diarrhoea and often lameness. These animals are immunotolerant (Ab -ve) but have a persistent viraemia (Virus +ve)	• Clinical signs • FA on frozen sections • Virus isolation: buffy coat • Serology: ELISA or VN on paired sera
CALF DIPHTHERIA **(Necrotic stomatitis)**	*Fusobacterium necrophorum*	Predisposing cause may be rough foodstuffs. Usually seen in calves under 6 months-old. Necrotic lesions occur in buccal cavity or laryngeal region (dyspnoea)	• Clinical signs • Gram-stained smear on necrotic debris • Culture if necessary

Table [illegible]. Infectious diseases of cattle *(continued)*

BUCCAL CAVITY: cattle

Disease	Agent(s)	Comments	Diagnosis
DERMATOPHILOSIS (Streptothricosis)	*Dermatophilus congolensis*	Lesions can occasionally occur on the tongue as a result of the animal licking skin lesions	• Clinical • Giemsa-stained or Gram-stained smear of scabs • Culture, if necessary
EPHEMERAL FEVER (3-day-sickness)	*Rhabdoviridae*	Signs include apparent pain in the throat region, accompanied by dysphagia and hypersalivation. Some animals have impaired swallowing reflexes and if death occurs it is often attributable to inhalation pneumonia. Other signs include fever, sudden drop in milk production and lameness, either constant or shifting. Usual course is 1-3 days. Milk production may be depressed for that lactation	• Usually based on clinical signs • Haematology: neutrophilia is a constant finding • Serology: VN for rising Ab titre • Virus isolation: TC or mouse inoculation
GLOSSOPLEGIA **Botulism** **Listeriosis** **Actinobacillosis**	 *Clostridium botulinum* *Listeria monocytogenes* *Actinobacillus lignieresii*	Partial or complete loss of function of the tongue may be peripheral or central in origin. The aetiology can be traumatic or infectious. The unilaterally affected tongue is deviated towards the non-affected side. The bilaterally affected tongue is limp and protrudes from relaxed jaws	• Diagnosis of the specific condition and removal of any predisposing causes
IBARAKI DISEASE (Kaeshi disease)	*Orbivirus* (*Reoviridae*) Distinct from BT virus	Occurs in Far East and South East Asia. Acute arthropod-borne disease characterized by fever, ulcerative stomatitis and dysphagia that leads to dehydration and emaciation. Seasonal, in late summer and autumn	• Histopathology • Virus isolation in TC or fertile eggs and identification by VN
INFECTIOUS BOVINE RHINOTRACHEITIS	Bovine herpesvirus 1	Systemic disease in neonatal calves with rhinitis, conjunctivitis, erosions of soft palate, bronchopneumonia, often encephalitis and high mortality. In young adults the URT form is characterised by inflamed nares and ulcers in the nasal mucosa. In severe cases lesions occur in pharynx, larynx and trachea.	• Clinical signs • FA for Ag on frozen sections • Virus isolation
MALIGNANT CATARRHAL FEVER	Gammaherpesvirus	Hyperaemia and diffuse, superficial necrosis of oral and nasal mucosa are constant findings in this disease	• Clinical signs • Histopathology
RABIES (pharyngeal paralysis)	*Lyssavirus* (*Rhabdoviridae*)	Pharyngeal paralysis is usually a sign of encephalitis and occurs in cattle with rabies. The animal is unable to swallow, salivation is noticed with gurgling noises from the pharynx	• History of rabies being endemic • Clinical signs • Histopathology of brain: Negri bodies • FA (brain) : viral antigen

Table 168. Infectious diseases of cattle. *(continued)*

BUCCAL CAVITY: cattle

Disease	Agent(s)	Comments	Diagnosis
RINDERPEST	*Morbillivirus* (*Paramyxoviridae*)	Erosive stomatitis and gastroenteritis are characteristic for this disease. There are punched-out lesions on gums, lips, tongue and hard palate	• History of endemic area and clinical signs • AGID or CFT on lymph node biopsy for Ag • Histopathology • Virus isolation • Serology: CFT, ELISA, VN, HAI
VESICULAR DISEASES **Foot-and-mouth disease (FMD)** **Vesicular stomatitis (VS)**	*Aphthovirus* (*Picornaviridae*) *Vesiculovirus* (*Rhabdoviridae*)	Vesicular lesions on tongue, buccal mucosa, teats (milking cows) and interdigital cleft. Lesions on feet and teats less constant in VS than in FMD	• ELISA or CFT for Ag in vesicular fluid • Virus isolation • Serology: ELISA, CFT, VN

GASTROINTESTINAL TRACT: cattle

Disease	Agent(s)	Comments	Diagnosis
ANTIBIOTIC-INDUCED DIARRHOEA	*Pseudomonas* sp., *Proteus* sp., or *Candida albicans*	Calves treated for diarrhoea with a prolonged course of oral antibiotics. Normal flora destroyed predisposing to a chronic diarrhoea with poor response to treatment and progressive weight loss	• History • Isolation of secondary invaders from faeces in heavy growth
BOVINE VIRAL DIARRHOEA (BVD) AND MUCOSAL DISEASE	*Pestivirus* (*Flaviviridae*)	BVD: Young cattle 6-24 months show mild depression, oculonasal discharge and occasionally shallow ulcers in buccal cavity. Diarrhoea occurs in susceptible herds. High morbidity but zero mortality	• FA: frozen sections • Virus isolation: buffy coat • Serology: four-fold rise in Ab titre
		Mucosal disease: Low morbidity/100 per cent mortality. BVD-immunotolerant, 6-24 month-old animals at risk (virus +ve/Ab -ve). Severe lameness (laminitis), profuse diarrhoea and buccal cavity lesions extending to intestines occur	• Herd history and clinical signs • FA on buffy coat or lymphocyte smears for virus (virus +ve) • Serology: VN or ELISA for antibody (Ab -ve) • Virus isolation: buffy coat
CLOSTRIDIAL ENTEROTOXAEMIA	*Clostridium perfringens* types B and C	Occurs in young well-nourished calves up to 10 days of age. Severe haemorrhagic enterotoxaemia with rapid death. Uncommon	• Gross pathology and histopathology • Gram-stain on mucosa: large numbers of thick Gram +ve rods • Mouse tests for toxin in small intestine
COLIBACILLOSIS AND COLISEPTICAEMIA	*Escherichia coli*	Neonates under 7 days of age. Colostral immunity determines survival. Acute profuse diarrhoea, dehydration and acidosis	• Isolation of *E. coli* • Enteropathogenicity tests (*see* Chapter 18)

Table 168. Infectious diseases of cattle. *(continued)*

GASTROINTESTINAL TRACT: cattle

Disease	Agent(s)	Comments	Diagnosis
CRYPTOSPORIDIOSIS	*Cryptosporidium parvum*	Outbreaks of diarrhoea in 5-35 day-old calves. Both affected and clinically normal calves shed large numbers of oocysts in faeces. Villous atrophy and enlargement of crypts occurs	• Safranin-methylene blue stain on faecal smears • Auramine-O technique on faecal smear
JOHNE'S DISEASE	*Mycobacterium paratuberculosis*	Seen in cattle over 2 years of age, although infection is acquired soon after birth. Not all infected cattle become clinical. Chronic disease with emaciation, profuse diarrhoea and eventual death. Corrugated and reddened ileocaecal valve area is characteristic	• ZN smear of rectal scraping or mucosa of ileocaecal valve area • Culture: Herrold's egg-yolk medium : up to 16 weeks incubation • Histopathology
MALIGNANT CATARRHAL FEVER (MCF)	Ovine herpesvirus 2 (OHV 2) or Alcelaphine herpesvirus 1 (AHV 1, Africa).	Low morbidity/high mortality. Corneal opacity, enlarged lymph nodes of head and neck, ragged erosions in buccal cavity. Terminal diarrhoea and encephalitis. Reservoirs are sheep (OHV 2) and wildebeest (AHV 1)	• Histopathology • Virus isolation from buffy coat cells: calf thyroid cell line for wildebeest derived-MCF only
RINDERPEST	*Morbillivirus* (*Paramyxoviridae*)	High morbidity/high mortality. Very contagious. Profuse diarrhoea, dehydration, weakness, buccal erosions ('punched-out' ulcers) extending into the intestinal tract, 'zebra-striping' of terminal large intestine	• History and clinical signs • Detection of Ag in tissue: CFT, AGID • Histopathology • Virus isolation • Serology: VN, ELISA, CFT
ROTAVIRUS AND CORONAVIRUS INFECTIONS	*Rotavirus* (*Reoviridae*) *Coronavirus* (*Coronaviridae*)	Neonates 5-21 days old. Explosive outbreaks of profuse watery diarrhoea. Extensive villous atrophy (most severe with coronavirus). Calves that recover may be unthrifty until villi regenerate	• EM (faeces) • ELISA for Ag capture • Virus isolation and identification
SALMONELLOSIS	*Salmonella* spp.	Animals of all ages are susceptible. Young calves often develop the septicaemic form of disease. May be stress-induced. Acute diarrhoea / dysentery and fever are present. Deaths can occur in young animals	• Culture for salmonellae
WINTER DYSENTERY ('Black scours')	*Coronavirus*	Once thought to be due to *Campylobacter jejuni*. Explosive outbreaks of diarrhoea/dysentery in mature housed cattle. Outbreak lasts 24 hours, is usually non-febrile and there are dark watery faeces with a fetid odour. High morbidity / low mortality	• Virus isolation • Virus detection in faeces: EM, passive haemagglutination • Serology: VN, HAI

Table 168. Infectious diseases of cattle. *(continued)*

LIVER: cattle

Disease	Agent(s)	Comments	Diagnosis
BACILLARY HAEMOGLOBINURIA	*Clostridium haemolyticum*	Ingested spores lodge in the liver. Migrating liver fluke cause tissue damage and provide conditions suitable for germination of spores. Sudden death or fever, abdominal pain, 'port-wine' urine and infarcts in liver. The infarcts are pale, raised and surrounded by a bluish-red zone. Disease affects cattle and occasionally sheep	• History of a liver fluke area • Clinical or postmortem findings • FA on smears from liver lesion
BOVINE LIVER ABSCESSES	*Fusobacterium necrophorum* and *Actinomyces pyogenes*	Usually no clinical signs but the lesions are discovered at slaughter. Most common in feedlot cattle. Similar lesions have been reported in pigs	• Pathology • Direct microscopy: Gram-stained smear • Culture of pathogens
FACIAL ECZEMA (mycotoxicosis)	Sporidesmin in spores of *Pithomyces chartarum* (mycotoxin)	Bile duct obstruction, liver fibrosis, jaundice and failure to excrete phylloerythrin results in photosensitization seen on unpigmented or bare skin	• Clinical signs • Spore count on pastures • Gross and histopathology of livers
LISTERIOSIS (SEPTICAEMIC VISCERAL)	*Listeria monocytogenes*	Occurs in many species of young animals and in birds. There is fever, anorexia, depression and death in 1-3 days. Necrotic foci are seen throughout the liver and other body organs at postmortem examination.	• Direct Gram-stained smears from lesions • Culture of pathogen
RIFT VALLEY FEVER	*Phlebovirus* (*Bunyaviridae*)	Hepatitis and high mortality occurs in lambs, kids and calves; severe disease and abortions in adult sheep and goats, but only mild or subclinical infections in cattle with a high percentage of abortions. Vector: mosquitoes	• History of endemic area • Clinical signs • Gross and histopathology • Virus isolation: TC or i/p mouse inoculation • Serology: VN, CFT, ELISA, or HAI

Table 168. Infectious diseases of cattle. *(continued)*

GENITAL SYSTEM: cattle

Disease	Agent(s)	Comments	Diagnosis
AKABANE DISEASE	*Bunyavirus* (*Bunyaviridae*)	Japan, Australia, South Africa and Israel. Cattle, sheep and goats. • Severe damage to foetus with death and abortion • Congenital abnormalities: hydranencephaly and arthrogryposis Solid immunity after infection Vector: mosquitoes	• Virus isolation: suckling mice or TC • Serology: VN
ANAPLASMOSIS **(Gall Sickness)**	*Anaplasma marginale*	Arthropod-transmitted disease of ruminants in tropics and subtropics. Clinical disease seen in introduced, non-immune, adult cattle: fever, anaemia, icterus (but not haemoglobinuria), weakness and abortion in pregnant animals. Death, recovery or chronic disease with emaciation may ensue	• History of endemic area • Giemsa-stained blood smears for rickettsiae • Serological tests for carriers and chronic cases
BLUETONGUE	*Orbivirus* (*Reoviridae*)	Sheep, deer and cattle are affected but only about 5 per cent of infected cattle show clinical signs. If cattle are infected during gestation:- • Abortion • Congenital abnormalities: cerebellar hypoplasia, arthrogryposis or hydranencephaly Vector: *Culicoides* spp.	• Isolation: i/v inoculation of 10-12 day embryonated eggs • Serology: ELISA, VN, AGID
BOVINE GENITAL CAMPYLOBACTERIOSIS **(Vibriosis)**	*Campylobacter fetus* ss. *venerealis.* (*C. fetus* ss. *fetus*)	Venereal infection, bulls are carriers. Non-immune cows suffer mild metritis, salpingitis and embryonic death with irregular cycles at 28-35 days. Self - limiting disease and natural immunity in 3-5 months with destruction of *Campylobacter.* Occasional carrier cow (vagina). Sporadic abortions with *C. fetus* ss. *fetus*	• Isolation of pathogen from abomasal contents of foetus • Direct microscopy using DCF or FA

Table 168. Infectious diseases of cattle. *(continued)*

GENITAL SYSTEM: cattle

Disease	Agent(s)	Comments	Diagnosis
BOVINE VIRAL DIARRHOEA (BVD)	*Pestivirus* (*Flaviviridae*)	Syndromes include: • Neonatal calves: immunosuppression • Young cattle 6-24 months: diarrhoea and erosions of buccal mucosa • Adult pregnant cows, depending on stage of gestation when infected: a. 50-100 days: foetal death and abortion or mummification b. 100-150 days: congenital defects in foetus c. Before 120 days of pregnancy (with non-CPE strain): immunotolerance (virus +ve/Ab -ve) and calf is at risk from mucosal disease when 6-24 months old d. Virus via coitus: fertilization failure and a 'repeat breeder' problem	• Virus isolation • FA: frozen sections (Ag +ve) • Serology: VN, IFA, ELISA (Ab -ve)
BRUCELLOSIS	*Brucella abortus* (*B. melitensis* and *B. suis*)	Abortion storms occur in non-immune herds. A cow usually only aborts once but remains infected and excretes brucellae at subsequent parturitions. Organisms can be excreted in milk. Infection is usually by oral route. Notifiable disease in many countries	• Serology: many tests used on a herd basis as part of a national eradication scheme • Isolation of brucellae • Direct microscopy: MZN-stained smears
CHLAMYDIAL ABORTION	*Chlamydia psittaci*	Similar to enzootic abortion of ewes	• Isolation • Impression smears from cotyledons • Serology: CFT, IFA, ELISA
EPIZOOTIC BOVINE ABORTION (Foothill Abortion)	*Borrelia* sp.?	Abortion or full-term weak calves. Vector is *Ornithodoros* tick. No signs of illness in cows but abortions in 10-90 per cent of susceptible cattle in third trimester. Incubation period 90-150 days. Cows usually abort only once and are normal at next pregnancy. Occurs in Western USA	• Gross and histopathology. • Microscopy: darkfield (foetal blood)

Table 168. Infectious diseases of cattle. *(continued)*

GENITAL SYSTEM: cattle			
Disease	**Agent(s)**	**Comments**	**Diagnosis**
INFECTIOUS BOVINE RHINOTRACHEITIS/ INFECTIOUS PUSTULAR VULVOVAGINITIS (IBR/IPV)	Bovine herpesvirus 1.	IBR: Incubation period 2-6 days. Syndromes include: • Young adults: respiratory disease ('red nose') • Abortions up to 90 days after infection with or without previous respiratory signs. Abortion most common between 4-7 months of pregnancy. Infertility is not a sequel of IBR. No gross lesions in foetus but microscopic foci of necrosis in many organs with IN inclusions • Neonates: diarrhoea and/or encephalitis IPV: venereal infection, localized in genitalia of both sexes. No viraemia occurs so abortion is not seen. Self-limiting infection. Many subclinical cases occur	• Histopathology (foetal tissues) • Virus isolation: vaginal and preputial swabs • FA: foetal tissues • Serology: VN or ELISA
LEPTOSPIROSIS	*Leptospira interrogans* serovars	Abortion storms common with some serovars but are sporadic with others. Abortion occurs 6-12 weeks after infection and is most common in 7th month of pregnancy. Other signs include infertility, weak calves and an agalactia syndrome. Carrier state common with leptospires excreted in urine	• Darkfield or FA microscopy on urine • Culture from urine or kidneys • Serology (herd basis): MAT, CFT, ELISA
LISTERIOSIS	*Listeria monocytogenes* (*L. ivanovii* : abortion in cattle and sheep only)	Syndromes include: • Visceral or septicaemic listeriosis in many young animals and birds • Neural listeriosis (circling disease) in cattle, sheep and goats • Abortion in cattle, sheep and goats. Usually sporadic and in late gestation. No systemic illness in dam and no infertility. Listeriae are shed in milk and uterine discharges for some months after infection. Minor outbreaks occur with silage-feeding to pregnant animals • Ocular form: a self-limiting iritis often with corneal opacity. Associated with silage feeding Circling disease and abortions do not usually occur together on same property. Focal necrosis of foetal liver occurs but may be masked by autolysis	• Isolation and identification of pathogen • Histopathology

Table 168. Infectious diseases of cattle. *(continued)*

GENITAL SYSTEM: cattle

Disease	Agent(s)	Comments	Diagnosis
MYCOTIC ABORTION	*Aspergillus fumigatus*, or *Mortierella wolfii*	Abortions sporadic and usually between 6-9 months of pregnancy. Characteristic findings: • 'Wooden' appearance of placenta as some maternal caruncles detach and adhere to cotyledons • Ringworm-like lesions on foetus: pathognomonic when present • Placenta often retained *M. wolfii* abortions (in 5 per cent cases) can be followed, within 48 hours by peracute pneumonia and death of cow	• Histopathology on cotyledons or foetal lesions • Isolation of fungal pathogen
RIFT VALLEY FEVER	*Phlebovirus* (*Bunyaviridae*)	Epidemics in cattle, sheep and goats in South and East Africa. • Hepatitis and high mortality in young animals • Severe disease and 90-100 per cent abortions in sheep and goats • Mild disease in cattle but 100 per cent abortion rate • Influenza-like disease in man Vector: mosquitoes. Reservoir: wild ruminants	• Histopathology: liver necrosis. • Virus isolation: lab. animal inoculation or TC • Serology: VN, CFT, ELISA, HAI
SALMONELLOSIS	*Salmonella* species (especially *S.dublin*)	Sporadic abortions and the cow may or may not show systemic illness. Many abortions may be seen in a herd where an outbreak of enteric disease has occurred. Faecal-oral transmission and carrier state is common. Stress can convert a carrier state to a clinical case	• Isolation of salmonellae from placenta or foetus
TICK-BORNE FEVER	*Ehrlichia (Cytoecetes) phagocytophila*	Seen in Western Europe and Finland. Relatively mild enzootic disease with dullness, fever, and immunosuppression. Abortions and stillbirths in cattle and sheep. Rickettsial organism has predilection for neutrophils. Vector: *Ixodes ricinus*	• Direct microscopy. Giemsa-stained blood smear: purple bodies 0.3-0.7μm
TRICHOMONIASIS	*Trichomonas foetus*	Early embryonic death (infertility) and occasionally abortions or pyometra. Venereal infection: bull carrier in prepuce and can infect 90 per cent of cows served. Foetus dies 50-100 days after conception and irregular oestrous cycles follow. Uterine discharge often purulent and contains large numbers of protozoan parasites (maximum numbers 3-7 days before oestrus)	• Direct microscopic examination of uterine discharges. Specimens should be kept warm or protozoan parasite will become non-motile • Culture

Table 168. Infectious diseases of cattle. *(continued)*

URINARY SYSTEM: cattle

Disease	Agent(s)	Comments	Diagnosis
ABSCESSES IN KIDNEYS	Pyogenic bacteria such as: streptococci, *Staphylococcus aureus*, *Actinomyces pyogenes*	If a bacteraemia occurs, these pyogenic bacteria can lodge and multiply in the kidneys and in other body organs	• Isolation of the pathogenic bacteria • Histopathology
BACILLARY HAEMOGLOBINURIA	*Clostridium haemolyticum*	Associated with liver damage due to migrating liver fluke. The urine is 'port-wine' coloured and foams when voided. Sudden deaths are common. On postmortem, the liver infarct is pathognomonic: raised, light in colour and outlined by a bluish-red zone of congestion	• History: liver fluke area • Clinical signs • Pathology • FA on smears from liver lesions • Isolation of pathogen • Demonstrate toxin in peritoneal cavity fluid
BOVINE LEUKOSIS	*Retroviridae*	Tumours (B lymphocyte infiltration) can occur in the kidneys and in other body organs	• History of the disease in the herd • Clinical signs • Histopathology (dead animals or biopsy) • Serology: AGID or ELISA (herd test)
LEPTOSPIROSIS	*Leptospira interrogans* serovar *pomona* and other serovars	Some serovars such as *pomona*, produce a haemolysin resulting in haemoglobinuria in calves and occasionally older animals. There is usually accompanying fever, icterus and anorexia	• Clinical signs • Urine for dark-field microscopy • FA technique on urinary deposits • Isolation of leptospires
PYELONEPHRITIS (BOVINE)	*Corynebacterium renale* group (Streptococci and other bacteria often present)	Disease of mature cows. Often precipitated by pregnancy and dystocia. Urine cloudy with blood clots in advanced cases. There is frequent micturition, uneasiness, hunched back and enlarged kidney may be felt on palpation	• History and clinical signs • Isolation of a member of the *C. renale* group
'WHITE SPOTTED KIDNEY' IN CALVES	*Escherichia coli*, *Leptospira interrogans* serovars or other bacteria	Focal interstitial nephritis following a bacteraemia or septicaemia. Often only detected at slaughter	• Isolation of the pathogens may be difficult at this stage in the disease • Fixed tissue for histopathology may assist in the diagnosis

Table 168. Infectious diseases of cattle. *(continued)*

EYES AND EARS: cattle

Disease	Agent(s)	Comments	Diagnosis
INFECTIOUS BOVINE KERATOCONJUNCTIVITIS (IBK)	*Moraxella bovis*	Predisposing causes are irritants, flies and sunlight. Acute disease with highest incidence in animals under 2 years of age. Initial signs are photophobia, blepharospasm, lacrimation and conjunctivitis. Ulcers, corneal oedema and opacity occur with vascularisation in severe cases. Healing stage involves granulation tissue projecting from the ulcer as a characteristic 'red-cone'. Condition resolves completely or leaves a white corneal scar	• History of outbreak • Clinical signs • Isolation of *M. bovis* from lacrimal secretions (within 2 hours of collection)
INFECTIOUS BOVINE RHINOTRACHEITIS	Bovine herpesvirus 1 (BHV 1)	Conjunctivitis occurs as part of the acute syndrome and in mild cases may be the only sign present. In acute and severe disease, signs include fever, depression, nasal discharge and inflamed nares ('red nose'). Ulcers develop in the nasal mucosa. There is dyspnoea, mouth breathing and excessive salivation	• Clinical signs • FA: frozen sections • Virus isolation
LISTERIOSIS (corneal opacity)	*Listeria monocytogenes*	Corneal opacity is often unilateral when it occurs in neural listeriosis. Listerial iritis can be associated with feeding of big bale silage. The condition may progress to corneal opacity and blindness. Systemic signs are usually absent	• History of silage feeding • Clinical signs • Attempted isolation of listeriae from eyes in ocular form
MALIGNANT CATARRHAL FEVER (corneal opacity)	Ovine herpesvirus 2 or Alcelaphine herpes-virus 1	Bilateral corneal opacity is a constant finding in this sporadic but usually fatal disease	• Clinical course of disease • Histopathology
MYCOPLASMAL CONJUNCTIVITIS	*Mycoplasma bovoculi*	Causes conjunctivitis and transient corneal opacity. A concurrent infection with *Moraxella bovis* may increase the severity of IBK	• Isolation of *M. bovoculi*

Table 168. Infectious diseases of cattle. *(continued)*

NERVOUS SYSTEM: cattle

Disease	Agent(s)	Comments	Diagnosis
ABSCESSES (SPINAL OR BRAIN)	*Actinomyces pyogenes*, *Fusobacterium necrophorum*, *Actinomyces bovis*, or *Mycobacterium bovis*	Usually occurs in young animals. May arise from direct trauma such as dehorning or as an extension of otitis media, paranasal infections or lesions of the meninges. There is rotation or deviation of neck, ataxia, circling, blindness or nystagmus in one eye	• Gross pathology • Histopathology • CSF: high neutrophil count • Isolation of pathogen from CSF or lesion
BOTULISM	*Clostridium botulinum*	History of cattle eating toxin-containing foods (baled silage, processed poultry litter or carrion), but wound infections (toxicoinfections) can occur. Toxin causes progressive muscular paralysis with dysphagia, recumbency and respiratory failure. There is no fever. Mild cases may recover	• Type of foodstuffs • Clinical signs • Demonstration of toxin in serum: mouse inoculation or ELISA
BOVINE LEUKOSIS	*Retroviridae*	Tumours can occur anywhere in the body including the brain or spinal canal. Clinical signs will depend on the site of the tumour.	• Serology: AGID or ELISA on a herd basis • Histopathology
BOVINE SPONGIFORM ENCEPHALOPATHY (BSE)	A prion (probably the scrapie agent)	Highest incidence in adults 3-6 years old. Long incubation period 2-8 years. Onset is insidious. There is apprehension, low head carriage, irritability and excessive ear movement. Later ataxia, falling and recumbency occur. Occasionally there is aggression towards other animals. Pruritis is very rare. Course usually 1-3 months	• Clinical signs • Age • Histopathology: astrocytosis, vacuolation of neurons • EM: scrapie-associated fibrils in brain tissue
CONGENITAL CNS LESIONS	Akabane disease virus	Hydranencephaly and/or arthrogryposis	• Serology: test dam for high antibody titres to the appropriate virus • Virus isolation
	Bovine viral diarrhoea virus	Hydranencephaly, hydrocephalus, cerebellar hypoplasia, cataract or arthrogryposis	
HEARTWATER	*Cowdria ruminantium* Vector: *Amblyomma* ticks.	Endemic in southern Africa, Malagasy and some West Indian islands. Clinical signs occur in non-immune animals: lip-licking, high-stepping, recumbency and death during a galloping convulsion. Course of the disease is 3-6 days	• History of endemic area • Giemsa-stained smears of cerebral grey matter to visualise rickettsiae

Table 168. Infectious diseases of cattle. *(continued)*

NERVOUS SYSTEM: cattle

Disease	Agent(s)	Comments	Diagnosis
INFECTIOUS BOVINE RHINOTRACHEITIS (IBR)	Bovine herpesvirus 1	Infection in neonatal calves can cause conjuntivitis, pneumonia and nervous signs such as excitement, tremor, ataxaxia and recumbency. A severe non-suppurative meningoencephalitis/myelitis with marked vascular cuffing and gliosis is present. A strain of BHV-1 has been isolated from encephalitis in adults	• Clinical signs • Histopathology: tissue damage and IBs • Virus isolation • FA: frozen sections
LISTERIOSIS	*Listeria monocytogenes*	Neural form: occurs in all ages but most common in adults. Signs include drooling, facial hypalgesia, head tilt and unilateral drooping ear. Loss of blink reflex can lead to keratitis and corneal ulceration. Ataxia, circling and occasionally mania and bellowing occur. Course is less than 14 days. Other syndromes are abortion, septicaemic (visceral) form in young animals, and ocular form	• History and clinical signs • Histopathology: microabscesses and perivascular cuffing in brain • Isolation of listeriae from brain (cold enrichment)
LOUPING ILL	*Flavivirus* (*Flaviviridae*)	The disease in cattle is usually mild and seen mainly in calves, as adults acquire an immunity in endemic areas. Signs include excitement, tremors, inco-ordination and ataxia. A non-suppurative meningoencephalitis is present mainly affecting the lower brain stem and cerebellum. Vector: *Ixodes ricinus*	• History of endemic area with vector present • Clinical signs • Histopathology • Virus isolation: TC or i/c inoculation of mice • Serology: HAI, AGID, CFT, VN, IFA
MALIGNANT CATARRHAL FEVER	Ovine herpesvirus 2 or Alcelaphine herpes-virus 1 (Africa)	Sporadic disease, usually in adults. Reservoir is sheep (OHV 2) or wildebeests (AHV 1). Generalised disease with fever, encrusted muzzle, diffuse erosions in buccal cavity, corneal opacity, cervical lymphadenopathy, deep depression, incoordination, head-pressing and eventually paralysis and death. Mortality 100 per cent. A non-suppurative encephalomyelitis is present	• History of contact with sheep or wildebeests • Clinical signs: sporadic cases and usually fatal • Histopathology • Virus isolation: AHV 1 but not OHV 2 from buffy coat
MENINGITIS (BACTERIAL)	*Staphylococcus aureus*, *Haemophilus somnus*, *Escherichia coli*, Streptococci or others	Seen in young calves and condition usually results from a bacteraemia following infection of the umbilicus. There is fever, neck rigidity, opisthotonos, nystagmus, extensor spasms, clonus and coma. Glairy thickening of meninges and congestion of vessels occurs	• Histopathology • Isolation of pathogen from CSF or meninges

Table 168. Infectious diseases of cattle. *(continued)*

NERVOUS SYSTEM: cattle

Disease	Agent(s)	Comments	Diagnosis
POLIO-ENCEPHALOMALACIA (Cerebrocortical necrosis)	Attributed to thiaminases produced by bacteria in the rumen	Thiamine deficiency. Animals are afebrile, have decreased mobility, blindness with active pupillary and corneal reflexes, recumbency with extensor spasms and 'paddling' can occur. Rapid response to thiamine early in condition	• Clinical signs • Swift response to thiamine • Histopathology: focal necrosis of cortex
PSEUDORABIES	Porcine herpesvirus 1	History of close contact with pigs or rats. Signs include intense focal pruritis often in flank area, dog - sitting position, bellowing, teeth grinding, salivation, pharyngeal paralysis but no aggression is seen. Death from respiratory or cardiac failure	• Virus isolation • FA on cryostats of brain tissue • History: association with pigs
RABIES	*Lyssavirus* (*Rhabdoviridae*)	The dumb form is most common in cattle with salivation, tenesmus, constipation, ataxia and paralysis. Rarely the furious form is seen with bellowing, mania and aggression. A non-suppurative encephalitis is present with Negri bodies in neurons of cerebellum and hippocampus	• History of dog or fox bites (vampire bats in South America) • Clinical signs • FA on brain • Histopathology of brain for Negri bodies
SPORADIC BOVINE ENCEPHALOMYELITIS (Buss disease)	*Chlamydia psittaci*	Described in USA, Europe, Japan, Australia and South Africa. Incubation period 6-31 days. Chlamydiae excreted in faeces and urine. Depression, fever, salivation and dyspnoea. Recovery can occur at this stage but CNS signs usually develop: stiff, staggering gait, circling and falling over small obstacles. Limbs become weaker and general paralysis follows. Low morbidity but mortality >50 per cent. All ages affected but most common in young animals. In chronic cases, serofibrinous exudates occur in body cavities. Course of disease usually 10-14 days	• Pathology: peritonitis • Isolation: yolk sac or TC • Serology: group antigen. Rising titre, CFT or ELISA
TETANUS	*Clostridium tetani*	Spores enter via traumatised tissue following dystocia, wounds, injections and umbilicus. Signs include muscle stiffness ("saw horse" stance), tremors, trismus, hyperaesthesia, raised tail-head, bloat, tetanic convulsions, opisthotonos and death from respiratory paralysis	• History: predisposing cause • Clinical signs • Gram-stained smear from deep wound if present

Table 168. Infectious diseases of cattle. *(continued)*

NERVOUS SYSTEM: cattle

Disease	Agent(s)	Comments	Diagnosis
THROMBOEMBOLIC MENINGOENCEPHALITIS (TEME)	*Haemophilus somnus*	Septicaemic form of infection ('sleeper syndrome') is seen most often in young feedlot cattle in autumn and winter. There is fever, stiffness, extension of head, lingual paralysis, ataxia, stupor, opisthotonos, occasional circling and blindness. Haemorrhagic infarcts occur in brain and retina	• Clinical signs • Gross pathology of brain • Histopathology • Isolation of *H. somnus*
TREMORGEN STAGGERS	*Penicillium* or *Aspergillus* spp.	Often seen in calves. There is stiffness, ataxia, trembling in large muscle masses, falling and convulsions if hurried. Animals should be moved slowly and gently from suspect pasture	• Clinical signs • Recovery on removal from suspect pasture
VERTEBRAL OSTEOMYELITIS	*Salmonella* spp. *Actinomyces bovis*	Signs dependent on site of lesion. Ataxia, hemiplegia or paraplegia can occur due to either pressure on spinal cord or extension to and inflammation of the spinal meninges	• Gross pathology • Isolation of pathogen

MUSCULOSKELETAL SYSTEM: cattle

Disease	Agent(s)	Comments	Diagnosis
BLACKLEG	*Clostridium chauvoei*	Sudden death usually occurs, especially if heart muscle is involved. Muscle masses of hind quarters commonly affected and age range is 3-24 months. The muscles are dry, dark, spongy with small gas bubbles and have a sweet, rancid odour. Crepitation can be felt. Usually an endogenous infection in cattle	• History: endemic area • Clinical signs • FA on muscle or bone marrow from a rib • Isolation of pathogen, if necessary
BOTULISM	*Clostridium botulinum*	Usually an intoxication from toxin-containing baled silage, processed poultry litter or carrion-eating. Less commonly wounds are contaminated by spores and toxicoinfectious botulism occurs. Signs include progressive weakness, tongue paralysis, inability to swallow, flaccid paralysis and death from cardiac or respiratory paralysis	• History of foodstuffs • Clinical signs • Demonstration of toxin in serum by mouse inoculation
CHLAMYDIAL POLYARTHRITIS	*Chlamydia psittaci*	Can involve all ages but calves 4-30 days of age are most severely affected. Lameness is pronounced but calves remain alert and will suck if aided. Limb joints are swollen and painful. There is no navel involvement	• Clinical signs • Cytological examination of joint fluid for elementary bodies or inclusions • Isolation of pathogen

Table 168. Infectious diseases of cattle. *(continued)*

MUSCULOSKELETAL SYSTEM: cattle

Disease	Agent(s)	Comments	Diagnosis
CONGENITAL DISEASES **Bluetongue** **Akabane** **Bovine viral diarrhoea**	 *Orbivirus*, *Bunyavirus*, *Pestivirus*	Congenital abnormalities can occur if each of these viruses infect cows at a critical point in gestation. In all three infections, common abnormalities in the calves include arthrogryposis and hydrocephaly	• Serology: detection of antibodies to each of these viruses in the dam • Virus isolation
EPHEMERAL FEVER	*Rhabdovirus*	Characterised by a sudden onset, biphasic fever, drop in milk production, depression, muscle stiffness and lameness. Usually recovery is dramatic and complete in 3 days ('three-day-sickness'). Vectors are mosquitoes and *Culicoides* spp.	• Clinical signs in an endemic region • Haematology: neutrophilia • Serology: VN (specific) or IFA • Virus isolation: TC or mouse inoculation
ERGOTISM (a mycotoxicosis)	*Claviceps purpurea*	Lameness is the first sign and occurs 2-6 weeks after ingestion of ergot alkaloids. All ages can be affected. There is tenderness and swelling in the fetlock and pastern joints, followed in about 1 week by loss of sensation and dry gangrene of the distal part of the limb	• History of pasture or foodstuffs contaminated by ergots • Clinical signs
FOOT ROT	*Fusobacterium necrophorum,* *Bacteroides melaninogenicus* and *Actinomyces pyogenes*	Infection is usually confined to one foot. There is acute lameness and paring of the horn of the sole liberates a foul-smelling exudate. Infection can spread to involve joints if not treated	• Clinical signs • Gram-stained smear from pus • Isolation of pathogens if necessary
JOINT-ILL	*Escherichia coli* *Staphylococcus aureus* Streptococci or others	Lodgement of pathogenic bacteria in joints following an omphalitis or a septicaemia. In the acute phase there is fever, severe lameness and swollen joints	• Clinical signs • Isolation of pathogen from aspirated joint fluid
LAMENESS (GENERALISED DISEASES) **Foot-and-mouth disease** **Vesicular stomatitis**	 *Aphthovirus*, *Vesiculovirus*, *Pestivirus*	In foot-and-mouth disease and vesicular stomatitis, lameness is due to interdigital vesicular lesions, and all four feet are usually affected. Lameness occurs in some animals with mucosal disease and is due to laminitis, coronitis or erosive lesions of skin in the interdigital cleft. All four feet are usually affected	• Tests for detection of viral Ag, virus isolation, and demonstration of antibodies to each virus
MALIGNANT OEDEMA	*Clostridium septicum,* *C. sordellii* or *C. novyi* type A	Animals of all ages affected. If alive, there is fever and soft swelling around a wound that spreads to muscle masses. The swelling pits on pressure and if muscle is incised the tissue is dark. Exudate and gas are present	• Clinical signs • FA in muscle or bone marrow from a rib • Isolation if necessary

Table 168. Infectious diseases of cattle. *(continued)*

MUSCULOSKELETAL SYSTEM: cattle			
Disease	**Agent(s)**	**Comments**	**Diagnosis**
MYCOPLASMAL ARTHRITIS	*Mycoplasma bovis*	Recognised most frequently in feed-lot cattle, 6-8 months of age. There is moderate fever, stiffness, lameness and progressive weight loss. Swelling of joints and distension of tendon sheaths occurs associated with fibrinous synovitis and synovial fluid effusions	• Clinical signs • Isolation of *M. bovis* from joint fluid (transport medium is required)
OSTEOMYELITIS	*Salmonella* spp., *Brucella abortus*, *Actinomyces bovis*	Vertebrae are often affected leading to pressure on the spinal cord or extension of infection to meninges. There is often ataxia and eventually hemiplegia or paraplegia depending on the site of the lesion. Lumpy jaw *(A. bovis)* is a specific disease affecting bone and soft tissue in the jaw region	• Clinical signs and X-ray examination • Pathology • Isolation of pathogen from CSF or lesion
TERMINAL DRY GANGRENE	*Salmonella dublin*	Ischaemic necrosis of tips of ears, tail and distal part of hind limbs can follow a few weeks after recovery from acute diarrhoeal disease. The condition characteristically occurs in young calves and is thought to be a localised form of disseminated intravascular coagulation	• History of previous illness • Differentiate from ergotism
TETANUS	*Clostridium tetani*	Tetanus may occur in cows following dystocia, calves can be infected via the umbilicus, castration or dehorning wounds, and occasionally all ages via deep wounds or injections. There is muscle stiffness, raised tail-head, bloat, tetanic spasms and opisthotonos	• History and clinical signs • Gram-stained smear of necrotic tissue deep in wound
THROMBOEMBOLIC MENINGOENCEPHALITIS	*Haemophilus somnus*	Bacterial colonisation of the meningeal vessels produces a thrombolic vasculitis leading to encephalitis and meningitis. There is fever, and with CNS involvement, motor and behavioural abnormalities develop such as stiffness, ataxia, stupor and opisthotonos	• Clinical signs • Pathology • Isolation of *H. somnus* from brain lesions

Table 168. Infectious diseases of cattle. *(continued)*

RESPIRATORY SYSTEM: cattle

Disease	Agent(s)	Comments	Diagnosis
CALF DIPHTHERIA	Trauma (coarse feed) and *Fusobacterium necrophorum*	Necrotic lesions in oral, pharyngeal or laryngeal mucosa. Fever, anorexia and salivation are seen and the condition can lead to pneumonia	• Gram-stained smear on scrapings from lesions • Isolation (anaerobic)
CONTAGIOUS BOVINE PLEUROPNEUMONIA (CBPP)	*Mycoplasma mycoides* ss. *mycoides* (small colony type)	**Acute**: 'Marbling' of lungs and a large volume of fluid in thorax. **Chronic**: Necrosis and walling-off of portion of lungs. These animals can be latent carriers or 'lungers' for up to 3 years. In time the lesion may break down and mycoplasmas are shed.	• Gross pathology and histopathology • Isolation from lung or pleural fluid • Serology: CFT and AGID
ENZOOTIC PNEUMONIA	Complex involves some or all of the following pathogens: Parainfluenzavirus 3 (PI3); bovine respiratory syncytial virus; bovine viral diarrhoea (BVD) virus; infectious bovine rhinotracheitis (IBR); *Mycoplasma bovis; M. dispar; Pasteurella haemolytica; Pasteurella multocida* and *Haemophilus somnus*	Predisposing causes are important in this disease complex and morbidity can reach 100 per cent. Mainly a problem of intensively reared calves 2-6 months of age. There is fever, depression, increased respiratory rate and coughing. Gradual recovery unless a severe bacterial pneumonia develops. Lesions usually in anteroventral portion of lungs	• Isolation and identification of pathogen(s). Pasteurellae are isolated in later stages, by then the isolation of viruses and mycoplasmas is difficult • Serology: for viruses
'FARMER'S LUNG' IN CATTLE	Sensitisation to spores of thermophilic actinomycetes such as: *Micropolyspora faeni*	Immediate type hypersensitivity. Condition due to mouldy hay fed in an enclosed area. Clinical signs of respiratory distress are seen at the end of winter. A few acute cases occur but the condition is usually chronic	• Usually only 1 or 2 acute cases which occur in cattle over 5 years old • Serology: AGID antibodies to *M. faeni*
HAEMORRHAGIC SEPTICAEMIA	*Pasteurella multocida* serotype B:2 (Asia) or E:2 (Africa)	An acute pasteurellosis characterized by a rapid course, oedematous swelling in the head-throat-brisket area, respiratory distress, swollen and haemorrhagic lymph nodes and subserous petechiation. Cattle and water buffaloes are susceptible	• History of endemic area • Isolation of pathogen from heart-blood, liver, spleen or lymph nodes • Serotype identification

Table 168. Infectious diseases of cattle. *(continued)*

RESPIRATORY SYSTEM: cattle

Disease	Agent(s)	Comments	Diagnosis
INFECTIOUS BOVINE RHINOTRACHEITIS (IBR)	Bovine herpesvirus 1	Most common in young cattle under stress such as in feedlots. Nasal discharge, 'red nose', ulcers of mucous membranes in nasal passages, conjunctivitis, mouth breathing. Recovery 4-5 days without secondary bacterial invasion. Abortions 3-4 weeks after respiratory disease in pregnant adults	• Virus isolation or FA: nasal and eye swabs or nasopharyngeal aspirate early in disease • FA: frozen sections from aborted foetuses • Histopathology (IBs transitory, occasionally seen) • Serology: VN or ELISA
INHALATION PNEUMONIA	Anaerobes from rumen.	Following milk fever or general anaesthesia	• History • PM findings
LUNG ABSCESSES	*Actinomyces pyogenes*	Occurs alone or as a complication of other respiratory diseases, especially in pasteurellosis of young cattle	• Direct microscopy • Isolation and identification
MYCOTIC PNEUMONIA	*Aspergillus fumigatus* or *Mortierella wolfii*	Usually chronic except for peracute and fatal *M. wolfii* pneumonia that occurs within 48 hours of abortion . The lungs are wet, oedematous with a large volume of fluid in thorax	• Histopathology • Isolation and identification of pathogenic fungi
SHIPPING FEVER **(Bovine pneumonic pasteurellosis)**	Parainfluenzavirus 3 (PI3); *Pasteurella haemolytica*; *Pasteurella multocida*; *Haemophilus somnus*	Aetiology involves stress + virus + bacteria. All ages can be affected and deaths are usually due to an overwhelming *P. haemolytica* infection. The disease varies from mild pneumonia to a fulminating bronchopneumonia	• Isolation and identification of bacteria • Serology for PI3: IFA, HAI, VN (rising Ab titre)
TUBERCULOSIS (BOVINE)	*Mycobacterium bovis*	Tubercles in lymph nodes, lungs and pleural cavity. 'Open cases' can create an aerosol of *M. bovis*	• Tuberculin test • ZN-smears on tissue • Isolation and identification of organism • Blood-based diagnostic tests currently being evaluated.

SKIN: cattle

Disease	Agent(s)	Comments	Diagnosis
ACTINOBACILLOSIS	*Actinobacillus lignieresii*	Classical disease is a granulomatous infection of the tongue: wooden or timber tongue. But the pathogen can cause granulomatous lesions of skin with purulent exudates from fistulae anywhere on body including the jaw area. Bone not involved and prognosis is usually good	• Microscopy of crushed granules from pus: Gram -ve rods • Culture: aerobic

Table 168. Infectious diseases of cattle. *(continued)*

SKIN: cattle

Disease	Agent(s)	Comments	Diagnosis
BACTERIAL ABSCESSES	*Staphylococcus aureus,* Streptococci, or *Actinomyces pyogenes*	Infection usually via abrasions and other trauma such as injections by a non-aseptic technique	• Gram-stained smear from pus. • Culture for pathogen.
BOVINE PAPILLOMATOSIS	Bovine Papillomaviruses types 1-6 (*Papovaviridae*)	Cutaneous warts (types 1, 2 and 3), teat warts (types 5 and 6) and type 4 occurs in bladder or intestines. Affects all ages but highest incidence in calves and yearlings. Type 1 and 2 warts range from small nodules to cauliflower-like growths. Most common on head and neck. Self-limiting condition	• Clinical signs • EM • Histopathology
BOVINE VIRAL DIARRHOEA IN DAM **(congenital alopecia)**	*Pestivirus* (*Flaviviridae*)	Infection of dam during mid-gestation. The dam may or may not show clinical signs. Congenital defects in calves infected with the virus are various, including cerebellar hypoplasia, cataracts and alopecia. Such calves if born alive may be immunotolerant and persistently infected (V+ve/Ab-ve)	• FA on buffy coat smear for virus, frozen sections • Serology: VN or ELISA • Virus isolation
DERMATOPHILOSIS **(Streptothricosis, Mycotic dermatitis)**	*Dermatophilus congolensis*	Reservoir often small foci on carrier animal. Condition most common in young animals. Predisposing causes include: wet conditions (lesions along backline), abrasions from vegetation (muzzle, face and limbs) and tick infestation (tick predilection sites). Exudative dermatitis with extensive scab formation occurs and scab comes away with tufts of hair leaving a raw bleeding surface	• Gram or Giemsa-stained smears from scab. Filamentous and branching with zoospores at least two across; 'tram-track' appearance • Culture: 10 per cent CO_2
FACIAL ECZEMA	Sporidesmin in spores of *Pithomyces chartarum* (mycotoxin)	Mycotoxicosis of grazing cattle and sheep. Fungus grows in pasture litter under moist, warm conditions. Liver damage and biliary obstruction occurs that restricts excretion of bile pigments and jaundice can occur. Failure to excrete phylloerythrin leads to photosensitization with lesions in non-pigmented skin including the udder and ears	• Spore counts on pasture • Gross pathology: liver • Histopathology
LUMPY JAW	*Actinomyces bovis*	Granulomatous lesions in jaw region with abscesses and fistulous tracts exuding pus. Bone is attacked and once rarefying osteitis becomes extensive the prognosis is poor	• Microscopy of crushed sulphur granules from pus: Gram +ve filaments • Culture: anaerobic

Table 168. Infectious diseases of cattle. *(continued)*

SKIN: cattle

Disease	Agent(s)	Comments	Diagnosis
LUMPY SKIN DISEASE	*Orthopoxvirus* (Neethling virus) (*Poxviridae*)	Limited to Africa. Nodules in skin all over body with general lymphadenitis, oedema of the limbs, nasal discharge and internal organs including lungs are involved. There is fever and anorexia. Mortality only 1-2 per cent but animals remain debilitated for long periods. Nodules ulcerate and heal slowly leaving scars and alopecia	• Histopathology on biopsy: IBs • EM: demonstration of poxvirus particles • Isolation: CAM or TC
PSEUDOLUMPY SKIN DISEASE	Bovine herpesvirus 2 (Allerton virus)	This syndrome seen mainly in Africa but can occur elsewhere. The virus also causes bovine ulcerative mammillitis in Europe and North America. Pseudolumpy skin disease is a comparatively mild disease with fever and skin nodules over the body. The nodules undergo necrosis with central depression but no scar. Mortality very low	• EM • Histopathology on biopsy: IN inclusion bodies • Virus isolation • Serology: VN, CFT, AGID, IFA
PSEUDORABIES ('Mad itch')	Porcine herpesvirus 1	Reservoir is usually pigs with Aujeszky's disease. Rats may take virus from farm to farm. Infection mainly by ingestion and less commonly by inhalation or via wounds (pig bites). Dominant sign is an intense pruritis; mainly flanks and hind limbs. Incessant licking, biting and rubbing so infected areas become abraded. Progressive involvement of CNS with frenzy and bellowing but not aggression. Death may occur within a few hours to a maximum of 6 days after first signs	• History of association with pigs • Clinical signs • Diagnosis of Aujeszky's disease in pigs • FA: frozen sections • Virus isolation
RINGWORM	*Trichophyton verrucosum*	Usually seen in calves or yearlings. The lesions are most common on the face, around the eyes and on the neck. They are circular and later develop a grey - white crust. Self-limiting disease	• Hairs in 10-20 per cent KOH • Culture at 27°C and 37°C, for 6 weeks
SKIN TUBERCULOSIS	Acid-fast bacterium (unnamed)	Chronic indurative nodules associated with the presence of acid-fast bacteria. Usually occurs on the lower limbs. The affected animals often react to the tuberculin test. The acid-fast bacteria have not yet been cultured	• Clinical signs • ZN-stained smears
VESICULAR DISEASES	Foot-and-mouth disease: (*Aphthovirus*, *Picornaviridae*) Vesicular stomatitis: (*Vesiculovirus*, *Rhabdoviridae*)	Vesicular lesions on muzzles, buccal cavity, interdigital cleft and on teats of milking animals	• Viral Ag detection : CFT, ELISA (vesicle fluid or epithelial material) • Virus isolation • Serology: VN, ELISA, CFT

Table 169. Infectious diseases of sheep (goats).

BUCCAL CAVITY: sheep (goats)

Disease	Agent(s)	Comments	Diagnosis
BLUETONGUE	*Orbivirus* (*Reoviridae*)	The highest losses are in growing lambs. There is depression, fever, oedema of lips, tongue and throat. The buccal mucosa is erythematous or cyanotic and erosions appear on the dental pad, tongue, gums and lips. The muzzle is usually encrusted. Stiffness and lameness occur	• History of endemic area • Virus isolation • Serology: VN, ELISA, AGID, modified CFT
CONTAGIOUS PUSTULAR DERMATITIS (ORF)	*Parapoxvirus* (*Poxviridae*)	Scab formation is usually restricted to the lips and nostrils but, as there is no colostral immunity, the disease can be severe in young lambs and kids. With stiff, sensitive lips and sometimes lesions in the buccal cavity, the young animals cannot suck or graze	• Clinical signs • EM for parapoxvirus particles • Histopathology on biopsy. • Virus isolation
FOOT-AND-MOUTH DISEASE	*Aphthovirus* (*Picornaviridae*)	Lameness, usually in all four feet due to vesicular lesions of the interdigital clefts, is the constant finding in sheep. Occasionally vesicular lesions occur in the buccal cavity	• Clinical signs • Viral Ag: CFT, ELISA on vesicular fluid • Virus isolation • Serology: ELISA, CFT, VN
PESTE DES PETITS RUMINANTS	*Morbillivirus* (*Paramyxoviridae*)	Highly contagious, systemic disease of sheep and goats (West Africa), although many infections are subclinical. Signs include fever, anorexia, necrotic stomatitis with gingivitis and diarrhoea. Mortality in goats can reach 95 per cent, sheep slightly less susceptible	• Gross and histopathology • AGID and CFT on lymph node biopsies for Ag • Virus isolation: it is distinguished from rinderpest virus using cross-neutralisation tests • Serology: VN, HAI, AGID
RINDERPEST	*Morbillivirus* (*Paramyxoviridae*)	Sheep and goats are susceptible to infection but disease is usually mild. A few large and serious outbreaks have been reported with signs similar to those in cattle: severe diarrhoea and shallow erosions of lips, dental pads and gums	• AGID and CFT on lymph node biopsies for Ag • Histopathology • Virus isolation • Serology: CFT, ELISA, VN, HAI
SHEEP POX (GOAT POX)	*Capripox* (*Poxviridae*)	The disease affects all ages. There is fever, rhinitis, anorexia and generalized pox eruptions on skin and mucosa of buccal cavity and pharynx within 1-2 days of first signs. Caseous nodules and catarrhal pneumonia occur in the lungs. Mortality varies from 5-50 per cent	• History of endemic area and clinical signs • EM • AGID for viral Ag • Histopathology • Virus isolation

Table 169. Infectious diseases of sheep (goats). *(continued)*

GASTROINTESTINAL TRACT: sheep (goats)

Disease	Agent(s)	Comments	Diagnosis
COLIBACILLOSIS AND COLISEPTICAEMIA	*Escherichia coli*	Most commonly seen in neonatal lambs in crowded lambing sheds. Predisposing causes: mismothering or other causes of insufficient colostrum. Acute diarrhoea, septicaemia and sudden deaths may occur	• History of predisposing causes • Isolation of *E. coli* • Tests for enterotoxigenicity
CRYPTOSPORIDIOSIS	*Cryptosporidium* sp.	Lambs 7-10 days of age are affected. There is dullness, anorexia and diarrhoea but no fever. Death can occur in 2-3 days and the survivors may be unthrifty. Not common in lambs	• Safranin-methylene blue stained smears of faeces • Auramine-O technique on faecal smears
JOHNE'S DISEASE	*Mycobacterium paratuberculosis*	Main sign is emaciation. As diarrhoea may not occur in sheep and goats, the disease is often not diagnosed. Mesenteric lymph nodes are best specimens for culture, rather than ileocaecal valve	• ZN stained smear of faeces, mucosal scrapings or lymph nodes • Culture from mesenteric lymph nodes
LAMB DYSENTERY	*Clostridium perfringens* type B	A clostridial enterotoxaemia with death a few hours after signs of dysentery, abdominal pain and continuous bleating. Seen in newborn lambs up to 3 weeks of age	• Clinical signs • Gross pathology and histopathology • Mouse neutralisation test for toxins in small intestine
RINDERPEST AND PESTE DES PETITS RUMINANTS	Closely related morbilliviruses *(Paramyxoviridae)*	Viruses have an affinity for lymphoid tissue and epithelium of intestines. There is fever, necrotic stomatitis and gastroenteritis. Infections can be subclinical in peste des petits ruminants, but not in rinderpest. Rinderpest vaccines protect against both diseases	• Pathology • Virus isolation • Detection of viral antigen in tissues: AGID and CFT • Serology: AGID, HAI
ROTAVIRUS INFECTION	*Rotavirus (Reoviridae)*	Occurs in neonatal to 4-week-old lambs. Intestinal villi are shortened and this affects lactose digestion. Usually recover if no complications such as a concurrent *E. coli* infection	• Electron microscopy on faeces • ELISA for Ag capture • Virus isolation and identification
SALMONELLOSIS	*Salmonella* spp.	The disease can be a problem in young lambs, adult ewes in late pregnancy or as a flock problem after mustering. Acute diarrhoea / dysentery, septicaemia and/or abortion may occur	• Clinical signs • Gross and histopathology • Isolation of salmonellae from faeces or bone marrow

Table 169. Infectious diseases of sheep (goats). *(continued)*

GASTROINTESTINAL TRACT: sheep (goats)

Disease	Agent(s)	Comments	Diagnosis
'WATERY MOUTH' OF LAMBS	Associated with *E. coli*	Occurs in lambs up to 3-days-old and almost always in those born in enclosed lambing pens. The lamb is dull and ceases to suck. There is saliva-drooling, abomasal tympany and a bloated appearance. Death occurs within 6-24 hours. Thought to be due to endotoxaemia	• Clinical signs • History of colostrum deprivation • Isolation of *E. coli* from body organs

LIVER: sheep (goats)

Disease	Agent(s)	Comments	Diagnosis
BACILLARY HAEMOGLOBINURIA	*Clostridium haemolyticum*	Less common in sheep than in cattle. Liver infarcts are pale, raised and surrounded by a bluish-red zone	• History of a liver fluke area • PM findings • FA on smears from liver lesion
FACIAL ECZEMA (MYCOTOXICOSIS)	Sporidesmin in spores of *Pithomyces chartarum*	Mycotoxicosis with bile duct obstruction, liver damage, jaundice and photosensitization (non-excretion of phylloerythrin). Photosensitisation seen on non-woolled areas of face and ears. There is erythema and oedema of affected skin. Livers are initially enlarged and icteric and later atrophied and fibrotic	• Spore count on pasture • Gross and histopathological appearance of livers
INFECTIOUS NECROTIC HEPATITIS (BLACK DISEASE)	*Clostridium novyi* type B	Endogenous infection with spores present in liver tissue. Migration of liver fluke larvae produces conditions suitable for spore germination and toxin production. Usually sudden death. Postmortem findings include greyish-yellow foci in liver, excess fluid in body cavities and extensive blood-stained oedema under the skin. The skin appears black (black disease)	• History of a liver fluke area • Clinical or postmortem signs • FA on smears from lesions • Demonstration of alpha toxin in body cavity fluid or liver lesion.
LISTERIOSIS (SEPTICAEMIC/ VISCERAL)	*Listeria monocytogenes*	Occurs in lambs and kids. There is fever, anorexia, depression and death in 1-3 days. Necrotic foci occur in the liver and other body organs	• Gram-stained smears from lesions • Culture of the pathogen
OVINE GENITAL CAMPYLOBACTERIOSIS ('VIBRIOSIS')	*Campylobacter fetus* subsp. *fetus* (*C. jejuni*)	Round (1-3 cm), grey necrotic foci in liver of aborted foetus are pathognomonic, if present. Abortion usually occurs in the last 8 weeks of pregnancy	• Characteristic lesions in foetus • Direct microscopy on foetal abomasal contents • Isolation from abomasal contents
RIFT VALLEY FEVER	*Phlebovirus* (*Bunyaviridae*)	South and East Africa: sheep, goats, cattle and man. Syndromes: • Hepatitis and high mortality in lambs, kids and calves • Severe disease and 90-100 per cent abortion in sheep and goats • Mild or subclinical in cows but 100 per cent abortions Vector: mosquitoes	• Histopathology: liver necrosis • Virus isolation: lab. animal or TC • Serology: VN, CFT,ELISA, HAI

Table 169. Infectious diseases of sheep (goats). *(continued)*

LIVER: sheep (goats)

Disease	Agent(s)	Comments	Diagnosis
WESSELSBRON DISEASE	*Flavivirus* (*Flaviviridae*)	Clinical disease and epidemiology (mosquito-borne) resemble Rift Valley Fever. Occurs in sub-Saharan Africa. Sheep are the most susceptible species, with signs of fever, hepatitis with jaundice, subcutaneous oedema and abortion. Mortality high in pregnant ewes and neonatal lambs. Cattle, horses and pigs are infected subclinically	• History of endemic area and clinical signs • Gross and histopathology • Virus isolation : i/c inoculation of mice • Serology: VN, CFT, HAI

GENITAL SYSTEM: sheep (goats)

Disease	Agent(s)	Comments	Diagnosis
AKABANE DISEASE	*Bunyavirus* (*Bunyaviridae*)	Occurs in Japan, Australia, South Africa and Israel. Teratogenic virus (congenital abnormalities). Severe damage to foetus leads to abortion, otherwise arthrogryposis and / or hydranencephaly occur. Vector: mosquitoes and gnats	• Virus isolation: suckling mice or TC • Serology: VN
BLUETONGUE	*Orbivirus* (*Reoviridae*)	Incubation period 6-8 days. Strains of virus vary greatly in virulence. Seasonal incidence dependent on vector. Acute disease: hyperaemia and swelling of mucosa of lips, tongue and dental pad, laminitis with severe lameness, fever and leukopenia. Abortion and congenital abnormalities can also occur. Vector: *Culicoides imicola*	• Virus isolation: i/v inoculation of 10-12 day eggs • Serology: ELISA, VN, AGID, modified CFT
BORDER DISEASE ('HAIRY SHAKER' DISEASE)	*Pestivirus* (*Flaviviridae*)	Infection in pregnant ewes: 1 Abortion and/or congenital abnormalities 2 Birth of 'hairy shaker' lambs, ataxia and muscle tremors due to defective myelination of CNS nerve fibres 3 Normal but immunotolerant lambs that can shed virus	• Virus isolation: blood clots or tissue • FA: frozen sections
BRUCELLOSIS (OVINE)	*Brucella ovis*	Unique to sheep. Transmission venereal, ram-ram or ram-ewe-ram. Infected rams shed brucellae in semen intermittently for 4 years or more. Syndromes are: • Epididymitis, orchitis and impaired fertility in rams • Occasionally placentitis and abortion in ewes	• Isolation of *B. ovis* from semen, foetus or placenta • Direct microscopy (MZN-stained smears): foetal abomasal contents or placenta • Serology: CFT
	Brucella abortus or *B. melitensis*	Ingestion main route but infection via vagina and conjunctiva can occur. Abortions usually in 4th month of pregnancy. Orchitis and arthritis occur rarely. Organisms shed in milk, placental fluids and placenta. Infection with *B. abortus* usually occurs due to contact with cattle. Goats are natural host of *B. melitensis*	• Direct microscopy (MZN-stained smears) • Isolation of brucellae

Table 169. Infectious diseases of sheep (goats). *(continued)*

GENITAL SYSTEM: sheep (goats)

Disease	Agent(s)	Comments	Diagnosis
ENZOOTIC ABORTION OF EWES (CHLAMYDIAL ABORTION)	*Chlamydia psittaci*	Scotland, Germany, USA and South Africa. Lambs may become infected soon after birth (oral) and infection remains latent until conception. The agent invades placenta and abortion occurs. Ewes affected in late pregnancy may not abort until the next gestation. No systemic effect on ewe but placentitis and thickening of intercotyledonary areas occurs. Solid flock immunity does not develop	• Isolation • Impression smears • Serology: CFT, IFA, ELISA
EPIDIDYMITIS (SUPPURATIVE)	*'Actinobacillosis seminis'*	An acute unilateral (occasionally bilateral) suppurative epididymitis of young rams. In advanced cases the contents of the scrotal sac consists largely of purulent material. Polyarthritis may also occur	• Clinical signs • Isolation and identification of the pathogen
LISTERIOSIS	*Listeria monocytogenes* (*L. ivanovii:* abortion only)	Syndromes include: • Visceral or septicaemic listeriosis in lambs and kids • Neural listeriosis (circling disease) • Abortions in cattle, sheep and goats (humans) Sporadic and late in gestation. No illness in dam except, rarely, a fatal septicaemia secondary to a metritis. Infection may be associated with silage feeding	• Isolation of listeriae from foetus or placenta • Histopathology
NAIROBI SHEEP DISEASE	*Nairovirus (Bunyaviridae)*	Central Africa: sheep and goats. Acute disease with fever, haemorrhagic gastroenteritis and nasal discharge. Mortality 30-90 per cent. Abortions in ewes. Vector: *Rhipicephalus* spp.	• Histopathology • Virus isolation: TC or mouse inoculation • AGID: detection of viral antigen • Serology: VN, CFT, IFA
OVINE GENITAL CAMPYLOBACTERIOSIS ('VIBRIOSIS')	*Campylobacter fetus* ss. *fetus* or *Campylobacter jejuni*	Oral route of transmission. Carrier animals main source of infection. Incubation period 10-50 days. After an outbreak of abortion there is flock immunity for 3 years but 10-70 per cent abortion rate in susceptible, introduced ewes. Abortion usually occurs in last 8 weeks of pregnancy. No illness in ewe and fertility normal at subsequent mating. Round (1-3 cm) grey necrotic foci can occur in foetal livers that are pathognomonic, if present.	• Gross pathology: foetal liver lesions • Direct microscopy: DCF or FA technique (foetal abomasal contents) • Isolation of pathogen

Table 169. Infectious diseases of sheep (goats). *(continued)*

GENITAL SYSTEM: sheep (goats)

Disease	Agent(s)	Comments	Diagnosis
Q-FEVER	*Coxiella burnetii*	Infection is usually subclinical in cattle, sheep and goats but anorexia and abortion can occur rarely in sheep and goats. The organism is shed in placental fluids at parturition and is excreted in milk. A biological cycle occurs in ticks. Humans are most commonly infected by aerosols and less commonly by ingestion of the pathogen in milk	• Serology: CFT, indirect FA • Giemsa-stained smears from placenta • Culture in fertile eggs. Laboratory culture is a human health hazard
RIFT VALLEY FEVER	*Phlebovirus (Bunyaviridae)*	South and East Africa: sheep, goats, cattle and man. Syndromes: • Hepatitis and high mortality in lambs, kids and calves • Severe disease and 90-100 per cent abortion in sheep and goats • Mild or subclinical in cows but 100 per cent abortions Vector: mosquitoes	• Histopathology: liver necrosis • Virus isolation: lab. animal or TC • Serology: VN, CFT, ELISA, HAI
SALMONELLOSIS	*Salmonella* spp.	Infection most common in sheep after mustering or during periods of drought and feed shortages. Abortions may occur with enteric infections but some serotypes can cause abortion without apparent clinical signs in ewes	• Isolation of salmonellae from foetus or placenta
TICK-BORNE FEVER	*Ehrlichia (Cytoecetes) phagocytophila*	Occasionally abortions and stillbirths in ewes. Occurs in Europe. Vector: *Ixodes ricinus*	• Direct microscopy: Giemsa-stained smears (blue coccobacilli in cytoplasm of neutrophils)
TOXOPLASMOSIS	*Toxoplasma gondii*	Congenital toxoplasmosis is a major cause of abortion in sheep (occasionally goats and pigs). Source of infection: oocytes in cat faeces on pasture or feed. Oocytes are resistant and survive for over 1 year. Lesions on foetal cotyledons are characteristic: white foci of necrosis up to 2 mm. Flock immunity develops after infection	• Gross lesions on cotyledons • Histopathology • Serology: IFA, IHA
WESSELSBRON DISEASE	*Flavivirus (Flaviviridae)*	Mosquito-borne infection of sheep, cattle and man in southern Africa. The virus is both teratogenic and abortigenic. Occasionally abortion storms in ewes and hepatitis in neonatal lambs may mimic Rift Valley fever	• Histopathology: IBs in hepatocytes • Virus isolation from liver or brain • Serology: ELISA, VN

Table 169. Infectious diseases of sheep (goats). *(continued)*

URINARY SYSTEM: sheep (goats)

Disease	Agent(s)	Comments	Diagnosis
KIDNEY ABSCESSES	*Staphylococcus aureus*, Streptococci, or *Actinomyces pyogenes*	In tick pyaemia of lambs (*S. aureus*), abscesses may occur in many body organs	• Tick pyaemia: rough grazing with tick infestation (*Ixodes* spp.) • Postmortem findings • Isolation and identification of bacterial pathogen
OVINE POSTHITIS ('PIZZLE ROT')	*Corynebacterium renale*	Condition is precipitated by high urinary urea. Ulcers around preputial orifice and a brown crust develops. Total occlusion of the preputial orifice can occur, if the condition is not treated	• Clinical findings • Isolation of principal pathogen
PULPY KIDNEY	*Clostridium perfringens* type D	One feature of this enterotoxaemia is the rapid autolysis of the kidneys after death	• History of unvaccinated animal and/or lush pasture • Clinical signs or sudden death • Demonstration of epsilon toxin in ileal contents • Histopathology: focal symmetrical encephalomalacia (pathognomonic when present)
'WHITE SPOTTED KIDNEY' IN LAMBS	*Escherichia coli*, *Leptospira interrogans* serovars or other bacteria	Focal interstitial nephritis. Often only discovered at slaughter	• Attempted isolation of the bacterial pathogen • Dark-field microscopy may reveal leptospires in kidneys • Histopathology

EYES AND EARS: sheep (goats)

Disease	Agent(s)	Comments	Diagnosis
CONTAGIOUS AGALACTIA (SHEEP AND GOATS)	*Mycoplasma agalactiae*	Ingestion is the main route of entry. There is a transient 'bacteraemia' and the mycoplasma localises in eyes, joints, lungs and pleura. Signs include conjunctivitis, fever, arthritis and mastitis. The eye lesions can vary from conjunctivitis to hypopyon and rarely perforation of the eyeball	• Clinical signs in an endemic region • Isolation of pathogen • Serology: ELISA, CFT

Table 169. Infectious diseases of sheep (goats). *(continued)*

EYES AND EARS: sheep (goats)

Disease	Agent(s)	Comments	Diagnosis
INFECTIOUS OVINE KERATOCONJUNCTIVITIS	*Mycoplasma conjunctivae (Acholeplasma oculi, Chlamydia psittaci)*	Initial signs are conjunctivitis, lacrimation, blepharospasm or blinking. The disease may progress to involve the cornea, usually at the periphery, but can be more extensive. Ulceration, if it occurs, is usually superficial and complete healing results	• Clinical signs • Isolation of pathogen(s). Swab rubbed vigorously over conjunctiva of early cases and placed in mycoplasmal and chlamydial transport medium

NERVOUS SYSTEM : sheep (goats)

Disease	Agent(s)	Comments	Diagnosis
ABSCESSES (BRAIN OR SPINAL)	*Staphylococcus aureus*	Most common as a sequel of tick pyaemia. Usually occurs in lambs under 1 year of age. Variety of clinical signs occur, depending on the location of the abscess and its size	• Gross and histopathology • Isolation of pathogen
BOTULISM	*Clostridium botulinum*	Sheep do not show the typical flaccid paralysis until the final stages of the intoxication. There is stiffness when walking, inco-ordination and some excitement in the early stages. The head tends to bob up and down when walking. In terminal stages there is abdominal breathing, limb paralysis and death	• History: feedstuffs • Clinical signs • Attempt demonstration of toxin from serum by mouse inoculation
CAPRINE ARTHRITIS-ENCEPHALITIS (CAE)	*Lentivirus, (Retroviridae)*	Three main syndromes: • Encephalomyelitis in kids 2-4 months of age. Kids show progressive wasting, tremors and a poor hair coat but are alert, afebrile and have a normal appetite. There is a progressive ascending paralysis with paresis in hind limbs seen first. Terminally, after a course of months, there is deviation of head and neck, paddling, paralysis and death • Arthritis and indurative mastitis in adults (2-9 years old) • Encephalitis in adults (1-5 years old) manifested as a slow progressive paralysis. Least common syndrome. It is often associated with interstitial pneumonia	• Histopathology: focal malacia of white matter and demyelination • Serology: ELISA, AGID • Virus isolation: buffy coat
COENUROSIS (STURDY OR GID)	*Taenia multiceps (Coenurus cerebralis)*	Eggs in canine faeces are ingested by sheep. Embryos from eggs pass from intestines to blood stream. Only those embryos that lodge in brain or spinal cord survive, where they mature and cause clinical signs in sheep (occasionally cattle). These include unilateral blindness, ataxia, tremors, deviation of head, circling, excitement and collapse. If the skull is palpated caudal to the horn buds, bone rarefaction may be detected. Surgery is possible in these cases	• Clinical signs: palpate skull • Necropsy findings

Table 169. Infectious diseases of sheep (goats). *(continued)*

NERVOUS SYSTEM: sheep (goats)

Disease	Agent(s)	Comments	Diagnosis
CONGENITAL CNS LESIONS	Akabane virus *(Bunyaviridae)*	Hydranencephaly and arthrogryposis (infected mid - pregnancy)	• Virus isolation • Serology: VN
	Border disease virus *(Flaviviridae)*	Hairy shaker lambs with excessively hairy, pigmented fleece and tremors due to defective myelination of CNS nerve fibres. A few recover but are carriers of virus	• Virus isolation. • FA: frozen sections
FOCAL SYMMETRICAL ENCEPHALOMALACIA	*Clostridium perfringens* type D	Occurs in pulpy kidney disease. The lambs are often found dead with the peracute form. In subacute intoxications with the epsilon toxin, there can be salivation, abdominal pain, excitement followed by depression, head-pressing, recumbency, paddling, opisthotonos and coma. Symmetrical haemorrhages and malacia are found in white matter of brain	• History: vaccination? • Histopathology • Mouse neutralisation: demonstration of epsilon toxin in SI contents
LISTERIOSIS	*Listeria monocytogenes*	Circling disease in sheep, goats and cattle that has a rapid course of 4-48 hours. Often associated with silage feeding. Head is held to side of lesion and circling occurs in one direction. There is fever, hemiplegia of facial muscles, ataxia, blindness, depression and paralysis. Other syndromes: abortion and septicaemic (visceral) listeriosis in young animals including lambs and kids	• Clinical and history • Histopathology • Isolation from brain (cold enrichment)
LOUPING ILL	*Flavivirus*, (*Flaviviridae*)	Seen in sheep under 2 years old in endemic areas as the older sheep are immune. Goats can excrete the virus in high titre in milk. Following a viraemia there can be a subclinical infection or CNS signs. The latter include fine muscular tremors, nervous nibbling, hind limb inco-ordination (louping gait), in severe cases collapse and death in 1-3 days. Non-suppurative polioencephalomyelitis is present. Vector: *Ixodes ricinus.* Reservoir: grouse.	• Histopathology: brain • Virus isolation: i/c inoculation of mice or TC • Serology: HAI, AGID, CFT, VN and IFA
POLIOENCEPHALO-MALACIA (CEREBROCORTICAL NECROSIS)	Bacterial thiaminases in rumen and intestines, or dietary thiaminases.	Sporadic occurrence in young sheep. Sudden onset of blindness, opisthotonos and convulsions. There is a rapid response to thiamine injections early in condition, Acute cerebral oedema and laminar necrosis of cerebral cortex occur. Condition seen in goats (2-36 months of age) and is associated with milk-replacer diets in kids and concentrate feeding in older goats	• Dietary causes? • Response rapid to thiamine injection in early cases • Histopathology

Table 169. Infectious diseases of sheep (goats). *(continued)*

NERVOUS SYSTEM: sheep (goats)

Disease	Agent(s)	Comments	Diagnosis
PSEUDORABIES	Porcine herpesvirus 1	Rare in sheep probably because of husbandry methods. Clinically the disease is similar to that in cattle with intense pruritis and a short course. Reported in goats housed with infected pigs. There may be no pruritis in goats. Goats lie down and get up frequently and there is plaintive crying, profuse sweating, spasms, paralysis and death, which is rapid	• History of association with pigs and diagnosis of Aujeszky's disease in pigs. • Virus isolation. • FA: cryostats of brain tissue.
RABIES	*Lyssavirus* (*Rhabdoviridae*)	In sheep the disease tends to occur in a number of animals at the same time, resulting from an attack on the flock by a rabid dog or fox. Incubation period is from 2 weeks to several months. It is clinically similar to that in cattle. Most affected sheep have dumb or paralytic rabies, they are quiet, anorexic, with salivation and muscle tremors. A minority show aggression and violent exertion. There is no excessive bleating. In goats, aggression and continuous bleating is more common	• FA for Ag in brain • Histopathology of brain: Negri bodies
SCRAPIE	A prion	Incubation period several months to 3 years and the course of disease averages about 6 months. In early stages there is a stilted gait, loss of condition but no fever and normal appetite. Intense pruritis develops, always bilateral and usually affecting rump, thighs and tail base. There is loss of fleece from these areas. Small stimuli cause the nibbling reflex, there are fine muscle tremors, incoordination and a staring gaze. Terminally there is extreme emaciation, sternal and then lateral recumbency and death	• Histopathology: astrocytosis, vacuolation of neurons • EM: scrapie-associated fibrils in brain tissue
TETANUS	*Clostridium tetani*	Entry is often via navel, castration and docking wounds, uterus after dystocia or deep puncture wounds. Average incubation period is 10-14 days, then stiffness of muscles and hyperaesthesia are seen. Mild stimuli cause sheep or goats to fall to the ground with tetanic spasms and opisthotonos. Mortality is about 80 per cent	• History • Clinical • Gram-stained smear from deep in wound
TREMORGEN STAGGERS	Tremorgen-producing *Penicillium, Aspergillus* spp. or other fungi.	Stiffness, ataxia and tremor in large muscle masses. The animals fall and have convulsions if chased or are excited. Recovery usually complete on removal from pasture, where the tremorgen-producing fungi are growing in the pasture litter	• Recovery when moved from suspect pasture

Table 169. Infectious diseases of sheep (goats). *(continued)*

NERVOUS SYSTEM: sheep (goats)

Disease	Agent(s)	Comments	Diagnosis
VISNA	Maedi-visna virus *Lentivirus,* *(Retroviridae)*	Visna is a wasting disease with meningoencephalitis and an incubation period of 2 years. There is an insidious onset with a progressive course, weight loss, paresis leading to paralysis and death. It is non-febrile and the course can be 1-2 years. Only sheep are affected	• Histopathology: brain • CSF: mononuclear cells 200/ml • Serology: ELISA, AGID

MUSCULOSKELETAL SYSTEM: sheep (goats)

Disease	Agent(s)	Comments	Diagnosis
BLACKLEG	*Clostridium chauvoei*	Occurs in all ages and usually follows trauma such as docking, castration or vulval damage from parturition. There can be extensive local lesions and severe lameness if limb muscles are involved. Fever, anorexia and depression occur and often rapid death	• History of trauma and clinical signs • FA on smears from affected tissue or bone marrow
BOTULISM	*Clostridium botulinum*	Sheep tend not to show typical flaccid paralysis until the final stages of the disease. Stiffness, ataxia and salivation occur and in terminal stages there is limb paralysis, abdominal breathing and rapid death	• History of contaminated foods or eating carcases of small rodents (protein or phosphorus deficiency) • Demonstration of toxin in serum
CAPRINE ARTHRITIS-ENCEPHALITIS (GOATS ONLY)	*Lentivirus* (*Retroviridae*)	Chronic arthritis is the most common sign in adult goats. Carpal joints often involved and lameness is usually intermittent	• Serology: ELISA and AGID tests (seropositive 4-5 months after infection) • Virus isolation: buffy coat
CHLAMYDIAL POLYARTHRITIS	*Chlamydia psittaci*	Can be common in lambs in feedlots and also at pasture. There is stiffness, lameness, depression, fever and conjunctivitis. Signs relating to lesions in other body systems may also occur. Morbidity rate is up to 80 per cent but mortality very low	• Clinical signs • Examination of joint fluid for elementary bodies or inclusions • Isolation of pathogen
CONTAGIOUS AGALACTIA (SHEEP AND GOATS)	*Mycoplasma agalactiae*	The main signs in this generalised disease include fever, conjunctivitis, arthritis and mastitis. Morbidity about 25 per cent but mortality is low	• Clinical signs in an endemic area • Isolation of pathogen • Serology: ELISA, CFT
CONTAGIOUS (VIRULENT) FOOT ROT	*Bacteroides nodosus,* *Fusobacterium necrophorum,* *Actinomyces pyogenes* and a motile anaerobe	Lameness present early when only other signs are warmth and tenderness of claw and mild inflammatory reaction in inner aspect of digit. Separation of sole occurs soon, first at the heel. There is necrosis of soft tissues under the horny sole and a little dark, watery, foetid pus is present between the layers of separating tissue. If untreated, involvement of deep tissues and joints occurs and claw becomes misshapen	• Clinical signs • Gram-stained smear of pus • Isolation of *B. nodosus* if necessary

Table 169. Infectious diseases of sheep (goats). *(continued)*

MUSCULOSKELETAL SYSTEM: sheep (goats)

Disease	Agent(s)	Comments	Diagnosis
ERYSIPELAS IN SHEEP	*Erysipelothrix rhusiopathiae*	***Polyarthritis in lambs:*** non-suppurative polyarthritis and signs appear about 14 days after birth (navel) or docking. Sudden lameness but minor swelling of affected joints. Can involve up to 50 per cent of the flock and chronic disease occurs in about 5 per cent of affected lambs.	• Clinical signs • Isolation of pathogen from aspirated joint fluid
		Post-dipping lameness (cellulitis/laminitis): infection from contaminated sheep-dip through skin abrasions. A cellulitis with extension to laminae of feet occurs but without joint involvement. Severe lameness with one or more feet affected. Legs are hot and slightly swollen often to the metacarpus. Recovery good with treatment	• History and clinical signs • Isolation of pathogen from aspirated joint fluid
FOOT ABSCESS	*Fusobacterium necrophorum* and *Actinomyces pyogenes*	Usually confined to one foot. Acute lameness is the first sign and affected claw is hot and painful. Paring of horn of sole liberates pus, otherwise the pus may break-out at the coronet	• Clinical signs • Gram-stained smear of pus
JOINT-ILL	*Staphylococcus aureus, Escherichia coli, Actinomyces pyogenes,* streptococci or others	Localisation of pathogens in joints following a navel infection or bacteraemia / septicaemia	• Clinical signs • Isolation of pathogen from aspirated joint fluid
LAMENESS (GENERALISED DISEASES)	Foot-and-mouth disease *(Aphthovirus)*	Acute lameness in all four feet. Vesicular lesions in interdigital cleft	• Tests for viral antigen detection • Serology for antibodies for virus, in each disease
	Bluetongue (*Orbivirus*)	In acute disease laminitis and coronitis, manifested by lameness and recumbency, occur in a few sheep. A dark purple band in the skin just above the coronet is diagnostic. In endemic areas milder disease can be seen such as "range stiffness in lambs" (Texas)	
MAEDI-VISNA (SHEEP ONLY)	*Lentivirus* (*Retroviridae*)	Arthritis may occur in visna together with meningo-encephalitis and wasting	• Pathology • Serology: AGID, ELISA
MALIGNANT OEDEMA	*Clostridium septicum* *C. novyi* type A	Entry is usually through wounds, signs appearing 12-48 hours after infection. There is a high fever, toxaemia, weakness, often lameness and a soft doughy swelling at site of infection. Course short (1-2 days) and is invariably fatal	• History of trauma and clinical signs • FA on smears from affected tissue (recently dead animals, as *C. septicum* is a rapid postmortem invader)

Table 169. Infectious diseases of sheep (goats). *(continued)*

Disease	Agent(s)	Comments	Diagnosis
MUSCULOSKELETAL SYSTEM: sheep (goats)			
MYCOPLASMAL ARTHRITIS (MAINLY IN GOATS)	*Mycoplasma capricolum* and some other spp.	Infection in goats results in septicaemia with resulting pneumonia and arthritis. It is manifested in 3-8 week-old kids by lameness, recumbency, diarrhoea and fever. *M. capricolum* is also a cause of arthritis in sheep	• Isolation of the mycoplasmas from joint fluid. Transport medium is desirable
OSTEOMYELITIS	*Actinomyces bovis* or *Corynebacterium pseudotuberculosis*	Non-specific infection by haematogenous route following omphalitis, castration or docking wounds. Not common in sheep	• Clinical and X-ray • Isolation of the pathogen
OVINE INTERDIGITAL DERMATITIS ('SCALD')	*Fusobacterium necrophorum* and *Actinomyces pyogenes*	One or all four feet can be affected. In mild cases, the interdigital skin is reddened, swollen and often covered by a moist film of necrotic material. In severe cases the skin is necrotic and eroded so that sensitive subcutaneous tissues are exposed. Lameness is present	• Clinical signs. • Gram-stained smear of necrotic material
PROLIFERATIVE DERMATITIS ('STRAWBERRY FOOT ROT')	*Dermatophilus congolensis*	Small heaped-up scabs appear from coronet to knee or hock. They enlarge to 3-5cm and become thick and wart-like. If scabs are removed, a bleeding, fleshy mass (resembling a strawberry) is seen surrounded by an ulcer. Later the ulcer becomes deeper with pus formation. No pruritis or lameness occurs unless lesions extend to interdigital space. Reported from Scotland and Australia	• Direct microscopy on scab material, using Giemsa or Gram stain • Isolation of pathogen if necessary
TETANUS	*Clostridium tetani*	Infection can be via umbilicus, uterus after dystocia or docking and castration wounds. The clinical signs are similar to those in other species. Course of disease is 3-4 days and mortality in young animals is high	• History and clinical signs • Gram-stained smear of necrotic material from wounds
RESPIRATORY SYSTEM : sheep (goats)			
CAPRINE ARTHRITIS-ENCEPHALITIS (CAE)	*Lentivirus (Retroviridae)*	'Slow' viral disease, transmitted in colostrum or milk. Syndromes include: • Encephalitis in kids 1-4 months-old • Arthritis and indurative mastitis in adults • Encephalitis and interstitial pneumonia in adults (least common) Only a small proportion of infected goats develop clinical disease	• Serology: AGID, ELISA • Virus isolation: buffy coat

Table 169. Infectious diseases of sheep (goats). *(continued)*

RESPIRATORY SYSTEM: sheep (goats)

Disease	Agent(s)	Comments	Diagnosis
CONTAGIOUS CAPRINE PLEUROPNEUMONIA (CCPP)	*Mycoplasma mycoides* ss. *capri* or *Mycoplasma* strain F38	A peracute, acute or chronic disease confined to goats. Similar disease to CBPP but the tendency to form necrotic sequestra (chronic form) is uncommon. Lesions may slowly resolve in surviving animals	• PM findings • Isolation and identification of the mycoplasmas
LUNG ABSCESSES	*Actinomyces pyogenes* or *Corynebacterium pseudotuberculosis*	May not cause clinical signs, frequently only discovered after death	• Gram-stained smear • Culture and identification
MAEDI-VISNA (PROGRESSIVE PNEUMONIA OF SHEEP)	*Lentivirus* (*Retroviridae*).	'Slow' viral disease that is transmitted in milk, colostrum or by aerosol. Long incubation period so rarely seen in sheep <2 years old and is most common in animals over 4 years old. Maedi is the respiratory form and lungs are firm and heavy and do not collapse when thorax is opened	• Gross pathology and histopathology • Serology: ELISA, AGID
PNEUMONIA COMPLEX	Complex involves one or more of the following pathogens: Parainfluenza virus 3 (PI3), *Mycoplasma ovipneumoniae*, *Pasteurella haemolytica*, *Pasteurella multocida*, *Chlamydia psittaci*	Viruses and mycoplasmas cause a relatively mild infection but serious disease with fatalities can occur if pasteurellae are involved as secondary invaders	• Isolation and identification of pathogens • Serology: PI3 (paired serum samples), HAI
PULMONARY ADENOMATOSIS JAAGSIEKTE ('DRIVING SICKNESS')	*Retroviridae*	'Slow' virus disease with a long incubation period. Signs occur in 3-4 year-old sheep. Lesions occur in the lungs and range from small nodules to extensive solid areas that are grey and flat with sharp demarcation. Copious froth in air passages is a characteristic finding	• Clinical signs: "wheel-barrow" test (elevation of hind limbs) • Histopathology • Detection of viral Ag and Ab in lung washings/exudate: ELISA, RIA
SEPTICAEMIC MYCOPLASMOSIS	*Mycoplasma mycoides* ss. *mycoides* (large colony type)	Clinical signs include septicaemia, pneumonia, polyarthritis and conjunctivitis	• PM findings • Isolation and identification of the mycoplasma

Table 169. Infectious diseases of sheep (goats). *(continued)*

SKIN: sheep (goats)

Disease	Agent(s)	Comments	Diagnosis
ACTINOBACILLOSIS (OVINE)	*Actinobacillus lignieresii*	Purulent disease of skin, lymph nodes, lungs and soft tissues of the head and neck in sheep. Most common in head region involving the jaw area	• Microscopy on granules in exudate • Culture: aerobic
BLUETONGUE	*Orbivirus, (Reoviridae)*	Incubation 6-8 days. Signs include hyperaemia and swelling of mucous membranes of lips, dental pads and tongue, cyanosis of tongue, coronitis, laminitis, fever, leukopenia, crusted muzzle and the lips bleed easily. Abortions and congenital abnormalities may occur. Many serotypes and mild and atypical forms of disease can be seen. Vector: *Culicoides* species. Reservoir: cattle and other ruminants	• Virus isolation from early cases • Serology: ELISA, VN, AGID, modified CFT
BORDER DISEASE ('HAIRY SHAKER' DISEASE)	*Pestivirus, (Flaviviridae)*	Dam infected in pregnancy with the border disease virus:- • Early pregnancy: foetal death • Later in pregnancy: congenital defects that involve skin, skeleton and/or CNS. 'Hairy shakers' have excessively hairy and pigmented fleeces, muscle tremors and ataxia. Nervous signs are due to defective myelination of nerve fibres. Some lambs can recover in 3-4 months but they may be immunotolerant carriers (V+ve/Ab-ve)	• Virus isolation: blood clots or tissue • FA: frozen sections
CASEOUS LYMPHADENITIS	*Corynebacterium pseudotuberculosis*	***Superficial form:*** abscessation and rupture of one or more superficial lymph nodes ***Internal form:*** characteristic abscesses in internal lymph nodes and occasionally in organs such as the lungs. These caseous lesions can bear a resemblance to those of tuberculosis. The internal form is often subclinical and the lesions are not discovered until slaughter	• Clinical signs in superficial form • Characteristic encapsulated, laminated abscesses containing greenish pus (postmortem examination in internal form) • Isolation and identification of the pathogen from pus samples
CONTAGIOUS PUSTULAR DERMATITIS (ORF)	*Parapoxvirus (Poxviridae)*	Primary lesions on skin of lips and muzzle. In lambs extension can occur to buccal cavity and in adults to other non-woolled areas such as udder and vulva. Initial papules develop into thick crusts, often friable and bleeding occurs if scabs are removed	• Clinical signs • EM on scab for parapoxvirus particles • Histopathology • Virus isolation

Table 169. Infectious diseases of sheep (goats). *(continued)*

SKIN: sheep (goats)

Disease	Agent(s)	Comments	Diagnosis
DERMATOPHILOSIS (STREPTOTHRICOSIS, MYCOTIC DERMATITIS)	*Dermatophilus congolensis*	***Lumpy wool:*** when woolled areas of body are involved. Scab material bound to wool fibres. Crusts usually on dorsal surface due to moisture from rain ***Strawberry foot rot:*** lesions occur on distal parts of limbs	• Direct microscopy on scab • Culture if microscopic results are equivocal
FACIAL ECZEMA	Sporidesmin in spores of *Pithomyces chartarum*	Mycotoxicosis with bile duct obstruction, liver damage, jaundice and photosensitization (non-excretion of phylloerythrin). Photosensitisation seen on non-woolled areas of face and ears. There is erythema and oedema of affected skin. Livers are initially enlarged and icteric and later atrophied and fibrotic	• Spore count on pasture • Gross and histopathology on livers
FOOT-AND-MOUTH DISEASE	*Aphthovirus, (Picornaviridae)*	Major sign is sudden and acute lameness in several sheep due to interdigital vesicular lesions. Rarely lesions on muzzle or in buccal cavity	• Viral antigen detection: ELISA, CFT (vesicular fluid or epithelial material) • Virus isolation. • Serology: ELISA, CFT, VN
RINGWORM	*Trichophyton verrucosum, Microsporum gypseum* or rarely other dermatophytes	Ringworm is rare in sheep and is usually on non-woolled areas of head or scrotum. High initial dose may be needed, such as housing lambs in sheds previously inhabited by infected calves	• Microscopy: skin scraping in 10 per cent KOH to demonstrate arthrospores • Culture and identification
SCRAPIE	A prion	Sheep and occasionally goats can be affected. Incubation period usually 2-3 years. There is pruritis, fine tremors of head and neck, nibbling movements of lips, weaving gait and staring eyes. Loss of weight and wool loss from rubbing occurs and eventually paralysis and death	• Histopathology: astrocytosis, vacuolation of neurons • EM: scrapie-associated fibrils in brain tissue
SHEEP (GOAT) POX	*Capripoxvirus (Poxviridae)*	Acute, severe, generalized disease with mortality rate 5 - 50 per cent. Widespread skin lesions as raised, circular plaques with red borders. Later necrosis and dark, hard scab formation occurs eventually leaving star-shaped scars and alopecia. Lungs and intestines are involved. Transmission: direct, aerosol or mechanical via biting insects.	• EM on lesions • AGID for viral Ag • Histopathology • Virus isolation

Table 169. Infectious diseases of sheep (goats). *(continued)*

SKIN: sheep (goats)

Disease	Agent(s)	Comments	Diagnosis
STAPHYLOCOCCAL DERMATITIS	*Staphylococcus aureus*	Pustular lesions forming scabs. Predisposing factors, usually present, such as insufficient trough space in concrete feeding troughs or thorny plants in vegetation. Lesions usually periorbital or on muzzle but can occur on other non-woolled areas	• Direct microscopy • Culture and identification
ULCERATIVE DERMATOSIS	Unclassified virus.	Characterised by destruction of epidermal and subcutaneous tissues with development of raw granulating ulcers on skin of lips, nares, feet, legs and male and female external genitalia. Two fairly distinct syndromes affecting (1) external genitalia or (2) other non-woolled areas. In males the glans penis is affected and the preputial orifice can become blocked. Transmission occurs at breeding time. Morbidity rate 15 - 20 per cent (up to 60 per cent) but mortality low	• Clinical signs • Histopathology • Differential diagnosis from orf

Table 170. Infectious diseases of pigs.

Disease	Agent(s)	Comments	Diagnosis
BUCCAL CAVITY: pigs			
NECROTIC STOMATITIS	*Fusobacterium necrophorum*	Necrotic lesions on tongue or in buccal cavity. The predisposing cause is usually trauma	• Clinical signs • Gram or DCF stained smear from necrotic material
VESICULAR DISEASES **Foot-and-mouth disease (FMD)** **Vesicular stomatitis (VS)** **Swine vesicular disease (SVD)** **Vesicular exanthema of swine(VES)**	*Aphthovirus* (*Picornaviridae*) *Vesiculovirus* (*Rhabdoviridae*) *Enterovirus* (*Picornaviridae*) *Calicivirus* (*Caliciviridae*)	Most constant sign in these vesicular diseases affecting pigs is sudden and acute lameness. The lameness is due to vesicular lesions in the interdigital cleft and lesions may or may not be present on the snout and in the buccal cavity	• Clinical signs • Specimens: vesicular fluid and epithelial flap • Demonstration of viral antigen: CFT, ELISA, VN • Virus isolation. • Serology: CFT, ELISA, VN
GASTROINTESTINAL TRACT: pigs			
AFRICAN SWINE FEVER	Unclassified DNA virus	In acute form there is fever, depression, erythema of skin, abortion in sows and a bloody diarrhoea. Mortality high. Post mortem findings include 'blackberry-jam spleen', 'blood-clot mesenteric lymph nodes', haemorrhages in many organs and oedema of gall bladder. In subacute disease the clinical signs and pathology are less marked with a mortality rate of about 5 per cent Vector: *Ornithodoros* ticks	• FA for antigen: frozen sections • Serology: IFA, ELISA • Virus isolation
CLASSICAL SWINE FEVER (HOG CHOLERA)	*Pestivirus* (*Flaviviridae*)	Fever, depression, erythema of skin and nervous signs early in disease. Diarrhoea can be bloody and submucosal haemorrhages widespread ('turkey-egg kidney', 'strawberry lymph nodes'). Spleen is not enlarged but often has infarcts and button ulcers occur in the colon. Pregnant sows abort. Virus strains vary in virulence. Mortality can be high but strains of reduced virulence now appear to be present in Europe	• FA or AGID for viral antigen detection in tissues • Virus isolation • Serology: VN

Table 170. Infectious diseases of pigs. *(continued)*

GASTROINTESTINAL TRACT: pigs

Disease	Agent(s)	Comments	Diagnosis
CLOSTRIDIAL ENTEROTOXAEMIA	*Clostridium perfringens* type C.	Very young piglets 1-4 days old. Claret red diarrhoea is characteristic. Necrosis, haemorrhage and gas bubbles occur in the wall of small intestine. There is blunting of villi. Mortality high and death of whole litter is common	• Pathology • Gram on mucosa of small intestine (recently dead pig): large numbers of clostridia are suggestive of the disease • Mouse neutralisation test
COLIBACILLOSIS	*Escherichia coli*	***Neonatal diarrhoea:*** Profuse watery diarrhoea and dehydration in 1-3-day old piglets. Death of whole litter can occur. There is gastric distention, watery contents in small intestine but villous atrophy is minimal. ***Weanling enteritis:*** Greyish diarrhoea with no blood occurs. There is anorexia, mild fever and loss of condition. Mortality less than 10 per cent but signs persist for some time. Most common of the post-weaning diarrhoeas	• Isolation of *E. coli* • Check for enterotoxigenicity: fimbrial antigens or enterotoxin
OEDEMA DISEASE	*E. coli* that produce the oedema disease enterotoxin (neurotoxin)	The toxin is absorbed from the gut and damages vascular endothelium. Usually seen in well-grown pigs and they are often found dead. If alive, oedema of forehead, eyelids and larynx (hoarse squeal) is observed. Diarrhoea and nervous signs such as ataxia and convulsions may occur. Oedema of stomach wall is the pathognomonic postmortem finding	• Clinical signs • Gross pathology and histopathology • Isolation of *E. coli* (often haemolytic)
PORCINE EPIDEMIC DIARRHOEA	*Coronavirus* (distinct from TGE virus)	Present in some European countries. Signs are most common in post-weaners and adults. Watery diarrhoea occurs with recovery after about 1 week. Mortality 1-3 per cent	• Clinical signs • FA technique on cryostat sections of small intestine • Serology: IFA or ELISA
PROLIFERATIVE HAEMORRHAGIC ENTEROPATHY (INTESTINAL ADENOMATOSIS)	*Campylobacter mucosalis*, *Campylobacter hyointestinalis* or some other *Campylobacter* sp.	Young adults show cutaneous pallor, weakness and black tarry faeces. High morbidity but mortality about 6 per cent. There is proliferation of small intestine mucosa with blood clots in lumen. Anaemia may result. Pigs 6-20 weeks old, if infected, can be subclinical or have watery diarrhoea with mucus and blood. Spontaneous recovery in most but chronic wasting in 1-2 per cent. Organisms are intracellular so faecal specimens are of no diagnostic use	• Stained smears of mucosa • Histopathology: intracellular *Campylobacter* • Isolation is difficult: specialised laboratories

Table 170. Infectious diseases of pigs. *(continued)*

GASTROINTESTINAL TRACT: pigs

Disease	Agent(s)	Comments	Diagnosis
ROTAVIRUS INFECTION	*Rotavirus (Reoviridae)*	Most common in sucking piglets about 2 weeks old. Watery diarrhoea of short duration (3-5 days) is seen. Vomiting if it occurs is less dramatic than in transmissible gastroenteritis. Mortality rare if uncomplicated	• Detection of virus in faeces: EM, ELISA • Virus isolation
SALMONELLOSIS	*Salmonella* spp. commonly *Salmonella choleraesuis* (hog cholera bacillus)	Disease can occur in any age group. In septicaemic form there is fever, depression and erythema of skin. Systemic involvement occurs with necrotic foci in liver, petechiated kidneys and swollen mesenteric lymph nodes. Mortality can be high	• Isolation of *Salmonella* sp. • Differential diagnosis from swine fever and African swine fever
SWINE DYSENTERY	*Serpulina (Treponema) hyodysenteriae*	Only diarrhoeal syndrome to have blood, mucus and necrotic debris in faeces. Only large intestine is involved and the pigs are afebrile. Usually all pigs in a pen are affected. Pigs become gaunt and dehydrated and mortality is 25 per cent in untreated cases	• Clinical signs and appearance of faeces • Histopathology: *Serpulina* in crypts (silver stain) • FA: mucosal smear • Isolation of pathogen
TRANSMISSIBLE GASTROENTERITIS (TGE)	*Coronavirus*	Profuse watery greenish-grey diarrhoea. Vomiting is a prominent sign. Severe villous atrophy but there is no blood in faeces and the pigs are afebrile. Mortality due to dehydration can be 100 per cent in piglets less than 7 days-old. All age groups can be affected in a non-immune herd	• Pathology: wash and examine jejunum and ileum: paper thin • FA: mucosal impression smears or frozen sections • Virus isolation • Serology: VN, ELISA

LIVER: pigs

Disease	Agent(s)	Comments	Diagnosis
LISTERIOSIS (SEPTICAEMIC/ VISCERAL)	*Listeria monocytogenes*	Infection can occur in young pigs. Usual course 1-3 days with fever, depression and death. Necrotic foci in liver and other body organs	• Gram-stained smears from lesions • Culture of pathogen
LIVER LESIONS (BACTERIAL)	A variety of bacteria including: *Erysipelothrix rhusiopathiae,* streptococci or staphylococci	Non-specific foci of infection can occur in the liver and other body organs following a bacteraemia or septicaemia, in pigs and other domestic animals	• Gram-stained smears from lesions • Culture of pathogen

Table 170. Infectious diseases of pigs. *(continued)*

LIVER: pigs			
Disease	**Agent(s)**	**Comments**	**Diagnosis**
PSEUDOTUBERCULOSIS	*Yersinia pseudotuberculosis*	Disease most common in guinea pigs, cage birds, captive deer and pigs. In classical infections there is emaciation, diarrhoea and death in 3-4 weeks. Abscesses in liver, spleen and other organs. Pigs may be infected by eating dead wild birds	• Gram-stained smears from lesions • Culture of pathogen

GENITAL SYSTEM: pigs			
AFRICAN SWINE FEVER	Unclassified DNA virus.	Endemic in Africa, Spain, Portugal and some West Indian islands. Acute disease: fever, erythema of skin, diarrhoea, and abortion in pregnant sows. Biological vector: *Ornithodoros* ticks	• FA: frozen sections • Serology: IFA, ELISA • Virus isolation
AUJESZKY'S DISEASE	Porcine herpesvirus 1 *(Alphaherpesvirinae)*	Transmission by contact or venereal. In susceptible herd:- • Neonates: severe generalized disease and death • Young adults: respiratory disease • Pregnant sows: SMEDI syndrome in 50 per cent of sows • Other animals (abnormal hosts) pseudorabies ('mad-itch')	• Virus isolation • FA: frozen sections or impression smears • Serology: ELISA, VN
BRUCELLOSIS (PORCINE)	*Brucella suis* *(Brucella abortus)*	*B. suis* is the main cause of brucellosis in pigs but sporadic abortion can occur with *B. abortus. B. suis* causes abortions in 50-80 per cent of pregnant sows in primary outbreaks, and reduction of fertility and orchitis in boars	• Direct microscopy: MZN-stained smears • Isolation • Serology: Agglutination tests or CFT
CLASSICAL SWINE FEVER (HOG CHOLERA)	*Pestivirus* *(Flaviviridae)*	A mild fever may be the only sign in sows but:- • Abortions, stillbirths and mummification • Congenital abnormalities in piglets • Immunotolerant piglets born that excrete virus	• FA or AGID for viral antigen detection in tissues • Virus isolation • Serology: VN
ENTEROVIRUS INFECTION (PORCINE)	Porcine enteroviruses 2-11. *(Picornaviridae)*	Viruses frequently isolated from normal pigs but also from aborted and stillborn foetuses. Disease producing status is uncertain. The term SMEDI viruses was given to these viruses but porcine parvovirus now considered to be more important in the SMEDI syndrome	• Serology: VN (pre-colostral sera) • Virus isolation

Table 170. Infectious diseases of pigs. *(continued)*

GENITAL SYSTEM: pigs

Disease	Agent(s)	Comments	Diagnosis
INCLUSION BODY RHINITIS	Porcine herpesvirus 2 *(Betaherpesvirinae)*	Subclinical disease in herds where endemic. In susceptible herds:- • Inclusion body rhinitis in piglets up to 10 weeks of age, after this subclinical infection • Generalised disease in piglets less than 2 weeks of age when infected *in utero*. Runting if they survive • Pregnant sows: sporadic abortions	• Virus isolation • Serology: IFA or ELISA
LEPTOSPIROSIS	*Leptospira interrogans* serovars	Subclinical infections in many pigs. Young pigs can shed leptospires in high concentrations in urine. Abortions in sows late in pregnancy or birth of weak piglets which is often the main sign of the disease	• Dark-field microscopy: urine of sow and young pigs on farm • Serology: MAT or CFT
MASTITIS-METRITIS-AGALACTIA (MMA) SYNDROME	Complex aetiology associated with coliform bacteria, hysteria and hormonal imbalance	Seen in young gilts after farrowing. There is depression and anorexia. If untreated the litter will die from starvation	• History and clinical signs • Isolation of *Escherichia coli* or other enterobacteria from milk and uterine discharges
PARVOVIRUS INFECTION (PORCINE)	Porcine parvovirus *(Parvoviridae)*	Virus replicates in small intestine. Oral or venereal transmission. On endemic farms subclinical disease occurs. In a susceptible herd or in introduced pigs the following syndromes may be evident: • Infection at mating: early embryonic death and return to service. • Sows infected in pregnancy: SMEDI syndrome. Occasional congenital abnormalities in piglets. • Low fertility in boars: virus in semen	• FA: frozen sections • Haemagglutination: detects viral haemagglutinin in foetal tissues • Virus isolation • Serology: ELISA or HAI
PORCINE REPRODUCTIVE AND RESPIRATORY SYNDROME (PRRS) ('BLUE-EARED' DISEASE)	Porcine arterivirus *(Togaviridae)* 'Lelystad virus'	Recorded in USA in 1987, Canada in 1988 and parts of Europe in 1991. Clinical signs variable and course of disease on a property is 1-3 months. Sows: fever, laboured breathing, "blue-ear" in 1 per cent, SMEDI syndrome. Piglets: death, weak piglets, increased susceptibility to secondary infections and haemorrhage easily. About 55 per cent survive to weaning (usual target is 82 per cent). Boars: temporary infertility in affected animals. Growers: fever, respiratory signs, 'blue-ear' in a small percentage, increase in prevalence of diseases endemic on property	• Clinical signs • Serological tests • Virus isolation

Table 170. Infectious diseases of pigs. *(continued)*

Disease	Agent(s)	Comments	Diagnosis
URINARY SYSTEM: pigs			
AFRICAN SWINE FEVER	Unclassified DNA virus	Petechial and ecchymotic haemorrhages in kidneys and other body organs in acute disease. Immune-complex glomerulonephritis in chronic disease	• FA for viral antigen • Serology IFA or ELISA (no neutralising antibodies are produced)
CLASSICAL SWINE FEVER (HOG CHOLERA)	*Pestivirus* *(Flaviviridae)*	Petechial haemorrhages through kidneys ('turkey egg kidneys') and elsewhere	• FA or AGID for viral antigen in tissues • Virus isolation • Serology: VN
PYELONEPHRITIS (PORCINE)	*Eubacterium suis*	Disease of sows. Boars are healthy carriers that transmit the infection venereally. Disease precipitated by pregnancy and parturition. Depression, anorexia, passage of blood-stained urine and pain during micturition are features of the disease	• Clinical signs • Culture from deposit of centrifuged urine (anaerobic)
SWINE ERYSIPELAS (KIDNEY INFARCTS)	*Erysipelothrix rhusiopathiae*	Virulent strains cause vascular damage, thrombus formation and metastatic emboli in various body organs, including kidneys, leading to infarcts	• History of swine erysipelas in herd • Isolation of the pathogen from lesions (chronic condition, may be unsuccessful)
'WHITE-SPOTTED KIDNEY'	*Leptospira interrogans* serovar *pomona*, *Escherichia coli* or other bacteria	Usually not discovered until slaughter. If leptospirosis, clinically normal young pigs on farm may be excreting leptospires in urine	• Mid-stream urine from live pigs. Darkfield for demonstration of leptospires • Attempted isolation of bacteria from lesions
EYES AND EARS: pigs			
'BLUE EARED' DISEASE	Porcine arterivirus	Cyanosis of the ears occurs in a small percentage of pigs in the porcine reproductive and respiratory syndrome	• Clinical signs • Serology
CONJUNCTIVITIS **Swine fever** **African swine fever** **Atrophic rhinitis**	 *Pestivirus* Unclassified DNA virus *Bordetella bronchiseptica*	Conjunctivitis is one sign in the disease syndromes	• Isolation of pathogen • Demonstration of Ag in tissues • Serology to detect Ab for viral diseases
GLASSER'S DISEASE	*Haemophilus parasuis*	Essentially a polyserositis of pigs up to 4 months of age. Cyanosis and oedematous thickening of the ears is considered almost pathognomonic	• Gross and histopathology • Isolation of pathogen

Table 170. Infectious diseases of pigs. *(continued)*

Disease	Agent(s)	Comments	Diagnosis
EYES AND EARS: pigs			
NECROTIC EAR SYNDROME OF PIGS	Trauma and secondary bacterial invasion	Condition is characterised by unilateral or bilateral necrosis of the pinnae of the ears, occurring sporadically in weaned or growing pigs. A septicaemia with septic arthritis and death are not uncommon sequelae	• Clinical signs • Isolation of secondary bacterial invader
OTITIS EXTERNA	Mixed non-specific flora	Otitis externa occurs in pigs but is much less common than in the dog. *Sarcoptes* mites may cause ear mange in pigs	• Clinical signs • Isolation of pathogen(s)
NERVOUS SYSTEM: pigs			
AFRICAN SWINE FEVER	Unclassified DNA virus	Clinically similar to swine fever but nervous signs much less marked, often limited to weakness and incoordination of hind limbs. Histologically an encephalitis, similar to that in swine fever, can be present	• FA technique for Ag (frozen sections) • Serology: IFA, ELISA • Virus isolation
BOTULISM	*Clostridium botulinum*	Reports of botulism in pigs are rare. Signs include staggering, limb weakness, vomiting and pupillary dilation. Terminally there is recumbency with flaccid paralysis and death from respiratory failure	• History: feedstuffs • Demonstration of toxin in serum by mouse inoculation
CLASSICAL SWINE FEVER (HOG CHOLERA)	*Pestivirus (Flaviviridae)*	Severe, contagious and generalised disease. CNS signs are usually seen early in the disease, these include high-stepping, incoordination, paddling of limbs and convulsions. Other signs are high fever, erythema of the skin, burrowing into bedding and piling on top of each other, vomiting, diarrhoea, swaying of hind limbs and muscle tremor. Course of disease is 5-7 days. A form of the disease has been recorded where CNS signs predominate such as tetanic then clonic convulsions with loud squealing, the mortality approaches 100 per cent. Occasionally BVD virus can cause signs almost identical to swine fever but usually confined to one litter. Virus strains causing subacute disease, with less severe clinical signs, are now common in Europe	• Virus isolation • FA technique or AGID for Ag • Serology: VN

Table 170. Infectious diseases of pigs. *(continued)*

NERVOUS SYSTEM: pigs

Disease	Agent(s)	Comments	Diagnosis
CONGENITAL CNS SIGNS	Swine fever virus (occasionally the closely related bovine viral diarrhoea virus)	Infection of pregnant sow: abortions, stillbirths, mummification and weak piglets (carriers). Congenital abnormalities: myoclonia congenita (congenital trembles) associated with cerebellar hypoplasia. Survivors may be immunotolerant carriers (Virus +ve/Ab -ve)	• Virus isolation • Serology: Ab titre in sow
ENCEPHALOMYOCARDITIS	*Cardiovirus* *(Picornaviridae)*	Natural hosts are rodents that excrete virus in faeces. Outbreaks of disease in pigs are associated with rodent plagues. Clinical signs in young pigs include depression, trembling and incoordination, but pigs are usually found dead. Mortality over 50 per cent. The ventral myocardium often has multiple, discrete, pale areas due to necrosis and lymphocytic infiltration	• Virus isolation: TC or mouse inoculation • Serology: VN, HAI • Histopathology
GLASSER'S DISEASE	*Haemophilus parasuis*	Meningitis can occur as part of the polyserositis in pigs up to 4 months of age	• Gross and histopathology • Isolation of pathogen
OEDEMA DISEASE	Strains of *Escherichia coli* producing oedema disease toxin	Predisposed to by changes in diet and occurs in well-grown, thrifty weaner and growing pigs. Characterised by oedema of eyelids, face and stomach wall, a hoarse squeal, stiffness, inco-ordination, blindness and complete flaccid paralysis. Course is 6-36 hours or pig may be found dead. Encephalomalacia occurs	• Gross pathology • Isolation of haemolytic *E. coli*
RABIES	*Lyssavirus* *(Rhabdoviridae)*	Clinical signs in pigs are very variable. May show excitement and a tendency to attack or dullness and inco-ordination. There can be twitching of the snout, rapid chewing movements, excessive salivation, walking backwards and clonic convulsions. Terminally there is paralysis and death	• FA technique for viral Ag in brain. • Histopathology: Negri bodies in brain
STREPTOCOCCAL MENINGITIS	*Streptococcus suis* type 2	Epidemic outbreaks of meningitis and septicaemia can occur in 5-10 week-old pigs. Depression, tremors, inco-ordination, blindness, opisthotonos, convulsions, paddling, fever, nystagmus and paralysis. There is thickening and congestion of the meninges, oedema of the brain and excess, purulent CSF	• CSF examination • Isolation and identification of *S. suis* type 2

Table 170. Infectious diseases of pigs. *(continued)*

NERVOUS SYSTEM: pigs

Disease	Agent(s)	Comments	Diagnosis
TALFAN/ TESCHEN	Porcine enterovirus 1 *(Picornaviridae)*	***Teschen*** (porcine enteroviral encephalomyelitis): virulent form of the disease but limited to Eastern Europe and Malagasy Republic: paralysis, coma and death occurs ***Talfan***: milder form and has a worldwide distribution. Young pigs are affected and there is ataxia and paresis, usually of the hind legs. Pigs are alert with normal appetite and recovery can occur. No gross lesions are seen. Microscopically: lesions in grey matter of brain stem, cerebellum and spinal cord. Non-suppurative encephalitis	• Virus isolation from brain tissue • Serology: VN, ELISA
TETANUS	*Clostridium tetani*	A high incidence of tetanus can occur after castration in a group of young pigs. Incubation period 1-3 weeks. Stiffness, muscle tremor, tail held stiffly, difficulty in eating or swallowing are seen. Later, hyperaesthesia occurs and mild stimuli will cause recumbency with tetanic spasms and opisthotonos	• History: castrations • Gram-stained smear from wound
VOMITING AND WASTING DISEASE	Haemagglutinating encephalomyelitis virus *(Coronaviridae)*	Most infections are inapparent, but susceptible pigs 4 days to 3 weeks-old may vomit, become anorexic, depressed, emaciated and a few develop CNS signs such as hyperaesthesia, jerking movements, posterior paralysis, lateral recumbency and death. No gross lesions are seen. Non-suppurative inflammation of brain stem and upper spinal cord occurs	• Virus isolation • Serology: VN, HAI

MUSCULOSKELETAL SYSTEM: pigs

Disease	Agent(s)	Comments	Diagnosis
ACTINOBACILLOSIS (PORCINE)	*Actinobacillus suis*	A fatal, acute septicaemia occurs in piglets aged 1-8 weeks, but the bacterium has been associated with arthritis, pneumonia and subcutaneous abscesses in older pigs	• Isolation and identification of *A. suis* from joints and other tissues

Table 170. Infectious diseases of pigs. *(continued)*

MUSCULOSKELETAL SYSTEM: pigs

Disease	Agent(s)	Comments	Diagnosis
AFRICAN SWINE FEVER (CHRONIC)	Unclassified DNA virus	Acute or peracute disease can be followed by the chronic form, or this may be the only syndrome seen in endemic areas. The chronic form is characterised by cutaneous ulcers, arthritis, pneumonia, pleuritis and pericarditis. The pigs are intermittently febrile, become emaciated with soft oedematous swellings over joints and under mandible. They may be persistently infected for life	• Acute disease: FA on frozen sections • Chronic disease: serology: IFA, ELISA
ATROPHIC RHINITIS (AR)	*Bordetella bronchiseptica* +/- AR$^+$ *Pasteurella multocida*	Atrophy of the turbinate bones, especially in piglets infected under 3 weeks of age. Signs include lacrimation, sneezing, and later twisting of snout	• Clinical signs • Examination of snouts at slaughter • Isolation of pathogen
BOTULISM	*Clostridium botulinum*	Rare in pigs. Signs include staggering, vomiting, pupillary dilation, flaccid paralysis and recumbency	• History of possible source of toxin • Clinical signs • Demonstration of toxin in serum
FOOT ROT	*Fusobacterium necrophorum, Actinomyces pyogenes,* streptococci and spirochaetes	Usually follows trauma to sole or wall of claw (rough abrasive floor). There is heat and pain when only moderate pressure is applied to the claw. Necrosis can extend between the sole and sensitive laminae with severe lameness	• History of rough concrete floor • Clinical signs • Gram-stained smear of necrotic material or exudate
GLASSER'S DISEASE (PORCINE POLYSEROSITIS)	*Haemophilus parasuis*	Outbreaks of disease in young pigs (3-16 weeks old). Manifested by acute polyarthritis, pleurisy, pericarditis and peritonitis. Usually occurs after stress such as chilling, transportation or weaning. There is a high fever, anorexia, dyspnoea and lameness with joints swollen and painful. Some pigs may die 2-5 days after first signs. Survivors often develop chronic arthritis	• History of stress • Clinical signs • Isolation of pathogen from joint fluid or affected tissues at PM • Gross and histopathology
JOINT-ILL	Streptococci, staphylococci, *Escherichia coli* or *Actinomyces pyogenes*	Usually an umbilical infection soon after birth leads to a bacteraemia with localisation of the bacterial pathogen in joints and other sites	• History of umbilical infection • Isolation of pathogen from aspirated joint fluid
MALIGNANT OEDEMA	*Clostridium septicum* and other gas gangrene clostridia	Soft, doughy swelling with marked local erythema and pain usually restricted to axillae, limbs and throat. The lesions are oedematous but there is usually no emphysema. Pigs are febrile, depressed, stiff and lame. Death often occurs within 1-2 days of first signs	• Clinical signs • FA on tissue or exudates • Isolation of pathogen if necessary

Table 170. Infectious diseases of pigs. *(continued)*

MUSCULOSKELETAL SYSTEM: pigs

Disease	Agent(s)	Comments	Diagnosis
MYCOPLASMAL POLYARTHRITIS	*Mycoplasma hyosynoviae*	Production of arthritis in growing pigs (12-24 weeks old). Sudden onset of acute lameness in one or more limbs but swelling of affected joints is minimal. Pigs are usually afebrile and recovery occurs in 3-10 days. A few animals may become permanently recumbent	• Clinical signs and age group affected • Isolation of pathogen from joints
MYCOPLASMAL POLYSEROSITIS	*Mycoplasma hyorhinis*	Produces arthritis and polyserositis in 1-8 week-old pigs. Transient fever, dyspnoea and acute lameness with moderate swelling in affected joints. Spontaneous recovery may occur in 1-2 weeks but often the affected pigs become unthrifty with chronic lameness	• Clinical signs and young age group • Isolation of pathogen from joints or serous surfaces
NECROTIC RHINITIS (BULLNOSE)	*Fusobacterium necrophorum*	Associated with trauma to nasal cavity such as placing a ring in a pig's nose. Lesions develop as a necrotic cellulitis of soft tissues of nose and face but may spread to involve bone	• History of injury and clinical signs • Gram-stained smear of necrotic material • Isolation of pathogen if necessary
OSTEOMYELITIS	Miscellaneous bacteria	Non-specific, sporadic infection, often associated with tail-biting in pigs	• Isolation of pathogen
STREPTOCOCCAL MENINGITIS AND ARTHRITIS	*Streptococcus suis* type 2	Outbreaks often occur in cold weather in intensively reared pigs aged 4-12 weeks. They show fever, incoordination, tremors, paddling movements, paralysis, convulsions and death. Polyarthritis occurs as part of a general septicaemia	• History of predisposing causes and clinical signs • Isolation of *S. suis* from tissues
SWINE ERYSIPELAS (POLYARTHRITIS)	*Erysipelothrix rhusiopathiae*	The acute forms of disease, septicaemia and urticarial (diamond disease) forms, if untreated, may lead to the chronic forms of polyarthritis and vegetative endocarditis. A synovitis occurs at first in the joints with a non-purulent effusion and then degenerative changes in the subendochondral bone, cartilages and ligaments	• History of erysipelas endemic on farm • Isolation of pathogen from joints
TETANUS	*Clostridium tetani*	Usually occurs in young pigs following castration. The signs are similar to those in other animals and the mortality rate can be high	• History of castration and clinical signs • Gram-stained smear of necrotic tissue from wound

Table 170. Infectious diseases of pigs. *(continued)*

MUSCULOSKELETAL SYSTEM: pigs

Disease	Agent(s)	Comments	Diagnosis
VESICULAR DISEASES	Foot-and-mouth disease *(Aphthovirus)*, Swine vesicular disease *(Enterovirus)*, Vesicular stomatitis *(Vesiculovirus)*, Vesicular exanthema of swine *(Calicivirus)*	Clinical signs in pigs are similar in all four diseases. Sudden lameness in a number of pigs in a group, and often, all four feet affected. The acute lameness is due to vesicular lesions in the interdigital cleft. Pigs less consistently develop vesicles on snout and in buccal cavity	• History and clinical signs • Specimens: vesicular fluid or epithelial flap • Ag detection: ELISA, CFT or VN • Virus isolation • Serology: ELISA, VN or CFT

RESPIRATORY SYSTEM : pigs

Disease	Agent(s)	Comments	Diagnosis
AFRICAN SWINE FEVER (CHRONIC)	Unclassified DNA virus	The chronic disease, in endemic areas, is characterized by cutaneous ulcers, pneumonia, pericarditis, pleuritis and arthritis. Vector: *Ornithodoros* ticks	• FA for antigen in frozen sections • Serology: IFA or ELISA
ATROPHIC RHINITIS (AR)	*Bordetella bronchiseptica* ± AR^+ strains of *Pasteurella multocida*	Signs at 3-8 weeks of age include lacrimation, sneezing and later twisting of snout may be seen. A primary or secondary pneumonia may occur. Disease most severe if young piglets under 3-weeks of age are infected	• Examination of snouts of pigs after slaughter • Culture and identification of pathogens
AUJESZKY'S DISEASE	Porcine herpesvirus 1 *(Alphaherpesvirinae)*	Generalised disease with encephalitis occurs in piglets. Respiratory signs are most common in weaned and growing pigs with sneezing, coughing and oculonasal discharge. Pregnant sows may abort	• Virus isolation • FA for antigen: frozen sections or brain/pharyngeal impression smears • Serology: VN, ELISA
GLASSER'S DISEASE	*Haemophilus parasuis*	Affects non-immune pigs usually up to 4 months of age. There is fever and polyserositis with serofibrinous pleurisy, pericarditis, septic arthritis and meningitis	• Gross and histopathology • Isolation of pathogen
INCLUSION BODY RHINITIS	Porcine herpesvirus 2 *(Betaherpesvirinae,* cytomegalovirus)	Rhinitis in pigs up to 10 weeks of age. Sneezing, coughing and oculonasal discharges are evident. The virus may occasionally cause foetal death and abortion	• Histopathology: large basophilic IBs in mucous glands of turbinates or exfoliated cells in nasal secretions • Serology: IFA or ELISA • Virus isolation

Table 170. Infectious diseases of pigs. *(continued)*

RESPIRATORY SYSTEM: pigs

Disease	Agent(s)	Comments	Diagnosis
MYCOPLASMAL PNEUMONIA ('VIRUS PNEUMONIA' OR 'ENZOOTIC PNEUMONIA')	*Mycoplasma hyopneumoniae* ± *Haemophilus parasuis*	All ages can be infected but lung lesions most common at 8-16 weeks old. Endemic in herds with sporadic flare-ups of disease. Pigs cough if roused. Mild disease on its own, but severe with secondary bacterial invaders	• Gross pathology: grey-purple lung lesions in apical and cardiac lobes • Histopathology • Culture and identification
NECROTIC RHINITIS ('BULL-NOSE')	*Fusobacterium necrophorum*	Predisposing causes include trauma to oral or nasal mucosa. There is swelling and deformity of facial area with snuffling, sneezing and a foul-smelling nasal discharge. Generally only 1 or 2 pigs in herd are affected at any one time	• Swelling of face and foul-smelling necrotic tissue • Gram-stained smear • Isolation: anaerobic
PASTEURELLOSIS	*Pasteurella multocida* ± *Mycoplasma* spp.	The condition is a bronchopneumonia, sometimes with pleuritis and pericarditis. Primary pasteurellosis is usually seen in pigs over 1 year old. Disease can become chronic with polyarthritis and chronic thoracic lesions	• Gross pathology • Isolation and identification of pathogens
PLEUROPNEUMONIA	*Actinobacillus pleuropneumoniae*	The disease is severe and contagious and usually seen in young pigs less than 6 months old. Rapid spread and mortality can be high if untreated. Severe respiratory distress, fever, and bloody, frothy nasal and oral discharges can be seen. Explosive outbreaks occur in non-immune herds but it is a chronic disease when endemic	• PM: pleuritic adhesions, lungs dark and swollen. Bloody fluid oozes from cut surfaces • Isolation and identification • Serology: CFT (herd basis)
PORCINE REPRODUCTIVE AND RESPIRATORY SYNDROME ('BLUE-EARED' DISEASE)	Porcine arterivirus *(Togaviridae)* 'Lelystad virus'	Respiratory disease occurs in sows, weaners and growers in this novel disease. However, the main economic losses are in disruption of the breeding programme and deaths of piglets	• Clinical findings • Differentiate from other respiratory diseases • Serology: effective tests being developed
SWINE INFLUENZA	Swine influenzavirus ± *Haemophilus parasuis*	The disease is highly contagious. Stress such as cold conditions is an important predisposing cause, and infections often occur in autumn and winter. Plum-coloured lesions can be seen in apical and intermediate lobes of lungs. The condition is accompanied by coughing, fever and muscular weakness	• Large number of pigs affected • Isolation of virus early in disease (nasal secretions during febrile period) • Serology: HAI test

Table 170. Infectious diseases of pigs. *(continued)*

SKIN: pigs

Disease	Agent(s)	Comments	Diagnosis
ANTHRAX	*Bacillus anthracis*	Subacute infection most common in pigs and characterized by localized, subcutaneous, oedematous swelling of ventral neck, thorax and shoulders. Swelling is secondary to a septicaemia. Many pigs make a gradual recovery but some die from asphyxia due to swelling of throat region or from toxaemia	• Polychrome methylene blue or Giemsa-stained smears on peritoneal fluid • Culture
CONTAGIOUS PYODERMA IN SUCKING PIGS	Streptococci or staphylococci	Pathogens enter through minor abrasions such as bites. Vesicles and pustules occur that rupture forming scabs. Involvement of hair follicles is common and can lead to acne and deeper, extensive lesions	• Gram-stained smears • Culture
EXUDATIVE EPIDERMITIS (GREASY PIG DISEASE)	*Staphylococcus hyicus*	Seen in sucking pigs and those recently weaned with high mortality in piglets under 10 days of age. There is marked cutaneous erythema with seborrhoea and greasy exudate. Severe dehydration and weakness occurs in young pigs	• Culture of *S. hyicus*
PITYRIASIS ROSEA	Unknown. Familial susceptibility suspected	Lesions very similar to those of ringworm. Seen in young pigs after weaning and several pigs in a litter can be affected. There are minimal systemic signs with spontaneous recovery in 4-8 weeks. Lesions are small red flat plaques enlarging to 2 cm or more, surrounded by a ring of erythema and coalescing to cover ventral abdomen.	• Differential diagnosis from ringworm
RINGWORM	*Microsporum nanum*	Occurs in fattening and mature pigs with high morbidity within a pen. The lesions are round, brownish with scabs and crusts. Usually there is no systemic reaction or pruritis (differentiate from pityriasis rosea)	• Microscopy: hairs in 10 per cent KOH • Culture: colony and macroconidia
SWINE ERYSIPELAS (URTICARIAL FORM)	*Erysipelothrix rhusiopathiae*	Usually seen in young adults. Small red spots develop into characteristic diamond lesions that are raised with an erythematous edge. Lesions coalesce and the skin may eventually slough. There is accompanying fever and systemic illness. Untreated cases can develop into the chronic forms of vegetative endocarditis or polyarthritis	• Characteristic skin lesions with systemic illness

Table 170. Infectious diseases of pigs. *(continued)*

SKIN: pigs

Disease	Agent(s)	Comments	Diagnosis
SWINEPOX	*Suipoxvirus* *(Poxviridae)*	Mild infection in young pigs transmitted by contact and pig lice. Papules, vesicles and circular red-brown scabs occur on belly, axillae and elsewhere on the body. Pigs continue to eat well with recovery in 3 weeks. Congenital infections can occur with lesions on the foetus. Vector: *Haematopinus suis* (mechanical)	• EM on scab • Histopathology on biopsy • FA: frozen sections • Virus isolation • Serology: AGID (screening test)
VESICULAR DISEASES	Foot-and-mouth disease *(Aphthovirus)* Vesicular stomatitis *(Vesiculovirus)* Swine vesicular disease *(Enterovirus)* Vesicular exanthema of swine *(Calicivirus)*	Main and cardinal clinical sign is acute lameness in several pigs in a group due to vesicles between the claws. Vesicular lesions on snout or in buccal cavity are inconstant	• Detection of viral antigen: CFT, ELISA, VN • Virus isolation • Serology: CFT, ELISA, VN

Table 171. Infectious diseases of horses.

BUCCAL CAVITY: horses

Disease	Agent(s)	Comments	Diagnosis
GLOSSOPLEGIA **Strangles** **Botulism** **Equine encephalomyelitides**	*Streptococcus equi* subsp. *equi* *Clostridium botulinum* *Alphaviruses* *(Togaviridae)*	Glossoplegia of central origin may accompany or follow such diseases as strangles, botulism or the equine encephalomyelitides. The unilaterally affected tongue is deviated towards the unaffected side and the bilaterally affected tongue is limp and often protrudes from the mouth	• Diagnosis of the specific condition involved
HORSE POX	*Poxvirus* similar to vaccinia and cowpox viruses.	The disease in horses takes two forms: 1. Infection of the pastern region and the virus may be one cause of 'grease' or 'greasy heel'. 2. Multiple lesions inside lips, on gums and tongue. Young animals may become ill and die	• EM • Virus isolation
PHARYNGEAL PARALYSIS **Botulism** **Guttural pouch infection** **Rabies**	*Clostridium botulinum* *Aspergillus fumigatus* *Lyssavirus* *(Rhabdoviridae)*	Paralysis due to *Aspergillus fumigatus* infection and erosion of the guttural pouch wall is relatively common. This condition is also seen in intoxications in the horse such as botulism and chronic lead poisoning	• Differentiate from oesophageal obstruction or foreign bodies wedged between the teeth
VESICULAR STOMATITIS	*Vesiculovirus*	The principal lesions in the horse are on the dorsum of the tongue and lips. There is a mild fever and champing of jaws and drooling of saliva. If foot lesions occur there is hyperaemia and ulceration of the coronary band	• Demonstration of viral Ag in vesicular fluid by CFT or EM • Virus isolation • Serology: CFT for rise in Ab titre

GASTROINTESTINAL TRACT: horses

Disease	Agent(s)	Comments	Diagnosis
ADENOVIRUS INFECTION	Equine adenovirus	Subclinical infection in normal foals but Arab foals with combined immunodeficiency disease succumb to adenovirus pneumonia at about 2 months of age and an enteric syndrome can sometimes be a feature of the disease	• Virus isolation: nasal and ocular discharges • IN inclusions in cells of lacrimal secretions • FA: virus in tissues

Table 171. Infectious diseases of horses. *(continued)*

GASTROINTESTINAL TRACT: horses

Disease	Agent(s)	Comments	Diagnosis
CHRONIC DIARRHOEA (IDIOPATHIC)	Certain antimicrobial drugs	Large doses and sometimes normal doses (parenteral route), of the tetracyclines, tylosin and lincomycin. Sudden onset of diarrhoea 3-4 days after antibiotic administration. Course short and mortality rate high. Post mortem findings include oedema of large intestinal wall, colitis and typhilitis	• History and clinical signs • Gross pathology • Differentiate from colitis-X
CHRONIC DIARRHOEAS (UNDIFFERENTIATED)	Aetiology: miscellaneous	Course 5-8 weeks but up to 1 year. There is weight loss and faeces are porridge-like to fluid in consistency. The appetite is maintained. Prognosis often poor. Possible causes include: • Massive strongyle larval migration • Granulomatous enteritis (tissue strongylosis?) • Avian tuberculosis • Chronic *Rhodococcus equi* infection • Chronic salmonellosis • *Eimeria* infections • Allergy to components in food • Neoplasia • Tetracycline- or lincomycin-induced diarrhoea	• Clinical and laboratory diagnosis to eliminate or confirm each of the possible causes
CLOSTRIDIAL ENTEROTOXAEMIA	*Clostridium perfringens* types B and C	Not common but is a very serious disease that occurs in first few days of life. Severe depression, abdominal pain, diarrhoea/dysentery and foals can die in a few hours. Usually afebrile as the disease is a toxaemia. Haemorrhagic enteritis seen on postmortem examination	• Gram-smear on mucosal scraping of small intestine from a recently dead foal (large numbers of clostridia) • Neutralisation test for toxin (mice)
COLIBACILLOSIS AND COLISEPTICAEMIA	*Escherichia coli*	Failure to ingest adequate colostrum is a prime determinant for the septicaemic or the enteric form of disease. Signs similar to that in calves. May account for 25 per cent of septicaemias in foals	• Isolation of *E. coli* • Enterotoxigenicity (pili antigens uncertain for equine strains)
COLITIS-X	Uncertain:- • *(E. coli)* endotoxic shock ? • Adreno-corticoid exhaustion ? • Other causes?	Adult horses may die within 24 hours of first signs. Profuse diarrhoea may or may not occur. Horses surviving for less than 3-4 hours often have normal faeces. Usually there is a history of stress. Depression, sweating, abdominal pain, skin cold and clammy, rapid pulse. Postmortem findings: oedema seen early and later haemorrhagic necrosis of walls of caecum and colon. Described in USA, Canada and Australia	• Clinical signs and post mortem findings • Differentiate from salmonellosis and tetracycline therapy

Table 171. Infectious diseases of horses. *(continued)*

GASTROINTESTINAL TRACT: horses

Disease	Agent(s)	Comments	Diagnosis
EQUINE INTESTINAL CLOSTRIDIOSIS	Associated with large numbers of *Clostridium perfringens* type A in gut	Highly fatal diarrhoea. Sudden onset of depression, profuse watery foetid diarrhoea. Circulatory failure and death can occur within 24 hours. Hyperaemia, oedema and haemorrhages of caecum and colon are characteristic. Associated with a protein-rich diet / low cellulose, plus stress	• Gross pathology • Mucosal smear from horse recently dead: large numbers of clostridia • Differentiate from colitis-X and grass sickness
NEONATAL SEPTICAEMIAS	*Actinobacillus equuli*, Group C Streptococci, *E. coli*, *Salmonella* spp. or *Klebsiella* spp.	In neonatal septicaemias death can be rapid. If the foal survives for a period then there is often diarrhoea in the terminal part of the illness	• Isolation and identification of pathogen
POTOMAC HORSE FEVER	*Ehrlichia risticii*	Disease is sporadic and seasonal. Not common in young horses. Mortality 5-30 per cent. Fever, paralytic ileus, and profuse watery diarrhoea (can be projectile). Post mortem: congestion, haemorrhage and mucosal erosion especially in caecum and colon, swollen mesenteric lymph nodes and subcutaneous oedema of abdominal wall. Vector: possibly a tick. Reservoir: rodents suspected	• Serology: ELISA and FA • Differentiate from salmonellosis and colitis-X
ROTAVIRUS INFECTION	*Rotavirus*, *(Reoviridae)*	Foals 5-35 days-old in high population density groups are at risk. There is profuse watery diarrhoea and dehydration is serious in neonates but recovery is uneventful in older foals. Villous atrophy occurs and villi return to normal about 7 days after onset of the diarrhoea	• Electron microscopy • Virus isolation and identification.
SALMONELLOSIS	*Salmonella* spp.	***Neonatal foals*** (up to 4 months of age): usually the septicaemic form occurs. Predisposing factors include overcrowding and build-up of salmonellae in foaling paddocks. Rapid onset with fever, anorexia and profuse diarrhoea if animals survive longer than 24-48 hours ***Adult horses:*** up to 50 per cent can be subclinical excretors of salmonellae. Clinical disease is predisposed to by stress (transportation, surgery, anaesthesia). There is high fever, complete anorexia, profuse diarrhoea, faeces may contain blood and have a foetid smell	• Culture for salmonellae

Table 171. Infectious diseases of horses. *(continued)*

Disease	Agent(s)	Comments	Diagnosis
GASTROINTESTINAL TRACT: horses			
SLEEPY FOAL DISEASE	*Actinobacillus equuli*	Foal may be infected *in utero* or via umbilicus. Can be an acute septicaemia and sudden death but if foal survives 24 hours then suppurative lesions occur in kidneys, joints and intestines. Illness seen a few hours after birth to 3 days of age. Fever, prostration, anorexia, diarrhoea and death usually within 24 hours of clinical signs	• Blood culture from live foal, lesions from dead animals • Cervical swab from mare (dam may be a carrier)
SUPPURATIVE BRONCHOPNEUMONIA	*Rhodococcus equi*	***Foals 2-4 months old*:** suppurative bronchopneumonia with gradual onset of signs. There is fever, weakness and anorexia. Diarrhoea may occur due to granulomatous colitis and mesenteric lymphadenitis from the foal swallowing infected sputum	• Radiography • Isolation of pathogen from transtracheal aspirates
LIVER: horses			
EQUINE RHINOPNEUMONITIS **(foetal infection)**	Equine herpesvirus 1 (sometimes equine herpesvirus 4)	Small necrotic foci in the liver of aborted foetuses	• Pathology: IN inclusions in liver • FA on frozen sections of liver
TYZZER'S DISEASE	*Bacillus piliformis*	Acute, fatal disease of laboratory rodents but also reported in wild and domestic animals. Infection in 1-6 week-old foals. Death can be sudden. The liver is enlarged, pale and mottled. Gastric or duodenal ulceration may also occur	• Gross pathology • Histopathology: Giemsa or Warthin-Starry to demonstrate pathogen within hepatocytes
GENITAL SYSTEM : horses			
CONTAGIOUS EQUINE METRITIS (CEM)	*Taylorella equigenitalis*	Specific and highly contagious, uterine infection of horses and donkeys. Stallion is the carrier in prepuce or penis (urethral fossa) and shows no clinical signs. Infection transmitted by coitus or via veterinary instruments. Mares have a profuse mucopurulent discharge 2-6 days after service or can be subclinical carriers. Low conception rate due to failure of egg implantation or early abortions	• Specimens (*see* Chapter 27) • Culture from swabs placed in Amies transport medium + charcoal • Serology: CF Abs detected 3-7 weeks post-infection

Table 171. Infectious diseases of horses. *(continued)*

GENITAL SYSTEM: horses

Disease	Agent(s)	Comments	Diagnosis
'DIRTY MARE SYNDROME' (NON-SPECIFIC METRITIS)	Group C streptococci, *Klebsiella pneumoniae, Actinobacillus equuli, Pseudomonas aeruginosa, E. coli* or opportunistic fungi	Vaginal discharge, often purulent, of uterine origin. Seen particularly at oestrus and when mare squats to urinate. Hind legs can be streaked with discharge in severe cases. Some of these pathogens may be spread by coitus. Occasionally the presence in the uterus of these organisms can cause a systemic reaction with fever, anorexia, depression and laminitis. If the infection becomes chronic there may be permanent infertility	• Specimens: Cervical swabs introduced through plastic sleeves. Swabs placed in transport medium, especially for streptococci • Culture and identification.
EQUINE COITAL EXANTHEMA	Equine herpesvirus 3	Acute but mild disease characterised by pustular and ulcerative lesions on vaginal mucosa, penis, prepuce and perineal region in the mare. Healing in 14 days with white depigmented permanent spots around vulva identifying potential carriers. No systemic signs or abortions. Decreased libido may occur in stallion with disruption of breeding schedule. Virus spread by coitus. Lesions seen 4-8 days after service	• EM: epithelial cells from the margins of the lesions • Virus isolation • Serology: epidemiological studies only
EQUINE VIRAL ARTERITIS (EVA)	*Arterivirus (Togaviridae)*	Mild or severe febrile disease with conjunctivitis, nasal discharge, palpebral oedema and dependent oedema of legs, udder and scrotum. Abortion in mares during the respiratory disease. Virus shed in nasal secretions for 8-10 days and virus persists in kidney with urinary shedding	• FA: frozen sections • Virus isolation from foetal lung and spleen • Serology: CFT, VN, AGID and IFA
EQUINE VIRAL RHINOPNEUMONITIS	Equine herpesvirus 1 (sometimes EHV 4)	Respiratory disease in foals and young adults. Abortions in mares 1-4 months after the outbreak of respiratory disease. Occasionally herpesvirus paralysis in adults and generalized disease in foals. Latent carrier state probable	• Histopathology of aborted foetus: pulmonary oedema, hepatic necrosis and petechiation of mucosae (IBs in hepatic cells) • FA: frozen sections • Virus isolation
LEPTOSPIROSIS	*Leptospira interrogans* serovars.	Serovars *pomona, hardjo, bratislava* and *icterohaemorrhagiae* have been isolated from aborted foetuses. Mild, transient illness in mare with anorexia, low grade fever and depression. Abortion occurs some weeks after the fever and periodic ophthalmia may be seen some months afterwards	• Leptospiral abortion in horses often goes undiagnosed. In an infected group of horses about 30 per cent will have positive antibody titres to leptospira

Table 171. Infectious diseases of horses. *(continued)*

Disease	Agent(s)	Comments	Diagnosis
GENITAL SYSTEM: horses			
SALMONELLOSIS	*Salmonella abortusequi*	This *Salmonella* species is now rare. Specific disease of equidae characterised by abortion in mares, testicular lesions in stallions and septicaemia and polyarthritis in foals	• Isolation and identification of the salmonella
SCIRRHOUS CORD	*Staphylococcus aureus* (botryomycosis) or *Actinomyces bovis*	Infection of the stump of the spermatic cord seen a few weeks after castration. May be accompanied by fever, toxaemia and lameness. The lesion is a mass of fibrous tissue interspersed with small abscess cavities and sinus tracts	• Specimens: biopsy or surgically removed tissue: • Gram-stained smear • Culture
URINARY SYSTEM: horses			
CYSTITIS **(often accompanied by urethritis)**	Streptococci, *Escherichia coli*, *Staphylococcus aureus* , *Proteus* species or *Klebsiella pneumoniae*	Frequent urination and voiding of small volumes of urine. Fever, inappetence and pain on urination may occur. Urine is cloudy and contains inflammatory cells and bacteria	• Clinical signs • Attempted isolation of pathogen from mid-stream urine
EQUINE INFECTIOUS ANAEMIA	*Lentivirus* *(Retroviridae)*	Immune-complex glomerulonephritis can occur in chronic disease. This may lead to renal failure and death	• History of endemic area • Clinical signs • Viral antigen in spleen: FA or ELISA • Serology: AGID (Coggins test) or ELISA
SLEEPY FOAL DISEASE	*Actinobacillus equuli*	Neonatal septicaemia. If foals survive for 2-3 days, abscesses form throughout kidneys. Joints may also be affected	• Clinical signs • Isolation and identification of pathogen from lesions • Histopathology
EYES AND EARS: horses			
CONJUNCTIVITIS (GENERALISED DISEASES)			
Equine viral arteritis	*Arterivirus* *(Togaviridae)*	Incubation period 1-8 days followed by fever, leukopenia, lacrimation, conjunctivitis, nasal discharge and depression. Photophobia and oedema of eyelids, conjunctiva, legs and ventral body wall can occur. Abortion in pregnant mares	• Clinical signs • Serology: ELISA, CFT

Table 171. Infectious diseases of horses. *(continued)*

EYES AND EARS: horses

Disease	Agent(s)	Comments	Diagnosis
African horse sickness	*Orbivirus* *(Reoviridae)*	In the cardiac form, conjunctivitis, oedema of eyelids and subcutaneous oedema of localised areas of head and neck with bulging of supraorbital fossa occurs. Dependent oedema appears terminally, secondary to cardiac insufficiency	• History of endemic area • Clinical signs • Serology: VN
EQUINE SARCOID	Equine papillomavirus *(Papovaviridae)*	Eyelids are a comparatively common site for sarcoids	• History and signs • Histopathology on biopsy
KERATOMYCOSIS	*Aspergillus fumigatus*	Infection of the cornea can occur following antibacterial and steroid treatment, suggesting immunosuppression and impaired resistance to colonisation as predisposing causes	• Clinical signs and history • Isolation of pathogen
PERIODIC OPHTHALMIA ('MOON BLINDNESS')	*Leptospira interrogans* serovars (especially *pomona*)	Strong presumptive evidence for leptospires as the aetiological agent. Recurrent episodes of inflammation of one or both eyes occurs, the iris and uveal tract are affected. The sequel may be blindness. Probably an immune-complex disease, as condition occurs several months after initial infection	• Clinical signs • Serology: MAT, CFT

NERVOUS SYSTEM: horses

Disease	Agent(s)	Comments	Diagnosis
ABSCESSES IN SPINAL CORD	*Staphylococcus aureus* and streptococci	Pressure on spinal cord and nervous signs will depend on the site of the abscess	• Pathology • Isolation of pathogen
BOTULISM	*Clostridium botulinum* type C_β	Onset 3-17 days after ingestion of feed containing the botulinal toxin. Signs include muscle tremor then stumbling, knuckling and ataxia with inability to lift the head. Animals in sternal recumbency. In some cases the tongue is paralysed with drooling of saliva and inability to chew. Death from respiratory failure 1-4 days after first signs	• History of suspect feed • Differentiate from other causes of motor paralysis • Acute cases: demonstration of toxin in serum by mouse inoculation

Table 171. Infectious diseases of horses. *(continued)*

NERVOUS SYSTEM: horses

Disease	Agent(s)	Comments	Diagnosis
EQUINE ENCEPHALOMYELITIDES	*Alphavirus, (Togaviridae)* Western equine encephalomyelitis virus (WEE), Eastern equine encephalomyelitis virus (EEE) and Venezuelan equine encephalomyelitis virus (VEE)	WEE and EEE have a wild bird/mosquito cycle with the horse (and man) as accidental hosts. Non-contagious as there is minimal viraemia in horses. Fever, drowsiness, paralysis of lips and pharynx, head-pressing, 'sawhorse' stance, incoordination, general paralysis and death may occur. Mortality 30 per cent for WEE and 90 per cent for EEE. VEE has a swamp rodent/mosquito cycle. The disease is contagious as horses have a significant viraemia. Peracute disease as for WEE and EEE but acute illness is without encephalitis and fever, pain, diarrhoea and weight loss occur. Mortality: peracute 80 per cent and acute 50 per cent	• History: seasonal • Histopathology • Serology: CFT, VN or HAI • Virus isolation when horse is febrile and before CNS signs
GUTTURAL POUCH INFECTION	*Aspergillus fumigatus* and/or Group C streptococci	Enlarged guttural pouch causes pressure on 7th cranial nerve. Horse is unable to prehend or chew and there are inappropriate movements of lips	• Clinical signs • Isolation of pathogen from biopsy
HERPESVIRUS PARALYSIS	Equine herpesvirus 1	A myeloencephalopathy that can accompany outbreaks of respiratory disease and abortions. Sudden onset of signs varying from mild ataxia to severe paralysis. Urine dribbling is a feature. The CSF is yellow and abnormal. The condition is seen in foals and adults. Vaccines do not always protect against this neurological form of the disease	• History of other herpesvirus syndromes • Isolation of virus from brain • Serology: HAI, CFT. VN (rise in antibody titres in in-contact horses)
LOUPING ILL	*Flavivirus, (Flaviviridae)*	Mainly an encephalomyelitis of sheep but rarely the horse can be infected. Signs include fever, abnormality of gait, convulsions and paralysis. Horses can develop a sufficiently high viraemia to infect blood-sucking insects. Vector: ticks. Reservoir: grouse	• History: endemic area • Histopathology: brain and brain stem • Serology: HAI and VN • Virus isolation: brain
MENINGITIS (BACTERIAL)	*Streptococcus equi* subsp. *equi*	Uncommon sequel to strangles with excitation, hyperaesthesia, rigidity of neck and terminal paralysis	• Pathology • Isolation of pathogen
RABIES	*Lyssavirus, (Rhabdoviridae)*	An ascending paralysis with hypersalivation, inco - ordination followed by paralysis and recumbency in 2-4 days. Death within 1 week. Occasionally horses can show extreme agitation and become aggressive. Self - inflicted injuries may also occur	• FA technique for Ag in brain • Histopathology for Negri bodies

Table 171. Infectious diseases of horses. *(continued)*

Disease	Agent(s)	Comments	Diagnosis
NERVOUS SYSTEM: horses			
RYE GRASS STAGGERS (MYCOTOXICOSIS)	*Acremonium loliae*	Ingestion of toxic metabolic products of fungus growing in rye grass pastures. Can affect cattle, sheep, horses and deer. Tremor, hypersensitivity, drunken gait and posterior paralysis occur. Animals recover if removed from affected pasture	• Recovery on removal from affected pasture • Histopathology: degeneration of Purkinje fibres in long-standing cases
SHAKER FOAL SYNDROME (TOXICO-INFECTIOUS BOTULISM)	Possibly *Clostridium botulinum* type A.	Reported from UK and USA. Sporadic disease in 3-8 week-old foals of either sex and in all breeds. There is a sudden onset of severe muscular weakness and foals become prostrate but are bright and alert. If foals are lifted, or try to rise, there is severe muscle tremor. Death occurs within 72 hours from respiratory failure. The disease has been reproduced experimentally with the toxin of *Cl. botulinum* type A	• As for botulism • In toxicoinfectious botulism it may be possible to isolate *Cl. botulinum* from the tissues
TETANUS	*Clostridium tetani*	Incubation period 1-3 weeks. Signs include prolapse of third eyelid, anxious expression, pricked ears, flared nostrils, paralysis of jaw muscles, urinary retention, stiffness of muscles with 'sawhorse' posture, tail held high, difficulty in walking with a tendency to fall and become recumbent. Tetanic convulsions are triggered by touch or sound and respiratory arrest may occur during a tetanic spasm. Death in 5-10 days. The longer the incubation period the better the prognosis and mild cases can recover	• History: vaccination? • Clinical signs • Gram-stained smear from deep wound (if present)
VERTEBRAL OSTEOMYELITIS	*Salmonella* spp.	Specific syndrome if cervical vertebrae 4-6 are affected with stumbling gait and stiffness. There is a reluctance to bend the neck so that the animals often have difficulty in grazing. Atrophy of cervical muscles occurs	• Pathology • Isolation of pathogen
MUSCULOSKELETAL SYSTEM: horses			
BOTULISM	*Clostridium botulinum*	Onset 3-17 days after ingestion of food containing toxin. Signs include muscle tremor, stumbling, knuckling, ataxia and inability to raise the head. Animals go down in sternal recumbency. Death from respiratory failure 1-4 days after first clinical signs	• History of food that might have contained toxin • Clinical signs • Demonstration of toxin in serum

Table 171. Infectious diseases of horses. *(continued)*

MUSCULOSKELETAL SYSTEM: horses

Disease	Agent(s)	Comments	Diagnosis
FISTULOUS WITHERS AND POLL EVIL	*Actinomyces bovis* or *Brucella abortus*	Bursitis. Trauma predisposes to the condition, including an ill-fitting saddle or blow to the poll	• Gram- or MZN-stained smears from aspirate or exudate • Isolation of pathogen • Serology for *Brucella abortus*
JOINT ILL (POLYARTHRITIS)	*Escherichia coli, Actinobacillus equuli, Salmonella* spp. or Group C streptococci	Young foal infected via the umbilicus or following a bacteraemia or septicaemia with localisation in the joints and in other tissues	• Clinical signs • Isolation of pathogen from aspirated joint fluid
LYMPHANGITIS	Aetiology includes: *Sporothrix schenckii* (sporothricosis), *Corynebacterium pseudotuberculosis* (ulcerative lymphangitis), *Histoplasma farciminosum* (African farcy) or *Pseudomonas mallei* (farcy)	Lameness may be present in any of these infections that cause lymphangitis, depending on the severity of the condition	• Direct microscopy • Culture • FA technique: *H. farciminosum* • Serology: farcy (CFT)
OSTEOMYELITIS	*Salmonella* spp. and other bacteria	Infection introduced by haematogenous route or by trauma. Lameness and local swelling are the major signs if limb bones are affected. Foals with polyarthritis may have osteomyelitis of the bones adjacent to the affected joints. Infection of vertebrae usually causes nervous signs	• Clinical signs and radiography • Isolation of pathogen from biopsy or tissue
TETANUS	*Clostridium tetani*	Entry of spores usually through puncture wounds in the hooves or penetrating wounds of muscle. Incubation period 1-3 weeks. Muscle stiffness is followed by prolapse of third eyelid, anxious expression, erect ears, dilated nostrils and an exaggerated response to normal stimuli. Later, muscle tetany increases and horse adopts a 'sawhorse' position. The tail may be deviated to one side. In a fatal infection there are tetanic convulsions and death from respiratory failure 5-10 days after first signs	• Vaccination history • Clinical signs • Gram-stained smear from wound for 'drumstick' sporing form

Table 171. Infectious diseases of horses. *(continued)*

Disease	Agent(s)	Comments	Diagnosis
MUSCULOSKELETAL SYSTEM: horses			
THRUSH OF THE FROG	*Fusobacterium necrophorum*	Degeneration of the frog with secondary bacterial infection. Predisposing factors include dirty, wet conditions and failure to clean the hooves regularly. The affected region is moist and contains a black, thick discharge with a foul odour. The condition is more common in the hind feet	• Clinical signs • Gram-stained smear of exudate
RESPIRATORY SYSTEM: horses			
ADENOVIRUS INFECTION	Equine adenovirus 1	Usually mild URT signs or asymptomatic in most young horses but Arab foals with severe combined immunodeficiency disease suffer a generalised disease with pneumonia that is usually fatal	• Virus isolation : nasal and ocular discharges • IN inclusions in cells from lacrimal secretions • FA: virus in tissues • Serology: VN and HAI; rising titres
AFRICAN HORSE SICKNESS	*Orbivirus (Reoviridae)*	Severe pulmonary form: fever, pulmonary oedema, hydrothorax with frothy fluid from nares in moribund horses. Mortality is over 95 per cent	• Virus isolation: TC or i/c inoculation of 2-6 day old mice • Serology: VN. CFT, HAI and AGID
CHRONIC OBSTRUCTIVE PULMONARY DISEASE	Spores of thermophilic actinomycete *Micropolyspora faeni* and dust mites in mouldy hay (hypersensitivity reaction)	Poor performance of thoroughbreds at exercise. There are degrees of coughing, dyspnoea and double respiratory effort	• Intradermal skin tests: test positive for dust mites and/or *Micropolyspora faeni*
EQUINE INFECTIOUS ANAEMIA (EIA)	*Lentivirus* (*Retroviridae*)	In acute primary infections there is fever, ocular and nasal discharges, subcutaneous oedema of ventral abdomen and legs, widespread serosal and mucosal haemorrhages, splenomegaly and hepatomegaly. Relapses are common but intermittent and last 3-5 days with fever, depression, anaemia, and ventral oedema. Emaciation and incoordination are progressive	• Histopathology • Serology: Coggins AGID or ELISA • Ag detection in leukocytes by FA

Table 171. Infectious diseases of horses. *(continued)*

RESPIRATORY SYSTEM: horses

Disease	Agent(s)	Comments	Diagnosis
EQUINE INFLUENZA ('THE COUGH')	Equine influenzaviruses (A/equi/1 and A/equi/2) *(Orthomyxoviridae)*	Usually seen in young horses under 2 years of age. Short incubation period of 1-3 days. Characterized by a persistent, strong, dry cough, nasal discharge, fever and weakness. In severe cases the virus causes myocarditis and pneumonia. A/equi/2 is the more pneumotropic of the two subtypes. Restriction of exercise for 2 weeks after abatement of signs is mandatory	• Virus isolation: nasopharyngeal swab in first few days of illness (transport medium essential) • Serology: HAI, VN. Paired sera for four-fold rise in Ab titre
EQUINE VIRAL ARTERITIS (EVA)	*Arterivirus* *(Togaviridae)*	The virus has a predilection for arteries. Fever, respiratory signs, oedema of legs and leucopenia occur. Abortion in 50-80 per cent of mares is concurrent with clinical signs	• Histopathology: lesions in arteries • FA: frozen sections • Virus isolation from nasal cavity and conjunctival sac • Serology: VN, CFT, AGID and IFA
EQUINE VIRAL RHINOPNEUMONITIS	Equine herpesvirus 4 occasionally equine herpesvirus 1	Latent carrier state probably exists and the infection is often inapparent in older horses. Respiratory disease occurs in foals and young adults. Mares may abort 1- 4 months after an inapparent infection	• Virus isolation early in disease (less than 72 hours post infection) • Serology: HAI, CFT, VN four-fold rise in antibody titre
GLANDERS (FARCY)	*Pseudomonas mallei*	Glanders has been eradicated from many countries and is now comparatively rare. It is a highly contagious disease of equidae although cats and man are susceptible. An acute form can occur with fever, respiratory signs, septicaemia and death in a few days. The chronic pulmonary form is a debilitating disease with tubercle-like nodules in the nasal cavity or lungs. Farcy is the chronic cutaneous form with ulcerating nodules along lymphatic vessels	• Mallein test • Isolation and identification of *P. mallei* • Serology: CFT, indirect HAI
GUTTURAL POUCH INFECTION	*Aspergillus fumigatus* and/or Group C streptococci	Nasal discharge, swelling or displacement of parotid gland, dysphagia and profuse epistaxis unrelated to exercise and occasionally unilateral facial paralysis	• Endoscopic examination: fungal plaques • Radiography: fluid in pouch • Biopsy: isolation and identification of pathogen

Table 171. Infectious diseases of horses. *(continued)*

RESPIRATORY SYSTEM: horses			
Disease	**Agent(s)**	**Comments**	**Diagnosis**
NASAL POLYPS OR NASAL GRANULOMAS	Usually caused by one of the following: *Aspergillus fumigatus, Cryptococcus neoformans, Rhinosporidium seeberi* or *Conidibolus coronatus*	Lesions cause interference with breathing and a nasal discharge, often unilateral occurs	• Demonstration of fungal elements microscopically and culturally from biopsy
PNEUMONIA (BACTERIAL)	Group C streptococci, *Actinobacillus equuli* or *Bordetella bronchiseptica*	Predisposing causes include a comparatively mild viral respiratory infection and / or stress. A secondary bacterial pneumonia can often be severe or fatal	• Isolation and identification of secondary invaders
RESPIRATORY INFECTIONS (MINOR VIRAL)	Equine rhinovirus, Equine parainfluenza virus 3 and Equine herpesvirus 2	Mild URT infection in young horses with nasal discharge and cough unless complicated by secondary bacterial invaders	• Isolation of viruses • Serology
STRANGLES	*Streptococcus equi* subsp. *equi*	Young horses are at greatest risk. There is fever, cough, purulent oculonasal discharge and abscesses (especially in submaxillary and pharyngeal lymph nodes). In bastard strangles, abscesses can develop in lungs and other body organs	• Isolation and identification of *S. equi.* subsp. *equi*
SUPPURATIVE BRONCHOPNEUMONIA	*Rhodococcus equi* (soil organism)	Foals 1-4 months old are affected. Infection is usually by inhalation leading to bronchopneumonia	• Radiography • Isolation and identification of *R. equi* from transtracheal aspirates

SKIN: horses			
CONTAGIOUS ACNE OF HORSES	*Corynebacterium pseudotuberculosis*	Uncommon. Development of pustules (1.0 - 2.5 cm) particularly in harness area. Pustules rupture and exude greenish pus and a crust forms. There is no pruritis but lesions can be painful to the touch. Healing in 1 week but there may be successive crops of pustules. Transmission via harness and grooming gear	• Gram-stained smear from pus • Culture

Table 171. Infectious diseases of horses. *(continued)*

SKIN: horses

Disease	Agent(s)	Comments	Diagnosis
DERMATOPHILOSIS (STREPTOTHRICOSIS)	*Dermatophilus congolensis*	Exudative dermatitis with extensive scab formation. Clumps of hairs come away with scab leaving an ovoid bleeding surface. Affected area of skin is rough and lumpy	• Gram or Giemsa-stained smears from scab • Culture if necessary under 10 per cent CO_2
EQUINE COITAL EXANTHEMA	Equine herpesvirus 3 *(Alphaherpesvirinae)*	Acute but mild disease characterized by pustular and ulcerative lesions on vaginal mucosa and perineal region in mares and on penis and prepuce of stallions. Healing occurs in about 14 days with white depigmented spots marking the site of lesions, that are particularly noticeable around vulval area in mares. The spots remain for life and identify potential carriers. It is a venereal disease and lesions are seen 4-8 days after coitus. No systemic signs or abortions occur	• Virus isolation • Serology: VN (demonstration of rising Ab titre)
EQUINE PAPILLOMATOSIS (EQUINE WARTS)	Equine papillomavirus *(Papovaviridae)*	Warts or papillomas are usually seen in young horses under 3 years of age. They are self-limiting and will regress in a few months. Warts are comparatively small. Individual growths often around the lips or on the face	• Histopathology • EM
EQUINE SARCOID	Papillomavirus *(Papovaviridae)*	All ages are susceptible. Sarcoids can persist for life, do not regress, can become large and tend to have a multiple base	• History and signs • Histopathology on biopsy
GREASY HEEL (SEBORRHOEA)	Attributed to a poxvirus. Parasitic (chorioptic mange) or secondary bacteria may complicate the infection	Often on hindlegs of horses and predisposed to by unsanitary conditions but can occur in well-managed horses. Thickening and greasiness of skin on back of pastern to coronary band. It can spread if not treated	• Gram-stained smear • Examination for mange mites • Culture for secondary bacteria
HORSE POX	*Orthopoxvirus (Poxviridae)*	Rare disease. ***Leg form:*** pustules and scabs on back of pastern with pain, lameness and a slight systemic reaction ***Buccal form:*** pustules on inside of lips that can spread over entire buccal mucosa. In rare cases lesions can occur on vulva or over entire body. Lesions heal in 2-4 weeks	• Electron microscopy

Table 171. Infectious diseases of horses. *(continued)*

SKIN: horses

Disease	Agent(s)	Comments	Diagnosis
LYMPHANGITIS			
Ulcerative lymphangitis	*Corynebacterium pseudotuberculosis*	Infection of skin wounds occurs with invasion of lymphatic vessels and abscesses along their course. Lesions on lower limbs with minimal lymph node involvement. Abscesses exude greenish pus and ulcerate. Heal in 1-2 weeks but fresh crops can occur for up to 12 months.	• Direct microscopy (exudate): pleomorphic Gram +ve rods • Culture and identification
Sporotrichosis	*Sporothrix schenckii*	Fungus ubiquitous but infections are sporadic. Raised cutaneous nodules occur along the lymphatics of lower limbs. Nodules ulcerate and discharge pale yellow pus. Infection mild and localised	• Direct microscopy (exudate): cigar-shaped yeast • Culture: dimorphic fungus
Farcy	*Pseudomonas mallei*	Now a rare condition. Subcutaneous nodules occur along lymphatics that ulcerate and discharge a sticky, honey-like pus. Ulcers heal in star-shaped scars and the lymphatics are fibrous and thickened. Lesions often present in hock area but can extend over body. Local lymph nodes are involved, there is debility and pneumonic lesions may also be present	• Direct microscopy (exudate): Gram -ve rods • Culture • Serology: CFT
Epizootic Lymphangitis (African Farcy)	*Histoplasma farciminosum*	Moveable nodules along superficial lymph vessels. The nodules ulcerate and discharge a thick, yellow, oily pus. There is cording of lymphatics and the local lymph nodes are swollen and hard. Lesions may also occur in lungs and nasal mucosa. It is considered a chronic and incurable disease	• Direct microscopy (exudate): oval, double-contoured yeast • FA technique
RINGWORM	*Microsporum gypseum, Trichophyton equinum* or *Microsporum canis*	Lesions (3 cm) with raised hair, alopecia and fine scab. May be sore to the touch and itchy. Regrowth of hair often with loss of pigmentation in 25-30 days. *M. gypseum* (from soil): back and sides of horse. *T. equinum* (from harness and grooming gear): in girth-strap area or widespread over body. *M. canis*: stable cats may be affected	• Direct microscopy on scab and hairs in 10 per cent KOH for arthrospores • Culture on Sabouraud agar • Identification: colonial morphology and macroconidia

Table 171. Infectious diseases of horses. *(continued)*

SKIN: horses			
Disease	**Agent(s)**	**Comments**	**Diagnosis**
STAPHYLOCOCCAL DERMATITIS	*Staphylococcus aureus* occasionally *Staphylococcus intermedius*	Quite common but sporadic. Lesions often under harness suggesting transmission by contact with contaminated harness. Small (5 mm) nodules then pustules appear that are painful to the touch and horse may not tolerate the harness	• Gram-stained smear • Culture
VESICULAR STOMATITIS	*Vesiculovirus (Rhabdoviridae)*	Disease once common in US military horses but now mostly seen in cattle and pigs. Vesicular lesions in horses usually on lips and tongue but can occur on udder of mares and prepuce of stallions. Vesicles rupture with rapid healing. Probably transmitted via biting insects	• Culture: fertile eggs • Serology: CFT and VN

Table 172. Infectious diseases of dogs.

BUCCAL CAVITY: dogs

Disease	Agent(s)	Comments	Diagnosis
CANINE ORAL PAPILLOMATOSIS	Canine papillomavirus *(Papovaviridae)*	Usually occurs in dogs under 1 year old. The condition can spread to all young dogs in kennels. Warts first appear on the lips but extend into the buccal cavity. Spread of lesions occurs for 4–6 weeks then there is spontaneous regression over several months	• Clinical signs • Histopathology on a biopsy
LEPTOSPIROSIS (CANINE)	*Leptospira interrogans* (often serovar *canicola* or *icterohaemorrhagiae*)	In the icteric form, oral mucous membranes may first show irregular haemorrhagic patches which later become dry and necrotic and slough in sections. A tenacious salivary secretion is present around the gums, at times blood-tinged	• Clinical signs • Darkfield microscopy on urine • Culture from urine • Serology: agglutination tests
MYCOTIC STOMATITIS	*Candida albicans*	The condition often follows prolonged antibiotic therapy in pups. It is a specific type of ulcerative stomatitis characterized by ulcers and soft, white, slightly elevated patches on the oral mucosa. The periphery is reddened and the lesions may coalesce	• History of antibiotic therapy • Clinical signs • Isolation of pathogen
TONSILLITIS	Beta-haemolytic streptococci	Comparatively common in dogs and may occur as a primary disease or secondary to infections in buccal cavity or pharynx. There may be fever, a short, soft cough followed by retching and expulsion of small amounts of mucus	• Clinical signs • Isolation of pathogen
ULCERO-MEMBRANOUS STOMATITIS **(Trench mouth)**	Commensals in mouth: spirochaetes and fusiform bacteria	The condition is usually secondary to other infections or deficiency diseases. At first there is reddening and swelling of gingival margins, that bleed easily. The lesions progress to ulceration and necrosis of gingivo-alveolar tissues with the formation of pseudomembranes	• History of predisposing causes • Clinical signs • Gram or DCF-smear of affected tissue

Table 172. Infectious diseases of dogs. *(continued)*

GASTROINTESTINAL TRACT: dogs

Disease	Agent(s)	Comments	Diagnosis
CANINE CORONAVIRUS INFECTION	Canine coronavirus	Highly contagious enteric disease of dogs marked by vomiting and diarrhoea. Dogs, foxes and coyotes are susceptible. The virus replicates in cats without clinical signs. Faecal-oral transmission. Dogs can excrete the virus for over 2 weeks. Signs similar to, but milder than, canine parvovirus infection. There is no leukopenia, faeces are liquid, may contain blood and mucus and have a foetid odour. Atrophy of villi with deepening of crypts occurs	• EM: faeces • Virus isolation • Serology: VN, IFA
CANINE HAEMORRHAGIC GASTROENTERITIS	Aetiology uncertain, possibly: *Escherichia coli* endotoxaemia or clostridial enterotoxaemia	Toy breeds particularly susceptible. An acute onset of intestinal haemorrhage occurs in mature dogs with collapse, bloody diarrhoea, rapid course and death in untreated cases. Faeces have a jam-like consistency and characteristic odour. Elevated packed cell volume (over 70 per cent in severe cases)	• Clinical findings • Haematology: elevated packed cell volume • Haemoglobin concentration
CANINE HERPESVIRUS INFECTION IN PUPS	Canine herpesvirus 1	Acute generalised disease and death in neonates less than 1 week-old. Soft yellow-green faeces, abdominal pain, anorexia and incessant crying occur. Pups are infected *in utero*, via birth canal or by aerosol after birth. Usually the whole litter is affected. Postmortem: haemorrhages in kidneys, liver and other organs. Bitch only aborts or has an affected litter once, as immunity is acquired. Venereal transmission may occur. Respiratory disease may be seen in young adults	• Pathology: IN inclusions in lungs, liver and kidneys • FA: frozen sections • Virus isolation • Serology: VN, IFA
CANINE HISTOPLASMOSIS	*Histoplasma capsulatum*	Sporadic occurrence worldwide and endemic in central USA. Chronic respiratory disease (lungs primary site) and ulceration of intestines with diarrhoea. The disease may be suspected in dogs with intractable cough, diarrhoea, emaciation and a lack of response to antibiotics	• Histopathology: small yeast in macrophages • Culture at 25° and 37°C • Serology: CFT

Table 172. Infectious diseases of dogs. *(continued)*

GASTROINTESTINAL TRACT: dogs

Disease	Agent(s)	Comments	Diagnosis
CANINE PARVOVIRUS INFECTION	Canine parvovirus	Stable virus spread in minute particles of faeces on fomites such as human shoes. There is no extended carrier state but virus is excreted in faeces for 5–8 weeks after illness. Targets are lymphoid tissue (immunodepression), bone marrow (leukopenia) and small intestine (necrosis + secondary Gram-ve bacterial overgrowth giving diarrhoea, vomiting and endotoxaemia). Infection *in utero* or in pups under 2 weeks-old can lead to generalized disease and death. Pups infected when 3–8 weeks-old suffer myocarditis (rare now as immunity has built up in dog population)	• EM on faeces in acute phase • Haematology: leukopenia • Histopathology • Virus isolation • Detection of viral antigen in faeces or tissues: FA, ELISA, HA • Serology: HAI, VN, ELISA
COLITIS (BACTERIAL)	*Salmonella* sp. or *Campylobacter jejuni*	**Acute:** vomiting, haemorrhagic mucoid diarrhoea causing dehydration **Chronic:** slow onset with tenesmus and scant watery blood-stained, mucoid faeces. Insidious weight loss, anaemia and a gaunt-look. Appetite is normal. Colon thick-walled, rubbery and lumen narrow. Mesenteric lymph nodes enlarged and firm	• Lab. examinations for bacteria or parasites • Histopathology on colon biopsy • Differentiate from parasites or foreign bodies
ENTERIC CAMPYLOBACTERIOSIS	*Campylobacter jejuni*	Causes severe diarrhoea in young animals and is most common in urban dogs. Faeces are watery, contain mucus and can be bile-streaked. Dog may be febrile, partially anorectic and occasional vomiting may occur. Diarrhoea usually lasts 3–7 days but intermittent diarrhoea may last 2 weeks to 2 months	• Isolation from faeces • Phase contrast on fresh faeces: large numbers of motile bacteria in acute phase of disease
ENTERITIS	Feature of: canine distemper, infectious canine hepatitis, canine parvovirus infection, canine coronavirus infection, salmonellosis, *Escherichia coli* infections in neonates	Diarrhoea is the outstanding sign. Can be accompanied by vomiting when anterior duodenum or stomach is involved and by tenesmus when inflammation extends into the colon. Faeces are liquid, foul-smelling and may be dark-green or black (bleeding high in small intestine) or blood-streaked (haemorrhage high in large intestine). Fever occurs if the cause is an infectious agent. Abdominal pain is a feature and dogs may stretch out on a cool floor or adopt a praying position	• Clinical examination • Lab. examination for bacteria • Differentiate from protozoan infections, poisons (especially heavy metals) or helminths

Table 172. Infectious diseases of dogs. *(continued)*

Disease	Agent(s)	Comments	Diagnosis
GASTROINTESTINAL TRACT: dogs			
PROTOTHECOSIS	*Prototheca zopfii, P. wickerhamii* (colourless algae)	Protothecosis in the dog is usually a widely disseminated disease. Bloody diarrhoea (intermittent or protracted) is the most common syndrome with weight loss and debility. CNS involvement (about 40 per cent of cases) can lead to depression, ataxia, circling or paresis. Eye involvement is common. Chronic skin lesions, characterised by ulcers with crusty exudates on trunk or extremities are seen in some cases	• Histopathology (PAS or silver stains) • Direct microscopy (CSF, urine) • Culture and identification
SALMON POISONING	*Neorickettsia helminthoeca*	The disease occurs along the Californian and Alaskan coasts. Rickettsiae are transmitted through various stages of a fluke in a snail-fish-dog cycle. The dog becomes infected by ingestion of salmon containing encysted metacercariae. Signs occur 5–9 days after ingestion and last 7–10 days before death (90 per cent of untreated cases). Fever, depression, anorexia, vomiting, bloody diarrhoea, dehydration, weight loss, lymphadenopathy, nasal and conjunctival discharge occur and the disease can resemble canine distemper	• Fluke ova in faeces is suggestive • Intracellular rickettsiae in aspirates from lymph nodes
LIVER: dogs			
INFECTIOUS CANINE HEPATITIS	Canine adenovirus 1	Common only in unvaccinated populations of dogs. Pups are most susceptible shortly after weaning and can become moribund within a few hours of first clinical signs. In the acute hepatic form there is severe hepatitis, oedema of the gallbladder, multifocal vasculitis and haemorrhage. All *Canidae* are susceptible. CNS signs are not common except in the fox (fox encephalitis). Hepatic cirrhosis is a common sequel	• Clinical signs • Pathology: IN inclusions in many tissues • FA on frozen sections • Virus isolation • Serology: VN, HAI
LEPTOSPIROSIS (CANINE)	*Leptospira interrogans* serovars *canicola* and *icterohaemorrhagiae*	The pathogens replicate in liver and kidneys where they produce degenerative changes. Icterus occurs due to liver damage and destruction of red cells. Serovar *icterohaemorrhagiae* often causes the acute haemorrhagic or icteric forms and *canicola* is associated with the icteric and uraemic syndromes	• Dark-field microscopy on recently collected urine • Isolation from urine or kidneys • Serology: agglutination tests

Table 172. Infectious diseases of dogs. *(continued)*

GENITAL SYSTEM: dogs

Disease	Agent(s)	Comments	Diagnosis
BALANOPOSTHITIS	Miscellaneous bacteria	Comparatively common in dogs as the preputial cavity offers an ideal environment for bacterial growth. Mucopurulent preputial discharge is seen but systemic signs only occur if deeper tissues of glans penis are invaded	• Bacterial culture and identification
BRUCELLOSIS (CANINE)	*Brucella canis* (*B. abortus*, *B. suis* or *B. melitensis* rarely)	Epidemics of abortion can occur in breeding kennels. Transmission: congenital, venereal and by ingestion. Abortions in last trimester, stillbirths or infertility are observed. Prolonged discharge often follows abortion and males may be sterile following an infection	• MZN-stained smears on aborted foetuses, discharge or semen • Culture for brucellae • Serology: agglutination test
CANINE HERPESVIRUS INFECTION	Canine herpesvirus	• Bitch: occasional abortions, stillbirths or vesicular vaginitis • Generalised disease in neonatal pups and death within 24 hours: one cause of the 'fading puppy syndrome' • Young adults: mild rhinitis Transmission: *in utero*, via birth canal, contact with saliva, nasal secretions or urine and venereal	• Pathology in pups: focal necrosis and petechiation occur in most organs. IN inclusions in lung, liver and kidney lesions • Virus isolation • FA: frozen sections • Serology: VN, IFA
CANINE PARVOVIRUS INFECTION	Canine parvovirus	Parvoviruses are known to cross the placental barrier. As modified live virus vaccine causes a viraemia it is unwise to use modified live vaccines in pregnant bitches as this might result in foetal infection or abortion	• History of vaccination of pregnant bitch • Virus isolation • Serology: HAI, VN, ELISA
'FADING PUPPY SYNDROME'	Canine herpesvirus (CHV) Bacterial septicaemias often caused by *Escherichia coli* and streptococci	The pups usually appear normal at birth but may acquire infection from dam's vagina or the environment soon after birth. There are sequential deaths in the litter, usually within 1 week of parturition	• Clinical signs • Histopathology • Bacteria: culture from vaginal discharge and from pups • CHV 1: FA, virus isolation and serology
METRITIS	*Escherichia coli*, streptococci or staphylococci	Usually infection occurs at time of parturition. Acute illness with fever, depression, purulent and foul-smelling discharge. Bitch may neglect the pups	• Clinical signs and examination • Radiography to check on retained foetus • Guarded swab from cervix for culture of bacteria

Table 172. Infectious diseases of dogs. *(continued)*

Disease	Agent(s)	Comments	Diagnosis
GENITAL SYSTEM: dogs			
ORCHITIS/ EPIDIDYMITIS	*Escherichia coli*, streptococci, staphylococci, or *Brucella canis*	Predisposing causes: trauma or concomitant posthitis or cystitis. There is a stilted gait or stance and licking of scrotum occurs. Swelling of scrotum with pain and fever are common findings	• Physical examination • Urinalysis and bacterial culture • Serology for *Brucella canis*
PROSTATIC ABSCESS	Miscellaneous bacteria	Abnormal gait or stance, pain, haematuria and sometimes bulging of the perineum can be present. Chronic peritonitis occurs if abscess drains into abdominal cavity. Abscess can be secondary to a urinary tract infection or may be blood-borne	• Rectal palpation and radiography • Rectal massage and collection of semen or exudate for culture
PYOMETRA	*Escherichia coli,* streptococci, staphylococci, *Pseudomonas* or *Proteus* spp.	Excess progesterone levels can result in endometrial growth and accumulation of secretions with subsequent bacterial infection. High progesterone levels also inhibit the leucocyte response. Exogenous oestrogen ('mis - mating shots') tend to stimulate progesterone and increase the risk of pyometra. Clinical signs depend on patency of cervix. Closed cervix: very serious with depression, anorexia, toxaemia, vomiting, polydipsia and polyuria. Partially open cervix: discharge continues for 4-8 weeks after oestrus with less severe clinical signs	• Physical examination • Radiography • Elevated WBC count • Culture of discharge or of uterine contents if surgically removed
VAGINITIS	Bacteria such as *Escherichia coli*, *Brucella canis* or streptococci	Vaginal discharge occurs but usually no systemic signs. Vaginitis may have developed secondarily to conformational abnormalities or vaginal foreign bodies	• Guarded swabs from anterior of vagina for culture. Significant if heavy growth of not more than 2 potential pathogens
URINARY SYSTEM: dogs			
CANINE HERPESVIRUS IN PUPS	Canine herpesvirus	One cause of the 'fading puppy syndrome': pups usually die within 1–2 days of first clinical signs. Focal renal haemorrhages are characteristic, although ecchymoses can also occur in other abdominal organs	• History of respiratory disease in adults • Clinical signs • Pathology: intranuclear IBs • FA on frozen sections • Isolation of virus • Serology: VN or ELISA, IFA for a high titre in adult dogs

Table 172. Infectious diseases of dogs. *(continued)*

URINARY SYSTEM: dogs

Disease	Agent(s)	Comments	Diagnosis
CATARRHAL CYSTITIS (CANINE DISTEMPER)	*Morbillivirus (Paramyxoviridae)*	This may be associated with canine distemper. Inclusion bodies are present in cells of the urinary bladder	• Clinical signs • Histopathology for tissue changes and IBs in bladder tissue • FA on conjunctival smears or frozen tissue sections of bladder, lung and intestine • Serology: VN, IFA
CYSTITIS AND URETHRITIS	*Escherichia coli*, staphylococci, *Proteus* species, *Klebsiella pneumoniae*	Frequent urination, haematuria and dysuria occurs. Bladder may be thickened and tender when palpated. There may be a predisposing cause in recurring infections	• Clinical signs • Direct microscopy on mid-stream urine. Gram-stained smear on one drop of urine. One bacterial cell or more per oil-immersion field is suggestive of a bacteriuria • Bacterial count on urine. Clinical bacteriuria suggested by 10^5 bacteria/ml or more • Isolation of pathogen from urine
INFECTIOUS CANINE HEPATITIS	Canine adenovirus 1	Immune-complex glomerulonephritis can occur in the chronic phase of the disease	• History and clinical signs • Histopathology for tissue changes and IBs in liver and kidney • FA technique on frozen tissues • Serology: HAI
LEPTOSPIROSIS (CANINE)	*Leptospira interrogans* serovars *canicola* and *icterohaemorrhagiae*	Acute or chronic interstitial nephritis may be present. A chronic condition with progressive renal failure and uraemia can occur 1–3 years after an acute episode of disease	• Dark-field microscopy and FA technique on urinary deposits • Isolation of leptospires from urine • Histopathology if death occurs • Serology: agglutination tests
PYELONEPHRITIS	*Escherichia coli*, staphylococci, *Proteus* species, *Klebsiella pneumoniae*	Systemic signs such as fever, anorexia, depression and vomiting may be present. Kidneys are painful when palpated. Chronic disease may lead to uraemia	• Investigation as for cystitis or urethritis

Table 172. Infectious diseases of dogs. *(continued)*

EYES AND EARS: dogs

Disease	Agent(s)	Comments	Diagnosis
BLINDNESS **Canine distemper**	*Morbillivirus (Paramyxoviridae)*	Rarely retinal or optic nerve damage	• Diagnosis of the generalised infection
Protothecosis	*Prototheca zopfii* *P. wickerhamii*	Rare disseminated algal infection but blindness occurs in over 50 per cent of cases	
Toxoplasmosis	*Toxoplasma gondii*	Blindness (retinochoroiditis) has been recorded in some infections	
Cryptococcosis	*Cryptococcus neoformans*	Eye involvement leading to chorioretinitis occurs but is more common in the cat	
'BLUE-EYE' **(Infectious canine hepatitis)**	Canine adenovirus 1 [sometimes canine adenovirus 2 (CAV 2)]	Transient opacity of the cornea can occur in the acute disease, as the only sign in subclinical disease and after vaccination with attenuated vaccine. It is an immune complex phenomenon and oedema and anterior uveitis are present. Afghan hounds are particularly susceptible to blue eye. The sequel can be glaucoma. CAV 2 virus can also cause blue eye but rarely	• Clinical signs and history of disease or vaccination • Serology: HAI, VN
CONJUNCTIVITIS **(Non-specific)**	Staphylococci, streptococci, *Bordetella bronchiseptica*	Predisposing causes include irritants, allergens or a concurrent viral infection such as canine distemper or 'kennel cough'. Purulent discharge indicates a secondary bacterial component	• Clinical signs • Isolation of a potential pathogen in heavy, pure, culture
OTITIS EXTERNA	One or more of the following are present: Staphylococci *Proteus* spp. *Pseudomonas aeruginosa* *Aspergillus fumigatus* *Malassezia pachydermatis*	Predisposing causes include faulty drainage (pendulous ears), foreign bodies, ear mites *(Otodectes cynotis)* or polyps in ear canal, constant wetting or hot, humid weather. Signs include violent shaking of head, scratching and rubbing ears, often with haematoma formation in pinna, dark purulent discharge and swelling and inflammation of the mucosa of ear canal	• Clinical signs • Isolation of one or more pathogens from exudate • Examination for ear mites and foreign bodies
OTITIS MEDIA AND OTITIS INTERNA	Non-specific, similar to aetiology of otitis externa	An inflammation of tympanic cavity as an extension of otitis externa or due to penetration of ear drum by a foreign body. Occurs in all species but most common in dog, cat, rabbit and pig. Otitis media can lead to otitis interna and may result in ataxia and deafness	• Clinical signs • Radiography (sclerotic changes in bone of tympanic bulla) • Isolation of pathogens from discharge

Table 172. Infectious diseases of dogs. *(continued)*

NERVOUS SYSTEM: dogs

Disease	Agent(s)	Comments	Diagnosis
ASPERGILLOSIS	*Aspergillus fumigatus*	In untreated nasal granulomas, *A. fumigatus* can erode cribriform plate and head tilt, ataxia or seizures may occur	• Histopathology and culture on biopsy or tissues after death
BOTULISM	*Clostridium botulinum*	Natural botulism is rare in dogs. Incubation period varies from hours to 6 days. There is progressive symmetrical ascending weakness from rear to forelimbs although the tail-wag is maintained. Eventually quadriplegia occurs. There is normal pain response but lack of withdrawal reflex. Pupillary dilation is characteristic. A diminished jaw tone is present. The course in dogs that recover is 24 hours to 14 days	• Demonstration of toxin in dog's serum or in suspect food: mouse neutralization test
CANINE DISTEMPER	*Morbillivirus (Paramyxoviridae)*	CNS signs seen 1–3 weeks after recovery from the generalised disease but also after mild ('kennel cough' syndrome) or inapparent infections. Signs include one or more of the following: myoclonus, ataxia, hyperaesthesia, cervical rigidity and seizures: petit mal ('chewing gum' syndrome) or grand mal convulsions	• Histopathology • FA technique for Ag in brain • History: typical distemper signs? • Virus isolation: TC or ferret inoculation with brain suspension • Serology: VN, IFA
CANINE EHRLICHIOSIS	*Ehrlichia canis*	• **Acute**: fever, enlarged spleen and lymph nodes, mucopurulent nasal discharge and loss of stamina. CNS signs referable to inflammation and bleeding into meninges and include hyperaesthesia, twitching of muscles and cranial nerve deficits. Most animals recover • **Chronic**: **Tropical canine pancytopenia**: an immunologically mediated sequel to the acute form. Mortality high and epistaxis, petechiation of skin and mucous membrane, haematuria, wasting, cerebellar ataxia, glomerulonephritis and renal failure occur Vector: *Rhipicephalus sanguineus*	• Giemsa or FA on blood smears for *E. canis* (present in monocytes) • Haematology • Serology: IFA

Table 172. Infectious diseases of dogs. *(continued)*

NERVOUS SYSTEM: dogs

Disease	Agent(s)	Comments	Diagnosis
CANINE HERPESVIRUS INFECTION	Canine herpesvirus	• Adult dogs: mild respiratory signs or sub-clinical. Venereal transmission can occur with minor vesicular lesions on genitalia. After infection the bitch develops an immunity so only the initial litter is affected • Neonates: infected *in utero* or shortly after birth (ingestion or inhalation): 'Fading puppy syndrome' (pups less than 2 weeks old): dullness, depression, incessant crying, soft faeces and abdominal pain. Terminal opisthotonos and seizures can occur but are not always apparent. A meningoencephalitis is present. The few pups that survive may show persistent CNS signs such as blindness and ataxia	• Virus isolation from adrenals, kidneys of pups or external genitalia of adults • FA: frozen sections • Histopathology • Serology: VN, IFA
CRYPTOCOCCOSIS	*Cryptococcus neoformans*	CNS and eyes are most commonly affected in dog. Head-tilt, nystagmus, facial paralysis, ataxia, circling, neck pain, dilated pupils, seizures, and general paresis can occur. Skin lesions are seen in about 25 per cent of cases. In some areas pigeon faeces can be a source of the yeast	• Gram or culture on CSF, exudates, aqueous or vitreous humour of eye • Histopathology
MENINGITIS OR BRAIN ABSCESSES	*Staphylococcus intermedius*, streptococci, *Pasteurella multocida*, *Nocardia asteroides*, disseminated fungal pathogens	Entry of pathogens: • Haematogenous: may enter subarachnoid space from extracranial foci • Arterial spread: bacteraemia and the outcome depends on virulence and underlying damage to CNS defences • Local: from paranasal sinuses or tympanic bullae Signs depend on site and severity of infection. Seizures can be caused by excessively high fever, forebrain oedema and increased intracranial pressure	• CSF evaluation: cytology biochemistry; Gram-stained smears; culture • Histopathology
OLD-DOG ENCEPHALITIS	Associated with canine distemper virus but pathophysiology is uncertain	Occurs in older adult dogs. There may not be a history of previous generalised canine distemper. Marked by encephalitic signs of ataxia, compulsive movements such as head-pressing, or continual pacing and an incoordinated high-stepping gait. Canine distemper viral antigen can be demonstrated in brain by FA technique. Convulsions and neuromuscular twitching do not seem to occur in the old-dog encephalitis syndrome	• History: canine distemper earlier in life • FA technique on brain tissue for Ag

Table 172. Infectious diseases of dogs. *(continued)*

NERVOUS SYSTEM: dogs

Disease	Agent(s)	Comments	Diagnosis
PROTOTHECOSIS	*Prototheca zopfii* or *P. wickerhamii* (colourless algae)	Usually a disseminated disease with bloody diarrhoea, weight loss, debility and blindness. Internal lesions are widespread as granular foci. In about 40 per cent of the generalised cases nervous signs are seen with ataxia, incoordination or paresis. Necrotic foci occur in the brain with numerous protothecal cells present in the lesions.	• Clinical signs: bloody diarrhoea and blindness • Histopathology: *Prototheca* spp. present in lesions • Culture: will grow on agar media
PSEUDORABIES	Porcine herpesvirus 1	Disease hyperacute and fatal in dogs, with a course of 48 hours or less. Intense pruritis most commonly occurs in the head region and self-mutilation results. Signs include hypersalivation, hyperaesthesia, convulsions, changes in behaviour and deficits in cranial nerve function such as head tilt, paresis of facial muscles and voice change. A few atypical cases reported where aggression was seen	• Virus isolation from brain or spleen (not always successful in dogs) • FA on brain stem • History of association with pigs or eating raw offal • Serology: VN
RABIES	*Lyssavirus (Rhabdoviridae)*	Incubation period averages 2–12 weeks but can be up to 6 months. Virus is excreted in saliva 1–14 days before clinical signs appear. Stages of disease are: • Prodromal (2–3 days) changes in temperament • Furious stage (1–7 days): dogs are aggressive, hyperactive and lose all fear of humans • Paralytic (dumb) stage within 10 days of first signs: dropped jaw, excessive salivation, choking noises and finally, generalised paralysis, coma and death	• Clinical signs • FA on brain tissue • Histopathology: Negri bodies in brain
TETANUS	*Clostridium tetani*	Although tetanus occurs in dogs and cats they are comparatively resistant to toxin with incubation period of up to 3 weeks. Localised tetanus can occur with stiffness of a muscle or of a limb. This may progress to generalised tetanus. Dogs tend to have a worried facial expression, erect ears and show stiffness when walking. Mild stimuli can lead to tetanic spasms and these continue until death occurs from respiratory arrest. *C. tetani* spores enter via wounds or following surgery	• History of wound or surgery • Differentiate from strychnine poisoning • Gram-stained smear of necrotic tissue if wound is present

Table 172. Infectious diseases of dogs. *(continued)*

Disease	Agent(s)	Comments	Diagnosis
NERVOUS SYSTEM: dogs			
TOXOPLASMOSIS	*Toxoplasma gondii*	Disseminated toxoplasmosis can occur in normal dogs that ingest high numbers of oocysts or bradyzoites or in neonates and immunocompromised animals. There is an affinity for lymphoid tissue, lungs, liver, heart and skeletal muscle. Fever, jaundice and muscle stiffness (myositis) are seen. CNS signs occur in chronic post-natal infections with cerebral inflammation, eye lesions, hind leg paresis and cerebellar ataxia. The disease is often associated with canine distemper due to immunosuppression by the virus	• Histopathology • Serology: IHA, IFA, ELISA
MUSCULOSKELETAL SYSTEM: dogs			
ARTHRITIS	Staphylococci, streptococci	Non-specific infection of one or more joints, either following a bacteraemia or traumatic injury	• Clinical signs • Culture of joint aspirates
COCCIDIOIDOMYCOSIS	*Coccidioides immitis*	Primary lesions in dogs are invariably in lungs. Dogs with disseminated disease have a chronic cough, cachexia, lameness, enlarged joints and fever. The fungus can cause an osteomyelitis	• History of visit to endemic area • Radiography • Clinical signs • Direct microscopy on aspirates or exudates for spherules • Culture should only be attempted using a biohazard cabinet
LYME DISEASE	*Borrelia burgdorferi*	Fever and arthritis involving the limb joints are the main signs. Lameness is recurrent and may progress to a chronic arthritis. Vector:*Ixodes* ticks	• History: endemic area • Clinical signs • Serology: ELISA or IFA • Demonstration of pathogen in joint fluid
NASAL GRANULOMA	*Aspergillus fumigatus*	Lesion usually in region of ventral maxilloturbinate bone. Unilateral nasal discharge, serous to mucopurulent and epistaxis occurs. The mucosa and underlying bone may be necrotic with loss of bone definition in radiographs	• Radiography • Clinical signs • Culture from biopsy

Table 172. Infectious diseases of dogs. *(continued)*

MUSCULOSKELETAL SYSTEM: dogs			
Disease	**Agent(s)**	**Comments**	**Diagnosis**
OSTEOMYELITIS / DISCOSPONDYLITIS	Aetiology includes: *Staphylococcus intermedius, Brucella canis, Nocardia* spp. and streptococci	Comparatively common condition in young to middle-age adult dogs, especially males and those of heavy breeds. Condition is often due to blood-borne bacterial emboli. Signs range from hyperaesthesia to severe paresis/paralysis	• Clinical signs • Radiography • Culture of bacterial pathogen
ROCKY MOUNTAIN SPOTTED FEVER	*Rickettsia rickettsii*	Acute illness with high fever, lymphadenopathy, oedema of limbs and sheath and scrotum of males. Haemorrhages may be seen in ocular, oral, genital mucosae and on non-pigmented skin. Vector: ticks	• History (endemic area) • Clinical signs • FA on skin biopsy • Serology: IFA, ELISA
TETANUS	*Clostridium tetani*	Localised (ascending) tetanus can occur in dogs with stiffness in one or more limbs. This may progress to generalised tetanus with a stiff gait, erect ears, worried expression and tetanic convulsions	• History of a wound or surgery • Differentiate from strychnine poisoning • Gram-stained smear of necrotic material if a wound is present
TOXOPLASMA MYOSITIS	*Toxoplasma gondii*	A condition that can occur in neonatal pups. There is muscle wasting, stiffness of limbs and progressive paresis or tetraparesis	• Clinical signs • Serological tests: IHA, IFA, ELISA

RESPIRATORY SYSTEM : dogs			
ACTINOMYCOSIS (CANINE)	*Actinomyces viscosus*	Clinical signs and lesions indistinguishable from canine nocardiosis. Granulomatous skin lesions and/or bloody fluid and granulomas in thoracic cavity	• Gram and MZN direct smears (*A. viscosus* MZN-ve and *N. asteroides* MZN+ve) • Culture • Histopathology
CANINE DISTEMPER	*Morbillivirus (Paramyxoviridae)*	Strains vary greatly in virulence. Clinical signs reflect the predilection sites: respiratory tract, intestinal tract, CNS and skin. *B. bronchiseptica* often responsible for purulent oculonasal discharges	• Intracytoplasmic IBs in smears from conjunctiva • Pathology • FA: impression smears from urinary bladder epithelium, cerebellum or lymph nodes • Virus isolation: TC or inoculation of ferrets • Serology: VN, IFA, ELISA (rising titres)

Table 172. Infectious diseases of dogs. *(continued)*

RESPIRATORY SYSTEM: dogs

Disease	Agent(s)	Comments	Diagnosis
CHRONIC RHINITIS	*Aspergillus fumigatus*	Chronic nasal discharge (often unilateral), sneezing and possibly epistaxis	• Radiography: turbinate bone destruction • Endoscopy: fungal plaques • Culture from biopsy not just from nasal discharge • Histopathology on biopsy
INFECTIOUS CANINE HEPATITIS (ICH)	Canine adenovirus 1	Acute or chronic hepatitis. Immune complex disease such as glomerulonephritis and 'blue-eye' occur. Other endothelial tissues such as lungs, lymph nodes and brain are often affected	• Histopathology • FA: on frozen sections • Serology: HAI, VN for rising titres • Virus isolation from nasal secretions, urine
'KENNEL COUGH' (CANINE INFECTIOUS TRACHEOBRONCHITIS)	Major agents: Canine parainfluenza 2, canine distemper virus, canine adenovirus 2, *Bordetella bronchiseptica*	The condition is a mild upper respiratory tract infection that is self-limiting. Coughing usually occurs for a few days but longer if complicated (laryngotracheitis) by secondary bacterial invaders. Most commonly seen in kennels where a number of young dogs are kept together	• Virus isolation early in the disease • Culture: check on primary or secondary bacterial pathogens • Serology (rising Ab titre)
MYCOTIC INFECTIONS		Sporadic, chronic diseases where the primary site of infection, in dogs, is usually the respiratory tract. Dissemination may occur	• Direct microscopy • Histopathology: PAS stain • Culture: 25°C and 37°C for the dimorphic fungi • Serology
Cryptococcosis	*Cryptococcus neoformans*	Chronic URT signs and possible CNS involvement	
Histoplasmosis	*Histoplasma capsulatum*	Chronic respiratory and intestinal signs	
North American Blastomycosis	*Blastomyces dermatitidis*	Primary respiratory involvement and dissemination to the skin	
Coccidioidomycosis	*Coccidioides immitis*	Chronic cough and dissemination to bones and other organs	
NOCARDIOSIS (CANINE)	*Nocardia asteroides*	Indistinguishable from canine actinomycosis on clinical grounds. Granulomatous skin lesions and/or bloody fluid and granulomas in thoracic cavity	• Gram and MZN direct smears (*A. viscosus* MZN- / *N. asteroides* MZN+) • Culture • Histopathology

Table 172. Infectious diseases of dogs. *(continued)*

Disease	Agent(s)	Comments	Diagnosis
RESPIRATORY SYSTEM: dogs			
TUBERCULOSIS (CANINE)	*Mycobacterium tuberculosis* *M. bovis*	Dogs are susceptible to both mycobacteria. Lymph node lesions can resemble carcinomas, while lung and liver lesions are exudative. Grey-white areas of bronchopneumonia often occur with cavitation. Fluid in thorax	• Direct ZN-stained smear • Culture • Typing of mycobacteria (tuberculin test unreliable in dog)
SKIN: dogs			
ALGAL PATHOGENS			
Protothecosis	*Prototheca zopfii* or *P. wickerhamii* (colourless algae)	Cutaneous lesions can occur as part of a disseminated disease (bloody diarrhoea, CNS signs and eye involvement). The skin lesions are chronic and characterised by ulcers with crusty exudates in the skin of the trunk and extremities	• Histopathology on biopsy • Direct microscopy • Culture and identification
BACTERIAL INFECTIONS (PRIMARY)		Occur in normal skin and a single pathogen is usually isolated. There is a characteristic lesion or pattern of lesions. Antimicrobial therapy is usually successful	
Cutaneous canine actinomycosis	*Actinomyces viscosus*	Identical syndromes to canine nocardiosis. Thoracic form is more common than granulomatous skin lesions	• Clinical signs • Microscopy: Gram/MZN (*Nocardia* MZN +ve and *A. viscosus* MZN -ve) • Culture
Cutaneous canine nocardiosis	*Nocardia asteroides*	• Cutaneous form: granulomatous skin lesions with discharging sinuses • Thoracic form: lung involvement and accumulation of purulent, bloody fluid in thorax	• Clinical signs • Microscopy: Gram/MZN (*Nocardia* MZN +ve and *A. viscosus* MZN -ve) • Culture
Dermatophilosis (streptothricosis)	*Dermatophilus congolensis*	Rare in dogs but natural infections have been described. Systemic signs are minimal. Lesions occur in hairy regions of body. There is dry adherent scab formation (entrapped in hair) and removal of scabs leaves an ulcerated area	• Gram or Giemsa-stained smear from scab • Culture if necessary

Table 172. Infectious diseases of dogs. *(continued)*

SKIN: dogs

Disease	Agent(s)	Comments	Diagnosis
Folliculitis	Bacterial and fungal pathogens or *Demodex* spp.	Infection of hair follicles may be localised or generalised. Most common in short-haired dogs. The characteristic lesion is a pustule or papule surrounding a hair shaft **Furunculosis:** A more severe form of folliculitis in which rupture of the follicle causes spread of infection into the dermal tissues **Carbuncle:** A localised area in which there is multiple furunculosis and consequent sinus formation. The lesion is circumscribed with hair loss. There is thickening of the skin and numerous sinus openings from which there is a purulent discharge	• Direct microscopy on aspirated fluid if possible or from exudates • Culture and identification
Staphylococcal pustular dermatitis	*Staphylococcus intermedius*	Generalised pustular dermatitis with multiple pustular lesions over the entire body of either puppies or adults	• Culture of aspirated fluid from pustules
BACTERIAL INFECTIONS (SECONDARY)		Occur in damaged skin and a mixed bacterial flora is usually isolated. The primary skin disorder and/or predisposing cause must be resolved before antimicrobial therapy will be effective.	
Pyodermas (Classification can be based on age, depth of infection or anatomical site of the lesion)	*Staphylococcus intermedius* or *S. aureus* are commonly involved	**Juvenile pyoderma** is seen in pups under 4 months of age. Most common in short-coated breeds particularly Labradors, Bassets and Beagles. The initial lesion may be allergic in type. There is oedema of the area followed by a deep pyoderma often of head and lips. It is accompanied by a marked cervical and submandibular lymphadenitis. Prognosis is poor. In cases that recover, healing of the lesions is usually accompanied by permanent, disfiguring, hyperpigmented and hairless scarring of the affected areas **Acute moist pyoderma** ('Hot spot' or pyotraumatic dermatitis): a superficial pyoderma secondary to self - inflicted trauma in response to pruritis (flea infestation or some mange infestations). There is hair loss, erythema, exudation, excoriation and crusting. **General deep pyoderma** is seen in adult dogs. There is involvement of an extensive area of the trunk with numerous discharging openings interconnected by sinus tracks, that run through the dermis over much of the affected skin. The condition may be secondary to demodectic mange.	• Investigation of underlying causes including parasitic infestation • Direct microscopy on aspirates or exudates • Culture of aspirate from intact pustules • Histopathology on a punch biopsy SPECIMENS • Aspirate from intact pustules with fine hypodermic needle (ideal specimen). Surface should first be disinfected with 70 per cent ethyl alcohol • Surface swab is a less useful specimen • Punch biopsy for histopathology.

Table 172. Infectious diseases of dogs. *(continued)*

SKIN: dogs

Disease	Agent(s)	Comments	Diagnosis
Pyodermas *(continued)*	*Staphylococcus intermedius* or *S. aureus* are commonly involved	**Interdigital pyoderma** has a multifactorial aetiology, often demodectic mange + secondary bacterial infection. Most common in the front feet with swelling, erythema, hyperpigmentation of skin, discharging sinus openings and crusting. **Nasal pyoderma** occurs on bridge of nose and is most common in long-nosed dogs, especially in dogs that tend to root around in the ground. There are erosions, exudation and crusting of skin. **Callus pyoderma:** infection of a callus or pressure point that tends to occur in heavy breeds of dog. **Pyoderma in moist areas** such as **lip fold pyoderma** (lower longitudinal lip folds), **facial fold** (between nose and eyes of brachycephalic breeds) and **tail fold pyoderma** (tailhead area). **Canine acne:** pustule plus secondary bacterial infection, usually in the chin region. Predisposing cause can be a hormonal disturbance (Cushing's syndrome). In younger animals the condition may resolve at puberty or can persist for life. There is often swelling and erythema of the chin and bacteria invade the sebaceous and apocrine glands	• Investigation of underlying causes including parasitic infestation • Direct microscopy on aspirates or exudates • Culture of aspirate from intact pustules • Histopathology on a punch biopsy SPECIMENS • Aspirate from intact pustules with fine hypodermic needle (ideal specimen). Surface should first be disinfected with 70 per cent ethyl alcohol • Surface swab is a less useful specimen • Punch biopsy for histopathology.
Interdigital Cyst (Interdigital granuloma)	*Staphylococcus intermedius* and others	Primarily a foreign body granuloma with the foreign body being fragments of implanted hair shaft. Secondary bacterial infection may occur. There is an inherited predisposition in certain breeds or individuals	• Histopathology on punch biopsy • Microscopy and culture from deep tissue
Staphylococcal Scalded Skin Syndrome (SSSS)	*Staphylococcus intermedius*	A generalised dermatitis with raised, reddish lesions. Within 24–48 hours the superficial skin layers separate from the underlying epidermis and there is a characteristic slipping or wrinkling of the superficial skin layer. This separated epidermal layer sloughs leaving an ulcerated area. Systemic reaction may or may not be present	• Blood culture at acute stage of the disease (no bacteria in epidermal bullae) • Histopathology on a biopsy
FUNGAL PATHOGENS			
Canine Cryptococcosis	*Cryptococcus neoformans*	Uncommon in dogs. Infection occurs by inhalation (high titre of yeast in pigeon faeces). Primary lesion in lungs with dissemination to the CNS and eyes. About 25% of infected dogs have skin lesions: ulceration of nose, lips, nail beds and buccal cavity	• Clinical signs • Microscopy: yeast with a thick capsule • Culture at 37°C • Histopathology

Table 172. Infectious diseases of dogs. *(continued)*

SKIN: dogs

Disease	Agent(s)	Comments	Diagnosis
FUNGAL PATHOGENS *(continued)* **Canine Sporotrichosis**	*Sporothrix schenckii*	Uncommon in dogs. It is usually a cutaneous infection with multiple, nodular, ulcerated lesions that form crusts and alopecia occurs. Limited to skin and usually involves head, trunk or limbs. Dissemination of infection is rare. The fungus is a saprophyte in soil rich in organic matter or on the bark of trees and wooden posts	• Microscopic examination of exudate: cigar-shaped yeast • Culture: 25 and 37°C. Dimorphic fungus
North American Blastomycosis	*Blastomyces dermatitidis*	Primary lesion in dogs is almost always in the lungs. Soil organism in certain areas of Africa, Eastern USA and in Central America. Skin lesions seen in about 40 per cent of dogs with disseminated disease. They are granulomatous with sinuses and exudates	• Clinical signs • Radiography for lung lesions • Microscopy: thick-walled yeast budding on a broad base • Culture at 25°C and 37°C
Ringworm	*Microsporum canis*	Classical ring-shaped localised lesions (most common); chronic generalised infection; infection of nail-beds (onychomycosis); or kerions	SPECIMENS: plucked hairs and skin scrapings from edge of lesion. • Clinical signs • Wood's lamp for *M. canis* and *M. distortum*. Apple-green fluorescence • Hairs in 10 per cent KOH for arthrospores • Culture: Sabouraud's agar at 27°C for up to 3 weeks • Identification: colonial morphology and macroconidia
	Microsporum distortum	Rare, probably a mutant of *M. canis.* It has distorted macroconidia	
	Microsporum gypseum	Uncommon in dogs (more common in horses). Recorded infecting noses of dogs fond of rooting in soil or in dogs with an immunodeficiency. Geophilic fungus	
	Trichophyton mentagrophytes	Often acquired from rodents. Lesions can be: • Localised on muzzles and front paws and legs • Generalised infection involving much of the body	
	Trichophyton erinacei	Seen on muzzles, front paws and legs of dog that worry hedgehogs. Spines of hedgehogs break skin and aid establishment of infection. *T. erinacei* can be regarded as a variant of *T. mentagrophytes*	

Table 172. Infectious diseases of dogs. *(continued)*

SKIN: dogs

Disease	Agent(s)	Comments	Diagnosis
VIRAL PATHOGENS			
Canine Distemper	*Morbillivirus* *(Paramyxoviridae)*	Localised lesions associated with canine distemper can indicate whether or not CNS signs will develop. **Hard pad:** hyperkeratosis of footpads and often nose. Strong correlation with the development of nervous signs. If the dog survives, the thickened skin of the footpads sloughs but the nose lesions are usually permanent. **Rash on abdomen in pups:** these pups rarely develop the nervous signs of canine distemper	• Clinical examination of skin lesions • History: previous signs of generalised canine distemper • Serology: VN, IFA
Canine Papillomatosis	Canine papillomaviruses *(Papovaviridae)*	**Canine oral papillomatosis:** most common form and seen in young dogs under 2 years of age. Appears to be contagious within a group of dogs but is self-limiting and dogs are not infected a second time. There are white cauliflower-like growths in buccal cavity and on lips where they can be pigmented. Surgical interference is only needed if there is difficulty in eating. Autogenous vaccines are said to be effective **Ocular and facial papillomatosis:** not common and can involve cornea, conjunctiva and eyelid margins. Recorded in dogs 6 months to 4 years of age. These warts may be caused by the oral papillomavirus **Cutaneous papillomatosis:** occurs in mature or older dogs and the papillomas do not regress spontaneously. Lesions are present anywhere on body and are thought to be caused by a different and distinct papillomavirus	• Gross appearance of lesions • Histopathology: cryosurgery biopsy • EM for viral particles

Table 173. Infectious diseases of cats.

BUCCAL CAVITY: cats			
Disease	**Agent(s)**	**Comments**	**Diagnosis**
FELINE CALICIVIRUS INFECTION	Feline calicivirus	Ulcerative stomatitis is a constant finding in this respiratory disease. Ulcers occur on the tongue and elsewhere in the buccal cavity. URT signs are usually present and occasionally those of pneumonia	• Clinical signs • Virus isolation and identification • Serology: VN, IFA
FELINE REOVIRUS INFECTION	*Reovirus*	This virus is thought to play a minor part in the feline respiratory complex. It may cause conjunctivitis, gingivitis, lacrimation and depression	• Virus isolation • Serology: VN to demonstrate rising Ab titre
FELINE VIRAL RHINOTRACHEITIS	Feline herpesvirus 1	Ulceration of tongue and buccal cavity mucosa is a less constant sign compared to calicivirus infection. The disease is severe in neonates and often causes a milder URT infection in older kittens and cats	• Histopathology or FA on conjunctival scrapings early in disease • Virus isolation • Serology: VN, IFA
GINGIVITIS (SECONDARY) **Feline immunodeficiency virus disease** **Feline leukaemia**	 *Lentivirus (Retroviridae)* *Retroviridae*	A gingivitis or ulceromembranous stomatitis in cats may be secondary to a deficiency, dental calculus or a chronic debilitating disease particularly due to feline immunodeficiency virus or feline leukaemia virus infection. Reddening and swelling of the gingival margins is followed by ulceration and necrosis of gingivo-alveolar tissue. There is a foul odour and brown slimy saliva. Lesions are caused by bacteria, normally commensals in the buccal cavity	• Investigation of a primary disease • Clinical signs • Gram-stained smear of necrotic debris for bacteria
MYCOTIC STOMATITIS	*Candida albicans*	Usually seen in kittens and often follows a prolonged course of antibiotic therapy. Characterised by ulceration and elevated, soft, white patches on the oral mucosa. The periphery of the lesions is usually reddened and the affected patches may coalesce	• History of antibiotic therapy • Clinical signs • Isolation of pathogen
GASTROINTESTINAL TRACT: cats			
ENTERITIS / COLITIS (BACTERIAL)	Diverse aetiology including Gram -ve bacteria and occasionally fungi	Non-specific diarrhoea can be predisposed to by an immature natural flora (neonates), or changes to the indigenous flora (prolonged antibacterial therapy) with subsequent overgrowth of opportunistic enteric bacteria	• Clinical signs • Isolation of the pathogen in heavy culture

Table 173. Infectious diseases of cats. *(continued)*

GASTROINTESTINAL TRACT: cats

Disease	Agent(s)	Comments	Diagnosis
FELINE IMMUNODEFICIENCY VIRUS DISEASE	*Lentivirus (Retroviridae)*	Infected cats tend to suffer intermittently or persistently from intestinal, respiratory, urinary tract or ear and eye infections	• Serology: ELISA or IFA
FELINE INFECTIOUS PERITONITIS (FIP)	Feline coronavirus *(Coronaviridae)*	Clinical signs are most common in 6 month to 2 year-old cats. Three syndromes are recognised: • **Effusive ('wet') FIP**: characterised by an accumulation of fibrin-rich fluid in the abdominal cavity with progressive, painless, enlargement of the abdomen. The course is 1-8 months and the prognosis is poor • **Noneffusive ('dry') FIP**: insidious onset with clinical signs reflective of pyogranulomatous involvement of many organs including lungs, liver, pancreas, brain and eyes. Weight loss, depression, anaemia and fever are constant signs. Course often protracted (over 1 year) • Combination of 'wet' and 'dry' forms: this is less common	• Clinical signs • Gross and histopathology • Serology: IFA, ELISA
FELINE LEUKAEMIA	*Retroviridae*	• A panleukopenia-like syndrome as a form of the non-neoplastic associated diseases can occur with secondary diarrhoea • A feline leukaemia virus-induced enteritis, as a distinct syndrome, has been described. Vomiting and diarrhoea are the predominant signs • Alimentary lymphoma can occur in older cats. Clinical signs include vomiting, diarrhoea and weight loss	• Detection of viral antigen: ELISA (serum) and FA (blood smears)
FELINE PANLEUKOPENIA	Feline parvovirus	Most infections are subclinical. Virus is shed from all body secretions during active infection and for 6 weeks after recovery. No extended carrier state occurs but virus is viable at room temperature for 1 year. Main targets are lymphoid tissue (immunodepression), small intestine (necrosis plus secondary Gram -ve bacterial overgrowth with vomiting, diarrhoea and endotoxaemia) and bone marrow (leukopenia). Fever and oculonasal discharges present. If infected *in utero* or before 2-3 weeks of age, cerebellar hypoplasia resulting in ataxia can occur	• Pathology. • Detection of virus in faeces in acute cases: EM, ELISA • Virus isolation • FA: frozen sections • Serology: HAI, VN

Table 173. Infectious diseases of cats. *(continued)*

Disease	Agent(s)	Comments	Diagnosis
GASTROINTESTINAL TRACT: cats			
HISTOPLASMOSIS (FELINE)	*Histoplasma capsulatum*	More common in dogs than in cats. Chronic intractable cough, diarrhoea and emaciation	• Histopathology: small yeast in macrophages
LIVER: cats			
FELINE INFECTIOUS PERITONITIS (FIP)	Feline coronavirus *(Coronaviridae)*	Granulomatous lesions in the liver may occur in the noneffusive ('dry') form of FIP. Signs include progressive jaundice and anorexia	• Gross pathology • Histopathology • FA for viral Ag • Serology: ELISA, IFA
GENITAL SYSTEM : cats			
FELINE INFECTIOUS PERITONITIS	Feline coronavirus	'Wet' form of disease in males: abdominal distention with viscid yellow fluid (peritonitis) and enlarged scrotum due to inflammation of the tunica vaginalis	• Clinical signs • Histopathology: pyogranuloma • Serology: IFA
FELINE LEUKAEMIA	*Retroviridae*	Foetal resorption or abortion (reproductive failure) can be a part of the non-neoplastic disease syndrome associated with the virus	• FA on blood smears and ELISA on serum to detect virus in the queen
FELINE PANLEUKOPENIA	Feline parvovirus	Result of infection: • *In utero* infection: death of embryos and abortion. Queen may not show systemic signs of illness • Infection of foetus in last 2 weeks of gestation or in first 2 weeks of life: cerebellar ataxia • Kittens over 3 weeks old or susceptible adults: feline panleukopenia	• Pathology • Detection of virus in foetal tissues: FA, ELISA • Virus isolation • Serology: HAI, VN
FELINE RHINOTRACHEITIS	Feline herpesvirus 1	Syndromes include: • Abortions in queens • Kittens and susceptible young adults: severe URT infection with destruction of turbinates in some cases • Corneal ulceration in some adult cats Infection in cats over 6 months of age is usually mild or inapparent	• Histopathology on foetal tissue • Virus isolation • Serology for Ab in dam: VN

Table 173. Infectious diseases of cats. *(continued)*

Disease	Agent(s)	Comments	Diagnosis
GENITAL SYSTEM: cats			
KITTEN MORTALITY COMPLEX	Aetiology includes: feline leukaemia virus, feline parvovirus, less certainly feline coronavirus and bacterial septicaemias in neonates	The complex has been divided into two main syndromes: • Reproductive failure: repeat breeding, abortions and stillbirths • Fading kitten syndrome: deaths of neonates	• Clinical signs • Isolation of bacterial pathogens • Serology for viral pathogens
PYOMETRA	Miscellaneous bacteria	Pathophysiology similar to that in bitch except that queens are induced ovulators requiring coitus before progesterone secretion occurs. This results in a lower frequency of pyometra in queens	• Clinical signs • Radiography to check for retained foetus • Guarded swab from cervix for culture
VAGINITIS	Miscellaneous bacteria	No discharge may be obvious because of the fastidious nature of cats with excessive licking of the vulva. Otherwise as for bitches	• Guarded swab from anterior of vagina for culture
URINARY SYSTEM : cats			
CYSTITIS AND URETHRITIS	*Escherichia coli*, *Staphylococcus* species, *Proteus* species or *Klebsiella pneumoniae*	Similar to the condition in dogs but it is comparatively uncommon in cats. Signs include voiding of small amounts of urine, dysuria and urine that is cloudy and may contain blood	• Clinical signs • Isolation of pathogen in heavy growth from urine
FELINE IMMUNODEFICIENCY VIRUS INFECTION	*Lentivirus* *(Retroviridae)*	Recurrent, bacterial urinary tract infections can be associated with the immunosuppression induced by this virus	• Serology: ELISA, IFA • Isolation of the secondary bacterial pathogens
FELINE INFECTIOUS PERITONITIS	Feline coronavirus	Pyogranulomatous lesions are commonly present in kidneys of cats with generalised, noneffusive FIP. Clinical signs of renal insufficiency may result	• Histopathology • Serology: IFA, ELISA
FELINE LEUKAEMIA	*Retroviridae*	Persistent high levels of circulating viral antigens can lead to the formation of immune complexes with resultant glomerulonephritis	• Detection of viral antigen in blood: ELISA, FA

Table 173. Infectious diseases of cats. *(continued)*

EYES AND EARS: cats

Disease	Agent(s)	Comments	Diagnosis
FELINE PNEUMONITIS	*Chlamydia psittaci*	Lesions are confined principally to the upper respiratory tract and conjunctival membranes. Affected mucosa is reddened, swollen and covered with exudate. Signs include fever, inappetence, oculonasal discharges, sneezing and coughing. Course lasts about 2 weeks with occasional relapses	• Clinical signs • Conjunctival scrapings for inclusions • Culture: TC or inoculation of yolk sac of fertile eggs
FELINE REOVIRUS INFECTION	Reovirus 3	Mild URT disease, significance in feline respiratory complex is still undetermined. There may be conjunctivitis, lacrimation, photophobia, gingivitis and depression	• Virus isolation • Serology: VN (rising Ab titre)
FELINE VIRAL RHINOTRACHEITIS	Feline herpesvirus 1	Severe URT infection in neonates with a purulent conjunctivitis, necrosis and resorption of turbinate bones and fever. Milder infection in older cats but corneal ulceration can occur in adults and pregnant queens may abort. The lungs are not usually involved but occasional ulcers in buccal cavity occur	• Histopathology (IBs early) • Virus isolation • Serology:VN
MYCOPLASMAL CONJUNCTIVITIS	*Mycoplasma felis*	Conjunctivitis with hypertrophy of conjunctival mucosa giving a red, velvet appearance	• Clinical signs • Isolation and identification of mycoplasma
OCULAR ABNORMALITIES		Local eye abnormalities can be due to trauma with secondary bacterial invasion. However, the eye can often be involved in systemic diseases. Occasionally the systemic manifestation remains occult	• Diagnosis of the primary disease in all cases
Cryptococcosis	*Cryptococcus neoformans*	Blindness as part of CNS syndrome	
Feline immuno-deficiency virus disease	Feline lentivirus (*Retroviridae*)	Anisocoria has been observed. More commonly there is a chronic ocular discharge as part of a persistent URT infection	
Feline infectious peritonitis (FIP)	Feline coronavirus	Ocular signs, most common in noneffusive FIP, are usually bilateral and present as an anterior uveitis with keratic precipitates ('mutton fat' deposits)	

Table 173. Infectious diseases of cats. *(continued)*

Disease	Agent(s)	Comments	Diagnosis
EYES AND EARS: cats			
OCULAR ABNORMALITIES			
Feline leukaemia (FeL)	*Retroviridae*	Anterior uveitis is the most common ocular syndrome in FeL. Paradoxical pupil movements can occur in healthy FeLV-infected cats. These are irregular periods of anisocoria, bilateral mydriasis or miosis without evidence of intraocular inflammation. Most of these cats eventually develop systemic signs of FeL	
Toxoplasmosis	*Toxoplasma gondii*	Ocular disease is more commonly seen in cats than in dogs and is often a posterior retino-choroiditis that can result in retinal detachment	
OTITIS EXTERNA, MEDIA AND INTERNA	Non-specific bacteria and fungi	Similar to conditions in dogs but are less common in cats	
NERVOUS SYSTEM : cats			
CONGENITAL CNS LESIONS	Feline parvovirus	Due to infection *in utero* or when neonate is under 2-3 weeks of age. Cerebellar hypoplasia occurs with consequent ataxia. The debility is permanent	• History of illness in dam (often subclinical) • Clinical signs in kitten • Pathology • FA for antigens
CRYPTOCOCCOSIS	*Cryptococcus neoformans*	Forms in cat:- • Respiratory: nasal granulomas and chronic nasal discharge • Cutaneous: firm nodules on face and head that ulcerate (30 per cent of cases) • CNS and ocular: seizures, ataxia and paresis. There is a dilated, unresponsive pupil and blindness due to retinal detachment or optic neuritis Pigeon faeces an be a source of infection	• Radiography to determine if nasal granuloma present • Gram-stained smear of exudates: budding yeast with thick capsule • Histopathology on biopsy material: capsulated yeast present in tissue
FELINE IMMUNODEFICIENCY VIRUS (FIV) INFECTION	*Lentivirus (Retroviridae)*	Neurological signs, including psychotic behaviour, dementia, facial twitching and seizures, can occur in infected cats. These signs are thought to be due to a direct effect of FIV involvement of the brain	• Serology: ELISA, IFA

Table 173. Infectious diseases of cats. *(continued)*

NERVOUS SYSTEM: cats

Disease	Agent(s)	Comments	Diagnosis
FELINE INFECTIOUS PERITONITIS (FIP)	Feline coronavirus *(Coronaviridae)*	Neurological sign may be seen in noneffusive FIP. Most frequent signs are posterior paresis or ataxia progressing to tetraparesis	• Gross and histopathology • Serology: IFA, ELISA
FELINE LEUKAEMIA/ SARCOMA COMPLEX	*Retroviridae*	Tumours, as part of the neoplastic syndrome or feline sarcoma virus activity, can cause CNS signs such as hind leg paralysis due to spinal cord compression	• Detection of virus in blood: ELISA or IFA
FELINE SPONGIFORM ENCEPHALOPATHY	Prion	A novel disease of cats with neurological signs similar to those seen in cattle with BSE. Reported in 5-12 year old cats. The main signs include ataxia, particularly of the hind limbs, becoming progressively worse and leading to a crouching stance, changes in temperament to timidity or aggression, hyperaesthesia, head-nodding, hypersalivation unrelated to feeding and abnormal grooming. Appetite and bodily condition remain normal	• Clinical signs • Electron microscopy on frontal lobe preparations for scrapie associated fibrils • Histopathology for neuronal vacuolation • Immunochemistry for PrP-related proteins
MENINGITIS (BACTERIAL)	Aetiology includes: *Staphylococcus aureus* and anaerobic bacteria	Signs include fever, hyperaesthesia, neck rigidity and ataxia	• CSF analysis and culture
PSEUDORABIES	Porcine herpesvirus 1	Cats with close direct or indirect contact with pigs or rats are at risk. The excitement stage is preceded by a period of sluggishness. There is salivation, persistent mewing and the cat resists being caught. Paresis or paralysis of the limbs may occur before death. The course is often rapid and pruritis may not be seen. Recovery in a few cases has been reported	• History: association with pigs and eating raw offal (pig or rat) • Histopathology • FA technique on brain for Ag • Virus isolation, (not always successful)
RABIES	*Lyssavirus (Rhabdoviridae)*	Cats often develop the furious form and attack suddenly, biting and scratching viciously at any moving object. The paralytic form follows on about the 5th day of illness with general paralysis and death. Mandibular and laryngeal paralysis is less common in the cat than in the dog	• FA on brain tissue • Histopathology: Negri bodies in brain (not always reliable in cats) • Virus isolation: i/c inoculation of mice or TC

Table 173. Infectious diseases of cats. *(continued)*

Disease	Agent(s)	Comments	Diagnosis
NERVOUS SYSTEM: cats			
TETANUS	*Clostridium tetani*	Cats are comparatively resistant to the toxin and incubation period is often up to 3 weeks. A deep and obvious wound is usually present in cats. Localised tetanus in one limb is not uncommon. Generalised tetanus is characterised by stiff legs and an outstretched tail. In fatal cases tetanic spasms leading to respiratory arrest occur	• Clinical signs • Gram-stained smear from deep in wound
TOXOPLASMOSIS	*Toxoplasma gondii*	Cats can act as definitive and intermediate host. They are usually healthy carriers but can develop either acute or chronic disease. Lesions in eye or CNS may occur. Signs include depression or hyperexcitability, tremors, seizures, paresis or paralysis	• Histopathology • Serology: IHA, IFA, ELISA
MUSCULOSKELETAL SYSTEM: cats			
ARTHRITIS (SUPPURATIVE)	Miscellaneous bacteria including *Pasteurella* spp., streptococci and anaerobic bacteria	Suppurative arthritis may occur as a sequel to a bite wound. The joint is swollen, hot and painful. The cat is often febrile	• Culture from aspirated joint fluid
CHRONIC PROGRESSIVE POLYARTHRITIS	Associated with feline syncytium-forming virus (FeSFV) *(Retroviridae)*	This condition is thought to be an immune-mediated disease. It affects males between 1.5-5 years of age. Lymphadenopathy, swollen joints and a stiff gait are constant findings. Feline leukaemia or feline immunodeficiency viruses may potentiate the ability of FeSFV to cause disease	• Clinical signs • Analysis of joint fluid • Serology: IFA, AGID
FELINE CALICIVIRUS INFECTION	Feline calicivirus	An atypical syndrome in 6-12 week-old kittens is attributable to this virus. There is transient muscle stiffness and joint pain. The affected kitten may resent being handled	• History of respiratory syndrome in other cats • Clinical signs • Serology: VN, IFA
FELINE VIRAL RHINOTRACHEITIS	Feline herpesvirus 1	Necrosis of turbinate bones can occur in severe infections in neonates. Signs include foul-smelling nasal discharge, conjunctivitis and bronchopneumonia	• Histopathology (IBs) or FA (virus) on conjunctival scrapings early in disease • Virus isolation from ocular or pharyngeal swabs • Serology: VN, IFA

Table 173. Infectious diseases of cats. *(continued)*

Disease	Agent(s)	Comments	Diagnosis
MUSCULOSKELETAL SYSTEM: cats			
OSTEOMYELITIS	Miscellaneous bacteria including staphylococci, streptococci, Gram -ve and anaerobic bacteria	Occurs most commonly as a sequel to open fractures or surgical repair of closed fractures. In acute cases there is heat, swelling and pain and a serous or purulent discharge from the wound. Chronic signs include soft tissue swelling, lameness and draining sinuses	• Culture and identification of bacterial pathogen
TETANUS	*Clostridium tetani*	Localised tetanus occurs comparatively commonly in cats because of their relative resistance to the toxin. Increased stiffness of a muscle or limb may be first noticed close to a wound site. Later the stiffness spreads leading to the characteristic stiff limbs and outstretched tail of generalised tetanus	• Clinical signs • Gram-stained smear from deep in wound
RESPIRATORY SYSTEM: cats			
CRYPTOCOCCOSIS (FELINE)	*Cryptococcus neoformans*	The main systems affected are URT, CNS (incoordination and blindness) and skin. The respiratory syndrome is the most common in cats. The signs are sneezing, snuffling and a unilateral or bilateral nasal discharge. These signs are related to granulomas in the nasal cavity. There is often a hard, subcutaneous swelling over the bridge of the nose ('Roman-nose' appearance). The regional lymph nodes are usually enlarged	• Microscopy: India ink (thick capsule) • Histopathology: PAS stain • Culture: Sabouraud agar
FELINE CALICIVIRUS INFECTION	Feline calicivirus	Continuous shedding of virus and carrier state probably for life. Ulcers in buccal cavity are pathognomonic. Other signs include pneumonia in susceptible cats, transient stiffness and joint pains in 6-12 weeks-old kittens and queens may abort. Some strains of virus cause an URT infection and ulcers only	• Clinical signs • Virus isolation • Serology: VN, IFA
FELINE IMMUNODEFICIENCY VIRUS DISEASE	*Lentivirus (Retroviridae)*	Recurrent or chronic respiratory and intestinal infections are common syndromes. Other conditions include gingivitis, stomatitis, emaciation, chronic skin disease, cystitis, lymphadenopathy, and neurological abnormalities	• Serology: ELISA and IFA

Table 173. Infectious diseases of cats. *(continued)*

RESPIRATORY SYSTEM: cats

Disease	Agent(s)	Comments	Diagnosis
FELINE INFECTIOUS ANAEMIA (HAEMOBARTONELLOSIS)	*Haemobartonella felis*	• Acute cases: fever, anaemia, jaundice, anorexia, weakness and splenomegaly occur • Chronic cases: these present with anaemia, weakness, depression and emaciation Dyspnoea is often seen, varying in severity with the degree of anaemia	• Clinical signs • Haematology • Blood smears stained with Giemsa or Wright's stain (erythrocytic bodies)
FELINE INFECTIOUS PERITONITIS (FIP)	Feline coronavirus *(Coronaviridae)*	Dyspnoea may occur in effusive FIP due to accumulation of fluid in the abdomen. A persistent cough, without noticeable dyspnoea, can be due to a pyogranulomatous pneumonia in noneffusive FIP. There is some evidence that some FIPV-infected cats develop mild respiratory disease (rhinitis and conjunctivitis) as a primary infection. Most of these cats recover to become healthy carriers of the virus	• Clinical signs • Gross and histopathology • Serology: IFA, ELISA
FELINE LEUKAEMIA	*Retroviridae*	May present as a panleukopenia-like disease as part of the non-neoplastic associated disease syndromes	• Detection of viral Ag in bloodstream: IFA or ELISA
FELINE PANLEUKOPENIA	Feline parvovirus	Virus is shed in faeces for 6 weeks after recovery from illness. Respiratory signs occur but main targets are lymphoid tissue, small intestine and bone marrow. *In utero* or neonatal (2-3 weeks) infections lead to cerebellar hypoplasia with ataxia in kittens	• Pathology • Detection of virus in faeces: EM, ELISA • Virus isolation • FA: frozen sections • Serology: HAI or VN
FELINE PNEUMONITIS	*Chlamydia psittaci*	Signs include purulent conjunctivitis, rhinitis with sneezing and coughing. Chronic infections with relapses are common	• Giemsa or MZN staining of smears from conjunctival scrapings • FA technique (conjunctival smears) • Culture: TC (McCoy cells) or yolk sac of fertile eggs
FELINE REOVIRUS INFECTION	*Reovirus*	Mild URT disease, significance in feline respiratory complex is still undetermined. The virus may cause conjunctivitis, lacrimation, photophobia, gingivitis and depression	• Virus isolation • Serology: VN or HAI to demonstrate rising Ab titre

Table 173. Infectious diseases of cats. *(continued)*

RESPIRATORY SYSTEM: cats			
Disease	**Agent(s)**	**Comments**	**Diagnosis**
FELINE VIRAL RHINOTRACHEITIS	Feline herpesvirus 1	Latent infections occur. Signs include purulent conjunctivitis, corneal ulcerations (adults), necrosis and resorption of nasal turbinates (neonates) and abortion in pregnant queens. Ulcers in buccal cavity are less common than in feline calicivirus infection	• Viral isolation from nasal, ocular or pharyngeal swabs • Histopathology: IBs early in disease • Serology: VN and IFA (rising Ab titres)
MYCOPLASMAL CONJUNCTIVITIS	*Mycoplasma felis*	Conjunctivitis with hypertrophy of conjunctival surface giving a deep-red, velvet appearance	• Isolation and serological identification of the mycoplasma
SKIN: cats			
ALGAL PATHOGENS			
Protothecosis	*Prototheca zopfii* or *P. wickerhamii*	Only the cutaneous form has been described in cats. Soft dermal mass on forehead or legs extending into underlying tissue. *Prototheca* in all stages of reproduction compose a large proportion of the granulomatous lesion	• Histopathology on biopsy for *Prototheca* in tissues • Culture: growth on agar media
BACTERIAL INFECTIONS (PRIMARY)		A single pathogen is usually isolated. A characteristic lesion or pattern of lesions is present	
Feline Leprosy	*Mycobacterium lepraemurium*	Infection probably acquired from rodents infected with the rat leprosy bacillus. The lesions consist of rapidly growing, painless, cutaneous nodules that are freely mobile. The lesions usually ulcerate and healing is slow	• ZN-stained smear from lesions: large numbers of acid-fast rods • No growth *in vitro*
***Mycobacterium smegmatis* Infection**	*Mycobacterium smegmatis*	This organism is usually a saprophyte. Infection has only been reported in cats. They appear to be exceptionally susceptible to atypical mycobacterial infections. There is thickening of the skin with numerous sinus openings some of which exude a purulent discharge. The skin of the ventral abdomen is most commonly affected	• ZN-stained smear • Histopathology on biopsy • Culture on Lowenstein-Jensen agar

Table 173. Infectious diseases of cats. *(continued)*

SKIN: cats

Disease	Agent(s)	Comments	Diagnosis
Nocardiosis (feline)	*Nocardia asteroides*	Infection is less common in the cat than in the dog. Nasal granulomas or pleuritis with pleural effusion can occur. Chronic, non-healing fistulous tracts associated with an underlying osteomyelitis have been described	• MZN and Gram-stained smears on pleural effusion • Culture: aerobic at 37°C for up to 4 days
Plague	*Yersinia pestis*	The disease is endemic in rodents in areas such as Southwest USA and is transmitted by fleas. Acute disease in cats is characterized by high fever, lymphadenopathy and abscessation of cervical and submandibular lymph nodes. The mortality is about 50 per cent. Infected cats and their fleas pose a health hazard for humans	Specimens: exudate from abscesses to Public Health Service laboratory • Identification of organism in exudates by FA • Culture and identification of pathogens
BACTERIAL INFECTIONS (SECONDARY)		Mixed bacterial flora is often isolated. The primary skin disorder or predisposing cause(s) must be resolved before antimicrobial therapy will be effective	
Cat Bite Abscesses (Cat fights)	*Pasteurella dagmatis* and other buccal cavity flora	One of the most common bacterial skin infections in cats (especially in tom cats). Many of these bacteria are part of the normal flora of the buccal cavity. Predilection sites for abscesses are root of tail, hindlegs, head and neck. The abscess may not be noticed until it ruptures. Abscesses in some sites (back of neck), where there is inadequate drainage may develop as chronic granulomas	• Clinical signs • Gram-stained smears on exudate • Culture
Pyoderma	Most common *Staphylococcus aureus* or *S. intermedius.*	Similar to the conditions in dogs but not as common. ***Acute moist pyoderma*** (pyotraumatic dermatitis) is secondary to self-inflicted trauma in response to pruritis (flea or mite infestations) ***Feline acne*** occurs on the point of the chin and results from comedone formation and subsequent bacterial infection involving sebaceous and apocrine glands	• Investigation of underlying causes • Direct microscopy on aspirates or exudates • Histopathology on punch biopsy • Bacterial culture of aspirates from intact pustules
FUNGAL PATHOGENS			
Cryptococcosis (feline)	*Cryptococcus neoformans*	Comparatively common infection in the cat. Systems affected are the upper respiratory tract, CNS and skin. In the cutaneous form, the skin of head and neck are most frequently affected. Multiple, firm nodules occur that tend to ulcerate with a raw, granular surface	• Clinical signs • Direct microscopy: yeast with large capsule • Culture at 37°C. Mucoid, yeast-like colonies

Table 173. Infectious diseases of cats. *(continued)*

SKIN: cats

Disease	Agent(s)	Comments	Diagnosis
Ringworm	*Microsporum canis*, other dermatophytes rare	Syndromes:- ***Subclinical:*** most common in young kittens but in adults a slight hair loss may be the only sign of infection. ***Classical:*** characteristic round lesions. ***Chronic generalised:*** almost total loss of hair with erythema and accumulations of sebum on skin surface	• Specimens: plucked hair and skin scrapings from the edge of the lesion • Wood's lamp for *M. canis* • Hairs in 10 per cent KOH for arthrospores • Culture and identification
Sporotrichosis (feline)	*Sporothrix schenckii*	Cutaneous lesions include shallow ulcers, granulomas, nodules and abscesses with draining sinuses. Disseminated disease is uncommon and may be related to immunodeficiency	• Cigar-shaped yeast in exudates • Culture: dimorphic fungus
NON-INFECTIOUS			
Pre-auricular hypotrichosis		Bilateral bald patches in front of the pinnae that are a natural condition in some cats. The skin is normal	• Differentiate from other causes of hypotrichosis or alopecia
VIRAL PATHOGENS			
Feline Poxvirus Infection	Cowpox virus in cats *(Poxiviridae)*	Distribution limited to parts of the UK and Scandinavia. A rodent reservoir is suspected. Initial lesion is at the site of a bite wound. A viraemia occurs with secondary and multiple skin lesions. The lesions are 3-15 mm, red, hairless or ulcerated and scabbed. They develop over several weeks. Lesions heal in 4-5 weeks with localised alopecia. The disease is mild unless the cat is immunosuppressed	• EM • Histopathology: biopsy • Virus isolation: TC or CAM of fertile eggs • Serology: VN
Feline Sarcoma Virus Infection	*Retroviridae*	Fibrosarcomas induced by this virus are usually rapidly growing causing multiple cutaneous or subcutaneous nodules that are locally invasive. Metastasis to other sites, such as lungs, often occurs	• Clinical signs • Detection of viral Ag of helper feline leukaemia virus: ELISA, FA

Table 174. Infectious diseases of domestic birds.

GENERALISED DISEASES: poultry

Disease	Agent(s)	Comments	Diagnosis
AVIAN SPIROCHAETOSIS	*Borrelia anserina*	Gallinaceous birds and water fowl. Occurs in tropical and temperate regions. Fever, leg weakness, listlessness, greenish diarrhoea and eventually complete paralysis can occur. PM: swollen spleen with haemorrhages. Heart and liver may be enlarged Vector: fowl ticks	• Darkfield or Giemsa-stained blood smears from birds with a bacteraemia • Serology: AGID • Inoculation of young susceptible birds
AVIAN TUBERCULOSIS	*Mycobacterium avium*	Many avian species. Chronic, granulomatous disease with a prolonged course. Infected birds with advanced lesions can excrete organisms in faeces. Progressive emaciation may be only sign. Granulomas in many body organs. In chicken lesions are most common in liver, spleen, bone marrow and gut. Lesions do not become calcified. This bacterium can sensitise cattle to the tuberculin and johnin skin tests. Very resistant bacterium: can survive in soil for up to 4 years	• Direct ZN-stained smears from lesions. Large numbers of acid-fast bacteria present in avian tuberculosis • Isolation: Lowenstein-Jensen medium • Histopathology • Avian tuberculin test, not completely reliable
COLIBACILLOSIS/ COLISEPTICAEMIA	*Escherichia coli*	Chickens and turkeys. Common invader especially after respiratory pathogens such as infectious bronchitis virus, mycoplasmas and Newcastle disease virus. ***Colisepticaemia:*** often follows a primary viral infection: pericarditis, perihepatitis and air sacculitis. Fibrinous exudates covering liver and other organs. Salpingitis and synovitis of hock joints commonly found in broiler birds ***Coligranuloma*** (Hjarre's disease): focal granulomas in caeca and elsewhere. It can appear very similar to avian tuberculosis ***Egg peritonitis:*** *E. coli* is isolated from most cases. Yolk mixed with exudate in abdominal cavity. Death due to a septicaemia ***Acute salpingitis:*** affected oviducts in laying birds are very enlarged and occupy much of abdominal cavity. *E. coli* usually isolated ***Omphalitis*** ('mushy-chick disease'): yolk sac of newly-hatched chicks infected. Distended abdomen	• Isolation of *E. coli* from lesions in heavy, pure culture • Identification by biochemical tests • Histopathology

Table 174. Infectious diseases of domestic birds. *(continued)*

GENERALISED DISEASES: poultry

Disease	Agent(s)	Comments	Diagnosis
FOWL CHOLERA	*Pasteurella multocida* (specific serotypes)	Chickens, turkeys, ducks, geese, swans and other birds are susceptible. ***Acute:*** sudden deaths or fever, depression, anorexia, ruffled feathers, diarrhoea, increased respiratory rate and rales. Haemorrhages throughout body organs and increased fluid in all cavities. Necrotic foci in liver ***Chronic:*** Swollen wattles and fibrinosuppurative exudate in sternal bursas and joints. Conjunctivitis and pharyngitis may occur	• Isolation on blood agar • Serotyping • Histopathology • Direct Giemsa-stained smear of blood in acute cases
FOWL PLAGUE	Virulent avian influenzaviruses (often subtypes H5 or H7)	Many avian species are susceptible. Generalised disease with high mortality. Diarrhoea, respiratory signs, cyanosis and oedema of wattles, head and comb, blood-stained oral and nasal discharges occur and there can be CNS involvement. Haemorrhages are present throughout body organs	• Isolation in fertile eggs (allantoic cavity) • Identification of virus by HAI • Serology: VN, ELISA, AGID
INFECTIOUS BURSAL DISEASE (GUMBORO DISEASE)	*Birnavirus* (*Birnaviridae*)	Chickens are the only important natural host. Transmission faecal-oral and possibly via egg. ***Chicks 0-2 weeks old:*** no clinical signs but 50 per cent incidence of immunodeficiency ***Chicks 3-6 weeks-old:*** depression, ruffled feathers, anorexia, diarrhoea, trembling and dehydration. Morbidity 100 per cent and mortality 20-30 per cent. Bursa enlarged (up to 5 times), striped and oedematous, but atrophied at time of death. All recovered chicks are immunodeficient and prone to other infections with poor vaccination takes ***Adults:*** unaffected but seroconvert and protect chicks, via antibodies in the egg, through critical period. As flock immunity develops little clinical disease is seen. Very resistant virus, hard to eliminate from a poultry farm	• Ag detection: FA, EM or AGID using bursal tissue • Isolation on CAM of fertile eggs • Histopathology

Table 174. Infectious diseases of domestic birds. *(continued)*

GENERALISED DISEASES: poultry

Disease	Agent(s)	Comments	Diagnosis
LYMPHOID LEUKOSIS	*Retroviridae*	Part of avian leukosis-sarcoma complex. Occurs naturally only in chickens. Peak mortality in 6-9 month-old birds. Egg transmission. Viruses can cause various types of tumours but lymphoid leukosis is most common. Transformation occurs in bursa 4-8 weeks after infection. Birds infected via egg or a few days after hatching are immunotolerant and persistently viraemic. About 3-20 per cent develop disease depending on genetic susceptibility. Infiltration of lymphoid cells into liver, spleen, kidneys and bursa. Birds depressed with pale comb prior to death. No vaccine available	• Gross pathology • Histopathology
MAREK'S DISEASE	*Gammaherpesvirinae* (*Herpesviridae*)	Chickens are the only important natural host. Peak mortality in 2-5 month-old birds. Replicates in feather follicles. No egg transmission. Four forms of the disease occur: • ***Visceral form***: this is becoming the most common and must be differentiated from lymphoid leukosis. Diffuse or focal infiltration of liver, spleen, kidney gonads, lungs and heart by lymphoblastic cells. Bursa not usually affected • ***Classical neural form:*** infiltration of the peripheral nerves: paralysis of legs and/or wings • ***Ocular form:*** grey iris and eccentric pupil • ***Cutaneous form:*** nodular tumours at feather follicle sites Vaccine available	• Histopathology • FA for Marek's tumour-specific antigen (MATSA) on surface of infected cells

Table 174. Infectious diseases of domestic birds. *(continued)*

GENERALISED DISEASES: poultry

Disease	Agent(s)	Comments	Diagnosis
NEWCASTLE DISEASE	Avian paramyxovirus 1 (*Paramyxoviridae*)	Many avian species are susceptible. Often introduced into countries via cage birds. Many strains: lentogenic (mild), mesogenic (moderate virulence), velogenic (virulent). ***Viscerotropic form:*** respiratory signs, diarrhoea, swelling and oedema around the eyes ***Neurotropic form:*** CNS signs of paresis of legs and wings, torticollis and tremors. Rapid spread with depression and fever. Mortality can be 100 per cent with velogenic strains. Haemorrhages may occur in larynx and throughout intestines. Necrotic foci found in intestines. Stable virus, survives several months in particles of faeces. Transmission: aerosols and transovarially. Virus is present in all exudates and body tissues	• Isolation in allantoic cavity of fertile eggs • Identification of virus by HAI • Serology: HAI test
PSITTACOSIS / ORNITHOSIS	*Chlamydia psittaci*	Wild and domestic birds. Disease is uncommon in food birds but is more common in parrots and non-psittacine cage birds. Clinical disease often activated by stress such as egg-laying or transportation in small cages. Nasal discharge, diarrhoea, dullness, weakness, inappetence and loss of weight occurs. Generalised disease with enlargement of liver and spleen, pericarditis, air-sacculitis, perihepatitis and peritonitis. Chlamydiae are excreted in respiratory discharges and faeces. Transmission: aerosols and airborne particles	• Impression smears: FA, Giemsa or MZN stains • Isolation in yolk sac of fertile eggs or in McCoy cell line
STAPHYLOCOCCAL INFECTIONS	*Staphylococcus aureus*	Many avian species. ***Synovitis:*** in growing birds. Large quantity of purulent exudate especially in hock joint. May be associated with a reovirus infection ***Osteomyelitis:*** lameness, erosion of cartilage over tibiotarsal condyles. Occurs in growing birds ***Septicaemia:*** sporadic infection in adult birds. Hepatic lesions and vegetative endocarditis (streptococci can cause similar lesions) ***Omphalitis*** ('mushy-chick disease'): can be caused by *E. coli* or staphylococci ***'Bumble-foot':*** infection distorting one or both feet (staphylococci or *Mycobacterium avium*). Most common in free-range chickens and in birds of prey	• Direct Gram-stained smear • Isolation from lesions • Coagulase test • Histopathology

Table 174. Infectious diseases of domestic birds. *(continued)*

GASTROINTESTINAL TRACT: poultry

Disease	Agent(s)	Comments	Diagnosis
NECROTIC ENTERITIS	Associated with *Clostridium perfringens* types A or C.	Chickens and other birds. Seen most commonly in broiler birds 3 weeks-old onwards. Mucosal surface of small intestine is pale due to necrosis of tips of villi. Later the mucosa fissures and sloughs irregularly. Most of the upper small intestine is involved. It is an acute enterotoxaemic disease with sudden deaths and explosive overall mortality. Seen especially in birds on litter (build-up of spores). Peak incidence at 3-12 weeks old. Diarrhoea may be seen but death usually occurs in a few hours	• Gram-stained scraping of small intestine mucosa: large numbers of Gram +ve rods • Clinical and PM findings • Histopathology
ROTAVIRUS INFECTION	Avian rotavirus (*Reoviridae*)	Chickens and turkeys. Characterized by enteritis and diarrhoea in young birds. Appetite is poor and dehydration occurs rapidly with mortality up to 50 per cent. Transmission is faecal-oral. No egg transmission reported. Dehydrated carcase and watery contents in gut may be seen. Subclinical infections can occur	• Electron microscopy on clarified faeces • Isolation in embryo liver cells: CPE. Identify by FA
SALMONELLOSIS	*Salmonella* spp.	***Fowl typhoid:*** *Salmonella gallinarum*. Egg transmitted but most common in growing and adult birds, although it can cause illness in chicks similar to that caused by *S. pullorum*. In adults, mortality can be high. There is enteritis and enlarged and congested liver and spleen ***Pullorum disease*** (Bacillary white diarrhoea): *Salmonella pullorum*. Seen in first few days of life to 2-3 weeks of age. Transmission by egg and contact. Chicks huddle, appear sleepy, anorexic and have white pasting around vent. They die or become carriers. Focal necrosis of liver or spleen occurs, nodules in lungs and heart, cheesy material in caeca and synovitis can be prominent ***Fowl paratyphoid:*** infection with many other *Salmonella* species. Most infections subclinical but there is high mortality in chicks a few weeks old with stress factors increasing susceptibility. Enlarged liver with or without focal areas of necrosis occurs. True egg transmission is rare but eggs can be contaminated from salmonellae in faeces that enter through the egg shell. Perpetuated by use of recycled chicken offal in poultry feeds	• Culture: lesions and heart blood • Serotyping of isolates • Biochemical tests to distinguish between *S. pullorum* and *S. gallinarum* • Histopathology

Table 174. Infectious diseases of domestic birds. *(continued)*

GASTROINTESTINAL TRACT: poultry			
Disease	**Agent(s)**	**Comments**	**Diagnosis**
THRUSH OF THE CROP	*Candida albicans*	Many bird species affected. Lesions are most common in crop: thickened mucosa with white raised diphtheritic membrane ('terry-towelling' effect). Mouth and oesophagus can also be affected. Birds are depressed and become emaciated. Condition commonly seen after extensive use of antibiotics	• Direct microscopy: Gram or 10 per cent KOH • Isolation and identification
ULCERATIVE ENTERITIS (QUAIL DISEASE)	*Clostridium colinum*	Quail, chicken and turkeys. Bobwhite quail most susceptible but can occur in 5-7 week-old Leghorn chickens or in turkey poults. Outbreaks tend to follow coccidiosis or infectious bursal disease. Explosive outbreaks in quail with 100 per cent deaths and chickens with mortality of 10 per cent without obvious illness. Faeces streaked with urates and lesions vary from enteritis to ulceration of gut (ulcers to 3 mm). Yellow foci occur in liver. Faecal-oral transmission. Survivors become carriers of the bacterium	• Gram-stained smears of blood, liver or spleen of septicaemic birds • Gross and histopathology • Anaerobic isolation of *C. colinum*
RESPIRATORY SYSTEM: poultry			
ASPERGILLOSIS	*Aspergillus fumigatus*	Chickens, turkeys and other birds such as penguins in zoos. ***Brooder pneumonia:*** chicks are infected by spores during hatching. Seen at 2-days of age. Gasping, anorexia, huddling together and increased thirst. From 10-50 per cent of chicks can be affected. Caseous focal lesions are seen mainly in lungs but can occur in other organs ***Generalised aspergillosis:*** all ages from 1 week of age. Infected via inhalation of spore-laden dust. Sporadic disease in older birds. Yellowish nodules occur in air sacs, lungs and in other body organs, including the brain	• Isolation from lesions on Sabouraud agar • Identification on:- a) Colonial morphology b) Fruiting heads • Histopathology: PAS or silver stains

Table 174. Infectious diseases of domestic birds. *(continued)*

RESPIRATORY SYSTEM: poultry

Disease	Agent(s)	Comments	Diagnosis
AVIAN INFLUENZA	Avian influenzaviruses. Many subtypes. (*Orthomyxoviridae*)	Many avian species affected. ***Mild form:*** mild respiratory signs and a slight drop in egg production ***Fowl plague:*** virulent subtypes with rapid spread, generalized disease and a high mortality in a short space of time. There is diarrhoea, respiratory signs, cyanosis of the face and swelling of the face and wattles. Haemorrhages through body organs and CNS can be involved. Mortality 40-100 per cent. Often brought into a country by migrating birds	• Isolation in allantoic cavity of fertile eggs • Identification by HAI • Serology: ELISA, VN, AGID
CHRONIC RESPIRATORY DISEASE (CRD) (CHICKENS) AND INFECTIOUS SINUSITIS (TURKEYS)	*Mycoplasma gallisepticum*	Chickens and turkeys. ***CRD:*** mild respiratory disease but severe with concurrent pathogens. Sneezing, coughing, rales, and nasal discharge. Morbidity high but mortality usually low ***Infectious Sinusitis*** in turkeys: swollen face, as sticky exudate fills infraorbital sinuses Egg transmission, aerosol and on human clothes	• Isolation on mycoplasma medium • Identification of microcolonies by FA or growth-inhibition tests • Serology: slide agglutination (stained Ag) or HAI
FOWL CORYZA	*Haemophilus paragallinarum*	Chickens, guinea fowls, turkeys and pheasants. Chickens can be infected from 4 weeks old and susceptibility increases with age. Mild to subacute disease on its own: depression, nasal discharge and swollen face because of oedema and filling of infraorbital sinuses with exudate. Occasionally the oedema extends to the wattles. Transmission: contact, aerosol and via water	• Isolation on chocolate agar under 10 per cent CO_2 • Identification on colonial morphology and biochemical tests
FOWL POX	*Avipoxvirus* (*Poxviridae*)	Chickens and turkeys. ***'Dry' form:*** scabby lesions on unfeathered parts of head, especially on combs and around eyes. ***'Wet' form:*** respiratory disease: caseous and diphtheritic lesions in larynx and trachea. Slow spread by pecking at lesions but epidemic if mosquitoes are plentiful	• Isolation on CAM of fertile eggs. Virus identified by FA, EM or neutralisation tests • Serology: gel-diffusion using ground scab as antigen, or VN in eggs

Table 174. Infectious diseases of domestic birds. *(continued)*

RESPIRATORY SYSTEM: poultry

Disease	Agent(s)	Comments	Diagnosis
INFECTIOUS BRONCHITIS (IB)	*Coronavirus* (*Coronaviridae*)	Chickens. ***Young birds*** less than 6 weeks old: severe respiratory disease. Can be complicated with *Escherichia coli* and mycoplasmas ***Broilers*** over 6 weeks old: immune complex glomerulonephritis ***Adults:*** mild or inapparent URT infection unless complicated by mycoplasmas, *E. coli* or other secondary invaders. Main signs: • Drop in egg production (20-25 per cent) • Misshapen and soft-shelled eggs Not egg transmitted	• FA: tracheal scraping • Isolation: allantoic cavity of fertile eggs. Dwarfing of chick embryos characteristic. Virus identification by VN test • Serology: VN, ELISA, gel diffusion • Differentiate from egg drop syndrome *(Aviadenovirus)*
INFECTIOUS LARYNGOTRACHEITIS (ILT)	Avian herpesvirus 1	Chickens and pheasants. Acute, highly contagious disease of most importance in laying birds. Dyspnoea, high-pitched squawk on expiration, coughing up blood. Caseous plugs and haemorrhages found in trachea and bronchi. Rapid drop in egg production that returns to normal on recovery. Mortality varies but can reach 50 per cent	• Isolation on CAM of fertile eggs • Identification: FA, EM or neutralisation tests • Serology: VN in eggs or in chicken cell lines
INFECTIOUS SYNOVITIS	*Mycoplasma synoviae*	Chickens and turkeys • Mild respiratory disease with some strains • Synovitis: chicken 4-6 weeks and turkeys 10-12 weeks of age. Swelling of hocks and foot pads. Yellow viscid exudate in bursa of keel and hock and wing joints • Air-sacculitis especially if infected concurrently with infectious bronchitis virus Egg transmission and via fomites (human clothes)	• Isolation on mycoplasma medium • Identification by FA or by growth-inhibition tests • Serology: slide agglutination test (stained antigen) or HAI test
NEWCASTLE DISEASE	Avian paramyxovirus 1	Many avian species susceptible. Respiratory signs of gasping and coughing may accompany but often precede other signs	• Isolation in allantoic cavity of fertile eggs • Identification of virus by HAI • Serology: HAI

Table 174. Infectious diseases of domestic birds. *(continued)*

Disease	Agent(s)	Comments	Diagnosis
SKIN: poultry			
FAVUS (AVIAN RINGWORM)	*Trichophyton gallinae*	Gallinaceous birds. Not common but can affect chickens and turkeys. White patchy overgrowths of comb and wattles that develop into thick white crusts (resembles fowl pox). In severe infections the feather follicles can be invaded with systemic signs of illness	• Direct microscopy of scab in 10 per cent KOH • Isolation on Sabouraud agar
NECROTIC DERMATITIS	*Clostridium septicum* (*C. perfringens,* staphylococci and streptococci can also be present)	Gangrenous necrosis with wet, inflamed skin. Sudden onset and birds rarely seen ill but death rate high. Most common in 4-16 week old chicken and occasionally in turkeys. Often predisposed to by infectious bursal disease or a previous staphylococcal infection. Mortality 10-60 per cent and affected birds die in 8-24 hours	• Clinical signs • Histopathology • FA for *C. septicum*
FOWL POX	*Avipoxvirus* *(Poxviridae)*	Chickens and turkeys. In the 'dry' form of the disease, scabby lesions occur on unfeathered parts of the head, especially on the comb and around the eyes	• Clinical signs • EM • Isolation on CAM of fertile eggs
NERVOUS SYSTEM : poultry			
BOTULISM (LIMBERNECK)	*Clostridium botulinum* type C most common	Poultry and wild water fowl. Growth of the bacterium occurs in suitable foodstuffs and invertebrate carcases (anaerobic) with production of toxin. Not all birds develop the classical 'limberneck' (paralysis of neck) but this is seen in swans, geese and ducks. A constant finding is the flaccid paralysis of legs and wings and death from respiratory paralysis. Some mildly affected birds recover if fed and kept away from the source of the neurotoxin. No specific lesions on necropsy. ***Toxicoinfectious botulism:*** reported in broiler flocks with 10 per cent mortality. Alteration in normal gut flora may predispose to this form of the disease	• Clinical signs • Demonstration of toxin in serum of acutely ill birds by mouse-neutralisation tests

Table 174. Infectious diseases of domestic birds. *(continued)*

NERVOUS SYSTEM: poultry

Disease	Agent(s)	Comments	Diagnosis
EPIDEMIC TREMOR (AVIAN ENCEPHALOMYELITIS)	*Enterovirus* (*Picornaviridae*)	Chicks, ducks, pheasants, turkeys and Japanese quail. Most common in 7-10 day-old chicks. Ataxia and fine tremors of head and neck, paralysis and death. Any birds that recover are brain-damaged. Mortality 50 per cent when virus first enters flock. Once endemic the chicks are protected for 21 days after hatching by maternal antibodies in eggs. Subclinical in adults except for a transient drop in egg production when first infected. Spread by faecal-oral route and via eggs for 1 month after infection	• Clinical signs • Histopathology • Isolation in yolk sac of fertile eggs (from AE-free flock). Allow eggs to hatch and observe chicks for clinical signs • FA for virus
LISTERIOSIS	*Listeria monocytogenes*	Chickens, turkeys, geese, pigeons and canaries. Septicaemia often with encephalitis in young birds. Sudden death is common but wasting before death may occur. Myocardial necrosis, pericarditis, focal hepatic necrosis and encephalitis (tremors, torticollis and circling) can be present. Listeria can survive in a chicken flock for 4 years or more without active disease	• PM: massive necrosis of myocardium and focal hepatic necrosis • Isolation from heart, liver or brain • Identification by biochemical tests
MAREK'S DISEASE	Gammaherpesvirus (*Herpesviridae*)	In the classical neural form of Marek's disease there is infiltration of peripheral nerves by lymphoblastic cells with resulting paralysis of legs and/or wings	• Histopathology: • FA for Marek's tumour-specific antigen (MATSA) on surface of infected cells
NEWCASTLE DISEASE	Avian paramyxovirus 1 (neurotropic strains) (*Paramyxoviridae*)	Nervous signs include drooping wings, leg paresis, twisting of head and neck ('star-gazing'), depression and total paralysis. Clonic spasms may be seen in moribund birds. CNS signs may accompany, but often follow, respiratory signs	• Isolation in allantoic cavity of fertile eggs • Identification of virus by HAI • Serology: HAI

Table 174. Infectious diseases of domestic birds. *(continued)*

MISCELLANEOUS DISEASES: poultry

Disease	Agent(s)	Comments	Diagnosis
AVIAN INFECTIOUS HEPATITIS	*Campylobacter jejuni*	Variety of birds and mammals can be infected. Typically the infection is subclinical and the bacterium is excreted in faeces. In clinical disease in chicken there is a major decline in egg production and severely affected birds lose weight, have shrivelled combs and are listless. There may be sudden deaths in acute cases. Only a few birds are affected at any one time. At PM there are haemorrhagic and necrotic lesions in the liver, ascites, hydropericardium, enlarged pale kidneys and a catarrhal enteritis. In younger birds heart lesions are more severe and constant	• Direct microscopy on bile • Isolation from bile or faeces
CHICK ANAEMIA VIRUS INFECTION (CAV)	Unclassified DNA virus	CAV not known to infect birds other than chickens. Chicks under 1 week of age, without maternal antibodies, develop disease. Characterized by anorexia, lethargy, anaemia, atrophy or hypoplasia of lymphoid organs, haemorrhages and increased mortality rates	Tentative on: • Clinical signs. • Gross and histopathology • Serology: VN, IFA, ELISA
EGG DROP SYNDROME	*Aviadenovirus* (*Adenoviridae*)	Natural hosts of virus are ducks and geese. Introduced to an unnatural host (chicken) probably via Marek's disease vaccine grown in duck tissue culture cells. Egg transmission occurs in chicken. The virus is latent in chicks until puberty and then they excrete virus in faeces with the production of antibodies. Lateral spread is slow. If virus enters a laying flock without immunity, loss of pigment in brown eggs is the first sign followed by soft-shelled or shell-less eggs. Production drops by 10-40 per cent but there is no effect on hatchability or on fertility. Repeated episodes of infection can occur with no specific signs of illness in the birds	• Serology on birds that are producing defective eggs: VN and HAI • Isolation difficult because of lack of signs and intermittent excretion. Fertile duck eggs or tissue culture may be used for isolation

Table 174. Infectious diseases of domestic birds. *(continued)*

Disease	Agent(s)	Comments	Diagnosis
MISCELLANEOUS DISEASES: poultry			
VIRAL ARTHRITIS	*Reovirus* (*Reoviridae*)	Chickens and turkeys. Reovirus ubiquitous in flocks. Some strains can cause viraemia and localize in joints: arthritis, tendonitis and synovitis. Egg transmitted. Any respiratory or intestinal signs are of short duration. The arthritic form is most common in 4-8 week-old broilers, the birds have a stilted gait, and rupture of the gastrocnemius can be associated with this infection. Morbidity rate 5-50 per cent	• Isolation in yolk sac of fertile eggs or in fowl cell lines • Serology: many birds positive early in the infection: gel-diffusion
INFECTIOUS DISEASES: turkeys			
ARIZONA INFECTION (PARACOLON INFECTION)	*Salmonella arizonae*	Disease most common in turkey poults. Bacterium can be isolated from other birds, mammals and reptiles. Signs and lesions generally similar to those caused by some other *Salmonella* spp. Faecal and egg transmission occurs. The infection tends to persist in a flock. Acute disease most common in 3-4 week-old age group. Birds are listless, develop diarrhoea and corneal opacity with blindness is common. Nervous signs, such as torticollis, ataxia, leg paralysis and convulsions can occur due to localisation of salmonellae in the brain. Mortality variable, being highest in young birds	• Isolation on selective media • Serotyping of isolates • Histopathology
BLUE COMB (TRANSMISSIBLE ENTERITIS OF TURKEYS)	*Coronavirus* (*Coronaviridae*)	Acute, highly contagious disease. Sudden onset with marked depression, anorexia, diarrhoea and weight loss. Morbidity and mortality approach 100 per cent in young poults. Weight loss in adult and growing birds may be economically more important . Poults chirp constantly and appear to be cold. The intestines are distended and contents watery. Diarrhoea is profuse in older birds. Cyanosis of head is common ('blue comb'). There is a drop in egg production and catarrhal enteritis with haemorrhages in viscera. Variable morbidity/mortality in adults	• Pathology • Virus isolation

Table 174. Infectious diseases of domestic birds. *(continued)*

INFECTIOUS DISEASES: turkeys

Disease	Agent(s)	Comments	Diagnosis
HAEMORRHAGIC ENTERITIS OF TURKEYS	*Aviadenovirus* (*Adenoviridae*)	Infection of the spleen in turkeys over 4 weeks-old with secondary intestinal haemorrhage. Acute onset with bloody droppings. Mortality is usually 5-10 per cent but is occasionally higher. Short course (2 weeks) but often complicated and prolonged by a concurrent *Escherichia coli* infection. The spleen is enlarged and mottled or can be shrunken and pale. Haemorrhages from tips of villi are characteristic	• Histopathology • Detection of viral antigen: AGID • Serology: AGID
MYCOPLASMA MELEAGRIDIS DISEASE	*Mycoplasma meleagridis*	Syndromes include: • Air-sacculitis in 1-day-old poults • Reduced hatchability (egg transmission and venereal) • Respiratory rales in 3-8 week-old birds • Hock joints and cervical vertebrae affected in growing birds	• Rapid plate agglutination test for antibodies • Isolation and identification of mycoplasma
PSEUDOTUBERCULOSIS	*Yersinia pseudotuberculosis*	The bacterium can infect many mammalian and avian species. Cage birds and turkey poults are particularly susceptible. Acute septicaemia and death within 24 hours (differentiate from listeriosis and salmonellosis in cage birds). Chronic disease with wasting, diarrhoea and local necrotic lesions in internal organs can occur but is less common	• Isolation on BA and MacConkey agar • Identify by biochemical tests
RETICULOENDOTHELIOSIS	*Retroviridae*	Turkeys, ducks, pheasants and quail. Chickens have antibodies to virus. Non-defective virus strains cause: • Acute neoplasia of lymphoreticular system • Runting syndrome characterised by abnormal feathering and atrophy of thymus and bursa • Chronic lymphomas a. Bursal lymphoma b. Non-bursal lymphoma Replication defective strain can cause acute reticulum cell neoplasia in chickens and turkeys. This is rare in the field	• Gross and histopathology • Virus isolation • Serology: VN, ELISA • Specific DNA probe

Table 174. Infectious diseases of domestic birds. *(continued)*

INFECTIOUS DISEASES: turkeys

Disease	Agent(s)	Comments	Diagnosis
TURKEY CORYZA	*Bordetella avium*	Resistant bacterium and remains viable in contaminated litter for long periods. Spread by aerosol and in drinking water. Disease is most pronounced in poults but broiler and layer chickens are also susceptible. There is mucus from nares, swelling of submaxillary region, mouth - breathing and sneezing. Morbidity 100 per cent but mortality low if disease is uncomplicated. However, concurrent *Escherichia coli* infection is common	• Isolation and identification
TURKEY ERYSIPELAS	*Erysipelothrix rhusiopathiae*	Mainly turkeys, less common in chickens, ducks and geese. Source of infection: uncooked meat or fish scraps. Disease most common in growing birds. They become listless, anorectic and have greenish diarrhoea. A red - purple swelling of caruncle and face is suggestive of this disease. Mortality can reach 40 per cent. Lesions: diffuse haemorrhages in abdominal, pectoral and femoral muscles and subserosal haemorrhages occur in viscera. Liver and spleen are enlarged and may have infarcts	• Isolation on BA • Identification by biochemical tests
TURKEY PARAMYXOVIRUS DISEASE	Avian paramyxovirus (*Paramyxoviridae*)	Turkeys are the only domestic birds known to be naturally infected. Mild respiratory signs may precede a drop in egg production with a high level of white-shelled eggs, reduced hatchability and fertility. Signs last about 5-6 weeks. Poults may show severe respiratory signs with swelling of the infraorbital sinuses and submandibular oedema	• Isolation in allantoic cavity of fertile eggs, early in disease • Serology: ELISA
TURKEY X DISEASE (AFLATOXICOSIS)	*Aspergillus flavus*	Peanut meal or cereals are most commonly implicated. Depression, anorexia, reduced growth rate, loss of condition, bruising, decreased egg production, fertility and hatchability occur. Ducklings and turkey poults are particularly susceptible. Ataxia, convulsions and opisthotonos can be seen. PM: ascites, visceral oedema and liver is pale with widespread necrosis	• Histopathology • Demonstration of aflatoxin in feedstuff

Table 174. Infectious diseases of domestic birds. *(continued)*

Disease	Agent(s)	Comments	Diagnosis
INFECTIOUS DISEASES: turkeys			
VIRAL HEPATITIS OF TURKEYS	*Aviadenovirus* (*Adenoviridae*)	Usually subclinical unless birds are stressed. In poults under 6 weeks of age the morbidity can be 100 per cent and mortality 10-15 per cent. In adults there is decreased production, fertility and hatchability. PM: focal necrosis of liver with haemorrhages and congestion superimposed on the degenerative changes. Generally the morbidity and mortality vary according to degree of stress	• Pathology • Virus isolation
INFECTIOUS DISEASES: ducks			
DUCK PLAGUE	Duck herpesvirus 1	Ducks, geese and swans. Sudden deaths of birds in good condition, sometimes still sitting on eggs. If seen before death: photophobia, anorexia, extreme thirst, ataxia, nasal discharge and bloody diarrhoea are noticed. High mortality. Haemorrhages and necrosis occur in internal organs	• Pathology: intranuclear IBs • FA technique • Virus isolation
NEW DUCK DISEASE	*Pasteurella anatipestifer*	Contagious and widely distributed disease. Primarily affects young ducks 2-7 weeks-old but waterfowl, chickens and turkeys can be affected. Ocular and nasal discharges, tremor of head and neck and incoordination are seen. Fibrinous exudate occurs in all body cavities and a fibrinous meningitis can be present	• Isolation on BA under 5-10 per cent CO_2 • Identification
VIRAL HEPATITIS OF DUCKS	*Enterovirus* (*Picornaviridae*)	Highly contagious with a short course, high mortality and characteristic liver lesions. Liver is pale red, slightly enlarged and covered by haemorrhagic foci up to 1cm in diameter. Spleen and kidneys may be enlarged. Disease seen in ducklings under 7 weeks-old. Mortality 95 per cent. Adults are infected but disease is subclinical. Waterfowl may be carriers of the virus	• Pathology • FA technique • Virus isolation

Table 174. Infectious diseases of domestic birds. *(continued)*

Disease	Agent(s)	Comments	Diagnosis
INFECTIOUS DISEASES: geese			
GOOSE PARVOVIRUS INFECTION (GOOSE VIRAL HEPATITIS, DERZSY'S DISEASE)	Parvovirus	Highly contagious disease of young geese under 4-weeks of age and Muscovy ducklings. • Goslings under 1 week of age: anorexia, prostration and death in 2-5 days. Survivors exhibit runting, loss of neck down and marked ascites • Older goslings: anorexia, weakness, nasal and ocular discharge and diarrhoea Lesions include serofibrinous pericarditis, perihepatitis and excess fluid in abdominal cavity	• Gross and histopathology • Isolation of virus in TC or fertile goose eggs (allantoic cavity). IBs in TC cells • Serology: VN, AGID
INFECTIOUS DISEASES: pigeons			
PARAMYXOVIRUS INFECTION OF PIGEONS	Avian paramyxovirus (*Paramyxoviridae*)	Recently recognized in Europe. Severe and rapidly spreading disease: anorexia, diarrhoea, conjunctivitis and paralysis of wings and legs. This pigeon paramyxovirus is lentogenic for chickens	• Virus isolation • Serology: HAI
PARATYPHOID ('DROPPED WING')	*Salmonella typhimurium*	As well as enteric disease, arthritis of the wing joints occurs quite commonly in young pigeons	• Isolation from synovial fluid in acute cases • Serology: slide agglutination
PIGEON HERPESVIRUS INFECTION	Pigeon herpesvirus	Most adult birds are latent, asymptomatic carriers. In acute disease, young susceptible birds show conjunctivitis, blocked nares and caruncles turn from white to yellow-grey. Chronic disease with sinusitis and dyspnoea occurs with secondary bacterial invasion	• Histopathology: IBs in URT and liver cells • Isolation in TC and identification of virus by FA technique • Serology: VN or IFA
TRICHOMONIASIS	*Trichomonas gallinae*	Most common in pigeons but outbreaks have occurred in chickens and turkeys. Pigeon parents can infect squabs via pigeon milk. Rapid course: small yellowish lesions on oral mucosa that grow and spread and can completely block the oesophagus. Death often occurs in 8-10 days. Some pigeons are healthy carriers with the organism in their throats	• Microscopic examination of wet mount for *Trichomonas* sp.

Appendix 1

Reagents and Stains

Biochemical Test Reagents

Decarboxylase broth base (Falkow's)

Peptone	5.0 g
Yeast extract	3.0 g
Glucose	1.0 g
Distilled water	1.0 litre
Bromocresol purple (0.2 per cent solution)	10.0 ml
L-arginine hydrochloride	5.0 g
or L-lysine hydrochloride	5.0 g
or L-ornithine hydrochloride	5.0 g

The solids are dissolved in the distilled water and the indicator solution is added. The pH is adjusted to 6.7 and the medium dispensed in 2–5 ml quantities in small bottles or tubes. The broth is sterilised at 115°C for 10 minutes.

Lead acetate paper strips for hydrogen sulphide detection

Filter paper strips (approximately 5 x 50 mm) are immersed in saturated lead acetate solution and dried in an oven at 70°C. These are suspended above a suitable medium and held in place by the cap or cotton wool plug.

Kovac's reagent for the detection of indole

p-dimethylamino-benzaldehyde	20.0 g
Iso-amyl alcohol	300.0 ml
Concentrated hydrochloric acid	100.0 ml

The aldehyde is dissolved in the alcohol with gentle heat. The acid is added after cooling. The reagent is stored in a dark bottle at 4°C.

Methyl red reagent

Methyl red	0.1 g
95 per cent ethyl alcohol	300.0 ml
Distilled water	200.0 ml

The methyl red is dissolved in the alcohol and the solution is diluted to a total volume of 500 ml with distilled water.

Nitrate Reduction Test

Nitrate broth (5 ml) is inoculated with the test bacterium and incubated at 37°C for 24 hours (rarely an incubation period of 5 days is necessary). This broth can be obtained commercially and contains 1 g potassium nitrate per litre.

Reagents A and B that test for the presence of nitrite:

Reagent A: 5 g alpha-naphthylamine in 1 litre 5N acetic acid.

Reagent B: 8 g sulphanilic acid in 1 litre 5N acetic acid.

To obtain 5N acetic acid, add 2 parts glacial acetic acid to 5 parts distilled water. Five drops of each reagent (A and B) are added to the nitrate broth. The test is shaken and read after 1-2 minutes.

Result: Colourless reaction indicates that no nitrite is present in the broth. Red colouration means that the nitrate in the test broth has been reduced to nitrite.

A pinch of zinc powder can be added to a test where a broth has remained colourless. If the broth then becomes a red colour, this means that nitrate has now been converted to nitrite and the test bacterium did not reduce the nitrate in the broth. If, however, the broth remains colourless, this indicates that the nitrate in the broth was reduced to nitrogen gas by the test bacterium.

Oxalic acid test papers for indole production (SIM medium)

A piece of filter paper is soaked in saturated oxalic acid solution. The paper is dried and cut into strips approximately 10 x 50 mm. The strips are suspended over the medium and held in place with the cap. Pink colouration indicates a positive reaction.

Potassium nitrate paper strips for nitrate reduction test

Filter paper strips (10 x 50 mm) are soaked in 40 per cent potassium nitrate solution. The strips are dried and autoclaved at 115°C for 10 minutes.

A strip is placed on the surface of a blood agar plate. The test bacterium is then stab inoculated into the agar about 20 mm from the paper strip. The plate is incubated at 37°C and examined after 4 and 24 hours incubation. A positive result is indicated by a wide zone of browning of the medium between the colony of the test bacterium and the strip. A negative reaction is denoted either by no reaction or by a very narrow zone of browning around the stab inoculation.

Test reagent for sodium hippurate hydrolysis (acid ferric chloride solution)

Ferric chloride	12.0 g
Concentrated hydrochloric acid	2.5 ml
Distilled water	100.0 ml

The acid is diluted with 75 ml of water and the ferric chloride is dissolved in the fluid by warming. The volume is adjusted to 100 ml with distilled water.

Microbiological Staining Solutions and Procedures

Preparation of staining solutions

The staining solutions can be prepared from dye powders but some of the staining reagents can be bought as solutions, such as strong carbol fuchsin solution and the reagents for the Gram stain.

The methanol (methyl alcohol), ethanol (ethyl alcohol), acetic acid and concentrated hydrochloric acid used in the formulae should be of Analar grade. After a staining solution has been prepared it should be filtered into a clean container before use. Commercially prepared solutions may occasionally require filtering if a deposit forms.

Blue-Black Ink (for staining fungal elements)

10 per cent potassium hydroxide (KOH)	1 part
Permanent blue-black ink	2 parts
Wetting agent solution	1 part

Wetting agent solution:

Sodium benzoate	0.15 g
Dioctyl sodium sulphosuccinate	0.85 g
Distilled water	100.0 ml

The fungi take up the dye (ink) selectively and the specimen is also satisfactorily cleared by the 10 per cent KOH.

Calcofluor White (for fungal elements)

Preparation of Solutions

- STOCK SOLUTION : 1 per cent (w/v) calcofluor white H2R (Polysciences, Inc.) or fluorescent brightener 28 (Sigma Chemical Co.). Dissolve 1.0 g powder in 100 ml of distilled water with gentle heat.
- WORKING SOLUTION : 0.1 per cent calcofluor white with 0.05 per cent Evans blue dye for a counter stain.

For use add 1 drop of 10 per cent KOH and 1 drop of calcofluor white working solution to the specimen on a microscope slide.

Uses for calcofluor white

- Hair and skin samples : 1 drop of 10 per cent KOH and 1 drop calcofluor white working solution (0.1 per cent) is used. The preparation is examined microscopically under excitation light of 500 nm.
- Exudates and fluids : 1 drop of calcofluor white (0.1 per cent) is emulsified with the specimen on a microscope slide and covered with a coverslip. The preparation is examined microscopically under excitation light of 500 nm.
- Histopathological sections : either paraffin-embedded or frozen sections can be used. Four or five drops of calcofluor white (0.1 per cent) are placed on unstained sections and allowed to stand for 1 minute. The preparation is rinsed with tap water and counter-stained with 0.05 per cent Evans blue for 1 minute to minimize background fluorescence. The preparation is examined microscopically under an excitation light of 500 nm.

Castaneda's Technique (for chlamydial elementary bodies)

- Impression smears are air-dried and fixed in mordant solution (formalin 100 ml and glacial acetic acid, 7.5 ml), followed by washing in tap water.
- The smear is stained for 10 minutes in formol blue solution that consists of 90 ml 0.15M phosphate buffer, pH 7.0, 10 ml Unna's blue (1 per cent azure II in methyl alcohol) and 5 ml formalin.
- The smear is washed thoroughly in tap water.
- The stained smear is then differentiated for 5-10 seconds in 0.25 per cent aqueous safranine. Chlamydial elementary bodies stain blue against a reddish background.

Dienes' Stain (for mycoplasmal microcolonies)

Methylene blue	2.4 g
Maltose	10.0 g
Azure II	1.25 g
Sodium chloride	0.25 g
Distilled water	100.0 ml

All the ingredients are dissolved in the distilled water.

Giemsa Stain

Giemsa stain reagents:

Giemsa powder	1.0 g
Glycerol	66.0 ml
Absolute methanol	66.0 ml

The Giemsa powder is dissolved in the glycerol at 55-60°C for about 2 hours. Methanol is added and mixed thoroughly.

The solution from the above formula and the commercially available solution is mixed with the buffer solution before use : 1 volume of stain to 9 volumes of buffer.

Buffer for Giemsa stain (pH 7.0):

Sodium phosphate, Na_2HPO_4 (anhydrous) M/15 solution (9.47 g/litre) or $Na_2HPO_4.2H_2O$, M/15 solution (11.87 g/litre)	61.1 ml
Potassium phosphate, KH_2PO_4 (anhydrous) M/15 solution (9.08 g/litre)	38.9 ml
Distilled water	900.0 ml

Gram Stain Reagents

Crystal violet:

Crystal violet	2.0 g
Ethanol 95 per cent (vol/vol)	20.0 ml
Ammonium oxalate	0.8 g
Distilled water	80.0 ml

The crystal violet is first dissolved in the ethanol, then the ammonium oxalate is dissolved in the distilled water. The two solutions are added together. To aid the dissolving process, both mixtures are agitated in a bath of hot water.

Gram's iodine (mordant):

Iodine crystals	1.0 g
Potassium iodide	2.0 g
Distilled water	200.0 ml

The iodine crystals and the potassium iodide are ground together in a mortar and the distilled water is added slowly. If necessary the mixture can be agitated in a bath of hot water to aid dissolution.

Decolourizer:

Ethanol 95 per cent (vol/vol)

Dilute carbol fuchsin (counterstain):

Concentrated carbol fuchsin (*see* ZN stain)	10.0 ml
Distilled water	90.0 ml

India Ink (for demonstrating capsules)

India ink (Pelican brand)	1.0 ml
Formalin (40 per cent)	9.0 ml

The solutions are mixed together for use.

Lactophenol Cotton Blue Stain (for staining fungal elements in wet-preparations)

Phenol crystals	20.0 g
Glycerin	40.0 ml
Lactic acid	20.0 ml
Distilled water	20.0 ml

The ingredients are dissolved by gently heating over a steam bath. When in solution, 0.05 g of cotton blue dye is added. The solution is mixed thoroughly.

Macchiavello's Method (for staining chlamydial elementary bodies)

Basic fuchsin solution:

Basic fuchsin (88-90 per cent dye content)	0.25 g
Distilled water	100.0 ml

Buffer to pH 7.4 with 4.0 ml of 0.15 M phosphate buffer. Filter through coarse filter paper directly onto the slide.

Staining method:

- The prepared smear is air-dried and gently heat-fixed.
- The slide is flooded with basic fuchsin for 4-6 minutes.
- The stain is poured off and the slide washed with tap water.
- The slide is dipped in 0.5 per cent citric acid solution and immediately removed and washed with tap water.
- The preparation is counterstained for 10 seconds with 1 per cent aqueous methylene blue and then

washed with tap water and air-dried.The chlamydial elementary bodies stain red and the background is blue.

The Methylene Blue Staining Procedure for Chlamydiae

- McCoy cell coverslip cultures are fixed in two changes of methanol for 3-5 minutes.
- Methanol is replaced with an aqueous solution of methylene blue (5 g/litre) and the preparation stained for no longer than 5-7 minutes.
- The stain is removed and the cultures are washed in tap water.
- The preparations are washed for 15 seconds in 0.025 per cent H_2SO_4 and followed rapidly by a rinse with tap water.
- Dehydration is carried out in acetone for 15 seconds and this is followed by clearing in xylene for 15 seconds and mounting in DPX. If the slides are examined under darkfield microscopy, the chlamydial inclusions will appear as refractile, yellow-green bodies surrounded by a halo.

Modified Ziehl-Neelsen Stain Reagents

Dilute carbol fuchsin: same formula as for the Gram stain

Acetic acid (decolourizer)

Concentrated acetic acid	1.0 ml
Distilled water	200.0 ml

Methylene blue (counterstain):

Methylene blue	8.0 g
Ethanol 95 per cent (v/v)	300.0 ml
Distilled water	1300.0 ml
Potassium hydroxide	0.13 g

If Loeffler's methylene blue is used, the potassium hydroxide can be omitted.

Nigrosin Staining Solution (for demonstrating capsules)

Nigrosin (granular)	10.0 g
Formalin 10 per cent	100.0 ml
or 1:10,000 merthiolate	100.0 ml

The solution is placed in a bath of boiling water for 30 minutes and then any solvent lost by evaporation is replaced. The solution should be filtered twice through double filter paper (Whatman No. 1).

Silver Stain for Flagella (West et al., 1977)

Solution A:

Saturated aqueous aluminium phosphate	25.0 ml
5 per cent aqueous ferric chloride	5.0 ml
10 per cent aqueous tannic acid	10.0 ml

Solution B:

(a) A solution of 100 ml of 5 per cent silver nitrate is prepared.

(b) Two to five drops of conc. ammonium hydroxide are added to 90 ml of the silver nitrate solution. A brown precipitate forms which dissolves as more base is added. Adding the ammonium hydroxide is discontinued just at the point of clearing.

(c) The procedure is then reversed and the remaining silver nitrate solution is added, one drop at a time, until the stain solution is cloudy.

(d) The solution, if stored in a dark bottle at room temperature, is stable for several months.

Staining procedure:

A loopful of culture is mixed with 3-5 ml of distilled water. One drop of this bacterial suspension is placed on a clean, grease-free, microscope slide and allowed to air-dry. The smear is flooded with solution A and left for 4 minutes. The slide is rinsed with distilled water and then flooded with solution B. The slide is heated gently using a low bunsen flame held under the staining rack. Steam should issue from the slide but the stain must not be allowed to boil. This heating is continued for 4 minutes then the preparation is rinsed with distilled water and dried in a slanted position.

Victoria Blue Stain (for Serpulina hyodysenteriae)

Victoria blue 4-R	0.5 g
Ethyl alcohol (absolute)	5.0 ml
Distilled water	95.0 ml

The dye is dissolved in the alcohol and then the distilled water is added and the solution mixed well.

Staining method: a smear of faeces or colonic mucosal scrapings is made on a microscope slide. It is fixed in 10 per cent formalin for 15 minutes and air-dried. The smear is flooded with the 0.5 per cent solution of Victoria blue 4-R and stained for 5 minutes. The slide is washed under a running tap, air-dried and examined microscopically under the oil-immersion objective.

Ziehl-Neelsen Stain Reagents

Concentrated carbol fuchsin

Basic fuchsin	1.0 g
Ethanol 95 per cent (v/v)	10.0 ml
Phenol	5.0 g
Distilled water	100.0 ml

The basic fuchsin is dissolved in the ethanol, the phenol is then dissolved in the distilled water and the two solutions mixed together. The solution is allowed to stand for a few days and then filtered into a clean container.

Acid-alcohol (decolourizer):

Concentrated hydrochloric acid	8.0 ml
Ethanol 95 per cent (v/v)	97.0 ml

The acid should be added to the alcohol.

Methylene blue (counterstain): same formula as in the modified Ziehl-Neelsen staining technique.

Brilliant green (alkaline): can be used as an alternative counterstain.

Brilliant green	1.0 g
Sodium hydroxide 0.01 per cent	100.0 ml

Reference

West, M., Burdash, N.M. and Freimuth, F. (1977). Simplified silver-plating stain for flagella, *Journal of Clinical Microbiology*, **6**: 414-419.

Appendix 2

Culture and Transport Media

Culture Media

Selective Medium for Brucella abortus

Columbia agar	42.5 g
Dextrose	1.0 g
Distilled water	1.0 litre

The medium is autoclaved at 121°C for 15 minutes and cooled to 50°C. Sterile horse serum (5 per cent) and the antibiotic supplement are then added.

Antibiotic supplement:

Polymyxin B	6.25 units/ml of medium
Bacitracin	25.0 units/ml
Nalidixic acid	5.0 µg/ml
Actidione	100.0 µg/ml

Selective Medium for Brucella ovis

Mixture A:

GC agar base (BBL)	7.2 g
Distilled water	100.0 ml

Mixture B:

Bovine haemoglobin powder	1.0 g
Distilled water	100.0 ml

Mixture A is heated in a boiling water bath to dissolve the agar and the ingredients of B are mixed to obtain a solution of haemoglobin. Each solution is sterilised by autoclaving at 121°C for 15 minutes and then allowed to cool to 50°C.

The contents of a vial of VCN inhibitor (BBL) is dissolved in 10 ml of sterile water and added to mixture A. Mixture B is then added to A and mixed well but foaming should be avoided. The medium is poured into Petri dishes.

The final concentration of antibiotics per ml of medium is: vancomycin 15 µg/ml, colistin 37.5 µg/ml and nystatin 62.5 units/ml.

Chocolate Agar

Columbia blood agar base	39 g
Distilled water	1.0 litre

The blood agar base is boiled to dissolve the base prior to autoclaving at 121°C for 15 minutes. After cooling to 50°C, 70 ml of sterile defibrinated sheep blood is added and the mixture is heated to 80°C in a water bath with constant gentle mixing until it turns brown. The medium is cooled to 50°C and poured into Petri dishes.

Clostridium chauvoei blood agar (Batty and Walker)

Broth base	94.0 ml
Liver extract (Oxoid)	3.0 g
Glucose	1.0 g
New Zealand agar	1.6 g
Defibrinated sheep blood	5.0 ml

The broth base, liver extract, glucose and agar are mixed together and the mixture autoclaved at 115°C for 10 minutes. The medium is cooled to 50°C and the blood added. The blood is mixed gently with the medium and poured into Petri dishes.

Clostridium novyi (types B and C) and C. haemolyticum medium (Moore)

Basal medium:

Neopeptone (Difco)	1.0 g
Yeast extract (Difco)	0.5 g
Proteolyzed liver	0.5 g
Glucose	1.0 g
New Zealand agar	2.0 g
Salts solution*	0.5 ml
Distilled water	100.0 ml

*Salts solution (g/litre):

$MgSO_4.7H_2O$	40.0 g
$MnSO_4$	2.0 g
$FeCl_3$ (anhydrous)	0.4 g
HCl (concentrated)	0.5 ml

The neopeptone, yeast extract, liver extract and glucose are dissolved in 50 ml of water. The salt solution is added and the pH adjusted to 7.6-7.8. The agar is dissolved in 50 ml of water and the two solutions are combined. This basal medium is dispensed in 18 ml volumes in 28 ml screw-capped bottles and is sterilised by autoclaving at 115°C for 10 minutes. It is cooled and stored at 4°C.

Preparation of the reducing solution:

Cysteine HCl	120.0 mg
Dithiothreitol	120.0 mg
Glutamine	60.0 mg
Distilled water	10.0 ml

The ingredients are dissolved in the water and the pH adjusted to 7.6-7.8. The solution is sterilised by filtration and must be freshly prepared before use.

Complete medium

An 18 ml volume of basal medium is melted and cooled to 50°C. Defibrinated horse blood (2 ml) and reducing solution (0.15 ml) are added to, and mixed gently with the basal medium. The plates should be poured immediately.

Milk agar for casein digestion

Skim milk	500.0 ml
Nutrient agar (double-strength)	500.0 ml

The skim milk is sterilised at 115°C for 10 minutes and then cooled to 50°C. Sterilized, double-strength nutrient agar is cooled to 50°C and added to the skim milk. The skim milk and agar are mixed gently and poured into Petri dishes. (*Corynebacterium renale* is used as the positive control)

Niger Seed (Birdseed) agar (Staib agar)

Pulverized *Guizotia abyssinicia* seed*	50.0 g
Agar	15.0 g
Glucose	1.0 g
Creatinine	1.0 g
KH_2PO_4	1.0 g

The seed is added to 100 ml of distilled water and ground in a blender. The mixture is boiled for half-an-hour and then strained through gauze to separate the seed from the extract. The volume of the seed extract is adjusted to 1,000 ml with distilled water. The agar and other ingredients are added and the mixture boiled gently until the agar is dissolved. The medium is autoclaved at 110°C for 20 minutes, cooled to 50°C and then poured into Petri dishes.

*Obtainable from Philadelphia Seed Co., Plymouth, Meeting, PA, USA.

Staib agar is available commercially: Mycoplate CR Roche (Staib agar) from Hoffmann-La Roche; Staib agar 08-172 (Remel) and Staib agar 97096 (BBL).

PLET agar (for Bacillus anthracis)

Heart infusion agar (Difco)	25.0 g
EDTA	0.3 g
Thallous acetate	0.04 g
Deionized distilled water	1000.0 ml

The ingredients are mixed and dissolved completely. The medium is sterilised by autoclaving at 121°C for 15 minutes and then cooled to 56°C. A filter sterilised solution of polymyxin B (30,000 units) and lysozyme (300,000 units) are added to the medium and mixed well before being poured into Petri dishes.

Smith-Baskerville Medium for Bordetella bronchiseptica (Smith and Baskerville, 1979)

Bacto peptone	20.0 g
Sodium chloride	5.0 g
Agar	15.0 g
Distilled water	857.0 ml

The basal medium is autoclaved at 121°C for 15 minutes and then cooled to 55°C. The following supplementary solutions are mixed together and added to the cooled agar medium :

Antimicrobial supplement (given as the final concentration in the medium):

Gentamicin	0.5 µg/ml
Penicillin	20.0 µg/ml
Furaltadone	20.0 µg/ml

Carbohydrate supplement (filter-sterilised):

Glucose (10 per cent)	100.0 ml
Lactose (10 per cent)	100.0 ml

Indicator solution (filter-sterilised):

Bromothymol blue (0.2 per cent)	40.0 ml

The bromothymol blue is first prepared as a 2.0 per cent stock solution :

Bromothymol blue	1.0 g
0.1 N NaOH	25.0 ml
Distilled water	475.0 ml

Taylorella equigenitalis medium (Timoney et al., 1982)

Chocolate agar is prepared with Eugon agar (BBL) and 5 per cent horse blood.
The chocolate agar is supplemented with :

5 per cent lysed horse blood	
Trimethoprim	1 µg/ml
Clindamycin	5 µg/ml
Amphotericin B	5 µg/ml

The addition of Isovitalex (BBL), 1 ml reconstituted per 100 ml of medium, is optional. The shelf-life of the medium is 4 weeks at 4°C.

Tween 80 Medium for the hydrolysis of Tween 80

Blood agar base (Oxoid)	40.0 g
Distilled water	1000.0 ml
10 per cent $CaCl_2$ (filtered)	10.0 ml
Tween 80	10.0 ml

The blood agar base is added to the distilled water and allowed to soak for 10 minutes. The mixture is heated gently to dissolve the agar and the other ingredients are added. The medium is autoclaved at 121°C for 20 minutes, cooled and poured into Petri dishes (*Corynebacterium cystitidis* is used as the positive control).

Skirrow Medium

(Suggested as a medium suitable for both *Campylobacter fetus* and *Brucella* spp. by Terzolo *et al.*, 1991).

Blood agar base No. 2 (Oxoid CM 271)	1000 ml
Lysed, defibrinated horse blood	50 ml
Vancomycin	10 mg
Polymyxin B	2500 units
Trimethoprim	5 mg

The autoclaved blood agar base is cooled to 50°C and the other ingredients are added to the sterile molten agar medium.

Stonebrinks Medium for Mycobacterium bovis

Ingredients:

Salt mixture:

Sodium pyruvate	5.0 g
Potassium dihydrogen orthophosphate	2.0 g
Distilled water	300.0 ml

Dye mixture:

Crystal violet	100.0 mg
Malachite green	800.0 mg
Distilled water	100.0 ml

Hens' eggs:

20 fresh eggs from domestic fowl that have not received antimicrobial agents either therapeutically or prophylactically.

Preparation of medium:

• The salt solution is added to the distilled water and mixed until completely dissolved. The pH is adjusted to 6.5 using a solution of sodium phosphate, dibasic (Na_2HPO_4). The solution is placed in a 2 litre flask.

• The dyes are added to the distilled water and stirred until completely dissolved. The solution is placed in a 100 ml bottle.

• Both solutions are sterilised by autoclaving at 121°C for 15 minutes.

• The eggs are washed in soapy water, rinsed in tap water and dried. They are then placed in a container and covered with 75 per cent isopropyl alcohol for 30 minutes. The eggs are air-dried on sterile cotton wool in a cabinet with ultraviolet strip-lighting. When dry, the eggs are cracked on the edge of a sterile beaker and the contents placed in a sterile blender jar. The contents of the eggs are homogenized gently on a blender.

• The dye mixture is added first to the sterile salt solution and then the homogenized egg mixture is added and the mixtures are combined thoroughly.

• The mixture is dispensed in 12 ml amounts into universal bottles.

• The bottles are slanted and the contents inspissated at 80°C until the medium has solidified (usually 40-60 minutes).

• The slants are cooled for 30 minutes and then incubated at 37°C for 24-48 hours to facilitate the absorption of moisture into the medium.

Ureaplasmas : Hayflick's Medium (modified)

Broth Medium:

Mycoplasma broth base (BBL)	2.1 g
Distilled water	66.0 ml
Horse serum	20.0 ml
Yeast extract 25 per cent (w/v)	10.0 ml
Urea 25 per cent (w/v)	1.0 ml
Phenol red 0.5 per cent	0.2 ml
Penicillin solution	500 units/ml

Final pH should be adjusted to 6.0 ± 0.2

Agar Medium:

For solid agar medium add 1.0 g Ionagar (No. 2) and omit the urea and phenol red.

Vitamin K-Haemin Supplement for non-sporing anaerobes

Stock haemin solution:

50 mg haemin is dissolved in 1 ml 1 N NaOH.

100 ml of distilled water is added and the solution is autoclaved at 121°C for 15 minutes.

Stock vitamin K solution

100 mg menadione (vitamin K) is dissolved in

20 ml of 95 per cent ethanol and the solution is sterilised by filtration.

For addition to media:

Sterile menadione solution	1.0 ml
Sterile haemin solution	100.0 ml

1 ml of the vitamin K-haemin solution is added to 100 ml of sterile medium.

Transport Media

Transport medium and transportation procedures for anaerobes

Modified Cary-Blair Medium

Cary and Blair medium(BBL)	2.5 g
1 per cent solution of $CaCl_2$	1.8 ml
Rasazurin solution 0.1 per cent w/v	0.1 ml
L-cysteine HC1	0.1 g
Distilled water	198.0 ml

All the ingredients, except the cysteine, are heated in a flask with glass beads until the agar has dissolved. The mixture is cooled in an oxygen-free atmosphere with carbon dioxide. The cysteine is then added and the pH adjusted to 8.4. The medium is dispensed in 10 ml aliquots in 16 x 125 mm tubes and stoppered with rubber bungs. The filling of the tubes must be carried out under oxygen-free nitrogen gas or in an anaerobic chamber. The tubes are sterilised by steaming for 15 minutes. Any pink colouration in the tubes indicates that the contents are not anaerobic.

Oxygen-free Swabs

These oxygen-free swabs are sterilised in a tube filled with oxygen-free nitrogen gas. When taking specimens, the swabs are removed from the tube, used quickly and then pushed deep into a pre-reduced, anaerobically sterilised tube of modified Cary-Blair medium (prepared as above).

Commercially available transport systems for anaerobes

Swab transport systems for anaerobic bacteria

Port-A-Cul (BBL and Carr-Scarborough Microbiologicals)

Anaerobic Culturette (Marion Scientific)

Transport for small pieces of tissue

Anaport System (Scott Laboratories)

Bio-Bag Type A (Marion Scientific)

Vacutainer Anaerobic Specimen Collector (Becton and Dickinson)

Transport Medium for Campylobacter fetus (Clark and Dufty, 1978)

Fresh bovine serum containing per ml:-

5-fluorouracil	300 μg
Polymyxin B sulphate	100 units
Brilliant green	50 μg
Nalidixic acid	3 μg
Cycloheximide	100 μg

The mixture is dispensed in 10 ml aliquots into 30 ml vaccine bottles and the rubber stoppers inserted. The bottles are placed in a boiling water bath for 2 minutes to allow the serum to solidify. The medium is stirred with a sterile glass rod when cool. An 18-gauge hypodermic needle is inserted through the rubber stoppers and the bottles are placed in as anaerobic jar under an atmosphere of 2.5 per cent oxygen, 10 per cent carbon dioxide and 87.5 per cent nitrogen. The needles must be removed immediately after opening the jar. The medium should be stored at 4°C for at least one week before use. The transport medium has a shelf-life of up to 3 months at 4°C.

Transport Medium for Chlamydiae (Spencer and Johnson, 1983)

Sucrose	74.6 g
KH_2PO_4	0.512 g
K_2HPO_4	1.237 g
L-glutamic acid	0.721 g
Phenol red	0.015 g
Vancomycin	100.0 mg
Nystatin	50.0 mg
Streptomycin	100.0 mg
Gentamicin	50.0 mg
Distilled water	1000.0 ml

The pH is adjusted to 7.0 with KOH and the medium is sterilised by filtration (0.22 μm membrane filter). Foetal calf serum (100 ml) is then added in an aseptic manner. The medium is dispensed aseptically in 4 ml amount and stored at -20°C.

Mycoplasmal Transport of Culture Medium

Transport Medium:

PPLO broth (Difco)	16.8 g
Distilled water	800.0 ml
Horse serum	200.0 ml
50 per cent yeast extract	10.0 ml

DNA (Sigma)	0.02 g
Penicillin	2000 units/ml medium
Thallium acetate	1:100,000

Swabs can be placed into 2 ml aliquots of this medium.

Culture Medium:
The medium is also suitable as a broth culture medium for *Mycoplasma* spp. To convert the broth medium to an agar culture medium, substitute 28 g PPLO agar (Difco) for the 16.8 g of PPLO broth base.

Transport Medium for Viral Specimens

The medium is essentially an isotonic solution containing protein, a buffer to control the pH and antimicrobial agents to control contaminating bacteria and fungi.

Hank's Balanced Salt Solution	1000.0 ml
Sodium bicarbonate	8.0 g
Bovine albumin	10.0 g
Phenol red (0.4 per cent)	5.0 ml
Penicillin	500 units/ml of medium
Streptomycin sulphate	500 μg/ml
Nystatin	50 units/ml

The transport medium is sterilised by filtration through a 0.22 μm membrane filter. It can be dispensed into sterile bijoux bottles in 4 ml amounts. The shelf-life is approximately 3 weeks at 4°C.

References

Clark, B.L. and Dufty, J.H. (1978). Transport medium for *Campylobacter fetus* subsp. *venerealis* in bovine preputial washings, *Australian Veterinary Journal*, **54**: 262.

Smith, I.M. and Baskerville, A.J. (1979). A selective medium facilitating the isolation and recognition of *Bordetella bronchiseptica* in pigs, *Research in Veterinary Science*, **27**: 187-192.

Spencer, W.N. and Johnson, F.W.A. (1983). Simple transport medium for the isolation of *Chlamydia psittaci* from clinical material, *Veterinary Record*, **113**: 535-536.

Terzolo, H.R. Paolicchi, F.A. Moreira, A.R. and Homse, A. (1991). Skirrow agar for simultaneous isolation of *Brucella* and *Campylobacter* species, *Veterinary Record*, **129**: 531-532.

Timoney, P.J., Shin, S.J. and Jacobson, R.H. (1982). Improved selective medium for isolation of the contagious equine metritis organism, *Veterinary Record*, **111**: 107-108.

Appendix 3

Product Suppliers for Diagnostic Microbiology

Addresses of manufacturers or suppliers mentioned in the text are listed in this appendix. The list is not comprehensive and is intended as a guide for those engaged in veterinary diagnostic microbiology. General, regional firms should be consulted as they often act as agents for many of the international manufacturers of microbiological products. Table 151 lists the sources of kit-set tests for diagnostic virology.

AB Biodisk (E test), Pyramidvagen 7, S-171 36 Solna, Sweden.

Abbott Laboratories, P.O. Box 152020, Irving, TX 75015, USA.

American Microscan, 1584 Enterprise Blvd., W. Sacramento, CA 95691, USA.

American Type Culture Collection, 12301 Parklawn Drive, Rockville, MA 20852-1776, USA.

Analytab Products, 200 Express St., Plainview, NY 11803, USA.

BBL Microbiology Systems, P.O. Box 243, Cockeysville, MD 21030, USA.

Becton and Dickinson Microbiology Systems, P.O. Box 243, Cockeysville, MD 21030, USA.

Beldico S.A., Rue Andre Feher 1, 5411 Marche-en-Famenne, Belgium.

bioMerieux S.A., 69280 Marcy-L'Etoile, France/API Unite Bacteriologie, B.P. 2,38390 La Balme-les-Grottes, France.

Capco Instruments, Dodeca, Fremont, California, USA.

Carr-Scarborough Microbiologicals, Inc., P.O. Box 1328, Stone Mountain, GA 30086, USA.

Central Veterinary Laboratories, New Haw, Addlestone, Surrey KT15 3NB, UK.

Difco Laboratories, Inc., P.O. Box 1058, Detroit, MI 48232, USA.

Flow Laboratories, Inc., 7655 Old Springhouse Rd., McLean, VA 22102, USA.

General Diagnostics, Morris Plains, NJ., USA.

Gibco Laboratories, 421 Merrimack St., Lawrence, MA 01843, USA.

Hoffmann-LaRoche, D-7889 Grenzach/Wyhlen, Germany.

Immuno-Mycologics Inc., P.O. Box 1151, Norman, OK 73070, USA.

Innovative Diagnostic Systems Inc., 3404 Oakcliff Rd., Suite C-1, Atlanta, GA 30340, USA.

Lab M, Topley House, Walsh Lane, Bury, BL9 6AU, UK

Marion Scientific Corp., 9233 Ward Parkway, Suite 350, Kansas City, MO 64114, USA.

Meridian Diagnostics, 3471 River Hills Dr., P.O. Box 44216, Cincinnati, OH 45244, USA.

Millipore Corporation, Ashby Rd., Bedford, MA 01730, USA.

Molecular Biosystems, Inc., 10030 Barnes Canyon Rd., San Diego, CA 92121, USA.

National Committee for Clinical Laboratory Standards (NCCLS), Villanova, Pennsylvania, USA.

Organon Teknika, 100 Akzo Ave., Durham, NC 27704, USA.

Oxoid (UK Head Office), Wade Rd., Basingstoke, Hampshire RG24 OPW, UK. **Unipath Co.** (Oxoid Division), 9200 Rumsey Rd., Columbia MD 21045, USA.

Pharmacia LKB Biotechnology, Inc., 800 Centennial Ave., Piscataway, NJ 08854-9932, USA.

Pitman-Moore, Inc., Washing Crossing, NJ 08560, USA.

Polysciences, Inc., 400 Valley Rd., Warrington, PA 18976, USA.

Remel, 12076 Santa Fe Dr., Lenexa, KS 66215, USA.

Rijks Institut, Vor de Volksgezondheid, Bilthoven, Netherlands.

Roche Diagnostic Systems, 340 Kingsland St., Nutley, NJ 07110, USA.

Roche Products Ltd., Welwyn Garden City, Hertfordshire, UK.

Scott Laboratories, Inc., 771 Main St., Fiskeville, RI 02823, USA.

Scott-Nolan Laboratories, 4958 Hammermill Rd., Tucker, GA 30084, USA.

Sigma Chemical Co. Ltd., Fancy Rd., Poole, Dorset BH17 7TG, UK **or** 3050 Spruce St., St. Louis, MO 63103, USA.

Vitek Systems, Inc., 595 Anglum Dr., Hazelwood, Missouri 63042, USA.

Wellcome Diagnostics, Dartford, England, DAI 5AH, UK or 3030 Cornwallis Rd., Research Triangle Park, NC 27709, USA.

Whitakker M.A. Bioproducts, P.O. Box 127, Biggs Ford Rd., Walkersville, MD 21793, USA.

Winthrop Laboratories, 90 Park Ave., New York, NY 10018, USA.

Index

Index

All numbers refer to page numbers. Those in **bold** refer to major entries.